UNDERSTANDING RHEOLOGY

Topics in Chemical Engineering: A Series of Textbooks and Monographs

Deen, *Analysis of Transport Phenomena*
Doraiswamy, *Organic Synthesis Engineering*
Floudas, *Nonlinear and Mixed Integer Optimization: Fundamentals and Applications*
Friedlander, *Smoke, Dust, and Haze, 2nd Edition*
Fuller, *Optical Rheometry of Complex Fluids*
Larson, *The Structure and Rheology of Complex Fluids*
Lauffenburger and Linderman, *Receptors: Models for Binding, Trafficking, and Signalling*
Morrison, *Understanding Rheology*
Ogunnaike and Ray, *Process Dynamics, Modeling, and Control*
Pearson, *Discrete-Time Dynamic Models*
Phan-Thien and Kim, *Microstructures in Elastic Media*
Pozrikidis, *An Introduction to Theoretical and Computational Fluid Dynamics*
Pozrikidis, *Numerical Computation in Science and Engineering*
Schmidt, *The Engineering of Chemical Reactions*
Varma and Morbidelli, *Mathematical Methods in Chemical Engineering*

UNDERSTANDING ᘊᘊᘊ RHEOLOGY

Faith A. Morrison
Michigan Technological University

New York   Oxford
OXFORD UNIVERSITY PRESS
2001

Oxford University Press

Oxford   New York
Athens   Auckland   Bangkok   Bogotá   Buenos Aires   Calcutta
Cape Town   Chennai   Dar es Salaam   Delhi   Florence   Hong Kong   Istanbul
Karachi   Kuala Lumpur   Madrid   Melbourne   Mexico City   Mumbai
Nairobi   Paris   São Paulo   Shanghai   Singapore   Taipei   Tokyo   Toronto   Warsaw

and associated companies in
Berlin   Ibadan

Published by Oxford University Press, Inc.
198 Madison Avenue, New York, New York, 10016
http://www.oup-usa.org

Oxford is a registered trademark of Oxford University Press

Library of Congress Cataloging-in-Publication Data

Morrison, Faith A.
   Understanding rheology / Faith A. Morrison.
      p.   cm.—(Topics in chemical engineering (Oxford University Press))
   Includes bibliographical references and index.
   ISBN 0-19-514166-0
   1. Rheology.   I. Title.   II. Series.

QC189.5 .M67   2000
531'.1134—dc21   00-037447

Printing (last digit): 1 3 5 7 9 8 6 4 2

Printed in the United States of America
on acid-free paper

Contents

CHAPTER 3
Newtonian Fluid Mechanics 59

CHAPTER 4
Standard Flows for Rheology 105

Preface

If you have picked up this book, it is either because you want to understand rheology or because you want to know what rheology is. To respond to the latter possibility, rheology is the study of deformation and flow. To be more precise, it is typically the study of the flow of complex fluids such as polymers, pastes, suspensions, and foods. Simpler fluids, such as water and air, have their own well-defined field, called fluid mechanics.

I am most interested in addressing those of you who have picked up this volume for the first reason—you would like to understand rheology. It would seem that learning about rheology is straightforward since there are many books that address the subject [61, 162, 26, 238]. In my experience with studying rheology, I find that most books assume an understanding of either mathematics or fluid mechanics that is greater than that which I possessed when I entered the field. In teaching the subject, I also have found that most of my students arrive in my class without these prerequisite skills.

Therefore what I set out to create was a workbook/textbook with which engineers, scientists, and others could teach themselves rheology. This book is aimed at the same time at the many technology professionals who end up having to teach themselves this subject on the job, as well as at advanced undergraduates interested in the subject. It is deliberate that this book is more detailed than the average monograph. I hope that this will be the kind of book that students talk about and recommend to their friends as the one book that is totally clear on the subject. As a result of trying to be clear on the mechanics of this subject, this book lacks breadth. I fully admit this flaw and invite accomplished rheologists to skip this book and to proceed immediately to the many fine texts that cover the field of rheology more completely [26, 27, 61, 238, 220, 162, 138]. My goal in this text is to make easier the entry of newcomers into the field of rheology.

This book is an outgrowth of a quarter-long course in introductory rheology I have taught 9 times to undergraduates and first-year graduate students in chemical engineering and mechanical engineering (with a few chemists thrown in) at Michigan Technological University. For 7 years the text for that course was *Dynamics of Polymeric Liquids*, volume 1, by Bird, Armstrong, and Hassager, Chapters 1–5, 10 [26]. The order of the topics addressed in the current text as well as the approach taken were strongly influenced by that book. For two years I used drafts of Chapters 1–8 in the classroom, and I devised many improvements as a result. I twice taught the material in Chapter 9 in a 10-week graduate course on advanced rheology. In that course the text I used was *Constitutive Equations for Polymer Melts and Solutions* [138], by Larson, along with supplementary material.

As I stated above, I have included in this book all the background material I consider necessary for an advanced engineering or science undergraduate to learn rheology. Depending on the reader or class using the text, some of this material may be omitted. Chapter 1 is intended to orient the reader, particularly a reader engaged in self-study of rheology. Chapter 2 may be assigned for self-study or review and Chapter 3 may be omitted for chemical and mechanical engineers (although I never omitted it in my classes). The next 5 chapters contain the core of the material on rheology, including standard flows for rheology, material functions, and experimental data. The data discussed in Chapter 6 predominately concern polymer systems, revealing my bias toward polymer rheology. For students with a background in polymer science this material may be familiar, and for those readers Chapter 6 may be skimmed or assigned for self-study. The next three chapters concern constitutive equations, moving from the simplest (Chapter 7), to more complex (Chapter 8), finishing with the genuinely advanced (Chapter 9). Chapter 10 is a stand-alone chapter on rheological measurements that, depending on the background of the reader, may be read out of sequence, for example first, or immediately after Chapters 3, 5, 6, or 8. Many problems are included with each chapter, including some challenging flow problems, marked with an asterisk. A solutions manual is available to instructors.

In a 10-week quarter I never covered more than Chapters 1–8 (up to generalized linear viscoelastic fluids) with a little bit of Chapter 10 squeezed in. In a 12- to 15-week semester course it would be appropriate to cover all of that material plus a few of the topics in Chapter 9 and to do a thorough treatment of the rheometry chapter, Chapter 10.

One comment on nomenclature is needed. I have departed from standard practice in using the symbol $\dot{\varsigma}(t)$ for shear rate in the definition of shear flow in Chapter 4. I have done this to emphasize $\dot{\varsigma}(t)$ is a function we *specify* in the context of a standard flow. While it is true that $\dot{\varsigma}(t) = \dot{\gamma}_{21}(t)$, this fact is presented as a conclusion, arrived at by applying the definition of the rate-of-deformation tensor $\dot{\underline{\underline{\gamma}}}$ to the velocity field of shear flow. I have found this approach to be helpful to students who are new to the subject. All symbols used are identified in Appendix A to aid the reader. In addition, Appendix B contains definitions of rheological terms, including many expressions not used in this text that the reader will encounter in the rheological literature.

I have included several appendixes intended to aid the reader's study of rheology including Appendix C, which contains detailed mathematical explanations and hints, Appendix D, which summarizes the predictions of the most common nonlinear constitutive equations, and Appendix E, an extensive background chapter on optics to help the newcomer to rheo-optics. The final appendix includes some supplementary data for problems in Chapters 7 and 8 (Appendix F).

I would like to thank Michigan Technological University for supporting this project, including the award of Faculty Development grants in 1997 and 1999. Parts of this text were written while I was on sabbatical at 3M Company, St. Paul, MN, and I am grateful for the support and environment provided by 3M. My thanks go to all my current and former students who helped this book grow within me; particular thanks go to Kathleen Barnes, Srinivas Uppuluri, and Parag Karmarkar; to Forbes Robertson who drafted early versions of some figures; and to several students who read an early draft and made suggestions, Debabrata Sarkar, Mike Sieroki, Bryan Vogt, and Tsung-Hsweh Wu. Several colleagues gave me very useful feedback on later drafts, and the text was much improved by having their input. Thanks go to Susan Muller, who gave extensive feedback and who also provided sources for some material, Ronald Larson, Wesley Burghardt, Jeff Giacomin, Gerry Fuller, Scott Chesna, Mike Solomon, Douglas Devans, Scott Norquist, Robert Ginn, Saad Khan, Madhukar Vable, Kathleen Feigl, Xina Quan, David Malkus, and Robert Kolkka. I owe a significant debt to my husband, Tomas Co, who encouraged me to write this and has long been my best friend and a tireless supporter. I greatly appreciate his help in clarifying some of the mathematics contained in the book.

For their love and encouragement, I would like to thank my family, especially my parents, Frances P. Morrison and Philip W. Morrison, my siblings and in-laws, and many friends, especially Susan Muller, Pushpalatha Murthy, David and Beth Odde, Karen Hubbard, Jim and Sally Brozzo, Connie Gettinger, Denise Lorson, Yannis Nikolaidis, and Selen Ciftci. I would also like to thank some inspirational math teachers, John Checkley and Doris Helms, who helped me grow my love of mathematics, and a talented English teacher, Judith Smullen, who taught me to write. Thanks also to Jeff Koberstein who taught me that research is fun and to Jay Benzinger for demonstrating how to teach.

This book is dedicated to the memory of Professor Davis W. Hubbard, mentor, colleague, friend, and, to this day, a source of great inspiration to me.

F. A. M.
Houghton, Michigan

CHAPTER 1

Introduction
How Much Do I Need to Learn about Rheology?

Rheology is the study of the deformation and flow of matter. This field is dominated by inquiry into the flow behavior of complex fluids such as polymers, foods, biological systems, slurries, suspensions, emulsions, pastes, and other compounds. The relationships between stress and deformation for these types of materials differ from Newton's law of viscosity, which describes the shear behavior for normal liquids. Newton's law is the only stress–deformation relationship considered in most introductory fluid-mechanics courses. Complex fluids also do not follow Hooke's law of elasticity, the relationship between stress and deformation that is used for metals and other elastic materials. Engineers and scientists must know something about rheology when neither Newton's law nor Hooke's law suffice to explain the fluid behavior they encounter in their work.

When one is exploring a new field, there is always the question, how deep should I go? Since rheology is a subject that can involve a great deal of mathematical and physical analysis, the reader is immediately faced with a choice between pursuing a qualitative or a quantitative understanding of rheology. In scanning the library shelf of books on rheology, you will find that there are books available that approach rheology from a mostly descriptive point of view (see [11], and for polymer viscoelasticity specifically see [75, 3]), as well as books that involve mathematical analysis without introducing tensor calculus [61], and books that employ vector/tensor calculus [61, 26, 162, 238, 220, 138]. The choice then is between a descriptive understanding of rheology and a more thorough, and hence mathematical, understanding.

To help the reader make this choice, we begin here with a discussion of the kinds of effects that distinguish rheologically interesting materials (non-Newtonian fluids) from the more conventional (Newtonian fluids). If after reading this chapter you believe that it is worth your while to invest some time in understanding rheology in mathematical detail, the rest of the text is laid out to assist you. In this text we follow an explicit, step-by-step approach, which will allow you to master the background material that you will need to study rheology, including vector and tensor mathematics, Newtonian fluid mechanics, rheological standard flows, and rheological material functions. By reading this text and taking the time to master each topic, you will find that understanding rheology is straightforward.

1.1 Shear-Thinning and Shear-Thickening

Viscosity is the most commonly sought after rheological quantity, and viscosity is a qualitatively different property for Newtonian and non-Newtonian fluids. Several devices that are used to measure viscosity are limited to use with Newtonian fluids, for example, the Cannon–Fenske and some versions of the Brookfield viscometer [36]; see Figure 1.1. There are devices suitable for both Newtonian and non-Newtonian fluids, but without an understanding of non-Newtonian effects, the measurements can be confusing. The situation discussed next demonstrates the difficulties encountered if non-Newtonian effects are neglected.

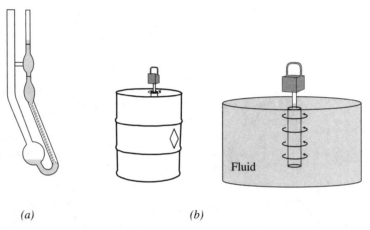

(a)   *(b)*

Figure 1.1 (*a*) Cannon–Fenske viscometer, which measures viscosity for Newtonian fluids using gravity as a driving force. The viscosity is related to the time interval for the fluid meniscus to pass from one mark on the tube to another. (*b*) Simple type of Brookfield viscometer, in which a mandrel is made to rotate in a drum of fluid. The viscosity of Newtonian fluids is related to the torque required to turn the mandrel.

EXAMPLE

Determine the viscosity of a solution used in a coating process by your employer, Acme Adhesive Associates. The information will be used to design process equipment.

SOLUTION

In the laboratory next to yours there is an instrument for measuring the viscosity of silicon oils and other Newtonian fluids (see Figure 1.2). The equation for the analysis of the flow is given in the operations manual:

$$\text{viscosity} = \frac{\pi P R^4}{8QL} \tag{1.1}$$

where R and L are the radius and length of a tube through which the solution flows in the viscometer ($R = 1.0$ mm and $L = 30$ mm are given in the manual), and P and Q are

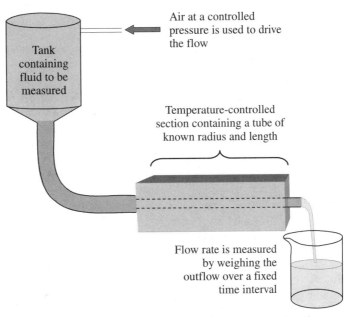

Air at a controlled
pressure is used to drive
the flow

Tank
containing
fluid to be
measured

Temperature-controlled
section containing a tube of
known radius and length

Flow rate is measured
by weighing the
outflow over a fixed
time interval

Figure 1.2 Rheometer that could be used for measuring viscosity as described in the example. Air is used to drive the fluid at a specified pressure. The flow rate of the fluid is measured by the pail-and-scale method.

the measured pressure drop and volumetric flow rate for the fluid of interest. You consult coworkers who have used the instrument and learn that for their fluids they usually use $P = 15$ psi because that gives a reasonable flow rate.

With the help of your coworkers, you carry out the measurement of flow rate for $P = 15$ psi and calculate the viscosity of your coating fluid to be 4.2 centipoise. On a hunch, you also measure flow rate for a driving pressure drop of $P = 20$ psi, and you find a viscosity of 0.1 centipoise! Repeated attempts verify that your results are accurate, and you conclude that the viscosity of the coating fluid depends on the driving pressure. You discuss your results with your coworkers, but they have never noticed a dependence of viscosity on driving pressure for the fluids they have examined.

You report viscosity as a function of driving pressure to your supervisor. The viscosity varies over two orders of magnitude, depending on the pressure. It is not clear how to proceed with the equipment design, since although the usual design equations take viscosity into account, no variation in viscosity is mentioned in the equations.

In this example we encounter the most common non-Newtonian effect, shear thinning. Shear thinning is the tendency of some materials to decrease in viscosity when they are driven to flow at high rates, such as by higher pressure drops. Some materials show the opposite effect, that is, they exhibit higher viscosity when they are made to flow at high rates; this is called shear-thickening. Both shear-thinning and shear-thickening can be modeled effectively by the equations discussed in Chapter 7. The solution to the design problem posed in the preceding example is to examine more thoroughly the rheological properties of the coating fluid and then to modify the design equations using the results. Rheological properties are measured using the techniques in Chapter 10, which are based on standard flows (Chapter 4) and standard material functions (Chapter 5).

1.2 Yield Stress

Yield stress is a complex rheological effect that we can easily observe in the kitchen. Yield is the tendency of a material to flow only when stresses are above a threshold stress. This is a non-Newtonian effect—Newtonian fluids will always flow when a stress is applied, no matter how small the stress. For example, honey is a food that has a high viscosity, but which is nonetheless Newtonian. When you serve yourself from a jar of honey, you disturb the flat surface of the honey. If you check the honey jar a small time later (maybe 2 minutes), the fluid surface will have returned to its original, level shape under the (small) force of gravity.

Mayonnaise, on the other hand, behaves quite differently (Figure 1.3). While digging out some mayonnaise from the jar to make a sandwich, you disturb the mayonnaise surface, much as we did when serving honey. Ten minutes later, the mayonnaise surface is still disturbed. Looking in the jar a week or even a year later, we see that the mayonnaise surface has the same shape—mayonnaise does not flow and form a level surface under gravity. Mayonnaise is a yield-stress fluid (discussed more in Chapter 7), which will flow easily if a high stress is imposed (as you spread it with a knife, for example), but which will sustain a small stress such as that due to gravity. When processing yield-stress fluids, which in addition to many foods include slurries, pastes, and paint, the existence of the yield stress means that larger stresses are required to cause fluid flow, and care must be taken to maintain the stress above the yield stress to sustain flow. One industrially important class of yield-stress fluids is road asphalt. Rheological measurements are used during its compounding to ensure that the resulting mixture will have properties that are known to be compatible with existing pavement, paving methods, and equipment. Designing such equipment requires detailed understanding of the rheological properties of asphalt.

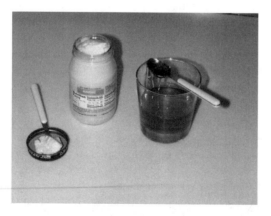

Figure 1.3 Mayonnaise is a yield-stress fluid. It can stay on the spoon indefinitely, and the surface of the mayonnaise in the jar also remains disturbed indefinitely. Honey is a Newtonian fluid. It will flow, even under the mild force imposed by gravity. After just seconds, the honey surface is level and smooth.

1.3 Elastic / Viscoelastic Effects

The most spectacular non-Newtonian effects are exhibited by polymers or their mixtures. These effects differ qualitatively from the behavior of Newtonian fluids, and we may be familiar with some of these effects from our experiences in the kitchen or from playing with popular toys like Silly Putty. We can appreciate these effects without studying rheology in mathematical detail. To understand the origin of the effects or to model a system that exhibits these effects, then once again we must seek a more detailed understanding of rheology.

1.3.1 THE WEISSENBERG EFFECT

Flour is a naturally occurring polymer, and concentrated mixtures of flour and water are non-Newtonian. A striking non-Newtonian effect can be demonstrated with a standard kitchen mixer. When water or other Newtonian fluids such as corn oil or syrup are mixed at high speeds, the fluid is flung away from of the mixing blades due to inertia; that is, the effect of accelerating the fluid particles in a circle is to make the fluid flow outward, toward the bowl walls (Figure 1.4a). This effect is described by Newton's first law, the principle that a body in motion tends to remain in motion unless an outside force acts upon it.

Figure 1.4 (*a*) A Newtonian fluid, such as a dilute mixture of flour, water, and food coloring, moves away from the mixing blades when it is stirred at a high rate. (*b*) A non-Newtonian, more concentrated flour–water dough climbs the mixing blades.

(*a*)

(*b*)

Non-Newtonian doughs formed of flour and water, however, do not flow to the outer walls of the bowl when stirred at high speeds; they climb the mixing blades (Figure 1.4*b*). This effect is called the Weissenberg effect or rod-climbing effect, and it cannot be explained through the relationships that govern Newtonian fluids. Coping with the Weissenberg effect is essential in the design and operation of many types of food-processing equipment, as well as in the processing of polymeric solutions. To predict the axial forces generated by the Weissenberg effect as a function of blade rotational speed, or to predict the shape of the free surface in flows of these types of materials, we must use rheological constitutive equations that correctly capture nonlinear effects.

1.3.2 FLUID MEMORY

The intriguing thing about Silly Putty and materials like it is that it is deformable, bounce-able, but it is ultimately a liquid. This latter fact is often discovered by unsuspecting parents when a child leaves a ball of Silly Putty on the carpet, and over time the liquid seeps into the carpet, making a mess that is difficult to clean up. That materials like Silly Putty are liquids can be seen clearly by allowing a ball of such a mixture to relax on a wide-mesh screen (Figure 1.5). The mixture flows through the screen over a matter of hours. Yet this liquid appears to hold its own shape like a solid if we examine it over a time period of seconds or minutes, and it has elasticity like a solid rubber ball, bouncing when thrown to the floor and partially retracting when pulled and then released.

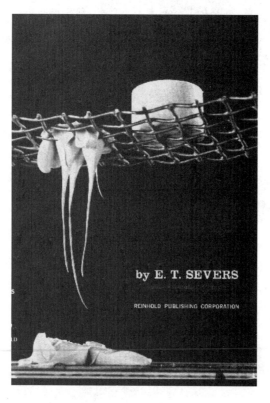

Figure 1.5 A ball of fluid resembling Silly Putty is placed on a mesh and allowed to flow, which proves that it is a liquid. The other material is a solid and does not flow.

Mixtures like Silly Putty are viscoelastic liquids, and the effects described are due to partial or fading memory. When Newtonian fluids like water or oil are subjected to stress, they deform (and flow), and when the stress is removed, the deformation stops instantly, and the liquid is at rest. Viscoelastic fluids subjected to stress also deform, but when stress is removed, the stress inside the viscoelastic fluid does not instantly vanish. For these types of fluids the internal molecular configuration of the fluid can sustain stress for some time. This time, called *relaxation time*, varies widely among materials. Moreover, because a viscoelastic fluid has internal stresses, the fluid will deform on its own even after the external stress has been removed. Thus, a viscoelastic fluid like Silly Putty, when deformed and released, will retract—deform in the opposite direction—producing a flow that reduces internal stress in the fluid.

Fading memory is exhibited by most long-chain polymers, and it impacts polymer manufacturing processes since trapped stresses (and frozen-in molecular orientation) can weaken or strengthen a part made of polymers (plastics). A qualitative understanding of fluid memory is helpful in handling polymer flows and in designing molds and other equipment in which polymers flow in the molten state. To do more in-depth analysis of polymer-processing operations, the study of more advanced constitutive equations is required (Chapter 9).

1.3.3 Die Swell

When plastic products are made by extrusion into air, so-called profile extrusion, the final product shape depends on orifice shape through which the polymer is forced, but it also depends on many details of the extrusion process [236]. This is because polymers exhibit die swell, that is, as the liquid polymer exits a die, the diameter of the liquid stream increases by up to an order of magnitude (Figure 1.6). Die swell is caused by the relaxation of extended

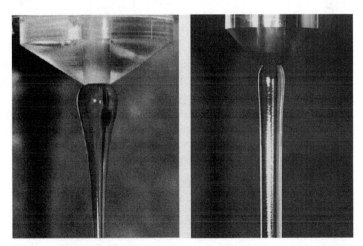

Figure 1.6 Partial memory of past configurations causes a viscoelastic fluid to swell as it exits from a die into air. The amount of swelling can be significant, as shown in these photos of aqueous solutions of polyvinyl alcohol and sodium borate. Newtonian fluids such as water actually contract slightly (∼15%) when exiting a die. *Photos:* Courtesy Dr. Andrew Kraynik.

polymer coils as the stress in a polymeric liquid reduces from the high, flow-producing stresses present within the die to the low stresses associated with the extruded stream moving through ambient air. Predictions of die-swell ratios can be made using advanced constitutive equations (such as those discussed in Chapter 9), and calculations using these constitutive equations can be used to design profile dies so that the desired extrudate shape is obtained.

1.4 Rheology as Spectroscopy

In many applications the rheological response of a material is important to the functioning or processing of the system. Rheology is often used to monitor material properties and can be used in quality control or as a way to categorize an unknown material [61]. For this application, a detailed mathematical understanding of rheology is not necessary.

EXAMPLE

After much searching and experimenting, your company, Fjord Automotive, has found the perfect resin for producing specialized clips that secure wires inside its automobiles. You are assigned to develop a product-testing procedure that will ensure that every batch of the resin that is purchased has the desired rheological properties. How will you address this?

SOLUTION

If the material were Newtonian, measuring its viscosity would be sufficient to characterize its flow properties. Because it is a plastic resin, however, it has both viscous properties and elastic properties. We therefore need to find a test that captures both viscous and elastic characteristics. One common way of characterizing polymeric materials is to measure their linear-viscoelastic moduli, $G'(\omega)$ and $G''(\omega)$ (Chapter 5). The $G'(\omega)$ and $G''(\omega)$ curve shapes are sensitive to changes in polymer molecular weight, molecular-weight distribution, chemical composition (in case the resin is a blend or a copolymer), and just about every other material variable. Ferry's book [75] gives a particularly complete discussion of the effects of many material characteristics on linear-viscoelastic behavior, and there is a shorter discussion of this topic in this text (Chapter 6). Measurements of these moduli are straightforward and are easily incorporated into a quality-control procedure. If keeping track of a high-strain property is more appropriate for a particular application (if flows are rapid, for example), steady shear viscosity and normal stresses could be measured, or the industry involved may develop its own industry-specific tests like the melt-flow index or a slump test (see Glossary). To use rheology in this manner does not require an extensive understanding of the details of rheological analysis, provided accurate data are obtained and compared.

EXAMPLE

In the previous example we proposed that the linear-viscoelastic moduli $G'(\omega)$ and $G''(\omega)$ be measured and monitored for quality control of a resin at Fjord Automotive. You have implemented this strategy. A sales representative from a new potential supplier has called and is proposing a replacement resin for your application. How can you evaluate the new, proposed resin?

SOLUTION

Although linear-viscoelastic properties are sensitive to details of material composition, it is not true that two materials with the same linear-viscoelastic properties will necessarily behave in the same way in other flows. The quality control proposed in the last example is effective because you are always testing similar materials. Under these circumstances, small changes in $G'(\omega)$ and $G''(\omega)$ can be interpreted in a fairly accurate way to reflect subtle structural changes.

When comparing two completely different resins for a possible application, however, you need to examine your application and determine which characteristics make the process or the product successful. You must then measure these characteristics for the two materials, test the two materials in the process, and decide. This more detailed analysis best involves an experienced rheologist with an in-depth understanding of the field.

In the two examples mentioned we saw that the amount of detailed understanding that is required in using rheology for materials analysis depends on the type of questions being asked. As with most technological questions, sometimes a cursory examination is sufficient, while at other times a more complete understanding is required.

1.5 Process Modeling

Modern computing techniques allow engineers to make detailed calculations of critical quantities such as the maximum temperature experienced by the heat-shield tiles on the U.S. space shuttle as it reenters the Earth's atmosphere. The accuracy of these calculations depends on the accuracy of tile material data, such as heat capacity and thermal conductivity, and also on the model that indicates the amount of viscous heating in the Earth's atmosphere generated by the rapidly moving space vehicle.

Similar methods can be applied to calculate flow fields, temperature fields, and stress fields in polymer-processing equipment such as injection molders, extruders, and blow molders. The accuracy of these calculations depends critically on selecting an appropriate non-Newtonian rheological model, as illustrated in the next example.

EXAMPLE

You are a new engineer who has been hired in a polymer-processing group. You have experience in running finite-element computer calculations on turbulent flows and in using other types of simulation packages. You are asked to learn a polymer-flow simulation package and to work with a team to solve problems in your company's molding operations. The task of producing computer simulations of complex systems is familiar to you, but to start the simulation you must choose a stress–deformation law to use in modeling your material. You have the following choices: power law, Cross model, Ellis model, truncated power law, linear viscoelastic fluid, upper convected Maxwell model, Oldroyd B model, and Leonov model. How will you choose?

SOLUTION

The choice of stress–deformation law or constitutive equation is critical to making accurate simulations of polymer-processing problems. If the problem involves a constant pressure gradient, then a generalized Newtonian model (power law, Cross, Ellis, truncated power law; see Chapter 7) suffices. If the flow rate is slowly changing, then the generalized linear-viscoelastic model may be appropriate (Chapter 8), whereas to obtain information about nonlinear effects such as normal stress effects, instabilities due to normal stresses, and die swell, a more complex model such as the upper convected Maxwell, Oldroyd B, or Leonov models (Chapter 9) is required. The effort needed to carry out the calculation increases if these nonlinear models are used. Also, the number of parameters that must be obtained from experimental data is higher for more complex models. We must consider these types of tradeoffs when making the decision.

If you are faced with a dilemma like the engineer in the preceding example, it will be worthwhile to learn about rheology. Although simulation packages greatly simplify the process of making process simulations, the accuracy of the results depends critically on the assumptions and model choices input to the program. To make these choices, you must understand rheology.

There are more examples of interesting, non-Newtonian phenomena shown in Figures 1.7 and 1.8. In summary, if we wish to measure rheological properties or to make engineering predictions involving the flow of non-Newtonian fluids, we must have a

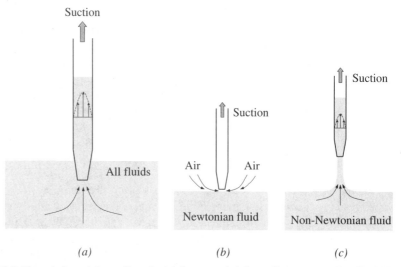

Figure 1.7 The tubeless-siphon effect. In (*a*) the normal siphon effect occurs when liquid is sucked into a tube held below the liquid surface. When the tube is raised above the liquid level, air is drawn into the tube, and the suction breaks (*b*). For some non-Newtonian fluids, however, when the tube is raised, the liquid continues to siphon, even as the tube is raised several centimeters above the liquid surface (*c*).

De = 4.629	De = 4.637	De = 4.645	De = 4.666	De = 4.779	De = 5.181	De = 5.231	De = 6.684
Re = 82.3	Re = 82.5	Re = 82.6	Re = 83.0	Re = 85.0	Re = 92.1	Re = 93.0	Re = 118.9

Figure 1.8 Photographs of an elastic instability occurring in a viscous fluid flowing in the gap between two cylinders; the inner cylinder is turning. In this flow geometry, called Taylor–Couette flow, there is an instability at high Reynolds numbers (Re) to stationary cells in the shape of doughnuts stacked in the gap. Here, the fluid has a high viscosity, making the Reynolds number small, well below the critical Reynolds number for the inertially driven Taylor instability, yet there is an instability in this flow as well [186]. Below each view of the whole flow cell there is a space–time plot of an axial slice through the image, with the axial dimension increasing from left to right and time increasing from top to bottom. At low Reynolds numbers and low Deborah numbers (De), a dimensionless group that accounts for the effects of elasticity, the vortices are stationary, and the space–time plots show vertical stripes. At higher De the vortices become nonaxisymmetric, with a well-defined azimuthal wavenumber. As De increases further, the vortices become increasingly disordered. *Photo:* Courtesy Dr. Susan Muller.

quantitative understanding of rheology. This book is designed to help the reader acquire a quantitative understanding of rheology through the presentation of background material, detailed derivations and explanations, practice exercises, additional problems at the ends of chapters, and extensive appendixes. One who masters the material in this text will be well prepared to take and interpret rheological measurements, tackle non-Newtonian simulation software, decipher the rheological literature, and deepen one's understanding of rheology through the study of more advanced texts [27, 138, 238, 220].

CHAPTER 2

Vector and Tensor Operations

We will focus our discussion of rheology on isothermal flows. The isothermal flow behavior of either conventional or complex fluids is determined by two physical laws, mass conservation and momentum conservation, plus the stress constitutive equation, a relationship that describes how a fluid responds to stress or to deformation. Mass conservation is a scalar equation, momentum conservation is a vector equation, and the constitutive equation is an equation of still higher mathematical complexity, a tensor equation.

To make fluid mechanics comprehensible to third-year chemical and mechanical engineering students, the vector and tensor nature of the subject is often given a light treatment in introductory courses. In these classes, emphasis lies with solving the vector momentum equation in the form of three scalar equations (conveniently tabulated in several coordinate systems). To relate the shear stress and the velocity gradient, the students incorporate the appropriate scalar component of the Newtonian constitutive equation, usually the 21-component, often called Newton's law of viscosity.

It is then straightforward to solve for the velocity and pressure fields and other quantities of interest. The scalar presentation of fluid mechanics works fine until, for example, shear-induced normal stresses are encountered or until one wishes to understand polymer die swell or memory effects. To describe such phenomena, more complex constitutive equations are required, and while these more complex constitutive equations may be expressed in scalar form, the scalar form will usually include six nontrivial equations. Furthermore, the forms of these six scalar equations will depend on the coordinate system in which the problem is written. This enormous increase in complexity can be understood and managed quite effectively if we employ the mathematical concept of a tensor. The tensor is thus a time-saving and simplifying device, and in studying rheology it is well worth the effort to learn tensor algebra. In fact, after taking the time to understand tensors, we will see that some aspects of Newtonian fluid mechanics become easier to understand and apply.

In this text we use tensor notation extensively. We assume no prior knowledge of tensors, however, and we begin in this chapter with a comprehensive review of scalars and vectors (Figure 2.1). This review is followed by the introduction of tensors and by the derivation of the conservation equations in vector/tensor format in Chapter 3. The review of Newtonian fluid mechanics in Chapter 3 provides an opportunity to work with tensors and with the conservation equations. The study of non-Newtonian constitutive equations is prefaced with three chapters that are needed to catalog non-Newtonian rheological behavior.

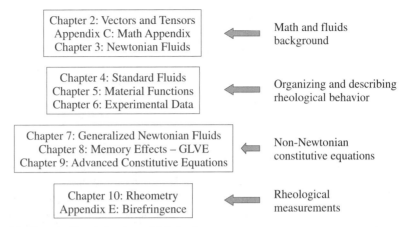

Figure 2.1 Organizational structure of this book.

In Chapter 4 we define and discuss the characteristics of standard flows used in rheology, and in Chapter 5 we define the material functions that are used to describe non-Newtonian behavior. We also provide in Chapter 6 a summary of the observed rheological behavior for many fluids. The rest of the text is dedicated to understanding and applying several simple non-Newtonian constitutive equations.

As stated before, first we will establish a common vocabulary of mathematics and fluid mechanics on which to build an in-depth understanding of rheology. There are some tools to help the reader in the appendixes, including a detailed table of nomenclature and a glossary. We begin, then, with scalars, vectors, and tensors. Readers familiar with vector and tensor analysis and Newtonian fluid mechanics may wish to skip ahead to Chapter 4.

2.1 Scalars

Scalars are quantities that have magnitude only. Examples of scalars include mass, energy, density, volume, and the number of cars in a parking lot. When we do ordinary arithmetic, we are dealing with scalars. Scalars may be constants, such as c, the speed of light ($c = 3.0 \times 10^{10}$ cm/s), or scalars may be variables, such as your height $h(t)$ over the course of your lifetime, which is a function of time t, or the density of an ideal gas $\rho(T, P)$, which is a function of temperature T and pressure P. The magnitude of a scalar has units associated with it since, for example, the numerical magnitude of your mass will be diffeent if it is expressed in kg or lbs.

Three scalars, for example, α, β, and ζ,[1] may be manipulated algebraically according to the following laws of scalar multiplication:

---

[1] Greek letters are used often in the text. They and all other symbols used are identified in Appendix A, Table of Nomenclature.

$$\left.\begin{array}{l} \text{Allowable algebraic} \\ \text{operations of} \\ \text{scalars with scalars} \end{array}\right\{ \begin{array}{ll} \text{commutative law} & \alpha\beta = \beta\alpha \\ \text{associative law} & (\alpha\beta)\zeta = \alpha(\beta\zeta) \\ \text{distributive law} & \alpha(\beta + \zeta) = \alpha\beta + \alpha\zeta \end{array}\right.$$

2.2 Vectors

Vectors are quantities that have both magnitude and direction. Examples of vectors that appear in fluid mechanics and rheology are velocity $\underline{v}$ and force $\underline{f}$. The velocity of a body is a vector because two properties are expressed: the speed at which the body is traveling (the magnitude of the velocity $|\underline{v}|$) and the direction in which the body is traveling. Likewise we can understand why force is a vector since to fully describe the force on, for example, a table (Figure 2.2), we must indicate both its magnitude $|\underline{f}|$ and the direction in which the force is applied. The same magnitude of force applied to the top of the table and to the side of the table will have different effects and must be treated differently. As with magnitudes of scalars, magnitudes of vectors have units associated with them.

In this text, most vectors will be distinguished from scalars by writing a single bar underneath vector quantities; vectors of unit length will be written without the underbar and with a caret (⌢) over the symbol for the vector, as will be discussed. An important vector property is that both the magnitude and the direction of a vector are independent of the coordinate system in which the vector is written. We will return to this property shortly.

Since a vector has two properties associated with it, we can examine these two properties separately. The magnitude of a vector is scalar valued. The magnitude a of a vector $\underline{a}$ is denoted as follows:

$$\text{Vector magnitude} \qquad \boxed{|\underline{a}| = a} \qquad (2.1)$$

The direction of a vector can be isolated by creating a new vector, such as $\hat{a}$, that points in the same direction as the original vector but has a magnitude of one. Called a *unit vector*, this is written as shown in Equation (2.2)

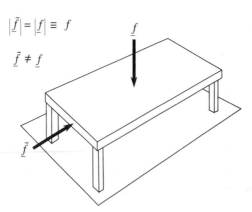

$$\left|\tilde{\underline{f}}\right| = \left|\underline{f}\right| \equiv f$$

$$\tilde{\underline{f}} \neq \underline{f}$$

Figure 2.2 Schematic representation of forces acting on a table. If the same magnitude of force f is applied in different directions, the vectors describing those forces differ too ($\underline{f}$, $\tilde{\underline{f}}$).

$$\hat{a} = \frac{a}{a} \qquad (2.2)$$

$$|\hat{a}| = 1 \qquad (2.3)$$

As stated before, we will distinguish unit vectors from general vectors with a caret ($\hat{}$). A special vector is the zero vector $\underline{0}$, which has zero magnitude and whose direction is unspecified.

2.2.1 VECTOR RULES OF ALGEBRA

The rules of algebra for vectors are not the usual laws of scalar arithmetic, since when manipulating two vectors both the magnitude and the direction must be taken into account. The rules for the addition and subtraction of vectors are reviewed in Figure 2.3.

The operation of multiplication with vectors takes on several forms since vectors may be multiplied by scalars or by other vectors. Each type of multiplication has its own rules associated with it. When a scalar (α) multiplies a vector ($\underline{a}$), it only affects the magnitude of the vector, leaving the direction unchanged,

$$\underline{b} = \alpha \underline{a} \qquad (2.4)$$

$$|\underline{b}| = |\alpha \underline{a}| = \alpha |\underline{a}| = \alpha a \qquad (2.5)$$

$$\hat{b} = \frac{\underline{b}}{|\underline{b}|} = \frac{\alpha \underline{a}}{\alpha a} = \hat{a} \qquad (2.6)$$

Since multiplication of a scalar with a vector only involves scalar quantities (the scalar α and the magnitude a), this type of multiplication has the same properties as scalar multiplication:

Allowable algebraic operations of scalars with vectors $\begin{cases} \text{commutative law} & \alpha \underline{a} = \underline{a} \alpha \\ \text{associative law} & (\alpha \underline{a})\beta = \alpha(\underline{a}\beta) \\ \text{distributive law} & \alpha(\underline{a} + \underline{b}) = \alpha \underline{a} + \alpha \underline{b} \end{cases}$

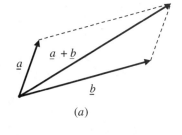

(a)

Figure 2.3 Pictorial representation of the addition and subtraction of two vectors.

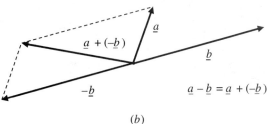

(b)

Two types of multiplication between vectors are the scalar product and the vector product. These are also called the *inner* (or *dot*) *product* and the *outer* (or *cross*) *product*. Their definitions are

Scalar product $\qquad \boxed{\underline{a} \cdot \underline{b} = ab \cos \psi}$ $\qquad$ (2.7)

Vector product $\qquad \boxed{\underline{a} \times \underline{b} = ab \sin \psi \, \hat{n}}$ $\qquad$ (2.8)

where $\hat{n}$ is a unit vector perpendicular to $\underline{a}$ and $\underline{b}$ subject to the right-hand rule (Figure 2.4), and ψ is the angle between $\underline{a}$ and $\underline{b}$. When a vector $\underline{b}$ is dotted with a unit vector $\hat{a}$, the scalar product yields the projection of $\underline{b}$ in the direction of the unit vector $\hat{a}$:

Projection of $\underline{b}$
in direction $\hat{a}$ $\qquad \boxed{\underline{b} \cdot \hat{a} = (b)(1) \cos \psi = b \cos \psi}$ $\qquad$ (2.9)

Also, when two vectors are perpendicular ($\psi = \pi/2$), the dot product is zero [$\cos (\pi/2) = 0$], and when two vectors are parallel ($\psi = 0$), the dot product is just the product of the magnitudes ($\cos 0 = 1$). The rules of algebra for the dot and cross products are:

Laws of algebra for
vector dot product
$\begin{cases} \text{commutative} & \underline{a} \cdot \underline{c} = \underline{c} \cdot \underline{a} \\ \text{associative} & \text{not possible} \\ \text{distributive} & \underline{a} \cdot (\underline{c} + \underline{w}) = \underline{a} \cdot \underline{c} + \underline{a} \cdot \underline{w} \end{cases}$

Laws of algebra for
vector cross product
$\begin{cases} \textit{not} \text{ commutative} & \underline{a} \times \underline{c} \neq \underline{c} \times \underline{a} \\ \text{associative} & (\underline{a} \times \underline{c}) \times \underline{w} = \underline{a} \times (\underline{c} \times \underline{w}) \\ \text{distributive} & \underline{a} \times (\underline{c} + \underline{w}) = \underline{a} \times \underline{c} + \underline{a} \times \underline{w} \end{cases}$

Performing the dot product is a convenient way to calculate the magnitude of a vector, as shown in Equations (2.10) and (2.11)

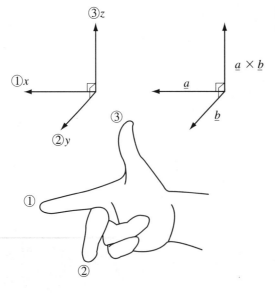

Figure 2.4 Definition of a right-handed coordinate system.

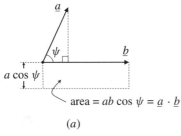

area $= ab \cos \psi = \underline{a} \cdot \underline{b}$

(a)

Figure 2.5 Pictorial representation of the multiplication of two vectors. (a) Scalar product. (b) Vector product.

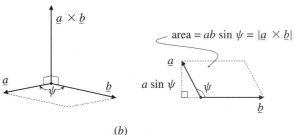

(b)

$$\underline{a} \cdot \underline{a} = a^2 \qquad (2.10)$$

and therefore

$$|\underline{a}| = \text{abs}\left(\sqrt{\underline{a} \cdot \underline{a}}\right) \qquad (2.11)$$

where abs( ) denotes the absolute value of the quantity in parentheses. By convention the magnitude of a vector is taken to be positive. The graphical interpretations of the two types of vector products are shown in Figure 2.5.

2.2.1.1 Coordinate Systems

When we introduced vectors we mentioned that an important vector property is that its magnitude and direction are independent of the coordinate system in which it is written. We now would like to elaborate on this concept, which will be important in understanding rheology.

Vectors, such as those that describe the forces on a body, exist independently of how we describe them mathematically. Imagine, for example, that you are leaning against a wall (Figure 2.6). Your hips are exerting a force on the wall. The vector direction in which you are exerting this force makes some angle ψ with the wall. This force has a magnitude (related to your weight and the angle ψ), and it has a direction (related to how exactly you are positioned with respect to the wall). All of this is true despite the fact that we have yet to describe the vector force with any type of mathematical expression.

If we wish to do a calculation involving the force you are exerting on the wall, we must translate that real force into a mathematical expression that we can use in our calculations. What we would typically do is to choose some reference coordinate system, probably composed of three mutually perpendicular axes, and to write down the coefficients of the force vector in the chosen coordinate system. If we write down all the forces in a problem in the same coordinate system, we can then solve the problem.

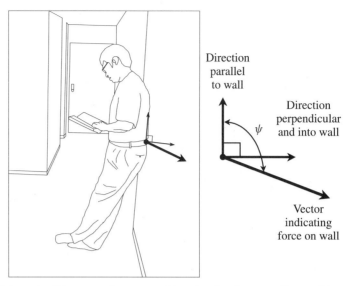

Figure 2.6 Schematic of the vector force exerted by a leaning figure, as discussed in text.

Now we ask, what is a coordinate system? Can any vectors be chosen as the bases of the coordinate system? Must they be mutually perpendicular? Must they be unit vectors? We will answer these questions by laying down two rules for coordinate systems, also called coordinate bases (Table 2.1).

The coordinate system with which we are most familiar is the Cartesian coordinate system, usually called $\hat{i}$, $\hat{j}$, and $\hat{k}$, or alternatively $\hat{e}_x$, $\hat{e}_y$, and $\hat{e}_z$ or $\hat{e}_1$, $\hat{e}_2$, and $\hat{e}_3$ (Figure 2.7). In the Cartesian coordinate system the basis vectors are three mutually perpendicular unit vectors (orthonormal basis vectors), and $\hat{i} = \hat{e}_x = \hat{e}_1$ points along the x-axis, $\hat{j} = \hat{e}_y = \hat{e}_2$ points along the y-axis, and $\hat{k} = \hat{e}_z = \hat{e}_3$ points along the z-axis. These basis vectors are constant and therefore point in the same direction at every point in space. Although the Cartesian system is the most commonly used coordinate system, we see from the rules listed in Table 2.1 that a coordinate system need not be composed of either unit vectors or mutually perpendicular basis vectors. We will use bases in which the basis vectors are not mutually perpendicular when we consider advanced rheological constitutive equations in Chapter 9.

One requirement of all coordinate systems is that the basis vectors be noncoplanar, i.e., that the three vectors not all lie in the same plane. This requirement can be understood by

TABLE 2.1
Rules for Coordinate Bases

1. In three-dimensional space, any vector may be expressed as a linear combination of three nonzero, noncoplanar vectors, which we will call basis vectors.

2. The choice of coordinate system is arbitrary. We usually choose the coordinate system to make the problem easier to solve. The coordinate system serves as a reference system, providing both units for magnitude and reference directions for vectors and other quantities.

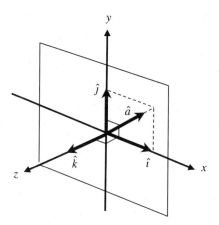

Figure 2.7 Cartesian coordinate system (x, y, z) and Cartesian basis vectors $(\hat{i}, \hat{j}, \hat{k})$. The vector $\hat{a}$ is in the xy-plane, as discussed in text.

imagining that we choose $\hat{i}$ and $\hat{j}$ as two of our basis vectors, and then as the third basis vector we choose a vector $\hat{a} = (1/\sqrt{2})\hat{i} + (1/\sqrt{2})\hat{j}$, which is parallel to the sum of $\hat{i}$ and $\hat{j}$ (Figure 2.7). Note that $\hat{a}$ is a unit vector and that all three proposed basis vectors lie in the same plane. The problem arises when we try to express a vector such as $\hat{k}$ in our chosen coordinate system. Since $\hat{k}$ is perpendicular to all three of the vectors in our chosen coordinate system, there is no combination of $\hat{i}$, $\hat{j}$, and $\hat{a}$ that will produce $\hat{k}$.

Mathematically the requirement that the three basis vectors be noncoplanar is the same as saying that the three vectors must be linearly independent. The requirement of being linearly independent means that the linear combination of the three vectors, here $\underline{a}$, $\underline{b}$, and $\underline{c}$, can be made to be zero, that is, the vectors can be added together with scalar coefficients $\alpha, \beta,$ and ζ, such that

$$\alpha\underline{a} + \beta\underline{b} + \zeta\underline{c} = 0 \tag{2.12}$$

if and only if $\alpha = \beta = \zeta = 0$. If scalars α, β, ζ can be found so that Equation (2.12) is satisfied but where one or more of these coefficients (α, β, ζ) is nonzero, then $\underline{a}$, $\underline{b}$, and $\underline{c}$ are linearly dependent, coplanar, and may not form a set of basis vectors.

Once a set of appropriate basis vectors is chosen (such as $\underline{a}$, $\underline{b}$, and $\underline{c}$), we know that any vector may be expressed as a linear combination of these three vectors. This means that for an arbitrary vector $\underline{v}$ we can find three scalars $\tilde{v}_1$, $\tilde{v}_2$, and $\tilde{v}_3$, such that

$$\underline{v} = \tilde{v}_1\underline{a} + \tilde{v}_2\underline{b} + \tilde{v}_3\underline{c} \tag{2.13}$$

Note that for the chosen basis of $\underline{a}$, $\underline{b}$, $\underline{c}$, the scalar coefficients $\tilde{v}_1$, $\tilde{v}_2$, $\tilde{v}_3$ are unique. Thus if we choose a different basis (oriented differently in space, or with different angles between the basis vectors, or composed of vectors with lengths different from those of the original basis vectors), different coefficients will be calculated. For example, for the orthonormal Cartesian basis $\hat{e}_1 = \hat{i}$, $\hat{e}_2 = \hat{j}$, $\hat{e}_3 = \hat{k}$, $\underline{v}$ can be written as

$$\underline{v} = v_1\hat{e}_1 + v_2\hat{e}_2 + v_3\hat{e}_3 \tag{2.14}$$

and in general, $v_1 \neq \tilde{v}_1$, $v_2 \neq \tilde{v}_2$, and $v_3 \neq \tilde{v}_3$.

EXAMPLE

Write the vector that describes the force $\underline{f}$ on the string on the right in Figure 2.8a. Express it in the three different coordinate systems: $\hat{e}_x, \hat{e}_y, \hat{e}_z$; $\hat{e}_{\bar{x}}, \hat{e}_{\bar{y}}, \hat{e}_{\bar{z}}$; and $\hat{e}_{\bar{x}}, \hat{e}_{\bar{y}}, \hat{e}_{\bar{z}}$.

SOLUTION

In the $\hat{e}_x, \hat{e}_y, \hat{e}_z$ coordinate system there are components of $\underline{f}$ in both the x- and y-directions. We can work these out in terms of the magnitude of the vector $\underline{f}$ using the usual trigonometric relations (Figure 2.9):

$$\underline{f} = f \cos \psi \, \hat{e}_x + f \sin \psi \, \hat{e}_y \tag{2.15}$$

The magnitude $f = |\underline{f}|$ can be found from a force balance between gravity affecting the mass m and the upward component of $\underline{f}$ for each string,

$$2f \sin \psi = mg \tag{2.16}$$

$$f = \frac{mg}{2 \sin \psi} \tag{2.17}$$

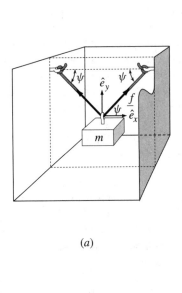

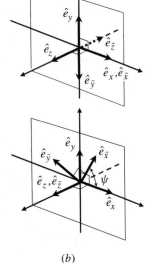

(a)

(b)

Figure 2.8 (a) Schematic for example problem showing a weight hanging between two walls. (b) Coordinate systems referred to in example.

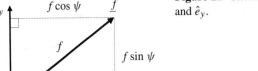

Figure 2.9 Relationship between $\underline{f}$ and the unit vectors $\hat{e}_x$ and $\hat{e}_y$.

Therefore $\underline{f}$ may be written as

$$\underline{f} = \frac{mg}{2\sin\psi}\left(\cos\psi\,\hat{e}_x + \sin\psi\,\hat{e}_y\right) \tag{2.18}$$

This same vector can be expressed in the $\hat{e}_{\bar{x}}, \hat{e}_{\bar{y}}, \hat{e}_{\bar{z}}$ and $\hat{e}_{\tilde{x}}, \hat{e}_{\tilde{y}}, \hat{e}_{\tilde{z}}$ coordinate systems by following the same procedure. Alternatively, we can write the new basis vectors in terms of the $\hat{e}_x, \hat{e}_y, \hat{e}_z$ basis vectors and substitute into Equation (2.18). For example, $\hat{e}_x, \hat{e}_y$, and $\hat{e}_z$ are related to $\hat{e}_{\bar{x}}, \hat{e}_{\bar{y}}$, and $\hat{e}_{\bar{z}}$ as follows:

$$\hat{e}_x = \hat{e}_{\bar{x}} \qquad \hat{e}_y = -\hat{e}_{\bar{y}} \qquad \hat{e}_z = -\hat{e}_{\bar{z}} \tag{2.19}$$

Using either procedure, the results for the two alternate coordinate systems are

$$\underline{f} = \frac{mg}{2\sin\psi}(\cos\psi\,\hat{e}_{\bar{x}} - \sin\psi\,\hat{e}_{\bar{y}}) \tag{2.20}$$

$$\underline{f} = \frac{mg}{2\sin\psi}\,\hat{e}_{\tilde{x}} \tag{2.21}$$

EXAMPLE

Express the vector force $\underline{f}$ from the previous example (Figure 2.8) in the two coordinate systems shown in Figure 2.10 (that is, with respect to the bases $\underline{a}, \underline{b}, \hat{e}_z$ and $\hat{e}_y, \hat{e}_{\bar{y}}, \hat{e}_z$). Note that $\underline{a}$ and $\underline{b}$ are not unit vectors, and $\hat{e}_y, \hat{e}_{\bar{y}}$, and $\hat{e}_z$ are not mutually perpendicular.

SOLUTION

In the previous example we found that $\underline{f}$ could be written in the $\hat{e}_x, \hat{e}_y, \hat{e}_z$ coordinate system as

$$\underline{f} = f\cos\psi\,\hat{e}_x + f\sin\psi\,\hat{e}_y \tag{2.22}$$

where $f = mg/(2\sin\psi)$. First we must express $\underline{f}$ using the following basis vectors, which are mutually orthogonal but not of unit length:

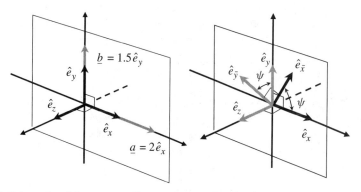

Figure 2.10 Schematic of the two coordinate systems referred to in example problem.

$$\underline{a} = 2\hat{e}_x, \qquad \underline{b} = 1.5\hat{e}_y, \qquad \hat{e}_z \tag{2.23}$$

Solving expressions (2.23) for $\hat{e}_x$ and $\hat{e}_y$ in terms of $\underline{a}$ and $\underline{b}$, we can substitute the results into Equation (2.22) and obtain the answer:

$$\underline{f} = f \cos \psi \, \hat{e}_x + f \sin \psi \, \hat{e}_y \tag{2.24}$$

$$= \left(\frac{f \cos \psi}{2} \right) \underline{a} + \left(\frac{f \sin \psi}{1.5} \right) \underline{b} \tag{2.25}$$

$$= \frac{mg}{2 \sin \psi} \left(\frac{\cos \psi}{2} \underline{a} + \frac{\sin \psi}{1.5} \underline{b} \right) \tag{2.26}$$

We can use the same technique to write $\underline{f}$ in terms of the second coordinate system, $\hat{e}_y, \hat{e}_{\bar{y}}, \hat{e}_z$ (see Figure 2.10), in which the basis vectors are all of unit length but not mutually orthogonal. First we must write $\hat{e}_{\bar{y}}$ in terms of $\hat{e}_x$ and $\hat{e}_y$. Then we will solve for $\hat{e}_x$ in terms of $\hat{e}_y$ and $\hat{e}_{\bar{y}}$, substitute the result into Equation (2.22), and simplify. Referring to Figure 2.10,

$$\hat{e}_{\bar{y}} = - \sin \psi \, \hat{e}_x + \cos \psi \, \hat{e}_y \tag{2.27}$$

Solving for $\hat{e}_x$ and substituting,

$$\hat{e}_x = - \frac{1}{\sin \psi} \left(\hat{e}_{\bar{y}} - \cos \psi \, \hat{e}_y \right) \tag{2.28}$$

$$\underline{f} = f \cos \psi \, \hat{e}_x + f \sin \psi \, \hat{e}_y \tag{2.29}$$

$$= -f \frac{\cos \psi}{\sin \psi} \hat{e}_{\bar{y}} + f \frac{\cos^2 \psi}{\sin \psi} \hat{e}_y + f \sin \psi \, \hat{e}_y \tag{2.30}$$

$$= \frac{mg}{2 \sin \psi} \left(- \cot \psi \, \hat{e}_{\bar{y}} + \sec \psi \, \hat{e}_y \right) \tag{2.31}$$

In the previous example, where we were working with orthonormal bases, it was easier to express vectors. The right angles in orthonormal bases allow us to relate vectors to their components directly, using trigonometric functions. When the basis vectors are not mutually orthogonal unit vectors (as in these last examples), we must do more work to get the final results. As we will see, vector multiplication is also much easier to carry out when the vectors are written with respect to orthonormal coordinate systems.

2.2.1.2 Vector Addition

We want to express vectors in a common coordinate system so that we can manipulate them. The advantage of expressing vectors in this way can be shown when we add two vectors. Consider the addition of two vectors $\underline{v}$ and $\underline{u}$ to produce $\underline{w}$. We may write each of these vectors with respect to the Cartesian coordinate system $\hat{e}_i$ ($i = 1, 2, 3$) as follows:

$$\underline{u} = u_1 \hat{e}_1 + u_2 \hat{e}_2 + u_3 \hat{e}_3 \tag{2.32}$$

$$\underline{v} = v_1 \hat{e}_1 + v_2 \hat{e}_2 + v_3 \hat{e}_3 \tag{2.33}$$

$$\underline{w} = w_1\hat{e}_1 + w_2\hat{e}_2 + w_3\hat{e}_3 \tag{2.34}$$

Adding $\underline{u}$ and $\underline{v}$ and factoring out the basis vectors yields

$$\underline{w} = \underline{u} + \underline{v} = (u_1 + v_1)\hat{e}_1 + (u_2 + v_2)\hat{e}_2 + (u_3 + v_3)\hat{e}_3 \tag{2.35}$$

Thus by comparing Equations 2.35 and 2.34 we see that the coefficients of the vector $\underline{w}$ are just the sums of the coefficients of the two other vectors:

$$w_1 = u_1 + v_1 \tag{2.36}$$

$$w_2 = u_2 + v_2 \tag{2.37}$$

$$w_3 = u_3 + v_3 \tag{2.38}$$

If we know what set of basis vectors we are using, it is a bit easier not to write the basis vectors each time. Thus the Cartesian version of the vector $\underline{v}$ can be written in one of two ways—as written in Equation (2.33) or by writing just the coefficients, v_1, v_2, and v_3 and understanding that the Cartesian coordinate system is being used. A convenient way of writing these coefficients is in matrix form:

$$\underline{v} = \begin{pmatrix} v_1 \\ v_2 \\ v_3 \end{pmatrix}_{123} = (\, v_1 \quad v_2 \quad v_3 \,)_{123} \tag{2.39}$$

We write the subscript 123 on the matrix version of $\underline{v}$ to remind us what coordinate system was used to define v_1, v_2, and v_3. Since we are using the matrix notation only to hold the vector coefficients, it is arbitrary whether we write these vectors as column or row vectors.

EXAMPLE

In the previous two examples we wrote the force on a string (see Figure 2.8) with respect to five different coordinate systems. Write each of these representations of the vector $\underline{f}$ in matrix notation.

SOLUTION

The five different representations of the vector $\underline{f}$ are

$$\underline{f} = \begin{pmatrix} \frac{mg \cot \psi}{2} \\ \frac{mg}{2} \\ 0 \end{pmatrix}_{xyz} \tag{2.40}$$

$$\underline{f} = \begin{pmatrix} \frac{mg \cot \psi}{2} \\ -\frac{mg}{2} \\ 0 \end{pmatrix}_{\bar{x}\bar{y}\bar{z}} \tag{2.41}$$

$$\underline{f} = \begin{pmatrix} \frac{mg}{2 \sin \psi} \\ 0 \\ 0 \end{pmatrix}_{\bar{x}\bar{y}\bar{z}} \tag{2.42}$$

$$\underline{f} = \begin{pmatrix} \frac{mg \cot \psi}{4} \\ \frac{mg}{3} \\ 0 \end{pmatrix}_{abz} \qquad (2.43)$$

$$\underline{f} = \begin{pmatrix} \frac{mg}{2 \sin^2 \psi} \\ \frac{-mg \cos \psi}{2 \sin^2 \psi} \\ 0 \end{pmatrix}_{y\bar{y}z} \qquad (2.44)$$

We see then that the coefficients associated with a vector will vary as the vector is expressed with respect to different coordinate systems. The magnitude and direction of the vector do not change, however. To completely describe a vector, both the identity of the basis vectors and the coefficients of the vector with respect to that basis are needed.

2.2.1.3 Vector Dot Product

Taking the dot product of two vectors is particularly easy when they are written with respect to the same orthonormal basis. For an example, we can take the dot product of the vectors $\underline{v}$ and $\underline{u}$:

$$\underline{v} \cdot \underline{u} = (v_1 \hat{e}_1 + v_2 \hat{e}_2 + v_3 \hat{e}_3) \cdot (u_1 \hat{e}_1 + u_2 \hat{e}_2 + u_3 \hat{e}_3) \qquad (2.45)$$

Using the distributive and commutative rules for the dot product we get

$$\underline{v} \cdot \underline{u} = v_1 u_1 \hat{e}_1 \cdot \hat{e}_1 + v_2 u_1 \hat{e}_2 \cdot \hat{e}_1 + v_3 u_1 \hat{e}_3 \cdot \hat{e}_1 + v_1 u_2 \hat{e}_1 \cdot \hat{e}_2$$
$$+ v_2 u_2 \hat{e}_2 \cdot \hat{e}_2 + v_3 u_2 \hat{e}_3 \cdot \hat{e}_2 + v_1 u_3 \hat{e}_1 \cdot \hat{e}_3 + v_2 u_3 \hat{e}_2 \cdot \hat{e}_3 + v_3 u_3 \hat{e}_3 \cdot \hat{e}_3 \qquad (2.46)$$

Because the basis vectors are orthonormal, however, when two like basis vectors are multiplied (e.g., $\hat{e}_1 \cdot \hat{e}_1$) the answer is one (cos $0 = 1$), whereas when two unlike unit vectors are multiplied (e.g., $\hat{e}_1 \cdot \hat{e}_2$), the answer is zero (cos $(\pi/2) = 0$). Thus expression (2.46) simplifies to

$$\underline{v} \cdot \underline{u} = v_1 u_1 + v_2 u_2 + v_3 u_3 \qquad (2.47)$$

This answer is a scalar, as required [see Equation (2.7)], that is, none of the basis vectors appears. We see that if we know two vectors with respect to the same orthonormal basis, we can easily calculate the dot product by multiplying the components term by term and adding them.

When we introduced the dot product of two vectors, we noted that the projection of a vector in a certain direction could be found by dotting the vector with a unit vector in the desired direction. For an orthonormal basis, the basis vectors themselves are unit vectors, and we can solve for the components of a vector with respect to the orthonormal basis by taking the following dot products:

$$v_1 = \underline{v} \cdot \hat{e}_1 \qquad (2.48)$$

$$v_2 = \underline{v} \cdot \hat{e}_2 \qquad (2.49)$$

$$v_3 = \underline{v} \cdot \hat{e}_3 \qquad (2.50)$$

This may also be confirmed by dotting Equation (2.33) with each of the unit vectors in turn and remembering that we are taking the three basis vectors $\hat{e}_i$ ($i = 1, 2, 3$) to be mutually perpendicular and of unit length. For example,

$$\hat{e}_1 \cdot \underline{v} = \hat{e}_1 \cdot (v_1\hat{e}_1 + v_2\hat{e}_2 + v_3\hat{e}_3) \tag{2.51}$$

$$= v_1 \tag{2.52}$$

EXAMPLE

Evaluate the dot product $\underline{f} \cdot \hat{e}_x$ for the vectors $\underline{f}$ and $\hat{e}_x$ shown in Figure 2.8. First use $\underline{f}$ expressed in the x, y, z coordinate system. Then calculate the same dot product first with $\underline{f}$ expressed in the $\bar{x}, \bar{y}, \bar{z}$ system and then with $\underline{f}$ expressed in the $y, \bar{y}, z$ system.

SOLUTION

In x, y, z-coordinates the dot product is straightforward to carry out since x, y, z is an orthonormal coordinate system and $\hat{e}_x$ is one of the basis vectors of this system:

$$\underline{f} \cdot \hat{e}_x = (f_x\hat{e}_x + f_y\hat{e}_y + f_z\hat{e}_z) \cdot \hat{e}_x \tag{2.53}$$

$$= f_x = \frac{mg \cot \psi}{2} \tag{2.54}$$

where f_x was obtained from Equation (2.40). Alternatively, we can write out the calculation using column-vector notation:

$$\underline{f} \cdot \hat{e}_x = \begin{pmatrix} \frac{mg \cot \psi}{2} \\ \frac{mg}{2} \\ 0 \end{pmatrix}_{xyz} \cdot \begin{pmatrix} 1 \\ 0 \\ 0 \end{pmatrix}_{xyz} \tag{2.55}$$

$$= \frac{mg \cot \psi}{2} \tag{2.56}$$

To carry out this calculation in $\bar{x}, \bar{y}, \bar{z}$ coordinates, we must first determine the coefficients of $\underline{f}$ and $\hat{e}_x$ in that system. Once the $\bar{x}, \bar{y}, \bar{z}$ coefficients are obtained, the procedure matches that employed before since $\bar{x}, \bar{y}, \bar{z}$ coordinates are orthonormal. The result is, of course, the same,

$$\underline{f} \cdot \hat{e}_x = (f_{\bar{x}}\hat{e}_{\bar{x}} + f_{\bar{y}}\hat{e}_{\bar{y}} + f_{\bar{z}}\hat{e}_{\bar{z}}) \cdot \hat{e}_x \tag{2.57}$$

$$= \begin{pmatrix} \frac{mg}{2\sin\psi} \\ 0 \\ 0 \end{pmatrix}_{\bar{x}\bar{y}\bar{z}} \cdot \begin{pmatrix} \cos\psi \\ -\sin\psi \\ 0 \end{pmatrix}_{\bar{x}\bar{y}\bar{z}} \tag{2.58}$$

$$= \frac{mg \cot \psi}{2} \tag{2.59}$$

The basis vectors of the $y, \bar{y}, z$ system are not mutually orthogonal. This complicates any calculation carried out in this coordinate system. First we write $\underline{f}$ in the $y, \bar{y}, z$

coordinate system, and then we carry out the dot product using the distributive rule of algebra,

$$f = f_y \hat{e}_y + f_{\bar{y}} \hat{e}_{\bar{y}} + f_z \hat{e}_z \tag{2.60}$$

$$f \cdot \hat{e}_x = (f_y \hat{e}_y + f_{\bar{y}} \hat{e}_{\bar{y}} + f_z \hat{e}_z) \cdot \hat{e}_x \tag{2.61}$$

$$= f_y (\hat{e}_y \cdot \hat{e}_x) + f_{\bar{y}} (\hat{e}_{\bar{y}} \cdot \hat{e}_x) + f_z (\hat{e}_z \cdot \hat{e}_x) \tag{2.62}$$

Since $\hat{e}_x$, $\hat{e}_y$, and $\hat{e}_z$ are mutually perpendicular, the first and last terms vanish. We are left with one dot product:

$$f \cdot \hat{e}_x = f_{\bar{y}} (\hat{e}_{\bar{y}} \cdot \hat{e}_x) \tag{2.63}$$

where $f_{\bar{y}}$ is the $\bar{y}$ coefficient of f when that vector is written in the y, $\bar{y}$, z coordinate system [see Equation (2.44)],

$$f_{\bar{y}} = -\frac{mg \cos \psi}{2 \sin^2 \psi} \tag{2.64}$$

We can evaluate the vector dot product on the right side of Equation (2.63) by recalling the definition of dot product (Equation (2.7)):

$$\hat{e}_{\bar{y}} \cdot \hat{e}_x = |\hat{e}_{\bar{y}}||\hat{e}_x| \cos \left(\begin{array}{c} \text{angle between} \\ \text{two vectors} \end{array} \right) \tag{2.65}$$

$$= (1)(1) \cos \left(\frac{\pi}{2} + \psi \right) = -\sin \psi \tag{2.66}$$

where we have used Figure 2.10 and a trigonometric identity to evaluate the cosine function. Substituting this result into Equation (2.63), we obtain the correct answer:

$$f \cdot \hat{e}_x = f_{\bar{y}} (\hat{e}_{\bar{y}} \cdot \hat{e}_x) \tag{2.67}$$

$$= \frac{mg \cos \psi}{2 \sin \psi} \tag{2.68}$$

This last example, which dealt with evaluating a dot product for a vector written with respect to a nonorthonormal basis, was simplified by the orthogonality of $\hat{e}_x$ with two of the basis vectors ($\hat{e}_y$ and $\hat{e}_z$). In the general case of calculating a dot product with vectors written with respect to nonorthonormal bases, there would be six independent dot products among the basis vectors to evaluate geometrically or trigonometrically [see Equation (2.46); recall that the dot product is commutative]. This is why we prefer to carry out vector calculations by writing the vectors with respect to orthonormal bases.

2.2.1.4 Vector Cross Product

Cross products may also be written simply using an orthonormal basis [239]. For the basis vectors, any cross products of like vectors (e.g., $\hat{e}_1 \times \hat{e}_1$) vanish since $\sin \psi = 0$ [Equation (2.8)]. For unlike vectors, following the right-hand rule (Figure 2.7) we can see that

$$\hat{e}_1 \times \hat{e}_2 = \hat{e}_3 \tag{2.69}$$

$$\hat{e}_2 \times \hat{e}_3 = \hat{e}_1 \tag{2.70}$$

$$\hat{e}_3 \times \hat{e}_1 = \hat{e}_2 \tag{2.71}$$

$$\hat{e}_3 \times \hat{e}_2 = -\hat{e}_1 \tag{2.72}$$

$$\hat{e}_2 \times \hat{e}_1 = -\hat{e}_3 \tag{2.73}$$

$$\hat{e}_1 \times \hat{e}_3 = -\hat{e}_2 \tag{2.74}$$

Notice that $\hat{e}_i \times \hat{e}_j = +\hat{e}_k$ when ijk are permutations of 123 that are produced by removing the last digit and placing it in front. This is called a *cyclic* permutation. Likewise $\hat{e}_i \times \hat{e}_j = -\hat{e}_k$ when ijk are cyclic permutations of 321.

For the cross product of arbitrary vectors we write each vector in terms of an orthonormal basis and carry out the individual cross products. Remember that the cross product of parallel vectors is zero.

$$\underline{v} \times \underline{u} = (v_1\hat{e}_1 + v_2\hat{e}_2 + v_3\hat{e}_3) \times (u_1\hat{e}_1 + u_2\hat{e}_2 + u_3\hat{e}_3) \tag{2.75}$$

$$= v_1u_1\hat{e}_1 \times \hat{e}_1 + v_1u_2\hat{e}_1 \times \hat{e}_2 + v_1u_3\hat{e}_1 \times \hat{e}_3 + v_2u_1\hat{e}_2 \times \hat{e}_1$$

$$+ v_2u_2\hat{e}_2 \times \hat{e}_2 + v_2u_3\hat{e}_2 \times \hat{e}_3 + v_3u_1\hat{e}_3 \times \hat{e}_1 + v_3u_2\hat{e}_3 \times \hat{e}_2 \tag{2.76}$$

$$+ v_3u_3\hat{e}_3 \times \hat{e}_3$$

$$= \hat{e}_1(v_2u_3 - v_3u_2) - \hat{e}_2(v_1u_3 - v_3u_1) + \hat{e}_3(v_1u_2 - v_2u_1) \tag{2.77}$$

This operation can be summarized using the matrix operation of taking a determinant, denoted by enclosing a square array in vertical lines and defined for a 3×3 matrix:

$$\det|Z| = \begin{vmatrix} Z_{11} & Z_{12} & Z_{13} \\ Z_{21} & Z_{22} & Z_{23} \\ Z_{31} & Z_{32} & Z_{33} \end{vmatrix} \tag{2.78}$$

$$= Z_{11}(Z_{22}Z_{33} - Z_{23}Z_{32}) - Z_{12}(Z_{21}Z_{33} - Z_{23}Z_{31})$$

$$+ Z_{13}(Z_{21}Z_{32} - Z_{22}Z_{31}) \tag{2.79}$$

To carry out the cross product of two arbitrary vectors, we construct a 3×3 matrix using the basis vectors and the coefficients of the vectors as follows:

$$\underline{v} \times \underline{u} = \begin{vmatrix} \hat{e}_1 & \hat{e}_2 & \hat{e}_3 \\ v_1 & v_2 & v_3 \\ u_1 & u_2 & u_3 \end{vmatrix} \tag{2.80}$$

$$= \hat{e}_1(v_2u_3 - v_3u_2) - \hat{e}_2(v_1u_3 - v_3u_1) + \hat{e}_3(v_1u_2 - v_2u_1) \tag{2.81}$$

$$= \begin{pmatrix} (v_2u_3 - v_3u_2) \\ (v_3u_1 - v_1u_3) \\ (v_1u_2 - v_2u_1) \end{pmatrix}_{123} \tag{2.82}$$

which is the same result as before. Again the subscript 123 denotes that this 3×1 array holds the coefficients of the vector $\underline{v} \times \underline{u}$ with respect to the basis vectors $\hat{e}_1$, $\hat{e}_2$, and $\hat{e}_3$.

When using this kind of mathematics in rheology we will often be generating even more complex expressions than those shown. To take advantage of the order in this chaos of letters and subscripts, we use a short-hand called *Einstein* notation.

2.2.2 VECTOR EINSTEIN NOTATION

Einstein notation, also called *summation convention*, is a way of writing the effects of operations on vectors in a compact and easier-to-read format. To use Einstein notation most effectively, the vectors must be written with respect to orthonormal basis vectors. We begin with the basic expression for a vector written with respect to an orthonormal basis:

$$\underline{v} = v_1\hat{e}_1 + v_2\hat{e}_2 + v_3\hat{e}_3 \tag{2.83}$$

This can be written more compactly as

$$\underline{v} = \sum_{i=1}^{3} v_i\hat{e}_i \tag{2.84}$$

A further simplification can be made by leaving out the summation sign and understanding that when an index (i in the example) is repeated, a summation from 1 to 3 over that index is understood,

$$\underline{v} = v_i\hat{e}_i \tag{2.85}$$

The power of Einstein notation is harnessed when expressing the results of the multiplication of vectors and, later, tensors. Consider the dot product of two vectors $\underline{v}$ and $\underline{u}$, which was carried out in detail in Equation (2.46). This becomes

$$\underline{v} \cdot \underline{u} = v_i\hat{e}_i \cdot u_j\hat{e}_j = v_iu_j\,\hat{e}_i \cdot \hat{e}_j \tag{2.86}$$

Remember that the summation signs over indices i and j are understood since these are repeated. If we expanded Equation (2.86) by reinstating the summations, the result would be

$$\underline{v} \cdot \underline{u} = \sum_{i=1}^{3}\sum_{j=1}^{3} v_iu_j\,\hat{e}_i \cdot \hat{e}_j \tag{2.87}$$

Also note that we used different indices for the two vectors. This is important since there are two summations in this expression—one for each vector. If we use the same index (e.g., i) for both summations, we incorrectly reduce the number of summations to one.

Now, to carry out the dot product it is helpful to use the Kronecker delta δ_{ij}:

$$\text{Kronecker delta} \qquad \boxed{\delta_{ij} \equiv \begin{cases} 1 & i = j \\ 0 & i \neq j \end{cases}} \tag{2.88}$$

This quantity expresses exactly the result of $\hat{e}_i \cdot \hat{e}_j$, where $\hat{e}_i$ and $\hat{e}_j$ are any of the three orthonormal basis vectors,

$$\hat{e}_i \cdot \hat{e}_j = \delta_{ij} \qquad (2.89)$$

Note that the subscripts i and j can take on any of the values 1, 2, or 3. When the Kronecker delta appears in an expression, it tells us that the two indices associated with the delta are now redundant as a result of some action. In the example outlined [Equation (2.89)], the indices are made redundant as a result of dot multiplication of orthonormal basis vectors. To simplify an Einstein expression in which the Kronecker delta appears, we replace the two subscript indices with a single index and drop the delta:

$$\underline{v} \cdot \underline{u} = v_i \hat{e}_i \cdot u_j \hat{e}_j \qquad (2.90)$$

$$= v_i u_j \, \hat{e}_i \cdot \hat{e}_j \qquad (2.91)$$

$$= v_i u_j \, \delta_{ij} \qquad (2.92)$$

$$= v_i u_i \qquad (2.93)$$

Note that the choice of the index i for the final scalar result of this operation is completely arbitrary. We could just as well have written $v_j u_j$ or $v_m u_m$. Remember that the indices serve to remind us that a summation is required, and they identify which terms change as the summation is performed; the specific letter used as the index is arbitrary. The final result in Equation (2.93) is the same result we derived in Equation (2.47):

$$\underline{v} \cdot \underline{u} = v_i u_i \qquad (2.94)$$

$$= \sum_{i=1}^{3} v_i u_i \qquad (2.95)$$

$$= v_1 u_1 + v_2 u_2 + v_3 u_3 \qquad (2.96)$$

The vector cross product ($\underline{v} \times \underline{u}$) can also be expressed in Einstein notation. To do this we must use a new expression, the epsilon permutation symbol ε_{ijk}:

Epsilon permutation symbol $\qquad \varepsilon_{ijk} \equiv \begin{cases} 1 & ijk = 123, \ 231, \ \text{or } 312 \\ -1 & ijk = 321, \ 213, \ \text{or } 132 \\ 0 & i = j, \ j = k, \ \text{or } k = i \end{cases} \qquad (2.97)$

The combinations of indices that give $+1$ are called even permutations of 123, and the combinations that give -1 are called odd permutations of 123. Using this function, the cross product can be written in Einstein notation as

$$\underline{v} \times \underline{u} = v_p \hat{e}_p \times u_s \hat{e}_s \qquad (2.98)$$

$$= v_p u_s \, \hat{e}_p \times \hat{e}_s \qquad (2.99)$$

$$= v_p u_s \varepsilon_{psj} \hat{e}_j \qquad (2.100)$$

Remember that summing of repeated indices (p, s, and j in this case) is assumed; thus there are three summations understood in Equation (2.100). The reader can verify that this result is the same as that obtained by the determinant method described in Section 2.2.1.4. As with all aspects of Einstein notation, this method of expressing the dot and cross products is limited to orthonormal bases.[2]

2.3 Tensors

Now that we have reviewed scalar and vector operations, we move on to the more complex quantities called *tensors*. A tensor is a mathematical entity related to vectors, but it is not easy to represent a tensor graphically. A tensor is better explained first by mathematical description and then by showing what it does.

A tensor is an ordered pair of coordinate directions. It is also called the *indeterminate vector product*. The simplest tensor is called a *dyad* or *dyadic product*, and it is written as two vectors side by side,

$$\text{Tensor} \qquad \boxed{\underline{\underline{A}} = \underline{a}\,\underline{b}} \qquad (2.101)$$

As you see, in our notation tensors will appear with two underbars. This indicates that they are of higher complexity, or order, than vectors. The tensors we are discussing are *second order* tensors, and we will have more to say about tensor order in Section 2.3.4. While scalars and vectors are physical entities (magnitude, magnitude and direction), tensors are *operators* (magnitude and two or more directions). We will discuss this in more detail after we familiarize ourselves with the algebraic rules for second-order tensors.

2.3.1 TENSOR RULES OF ALGEBRA

To understand tensors,[3] we must first know their rules of algebra. The indeterminate vector product that forms a tensor is not commutative, although it is associative and distributive:

$$\text{Laws of algebra for indeterminate vector product} \begin{cases} \textit{not } \text{commutative} & \underline{a}\,\underline{b} \neq \underline{b}\,\underline{a} \\ \text{associative} & (\underline{a}\,\underline{b})\underline{c} = \underline{a}(\underline{b}\,\underline{c}) \\ \text{distributive} & \underline{a}(\underline{b} + \underline{c}) = \underline{a}\,\underline{b} + \underline{a}\,\underline{c} \\ & (\underline{a} + \underline{b})(\underline{c} + \underline{d}) = \underline{a}\,\underline{c} + \underline{a}\,\underline{d} + \underline{b}\,\underline{c} + \underline{b}\,\underline{d} \end{cases}$$

Scalar multiplication of a tensor follows the same rules as scalar multiplication of a vector, such as $\alpha\underline{a}\,\underline{b} = \underline{a}(\alpha\underline{b}) = (\alpha\underline{a})\underline{b}$.

2.3.1.1 Tensor Addition

Adding and subtracting tensors is also possible, but unlike the case of adding vectors, it is not possible to graphically illustrate these. As we will show, the easiest way to carry out the addition of two tensors is to write them with respect to a common basis and to collect terms, as we did when adding two vectors.

---

[2] See Chapter 9 for a discussion of nonorthonormal bases.

[3] Throughout the text we will use the term tensor to mean second-order tensor.

Recall that any vector may be expressed as the linear combination of any three basis vectors. If we express two vectors this way and then take the indeterminate vector product to make a tensor (following the rules of algebra outlined earlier), we see that to express a tensor in the most general form we need a linear combination of nine pairs of coordinate basis vectors:

$$\underline{A} = \underline{u}\,\underline{v} \tag{2.102}$$

$$= (u_1\hat{e}_1 + u_2\hat{e}_2 + u_3\hat{e}_3)(v_1\hat{e}_1 + v_2\hat{e}_2 + v_3\hat{e}_3) \tag{2.103}$$

$$= u_1v_1\hat{e}_1\hat{e}_1 + u_1v_2\hat{e}_1\hat{e}_2 + u_1v_3\hat{e}_1\hat{e}_3 + u_2v_1\hat{e}_2\hat{e}_1$$

$$\quad + u_2v_2\hat{e}_2\hat{e}_2 + u_2v_3\hat{e}_2\hat{e}_3 + u_3v_1\hat{e}_3\hat{e}_1 + u_3v_2\hat{e}_3\hat{e}_2 + u_3v_3\hat{e}_3\hat{e}_3 \tag{2.104}$$

where we have used an orthonormal basis for convenience. Since the indeterminate vector product is not commutative, the coefficients of the terms $\hat{e}_1\hat{e}_2$ and $\hat{e}_2\hat{e}_1$ cannot be combined and must remain distinct. Thus, for any chosen coordinate system we can always express a tensor as a linear combination of nine ordered pairs of basis vectors, as outlined before and shown here:

$$\underline{\underline{A}} = A_{11}\hat{e}_1\hat{e}_1 + A_{12}\hat{e}_1\hat{e}_2 + A_{13}\hat{e}_1\hat{e}_3 + A_{21}\hat{e}_2\hat{e}_1 + A_{22}\hat{e}_2\hat{e}_2$$

$$\quad + A_{23}\hat{e}_2\hat{e}_3 + A_{31}\hat{e}_3\hat{e}_1 + A_{32}\hat{e}_3\hat{e}_2 + A_{33}\hat{e}_3\hat{e}_3 \tag{2.105}$$

To add two tensors we write each out with respect to the same basis vectors and add, grouping like terms:

$$\underline{\underline{C}} = \underline{\underline{A}} + \underline{\underline{B}} \tag{2.106}$$

$$= A_{11}\hat{e}_1\hat{e}_1 + A_{12}\hat{e}_1\hat{e}_2 + A_{13}\hat{e}_1\hat{e}_3 + A_{21}\hat{e}_2\hat{e}_1 + A_{22}\hat{e}_2\hat{e}_2$$

$$\quad + A_{23}\hat{e}_2\hat{e}_3 + A_{31}\hat{e}_3\hat{e}_1 + A_{32}\hat{e}_3\hat{e}_2 + A_{33}\hat{e}_3\hat{e}_3$$

$$\quad + B_{11}\hat{e}_1\hat{e}_1 + B_{12}\hat{e}_1\hat{e}_2 + B_{13}\hat{e}_1\hat{e}_3 + B_{21}\hat{e}_2\hat{e}_1 + B_{22}\hat{e}_2\hat{e}_2$$

$$\quad + B_{23}\hat{e}_2\hat{e}_3 + B_{31}\hat{e}_3\hat{e}_1 + B_{32}\hat{e}_3\hat{e}_2 + B_{33}\hat{e}_3\hat{e}_3 \tag{2.107}$$

$$= (A_{11} + B_{11})\hat{e}_1\hat{e}_1 + (A_{12} + B_{12})\hat{e}_1\hat{e}_2 + (A_{13} + B_{13})\hat{e}_1\hat{e}_3$$

$$\quad + (A_{21} + B_{21})\hat{e}_2\hat{e}_1 + (A_{22} + B_{22})\hat{e}_2\hat{e}_2 + (A_{23} + B_{23})\hat{e}_2\hat{e}_3$$

$$\quad + (A_{31} + B_{31})\hat{e}_3\hat{e}_1 + (A_{32} + B_{32})\hat{e}_3\hat{e}_2 + (A_{33} + B_{33})\hat{e}_3\hat{e}_3 \tag{2.108}$$

As was the case when adding vectors, when two tensors are expressed with respect to the same coordinate system, they may be added by simply adding the appropriate coefficients together.

Nine coordinates are unwieldy, and it is common to write the nine coefficients of a tensor in matrix form:

$$\underline{\underline{A}} = \underline{u}\,\underline{v} \tag{2.109}$$

$$= u_1v_1\hat{e}_1\hat{e}_1 + u_1v_2\hat{e}_1\hat{e}_2 + u_1v_3\hat{e}_1\hat{e}_3 + u_2v_1\hat{e}_2\hat{e}_1$$

$$\quad + u_2v_2\hat{e}_2\hat{e}_2 + u_2v_3\hat{e}_2\hat{e}_3 + u_3v_1\hat{e}_3\hat{e}_1 + u_3v_2\hat{e}_3\hat{e}_2$$

$$\quad + u_3v_3\hat{e}_3\hat{e}_3 \tag{2.110}$$

$$= \begin{pmatrix} u_1 v_1 & u_1 v_2 & u_1 v_3 \\ u_2 v_1 & u_2 v_2 & u_2 v_3 \\ u_3 v_1 & u_3 v_2 & u_3 v_3 \end{pmatrix}_{123} \tag{2.111}$$

More generally,

$$\underline{\underline{A}} = \begin{pmatrix} A_{11} & A_{12} & A_{13} \\ A_{21} & A_{22} & A_{23} \\ A_{31} & A_{32} & A_{33} \end{pmatrix}_{123} \tag{2.112}$$

where the subscripts of each coefficient indicate which two basis vectors are associated with the scalar coefficient and in what order. Because tensors are made of ordered pairs, A_{12} will not generally be equal to A_{21}, and so on. Further, in the convention used here and followed throughout the text, the first index indicates the row in which the coefficient is placed, and the second index indicates the column.

2.3.1.2 Tensor Dot Product

There exists a dot product between two dyads. This is carried out by dotting the two vectors that are closest together:

Tensor dot product

$$\boxed{\begin{aligned} \underline{a}\,\underline{b} \cdot \underline{c}\,\underline{d} &= \underline{a}(\underline{b} \cdot \underline{c})\,\underline{d} \\ &= (\underline{b} \cdot \underline{c})\underline{a}\,\underline{d} \end{aligned}} \tag{2.113}$$

Since scalar multiplication is commutative and $(\underline{b} \cdot \underline{c})$ is a scalar, we can move this quantity around to the front, as shown in Equation (2.113). We can also see from the example that the dot product of two tensors is a tensor, but the overall magnitude of the resulting tensor differs from the magnitude of either of the original tensors (since the magnitude now involves $\underline{b} \cdot \underline{c}$), and only certain vector directions ($\underline{a}$ and $\underline{d}$ in the example) are preserved [the directions of $\underline{b}$ and $\underline{c}$ do not appear in Equation (2.113)].

Similarly we may dot a vector with a tensor:

$$\underline{a} \cdot \underline{b}\,\underline{c} = (\underline{a} \cdot \underline{b})\underline{c} \equiv \underline{w} \tag{2.114}$$

The result $\underline{w}$ is a vector pointing in a direction that was part of the original tensor (parallel to $\underline{c}$), but the magnitude of $\underline{w}$ differs from the magnitudes of any of the original vectors ($\underline{a}$, $\underline{b}$, and $\underline{c}$ above). Neither the dot product of two tensors nor the dot product of a vector with a tensor is commutative. Both are associative and distributive, however,

Laws of algebra for tensor dot product

$$\begin{cases} \textit{not} \text{ commutative} & \underline{a}\,\underline{b} \cdot \underline{c}\,\underline{d} \neq \underline{c}\,\underline{d} \cdot \underline{a}\,\underline{b} \\ & \underline{\underline{A}} \cdot \underline{\underline{B}} \neq \underline{\underline{B}} \cdot \underline{\underline{A}} \\ \text{associative} & (\underline{a}\,\underline{b} \cdot \underline{c}\,\underline{d}) \cdot \underline{f}\,\underline{g} = \underline{a}\,\underline{b} \cdot (\underline{c}\,\underline{d} \cdot \underline{f}\,\underline{g}) \\ & (\underline{\underline{A}} \cdot \underline{\underline{B}}) \cdot \underline{\underline{C}} = \underline{\underline{A}} \cdot (\underline{\underline{B}} \cdot \underline{\underline{C}}) \\ \text{distributive} & \underline{a}\,\underline{b} \cdot (\underline{c}\,\underline{m} + \underline{n}\,\underline{w}) = (\underline{a}\,\underline{b} \cdot \underline{c}\,\underline{m}) + (\underline{a}\,\underline{b} \cdot \underline{n}\,\underline{w}) \\ & \underline{\underline{A}} \cdot (\underline{\underline{D}} + \underline{\underline{M}}) = \underline{\underline{A}} \cdot \underline{\underline{D}} + \underline{\underline{A}} \cdot \underline{\underline{M}} \end{cases}$$

$$\text{Laws of algebra for vector dot product with a tensor} \begin{cases} \textit{not} \text{ commutative} & \underline{b} \cdot \underline{c}\,\underline{d} \neq \underline{c}\,\underline{d} \cdot \underline{b} \\ & \underline{b} \cdot \underline{\underline{M}} \neq \underline{\underline{M}} \cdot \underline{b} \\ \text{associative} & \underline{a} \cdot (\underline{c}\,\underline{d} \cdot \underline{w}) = (\underline{a} \cdot \underline{c}\,\underline{d}) \cdot \underline{w} \\ & \underline{a} \cdot (\underline{\underline{B}} \cdot \underline{w}) = (\underline{a} \cdot \underline{\underline{B}}) \cdot \underline{w} \\ \text{distributive} & \underline{d} \cdot (\underline{c}\,\underline{m} + \underline{n}\,\underline{w}) = (\underline{d} \cdot \underline{c}\,\underline{m}) + (\underline{d} \cdot \underline{n}\,\underline{w}) \\ & \underline{d} \cdot (\underline{\underline{A}} + \underline{\underline{C}}) = \underline{d} \cdot \underline{\underline{A}} + \underline{d} \cdot \underline{\underline{C}} \end{cases}$$

In the previous section we started to use matrix notation for writing tensor and vector components. It may seem like shaky ground to begin to use matrix notation for these new entities called tensors. After all, in linear algebra courses, properties and techniques associated with matrices are taught, and we have yet to show whether these properties and techniques are appropriate for matrices composed of vector and tensor coefficients.

To address this concern, consider the multiplication of a vector $\underline{v}$ and a tensor $\underline{\underline{A}}$. To calculate the new vector that results from this multiplication, we first write out the two quantities in terms of their coefficients with respect to an orthonormal basis, $\hat{e}_1, \hat{e}_2, \hat{e}_3$:

$$\underline{v} \cdot \underline{\underline{A}} = (v_1\hat{e}_1 + v_2\hat{e}_2 + v_3\hat{e}_3) \cdot (A_{11}\hat{e}_1\hat{e}_1 + A_{12}\hat{e}_1\hat{e}_2 + A_{13}\hat{e}_1\hat{e}_3 +$$

$$A_{21}\hat{e}_2\hat{e}_1 + A_{22}\hat{e}_2\hat{e}_2 + A_{23}\hat{e}_2\hat{e}_3 + A_{31}\hat{e}_3\hat{e}_1 + A_{32}\hat{e}_3\hat{e}_2 + A_{33}\hat{e}_3\hat{e}_3) \quad (2.115)$$

Now we use the distributive rule of the dot product to multiply. Recall that every time the indices of the orthonormal basis vectors match, their dot product is one (e.g., $\hat{e}_1 \cdot \hat{e}_1 = 1$); when the indices of two dotting basis vectors differ, their dot product is zero (e.g., $\hat{e}_1 \cdot \hat{e}_2 = 0$). This allows us to simplify this complex expression. Recall that we will dot the unit vector from $\underline{v}$ with the first (leftmost) unit vectors in the tensor dyads.

$$\underline{v} \cdot \underline{\underline{A}} = v_1(A_{11}\hat{e}_1 + A_{12}\hat{e}_2 + A_{13}\hat{e}_3) + v_2(A_{21}\hat{e}_1 + A_{22}\hat{e}_2 + A_{23}\hat{e}_3)$$

$$+ v_3(A_{31}\hat{e}_1 + A_{32}\hat{e}_2 + A_{33}\hat{e}_3) \quad (2.116)$$

$$= (v_1 A_{11} + v_2 A_{21} + v_3 A_{31})\hat{e}_1 + (v_1 A_{12} + v_2 A_{22} + v_3 A_{32})\hat{e}_2$$

$$+ (v_1 A_{13} + v_2 A_{23} + v_3 A_{33})\hat{e}_3 \quad (2.117)$$

This final expression follows the rules of matrix multiplication if we write $\underline{v}$ and $\underline{\underline{A}}$ as follows:

$$\underline{w} = \underline{v} \cdot \underline{\underline{A}} \quad (2.118)$$

$$= \begin{pmatrix} v_1 & v_2 & v_3 \end{pmatrix}_{123} \cdot \begin{pmatrix} A_{11} & A_{12} & A_{13} \\ A_{21} & A_{22} & A_{23} \\ A_{31} & A_{32} & A_{33} \end{pmatrix}_{123} \quad (2.119)$$

$$= \begin{pmatrix} w_1 & w_2 & w_3 \end{pmatrix}_{123} \quad (2.120)$$

where

$$w_1 = v_1 A_{11} + v_2 A_{21} + v_3 A_{31} \quad (2.121)$$

$$w_2 = v_1 A_{12} + v_2 A_{22} + v_3 A_{32} \tag{2.122}$$

$$w_3 = v_1 A_{13} + v_2 A_{23} + v_3 A_{33} \tag{2.123}$$

Thus, using matrix algebra to carry out the dot product on components of vectors with tensors written with respect to orthonormal bases is correct. Similarly, we can show that the dot product of two tensors also follows the rules of matrix multiplication.

2.3.1.3 Tensor Scalar Product

There is a scalar product of two tensors, which is defined as follows:

Tensor
scalar product
$$\boxed{\underline{a}\,\underline{b} : \underline{c}\,\underline{d} = (\underline{b} \cdot \underline{c})(\underline{a} \cdot \underline{d})} \tag{2.124}$$

This amounts to dotting the two closest vectors (the inner pair) and then dotting the two remaining vectors (the outer pair),

$$\underline{a}\ \overbrace{\underline{b} \cdot \underline{c}}^{b \cdot c}\ \underline{d}$$
$$\underbrace{}_{a \cdot d}$$

The rules of algebra for the scalar product of two tensors are summarized as follows:

Laws of algebra $\left\{ \begin{array}{ll} \text{commutative} & \underline{a}\,\underline{w} : \underline{n}\,\underline{d} = \underline{n}\,\underline{d} : \underline{a}\,\underline{w} \\ \text{associative} & \text{not possible} \\ \text{distributive} & \underline{b}\,\underline{a} : (\underline{m}\,\underline{n} + \underline{w}\,\underline{d}) = \underline{b}\,\underline{a} : \underline{m}\,\underline{n} + \underline{b}\,\underline{a} : \underline{w}\,\underline{d} \end{array} \right.$

for tensor
scalar product

2.3.2 TENSOR EINSTEIN NOTATION

The Einstein summation convention can be used to simplify the notation that goes along with tensor multiplication. A tensor in the summation notation requires a double sum:

$$\underline{\underline{A}} = \sum_{i=1}^{3} \sum_{j=1}^{3} A_{ij} \hat{e}_i \hat{e}_j \tag{2.125}$$

In Einstein notation this becomes

$$\underline{\underline{A}} = A_{ij} \hat{e}_i \hat{e}_j \tag{2.126}$$

Using i and j as the dummy indices is completely arbitrary. Each of the following expressions is equivalent:

$$\underline{\underline{A}} = A_{ij} \hat{e}_i \hat{e}_j = A_{mp} \hat{e}_m \hat{e}_p = A_{rs} \hat{e}_r \hat{e}_s \tag{2.127}$$

What is important in these expressions is that the first index on the symbol A matches the index on the first unit vector; the second index on the symbol A matches that on the second unit vector. An example of a different tensor, related to $\underline{\underline{A}}$, is the transpose of $\underline{\underline{A}}$, written as $\underline{\underline{A}}^T$. This is a tensor that has the same coefficients as $\underline{\underline{A}}$, but they are associated with different ordered pairs of basis vectors. In coefficient matrix notation, $\underline{\underline{A}}^T$ is the mirror image of $\underline{\underline{A}}$ across the main diagonal, that is, to obtain the matrix of coefficients of $\underline{\underline{A}}^T$, interchange the rows and the columns of $\underline{\underline{A}}$:

$$\underline{\underline{A}} = \begin{pmatrix} A_{11} & A_{12} & A_{13} \\ A_{21} & A_{22} & A_{23} \\ A_{31} & A_{32} & A_{33} \end{pmatrix}_{123} \tag{2.128}$$

$$\underline{\underline{A}}^T = \begin{pmatrix} A_{11} & A_{21} & A_{31} \\ A_{12} & A_{22} & A_{32} \\ A_{13} & A_{23} & A_{33} \end{pmatrix}_{123} \tag{2.129}$$

In the Einstein expression for $\underline{\underline{A}}^T$, the first index of A_{pk} is associated with the *second* basis vector in the basis vector dyad:

$$\underline{\underline{A}} = A_{pk}\hat{e}_p\hat{e}_k \tag{2.130}$$

$$\underline{\underline{A}}^T = A_{pk}\hat{e}_k\hat{e}_p \tag{2.131}$$

Again, since the letters used to indicate the implicit summations (in this case p and k) are arbitrary, $\underline{\underline{A}}^T$ can be written any number of ways. In particular, note the difference between the first two examples:

$$\underline{\underline{A}}^T = A_{ji}\hat{e}_i\hat{e}_j = A_{ij}\hat{e}_j\hat{e}_i = A_{sr}\hat{e}_r\hat{e}_s \tag{2.132}$$

Since tensors are written in terms of vectors, the methods outlined earlier for vector multiplication with the Einstein convention works just as well for the multiplication of two tensors or the multiplication of a vector with a tensor. When preparing to multiply tensors in Einstein notation, all different letters (e.g., i, j, p, or k) must be used for indexing the implicit summations.

$$\underline{\underline{A}} \cdot \underline{\underline{B}} = A_{ij}\hat{e}_i\hat{e}_j \cdot B_{pk}\hat{e}_p\hat{e}_k \tag{2.133}$$

$$= A_{ij}B_{pk}\,\hat{e}_i\hat{e}_j \cdot \hat{e}_p\hat{e}_k \tag{2.134}$$

$$= A_{ij}B_{pk}\,\hat{e}_i\delta_{jp}\hat{e}_k \tag{2.135}$$

$$= A_{ip}B_{pk}\,\hat{e}_i\hat{e}_k \tag{2.136}$$

The Kronecker delta in Equation (2.135) tells us that one index, j or p, is redundant. We therefore replace all of the j's with p's (we could have replaced p's with j's or both p and j with a third letter) to arrive at Equation (2.136).

In the final result there are three summations, two of which involve the unit vectors, and one that does not. To clarify the answer obtained, we can carry out the summation that does not involve the unit vectors, the summation over p:

$$A_{ip}B_{pk}\hat{e}_i\hat{e}_k = \sum_{i=1}^{3}\sum_{p=1}^{3}\sum_{k=1}^{3} A_{ip}B_{pk}\hat{e}_i\hat{e}_k \tag{2.137}$$

$$= \sum_{i=1}^{3} \sum_{k=1}^{3} (A_{i1} B_{1k} + A_{i2} B_{2k} + A_{i3} B_{3k}) \hat{e}_i \hat{e}_k \qquad (2.138)$$

Now the result of the multiplication of $\underline{A}$ and $\underline{\underline{B}}$ looks more like a usual tensor. Each coefficient term $(A_{i1} B_{1k} + A_{i2} B_{2k} + A_{i3} B_{3k})$ contains only two unknown subscripts that are associated with the basis dyad $(\hat{e}_i \hat{e}_k)$. This coefficient expression is more complicated than a basic tensor in that the coefficients each consist of the sum of three scalars, but we can see that the tensor is still composed of a double sum of scalar coefficients multiplying nine unique pairs of basis vectors.

Multiplication of a vector with a tensor can be carried out no matter whether the vector comes first $(\underline{v} \cdot \underline{\underline{A}})$ or the tensor comes first $(\underline{\underline{A}} \cdot \underline{v})$, but in general the answers in these two cases differ:

$$\underline{v} \cdot \underline{\underline{A}} = v_i \hat{e}_i \cdot A_{rs} \hat{e}_r \hat{e}_s \qquad (2.139)$$

$$= v_i A_{rs} \hat{e}_i \cdot \hat{e}_r \hat{e}_s \qquad (2.140)$$

$$= v_i A_{rs} \delta_{ir} \hat{e}_s \qquad (2.141)$$

$$= v_r A_{rs} \hat{e}_s \qquad (2.142)$$

$$\underline{\underline{A}} \cdot \underline{v} = A_{mp} \hat{e}_m \hat{e}_p \cdot v_j \hat{e}_j \qquad (2.143)$$

$$= A_{mp} v_j \hat{e}_m \hat{e}_p \cdot \hat{e}_j \qquad (2.144)$$

$$= A_{mp} v_j \hat{e}_m \delta_{pj} \qquad (2.145)$$

$$= A_{mp} v_p \hat{e}_m \qquad (2.146)$$

$$= v_p A_{mp} \hat{e}_m \qquad (2.147)$$

To convince yourself that these two answers differ, note that the index that is on the surviving unit vector is not in the same place on the tensor coefficient A_{ij} in the two final expressions.

EXAMPLE

What is the 2-component of $\underline{\underline{A}} \cdot \underline{v}$?

SOLUTION

From the example in the text we know that $\underline{\underline{A}} \cdot \underline{v} = v_p A_{mp} \hat{e}_m$. Expanding that into summation and then into vector matrix notation, we obtain

$$\underline{\underline{A}} \cdot \underline{v} = \sum_{p=1}^{3} \sum_{m=1}^{3} v_p A_{mp} \hat{e}_m \qquad (2.148)$$

$$= \sum_{p=1}^{3} v_p A_{1p} \hat{e}_1 + \sum_{p=1}^{3} v_p A_{2p} \hat{e}_2 + \sum_{p=1}^{3} v_p A_{3p} \hat{e}_3 \qquad (2.149)$$

$$= \begin{pmatrix} \sum_{p=1}^{3} v_p A_{1p} \\ \sum_{p=1}^{3} v_p A_{2p} \\ \sum_{p=1}^{3} v_p A_{3p} \end{pmatrix}_{123} \tag{2.150}$$

$$= \begin{pmatrix} v_1 A_{11} + v_2 A_{12} + v_3 A_{13} \\ v_1 A_{21} + v_2 A_{22} + v_3 A_{23} \\ v_1 A_{31} + v_2 A_{32} + v_3 A_{33} \end{pmatrix}_{123} \tag{2.151}$$

The 2-component of $\underline{\underline{A}} \cdot \underline{v}$ is the coefficient of $\hat{e}_2$ in Equation (2.151), that is, $v_1 A_{21} + v_2 A_{22} + v_3 A_{23}$. Equation (2.151) is a vector (note the $+$ signs between the terms). The three scalar components of this vector are long and spread out; do not confuse such vectors with tensors.

2.3.3 LINEAR VECTOR FUNCTIONS

We have been concerned, thus far, with familiarizing ourselves with tensors and with tensor algebra. We saved the discussion of what tensors are for until now. In rheology we use tensors because they are a convenient way to express linear vector functions. Linear vector functions arise naturally in the equations describing physical quantities such as linear and angular momentum, light traversing a medium, and stress in a body. We will now show how a tensor expresses a linear vector function through a simple calculation.

We are familiar with functions such as $y = f(x)$. This is a scalar function because it takes a scalar variable x and transforms it into a scalar variable y. A vector function, for example, $f(\underline{b})$, behaves analogously, transforming a vector $\underline{b}$ to another vector $\underline{a}$,

$$\underline{a} = f(\underline{b}) \tag{2.152}$$

A function, as described, is a general transformation. We can further qualify the type of function we are talking about by describing its mathematical properties. An important type of function is a linear function. A function is linear if for all vectors $\underline{a}, \underline{b}$ and scalars α, the following properties hold:

Definition of a linear function

$$\boxed{\begin{array}{c} f(\underline{a} + \underline{b}) = f(\underline{a}) + f(\underline{b}) \\[4pt] f(\alpha \underline{a}) = \alpha f(\underline{a}) \end{array}} \tag{2.153}$$

To show that a tensor embodies the properties of a linear vector function, consider the function $f(\underline{b})$ and expand the vector $\underline{b}$ with respect to an orthonormal basis,

$$\underline{a} = f(\underline{b}) \tag{2.154}$$

$$= f(b_1 \hat{e}_1 + b_2 \hat{e}_2 + b_3 \hat{e}_3) \tag{2.155}$$

Since f is a linear function, we can expand Equation (2.155) as follows:

$$\underline{a} = f(b_1 \hat{e}_1 + b_2 \hat{e}_2 + b_3 \hat{e}_3) \tag{2.156}$$

$$= b_1 f(\hat{e}_1) + b_2 f(\hat{e}_2) + b_3 f(\hat{e}_3) \tag{2.157}$$

We know that when f operates on a vector, a new vector is produced. Thus each of the expressions $f(\hat{e}_1)$, $f(\hat{e}_2)$, and $f(\hat{e}_3)$ is a vector. We do not know what these vectors are, but we can call them $\underline{v}$, $\underline{u}$, and $\underline{w}$,

$$\underline{v} \equiv f(\hat{e}_1) \tag{2.158}$$

$$\underline{u} \equiv f(\hat{e}_2) \tag{2.159}$$

$$\underline{w} \equiv f(\hat{e}_3) \tag{2.160}$$

Thus,

$$\underline{a} = b_1 \underline{v} + b_2 \underline{u} + b_3 \underline{w} \tag{2.161}$$

$$= \underline{v} b_1 + \underline{u} b_2 + \underline{w} b_3 \tag{2.162}$$

where we have used the commutative rule of scalar multiplication in writing the second expression.

The three scalars b_1, b_2, and b_3 are just the coefficients of $\underline{b}$ with respect to the orthonormal basis $\hat{e}_1$, $\hat{e}_2$, $\hat{e}_3$, and thus we can write them as [see Equations (2.48)–(2.50)]

$$b_1 = \hat{e}_1 \cdot \underline{b} \tag{2.163}$$

$$b_2 = \hat{e}_2 \cdot \underline{b} \tag{2.164}$$

$$b_3 = \hat{e}_3 \cdot \underline{b} \tag{2.165}$$

Substituting these expressions into Equation (2.162) gives

$$\underline{a} = \underline{v}\,\hat{e}_1 \cdot \underline{b} + \underline{u}\,\hat{e}_2 \cdot \underline{b} + \underline{w}\,\hat{e}_3 \cdot \underline{b} \tag{2.166}$$

Factoring out $\underline{b}$ by the distributive rule of the dot product gives us

$$\underline{a} = (\underline{v}\,\hat{e}_1 + \underline{u}\,\hat{e}_2 + \underline{w}\,\hat{e}_3) \cdot \underline{b} \tag{2.167}$$

The expression in parentheses is the sum of three dyadic products and thus is a tensor. If we call this sum $\underline{\underline{M}}$, then our final result is

$$\underline{\underline{M}} \equiv \underline{v}\,\hat{e}_1 + \underline{u}\,\hat{e}_2 + \underline{w}\,\hat{e}_3 \tag{2.168}$$

$$\underline{a} = f(\underline{b}) = \underline{\underline{M}} \cdot \underline{b} \tag{2.169}$$

We see that the linear vector function f acting on the vector $\underline{b}$ is the equivalent of dotting a tensor $\underline{\underline{M}}$ with $\underline{b}$. We will often use tensors in this manner, taking advantage of Einstein notation to simplify the calculations.

2.3.4 ASSOCIATED DEFINITIONS

As we study rheology we will have use for some specialized definitions that relate to tensors. We have already encountered the transpose of a tensor. Some other definitions are summarized next.

Identity tensor $\underline{\underline{I}}$:

$$\underline{\underline{I}} \equiv \hat{e}_1\hat{e}_1 + \hat{e}_2\hat{e}_2 + \hat{e}_3\hat{e}_3 \tag{2.170}$$

$$= \hat{e}_i\hat{e}_i \tag{2.171}$$

$$= \begin{pmatrix} 1 & 0 & 0 \\ 0 & 1 & 0 \\ 0 & 0 & 1 \end{pmatrix}_{123} \tag{2.172}$$

This tensor has the same properties as the identity matrix that it resembles. For example, $\underline{\underline{I}} \cdot \underline{v} = \underline{v} \cdot \underline{\underline{I}} = \underline{v}$ and $\underline{\underline{I}} \cdot \underline{\underline{B}} = \underline{\underline{B}} \cdot \underline{\underline{I}} = \underline{\underline{B}}$, where $\underline{v}$ is any vector and $\underline{\underline{B}}$ is any tensor. $\underline{\underline{I}}$ is written as in Equation (2.172) for any orthonormal basis. Any tensor proportional to $\underline{\underline{I}}$ is an isotropic tensor. The linear vector function represented by an isotropic tensor has the same effect in all directions.

Zero tensor $\underline{\underline{0}}$: The zero tensor is a quantity of tensor order that has all coefficients equal to zero in any coordinate system. It is a linear vector function that transforms any vector to the zero vector

$$\underline{\underline{0}} \equiv \begin{pmatrix} 0 & 0 & 0 \\ 0 & 0 & 0 \\ 0 & 0 & 0 \end{pmatrix} \tag{2.173}$$

Magnitude of a tensor $|\underline{\underline{A}}|$:

$$|\underline{\underline{A}}| \equiv +\sqrt{\frac{\underline{\underline{A}} : \underline{\underline{A}}}{2}} \tag{2.174}$$

The magnitude of a tensor is a scalar that is associated with a tensor. The value of the magnitude does not depend on the coordinate system in which the tensor is written.

Symmetric and antisymmetric tensors: A tensor is said to be symmetric if

$$\underline{\underline{A}} = \underline{\underline{A}}^T \tag{2.175}$$

In Einstein notation this means that $A_{sm} = A_{ms}$. An example of the matrix of coefficients of a symmetric tensor is

$$\begin{pmatrix} 1 & 2 & 3 \\ 2 & 4 & 5 \\ 3 & 5 & 6 \end{pmatrix}_{123} \tag{2.176}$$

A tensor is said to be antisymmetric if

$$\underline{\underline{A}} = -\underline{\underline{A}}^T \tag{2.177}$$

In Einstein notation this means that $A_{sm} = -A_{ms}$. An example of the matrix of coefficients of an antisymmetric tensor is

$$\begin{pmatrix} 0 & 2 & 3 \\ -2 & 0 & 4 \\ -3 & -4 & 0 \end{pmatrix}_{123} \tag{2.178}$$

The diagonal elements of an antisymmetric tensor are always zeros.

Invariants of a tensor: Tensors of the type we have been discussing so far have three scalar quantities associated with them that are independent of the coordinate system. These are called the invariants of the tensor. Combinations of the three invariants are also invariant to change in a coordinate system, and therefore how the three invariants are defined is not unique (see Appendix C.6). The definitions of the tensor invariants that we will use are shown here for a tensor $\underline{\underline{B}}$ [26]. These definitions in terms of tensor coefficients are only valid when the tensor is written in an orthonormal coordinate system. Tensor invariants are

$$I_{\underline{\underline{B}}} \equiv \sum_{i=1}^{3} B_{ii} \tag{2.179}$$

$$II_{\underline{\underline{B}}} \equiv \sum_{i=1}^{3} \sum_{j=1}^{3} B_{ij} B_{ji} = \underline{\underline{B}} : \underline{\underline{B}} \tag{2.180}$$

$$III_{\underline{\underline{B}}} \equiv \sum_{i=1}^{3} \sum_{j=1}^{3} \sum_{k=1}^{3} B_{ij} B_{jk} B_{ki} \tag{2.181}$$

The magnitude $|\underline{\underline{B}}|$ of a tensor (defined previously) is equal to $+\sqrt{II_{\underline{\underline{B}}}/2}$.

Trace of a tensor: The trace of a tensor, written trace($\underline{\underline{A}}$), is the sum of the diagonal elements,

$$\underline{\underline{A}} = A_{pj} \hat{e}_p \hat{e}_j \tag{2.182}$$

$$= \begin{pmatrix} A_{11} & A_{12} & A_{13} \\ A_{21} & A_{22} & A_{23} \\ A_{31} & A_{32} & A_{33} \end{pmatrix}_{123} \tag{2.183}$$

$$\text{trace}(\underline{\underline{A}}) = A_{mm} \tag{2.184}$$

$$= A_{11} + A_{22} + A_{33} \tag{2.185}$$

The first invariant, defined by Equation (2.179), is the trace of the tensor written with respect to orthonormal coordinates. The second and third invariants may also be written as traces:

$$II_{\underline{\underline{B}}} = \text{trace}(\underline{\underline{B}} \cdot \underline{\underline{B}}) \tag{2.186}$$

$$III_{\underline{\underline{B}}} = \text{trace}(\underline{\underline{B}} \cdot \underline{\underline{B}} \cdot \underline{\underline{B}}) \tag{2.187}$$

Order of a tensor: The types of tensors we have been dealing with so far are called *second-order* tensors. Second-order tensors are formed by the indeterminate vector product of two vectors. Higher order tensors may be formed by taking the indeterminate vector product of more than two tensors,

$$\text{third-order tensor } \underline{v}\,\underline{u}\,\underline{w}$$

$$\text{fourth-order tensor } \underline{v}\,\underline{u}\,\underline{w}\,\underline{b}$$

In addition, a vector may be considered to be a first-order tensor, and a scalar may be considered to be a zero-order tensor. The number of components in three-dimensional space required to express a tensor depends on the order v, as summarized in Table 2.2. The order of a mathematical quantity is important to know when performing algebraic manipulations. Since scalars have magnitude only, while vectors denote magnitude and direction, scalars cannot equal vectors. Likewise, vectors cannot equal tensors, which are of higher order. When writing an equation, the rule is that each term must be of the same order. Examples of scalar, vector, and tensor equations used in engineering and physics are

$$\text{Scalar equation} \quad Q = mC_p(T_1 - T_2) \quad \begin{cases} Q = \text{ heat transferred} \\ m = \text{ mass} \\ C_p = \text{ heat capacity} \\ T_1, T_2 = \text{ temperatures} \end{cases}$$

$$\text{Scalar equation} \quad f_1 = \hat{e}_1 \cdot \underline{f} \quad \begin{cases} \underline{f} = \text{ force vector} \\ f_1 = \text{ scalar component of } \underline{f} \\ \hat{e}_1 = \text{ unit vector} \end{cases}$$

$$\text{Vector equation} \quad \underline{f} = m\underline{a} \quad \begin{cases} \underline{f} = \text{ force vector} \\ m = \text{ mass} \\ \underline{a} = \text{ acceleration vector} \end{cases}$$

$$\begin{array}{l} \text{Vector equation} \\ \text{(see Appendix E)} \end{array} \quad \underline{D} = \underline{\underline{\epsilon}} \cdot \underline{E} \quad \begin{cases} \underline{D} = \text{ electric displacement vector} \\ \underline{\underline{\epsilon}} = \text{ dielectric tensor} \\ \underline{E} = \text{ electric field vector} \end{cases}$$

$$\begin{array}{l} \text{Tensor equation} \\ \text{(see Chapter 3)} \end{array} \quad \underline{\underline{\tau}} = -\mu\underline{\underline{\dot{\gamma}}} \quad \begin{cases} \underline{\underline{\tau}} = \text{ stress tensor} \\ \mu = \text{ Newtonian viscosity} \\ \underline{\underline{\dot{\gamma}}} = \text{ rate-of-deformation tensor} \end{cases}$$

The net effect of vector–tensor operations on the order of an expression is summarized in Table 2.3 [28].

Inverse of a tensor $\underline{\underline{A}}^{-1}$: The inverse $\underline{\underline{A}}^{-1}$ of a tensor $\underline{\underline{A}}$ is a tensor that when dot multiplied by $\underline{\underline{A}}$ gives the identity tensor $\underline{\underline{I}}$,

$$\underline{\underline{A}} \cdot \underline{\underline{A}}^{-1} = \underline{\underline{A}}^{-1} \cdot \underline{\underline{A}} = \underline{\underline{I}} \tag{2.188}$$

TABLE 2.2
Summary of the Orders of Vector and Tensor Quantities and Their Properties

Order ν	Name	Number of Associated Directions	Number of Components	Examples
0	scalar	0	3^0	Mass, energy, temperature
1	vector	1	3^1	Velocity, force, electric field
2	2nd-order tensor	2	3^2	Stress, deformation
3	3rd-order tensor	3	3^3	Gradient of stress
ν	νth-order tensor	ν	3^ν	

TABLE 2.3
Summary of the Effect of Various Operations on the Order of an Expression

Operation	Order of Result	Example
No symbol	$\sum_{orders}$	$\alpha \underline{\underline{B}}$, order $= 2$
$\times$	$\sum_{orders} - 1$	$\underline{w} \times \underline{\underline{C}}$, order $= 2$
$\cdot$	$\sum_{orders} - 2$	$\underline{u} \cdot \underline{\underline{A}}$, order $= 1$
:	$\sum_{orders} - 4$	$\underline{\underline{B}} : \underline{\underline{C}}$, order $= 0$

Notes: $\sum_{orders}$ is the summation of the orders of the quantities in the expression.
Source: After [28].

An inverse does not exist for a tensor whose determinant is zero. It is straightforward to show that the determinant of $\underline{\underline{A}}$ [written $\det|\underline{\underline{A}}|$ and defined in Equation (2.79)] is related to the tensor invariants of $\underline{\underline{A}}$ as follows:

$$\det|\underline{\underline{A}}| = \frac{1}{6}\left(I_{\underline{\underline{A}}}^2 - 3I_{\underline{\underline{A}}}II_{\underline{\underline{A}}} + 2III_{\underline{\underline{A}}}\right) \tag{2.189}$$

The determinant of a tensor is invariant to any coordinate transformation.

2.4 Differential Operations with Vectors and Tensors

The rheologically important equations of conservation of mass and momentum are differential equations, and thus we must learn how to differentiate vector and tensor quantities. In vector and tensor notation, differentiation in physical space (three dimensions) is handled by the vector differential operator ∇, called *del* or *nabla*. In this section we will cover the operation of ∇ on scalars, vectors, and tensors.

To calculate the derivative of a vector or tensor we must first express the quantity with respect to a basis. Differentiation is then carried out by having a differential operator, for example, $\partial/\partial y$, act on each term, including the basis vectors. For example, if the chosen basis is the arbitrary basis $\tilde{\underline{e}}_1, \tilde{\underline{e}}_2, \tilde{\underline{e}}_3$ (not necessarily orthonormal or constant in space), we can express a vector $\underline{v}$ as

$$\underline{v} = \tilde{v}_1\tilde{\underline{e}}_1 + \tilde{v}_2\tilde{\underline{e}}_2 + \tilde{v}_3\tilde{\underline{e}}_3 \tag{2.190}$$

The y-derivative of $\underline{v}$ is thus

$$\frac{\partial \underline{v}}{\partial y} = \frac{\partial}{\partial y}(\tilde{v}_1\tilde{\underline{e}}_1 + \tilde{v}_2\tilde{\underline{e}}_2 + \tilde{v}_3\tilde{\underline{e}}_3) \tag{2.191}$$

$$= \frac{\partial}{\partial y}(\tilde{v}_1\tilde{\underline{e}}_1) + \frac{\partial}{\partial y}(\tilde{v}_2\tilde{\underline{e}}_2) + \frac{\partial}{\partial y}(\tilde{v}_3\tilde{\underline{e}}_3) \tag{2.192}$$

$$= \tilde{v}_1\frac{\partial \tilde{\underline{e}}_1}{\partial y} + \tilde{\underline{e}}_1\frac{\partial \tilde{v}_1}{\partial y} + \tilde{v}_2\frac{\partial \tilde{\underline{e}}_2}{\partial y} + \tilde{\underline{e}}_2\frac{\partial \tilde{v}_2}{\partial y} + \tilde{v}_3\frac{\partial \tilde{\underline{e}}_3}{\partial y} + \tilde{\underline{e}}_3\frac{\partial \tilde{v}_3}{\partial y} \tag{2.193}$$

Note that we used the product rule of differentiation in obtaining Equation (2.193). This complex situation is simplified if for the basis vectors $\tilde{\underline{e}}_i$ we choose to use the Cartesian coordinate system $\hat{e}_x$, $\hat{e}_y$, $\hat{e}_z$. In Cartesian coordinates, the basis vectors are constant in length and fixed in direction, and with this choice the terms in Equation (2.193) involving differentiation of the basis vectors $\tilde{\underline{e}}_i$ are zero; thus half of the terms disappear.

Since vector and tensor quantities are independent of the coordinate system, any vector or tensor quantity derived in Cartesian coordinates is valid when properly expressed in any other coordinate system. Thus, when deriving general expressions, it is most convenient to represent vectors and tensors in Cartesian coordinates. Limiting ourselves to spatially homogeneous (the directions of the unit vectors do not vary with position), orthonormal basis vectors (the Cartesian system) allows us to use Einstein notation for differential operations, as we shall see. This is a distinct advantage. There are times when coordinate systems other than the spatially homogeneous Cartesian system are useful, and we will discuss two such coordinate systems (cylindrical and spherical) in the next section. In addition, there are times when nonorthonormal bases are preferred to orthonormal systems. This will be discussed in Chapter 9. Remember that the choice of coordinate system is simply one of convenience, since vector and tensor expressions are independent of the coordinate system.

In Cartesian coordinates ($x = x_1$, $y = x_2$, $z = x_3$) the spatial differentiation operator ∇ is defined as

$$\nabla \equiv \hat{e}_1\frac{\partial}{\partial x_1} + \hat{e}_2\frac{\partial}{\partial x_2} + \hat{e}_3\frac{\partial}{\partial x_3} \tag{2.194}$$

In Einstein notation this becomes

$$\nabla = \hat{e}_i\frac{\partial}{\partial x_i} \tag{2.195}$$

∇ is a vector operator, not a vector. This means that it has the same order as a vector, but it cannot stand alone. We cannot sketch it on a set of axes, and it does not have a magnitude in the usual sense. Also, although ∇ is of vector order, convention omits the underbar from this symbol.

Since ∇ is an operator, it must operate on something. ∇ may operate on scalars, vectors, or tensors of any order. When ∇ operates on a scalar, it produces a vector,

$$\nabla\alpha = \left(\hat{e}_1\frac{\partial}{\partial x_1} + \hat{e}_2\frac{\partial}{\partial x_2} + \hat{e}_3\frac{\partial}{\partial x_3}\right)\alpha \tag{2.196}$$

$$= \hat{e}_1\frac{\partial\alpha}{\partial x_1} + \hat{e}_2\frac{\partial\alpha}{\partial x_2} + \hat{e}_3\frac{\partial\alpha}{\partial x_3} \tag{2.197}$$

$$= \hat{e}_i \frac{\partial \alpha}{\partial x_i} \tag{2.198}$$

$$= \begin{pmatrix} \frac{\partial \alpha}{\partial x_1} \\ \frac{\partial \alpha}{\partial x_2} \\ \frac{\partial \alpha}{\partial x_3} \end{pmatrix}_{123} \tag{2.199}$$

The vector it produces, $\nabla \alpha$, is called the *gradient* of the scalar quantity α.

We pause here to clarify two terms we have used, scalars and constants. Scalars are quantities that are of order zero. They convey magnitude only. They may be variables, however, such as the distance $x(t)$ between two moving objects or the temperature $T(x, y, z)$ at various positions in a room with a fireplace. Multiplication by scalars follows the rules outlined earlier, namely, it is commutative, associative, and distributive. When combined with a ∇ operator, however, the position of a scalar is quite important. If the position of a scalar variable is moved with respect to the ∇ operator, the meaning of the expression has changed. We can summarize some of this by pointing out the following rules with respect to ∇ operating on scalars α and ζ:

$$\left.\begin{array}{c} \text{Laws of algebra} \\ \text{for } \nabla \text{ operating} \\ \text{on scalars} \end{array}\right\{ \begin{array}{ll} \textit{not} \text{ commutative} & \nabla \alpha \neq \alpha \nabla \\ \textit{not} \text{ associative} & \nabla(\zeta \alpha) \neq (\nabla \zeta)\alpha \\ \text{distributive} & \nabla(\zeta + \alpha) = \nabla \zeta + \nabla \alpha \end{array}$$

The first limitation, that ∇ is not commutative, relates to the fact that ∇ is an operator: $\nabla \alpha$ is a vector whereas $\alpha \nabla$ is an operator, and they cannot be equal. The second limitation reflects the rule that the differentiation operator $(\partial/\partial x)$ acts on all quantities to its right until a plus, minus, equals sign, or bracket $((), \{\}, [])$ is reached. Thus, expressions of the type $\partial(\zeta \alpha)/\partial x$ must be expanded using the usual product rule of differentiation, and ∇ is not associative,

$$\frac{\partial(\zeta \alpha)}{\partial x} = \zeta \frac{\partial \alpha}{\partial x} + \alpha \frac{\partial \zeta}{\partial x} \tag{2.200}$$

The term "constant" is sometimes confused with the word "scalar." Constant is a word that describes a quantity that does not change. Scalars may be constant [as in the speed of light c $(= 3 \times 10^8$ m/s) or the number of cars sold last year worldwide], vectors may be constant (as in the Cartesian coordinate basis vectors $\hat{e}_x$, $\hat{e}_y$, and $\hat{e}_z$), and tensors may be constant (as in the isotropic pressure 2 m below the surface of the ocean). The issue of constancy only comes up now because we are dealing with the change operator ∇. Constants may be positioned arbitrarily with respect to a differential operator since they do not change.

Another thing to notice about the ∇ operator is that it increases the order of the expression on which it acts. We saw that when ∇ operates on a scalar, a vector results. We will now see that when ∇ operates on a vector, it yields a second-order tensor, and when ∇ operates on a second-order tensor, a third-order tensor results. Note that since the Cartesian basis vectors $\hat{e}_1$, $\hat{e}_2$, $\hat{e}_3$ are constant (do not vary with x_1, x_2, x_3), it does not matter where they are written with respect to the differentiation operator $\partial/\partial x_j$ in the Einstein summation convention. However, the order of the unit vectors in the final expression [as shown in Equation (2.203)] and in the original expansion into Einstein notation must match.

$$\nabla\underline{w} = \hat{e}_p \frac{\partial}{\partial x_p}(w_k\hat{e}_k) = \hat{e}_p\frac{\partial(w_k\hat{e}_k)}{\partial x_p} \tag{2.201}$$

$$= \hat{e}_p\,\hat{e}_k\,\frac{\partial w_k}{\partial x_p} \tag{2.202}$$

$$= \frac{\partial w_k}{\partial x_p}\hat{e}_p\,\hat{e}_k \tag{2.203}$$

$$= \begin{pmatrix} \frac{\partial w_1}{\partial x_1} & \frac{\partial w_2}{\partial x_1} & \frac{\partial w_3}{\partial x_1} \\ \frac{\partial w_1}{\partial x_2} & \frac{\partial w_2}{\partial x_2} & \frac{\partial w_3}{\partial x_2} \\ \frac{\partial w_1}{\partial x_3} & \frac{\partial w_2}{\partial x_3} & \frac{\partial w_3}{\partial x_3} \end{pmatrix}_{123} \tag{2.204}$$

$$\nabla\underline{\underline{B}} = \hat{e}_i\frac{\partial}{\partial x_i}(B_{rs}\hat{e}_r\hat{e}_s) = \hat{e}_i\frac{\partial(B_{rs}\hat{e}_r\hat{e}_s)}{\partial x_i} \tag{2.205}$$

$$= \hat{e}_i\hat{e}_r\hat{e}_s\frac{\partial B_{rs}}{\partial x_i} \tag{2.206}$$

$$= \frac{\partial B_{rs}}{\partial x_i}\hat{e}_i\hat{e}_r\hat{e}_s \tag{2.207}$$

There are 27 components (3^3) associated with the third-order tensor $\nabla\underline{\underline{B}}$, and we will not list them here. This is left as an exercise for the interested reader. $\nabla\underline{w}$ is called the gradient of the vector $\underline{w}$,[4] and $\nabla\underline{\underline{B}}$ is called the gradient of the tensor $\underline{\underline{B}}$.[5]

The rules of algebra for ∇ operating on nonconstant scalars (∇ is not commutative, not associative, but is distributive) also hold for nonconstant vectors and tensors as outlined:

Laws of algebra for ∇ operating on vectors and tensors $\left\{\begin{array}{l} \textit{not}\ \text{commutative}\quad \nabla\underline{w} \neq \underline{w}\nabla \\ \qquad\qquad\qquad\ \nabla\underline{\underline{B}} \neq \underline{\underline{B}}\nabla \\ \textit{not}\ \text{associative}\quad \nabla(\underline{a}\,\underline{b}) \neq (\nabla\underline{a})\underline{b} \\ \qquad\qquad\qquad\ \nabla(\underline{a}\cdot\underline{b}) \neq (\nabla\underline{a})\cdot\underline{b} \\ \qquad\qquad\qquad\ \nabla(\underline{a}\times\underline{b}) \neq (\nabla\underline{a})\times\underline{b} \\ \qquad\qquad\qquad\ \nabla(\underline{\underline{B}}\,\underline{\underline{C}}) \neq (\nabla\underline{\underline{B}})\underline{\underline{C}} \\ \qquad\qquad\qquad\ \nabla(\underline{\underline{B}}\cdot\underline{\underline{C}}) \neq (\nabla\underline{\underline{B}})\cdot\underline{\underline{C}} \\ \text{distributive}\quad \nabla(\underline{w}+\underline{b}) = \nabla\underline{w}+\nabla\underline{b} \\ \qquad\qquad\qquad\ \nabla(\underline{\underline{B}}+\underline{\underline{C}}) = \nabla\underline{\underline{B}}+\nabla\underline{\underline{C}} \end{array}\right.$

---

[4] Note that in the convention we follow [138, 26] the unit vector that accompanies ∇ is the first unit vector in the tensor $\nabla\underline{w}$, and the unit vector from the vector being operated upon is the second unit vector. Thus $\nabla\underline{w} = (\partial w_j/\partial x_i)\hat{e}_i\hat{e}_j$. The opposite convention is also in wide use, namely, $\tilde{\nabla}\underline{w} = (\partial w_i/\partial x_j)\hat{e}_i\hat{e}_j$ [162, 9, 205, 166, 61, 238, 179]. When reading other texts it is important to check which convention is in use (see Tables D.1 and D.2).

[5] The fact that $\nabla\underline{w}$ and $\nabla\underline{\underline{B}}$ are tensors, that is, frame-invariant quantities that express linear vector (or tensor) functions, should not simply be assumed. By examining these quantities under the action of a change in basis, however, both of these gradients as well as the gradients of all higher order tensors can be shown to be tensors [7].

A second type of differential operation is performed when ∇ is dot-multiplied with a vector or a tensor. This operator, the divergence $\nabla\cdot$, lowers by one the order of the entity on which it operates. Since the order of a scalar is already zero, one cannot take the divergence of a scalar. The following operations are defined.

Divergence of a vector:

$$\nabla \cdot \underline{w} = \frac{\partial}{\partial x_i} \hat{e}_i \cdot w_m \hat{e}_m \tag{2.208}$$

$$= \hat{e}_i \cdot \hat{e}_m \frac{\partial w_m}{\partial x_i} \tag{2.209}$$

$$= \delta_{im} \frac{\partial w_m}{\partial x_i} \tag{2.210}$$

$$= \frac{\partial w_m}{\partial x_m} \tag{2.211}$$

$$= \frac{\partial w_1}{\partial x_1} + \frac{\partial w_2}{\partial x_2} + \frac{\partial w_3}{\partial x_3} \tag{2.212}$$

The result is a scalar [no unit vectors present in Equation (2.211)].

Divergence of a tensor:

$$\nabla \cdot \underline{\underline{B}} = \frac{\partial}{\partial x_p} \hat{e}_p \cdot B_{mn} \hat{e}_m \hat{e}_n \tag{2.213}$$

$$= \hat{e}_p \cdot \hat{e}_m \hat{e}_n \frac{\partial B_{mn}}{\partial x_p} \tag{2.214}$$

$$= \delta_{pm} \hat{e}_n \frac{\partial B_{mn}}{\partial x_p} \tag{2.215}$$

$$= \frac{\partial B_{pn}}{\partial x_p} \hat{e}_n \tag{2.216}$$

$$= \begin{pmatrix} \frac{\partial B_{p1}}{\partial x_p} \\ \frac{\partial B_{p2}}{\partial x_p} \\ \frac{\partial B_{p3}}{\partial x_p} \end{pmatrix}_{123} = \begin{pmatrix} \sum_{p=1}^{3} \frac{\partial B_{p1}}{\partial x_p} \\ \sum_{p=1}^{3} \frac{\partial B_{p2}}{\partial x_p} \\ \sum_{p=1}^{3} \frac{\partial B_{p3}}{\partial x_p} \end{pmatrix}_{123} \tag{2.217}$$

$$= \begin{pmatrix} \frac{\partial B_{11}}{\partial x_1} + \frac{\partial B_{21}}{\partial x_2} + \frac{\partial B_{31}}{\partial x_3} \\ \frac{\partial B_{12}}{\partial x_1} + \frac{\partial B_{22}}{\partial x_2} + \frac{\partial B_{32}}{\partial x_3} \\ \frac{\partial B_{13}}{\partial x_1} + \frac{\partial B_{23}}{\partial x_2} + \frac{\partial B_{33}}{\partial x_3} \end{pmatrix}_{123} \tag{2.218}$$

The result is a vector [one unit vector is present in Equation (2.216)]. The rules of algebra for the operation of the divergence $\nabla\cdot$ on vectors and tensors can be deduced by writing the expression of interest in Einstein notation and following the rules of algebra for the operation of the differentiation operator $\partial/\partial x_p$ on scalars and vectors.

One final differential operation is the Laplacian, $\nabla \cdot \nabla$ or ∇^2. This operation leaves the order of its object unchanged, and thus we may take the Laplacian of scalars, vectors, and tensors as shown next.

Laplacian of a scalar:

$$\nabla \cdot \nabla \alpha = \frac{\partial}{\partial x_k} \hat{e}_k \cdot \frac{\partial}{\partial x_m} \hat{e}_m \alpha \tag{2.219}$$

$$= \hat{e}_k \cdot \hat{e}_m \frac{\partial}{\partial x_k} \frac{\partial \alpha}{\partial x_m} = \delta_{km} \frac{\partial}{\partial x_k} \frac{\partial \alpha}{\partial x_m} \tag{2.220}$$

$$= \frac{\partial}{\partial x_k} \frac{\partial \alpha}{\partial x_k} = \frac{\partial^2 \alpha}{\partial x_k^2} \tag{2.221}$$

$$= \frac{\partial^2 \alpha}{\partial x_1^2} + \frac{\partial^2 \alpha}{\partial x_2^2} + \frac{\partial^2 \alpha}{\partial x_3^2} \tag{2.222}$$

The result is a scalar. Although k appears only once in Equation (2.221), it is a repeated subscript with respect to Einstein notation since we have used the usual shorthand notation $\frac{\partial}{\partial x_k} \frac{\partial}{\partial x_k} = \left(\frac{\partial^2}{\partial x_k^2} \right)$ for twice differentiating with respect to x_k.

Laplacian of a vector:

$$\nabla \cdot \nabla \underline{w} = \frac{\partial}{\partial x_k} \hat{e}_k \cdot \frac{\partial}{\partial x_m} \hat{e}_m \, w_j \hat{e}_j \tag{2.223}$$

$$= \hat{e}_k \cdot \hat{e}_m \, \hat{e}_j \frac{\partial}{\partial x_k} \frac{\partial w_j}{\partial x_m} = \delta_{km} \hat{e}_j \frac{\partial}{\partial x_k} \frac{\partial w_j}{\partial x_m} \tag{2.224}$$

$$= \frac{\partial}{\partial x_k} \frac{\partial w_j}{\partial x_k} \, \hat{e}_j = \frac{\partial^2 w_j}{\partial x_k^2} \, \hat{e}_j \tag{2.225}$$

$$= \begin{pmatrix} \frac{\partial^2 w_1}{\partial x_1^2} + \frac{\partial^2 w_1}{\partial x_2^2} + \frac{\partial^2 w_1}{\partial x_3^2} \\ \frac{\partial^2 w_2}{\partial x_1^2} + \frac{\partial^2 w_2}{\partial x_2^2} + \frac{\partial^2 w_2}{\partial x_3^2} \\ \frac{\partial^2 w_3}{\partial x_1^2} + \frac{\partial^2 w_3}{\partial x_2^2} + \frac{\partial^2 w_3}{\partial x_3^2} \end{pmatrix}_{123} \tag{2.226}$$

The result is a vector. The same procedure may be followed to determine the expression for the Laplacian of a tensor.

Correctly identifying the quantities on which ∇ operates is an important issue, and the rules are worth repeating. The differentiation operator $\partial/\partial x_i$ acts on all quantities to its right until a plus, minus, equals sign, or bracket $((), \{\}, [])$ is reached. To show how this property affects terms in a vector/tensor expression, we now give an example.

EXAMPLE

What is $\nabla \cdot \underline{a} \, \underline{b}$?

SOLUTION

We begin with Einstein notation:

$$\nabla \cdot \underline{a}\,\underline{b} = \frac{\partial}{\partial x_m}\hat{e}_m \cdot a_p\hat{e}_p\, b_n\hat{e}_n \qquad (2.227)$$

Here, as always, we use different indices for the various implied summations.

Since the coefficients of both $\underline{a}$ and $\underline{b}$ are to the right of the ∇ operator, they are both acted upon by its differentiation action. The Cartesian unit vectors are also affected, but these are constant.

$$\nabla \cdot \underline{a}\,\underline{b} = \frac{\partial}{\partial x_m}\hat{e}_m \cdot \left(a_p\hat{e}_p\, b_n\hat{e}_n\right) \qquad (2.228)$$

$$= \hat{e}_m \cdot \hat{e}_p\,\hat{e}_n\frac{\partial(a_p\, b_n)}{\partial x_m} = \delta_{mp}\,\hat{e}_n\frac{\partial(a_p\, b_n)}{\partial x_m} \qquad (2.229)$$

$$= \hat{e}_n\frac{\partial(a_m\, b_n)}{\partial x_m} \qquad (2.230)$$

To further expand this expression, we use the product rule of differentiation on the quantity in parentheses,

$$\nabla \cdot \underline{a}\,\underline{b} = \hat{e}_n\frac{\partial(a_m\, b_n)}{\partial x_m} \qquad (2.231)$$

$$= \hat{e}_n\left(a_m\frac{\partial b_n}{\partial x_m} + b_n\frac{\partial a_m}{\partial x_m}\right) \qquad (2.232)$$

This is as far as this expression may be expanded. One can write these two terms in vector (also called *Gibbs*) notation,

$$\nabla \cdot \underline{a}\,\underline{b} = a_m\frac{\partial b_n}{\partial x_m}\hat{e}_n + \frac{\partial a_m}{\partial x_m}\, b_n\hat{e}_n \qquad (2.233)$$

$$= \underline{a} \cdot \nabla\underline{b} + (\nabla \cdot \underline{a})\underline{b} \qquad (2.234)$$

The equivalence of the last two equations may be verified by working backward from Equation (2.234). If the differentiation of the product had not been carried out correctly, the first term on the right-hand side would have been omitted.

2.5 Curvilinear Coordinates

Until now we have used (almost exclusively) the Cartesian coordinate system to express vectors and tensors with respect to scalar coordinates. Since vector and tensor quantities are independent of the coordinate system, we have chosen to use the Cartesian system, which is orthonormal and constant in space, to derive vector/tensor relations. This choice allows us to use Einstein notation to keep track of vector/tensor operations. The Cartesian system is also a natural choice when solving flow problems if the flow boundaries are straight lines, that is, if the boundaries coincide with coordinate surfaces (e.g., at $x = B$, $v = V$). When the boundaries are curved, however, as, for example, in flow in a pipe or flow around a falling sphere, it is mathematically awkward to use the Cartesian system. To solve problems with cylindrical and spherical symmetry we will choose to use coordinate systems that share these symmetries (see Figures 2.11 and 2.12).

The cylindrical and spherical coordinate systems, jointly called curvilinear coordinate systems, allow for considerable simplification of the analysis of problems for which the boundaries are coordinate surfaces of these systems, that is, problems for which the boundaries are cylindrical or spherical (see Chapter 3 for worked-out examples). The disadvantage of these systems, however, is that for both the cylindrical and the spherical coordinate systems the basis vectors vary with position, as we will demonstrate in the next section. Hence, the differential operator ∇ is more complicated in curvilinear systems, and care must be taken to use the correct form of vector and tensor quantities involving ∇. Also, Einstein notation is inconvenient to use in curvilinear systems when ∇ is involved, since ∇ must be made to operate on the (spatially varying) basis vectors as well as on the vector and tensor coefficients. We will discuss both of these concerns in the next section.

2.5.1 CYLINDRICAL COORDINATE SYSTEM

The coordinates and unit vectors that are used for the cylindrical and spherical coordinate systems are shown in Figures 2.11 and 2.12. In cylindrical coordinates (Figure 2.12a), the three basis vectors are $\hat{e}_r$, $\hat{e}_\theta$, and $\hat{e}_z$. The vector $\hat{e}_z$ is the same as the vector of the same name in the Cartesian coordinate system. The vector $\hat{e}_r$ is a vector that is perpendicular to the z-axis and makes an angle θ with the positive x-axis of the Cartesian system. The last vector, $\hat{e}_\theta$, is perpendicular to $\hat{e}_r$, resides in an xy-plane of the Cartesian system, and points in the direction counterclockwise to the x-axis, that is, in the direction of increasing θ. Note that this is an orthonormal basis system.

In cylindrical coordinates the vector $\hat{e}_z$ remains constant in direction and in magnitude no matter what point in space is being considered. Both $\hat{e}_r$ and $\hat{e}_\theta$ vary with position, however. To convince yourself of this, consider two points in the xy-plane, $(1, 1, 0)$ and $(-1, -1, 0)$ (Figure 2.13). We can write the vectors $\hat{e}_r$ and $\hat{e}_\theta$ with respect to Cartesian coordinates

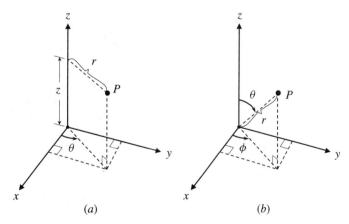

(a)   (b)

Figure 2.11 Geometries of (a) cylindrical and (b) spherical coordinate systems.

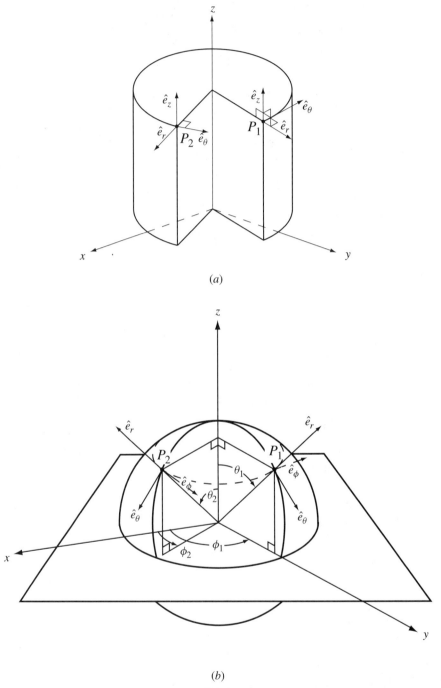

(a)

(b)

Figure 2.12 Pictorial representation of the basis vectors associated with (a) cylindrical and (b) spherical coordinate systems.

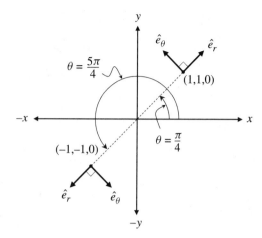

Figure 2.13 Variation of the basis vectors $\hat{e}_\theta$ and $\hat{e}_r$ in cylindrical coordinates.

for each point. The vector $\hat{e}_r$ is a unit vector pointing in the direction of increasing r, that is, in the direction of a line from the origin to the point of interest. The unit vector $\hat{e}_\theta$ points from the point of interest in the direction of increasing θ. For the point $(1,1,0)$, $\hat{e}_r$ and $\hat{e}_\theta$ are

$$\hat{e}_r\big|_{(110)} = \begin{pmatrix} \frac{1}{\sqrt{2}} \\ \frac{1}{\sqrt{2}} \\ 0 \end{pmatrix}_{xyz} \tag{2.235}$$

$$\hat{e}_\theta\big|_{(110)} = \begin{pmatrix} -\frac{1}{\sqrt{2}} \\ \frac{1}{\sqrt{2}} \\ 0 \end{pmatrix}_{xyz} \tag{2.236}$$

where the subscript xyz emphasizes that the vectors are written in the Cartesian, x, y, z, coordinate system. For $(-1,-1,0)$, $\hat{e}_r$ and $\hat{e}_\theta$ are

$$\hat{e}_r\big|_{(-1,-1,0)} = \begin{pmatrix} -\frac{\sqrt{2}}{2} \\ -\frac{\sqrt{2}}{2} \\ 0 \end{pmatrix}_{xyz} \tag{2.237}$$

$$\hat{e}_\theta\big|_{(-1,-1,0)} = \begin{pmatrix} \frac{\sqrt{2}}{2} \\ -\frac{\sqrt{2}}{2} \\ 0 \end{pmatrix}_{xyz} \tag{2.238}$$

The vectors $\hat{e}_r$ and $\hat{e}_\theta$ clearly differ at the two points. For an arbitrary point at coordinates x, y, z or r, θ, z, the cylindrical basis vectors are related to the Cartesian basis vectors as follows:

$$\hat{e}_r = \cos\theta\,\hat{e}_x + \sin\theta\,\hat{e}_y \tag{2.239}$$

$$\hat{e}_\theta = -\sin\theta\,\hat{e}_x + \cos\theta\,\hat{e}_y \tag{2.240}$$

$$\hat{e}_z = \hat{e}_z \tag{2.241}$$

The cylindrical coordinate variables r, θ, and z are related to the Cartesian coordinate variables x, y, and z as follows:

$$x = r\cos\theta \tag{2.242}$$

$$y = r\sin\theta \tag{2.243}$$

$$z = z \tag{2.244}$$

The fact that the cylindrical basis vectors vary with position impacts the use of the ∇ operator in this coordinate system. The ∇ operator in cylindrical coordinates may be derived from the Cartesian expression by making use of the chain rule ([239] and Chapter 9) and the relations between x, y, and z and r, θ, and z. The result is

$$\nabla = \hat{e}_r \frac{\partial}{\partial r} + \hat{e}_\theta \frac{1}{r}\frac{\partial}{\partial \theta} + \hat{e}_z \frac{\partial}{\partial z} \tag{2.245}$$

We will now operate ∇ on a vector $\underline{v}$, writing both ∇ and the vector $\underline{v}$ in cylindrical coordinates and following the rules of algebra outlined earlier in this chapter,

$$\nabla\underline{v} = \left(\hat{e}_r \frac{\partial}{\partial r} + \hat{e}_\theta \frac{1}{r}\frac{\partial}{\partial \theta} + \hat{e}_z \frac{\partial}{\partial z}\right)(v_r\hat{e}_r + v_\theta\hat{e}_\theta + v_z\hat{e}_z) \tag{2.246}$$

$$= \hat{e}_r \frac{\partial}{\partial r}(v_r\hat{e}_r + v_\theta\hat{e}_\theta + v_z\hat{e}_z) + \hat{e}_\theta \frac{1}{r}\frac{\partial}{\partial \theta}(v_r\hat{e}_r + v_\theta\hat{e}_\theta + v_z\hat{e}_z) \tag{2.247}$$

$$+ \hat{e}_z \frac{\partial}{\partial z}(v_r\hat{e}_r + v_\theta\hat{e}_\theta + v_z\hat{e}_z)$$

$$= \hat{e}_r \left[\frac{\partial(v_r\hat{e}_r)}{\partial r} + \frac{\partial(v_\theta\hat{e}_\theta)}{\partial r} + \frac{\partial(v_z\hat{e}_z)}{\partial r}\right] + \hat{e}_\theta \frac{1}{r}\left[\frac{\partial(v_r\hat{e}_r)}{\partial \theta} + \frac{\partial(v_\theta\hat{e}_\theta)}{\partial \theta} + \frac{\partial(v_z\hat{e}_z)}{\partial \theta}\right]$$

$$+ \hat{e}_z \left[\frac{\partial(v_r\hat{e}_r)}{\partial z} + \frac{\partial(v_\theta\hat{e}_\theta)}{\partial z} + \frac{\partial(v_z\hat{e}_z)}{\partial z}\right] \tag{2.248}$$

Each of the derivatives operates on a product of two quantities that vary in space, the coefficient of $\underline{v}$ and the basis vector. When we operated in the Cartesian coordinate system, since the basis vectors are not a function of position, they could be removed from the differentiation. For the cylindrical (and spherical) coordinate systems, which have variable unit vectors, this is not possible. One correct way to calculate $\nabla\underline{v}$ in cylindrical coordinates is to write r, θ, z, $\hat{e}_r$, $\hat{e}_\theta$, and $\hat{e}_z$ with respect to the constant Cartesian system and then carry out the appropriate differentiations. The results are shown in Table C.7 in Appendix C.2.

2.5.2 Spherical Coordinate System

In the spherical coordinate system, all three basis vectors vary with position (Figure 2.12b). The three unit vectors are $\hat{e}_r$, $\hat{e}_\theta$, and $\hat{e}_\phi$. The vector $\hat{e}_r$ emits radially from the origin toward a point of interest. The vector $\hat{e}_\theta$, which lies in the plane formed by the point of interest and the Cartesian z-direction, is perpendicular to $\hat{e}_r$, and points in the direction that rotates away from the positive z-axis. The vector $\hat{e}_\phi$ lies in a Cartesian xy-plane, is perpendicular to the projection of $\hat{e}_r$ in the Cartesian xy-plane, and points counterclockwise from the x-axis. Note that the definitions of $\hat{e}_r$ and $\hat{e}_\theta$ in cylindrical and spherical coordinates differ.

The spherical coordinate variables r, θ, and ϕ and basis vectors $\hat{e}_r, \hat{e}_\theta$, and $\hat{e}_\phi$ are related to their Cartesian counterparts as follows:

$$x = r \sin \theta \cos \phi \tag{2.249}$$

$$y = r \sin \theta \sin \phi \tag{2.250}$$

$$z = r \cos \theta \tag{2.251}$$

$$\hat{e}_r = (\sin \theta \cos \phi)\hat{e}_x + (\sin \theta \sin \phi)\hat{e}_y + (\cos \theta)\hat{e}_z \tag{2.252}$$

$$\hat{e}_\theta = (\cos \theta \cos \phi)\hat{e}_x + (\cos \theta \sin \phi)\hat{e}_y + (-\sin \theta)\hat{e}_z \tag{2.253}$$

$$\hat{e}_\phi = (-\sin \phi)\hat{e}_x + (\cos \phi)\hat{e}_y \tag{2.254}$$

The ∇ operator for spherical coordinates is also calculated from the chain rule (see Chapter 9),

$$\nabla = \hat{e}_r \frac{\partial}{\partial r} + \hat{e}_\theta \frac{1}{r} \frac{\partial}{\partial \theta} + \hat{e}_\phi \frac{1}{r \sin \theta} \frac{\partial}{\partial \phi} \tag{2.255}$$

The extra difficulty caused by using the curvilinear coordinates is offset by the mathematical simplifications that result when cylindrically or spherically symmetric flow problems are expressed in these coordinate systems. To simplify the use of these coordinate systems, the effects of the ∇ operator on scalars, vectors, and tensors in the cylindrical and spherical coordinate systems are summarized in Tables C.7 and C.8 in Appendix C.2. We will often refer to these tables when we solve Newtonian and non-Newtonian flow problems in curvilinear coordinates.

2.6 Vector and Tensor Integral Theorems

Upon completion of this chapter on mathematics review, we will move on to define and to derive the governing equations of Newtonian and non-Newtonian fluid mechanics. We will need some theorems and formulas from vector mathematics, and these are presented here without proof. The reader is directed to textbooks on advanced mathematics for a more detailed discussion of these subjects [100].

2.6.1 GAUSS–OSTROGRADSKII DIVERGENCE THEOREM

The Gauss–Ostrogradskii divergence theorem[6] relates the change of a vector property $\underline{b}$, taking place in a closed volume V, with the flux of that property through the surface S that encloses V [100, 7] (Figure 2.14). The theorem is shown in Equation (2.256)

---

[6] Also known as Green's theorem or simply as the divergence theorem; see Aris [7].

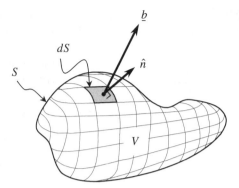

Figure 2.14 Arbitrary volume V enclosed by a surface area S. Each particular piece of the surface dS is characterized by the direction of its unit normal $\hat{n}$. $\underline{b}$ represents any vector property associated with dS.

Gauss–Ostrogradskii
divergence theorem

$$\int_V \nabla \cdot \underline{b} \, dV = \int_S \hat{n} \cdot \underline{b} \, dS \qquad (2.256)$$

where $\hat{n}$ is the outwardly pointing unit normal of the differential surface element dS (Figure 2.14). The volume V is not necessarily constant in time. Use of the Gauss–Ostrogradskii divergence theorem allows us to convert an integral over a volume into a surface integral (or vice versa) without loss of information. This is handy when it is more intuitive to write part of an expression as a surface integral but all other terms of the equation as volume integrals.

2.6.2 LEIBNITZ FORMULA

The Leibnitz formula interprets for us the effect of differentiating an integral [100]. Most simple integrals encountered by beginning students involve integrals over fixed limits. For example, a quantity $J(x, t)$ is defined as the following integral:

$$J(x, t) = \int_\alpha^\beta f(x, t) \, dx \qquad (2.257)$$

where α and β are constants, and f is a function of x and t. When it is desired to take the derivative of J, the procedure is straightforward:

$$\frac{dJ(x, t)}{dt} = \frac{d}{dt} \left[\int_\alpha^\beta f(x, t) \, dx \right] \qquad (2.258)$$

$$= \int_\alpha^\beta \frac{\partial f(x, t)}{\partial t} \, dx \qquad (2.259)$$

This is actually a simplified version of the Leibnitz formula, which tells us how to carry out this differentiation if the limits are not constant but rather are functions of t. The Leibnitz formula is given.

$$J(x, t) = \int_{\alpha(t)}^{\beta(t)} f(x, t)\, dx \tag{2.260}$$

$$\frac{dJ(x, t)}{dt} = \frac{d}{dt}\left[\int_{\alpha(t)}^{\beta(t)} f(x, t)\, dx\right] \tag{2.261}$$

Leibnitz formula
(single integral)
$$\boxed{\frac{dJ(x, t)}{dt} = \int_{\alpha(t)}^{\beta(t)} \frac{\partial f(x, t)}{\partial t}\, dx + f(\beta, t)\frac{d\beta}{dt} - f(\alpha, t)\frac{d\alpha}{dt}} \tag{2.262}$$

There are two versions of the Leibnitz formula that we will use, the preceding one for differentiating single integrals, and a second version for differentiating in three dimensions.[7] For a time-varying volume $V(t)$, enclosed by the moving surface $S(t)$, J is defined as

$$J(x, y, z, t) = \int_{V(t)} f(x, y, z, t)\, dV \tag{2.263}$$

The derivative of J is given by the three-dimensional Leibnitz formula:

Leibnitz formula
(volume integral)
$$\boxed{\frac{dJ}{dt} = \int_{V(t)} \frac{\partial f}{\partial t}\, dV + \int_{S(t)} f\, (v_{\text{surface}} \cdot \hat{n})\, dS} \tag{2.264}$$

The meaning of $\hat{n}$ is the same as in the Gauss–Ostrogradskii divergence theorem (see Figure 2.14), and v_{surface} is the velocity of the surface element dS (the surface is moving). If the volume is fixed in space, the second term goes to zero because v_{surface} is zero.

2.6.3 SUBSTANTIAL DERIVATIVE

In fluid mechanics and rheology, we often deal with properties that vary in space and also change with time. Thus we must consider the differentials of multivariable functions.

Consider such a multivariable function $f(x_1, x_2, x_3, t)$ associated with a particle of fluid, where x_1, x_2, and x_3 are the three spatial coordinates and t is time. This might be, for example, the density of flowing material as a function of time and position. We know that the differential of f is given by

$$df = \frac{\partial f}{\partial t}dt + \frac{\partial f}{\partial x_1}dx_1 + \frac{\partial f}{\partial x_2}dx_2 + \frac{\partial f}{\partial x_3}dx_3 \tag{2.265}$$

If we divide this expression through by dt we get

$$\frac{df}{dt} = \frac{\partial f}{\partial t} + \frac{\partial f}{\partial x_1}\frac{dx_1}{dt} + \frac{\partial f}{\partial x_2}\frac{dx_2}{dt} + \frac{\partial f}{\partial x_3}\frac{dx_3}{dt} \tag{2.266}$$

The terms dx_i/dt are just the velocity components of the particle v_i:

---

[7] Also known as Reynolds' transport theorem [7,148].

$$\frac{df}{dt} = \frac{\partial f}{\partial t} + \frac{\partial f}{\partial x_1}v_1 + \frac{\partial f}{\partial x_2}v_2 + \frac{\partial f}{\partial x_3}v_3 \tag{2.267}$$

Equation (2.267) may be written in Einstein notation and in vector (Gibbs) notation as follows:

$$\frac{df}{dt} = \frac{\partial f}{\partial t} + \frac{\partial f}{\partial x_i}v_i \tag{2.268}$$

Substantial
derivative

$$\boxed{\frac{df}{dt} = \frac{Df}{Dt} = \frac{\partial f}{\partial t} + \underline{v} \cdot \nabla f} \tag{2.269}$$

This expression is called the *substantial derivative* and is often written as Df/Dt. It indicates the rate of change of the function f as observed from a particle of fluid moving with velocity $\underline{v}$.

The mathematical techniques discussed in this chapter will be used extensively throughout the text. In the next chapter we will apply them to deriving conservation equations for mass and momentum.

2.7 PROBLEMS

2.1   What is the magnitude of $\underline{a} = \hat{i} + \hat{j} = \hat{e}_1 + \hat{e}_2$?

2.2   Show that when a vector is dotted with an arbitrary unit vector, the scalar product yields the projection of the vector in the direction of the unit vector.

2.3   What is the unit vector parallel to $\underline{v} = \begin{pmatrix} 2 \\ 3 \\ 6 \end{pmatrix}_{xyz}$?

2.4   What vector goes between the points $(1, 0, 3)$ and $(0, 2, 1)$?

2.5   What is a unit vector perpendicular to $\underline{v} = \begin{pmatrix} 1 \\ 3 \\ 6 \end{pmatrix}_{xyz}$?

2.6   Show that $\underline{u} = a\hat{i} + b\hat{j} + c\hat{k}$ is perpendicular to the plane $ax + by + cz = \alpha$.

2.7   For a general vector $\underline{v}$,

$$\underline{v} = v_1\hat{e}_1 + v_2\hat{e}_2 + v_3\hat{e}_3$$

show that

$$v_1 = \underline{v} \cdot \hat{e}_1$$
$$v_2 = \underline{v} \cdot \hat{e}_2$$
$$v_3 = \underline{v} \cdot \hat{e}_3$$

2.8   Do the following vectors form a basis? If yes, write the vector $\underline{w} = 2\hat{e}_x + 3\hat{e}_y + \hat{e}_z$ in the basis. Prove your answers.

(a) $\begin{pmatrix} 2 \\ 1 \\ 0 \end{pmatrix}_{xyz}$, $\begin{pmatrix} 0 \\ 1 \\ 4 \end{pmatrix}_{xyz}$, $\begin{pmatrix} 1 \\ 1 \\ 2 \end{pmatrix}_{xyz}$

(b) $\begin{pmatrix} 1 \\ 1 \\ 0 \end{pmatrix}_{xyz}$, $\begin{pmatrix} 2 \\ 1 \\ 0 \end{pmatrix}_{xyz}$, $\begin{pmatrix} 1 \\ 0 \\ 1 \end{pmatrix}_{xyz}$

2.9   (a) Show that the vectors $\underline{u}$, $\underline{v}$, and $\underline{w}$ form a basis:

$$\underline{u} = \begin{pmatrix} 1 \\ 1 \\ 0 \end{pmatrix}_{xyz}, \quad \underline{v} = \begin{pmatrix} 0 \\ 2 \\ 3 \end{pmatrix}_{xyz}, \quad \underline{w} = \begin{pmatrix} 1 \\ -2 \\ 3 \end{pmatrix}_{xyz}$$

(b) Write the vector $\underline{t}$ with respect to the basis $\underline{u}$, $\underline{v}$, $\underline{w}$:

$$\underline{t} = \begin{pmatrix} -1 \\ 0 \\ -2 \end{pmatrix}_{xyz}$$

2.10   What is the difference between a 3 by 3 matrix and a 2nd-order tensor?

2.11   In Cartesian coordinates $(\hat{e}_1, \hat{e}_2, \hat{e}_3)$ the coefficients of a tensor $\underline{\underline{A}}$ are given by

$$\underline{\underline{A}} = \begin{pmatrix} 3 & 0 & 0 \\ 0 & 2 & 0 \\ 0 & 0 & 1 \end{pmatrix}_{123}$$

What are the coefficients of $\underline{\underline{A}}$ written with respect to the basis vectors $\underline{a}$, $\underline{b}$, and $\underline{c}$ given below?

$$\underline{a} = \frac{1}{2}\hat{e}_1 + \frac{1}{2}\hat{e}_2$$

$$\underline{b} = \frac{1}{2}\hat{e}_1 - \frac{1}{2}\hat{e}_2$$

$$\underline{c} = \hat{e}_3$$

2.12 How are $\underline{a} \times \underline{b}$ and $\underline{b} \times \underline{a}$ related?

2.13 How can we simplify $(\underline{\underline{B}}^T)^T$? Use Einstein notation.

2.14 Express $\underline{\underline{A}} \cdot \underline{\underline{B}}$ in Einstein notation.

2.15 Expand $(\underline{a}\,\underline{b})^T$ using Einstein notation.

2.16 What is the component of $(\underline{v} \cdot \underline{\underline{A}} \cdot \underline{b}\,\underline{c})$ in the 2-direction?

2.17 Using Einstein notation, show that $\underline{\underline{A}} \cdot \underline{\underline{A}}^T$ is symmetric.

2.18 Using Einstein notation, show that $\underline{\underline{A}}^T + \underline{\underline{B}}^T = (\underline{\underline{A}} + \underline{\underline{B}})^T$.

2.19 Using Einstein notation, show that $(\underline{\underline{A}} \cdot \underline{\underline{B}} \cdot \underline{\underline{C}})^T = \underline{\underline{C}}^T \cdot \underline{\underline{B}}^T \cdot \underline{\underline{A}}^T$.

2.20 Using Einstein notation, show that the tensor $\underline{\underline{A}} + \underline{\underline{A}}^T$ is symmetric. Show that $\underline{\underline{A}} - \underline{\underline{A}}^T$ is antisymmetric.

2.21 Does $\underline{a} \cdot \underline{\underline{B}} = \underline{\underline{B}} \cdot \underline{a}$ in general? Show why or why not using Einstein notation.

2.22 Show that $\underline{\underline{A}} : \underline{\underline{A}}^T > 0$. Use Einstein notation.

2.23 The magnitude of a tensor $\underline{\underline{A}}$ is defined by

$$\left|\underline{\underline{A}}\right| = \text{abs}\left(\sqrt{\frac{\underline{\underline{A}} : \underline{\underline{A}}}{2}}\right)$$

What is the magnitude of the tensor $\underline{\underline{A}}$ given below?

$$\underline{\underline{A}} = 5\hat{e}_1\hat{e}_1 + 3\hat{e}_1\hat{e}_2 - 3\hat{e}_1\hat{e}_3 - \hat{e}_2\hat{e}_1 - \hat{e}_2\hat{e}_2 + 2\hat{e}_2\hat{e}_3 - 3\hat{e}_3\hat{e}_1$$

2.24 What is the Laplacian of a tensor $\underline{\underline{B}}$ in Einstein notation? What is the order of this quantity?

2.25 What are the rules of algebra for taking the divergence of a vector (e.g., $\nabla \cdot \underline{w}$) and for taking the divergence of a tensor (e.g., $\nabla \cdot \underline{\underline{B}}$)? Show that your rules hold by working out the expressions in Einstein notation.

2.26 Prove that the following equality holds:

$$\frac{\partial f}{\partial t} + \frac{\partial f}{\partial x_i}v_i = \frac{\partial f}{\partial t} + \underline{v} \cdot \nabla f$$

2.27 Using Einstein notation, show that for a symmetric tensor $\underline{\underline{A}}$:

$$\underline{\underline{A}} : \nabla\underline{v} = \nabla \cdot (\underline{\underline{A}} \cdot \underline{v}) - \underline{v} \cdot (\nabla \cdot \underline{\underline{A}})$$

2.28 For the vector $\underline{x} = x_1\hat{e}_1 + x_2\hat{e}_2 + x_3\hat{e}_3$, what is $\nabla(\underline{x} \cdot \underline{x})$? Write your final answer in Gibbs notation.

2.29 Prove that the following equality holds:

$$\nabla \cdot \underline{a}\,\underline{b} = \underline{a} \cdot \nabla\underline{b} + (\nabla \cdot \underline{a})\underline{b}$$

2.30 For $\underline{v} = \begin{pmatrix} ax_2 \\ bx_1 + x_2^2 \\ cx_3 \end{pmatrix}_{123}$, what is $\nabla \cdot \underline{v}$?

2.31 Expand $\nabla \cdot (\alpha\underline{v})$, where α is a scalar but is not constant. Write your answers in Einstein notation and vector form.

2.32 Using Einstein notation, show that $\nabla \cdot \underline{v}\,\underline{w} = \underline{v} \cdot \nabla\underline{w} + \underline{w}(\nabla \cdot \underline{v})$.

2.33 What is $\nabla \cdot a\underline{\underline{I}}$? Express your answer in Einstein notation and in vector form.

2.34 Simplify the expression $\underline{\underline{I}} : \nabla\underline{v}$.

2.35 What is the x_2-component of $\nabla \cdot \nabla\underline{v}$?

2.36 The trace of a tensor is the sum of the elements on the diagonal, as shown below:

$$\underline{\underline{A}} = \begin{pmatrix} a & b & c \\ d & e & f \\ g & h & i \end{pmatrix}_{xyz}$$

$$\text{trace}\left(\underline{\underline{A}}\right) = A_{11} + A_{22} + A_{33} = a + e + i$$

Show that the trace of $\nabla\underline{v}$ is equal to $\nabla \cdot \underline{v}$.

2.37 What are the orders of the following quantities? Prove your answers using Einstein notation.

(a) $\underline{\underline{A}} : \underline{\underline{B}}$

(b) $(\underline{\underline{A}} \cdot \underline{b}) \cdot \underline{v}$

(c) $\underline{a}\,\underline{b} \cdot \underline{\underline{C}}$

(d) $\nabla^2\underline{\underline{A}}$

(e) $(\underline{a} \cdot \underline{\underline{B}}) \times \underline{c}$

2.38 Are the following equations valid? Explain.

(a) $\underline{a} \cdot \underline{b} + \lambda \underline{b} = \underline{v}$

(b) $\underline{a}\,\underline{b} + \nabla \underline{b} = \underline{\underline{C}}$

(c) $\underline{a} : \underline{b} + \nabla \cdot \underline{b} = \alpha$

2.39 Calculate the invariants of the tensor $\underline{\underline{B}} =$
$$\begin{pmatrix} 1 & 0 & 3 \\ 2 & 1 & -1 \\ 1 & 4 & 0 \end{pmatrix}_{xyz}.$$

2.40 Show that

$$I_{\underline{\underline{B}}} \equiv \sum_{i=1}^{3} \sum_{j=1}^{3} B_{ij} B_{ji} = \text{trace}(\underline{\underline{B}} \cdot \underline{\underline{B}})$$

2.41 Show that

$$III_{\underline{\underline{B}}} \equiv \sum_{i=1}^{3} \sum_{j=1}^{3} \sum_{m=1}^{3} B_{ij} B_{jm} B_{mi} = \text{trace}(\underline{\underline{B}} \cdot \underline{\underline{B}} \cdot \underline{\underline{B}})$$

2.42 Show that the cross product written as

$$\underline{v} \times \underline{u} = v_p u_s \varepsilon_{psj} \hat{e}_j$$

is equivalent to the cross product carried out with the determinant method [Equation (2.80)].

2.43 Show that the θ-derivative of the cylindrical unit vector e_r is given by $\partial \hat{e}_r / \partial \theta = \hat{e}_\theta$.

2.44 What is $\partial \hat{e}_\theta / \partial \theta$ equal to in the cylindrical coordinate system? Derive your answer.

2.45 Calculate $\partial \hat{e}_r / \partial \phi$, $\partial \hat{e}_\theta / \partial \phi$, and $\partial \hat{e}_\phi / \partial \phi$ for the spherical coordinate system. Your final answer will be in terms of $\hat{e}_r$, $\hat{e}_\theta$, and $\hat{e}_\phi$.

2.46 For the points listed below in Cartesian coordinates x_1, x_2, x_3, write the cylindrical system unit vectors at that point, $\hat{e}_r$, $\hat{e}_\theta$, and $\hat{e}_z$. Express the unit vectors with respect to the 123 Cartesian coordinate system. Sketch the results in the $x_1 x_2$-plane.

(a) $\begin{pmatrix} 1 \\ 2 \\ 0 \end{pmatrix}_{123}$

(b) $\begin{pmatrix} 0 \\ 1 \\ 0 \end{pmatrix}_{123}$

2.47 Consider the steady flow of a fluid of density ρ in which the velocity at every point is given by the vector field $\underline{v}(x, y, z)$. What is the mass flow rate (mass/time) through a surface of area A located at point P? The unit normal to the surface considered is given by the unit vector $\hat{n}$ (Figure 2.15).

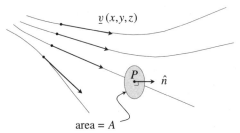

Figure 2.15 Flow considered in Problem 2.47.

CHAPTER 3

Newtonian Fluid Mechanics

It is now helpful to recall the big picture of what we are trying to accomplish. The purpose of this text is to help the reader to understand rheology. Our goal is to be able to take knowledge of material properties and interest in a flow situation and to be able to predict stresses, strains, velocities, or any other variable of interest that will result from the ensuing flow or deformation. Once material properties and the flow situation are supplied, certain stresses, strains, and so on, will be produced because some physical laws are known to hold when matter flows. The physical laws form mathematical constraints on the variables in the problem and allow only particular solutions, that is, once we decide on the material and the flow situation, a particular, nonarbitrary value of the stress, for example, will be produced.

The two physical laws governing the isothermal deformation of matter are the law of conservation of mass and the law of conservation of linear momentum.[1] To obtain the conservation equations in a form compatible with the mathematics we have discussed so far, we will derive equations for both of these laws. These two equations are sometimes called the equations of change. After this we will introduce the Newtonian constitutive equation, which captures, mathematically, how simple fluids respond to stresses and deformation. We spend the latter half of this chapter solving flow problems for Newtonian fluids.

The goal of this chapter, then, is to introduce Newtonian fluid mechanics as a stepping-off point for the study of non-Newtonian fluid mechanics in the remainder of the text. If you are already familiar with Newtonian fluid mechanics, you may wish to skip this chapter.

We begin now by deriving the equations of change.

3.1 Conservation of Mass

The usual engineering problem-solving procedure for applying the principle of conservation of mass is to choose a system (a mixer, for example), identify the streams passing into and out of the system, and set up an equation where

$$\text{mass in } - \text{ mass out } = 0$$

---

[1] A third conservation law is conservation of energy, and it is essential in solving nonisothermal problems; see [26]. A fourth conservation law, conservation of angular momentum, will be invoked in our discussion of the stress tensor.

We will use this procedure to derive a microscopic mass-balance equation for a tiny piece of fluid.

To solve flow problems, we need a mass conservation equation that is applicable at any point in a flowing fluid. Therefore we choose as our system an arbitrary, fixed volume V, located in a flowing stream (Figure 3.1). The surface of V is of arbitrary shape (unspecified), and it has a total surface area we will call S. The volume V is chosen to be fixed in space, and its size does not change, that is, both V and S are constant.

Since V is located in a flowing stream, mass is passing through it, and the total amount of mass enclosed by V may be changing. The latter statement may cause some confusion if you think in terms of water or of some other incompressible fluid—a fluid with constant density. The change in mass in a fixed volume V comes from changing density, as in a gas, for example.

Our next task is to write the net amount of mass leaving the volume V. Since the specifics of our volume element are arbitrary and unspecified, we must write the mass balance in terms of an integral in the most general terms. Consider a differential surface element dS, where the local velocity through dS is equal to $\underline{v}$ (see Figure 3.1). The velocity vector, which indicates the fluid velocity in the vicinity of dS, does not necessarily point such that it is perpendicular to dS. The vector $\underline{v}$ may be resolved into two components, one that runs parallel to dS, which does not contribute to flow through dS, and another component perpendicular to dS, which causes fluid to pass through dS. To calculate the volumetric flow through dS we must isolate only the component normal to dS. To do this we use the dot product and the unit normal to dS, which we will call $\hat{n}$,

$$\begin{pmatrix} \text{local volumetric} \\ \text{rate of flow} \\ \text{through } dS \\ \text{due to } \underline{v} \end{pmatrix} = \hat{n} \cdot \underline{v}\, dS \qquad (3.1)$$

We choose $\hat{n}$ to be the outwardly pointing normal, and thus expression (3.1) is positive for outward flow and negative for inward flow.

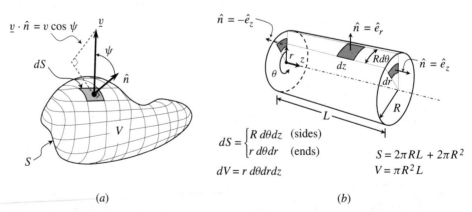

$$\underline{v} \cdot \hat{n} = v \cos \psi$$

$$dS = \begin{cases} R\, d\theta\, dz & \text{(sides)} \\ r\, d\theta\, dr & \text{(ends)} \end{cases}$$

$$dV = r\, d\theta\, dr\, dz$$

$$S = 2\pi R L + 2\pi R^2$$

$$V = \pi R^2 L$$

(a) (b)

Figure 3.1 (a) Schematic of an arbitrary volume, V, used for deriving the equations of change. (b) A particular choice for V, a cylindrical shape.

To calculate the local mass flow rate, we must incorporate the density ρ into this expression,

$$\begin{pmatrix} \text{local mass} \\ \text{rate of flow} \\ \text{out through } dS \\ \text{due to } \underline{v} \end{pmatrix} = \rho(\hat{n} \cdot \underline{v}) \, dS \tag{3.2}$$

$$= \hat{n} \cdot (\rho\underline{v}) \, dS \tag{3.3}$$

The net rate of outward mass flow is then calculated by integrating this expression over the entire surface S that bounds V,

$$\begin{pmatrix} \text{net mass rate} \\ \text{of outward flow} \\ \text{through } S \end{pmatrix} = \int_S \hat{n} \cdot (\rho\underline{v}) \, dS \tag{3.4}$$

For non-steady-state conditions, a net outflux of mass results in a decrease in mass in V. The mass in a small differential volume element of V can be written as $\rho \, dV$, and the total mass in V is the integral of this quantity over V. Thus, the net decrease in mass in V is given by

$$\begin{pmatrix} \text{net decrease} \\ \text{in mass in } V \end{pmatrix} = -\frac{d}{dt}\left(\int_V \rho \, dV \right) \tag{3.5}$$

The negative sign is added to make the expression reflect the decrease rather than the increase in mass.

Putting these two expressions together we get the mass conservation equation,

$$\begin{pmatrix} \text{net decrease} \\ \text{in mass in } V \end{pmatrix} = \begin{pmatrix} \text{net mass rate} \\ \text{of outward flow} \\ \text{through } S \end{pmatrix} \tag{3.6}$$

$$-\frac{d}{dt}\left(\int_V \rho \, dV \right) = \int_S \hat{n} \cdot (\rho\underline{v}) \, dS \tag{3.7}$$

To simplify this, we must change both terms to be integrals over V. We can apply the Gauss–Ostrogradskii divergence theorem to the surface integral to accomplish this (see Chapter 2),

$$0 = \frac{d}{dt}\left(\int_V \rho \, dV \right) + \int_S \hat{n} \cdot (\rho\underline{v}) \, dS \tag{3.8}$$

$$= \frac{d}{dt}\left(\int_V \rho \, dV \right) + \int_V \nabla \cdot (\rho\underline{v}) \, dV \tag{3.9}$$

We want to move both integrals under the same integral sign. First we use the Leibnitz rule to differentiate the integral in the first term. Since V is fixed, $\underline{v}_{\text{surface}} = 0$, and the second term of the Leibnitz formula [Equation (2.264)] is zero,

$$0 = \frac{d}{dt}\left(\int_V \rho \, dV \right) + \int_V \nabla \cdot (\rho\underline{v}) \, dV \tag{3.10}$$

$$= \int_V \frac{\partial \rho}{\partial t} \, dV + \int_V \nabla \cdot (\rho \underline{v}) \, dV \tag{3.11}$$

Finally we combine the two terms under a common integral sign,

$$0 = \int_V \left[\frac{\partial \rho}{\partial t} + \nabla \cdot (\rho \underline{v}) \right] dV \tag{3.12}$$

We have arrived at an equation for the conservation of mass over an arbitrary volume V in our flowing stream. Since that volume is arbitrary, however, this equation must hold over every volume we choose. The only way that this can be true is if the expression within the integral sign is zero at every position in space. Note that it is unusual that one may conclude that an integral being zero implies that the integrand must therefore be zero everywhere. To convince yourself of this, consider the function $y = \sin x$. While we know that

$$\int_0^{2\pi} \sin x \, dx = 0 \tag{3.13}$$

we also know that $\sin x \neq 0$ at every point. What makes this integral equal to zero is the choice of limits. In our derivation, the limits of the integral are the boundaries of V, which are arbitrary. It is the arbitrariness of V that implies that the integrand must be zero at every point.

We have arrived at the following scalar differential equation for the mass balance:

$$
\begin{array}{cc}
\text{Continuity} \\
\text{equation}
\end{array}
\qquad
\boxed{0 = \frac{\partial \rho}{\partial t} + \nabla \cdot (\rho \underline{v})}
\tag{3.14}
$$

which holds at every point in a flowing fluid. This equation is known as the *continuity equation*, and it expresses the physical law that mass is conserved. This equation can be worked out in Cartesian components by using the Einstein summation convention. To write the continuity equation in cylindrical and spherical components, we can evaluate $\nabla \cdot \rho \underline{v}$ using expressions in Tables C.7 and C.8 of Appendix C.2.

3.2 Conservation of Momentum

To derive a vector equation that expresses the physical law of linear momentum conservation in a flowing fluid, we can follow a procedure resembling that used in the last section to derive the continuity equation. The law of momentum conservation is Newton's second law of motion, which is commonly

$$\underline{f} = m\underline{a} \tag{3.15}$$

where $\underline{f}$ is the force on a body of mass m, and $\underline{a}$ is the acceleration experienced by the mass. When there are multiple forces $\underline{f}_{(i)}$, and the mass of the body may be changing, the more complete way of writing Newton's law of motion is

$$\sum_{\substack{\text{all forces } i \\ \text{on body}}} \underline{f}_{(i)} = \frac{d(m\underline{v})}{dt} \tag{3.16}$$

When the mass is constant and there is a single force, the original expression is recovered.

Newton's second law tells us that forces bring about changes in momentum, that is, forces are a type of momentum flux. We can rearrange Newton's law as

$$\underline{0} = \sum_{\substack{\text{all changes in} \\ \text{momentum } i}} \frac{d(m\underline{v})_i}{dt} \tag{3.17}$$

where the forces are included in the terms accounting for changes in momentum. This way of writing Newton's second law emphasizes that momentum is conserved or that there is no net loss or net gain of momentum in a chosen system, only exchanges of momentum between different parts of the system.

For us to apply this principle to a portion of a flow we must account for all of the forces that act on the flow, and we must account for any types of momentum flow that occur due to the transfer of mass through the boundaries of a chosen system in the flow. Since momentum is conserved, the net effect of these momentum flows and forces will be to change the state of momentum of the system.

Momentum balance equation:

$$\begin{pmatrix} \text{rate of } \textit{increase} \\ \text{of momentum in a} \\ \text{fixed volume } V \end{pmatrix} = \begin{pmatrix} \text{net } \textit{inward} \\ \text{flow of} \\ \text{momentum} \end{pmatrix} + \begin{pmatrix} \text{net force} \\ \text{acting } \textit{on} \\ \text{volume} \end{pmatrix} \tag{3.18}$$

$$\begin{pmatrix} \text{rate of } \textit{decrease} \\ \text{of momentum in a} \\ \text{fixed volume } V \end{pmatrix} = \begin{pmatrix} \text{net } \textit{outward} \\ \text{flow of} \\ \text{momentum} \end{pmatrix} - \begin{pmatrix} \text{net force} \\ \text{acting } \textit{on} \\ \text{volume} \end{pmatrix} \tag{3.19}$$

The left side of the momentum balance equation is straightforward as it follows the pattern of the left side of the mass balance:

$$\begin{pmatrix} \text{rate of } \textit{decrease} \\ \text{of momentum in a} \\ \text{fixed volume } V \end{pmatrix} = -\frac{d}{dt}\left(\int_V \rho\underline{v}\, dV \right) \tag{3.20}$$

$$= -\int_V \frac{\partial(\rho\underline{v})}{\partial t}\, dV \tag{3.21}$$

We have used the Leibnitz rule in going from the first equation to the second. Again, since V is fixed in space, $\underline{v}_{\text{surface}} = 0$, the second term of the Leibnitz rule [Equation (2.264)] is zero.

There are three contributions to the right side of the momentum balance [Equation (3.19)] in a flowing fluid: momentum flow by convection and two types of forces, molecular and body forces. We will account for how each affects our volume V and then combine them in Equation (3.19) to obtain the momentum balance for an arbitrary volume in a flowing fluid.

3.2.1 MOMENTUM FLUX BY CONVECTION

In accounting for the net convection of momentum out of V, we will be following the same procedure used when we wrote the convection of mass out of V. Consider a differential surface element dS, where the local fluid velocity through dS is $\underline{v}$. Again, the velocity vector $\underline{v}$ does not necessarily point such that it is perpendicular to dS. The local rate of flow of momentum due to $\underline{v}$ equals

$$\begin{pmatrix} \text{local rate of flow} \\ \text{of momentum through} \\ dS \text{ due to } \underline{v} \end{pmatrix} = \begin{pmatrix} \text{momentum} \\ \overline{\text{volume flow}} \end{pmatrix} \begin{pmatrix} \text{volume} \\ \overline{\text{time}} \end{pmatrix} \qquad (3.22)$$

$$= (\rho\underline{v})(\hat{n} \cdot \underline{v}\, dS) \qquad (3.23)$$

where on the right side the first term in parentheses is the momentum per unit volume, and the second term is the volumetric flow rate through dS. Since ρ and $\hat{n} \cdot \underline{v}$ are scalars, we can rearrange this expression to be

$$\begin{pmatrix} \text{local rate of flow} \\ \text{of momentum through} \\ dS \text{ due to } \underline{v} \end{pmatrix} = \hat{n} \cdot (\rho\underline{v}\,\underline{v})\, dS \qquad (3.24)$$

The dyad $\underline{v}\,\underline{v}$ is an indeterminate vector product, that is, tensor, and it has appeared naturally in the development of the momentum balance. The net rate of outward flow of momentum is then calculated by integrating this expression over the entire surface S,

$$\begin{pmatrix} \text{net rate of outward flow} \\ \text{of momentum from } V \\ \text{due to flux through } S \end{pmatrix} = \int_S \hat{n} \cdot (\rho\underline{v}\,\underline{v})\, dS \qquad (3.25)$$

Recall that this is the net outward flow because we choose to use the outwardly pointing normal $\hat{n}$. Using the divergence theorem to convert this to a volume integral we obtain

$$\begin{pmatrix} \text{net rate of outward flow} \\ \text{of momentum from } V \\ \text{due to flux through } S \end{pmatrix} = \int_V \nabla \cdot (\rho\underline{v}\,\underline{v})\, dV \qquad (3.26)$$

This expression will be substituted into Equation (3.19).

3.2.2 MOMENTUM FLUX BY MOLECULAR FORCES

The second term in the general momentum balance [Equation (3.19)] has to do with forces on V, and we break these into two types: molecular forces, which occur because of the collisions or interactions among molecules and parts of molecules, and body forces, which are external forces due to, for example, gravity or an electromagnetic field. First, we examine molecular forces.

The nature of molecular forces varies greatly with the type of material being studied. In fact, the complexity of rheology is due in large part to our limited understanding of molecular forces. We seek here to derive a general expression that will allow us to describe molecular forces, and thus to account for their effects in the momentum balance, without having to specify the origin of these forces.

We begin with our arbitrary volume V and examine the small surface element dS centered around a point on the surface (Figure 3.1). The surface element dS has an outward unit normal vector $\hat{n}$, which has already been discussed. The quantity we are trying to express is the force on the surface dS due to molecular forces. We will call this force $\underline{f}$, which we can write in Cartesian coordinates as follows:

$$\underline{f} = f_i \hat{e}_i = f_1 \hat{e}_1 + f_2 \hat{e}_2 + f_3 \hat{e}_3 \tag{3.27}$$

We would now like to calculate the coefficients of $\underline{f}$ in terms of a general expression that keeps track of the state of stress in the fluid at every point. Recall that stress is force per unit area. If we can write a vector expression for the stress on dS, we can calculate $\underline{f}$ from the relation

$$\underline{f} = \left(\begin{array}{c} \text{stress on } dS \\ \text{at point } P \end{array} \right) dS \tag{3.28}$$

Our first task is therefore to examine the state of stress at an arbitrary point. We will use the Cartesian coordinates $\hat{e}_1, \hat{e}_2, \hat{e}_3$.

To describe the state of stress in a fluid at an arbitrary point consider three mutually perpendicular planes passing through a point P, as shown in Figure 3.2. On each plane there is a stress (force per unit area). The stress on each plane can be described by a vector. This vector will not, in general, point in any special direction; for instance, it will not necessarily be perpendicular to the plane. To derive a general expression for the stress at P, we will make no assumptions about the directions of the stress vectors on the three planes.

The three stress vectors that describe the state of stress at P will be called $\underline{a}$, $\underline{b}$, and $\underline{c}$ and will be taken to act on surfaces whose unit normal vectors are $\hat{e}_1, \hat{e}_2$, and $\hat{e}_3$, respectively. We begin by examining $\underline{a}$ in the Cartesian coordinate system:

$$\underline{a} = a_1 \hat{e}_1 + a_2 \hat{e}_2 + a_3 \hat{e}_3 \tag{3.29}$$

Note that a_1 is the stress on a 1-plane (i.e., on a plane whose unit normal is $\hat{e}_1$) in the 1-direction. We now define a scalar quantity Π_{11} to be equal to a_1. By writing the coefficients of $\underline{a}$ (and subsequently $\underline{b}$ and $\underline{c}$) in terms of these double-subscripted scalars, we can organize all of the different stress components that exist at point P:

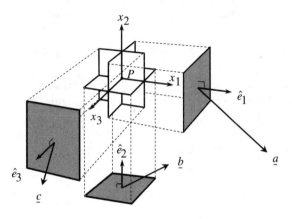

Figure 3.2 Schematic of the state of stress at point P in a flowing system. The vectors shown indicate the stresses on three mutually perpendicular planes passing through P.

$$\Pi_{11} = a_1 = \begin{pmatrix} \text{stress at point } P \\ \text{on a 1-surface} \\ \text{in the 1-direction} \end{pmatrix} \tag{3.30}$$

$$\Pi_{12} = a_2 = \begin{pmatrix} \text{stress at point } P \\ \text{on a 1-surface} \\ \text{in the 2-direction} \end{pmatrix} \tag{3.31}$$

$$\Pi_{13} = a_3 = \begin{pmatrix} \text{stress at point } P \\ \text{on a 1-surface} \\ \text{in the 3-direction} \end{pmatrix} \tag{3.32}$$

$$\underline{a} = (\text{stress on a 1-surface})$$

$$= \Pi_{11}\hat{e}_1 + \Pi_{12}\hat{e}_2 + \Pi_{13}\hat{e}_3 \tag{3.33}$$

$$= \Pi_{1i}\hat{e}_i \tag{3.34}$$

Following the same logic for $\underline{b}$ and $\underline{c}$ we obtain:

$$\Pi_{21} = b_1 = \begin{pmatrix} \text{stress at point } P \\ \text{on a 2-surface} \\ \text{in the 1-direction} \end{pmatrix} \tag{3.35}$$

$$\Pi_{22} = b_2 = \begin{pmatrix} \text{stress at point } P \\ \text{on a 2-surface} \\ \text{in the 2-direction} \end{pmatrix} \tag{3.36}$$

$$\Pi_{23} = b_3 = \begin{pmatrix} \text{stress at point } P \\ \text{on a 2-surface} \\ \text{in the 3-direction} \end{pmatrix} \tag{3.37}$$

$$\underline{b} = (\text{stress on a 2-surface})$$

$$= \Pi_{21}\hat{e}_1 + \Pi_{22}\hat{e}_2 + \Pi_{23}\hat{e}_3 \tag{3.38}$$

$$= \Pi_{2i}\hat{e}_i \tag{3.39}$$

$$\underline{c} = (\text{stress on a 3-surface})$$

$$= \Pi_{31}\hat{e}_1 + \Pi_{32}\hat{e}_2 + \Pi_{33}\hat{e}_3 \tag{3.40}$$

$$= \Pi_{3i}\hat{e}_i \tag{3.41}$$

In general then, Π_{ik} is the stress on an i-plane in the k-direction. There are nine stress quantities Π_{ik}.

We turn now to the question of calculating the stress vector on dS, an arbitrarily chosen differential piece of the surface enclosing V, in terms of the nine stress quantities we have defined and organized. Since dS is oriented arbitrarily in space, we need to find a way to relate the stress on dS with the stresses on the three mutually perpendicular planes we considered earlier. We defined the vector $\underline{f}$, which is the force (not stress) on dS:

$$\underline{f} = f_1\hat{e}_1 + f_2\hat{e}_2 + f_3\hat{e}_3 \tag{3.42}$$

The coefficient of $\hat{e}_1$ is f_1, which is the force on dS in the 1-direction. Examining the Π_{ik} we see that there are three Π_{ik} that describe stresses in the 1-direction:

$$\Pi_{11} = \begin{pmatrix} \text{stress at point } P \\ \text{on a 1-surface} \\ \textit{in the 1-direction} \end{pmatrix} \tag{3.43}$$

$$\Pi_{21} = \begin{pmatrix} \text{stress at point } P \\ \text{on a 2-surface} \\ \textit{in the 1-direction} \end{pmatrix} \tag{3.44}$$

$$\Pi_{31} = \begin{pmatrix} \text{stress at point } P \\ \text{on a 3-surface} \\ \textit{in the 1-direction} \end{pmatrix} \tag{3.45}$$

These three quantities all contribute to the final stress on dS directed along $\hat{e}_1$ (Figure 3.3). The areas over which each of these stresses acts differ, however. Π_{11} is the stress on a 1-surface in the 1-direction, and thus it acts on the projection of dS in the 1-direction. Π_{21} is the stress on a 2-surface in the 1-direction, and thus it acts on the projection of dS in the 2-direction. The projection in a coordinate direction $\hat{e}_i$ of a surface of area A, with unit normal vector $\hat{n}$, is given by (see Appendix C.3)

$$\begin{pmatrix} \text{projection of } A \\ \text{in direction of } \hat{e}_i \end{pmatrix} = \hat{n} \cdot \hat{e}_i \, A \tag{3.46}$$

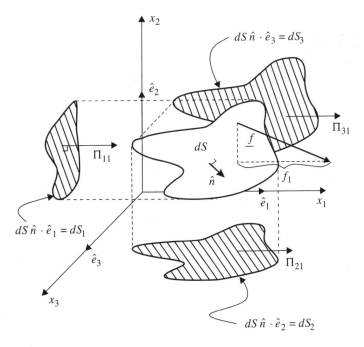

$$f_1 = \Pi_{11} dS_1 + \Pi_{21} dS_2 + \Pi_{31} dS_3$$

Figure 3.3 Effect of stresses in the $\hat{e}_1$-direction as acting on dS.

Thus, the area over which Π_{11} acts is $\hat{n} \cdot \hat{e}_1 \, dS$, and likewise the areas over which Π_{21} and Π_{31} act are $\hat{n} \cdot \hat{e}_2 \, dS$ and $\hat{n} \cdot \hat{e}_3 \, dS$, respectively. Now we can write an expression for f_1 in terms of the Π_{i1}:

$$\begin{pmatrix} \text{force on } dS \\ \text{in the} \\ \hat{e}_1\text{-direction} \end{pmatrix} = \sum_{i=1}^{3} \left[\begin{pmatrix} \text{stress acting} \\ \text{in the } \hat{e}_1\text{-} \\ \text{direction on an} \\ i\text{-surface} \end{pmatrix} \begin{pmatrix} \text{area of an} \\ i\text{-surface} \end{pmatrix} \right] \tag{3.47}$$

$$f_1 = \Pi_{11}(\hat{n} \cdot \hat{e}_1) \, dS + \Pi_{21}(\hat{n} \cdot \hat{e}_2) \, dS + \Pi_{31}(\hat{n} \cdot \hat{e}_3) \, dS \tag{3.48}$$

This can be simplified and rearranged as:

$$f_1 = dS \left(\hat{n} \cdot \hat{e}_1 \Pi_{11} + \hat{n} \cdot \hat{e}_2 \Pi_{21} + \hat{n} \cdot \hat{e}_3 \Pi_{31} \right) \tag{3.49}$$

$$= dS \, \hat{n} \cdot \left(\hat{e}_1 \Pi_{11} + \hat{e}_2 \Pi_{21} + \hat{e}_3 \Pi_{31} \right) \tag{3.50}$$

We can follow the same logic to arrive at expressions for f_2, the force on dS in the 2-direction, and f_3, the force on dS in the 3-direction:

$$f_2 = dS \, \hat{n} \cdot \left(\hat{e}_1 \Pi_{12} + \hat{e}_2 \Pi_{22} + \hat{e}_3 \Pi_{32} \right) \tag{3.51}$$

$$f_3 = dS \, \hat{n} \cdot \left(\hat{e}_1 \Pi_{13} + \hat{e}_2 \Pi_{23} + \hat{e}_3 \Pi_{33} \right) \tag{3.52}$$

We can now combine these expressions for the scalar coefficients of $\underline{f}$ and write the complete expression for $\underline{f}$ [Equation (3.42)]:

$$\underline{f} = f_1 \hat{e}_1 + f_2 \hat{e}_2 + f_3 \hat{e}_3 \tag{3.53}$$

$$= dS \, \hat{n} \cdot \left(\hat{e}_1 \Pi_{11} + \hat{e}_2 \Pi_{21} + \hat{e}_3 \Pi_{31} \right) e_1$$
$$+ dS \, \hat{n} \cdot \left(\hat{e}_1 \Pi_{12} + \hat{e}_2 \Pi_{22} + \hat{e}_3 \Pi_{32} \right) e_2$$
$$+ dS \, \hat{n} \cdot \left(\hat{e}_1 \Pi_{13} + \hat{e}_2 \Pi_{23} + \hat{e}_3 \Pi_{33} \right) e_3 \tag{3.54}$$

$$= dS \, \hat{n} \cdot \left(\hat{e}_1 \hat{e}_1 \Pi_{11} + \hat{e}_2 \hat{e}_1 \Pi_{21} + \hat{e}_3 \hat{e}_1 \Pi_{31} \right.$$
$$+ \hat{e}_1 \hat{e}_2 \Pi_{12} + \hat{e}_2 \hat{e}_2 \Pi_{22} + \hat{e}_3 \hat{e}_2 \Pi_{32}$$
$$\left. + \hat{e}_1 \hat{e}_3 \Pi_{13} + \hat{e}_2 \hat{e}_3 \Pi_{23} + \hat{e}_3 \hat{e}_3 \Pi_{33} \right) \tag{3.55}$$

This final expression is a scalar (dS) multiplied by the dot product of a vector ($\hat{n}$) with a tensor ($\underline{\underline{\Pi}} \equiv$ sum of dyads). The terms of $\underline{\underline{\Pi}}$ are not in the usual order we take when we write a tensor in matrix form, but all the terms are there. Rearranging, we obtain

$$\underline{f} = dS \, \hat{n} \cdot \left(\hat{e}_1 \hat{e}_1 \Pi_{11} + \hat{e}_1 \hat{e}_2 \Pi_{12} + \hat{e}_1 \hat{e}_3 \Pi_{13} \right.$$
$$+ \hat{e}_2 \hat{e}_1 \Pi_{21} + \hat{e}_2 \hat{e}_2 \Pi_{22} + \hat{e}_2 \hat{e}_3 \Pi_{23}$$
$$\left. + \hat{e}_3 \hat{e}_1 \Pi_{31} + \hat{e}_3 \hat{e}_2 \Pi_{32} + \hat{e}_3 \hat{e}_3 \Pi_{33} \right) \tag{3.56}$$

$$= dS \, \hat{n} \cdot \begin{pmatrix} \Pi_{11} & \Pi_{12} & \Pi_{13} \\ \Pi_{21} & \Pi_{22} & \Pi_{23} \\ \Pi_{31} & \Pi_{32} & \Pi_{33} \end{pmatrix}_{123} \tag{3.57}$$

$$\text{Force on a surface} \atop \text{with unit normal } \hat{n} \qquad \boxed{\underline{f} = dS\,\hat{n}\cdot\underline{\underline{\Pi}}} \qquad (3.58)$$

The tensor $\underline{\underline{\Pi}}$ is called the *total stress tensor*. As we have shown, $\underline{\underline{\Pi}}$ contains all the information about the state of stress at a point. To calculate the force acting on a particular surface, we must simply dot the unit normal to the surface with $\underline{\underline{\Pi}}$ and multiply by the area of the surface [Equation (3.58)]. Comparing the preceding with Equation (3.28) we see in a fluid whose state of stress is characterized by the total stress tensor $\underline{\underline{\Pi}}$ that the stress vector acting on an area dS with normal $\hat{n}$ is given by $\hat{n}\cdot\underline{\underline{\Pi}}$.

It should be clear now that tensors are useful in fluid mechanics and rheology. The final expression, Equation (3.58), is simple and nicely does the bookeeping associated with expressing the state of stress of a fluid at a point. Experimental studies on $\underline{\underline{\Pi}}$ have shown that it is symmetric for the vast majority of fluids, and it can be shown rigorously to be symmetric for nonpolar fluids [7] or for fluids with no body moments[2] to couple stresses [247, 148]. We will always take $\underline{\underline{\Pi}}$ to be symmetric.

One last unresolved issue is the sign of the stress tensor $\underline{\underline{\Pi}}$. Note that $\hat{n}\cdot\underline{\underline{\Pi}}$ gives the stress vector on a surface. Surfaces have two sides, however, and we need to choose a convention for whether the stress is positive in the $\hat{e}_1$-, $\hat{e}_2$-, $\hat{e}_3$-directions or positive in the $-\hat{e}_1$-, $-\hat{e}_2$-, $-\hat{e}_3$-directions. In this text we follow Bird et al. [28, 26] and choose that $\Pi_{ik} > 0$ for forces in the positive $\hat{e}_k$-direction. This convention implies that the forces generated by our volume V will be positive, and forces acting *on* V will be negative. This convention affects any expression that contains $\underline{\underline{\Pi}}$.[3]

We return now to our original goal, which was to write the term in the momentum balance that accounts for the effects of molecular forces on the volume V. We now have an expression for the molecular forces on dS. It remains only to integrate over S:

---

[2] Body moments are torques experienced by particles of fluid due to some intrinsic property of the fluid, that is, not due to the usual body forces (gravity) or surface forces (molecular action at the surface of the particle). Ferrofluids, which are suspensions of magnetic particles, experience body moments when they flow in the presence of a magnetic field [148]. For these fluids it is inappropriate to assume that the stress tensor is symmetric.

[3] A warning to the reader: most texts in mechanical engineering take the opposite convention [179]. The choice is arbitrary, and there are good reasons for both choices. We choose $\Pi_{ik} > 0$ for forces in the positive $\hat{e}_k$-direction so that Newton's law of viscosity takes this form:

$$\tau_{ki} = -\mu\frac{\partial v_i}{\partial x_k} \qquad (3.59)$$

For the opposite convention, the minus sign does not appear, and Newton's law of viscosity reads that positive momentum flux occurs up a gradient in velocity. This is not a desirable convention, since it does not match the other transport laws, Fourier's law (positive heat flux is down a temperature gradient) and Fick's law (positive mass flux occurs down a concentration gradient). We refer you to Bird et al. [28] for a complete discussion of their reasons for choosing this convention. Also, in some texts [179] the meaning of the subscripts of $\underline{\underline{\Pi}}$ are reversed from what we are using, that is, in some texts Π_{rs} is the stress on an s-plane in the r-direction. The latter is not as common as the one we are using, but when reading other sources it is important to note which convention is being followed.

$$\left(\begin{array}{c}\text{molecular force}\\ \text{acting } on\\ \text{surface } dS\end{array}\right) = -\hat{n} \cdot \underline{\underline{\Pi}} \, dS \tag{3.60}$$

$$\left(\begin{array}{c}\text{net molecular force}\\ \text{acting on}\\ \text{total surface } S\end{array}\right) = -\int_S \hat{n} \cdot \underline{\underline{\Pi}} \, dS \tag{3.61}$$

$$\left(\begin{array}{c}\text{net molecular force}\\ \text{acting on}\\ \text{total volume } V\end{array}\right) = -\int_V \nabla \cdot \underline{\underline{\Pi}} \, dV \tag{3.62}$$

We used the Gauss–Ostrogradskii divergence theorem in arriving at the last equation. The negative sign is required due to our convention for the sign of the total stress tensor.

3.2.3 MOMENTUM FLUX BY BODY FORCES

The final contribution to the momentum balance comes from body forces; these are due to external fields. The only body force we will be considering is generated by the gravity field g, but other types of fields such as electric and magnetic fields may have an influence in flows of special materials such as polar fluids [7] and charged suspensions [29].

Gravity acts on the mass in dV, and thus the force due to gravity is (mass)(acceleration) $= (\rho \, dV)(g)$, where ρ is the fluid density. The total effect of the gravity body force is therefore

$$\left(\begin{array}{c}\text{total force on}\\ V \text{ due to body}\\ \text{forces}\end{array}\right) = \int_V \rho \underline{g} \, dV \tag{3.63}$$

3.2.4 EQUATION OF MOTION

We now have all the pieces of the momentum balance, and we can assemble the final equation:

$$\left(\begin{array}{c}\text{rate of decrease}\\ \text{of momentum in a}\\ \text{fixed volume } V\end{array}\right) = \left(\begin{array}{c}\text{net outward}\\ \text{flux of}\\ \text{momentum}\end{array}\right) - \left(\begin{array}{c}\text{net force}\\ \text{acting on}\\ \text{volume}\end{array}\right) \tag{3.64}$$

$$-\int_V \frac{\partial(\rho\underline{v})}{\partial t} \, dV = \int_V \nabla \cdot (\rho\underline{v}\,\underline{v}) \, dV + \int_V \nabla \cdot \underline{\underline{\Pi}} \, dV - \int_V \rho\underline{g} \, dV \tag{3.65}$$

$$\underline{0} = \int_V \left[-\frac{\partial(\rho\underline{v})}{\partial t} - \nabla \cdot (\rho\underline{v}\,\underline{v}) - \nabla \cdot \underline{\underline{\Pi}} + \rho\underline{g} \right] dV \tag{3.66}$$

The same arguments used earlier to derive the continuity equation apply here, namely, since the volume V is arbitrary, this integral over V can be zero for all possible choices of V if and only if the integrand itself is zero. Thus we arrive at the equation of motion, which expresses conservation of linear momentum in a flowing fluid:

$$\text{Equation of motion} \qquad \boxed{\frac{\partial(\rho\underline{v})}{\partial t} = -\nabla \cdot (\rho\underline{v}\,\underline{v}) - \nabla \cdot \underline{\underline{\Pi}} + \rho\underline{g}} \qquad (3.67)$$

This equation is valid for compressible and incompressible fluids of all types, Newtonian and non-Newtonian.

We can use the continuity equation (mass conservation equation) to simplify the equation of motion. First we expand the first term on the right side using Einstein notation:

$$\nabla \cdot (\rho\underline{v}\,\underline{v}) = \frac{\partial}{\partial x_m}\hat{e}_m \cdot (\rho\,v_j\hat{e}_j\,v_p\hat{e}_p) \qquad (3.68)$$

$$= \frac{\partial(\rho v_j v_p)}{\partial x_m}\hat{e}_m \cdot \hat{e}_j\hat{e}_p = \frac{\partial(\rho v_j v_p)}{\partial x_m}\delta_{mj}\hat{e}_p \qquad (3.69)$$

$$= \frac{\partial(\rho v_m v_p)}{\partial x_m}\hat{e}_p \qquad (3.70)$$

To further expand this expression we must use the product rule to differentiate the terms in parentheses. Since we are considering a general fluid, ρ is not necessarily constant, and we must treat ρ and the two vector coefficients as variables:

$$\nabla \cdot (\rho\underline{v}\,\underline{v}) = \frac{\partial(\rho v_m v_p)}{\partial x_m}\hat{e}_p \qquad (3.71)$$

$$= \left\{ \rho\left[\frac{\partial(v_m v_p)}{\partial x_m}\right] + v_m v_p \frac{\partial\rho}{\partial x_m} \right\}\hat{e}_p \qquad (3.72)$$

$$= \rho v_m \frac{\partial v_p}{\partial x_m}\hat{e}_p + \rho v_p \frac{\partial v_m}{\partial x_m}\hat{e}_p + v_m v_p \frac{\partial\rho}{\partial x_m}\hat{e}_p \qquad (3.73)$$

$$= \rho v_m \frac{\partial v_p}{\partial x_m}\hat{e}_p + \rho(v_p\hat{e}_p)(\nabla \cdot \underline{v}) + (v_p\hat{e}_p)v_m \frac{\partial\rho}{\partial x_m} \qquad (3.74)$$

The quantity $v_p\hat{e}_p$ is just the vector $\underline{v}$, and the first and third terms in Equation (3.74) may be written as follows (verify using Einstein notation):

$$\rho v_m \frac{\partial v_p}{\partial x_m}\hat{e}_p = \rho\underline{v} \cdot \nabla\underline{v} \qquad (3.75)$$

$$v_p\hat{e}_p v_m \frac{\partial\rho}{\partial x_m} = \underline{v}(\underline{v} \cdot \nabla\rho) \qquad (3.76)$$

If we also expand the time derivative on the left side of Equation (3.67) (using Einstein notation and then reverting to Gibbs notation after differentiating) and combine that equation with Equations (3.74), (3.75), and (3.76), the equation of motion becomes

$$\rho\frac{\partial\underline{v}}{\partial t} + \underline{v}\frac{\partial\rho}{\partial t} = -\rho\underline{v} \cdot \nabla\underline{v} - \underline{v}\rho(\nabla \cdot \underline{v}) - \underline{v}(\underline{v} \cdot \nabla\rho) - \nabla \cdot \underline{\underline{\Pi}} + \rho\underline{g} \qquad (3.77)$$

$$\rho\left(\frac{\partial\underline{v}}{\partial t} + \underline{v} \cdot \nabla\underline{v}\right) = -\underline{v}\left[\frac{\partial\rho}{\partial t} + \rho(\nabla \cdot \underline{v}) + \underline{v} \cdot \nabla\rho\right] - \nabla \cdot \underline{\underline{\Pi}} + \rho\underline{g} \qquad (3.78)$$

If we look back at the continuity equation [Equation (3.14)] and expand the term $\nabla \cdot (\rho \underline{v})$, we can see that the terms in square brackets in Equation (3.78) sum to zero due to mass conservation. Thus, the microscopic momentum equation for general fluids becomes

Equation of motion
(compressible or
incompressible fluids)

$$\rho \left(\frac{\partial \underline{v}}{\partial t} + \underline{v} \cdot \nabla \underline{v} \right) = -\nabla \cdot \underline{\underline{\Pi}} + \rho \underline{g} \qquad (3.79)$$

and

$$\rho \frac{D\underline{v}}{Dt} = -\nabla \cdot \underline{\underline{\Pi}} + \rho \underline{g} \qquad (3.80)$$

3.3 Newtonian Constitutive Equation

The equation of motion is a powerful tool. It must hold at every point in a flowing fluid since momentum is conserved at every point. Since we have said almost nothing about the nature of the molecular forces described by the stress tensor $\underline{\underline{\Pi}}$, the equation is general. To actually use the equation of motion, however, we must specify $\underline{\underline{\Pi}}$.

There are two major contributions to the total stress tensor $\underline{\underline{\Pi}}$: the thermodynamic pressure p and a second portion that originates in the deformation of the fluid. The thermodynamic pressure is familiar to us from the study of chemistry, and it is given by an equation of state such as the ideal gas law:

Ideal gas
law

$$p\hat{V} = RT \qquad (3.81)$$

where $\hat{V}$ is the specific volume (volume/mole), R is the ideal gas constant, and T is the temperature. For nonideal gases more complex functions of the same variables give thermodynamic pressure. Thermodynamic pressure is an isotropic force, that is, it is a force that acts equally in every direction. Also, the pressure acts only normally (perpendicularly) to a surface. These two properties of pressure, isotropy and perpendicular action, can be captured mathematically by writing the pressure portion of $\underline{\underline{\Pi}}$ as a tensor proportional to the identity tensor $\underline{\underline{I}}$:

$$\begin{pmatrix} \text{pressure} \\ \text{contribution} \\ \text{to } \underline{\underline{\Pi}} \end{pmatrix} = \begin{pmatrix} p & 0 & 0 \\ 0 & p & 0 \\ 0 & 0 & p \end{pmatrix}_{123} = p\underline{\underline{I}} \qquad (3.82)$$

The pressure tensor has the same form in all orthonormal bases.

The portion of $\underline{\underline{\Pi}}$ that is not thermodynamic pressure is $\underline{\underline{\tau}}$, called the *extra stress tensor*:

Extra stress
tensor

$$\underline{\underline{\tau}} = \underline{\underline{\Pi}} - p\underline{\underline{I}} \qquad (3.83)$$

Since both $\underline{\underline{\Pi}}$ and $p\underline{\underline{I}}$ are symmetric tensors, the extra stress tensor $\underline{\underline{\tau}}$ is also symmetric. The tensor $\underline{\underline{\tau}}$ contains the contributions to stress that result from fluid deformation. When the fluid is at rest, $\underline{\underline{\Pi}}$ becomes $p\underline{\underline{I}}$, the hydrostatic pressure.

An equation that specifies $\underline{\underline{\tau}}$ for a fluid is called a *stress constitutive equation* for that fluid. The particular nature of the stresses described by $\underline{\underline{\tau}}$ depends on the type of fluid being studied. A great deal of current research in rheology is focused on developing accurate stress constitutive equations for non-Newtonian fluids [138]. The constitutive equation expresses the molecular stresses generated in the flow in terms of kinetic variables such as velocities, strains, and derivatives of velocities and strains. Once a constitutive equation is decided upon, $\underline{\underline{\tau}} = \underline{\underline{\tau}}(\underline{v}, \nabla\underline{v}$, etc.), it may be inserted into the equation for $\underline{\underline{\Pi}}$ [Equation 3.83)] and subsequently into the equation of motion [Equation (3.67)] (Figure 3.4). The equation of motion may then be solved, along with the continuity equation, for the velocity field or for other flow variables. Note also that since $\underline{\underline{\tau}}$ is a tensor, constitutive equations are tensor equations.

In this section we will introduce you to the simplest constitutive equation, that for a Newtonian fluid. There are two versions of the Newtonian constitutive equation, one for compressible and a second for incompressible fluids. Next we will show how the equation of motion simplifies for an incompressible Newtonian fluid, and finally we demonstrate how to solve Newtonian flow problems. Although most readers will already be familiar with Newtonian fluid mechanics, we use the Newtonian case to provide the reader with practice in dealing with the tensor nature of continuum mechanics. This will help us later when we deal with non-Newtonian fluid mechanics, that is, rheology.

3.3.1 COMPRESSIBLE FLUIDS

The Newtonian constitutive equation was originally an empirical equation when the basic principles were described by Newton in 1687 [190]. An empirical equation is one that is deduced from experimental observations rather than being derived from a fundamental principle. Newton conducted experiments in sliding (shear) flow on incompressible fluids and found that the shear stress τ_{21} was directly proportional to the gradient of velocity (Figure 3.5).

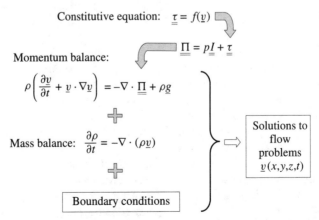

Constitutive equation: $\underline{\underline{\tau}} = f(\underline{v})$

$\underline{\underline{\Pi}} = p\underline{\underline{I}} + \underline{\underline{\tau}}$

Momentum balance:

$$\rho\left(\frac{\partial \underline{v}}{\partial t} + \underline{v} \cdot \nabla\underline{v}\right) = -\nabla \cdot \underline{\underline{\Pi}} + \rho\underline{g}$$

Mass balance: $\frac{\partial \rho}{\partial t} = -\nabla \cdot (\rho\underline{v})$

Boundary conditions

Solutions to flow problems $\underline{v}(x,y,z,t)$

Figure 3.4 Schematic of how the equations derived for momentum and mass balance combine with the constitutive equation to allow us to solve flow problems.

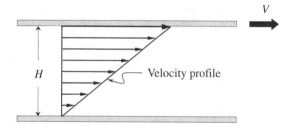

Newton conducted experiments on various fluids in shear flow by confining the fluids between two plates and measuring the amount of force required to move the upper plate at various speeds. The relationship he observed between shear stress (force/area) and shear rate (velocity/gap) is called *Newton's law of viscosity.*

$$\frac{\text{shear force}}{\text{area}} = \frac{V}{H} \tag{3.84}$$

Newton's law of viscosity

$$\tau_{21} = -\mu \frac{dv_1}{dx_2} \tag{3.85}$$

This scalar equation gives the relationship between shear stress and velocity in one particular kind of flow. A constitutive equation, however, is a complete expression of how a fluid responds to any kind of flow. The complete Newtonian constitutive equation was deduced later, and modern molecular modeling work [110] also gives the same result [107]. The Newtonian constitutive equation for compressible fluids is [7]

Newtonian constitutive equation (compressible)

$$\underline{\underline{\tau}} = -\mu \left[\nabla \underline{v} + (\nabla \underline{v})^T \right] + \left(\frac{2}{3}\mu - \kappa \right) (\nabla \cdot \underline{v}) \underline{\underline{I}} \tag{3.86}$$

where μ is the Newtonian shear viscosity and κ is the dilatational viscosity. The shear viscosity is the coefficient that describes the resistance of a fluid to sliding motion, and this is the primary material parameter with which we will be concerned in this chapter. The dilatational or bulk viscosity κ is a coefficient that describes an isotropic contribution to stress that is generated when the density of a fluid changes upon deformation. When the density changes, a stress is produced in all directions that is due to the density change. Since the density change comes about because of the flow, this contribution to the stress is zero when the flow stops. We can see that this requirement is reflected in Equation (3.86) since the term with κ goes to zero when $\underline{v}$ goes to zero. The dilatational viscosity is only of concern when suspensions and polyatomic gases are being considered [13] ($\kappa = 0$ for an ideal monatomic gas).

The Newtonian constitutive equation may seem like a handful, but there are some straightforward insights we can get from it. Note that the Newtonian constitutive equation is a tensor equation, as required. It consists of two parts, both symmetric. The first term has been made explicitly symmetric by the addition of the tensor $\nabla \underline{v}$ to its transpose before multiplying it by the viscosity (verify using Einstein notation). The second term is an additional isotropic stress that affects the diagonal of the stress tensor. This term, which accounts for stresses generated by bulk compressibility, is also symmetric. Both terms depend on $\underline{v}$, as expected, since $\underline{\underline{\tau}}$ is the part of $\underline{\underline{\Pi}}$ that depends on the flow field.

As we stated earlier, most known fluids generate a symmetric stress tensor. Although there are some specialty fluids described in the research literature that are characterized by nonsymmetric stress tensors [55], we will always take $\underline{\underline{\tau}}$ to be symmetric.

3.3.2 INCOMPRESSIBLE FLUIDS

For a fluid of constant density (an incompressible fluid, that is, ρ is not a function of space or time), the equation of conservation of mass becomes

$$\frac{\partial \rho}{\partial t} + \nabla \cdot (\rho \underline{v}) = 0 \qquad (3.87)$$

Continuity equation
for incompressible liquids
$$\boxed{\nabla \cdot \underline{v} = 0} \qquad (3.88)$$

This equation allows us to simplify the general expression for the Newtonian constitutive equation [Equation (3.86)]. Thus, for an incompressible Newtonian fluid,

$$\underline{\underline{\tau}} = -\mu \left[\nabla \underline{v} + (\nabla \underline{v})^T \right] \qquad (3.89)$$

The tensor in square brackets is called the rate-of-strain tensor and is given the symbol $\underline{\underline{\dot{\gamma}}}$,

Rate-of-strain
tensor
$$\boxed{\underline{\underline{\dot{\gamma}}} \equiv \nabla \underline{v} + (\nabla \underline{v})^T} \qquad (3.90)$$

Using this notation, the constitutive equation for an incompressible Newtonian fluid is

Newtonian
constitutive equation
(incompressible)
$$\boxed{\underline{\underline{\tau}} = -\mu \underline{\underline{\dot{\gamma}}}} \qquad (3.91)$$

Equation (3.91) applies to *all* flow situations, and, as we will see in this chapter and later in Chapter 4, it contains Newton's law of viscosity as a special case when shear flow is considered. We will demonstrate the use of Equation (3.91) in the later sections of this chapter.

3.4 Navier–Stokes Equation

Most of the terms in the equation of motion are composed of expressions involving the quantities $\underline{v}$, ρ, ∇, and P, all of which are straightforward to relate to flow problems. The exception is the term $\nabla \cdot \underline{\underline{\Pi}}$, which contains the unknown linear vector function $\underline{\underline{\Pi}}$, the stress tensor. With a constitutive equation in hand, we now know the total stress tensor, $\underline{\underline{\Pi}} = p\underline{\underline{I}} + \underline{\underline{\tau}}$, for one class of materials, Newtonian fluids. Thus, for the two Newtonian cases outlined (compressible and incompressible), we can simplify the equation of motion and use it to solve flow problems.

The equation of motion (microscopic conservation of momentum equation) is

$$\rho \left(\frac{\partial \underline{v}}{\partial t} + \underline{v} \cdot \nabla \underline{v} \right) = -\nabla \cdot \underline{\underline{\Pi}} + \rho \underline{g} \qquad (3.92)$$

For $\underline{\underline{\Pi}}$ we now substitute $\underline{\underline{\Pi}} = p\underline{\underline{I}} + \underline{\underline{\tau}}$,

$$\rho\left(\frac{\partial \underline{v}}{\partial t} + \underline{v} \cdot \nabla \underline{v}\right) = -\nabla \cdot (p\underline{\underline{I}} + \underline{\underline{\tau}}) + \rho\underline{g} \tag{3.93}$$

Examining the first term on the right side,

$$-\nabla \cdot (p\underline{\underline{I}} + \underline{\underline{\tau}}) = -\nabla \cdot p\underline{\underline{I}} - \nabla \cdot \underline{\underline{\tau}} \tag{3.94}$$

$$= -\nabla p - \nabla \cdot \underline{\underline{\tau}} \tag{3.95}$$

The equality between $\nabla \cdot p\underline{\underline{I}}$ and ∇p can be verified in a straightforward manner using Einstein notation. Substituting this into the equation of motion, we obtain

Equation of motion
(in terms of $\underline{\underline{\tau}}$)

$$\boxed{\rho\left(\frac{\partial \underline{v}}{\partial t} + \underline{v} \cdot \nabla \underline{v}\right) = -\nabla p - \nabla \cdot \underline{\underline{\tau}} + \rho\underline{g}} \tag{3.96}$$

$$\rho\frac{D\underline{v}}{Dt} = -\nabla p - \nabla \cdot \underline{\underline{\tau}} + \rho\underline{g} \tag{3.97}$$

For Newtonian fluids we can proceed further since we know the constitutive equation. We will confine our discussion to incompressible fluids, that is, those for which $\underline{\underline{\tau}} = -\mu\underline{\underline{\dot{\gamma}}}$. The expression $-\nabla \cdot \underline{\underline{\tau}}$ then becomes

$$-\nabla \cdot \underline{\underline{\tau}} = \mu\nabla \cdot \underline{\underline{\dot{\gamma}}} \tag{3.98}$$

$$= \mu\nabla \cdot [\nabla\underline{v} + (\nabla\underline{v})^T] \tag{3.99}$$

$$= \mu\nabla^2\underline{v} + \mu\nabla \cdot (\nabla\underline{v})^T \tag{3.100}$$

The second term on the right of Equation (3.100) can be simplified using Einstein notation:

$$\nabla \cdot (\nabla\underline{v})^T = \frac{\partial}{\partial x_j}\hat{e}_j \cdot \left(\frac{\partial}{\partial x_p}\hat{e}_p \, v_m\hat{e}_m\right)^T \tag{3.101}$$

$$= \frac{\partial}{\partial x_j}\hat{e}_j \cdot \frac{\partial}{\partial x_p}v_m\,\hat{e}_m\hat{e}_p = \frac{\partial}{\partial x_j}\frac{\partial v_m}{\partial x_p}\hat{e}_j \cdot \hat{e}_m\hat{e}_p = \frac{\partial}{\partial x_j}\frac{\partial v_m}{\partial x_p}\delta_{jm}\hat{e}_p \tag{3.102}$$

$$= \frac{\partial}{\partial x_m}\frac{\partial v_m}{\partial x_p}\hat{e}_p = \frac{\partial}{\partial x_p}\left(\frac{\partial v_m}{\partial x_m}\right)\hat{e}_p = \frac{\partial}{\partial x_p}(\nabla \cdot \underline{v})\hat{e}_p \tag{3.103}$$

$$= \underline{0}$$

where we have used the continuity equation for incompressible fluids ($\nabla \cdot \underline{v} = 0$) in the final step. Returning to the equation of motion and incorporating Equation (3.100), we obtain

Navier–Stokes
equation

$$\boxed{\rho\left(\frac{\partial \underline{v}}{\partial t} + \underline{v} \cdot \nabla \underline{v}\right) = -\nabla p + \mu\nabla^2\underline{v} + \rho\underline{g}} \tag{3.104}$$

$$\rho\frac{D\underline{v}}{Dt} = -\nabla p + \mu\nabla^2\underline{v} + \rho\underline{g} \tag{3.105}$$

This is the well-known Navier–Stokes equation. It is the microscopic momentum balance (equation of motion) for an incompressible Newtonian fluid. We used the continuity equation for an incompressible fluid in the derivation of the Navier–Stokes equation, and this explains the restriction of this equation to incompressible fluids. For incompressible, non-Newtonian fluids or compressible fluids Equation (3.96) is appropriate, and the constitutive equation must be known to evaluate $\underline{\underline{\tau}}$.

We have derived the two physical laws that govern fluid mechanics, conservation of mass (the continuity equation) and conservation of momentum (the equation of motion). Now we wish to apply them to flow situations to calculate flow patterns.

We close out the chapter with several example flow problems. This will conclude our review of vector/tensor mathematics and Newtonian fluid mechanics. After this we will move on to defining standard flows for examining non-Newtonian fluids in Chapter 4.

3.5 Flow Problems: Incompressible Newtonian Fluids

3.5.1 DRAG FLOW BETWEEN INFINITE PARALLEL PLATES

An incompressible Newtonian liquid is confined between two infinitely wide, parallel plates, separated by a gap of H. The top plate moves at a velocity V in the x_1-direction. Calculate the velocity profile, the flow rate per unit width, and the stress tensor $\underline{\underline{\tau}}$. Assume that the flow is fully developed and at steady state.

This is the classic problem of drag flow. First we choose our coordinate system. As shown in Figure 3.6, we have chosen a Cartesian coordinate system, $\hat{e}_1$, $\hat{e}_2$, $\hat{e}_3$, with flow occurring in the e_1-direction, and $x_2 = 0$ at the bottom of the channel. To solve for the velocity profile, we must apply the two conservation equations (mass, momentum) for incompressible Newtonian fluids.

First we examine mass conservation:

$$0 = \nabla \cdot \underline{v} \tag{3.106}$$

$$= \frac{\partial v_p}{\partial x_p} \tag{3.107}$$

$$= \frac{\partial v_1}{\partial x_1} + \frac{\partial v_2}{\partial x_2} + \frac{\partial v_3}{\partial x_3} \tag{3.108}$$

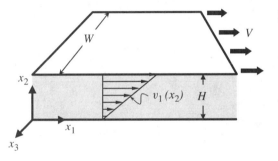

Figure 3.6 Drag flow in an infinite slit.

Since the flow is only in the $\hat{e}_1$-direction, the $\hat{e}_2$- and $\hat{e}_3$-components of $\underline{v}$ are zero:

$$\underline{v} = \begin{pmatrix} v_1 \\ 0 \\ 0 \end{pmatrix}_{123} \tag{3.109}$$

Thus, the continuity equation gives us

$$\frac{\partial v_1}{\partial x_1} = 0 \tag{3.110}$$

In words this means that in an incompressible fluid, for a flow to take place where there is only motion in the $\hat{e}_1$-direction ($v_2 = v_3 = 0$), the value of the velocity in the 1-direction must not change with the coordinate in the flow direction (x_1).

Momentum conservation is given by the equation of motion, which for an incompressible Newtonian fluid is

$$\rho \left(\frac{\partial \underline{v}}{\partial t} + \underline{v} \cdot \nabla \underline{v} \right) = -\nabla p + \mu \nabla^2 \underline{v} + \rho \underline{g} \tag{3.111}$$

To proceed further we must write the Navier–Stokes equation in component form with respect to the chosen coordinate system. Since we have chosen a Cartesian coordinate system, we will use Einstein notation, working term by term.

$$\rho \frac{\partial \underline{v}}{\partial t} = \begin{pmatrix} \rho \frac{\partial v_1}{\partial t} \\ \rho \frac{\partial v_2}{\partial t} \\ \rho \frac{\partial v_3}{\partial t} \end{pmatrix}_{123} \tag{3.112}$$

$$\rho \underline{v} \cdot \nabla \underline{v} = \rho v_p \hat{e}_p \cdot \frac{\partial}{\partial x_k} \hat{e}_k \, v_m \hat{e}_m \tag{3.113}$$

$$= \rho v_p \frac{\partial v_m}{\partial x_p} \hat{e}_m \tag{3.114}$$

$$= \rho \sum_{p=1}^{3} v_p \frac{\partial v_m}{\partial x_p} \hat{e}_m \tag{3.115}$$

$$= \rho \begin{pmatrix} v_1 \frac{\partial v_1}{\partial x_1} + v_2 \frac{\partial v_1}{\partial x_2} + v_3 \frac{\partial v_1}{\partial x_3} \\ v_1 \frac{\partial v_2}{\partial x_1} + v_2 \frac{\partial v_2}{\partial x_2} + v_3 \frac{\partial v_2}{\partial x_3} \\ v_1 \frac{\partial v_3}{\partial x_1} + v_2 \frac{\partial v_3}{\partial x_2} + v_3 \frac{\partial v_3}{\partial x_3} \end{pmatrix}_{123} \tag{3.116}$$

$$-\nabla p = \begin{pmatrix} -\frac{\partial p}{\partial x_1} \\ -\frac{\partial p}{\partial x_2} \\ -\frac{\partial p}{\partial x_3} \end{pmatrix}_{123} \tag{3.117}$$

$$\mu \nabla^2 \underline{v} = \mu \frac{\partial}{\partial x_m} \hat{e}_m \cdot \frac{\partial}{\partial x_j} \hat{e}_j \, v_s \hat{e}_s \tag{3.118}$$

$$= \mu \frac{\partial}{\partial x_m} \frac{\partial v_s}{\partial x_m} \hat{e}_s \tag{3.119}$$

$$= \begin{pmatrix} \mu \frac{\partial^2 v_1}{\partial x_1^2} + \mu \frac{\partial^2 v_1}{\partial x_2^2} + \mu \frac{\partial^2 v_1}{\partial x_3^2} \\ \mu \frac{\partial^2 v_2}{\partial x_1^2} + \mu \frac{\partial^2 v_2}{\partial x_2^2} + \mu \frac{\partial^2 v_2}{\partial x_3^2} \\ \mu \frac{\partial^2 v_3}{\partial x_1^2} + \mu \frac{\partial^2 v_3}{\partial x_2^2} + \mu \frac{\partial^2 v_3}{\partial x_3^2} \end{pmatrix}_{123} \tag{3.120}$$

$$\rho \underline{g} = \begin{pmatrix} \rho g_1 \\ \rho g_2 \\ \rho g_3 \end{pmatrix}_{123} \tag{3.121}$$

We know several things that can be used to simplify these terms. We know that $v_2 = v_3 = 0$, and we know from the continuity equation that $\partial v_1 / \partial x_1 = 0$. We can now cancel all terms involving v_2, v_3, or spatial derivatives of v_1 with respect to x_1. The terms remaining are

$$\rho \frac{\partial \underline{v}}{\partial t} = \begin{pmatrix} \rho \frac{\partial v_1}{\partial t} \\ 0 \\ 0 \end{pmatrix}_{123} \tag{3.122}$$

$$\rho \underline{v} \cdot \nabla \underline{v} = \begin{pmatrix} 0 \\ 0 \\ 0 \end{pmatrix}_{123} \tag{3.123}$$

$$-\nabla p = \begin{pmatrix} -\frac{\partial p}{\partial x_1} \\ -\frac{\partial p}{\partial x_2} \\ -\frac{\partial p}{\partial x_3} \end{pmatrix}_{123} \tag{3.124}$$

$$\mu \nabla^2 \underline{v} = \begin{pmatrix} \mu \frac{\partial^2 v_1}{\partial x_2^2} + \mu \frac{\partial^2 v_1}{\partial x_3^2} \\ 0 \\ 0 \end{pmatrix}_{123} \tag{3.125}$$

$$\rho \underline{g} = \begin{pmatrix} 0 \\ -\rho g \\ 0 \end{pmatrix}_{123} \tag{3.126}$$

We have taken gravity to be pointing in the negative x_2-direction.

Putting these together, we obtain for the equation of motion for this problem (all terms written in the 123 coordinate system)

$$\begin{pmatrix} \rho \frac{\partial v_1}{\partial t} \\ 0 \\ 0 \end{pmatrix} + \begin{pmatrix} 0 \\ 0 \\ 0 \end{pmatrix} = \begin{pmatrix} -\frac{\partial p}{\partial x_1} \\ -\frac{\partial p}{\partial x_2} \\ -\frac{\partial p}{\partial x_3} \end{pmatrix} + \begin{pmatrix} \mu \left(\frac{\partial^2 v_1}{\partial x_2^2} + \frac{\partial^2 v_1}{\partial x_3^2} \right) \\ 0 \\ 0 \end{pmatrix} + \begin{pmatrix} 0 \\ -\rho g \\ 0 \end{pmatrix} \tag{3.127}$$

Written this way, we can see that the Navier–Stokes equation, since it is a vector equation, can be thought of as three equations. The coefficients of $\hat{e}_1$ form one equation, and the coefficients of $\hat{e}_2$ and $\hat{e}_3$ form two other equations. The equation formed by the coefficients of $\hat{e}_3$ is particularly simple:

$$-\frac{\partial p}{\partial x_3} = 0 \qquad (3.128)$$

indicating that there is no variation of pressure in the x_3-direction. The equation tells us that this conclusion is required by conservation of momentum, given the assumptions we have made in the problem so far. The $\hat{e}_2$-component of the Navier–Stokes equation is

$$\frac{\partial p}{\partial x_2} = -\rho g \qquad (3.129)$$

We see that the only pressure gradient in the $\hat{e}_2$-direction is due to the weight of the fluid.

The component of the Navier–Stokes equation that tells us about the motion of the fluid is the $\hat{e}_1$-component.

$$\rho \frac{\partial v_1}{\partial t} = -\frac{\partial p}{\partial x_1} + \mu \left(\frac{\partial^2 v_1}{\partial x_2^2} + \frac{\partial^2 v_1}{\partial x_3^2} \right) \qquad (3.130)$$

We can simplify this expression a bit more by noting that because the flow is at steady state, the time derivative on the left side is zero. Also, because the plates are infinite in width, we can assume that there is no variation of any properties in the $\hat{e}_3$-direction. Finally, there is no imposed pressure gradient in the 1-direction, and therefore $\partial p / \partial x_1 = 0$. Thus, we obtain

$$0 = \mu \frac{\partial^2 v_1}{\partial x_2^2} \qquad (3.131)$$

and the solution is

$$v_1 = C_1 x_2 + C_2 \qquad (3.132)$$

where C_1 and C_2 are constants of integration. To evaluate C_1 and C_2 we need boundary conditions. We can assume that at the boundaries the velocity of the solid walls and of the fluid match. This is the no-slip boundary condition [28]. Writing this condition at both walls gives us the needed boundary conditions on velocity:

$$x_2 = 0 \qquad v_1 = 0 \qquad (3.133)$$

$$x_2 = H \qquad v_1 = V \qquad (3.134)$$

The solution for v_1 is then (Figure 3.7)

$$\boxed{v_1 = \frac{V x_2}{H}} \qquad (3.135)$$

To obtain the flow rate Q, we must integrate the expression for velocity over the cross-sectional area.

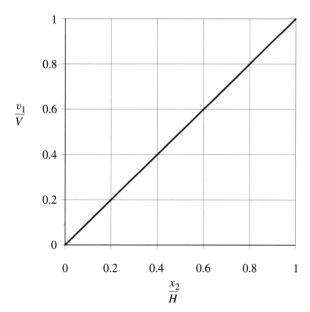

Figure 3.7 Velocity profile calculated for the drag flow of a Newtonian fluid confined between two infinite sliding plates.

$$dQ = v_1 \, dA \tag{3.136}$$

$$Q = \int_A v_1 \, dA \tag{3.137}$$

Taking the slit width to be W, the differential area perpendicular to the velocity at x_2 is just $dA = W d x_2$, and we obtain

$$Q = W \int_0^H v_1(x_2) \, dx_2 \tag{3.138}$$

$$= \frac{WVH}{2} \tag{3.139}$$

The flow rate per unit width is

$$\text{Flow rate,} \atop \text{drag flow} \qquad \boxed{\frac{Q}{W} = \frac{VH}{2}} \tag{3.140}$$

The average velocity can be calculated as the flow rate divided by the cross-sectional area:

$$v_{\text{av}} = \frac{Q}{A} = \frac{\displaystyle\int_A v_1 \, dA}{\displaystyle\int_A dA} \tag{3.141}$$

We can now go back and examine the stress tensor for this flow. Since we are considering an incompressible Newtonian fluid, $\underline{\underline{\tau}} = -\mu \underline{\underline{\dot{\gamma}}}$,

$$\underline{\underline{\tau}} = -\mu \underline{\underline{\dot{\gamma}}} \tag{3.142}$$

$$= -\mu \left[\nabla \underline{v} + (\nabla \underline{v})^T \right] \tag{3.143}$$

We have solved for $\underline{v}$, and we can therefore evaluate $\underline{\underline{\tau}}$:

$$\underline{v} = \begin{pmatrix} v_1 \\ 0 \\ 0 \end{pmatrix}_{123} = \begin{pmatrix} \frac{Vx_2}{H} \\ 0 \\ 0 \end{pmatrix}_{123} \tag{3.144}$$

$$\nabla \underline{v} = \begin{pmatrix} \frac{\partial v_1}{\partial x_1} & \frac{\partial v_2}{\partial x_1} & \frac{\partial v_3}{\partial x_1} \\ \frac{\partial v_1}{\partial x_2} & \frac{\partial v_2}{\partial x_2} & \frac{\partial v_3}{\partial x_2} \\ \frac{\partial v_1}{\partial x_3} & \frac{\partial v_2}{\partial x_3} & \frac{\partial v_3}{\partial x_3} \end{pmatrix}_{123} \tag{3.145}$$

$$= \begin{pmatrix} 0 & 0 & 0 \\ \frac{\partial v_1}{\partial x_2} & 0 & 0 \\ 0 & 0 & 0 \end{pmatrix}_{123} = \begin{pmatrix} 0 & 0 & 0 \\ \frac{V}{H} & 0 & 0 \\ 0 & 0 & 0 \end{pmatrix}_{123} \tag{3.146}$$

$$\underline{\underline{\dot{\gamma}}} = \nabla \underline{v} + (\nabla \underline{v})^T \tag{3.147}$$

$$= \begin{pmatrix} 0 & \frac{dv_1}{dx_2} & 0 \\ \frac{dv_1}{dx_2} & 0 & 0 \\ 0 & 0 & 0 \end{pmatrix}_{123} = \begin{pmatrix} 0 & \frac{V}{H} & 0 \\ \frac{V}{H} & 0 & 0 \\ 0 & 0 & 0 \end{pmatrix}_{123} \tag{3.148}$$

$$\underline{\underline{\tau}} = \begin{pmatrix} 0 & -\mu \frac{dv_1}{dx_2} & 0 \\ -\mu \frac{dv_1}{dx_2} & 0 & 0 \\ 0 & 0 & 0 \end{pmatrix}_{123} = \begin{pmatrix} 0 & -\mu \frac{V}{H} & 0 \\ -\mu \frac{V}{H} & 0 & 0 \\ 0 & 0 & 0 \end{pmatrix}_{123} \tag{3.149}$$

Notice in the equation for $\underline{\underline{\tau}}$ that there are only two nonzero components, $\tau_{21} = \tau_{12} = -\mu dv_1/dx_2$. This is just Newton's law of viscosity. We can see from this example that for incompressible Newtonian fluids, one simple scalar equation (Newton's law of viscosity) is adequate for describing stress in a simple shearing flow such as drag flow. The tensor version of the constitutive equation gives additional information, however. The constitutive equation tells us explicitly that no normal stresses are predicted ($\tau_{11} = \tau_{22} = \tau_{33} = 0$) and that the only shear stresses are τ_{21} and τ_{12}, which are equal ($\tau_{13} = \tau_{31} = \tau_{23} = \tau_{32} = 0$). We will see in Chapter 5 that shear normal stresses are nonzero for many non-Newtonian fluids.

In this example we followed a procedure that can be used quite generally to solve for velocity fields for Newtonian fluids. We can write the procedure as follows.

Problem-Solving Procedure

1. Sketch the problem and locate the flow domian.

2. Choose a coordinate system. The choice of coordinate system should be made so that the velocity vector and the boundary conditions are simplified. Locate and orient the coordinate system conveniently within the flow domain.

3. Apply the continuity equation (scalar equation) in the chosen coordinate system and simplify.

4. Apply the equation of motion (vector equation) and simplify.

5. Solve the resulting differential equation.

6. Write down the boundary conditions and solve for the unknown constants of integration.

7. Solve for the velocity field and for the pressure field.

8. Calculate $\underline{\underline{\tau}}$, v_{av}, or Q if desired.

An important step in the procedure described is to identify the boundary conditions. This step is sometimes a challenge when problems of this sort are first attempted. The number of boundary conditions that are used in fluid mechanics and rheology is relatively small, however, and we list the most common ones here.

Common Boundary Conditions in Fluid Mechanics

1. *No slip at the wall.* This boundary condition says that the fluid in contact with a wall will have the same velocity as the wall. Often the walls are not moving, so the fluid velocity is zero. In drag flows, like the previous example, the velocity of one wall is finite, and in that case the fluid velocity is equal to the wall velocity:

$$v_p\big|_{\text{at boundary}} = V_{\text{wall}} \tag{3.150}$$

2. *Symmetry.* In some flows there is a plane of symmetry. Since the velocity field is the same on either side of the plane of symmetry, the velocity must go through a minimum or a maximum at the plane of symmetry. Thus, the boundary condition to use is that the first derivative of the velocity is zero at the plane of symmetry:

$$\frac{\partial v_p}{\partial x_m}\bigg|_{\text{at boundary}} = 0 \tag{3.151}$$

3. *Stress continuity.* When a fluid forms one of the flow boundaries, the shear stress is continuous from one fluid to another. Thus for a viscous fluid in contact with an inviscid (zero or very low viscosity fluid), this means that at the boundary, the shear stress in the viscous fluid is the same as the shear stress in the inviscid fluid. Since the inviscid fluid can support no shear stress (zero viscosity), this means that the shear stress is zero at this interface. The boundary condition between a fluid such as a polymer and air, for example, would be that the shear stress in the polymer at the interface would be zero:

$$\tau_{jk}\big|_{\text{at boundary}} = 0 \tag{3.152}$$

Alternatively if two viscous fluids meet and form a flow boundary, as in coextrusion for example, this same boundary condition would require that the shear stress in one fluid equal the shear stress in the other at the boundary:

$$\tau_{jk}(\text{fluid 1})\big|_{\text{at boundary}} = \tau_{jk}(\text{fluid 2})\big|_{\text{at boundary}} \tag{3.153}$$

4. *Velocity continuity.* When a fluid forms one of the boundaries of the flow as described, the velocity is also continuous from one fluid to another:

$$v_p(\text{fluid 1})\big|_{\text{at boundary}} = v_p(\text{fluid 2})\big|_{\text{at boundary}} \qquad (3.154)$$

5. *Finite velocity and stress.* Occasionally an expression is derived that predicts infinite velocities or stresses at a point. An example would be an equation that includes $1/r$, for a flow domain where $r = 0$ is included. A possible boundary condition to use in this instance is the requirement that the velocity or the stress be finite throughout the flow domain. This boundary condition appears occasionally in flows with cylindrical symmetry.

Other flow boundary conditions can be found in standard texts on fluid mechanics [13, 28, 148].

In the next section we use the symmetry boundary condition, and, knowing that this boundary condition applies, we choose our coordinate system to make best use of this boundary condition. This problem is solvable with other choices of coordinate system, although the algebra is a bit more complex.

3.5.2 POISEUILLE FLOW BETWEEN INFINITE PARALLEL PLATES

Calculate the velocity profile, flow rate, and stress tensor $\underline{\underline{\tau}}$ for pressure-driven flow of an incompressible Newtonian liquid between two infinitely wide, parallel plates, separated by a gap of $2H$. The pressure at an upstream point is P_0, and at a point a distance L downstream, the pressure is P_L. Assume that the flow between these two points is fully developed and at steady state.

This is the classic problem of Poiseuille flow, and to solve it we follow the same procedure as in the previous example. First we choose our coordinate system. As shown in Figure 3.8, we will use a Cartesian coordinate system, $\hat{e}_1, \hat{e}_2, \hat{e}_3$, with flow occurring in the e_1-direction, and $x_2 = 0$ at the center of the channel. We choose the zero location of the x_1-axis such that at $x_1 = 0$, $p = P_0$, and at $x_1 = L$, $p = P_L$. To solve for the velocity profile, we must apply the two conservation equations (mass, momentum) for incompressible Newtonian fluids.

We begin with mass conservation:

$$0 = \nabla \cdot \underline{v} \qquad (3.155)$$

$$= \frac{\partial v_1}{\partial x_1} + \frac{\partial v_2}{\partial x_2} + \frac{\partial v_3}{\partial x_3} \qquad (3.156)$$

Since the flow is only in the $\hat{e}_1$-direction, the $\hat{e}_2$- and $\hat{e}_3$-components of $\underline{v}$ are zero:

Figure 3.8 Poiseuille flow in an infinite slit.

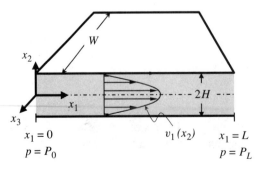

$$\underline{v} = \begin{pmatrix} v_1 \\ 0 \\ 0 \end{pmatrix}_{123} \tag{3.157}$$

The continuity equation then gives us the same result as in the case of drag flow:

$$\frac{\partial v_1}{\partial x_1} = 0 \tag{3.158}$$

Momentum conservation is given by the equation of motion for an incompressible Newtonian fluid, the Navier–Stokes equation:

$$\rho \left(\frac{\partial \underline{v}}{\partial t} + \underline{v} \cdot \nabla \underline{v} \right) = -\nabla p + \mu \nabla^2 \underline{v} + \rho \underline{g} \tag{3.159}$$

We now simplify this using the given information. We begin by writing each term out in component form in the chosen coordinate system:

$$\rho \frac{\partial \underline{v}}{\partial t} = \begin{pmatrix} \rho \frac{\partial v_1}{\partial t} \\ \rho \frac{\partial v_2}{\partial t} \\ \rho \frac{\partial v_3}{\partial t} \end{pmatrix}_{123} \tag{3.160}$$

$$\rho \underline{v} \cdot \nabla \underline{v} = \rho \begin{pmatrix} v_1 \frac{\partial v_1}{\partial x_1} + v_2 \frac{\partial v_1}{\partial x_2} + v_3 \frac{\partial v_1}{\partial x_3} \\ v_1 \frac{\partial v_2}{\partial x_1} + v_2 \frac{\partial v_2}{\partial x_2} + v_3 \frac{\partial v_2}{\partial x_3} \\ v_1 \frac{\partial v_3}{\partial x_1} + v_2 \frac{\partial v_3}{\partial x_2} + v_3 \frac{\partial v_3}{\partial x_3} \end{pmatrix}_{123} \tag{3.161}$$

$$-\nabla p = \begin{pmatrix} -\frac{\partial p}{\partial x_1} \\ -\frac{\partial p}{\partial x_2} \\ -\frac{\partial p}{\partial x_3} \end{pmatrix}_{123} \tag{3.162}$$

$$\mu \nabla^2 \underline{v} = \begin{pmatrix} \mu \frac{\partial^2 v_1}{\partial x_1^2} + \mu \frac{\partial^2 v_1}{\partial x_2^2} + \mu \frac{\partial^2 v_1}{\partial x_3^2} \\ \mu \frac{\partial^2 v_2}{\partial x_1^2} + \mu \frac{\partial^2 v_2}{\partial x_2^2} + \mu \frac{\partial^2 v_2}{\partial x_3^2} \\ \mu \frac{\partial^2 v_3}{\partial x_1^2} + \mu \frac{\partial^2 v_3}{\partial x_2^2} + \mu \frac{\partial^2 v_3}{\partial x_3^2} \end{pmatrix}_{123} \tag{3.163}$$

$$\rho \underline{g} = \begin{pmatrix} \rho g_1 \\ \rho g_2 \\ \rho g_3 \end{pmatrix}_{123} \tag{3.164}$$

We can now cancel all terms involving v_2, v_3, or spatial derivatives of v_1 with respect to x_1. Putting the terms together, we obtain for this problem (all terms written in the 123 coordinate system)

$$\begin{pmatrix} \rho \frac{\partial v_1}{\partial t} \\ 0 \\ 0 \end{pmatrix} + \begin{pmatrix} 0 \\ 0 \\ 0 \end{pmatrix} = \begin{pmatrix} -\frac{\partial p}{\partial x_1} \\ -\frac{\partial p}{\partial x_2} \\ -\frac{\partial p}{\partial x_3} \end{pmatrix} + \begin{pmatrix} \mu \left(\frac{\partial^2 v_1}{\partial x_2^2} + \frac{\partial^2 v_1}{\partial x_3^2} \right) \\ 0 \\ 0 \end{pmatrix} + \begin{pmatrix} 0 \\ -\rho g \\ 0 \end{pmatrix} \tag{3.165}$$

The equation formed by the coefficients of $\hat{e}_3$ is simple:

$$-\frac{\partial p}{\partial x_3} = 0 \tag{3.166}$$

indicating that there is no variation of pressure in the x_3-direction. The $\hat{e}_2$-component of the Navier–Stokes equation is

$$\frac{\partial p}{\partial x_2} = -\rho g \tag{3.167}$$

We see, as was true in the previous example of drag flow, that the only pressure gradient in the $\hat{e}_2$-direction is due to the weight of the fluid.

The component of the Navier–Stokes equation that tells us about the flow is the $\hat{e}_1$-component:

$$\rho \frac{\partial v_1}{\partial t} = -\frac{\partial p}{\partial x_1} + \mu \left(\frac{\partial^2 v_1}{\partial x_2^2} + \frac{\partial^2 v_1}{\partial x_3^2} \right) \tag{3.168}$$

We can simplify this expression a bit more by noting that because the flow is at steady state, the time derivative on the left side is zero. Also, because the plates are infinite in width, we can assume that there is no variation of any properties in the $\hat{e}_3$-direction. Finally we obtain

$$0 = -\frac{\partial p(x_1, x_2)}{\partial x_1} + \mu \frac{\partial^2 v_1}{\partial x_2^2} \tag{3.169}$$

This equation is difficult to solve since the pressure field is two-dimensional, $p = p(x_1, x_2)$. The variation of p in the x_2-direction, however, is extremely slight, since it is only caused by gravity, and the thickness being considered $(2H)$ is small. Thus, if we neglect the variation of pressure in the x_2-direction, that is, neglect gravity, p is a function of x_1 only, and we can solve Equation (3.169) by separation of variables, as we will now show.

If we examine the simplified x_3-component of the Navier–Stokes equation,

$$\mu \frac{\partial^2 v_1(x_2)}{\partial x_2^2} = \frac{dp(x_1)}{dx_1} \tag{3.170}$$

we see that the left side is only a function of x_2, $(\partial v_1 / \partial x_1 = 0$ from continuity and we have neglected x_3-variations in v_1) and the right side is only a function of x_1 (pressure is not a function of x_3 from the 3-component of the equation of motion, and we have neglected the x_2-variation due to gravity). Since the two sides of this equation equate for all values of these independently changing variables, each side must be independently equal to a constant that we will call λ [34]:

$$\frac{dp}{dx_1} = \lambda \tag{3.171}$$

$$\mu \frac{d^2 v_1}{dx_2^2} = \lambda \tag{3.172}$$

Note that we have changed partial to regular derivatives in the velocity expression since $v_1 = v_1(x_2)$ only. These two differential equations can now be solved independently using the appropriate boundary conditions. On pressure, the problem statement and our choice of coordinate system give us the boundary conditions at $x_1 = 0$ and $x_1 = L$:

$$x_1 = 0 \qquad p = P_0 \tag{3.173}$$

$$x_1 = L \qquad p = P_L \tag{3.174}$$

We can use one of these to solve for the integration constant and the the other to solve for *lambda*. On velocity, we can assume that the velocity of the solid walls and the velocity of the fluid match at the boundaries. This is the no-slip boundary condition described earlier [28]. Further, halfway between the planes is a plane of symmetry, which means that v_1 must go through a maximum or a minimum at this plane, that is, the derivative of v_1 with respect to x_2 must be zero at this plane. We can choose any two of these conditions to evaluate the two integration constants that arise when we solve Equation (3.172).

$$x_2 = 0 \qquad \frac{dv_1}{dx_2} = 0 \tag{3.175}$$

$$x_2 = \pm H \qquad v_1 = 0 \tag{3.176}$$

The symmetry boundary condition is particularly desirable to use since it simplifies the evaluation of the integration constants. The choice of coordinate system with $x_2 = 0$ at the centerline of the channel is well matched with the boundary conditions. The resulting solutions are (Figure 3.9)

$$\boxed{p = -\frac{(P_0 - P_L)}{L} x_1 + P_0} \tag{3.177}$$

$$\boxed{v_1 = \frac{H^2(P_0 - P_L)}{2\mu L} \left[1 - \left(\frac{x_2}{H}\right)^2 \right]} \tag{3.178}$$

To obtain the volumetric flow rate Q, we must integrate the expression for velocity. Taking the slit width to be W, we obtain

$$Q = W \int_{-H}^{H} v_1(x_2) \, dx_2 \tag{3.179}$$

$$= 2W \int_{0}^{H} v_1(x_2) \, dx_2 \tag{3.180}$$

Substituting $v_1(x_2)$ from Equation (3.178) yields the result for flow rate per unit width:

Flow rate,
Poiseuille flow
in a slit

$$\boxed{\frac{Q}{W} = \frac{2H^3(P_0 - P_L)}{3\mu L}} \tag{3.181}$$

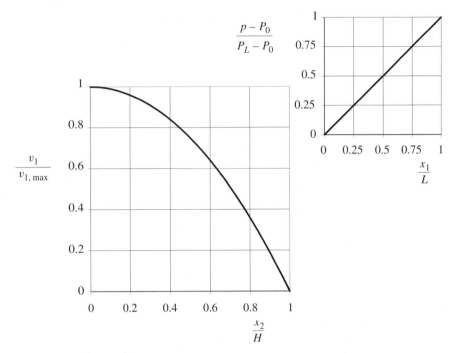

Figure 3.9 Velocity and pressure profiles calculated for Poiseuille flow (pressure-driven flow) of a Newtonian fluid in a slit.

Since we now know $v_1(x_2)$, we can calculate the stress $\underline{\underline{\tau}}$ from the constitutive equation:

$$\underline{v} = \begin{pmatrix} v_1 \\ 0 \\ 0 \end{pmatrix}_{123} = \begin{pmatrix} \frac{H^2(P_0 - P_L)}{2\mu L}\left[1 - \left(\frac{x_2}{H}\right)^2\right] \\ 0 \\ 0 \end{pmatrix}_{123} \tag{3.182}$$

$$\nabla \underline{v} = \begin{pmatrix} \frac{\partial v_1}{\partial x_1} & \frac{\partial v_2}{\partial x_1} & \frac{\partial v_3}{\partial x_1} \\ \frac{\partial v_1}{\partial x_2} & \frac{\partial v_2}{\partial x_2} & \frac{\partial v_3}{\partial x_2} \\ \frac{\partial v_1}{\partial x_3} & \frac{\partial v_2}{\partial x_3} & \frac{\partial v_3}{\partial x_3} \end{pmatrix}_{123} \tag{3.183}$$

$$= \begin{pmatrix} 0 & 0 & 0 \\ \frac{\partial v_1}{\partial x_2} & 0 & 0 \\ 0 & 0 & 0 \end{pmatrix}_{123} = \begin{pmatrix} 0 & 0 & 0 \\ -\frac{(P_0 - P_L)x_2}{\mu L} & 0 & 0 \\ 0 & 0 & 0 \end{pmatrix}_{123} \tag{3.184}$$

$$\underline{\underline{\dot{\gamma}}} = \nabla \underline{v} + (\nabla \underline{v})^T \tag{3.185}$$

$$= \begin{pmatrix} 0 & \frac{dv_1}{dx_2} & 0 \\ \frac{dv_1}{dx_2} & 0 & 0 \\ 0 & 0 & 0 \end{pmatrix}_{123} = \begin{pmatrix} 0 & -\frac{(P_0 - P_L)x_2}{\mu L} & 0 \\ -\frac{(P_0 - P_L)x_2}{\mu L} & 0 & 0 \\ 0 & 0 & 0 \end{pmatrix}_{123} \tag{3.186}$$

$$\underline{\underline{\tau}} = \begin{pmatrix} 0 & -\mu\frac{dv_1}{dx_2} & 0 \\ -\mu\frac{dv_1}{dx_2} & 0 & 0 \\ 0 & 0 & 0 \end{pmatrix}_{123} = \begin{pmatrix} 0 & \frac{(P_0-P_L)x_2}{L} & 0 \\ \frac{(P_0-P_L)x_2}{L} & 0 & 0 \\ 0 & 0 & 0 \end{pmatrix}_{123} \qquad (3.187)$$

We see that for this flow most of the coefficients of $\underline{\underline{\tau}}$ are zero, except $\tau_{21} = \tau_{12} = (P_0 - P_L)x_2/L$. The nonzero terms are the shear stresses, and these vary linearly with x_2. Also τ_{21} is positive, indicating that momentum moves in the positive x_2-direction, that is, down the velocity gradient, as it should, given our convention on the sign of stress. A quantity that could be measured in this flow is the shear stress at the wall. This is given by

$$\boxed{\tau_{21}(H) = \frac{(P_0 - P_L)H}{L}} \qquad (3.188)$$

The next section concerns the same type of flow (pressure-driven), but now it is carried out in a cylindrical geometry.

3.5.3 POISEUILLE FLOW IN A TUBE

Calculate the velocity profile, flow rate, and stress tensor $\underline{\underline{\tau}}$ for downward, pressure-driven flow of an incompressible Newtonian liquid in a tube of circular cross section. The pressure at an upstream point is P_0, and at a point a distance L downstream the pressure is P_L. Assume that the flow between these two points is fully developed and at steady state.

Figure 3.10 Poiseuille flow in a tube.

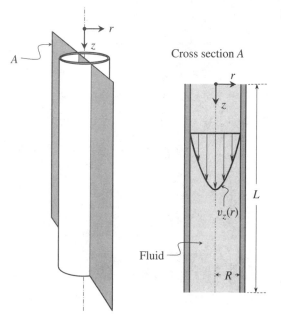

Since the tube is round, we will be working in cylindrical coordinates. To get all the terms right, we will refer to Table C.7 in Appendix C.2 to translate the vector equations into scalar component notation.

The mass conservation equation is

$$0 = \nabla \cdot \underline{v} \tag{3.189}$$

$$= \frac{1}{r}\frac{\partial (r v_r)}{\partial r} + \frac{1}{r}\frac{\partial v_\theta}{\partial \theta} + \frac{\partial v_z}{\partial z} \tag{3.190}$$

Since the flow is only in the z-direction, the r- and θ-components of $\underline{v}$ are zero:

$$\underline{v} = \begin{pmatrix} v_r \\ v_\theta \\ v_z \end{pmatrix}_{r\theta z} = \begin{pmatrix} 0 \\ 0 \\ v_z \end{pmatrix}_{r\theta z} \tag{3.191}$$

Thus, the continuity equation gives us the result

$$\frac{\partial v_z}{\partial z} = 0 \tag{3.192}$$

The equation of motion for an incompressible Newtonian fluid is the Navier–Stokes equation, derived in this chapter,

$$\rho \left(\frac{\partial \underline{v}}{\partial t} + \underline{v} \cdot \nabla \underline{v} \right) = -\nabla p + \mu \nabla^2 \underline{v} + \rho \underline{g} \tag{3.193}$$

In cylindrical coordinates the terms become (see Table C.7)

$$\rho \frac{\partial \underline{v}}{\partial t} = \begin{pmatrix} \rho \frac{\partial v_r}{\partial t} \\ \rho \frac{\partial v_\theta}{\partial t} \\ \rho \frac{\partial v_z}{\partial t} \end{pmatrix}_{r\theta z} \tag{3.194}$$

$$\rho \underline{v} \cdot \nabla \underline{v} = \rho \begin{pmatrix} v_r \frac{\partial v_r}{\partial r} + v_\theta \left(\frac{1}{r}\frac{\partial v_r}{\partial \theta} - \frac{v_\theta}{r} \right) + v_z \frac{\partial v_r}{\partial z} \\ v_r \frac{\partial v_\theta}{\partial r} + v_\theta \left(\frac{1}{r}\frac{\partial v_\theta}{\partial \theta} + \frac{v_r}{r} \right) + v_z \frac{\partial v_\theta}{\partial z} \\ v_r \frac{\partial v_z}{\partial r} + v_\theta \left(\frac{1}{r}\frac{\partial v_z}{\partial \theta} \right) + v_z \frac{\partial v_z}{\partial z} \end{pmatrix}_{r\theta z} \tag{3.195}$$

$$-\nabla p = \begin{pmatrix} -\frac{\partial p}{\partial r} \\ -\frac{1}{r}\frac{\partial p}{\partial \theta} \\ -\frac{\partial p}{\partial z} \end{pmatrix}_{r\theta z} \tag{3.196}$$

$$\mu \nabla^2 \underline{v} = \begin{pmatrix} \mu \frac{\partial}{\partial r}\left[\frac{1}{r}\frac{\partial}{\partial r}(r v_r) \right] + \mu \frac{1}{r^2}\frac{\partial^2 v_r}{\partial \theta^2} + \mu \frac{\partial^2 v_r}{\partial z^2} - \frac{2\mu}{r^2}\frac{\partial v_\theta}{\partial \theta} \\ \mu \frac{\partial}{\partial r}\left[\frac{1}{r}\frac{\partial}{\partial r}(r v_\theta) \right] + \mu \frac{1}{r^2}\frac{\partial^2 v_\theta}{\partial \theta^2} + \mu \frac{\partial^2 v_\theta}{\partial z^2} + \frac{2\mu}{r^2}\frac{\partial v_r}{\partial \theta} \\ \mu \frac{1}{r}\frac{\partial}{\partial r}\left(r \frac{\partial v_z}{\partial r} \right) + \mu \frac{1}{r^2}\frac{\partial^2 v_z}{\partial \theta^2} + \mu \frac{\partial^2 v_z}{\partial z^2} \end{pmatrix}_{r\theta z} \tag{3.197}$$

$$\rho \underline{g} = \begin{pmatrix} \rho g_r \\ \rho g_\theta \\ \rho g_z \end{pmatrix}_{r\theta z} \tag{3.198}$$

Substituting in what we already know about $\underline{v}$, we obtain

$$\rho \frac{\partial \underline{v}}{\partial t} = \begin{pmatrix} 0 \\ 0 \\ \rho \frac{\partial v_z}{\partial t} \end{pmatrix}_{r\theta z} \tag{3.199}$$

$$\rho \underline{v} \cdot \nabla \underline{v} = \begin{pmatrix} 0 \\ 0 \\ 0 \end{pmatrix}_{r\theta z} \tag{3.200}$$

$$-\nabla p = \begin{pmatrix} -\frac{\partial p}{\partial r} \\ -\frac{1}{r}\frac{\partial p}{\partial \theta} \\ -\frac{\partial p}{\partial z} \end{pmatrix}_{r\theta z} \tag{3.201}$$

$$\mu \nabla^2 \underline{v} = \begin{pmatrix} 0 \\ 0 \\ \frac{\mu}{r}\frac{\partial}{\partial r}\left(r\frac{\partial v_z}{\partial r}\right) + \frac{\mu}{r^2}\frac{\partial^2 v_z}{\partial \theta^2} \end{pmatrix}_{r\theta z} \tag{3.202}$$

$$\rho \underline{g} = \begin{pmatrix} 0 \\ 0 \\ \rho g \end{pmatrix}_{r\theta z} \tag{3.203}$$

Gravity is in the flow direction. Combining the terms we obtain (all terms are written in the r, θ, z coordinate system)

$$\begin{pmatrix} 0 \\ 0 \\ \rho \frac{\partial v_z}{\partial t} \end{pmatrix} + \begin{pmatrix} 0 \\ 0 \\ 0 \end{pmatrix} = \begin{pmatrix} -\frac{\partial p}{\partial r} \\ -\frac{1}{r}\frac{\partial p}{\partial \theta} \\ -\frac{\partial p}{\partial z} \end{pmatrix} + \begin{pmatrix} 0 \\ 0 \\ \frac{\mu}{r}\frac{\partial}{\partial r}\left(r\frac{\partial v_z}{\partial r}\right) + \frac{\mu}{r^2}\frac{\partial^2 v_z}{\partial \theta^2} \end{pmatrix} + \begin{pmatrix} 0 \\ 0 \\ \rho g \end{pmatrix} \tag{3.204}$$

In addition, since the problem asks for the steady-state solution, $\partial v_z / \partial t = 0$.
    Examining the r- and θ-components of the Navier–Stokes equation, we see that

$$\frac{\partial p}{\partial r} = 0 \tag{3.205}$$

$$\frac{\partial p}{\partial \theta} = 0 \tag{3.206}$$

This tells us that the pressure is only a function of z. The z-component of the Navier–Stokes equation is

$$\frac{dp}{dz} = \frac{\mu}{r}\frac{\partial}{\partial r}\left(r\frac{\partial v_z}{\partial r}\right) + \frac{\mu}{r^2}\frac{\partial^2 v_z}{\partial \theta^2} + \rho g \tag{3.207}$$

This equation contains θ- and r-derivatives of v_z; v_z definitely varies with r since it is zero at $r = R$ and nonzero in the center ($r = 0$). Now we examine the possibility of θ variation of v_z. We see from Equation (3.206) that the pressure does not vary in the θ-direction. Although we have not found any restriction on the θ variation of the velocity v_z, with no flow in the θ-direction and no pressure variation, it is reasonable to assume that there are no variations of v_z in the θ-direction, that is, the flow should be symmetric with respect to θ. On these physical arguments, we will take $\partial v_z / \partial \theta = 0$. Thus Equation (3.207) simplifies to

$$\frac{dp(z)}{dz} - \rho g = \frac{\mu}{r} \frac{\partial}{\partial r} \left[r \frac{\partial v_z(r)}{\partial r} \right] \tag{3.208}$$

Since by continuity v_z is not a function of z, and by the preceding assumption it is also not a function of θ, v_z is a function of r alone. From the discussion concerning the r- and θ-components of the Navier–Stokes equation we found that $p = p(z)$. Thus, the left-hand side of Equation (3.208) is a function of z only, the right-hand side is a function of r only, and the differential equation is separable. We can solve Equation (3.208) following the same method as used on the example in Section 3.5.2. For simplicity we will combine the pressure and gravity terms as follows:

$$\frac{dp}{dz} - \rho g = \frac{\mu}{r} \frac{d}{dr} \left(r \frac{dv_z}{dr} \right) \tag{3.209}$$

$$\frac{d\mathcal{P}}{dz} = \frac{\mu}{r} \frac{d}{dr} \left(r \frac{dv_z}{dr} \right) \tag{3.210}$$

where $\mathcal{P} \equiv p - \rho g z$. $\mathcal{P}$ is called the *equivalent pressure* (see Glossary).

The boundary conditions for this problem are analogous to those in the slit problem:

$$z = 0 \qquad p = P_0 \tag{3.211}$$

$$z = L \qquad p = P_L \tag{3.212}$$

$$r = 0 \qquad \frac{dv_z}{dr} = 0 \tag{3.213}$$

$$r = R \qquad v_z = 0 \tag{3.214}$$

The resulting pressure and velocity profiles are (Figure 3.11)

$$\boxed{\mathcal{P} = -\frac{\mathcal{P}_0 - \mathcal{P}_L}{z} + \mathcal{P}_0} \tag{3.215}$$

$$\boxed{v_z = \frac{(\mathcal{P}_0 - \mathcal{P}_L)R^2}{4\mu L} \left[1 - \left(\frac{r}{R} \right)^2 \right]} \tag{3.216}$$

In nondimensional form, the pressure and velocity curves for this problem have the same shapes as those for the previous example, Poiseuille flow in a slit (Figure 3.9).

The solution for the flow rate Q is the well-known Hagen–Poiseuille law:

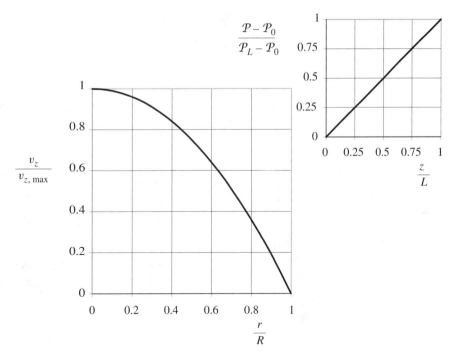

Figure 3.11 Velocity and pressure profiles calculated for Poiseuille flow (pressure-driven flow) of a Newtonian fluid in a tube.

$$Q = \int_A v_z \, dA \tag{3.217}$$

$$= \int_0^R \int_0^{2\pi} v_z(r) r \, d\theta \, dr \tag{3.218}$$

Substituting $v_z(r)$ from Equation (3.216) we obtain

Hagen–Poiseuille law $\qquad \boxed{Q = \dfrac{\pi(\mathcal{P}_0 - \mathcal{P}_L)R^4}{8\mu L}} \tag{3.219}$

As with the previous example problems, now that we know $\underline{v}$, we can calculate $\underline{\underline{\tau}} = -\mu \underline{\underline{\dot{\gamma}}}$:

$$\underline{v} = \begin{pmatrix} v_r \\ v_\theta \\ v_z \end{pmatrix}_{r\theta z} = \begin{pmatrix} 0 \\ 0 \\ v_z \end{pmatrix}_{r\theta z} = \begin{pmatrix} 0 \\ 0 \\ \frac{(\mathcal{P}_0 - \mathcal{P}_L)R^2}{4\mu L}\left[1 - \left(\frac{r}{R}\right)^2\right] \end{pmatrix}_{r\theta z} \tag{3.220}$$

$$\nabla \underline{v} = \begin{pmatrix} \frac{\partial v_r}{\partial r} & \frac{\partial v_\theta}{\partial r} & \frac{\partial v_z}{\partial r} \\ \frac{1}{r}\frac{\partial v_r}{\partial \theta} - \frac{v_\theta}{r} & \frac{1}{r}\frac{\partial v_\theta}{\partial \theta} + \frac{v_r}{r} & \frac{1}{r}\frac{\partial v_z}{\partial \theta} \\ \frac{\partial v_r}{\partial z} & \frac{\partial v_\theta}{\partial z} & \frac{\partial v_z}{\partial z} \end{pmatrix}_{r\theta z} \tag{3.221}$$

$$= \begin{pmatrix} 0 & 0 & \frac{\partial v_z}{\partial r} \\ 0 & 0 & 0 \\ 0 & 0 & 0 \end{pmatrix}_{r\theta z} = \begin{pmatrix} 0 & 0 & -\frac{(P_0-P_L)r}{2\mu L} \\ 0 & 0 & 0 \\ 0 & 0 & 0 \end{pmatrix}_{r\theta z} \tag{3.222}$$

$$\dot{\underline{\underline{\gamma}}} = \nabla\underline{v} + (\nabla\underline{v})^T \tag{3.223}$$

$$= \begin{pmatrix} 0 & 0 & \frac{\partial v_z}{\partial r} \\ 0 & 0 & 0 \\ \frac{\partial v_z}{\partial r} & 0 & 0 \end{pmatrix}_{r\theta z} = \begin{pmatrix} 0 & 0 & -\frac{(P_0-P_L)r}{2\mu L} \\ 0 & 0 & 0 \\ -\frac{(P_0-P_L)r}{2\mu L} & 0 & 0 \end{pmatrix}_{r\theta z} \tag{3.224}$$

$$\underline{\underline{\tau}} = \begin{pmatrix} 0 & 0 & -\mu\frac{\partial v_z}{\partial r} \\ 0 & 0 & 0 \\ -\mu\frac{\partial v_z}{\partial r} & 0 & 0 \end{pmatrix}_{r\theta z} = \begin{pmatrix} 0 & 0 & \frac{(P_0-P_L)r}{2L} \\ 0 & 0 & 0 \\ \frac{(P_0-P_L)r}{2L} & 0 & 0 \end{pmatrix}_{r\theta z} \tag{3.225}$$

Once again the only nonzero stress is the shear stress, $\tau_{rz} = \tau_{zr}$, and the shear stress is positive, indicating that the flux of momentum is in the positive r-direction as expected (down the velocity gradient). In this flow the quantity that is measurable is the shear stress at the wall given by (see also Chapter 10)

$$\boxed{\tau_{rz}(R) = \frac{(P_0 - P_L)R}{2L}} \tag{3.226}$$

We will discuss this flow for general fluids in Chapter 10.

Although this flow was solved in cylindrical rather than Cartesian (x_1, x_2, x_3) coordinates, we see that the shear stress is still given by an expression having the same form as Newton's law of viscosity adapted for the cylindrical coordinate system:

$$\tau_{rz} = -\mu\frac{\partial v_z}{\partial r} \tag{3.227}$$

For the three examples discussed thus far, $\underline{v} \cdot \nabla\underline{v}$ has been zero. This will always be the case for unidirectional flow, as can be shown using Einstein notation. For flows such as rotational flows, where the velocity varies in direction, the quantity $\underline{v} \cdot \nabla\underline{v}$ will not vanish, as we see in the next section.

3.5.4 TORSIONAL FLOW BETWEEN PARALLEL PLATES

In a torsional parallel-plate viscometer (Figure 3.12) a fluid is placed between two round plates, and one plate is rotated. For an incompressible Newtonian fluid, calculate the velocity field, the stress tensor $\underline{\underline{\tau}}$, and the torque required to turn the plate at steady state at an angular velocity Ω. To simplify the problem, assume that $v_\theta(r, \theta, z) = zf(r)$, where $f(r)$ is an unknown function of r for which we will solve.

Torsional flow between parallel plates is used in some types of rheometers to measure viscosity, as discussed in Chapter 10. Here we solve for the Newtonian case. This problem is worked in cylindrical coordinates, and we again use Table C.7 to evaluate the terms in the continuity equation and the equation of motion.

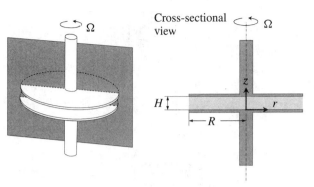

Figure 3.12 Torsional flow between parallel plates.

The equation of mass conservation is:

$$0 = \nabla \cdot \underline{v} \tag{3.228}$$

$$= \frac{1}{r}\frac{\partial(r v_r)}{\partial r} + \frac{1}{r}\frac{\partial v_\theta}{\partial \theta} + \frac{\partial v_z}{\partial z} \tag{3.229}$$

The flow is only in the θ-direction, and therefore the r- and z-components of $\underline{v}$ are zero:

$$\underline{v} = \begin{pmatrix} v_r \\ v_\theta \\ v_z \end{pmatrix}_{r\theta z} = \begin{pmatrix} 0 \\ v_\theta \\ 0 \end{pmatrix}_{r\theta z} \tag{3.230}$$

The continuity equation therefore gives

$$\frac{1}{r}\frac{\partial v_\theta}{\partial \theta} = 0 \tag{3.231}$$

The equation of motion for an incompressible Newtonian fluid is the Navier–Stokes equation:

$$\rho\left(\frac{\partial \underline{v}}{\partial t} + \underline{v} \cdot \nabla \underline{v}\right) = -\nabla p + \mu \nabla^2 \underline{v} + \rho \underline{g} \tag{3.232}$$

The terms of the Navier–Stokes equation were written in cylindrical coordinates in Section 3.5.3 [Equations (3.194)–(3.198)]. Substituting what we know already about $\underline{v}$ from this problem, we obtain

$$\rho\frac{\partial \underline{v}}{\partial t} = \begin{pmatrix} 0 \\ \rho\frac{\partial v_\theta}{\partial t} \\ 0 \end{pmatrix}_{r\theta z} \tag{3.233}$$

$$\rho\underline{v} \cdot \nabla\underline{v} = \rho \begin{pmatrix} -\frac{v_\theta^2}{r} \\ 0 \\ 0 \end{pmatrix}_{r\theta z} \tag{3.234}$$

$$-\nabla p = \begin{pmatrix} -\frac{\partial p}{\partial r} \\ -\frac{1}{r}\frac{\partial p}{\partial \theta} \\ -\frac{\partial p}{\partial z} \end{pmatrix}_{r\theta z} \tag{3.235}$$

$$\mu\nabla^2\underline{v} = \begin{pmatrix} 0 \\ \mu\frac{\partial}{\partial r}\left[\frac{1}{r}\frac{\partial}{\partial r}(rv_\theta)\right] + \mu\frac{\partial^2 v_\theta}{\partial z^2} \\ 0 \end{pmatrix}_{r\theta z} \tag{3.236}$$

$$\rho\underline{g} = \begin{pmatrix} 0 \\ 0 \\ -\rho g \end{pmatrix}_{r\theta z} \tag{3.237}$$

Note that gravity is in the negative z-direction. The flow should be symmetric in the θ-direction, and therefore terms with θ-derivatives will also be zero. With this assumption and putting the terms together we obtain (all terms are written in the r, θ, z coordinate system)

$$\begin{pmatrix} 0 \\ \rho\frac{\partial v_\theta}{\partial t} \\ 0 \end{pmatrix} + \begin{pmatrix} -\frac{\rho v_\theta^2}{r} \\ 0 \\ 0 \end{pmatrix} = \begin{pmatrix} -\frac{\partial p}{\partial r} \\ 0 \\ -\frac{\partial p}{\partial z} \end{pmatrix} + \begin{pmatrix} 0 \\ \mu\frac{\partial}{\partial r}\left[\frac{1}{r}\frac{\partial}{\partial r}(rv_\theta)\right] + \mu\frac{\partial^2 v_\theta}{\partial z^2} \\ 0 \end{pmatrix} + \begin{pmatrix} 0 \\ 0 \\ -\rho g \end{pmatrix} \tag{3.238}$$

Since the problem asks for us to calculate the steady-state solution, $\partial v_\theta/\partial t = 0$. Note that $\underline{v} \cdot \nabla\underline{v}$ is not zero in this flow since the flow is not unidirectional.

The z-component of the Navier–Stokes equation is simple and tells us that the pressure gradient in the z-direction is due to gravity:

$$\frac{\partial p}{\partial z} = -\rho g \tag{3.239}$$

The r-component of the equation of motion indicates that there is a radial pressure gradient due to centrifugal force:

$$\frac{\partial p}{\partial r} = \frac{\rho v_\theta^2}{r} \tag{3.240}$$

The θ-component of the Navier–Stokes equation tells us about the torsional flow:

$$\mu\frac{\partial}{\partial r}\left[\frac{1}{r}\frac{\partial}{\partial r}(rv_\theta)\right] + \mu\frac{\partial^2 v_\theta}{\partial z^2} = 0 \tag{3.241}$$

The velocity v_θ in this example is a function of both r and z, as can be seen by considering the values of velocity at various positions between the plates: at the bottom and top plates ($z = 0$, H) the velocity is 0 and nonzero; thus v_θ is a function of z. At the radial center and rim ($r = 0$, R) the velocity is zero and nonzero, and again we can conclude that v_θ is a function of r. Because $v_\theta = v_\theta(r, z)$ we cannot solve Equation (3.241) with the same separation of variables technique we used in the last two examples. Conveniently we were told in the problem statement to assume $v_\theta = zf(r)$. With this information, Equation (3.241) simplifies:

$$\frac{d}{dr}\left[\frac{1}{r}\frac{d(rf)}{dr}\right] = 0 \tag{3.242}$$

which we can solve to obtain

$$f(r) = \frac{C_1 r}{2} + \frac{C_2}{r} \tag{3.243}$$

where C_1 and C_2 are the usual constants of integration.

The boundary conditions for this problem are no slip at the plates and finite velocity everywhere:

$$z = 0 \qquad v_\theta = 0 \tag{3.244}$$

$$z = H \qquad v_\theta = r\Omega \tag{3.245}$$

$$\text{for all } z, \text{ at } r = 0 \qquad v_\theta = 0 \tag{3.246}$$

$$\text{for all } r \qquad v_\theta = \text{finite} \tag{3.247}$$

This is too many boundary conditions for the simple ordinary differential equation we are trying to solve, but these conditions were used in formulating the guess $[v_\theta = zf(r)]$ that we used to solve the problem. As a result, there are no contradictions between these boundary conditions and our solution. Writing $v_\theta = zf(r)$, using Equation (3.243), and applying the boundary conditions, we obtain the velocity field:

$$f(r) = \frac{r\Omega}{H} \tag{3.248}$$

$$\boxed{v_\theta = \frac{zr\Omega}{H}} \tag{3.249}$$

To calculate the torque $\mathcal{T}$ required to turn the plate, we apply the definition of torque and integrate.[4]

$$\mathcal{T} = (\text{lever arm})(\text{force}) \tag{3.250}$$

$$= \int_0^R r\,[-\tau_{z\theta}(r)](2\pi r\,dr) \tag{3.251}$$

The stress $\tau_{z\theta}$ in Equation (3.251) is calculated from the constitutive equation.

$$\underline{\underline{\tau}} = -\mu\underline{\underline{\dot{\gamma}}} = -\mu[\nabla\underline{v} + (\nabla\underline{v})^T] \tag{3.252}$$

---

[4] Note that our sign convention for stress requires that stress be positive when it is transporting momentum from high-velocity regions to low-velocity regions. Because the top plate is causing a flux of momentum in the negative z-direction, the stress $\tau_{z\theta}$ is negative. To obtain a positive torque, a negative sign is included.

$$\nabla \underline{v} = \begin{pmatrix} \frac{\partial v_r}{\partial r} & \frac{\partial v_\theta}{\partial r} & \frac{\partial v_z}{\partial r} \\ \frac{1}{r}\frac{\partial v_r}{\partial \theta} - \frac{v_\theta}{r} & \frac{1}{r}\frac{\partial v_\theta}{\partial \theta} + \frac{v_r}{r} & \frac{1}{r}\frac{\partial v_z}{\partial \theta} \\ \frac{\partial v_r}{\partial z} & \frac{\partial v_\theta}{\partial z} & \frac{\partial v_z}{\partial z} \end{pmatrix}_{r\theta z} \tag{3.253}$$

$$= \begin{pmatrix} 0 & \frac{\partial v_\theta}{\partial r} & 0 \\ -\frac{v_\theta}{r} & 0 & 0 \\ 0 & \frac{\partial v_\theta}{\partial z} & 0 \end{pmatrix}_{r\theta z} \tag{3.254}$$

$$\dot{\underline{\gamma}} = \begin{pmatrix} 0 & \frac{\partial v_\theta}{\partial r} - \frac{v_\theta}{r} & 0 \\ \frac{\partial v_\theta}{\partial r} - \frac{v_\theta}{r} & 0 & \frac{\partial v_\theta}{\partial z} \\ 0 & \frac{\partial v_\theta}{\partial z} & 0 \end{pmatrix}_{r\theta z} \tag{3.255}$$

$$\underline{\tau} = \begin{pmatrix} 0 & -\mu\left(\frac{\partial v_\theta}{\partial r} - \frac{v_\theta}{r}\right) & 0 \\ -\mu\left(\frac{\partial v_\theta}{\partial r} - \frac{v_\theta}{r}\right) & 0 & -\mu\frac{\partial v_\theta}{\partial z} \\ 0 & -\mu\frac{\partial v_\theta}{\partial z} & 0 \end{pmatrix}_{r\theta z} \tag{3.256}$$

From the preceding and the solution for $v_\theta(r, z)$ [Equation (3.249)] we can see that

$$\underline{\tau} = \begin{pmatrix} 0 & 0 & 0 \\ 0 & 0 & -\frac{\mu r \Omega}{H} \\ 0 & -\frac{\mu r \Omega}{H} & 0 \end{pmatrix}_{r\theta z} \tag{3.257}$$

and substituting $\tau_{z\theta}$ into Equation (3.251) and integrating, we obtain

$$\boxed{\mathcal{T} = \frac{\pi R^4 \mu \Omega}{2H}} \tag{3.258}$$

We will discuss this flow for general fluids in Chapter 10.

From these examples we see how mass and momentum conservation are used to solve Newtonian flow problems. The material information needed in these solutions is limited to two scalar constants, the density ρ and the viscosity μ. Non-Newtonian behavior is considerably more complex, as we will discuss beginning in the next chapter.

3.6 PROBLEMS

3.1 What are the mass flow rates through each of the surfaces described below and shown in Figure 3.13?

(a) A circular surface of area A with unit normal $\hat{n} = 1/\sqrt{2}(\hat{e}_1 + \hat{e}_2)$.

(b) A circular surface of area A with unit normal $\hat{n} = \hat{e}_2$.

(c) A circular surface of area A with unit normal $\hat{n} = \hat{e}_1$.

(d) A hemisphere whose cross-sectional area is A with unit normal $\hat{n} = \hat{e}_1$.

(e) A sphere whose cross-sectional area is A, that is, its radius is $\sqrt{A/\pi}$.

3.2 What does the continuity equation reduce to for incompressible fluids?

3.3 Using Einstein notation, rearrange the continuity

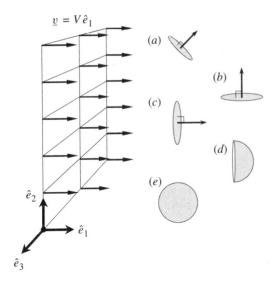

$$\underline{v} = V\hat{e}_1$$

Figure 3.13 Surfaces for Problem 3.1.

equation so that it contains the substantial derivative term $D\rho/Dt$. Do not assume constant density.

3.4 Write the continuity equation using cylindrical and spherical coordinates.

3.5 What assumptions are necessary to derive the Navier–Stokes equation from the general momentum balance?

3.6 Starting from the vector equation given below and using Einstein notation, work out the x_1-component of the Navier–Stokes equation.

$$\rho\left(\frac{\partial \underline{v}}{\partial t} + \underline{v} \cdot \nabla \underline{v}\right) = -\nabla p + \mu \nabla^2 \underline{v} + \rho \underline{g}$$

3.7 Show that the equation of motion for *compressible* Newtonian fluids is given by [64]

$$\rho\left(\frac{\partial \underline{v}}{\partial t} + \underline{v} \cdot \nabla \underline{v}\right) = -\nabla p + \mu \nabla^2 \underline{v}$$
$$+ \left(\frac{1}{3}\mu + \kappa\right)\nabla(\nabla \cdot \underline{v}) + \rho \underline{g}$$

where ρ is the density, μ is the viscosity, and κ is the dilatational viscosity. The equivalent pressure $\mathcal{P}$ is defined as the sum of pressure p and the potential energy due to gravity, $\Phi = \rho g h$, where h is the height of a fluid particle above a reference plane (see glossary).

$$\mathcal{P} = p + \rho g h$$

An alternative definition of equivalent pressure useful for compressible fluids is given as [64]:

$$\tilde{\mathcal{P}} \equiv p + \rho g h - \left(\frac{1}{3}\mu + \kappa\right)(\nabla \cdot \underline{v})$$

(a) How does the Navier–Stokes equation for compressible fluids simplify if $\tilde{\mathcal{P}}$ is incorporated?

(b) For *incompressible fluids*, how are $\mathcal{P}$ and $\tilde{\mathcal{P}}$ related?

3.8 Write velocity or stress boundary conditions for the flows depicted in Figure 3.14.

3.9 Show that the solutions for the pressure distribution and velocity distribution for Poiseuille flow in a slit [Equations (3.177) and (3.178)] result from the solution (with the appropriate boundary conditions) to the differential equation (3.170):

$$\mu \frac{\partial^2 v_1}{\partial x_2^2} = \frac{dp}{dx_1}$$

3.10 The Hagen–Poiseuille law gives the flow rate of an incompressible Newtonian fluid in pressure-driven flow in a pipe. For a horizontal pipe this law is

$$Q = \frac{\pi(P_0 - P_L)R^4}{8\mu L}$$

Why is density ρ absent from this expression?

3.11 Using a computer, plot the velocity profile for Poiseuille flow of a Newtonian fluid in a tube. Calculate the volumetric flow rate.

3.12 (a) Show that $\underline{v} \cdot \nabla \underline{v} = 0$ for unidirectional flow (flow in a straight line, no curves) of an incompressible fluid.

(b) Give an example of a flow for which $\underline{v} \cdot \nabla \underline{v}$ is not zero. Show that this is the case.

3.13 In the text we said that thermodynamic pressure could be cast in tensor form by writing $p\underline{I}$, where p is the pressure and $\underline{I}$ is the identity tensor. Show that this expression correctly reflects the fact that thermodynamic pressure acts with the same magnitude in every direction and acts normally (perpendicularly) to any surface.

3.14 **Flow Problem: Flow of a Newtonian fluid down an inclined plane.**

(a) Find the velocity profile at steady state for an incompressible Newtonian fluid flowing down an inclined plane. The upper surface of the fluid is exposed to air. The fluid film has a constant

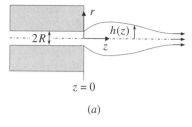

$z = 0$

(a)

Infinite fluid

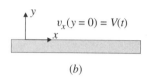

(b)

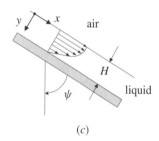

(c)

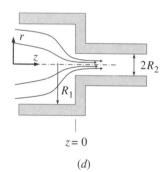

$z = 0$

(d)

Figure 3.14 Flows for Problem 3.8.

height H at steady state. The plane makes an angle ψ with the vertical. Be sure to explain the fate of each term of the Navier–Stokes equation.

(b) Calculate the maximum velocity, the average velocity, and the flow rate of the fluid.

(c) Plot the velocity profile.

3.15 Flow Problem: Pressure-driven flow of a Newtonian fluid in a slit that is tilted upward. Calculate the velocity profile and flow rate for pressure-driven flow of an incompressible Newtonian liquid between

two infinitely wide parallel plates separated by a gap of $2H$. The slit is inclined to the horizontal by an angle α. The pressure at an upstream point is P_0, and at a point a distance L downstream the pressure is P_L. Assume that the flow between the plates is fully developed and at steady state.

3.16 Flow Problem: Combined pressure and drag of a Newtonian fluid in a slit. Calculate the velocity profile and flow rate for combined pressure-driven and drag flow of an incompressible Newtonian liquid between two infinitely wide parallel plates separated by a gap of $2H$. The pressure at an upstream point is P_0, and at a point a distance L downstream the pressure is P_L. The upper plate is driven such that its velocity is V, and the lower plate is stationary. Assume that the flow between the plates is fully developed and at steady state.

3.17 Flow Problem: Drag flow of a Newtonian fluid in a slit that is tilted with respect to horizontal. Calculate the velocity profile and flow rate for drag flows of an incompressible Newtonian liquid between two infinitely wide, tilted parallel plates separated by a gap H as shown in Figure 3.15. The upper plate is driven such that its velocity is V, and the lower plate is stationary. Assume that the flow between the plates is well developed and at steady state.

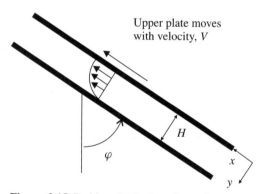

Upper plate moves with velocity, V

Figure 3.15 Problem 3.17: drag flow of a Newtonian fluid in a slit that is tilted.

3.18 Flow Problem: Tangential annular flow of a Newtonian fluid. Calculate the velocity and pressure profiles for tangential annular flow of an incompressible Newtonian fluid between concentric cylinders when the inner cylinder is turning (Figure 3.16). The pressure at the inner cylinder is P_0. Also calculate the torque required to turn the cylinder. The outer

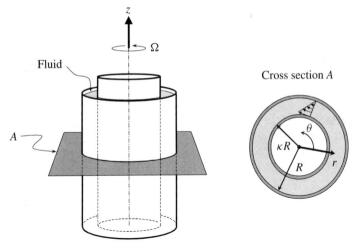

Figure 3.16 Problem 3.18: tangential annular flow of a Newtonian fluid.

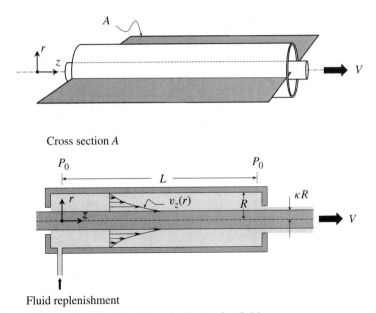

Figure 3.17 Problem 3.19: axial annular drag flow of a Newtonian fluid.

cylinder has a radius R, and the inner cylinder has a radius κR and is moving with an angular velocity Ω. The gap between the two cylinders is small.

3.19 Flow Problem: Axial annular drag flow of a Newtonian fluid. Axial annular drag flow is shown schematically in Figure 3.17. The inner cylinder is moving in the z-direction with a velocity V, and there is no pressure gradient in the z-direction. Calculate the steady-state velocity profile and the corresponding flow rate for an incompressible Newtonian fluid.

3.20 Flow Problem: Radial flow of a Newtonian fluid between parallel plates. Calculate the velocity and pressure profiles for an incompressible Newtonian fluid flowing radially outward from between parallel disks. The flow is slow and steady. Neglect the complications at the center of the flow where the fluid enters the parallel-plate region (Figure 3.18). The pressure at $r = \kappa R$ is $P_{\kappa R}$ and at $r = R$, $P = P_R$.

3.21 Flow Problem: Squeeze flow of a Newtonian fluid.

(a) Calculate the velocity field $\underline{v}$ and the pressure field p for an incompressible Newtonian fluid in slow squeeze flow between two circular, parallel plates separated by a gap of $2h$ (Figure 3.19). The upper and lower plates are approaching one another at speed V. The pressure at the edges of the plates is atmospheric, and the flow is two-dimensional (two nonzero components of velocity).

(b) Calculate the force on each plate required to maintain the motion.

(c) Calculate the plate separation as a function of time $h(t)$ if the applied force is constant.

3.22 *Flow Problem: Start-up flow of a plate in a semi-infinite Newtonian fluid [28].

A semi-infinite incompressible Newtonian fluid at rest is bounded on one side by an infinite plate (Figure 3.20). At $t = 0$ the plate is set in motion at a constant velocity V. Calculate the velocity field in the fluid as a function of time. Plot your results as velocity as a function of distance for several values of time.

3.23 *Flow Problem: Oscillating plate in a semi-infinite Newtonian fluid.

An incompressible Newtonian fluid of viscosity μ and density ρ is bounded on one side by an infinite flat plate [26] at $y = 0$. The flate plate is oscillating in the x-direction with a frequency ω and a velocity $v_x(y = 0) = V \cos \omega t = \Re\{V e^{i\omega t}\}$. What is the steady-state velocity profile in the fluid? Solution steps:

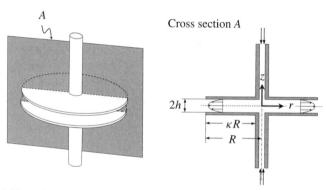

Figure 3.18 Problem 3.20: radial flow of a Newtonian fluid from between parallel disks.

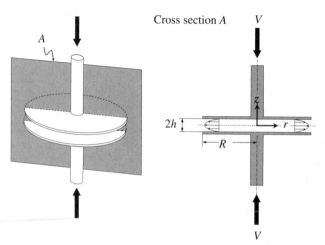

Figure 3.19 Problem 3.21: squeeze flow of a Newtonian fluid between parallel disks.

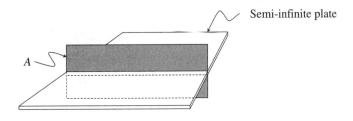

Cross section A

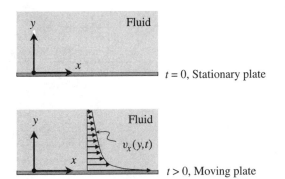

Figure 3.20 Problem 3.22: start-up flow of a plate in a semi-infinite Newtonian fluid.

(a) Show how the Navier–Stokes equation simplifies to a simple, scalar, nonseparable differential equation. Indicate the boundary conditions on y.

(b) Postulate a solution of the form $v_x(y, t) = \Re\{v^*(y)e^{i\omega t}\}$ and obtain an equation for v^*. Using the appropriate boundary conditions, solve for $v^*(y)$. *Answer:* $v^*(y) = Ve^{-(1+i)\sqrt{\frac{\omega\rho}{2\mu}}y}$.

(c) Calculate the velocity $v_x(y, t)$ and plot your results for $\omega t = \frac{\pi}{3}, \frac{2\pi}{3}, \pi, \frac{4\pi}{3}, \frac{5\pi}{3}, 2\pi$.

(d) (Messy algebra) Consider now that the fluid is bound at $y = H$ by a second stationary wall. What are the new boundary conditions? Using the new boundary conditions show that for $\sqrt{\rho\omega/2\mu}\,H \ll 1$ the velocity profile is nearly linear. *Hint:* $e^x = 1 + x + \cdots$, $e^{-x} = 1 - x + \cdots$.

3.24 *Flow Problem: Helical flow of a Newtonian fluid.** Calculate the steady-state velocity profile for an incompressible Newtonian fluid subjected to combined tangential annular flow and axial pressure-driven flow. The fluid is confined between concentric cylinders and the inner cylinder is turning (Figure 3.21). The axial pressure gradient is constant.

Also calculate the torque required to turn the cylinder. The outer cylinder has a radius R, and the inner cylinder has a radius κR and is moving with an angular velocity Ω. The gap between the two cylinders is small. Plot your answers.

3.25 *Flow Problem: Poiseuille flow in a rectangular duct.* Calculate the velocity field and flow rate for steady, well-developed, pressure-driven flow in a duct of rectangular cross section (Poiseuille flow in a duct; see Figure 3.22). The duct is not wide, and therefore variations in the y-direction may not be neglected. Thus the velocity is two-dimensional, that is, $v_z = v_z(x, y)$. Upstream the pressure is P_0 while a distance L downstream the pressure is P_L. Solution steps:

(a) Show how the Navier–Stokes equation simplifies to a simple, scalar, separable differential equation. Indicate the boundary conditions.

(b) Solve for the pressure profile.

(c) Solve for the velocity $v_z(x, y)$. *Note:* The solution is a series involving hyperbolic trigonometric functions [263].

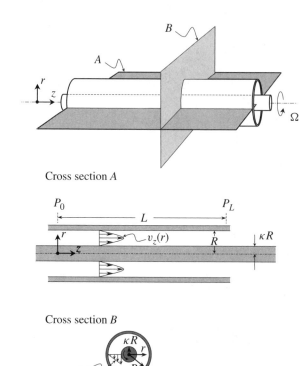

Figure 3.21 Problem 3.24: helical flow of a Newtonian fluid.

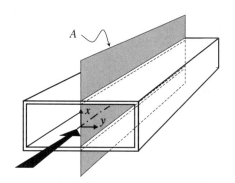

Figure 3.22 Problem 3.25: Poiseuille flow in a rectangular duct.

CHAPTER 4

Standard Flows for Rheology

Having established a common background of mathematics and Newtonian fluid mechanics, we now embark on our true subject, rheology.

We cannot study non-Newtonian constitutive relations until we are familiar with the non-Newtonian effects that these equations are trying to describe. Experimental rheologists have recorded a vast literature of observations on non-Newtonian fluids, organized around a small set of standard flows. We begin in this chapter by describing the standard flows that are commonly used to study non-Newtonian fluids. In Chapter 6 we devote some space to a brief catalog of experimental observations. To do this we define material functions in Chapter 5 that can be measured by subjecting a fluid to a standard flow or that can be predicted from a constitutive equation. After these background chapters, we move on to non-Newtonian constitutive equations in Chapters 7 through 9.

4.1 Introduction

How do we classify a material as non-Newtonian? Generally speaking, any material that exhibits behavior not predicted by the Newtonian constitutive equation, Equation (3.86) or Equation (3.91), is non-Newtonian. This approach only tells us what the fluid is not, however. To understand a fluid, we must concentrate on what it is, that is, how it behaves.

To describe a fluid's rheology, we must poke it and prod it in a variety of ways and see how it responds, to see what stresses are generated and how the material flows. One approach would be to specify a deformation to impose on the fluid and then measure the stresses generated by the flowing fluid. For example, we could pull on a polymer sample and measure the force required to stretch it at a certain rate (Figure 4.1a). Alternatively we could rapidly pull a sample to a specified length and measure the time-dependent force needed to hold the deformed sample in that position. A second approach is to impose a stress on a fluid and measure the velocity or deformation fields that are produced as a result. For example, we could hang a weight on a polymer sample and measure the change in sample length with time, or we could rotate a rod in a cup of fluid at a certain torque level and count how many turns per minute the rod makes as a function of torque (Figure 4.1b). Any of these approaches will yield information about the flow behavior (and hence the constitutive relationship) for the fluid. We could then compare how different materials respond to the imposed flows and organize fluids by their types of responses.

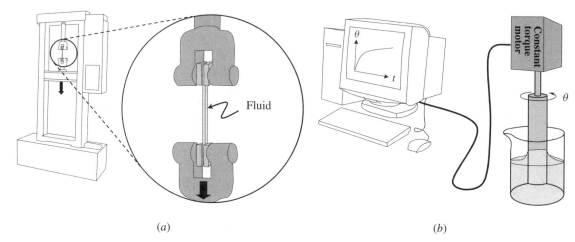

Fluid

(a)   (b)

Figure 4.1 Schematics of the kinds of tests that could yield rheological information about non-Newtonian fluids. (a) Tensile-testing machine. (b) Rotational viscometer.

The choice of the flow situation (the kinematics) used for probing non-Newtonian behavior is arbitrary, and the number of choices is infinite. If everyone interested in rheology were to choose different flows, however, the advancement of the study of rheology would be slow, since researchers could not compare their results. The rheological community has therefore settled on a small number of standard flows. The choice of standard flows is driven by two considerations: first, the flow should be sufficiently simple that the missing ingredient–either the velocity profile or the stress field–can be readily calculated with proposed constitutive equations; second, the flow should be realizable by experimentalists.

These two requirements are often at odds. A test that is easy for an experimentalist to perform (stretching a piece of polymer film, for instance) can be extremely complicated to describe mathematically (e.g., for a tensile experiment, the effects due to the clamps holding the sample at the ends are very difficult to model accurately). Conversely, a test that computationalists favor because it differentiates constitutive equations that are otherwise similar (steady stretching or elongational flow, described in Section 4.3.1) may be nearly impossible to achieve experimentally (e.g., in steady stretching flow, fluid elements separate exponentially in time, and this is very difficult to accomplish experimentally; see Chapter 10).

The two flows described in this chapter, shear and elongation, may be thought of as the classic flows used in rheological measurements. The list of standard flows is not closed–new ones are proposed on a regular basis. Whether or not a flow becomes a standard is based entirely on whether a large number of rheologists use it in calculations or in experiments.

4.2 Simple Shear Flow

Shear flow is the most common type of flow discussed in rheology. Figure 4.2 shows a two-dimensional schematic of the velocity profile in simple shear flow. In this flow, layers of fluid slide past each other and do not mix. The flow is rectilinear, and the velocity only varies in one direction, the direction x_2 in this diagram. Particle path lines in simple shear flow are straight

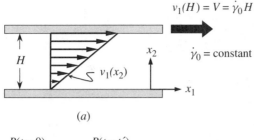

$v_1(H) = V = \dot\gamma_0 H$

$\dot\gamma_0 = \text{constant}$

Figure 4.2 Flow field in simple shear. (a) Velocity profile. (b) Particle path lines.

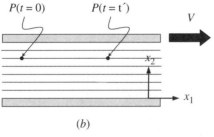

$P(t = 0)$ $P(t = t')$

parallel lines. Simple shear flow can be produced by sandwiching material between two parallel plates and then causing one plate to move at a constant velocity in some unchanging direction. Recall the drag flow example in Section 3.5.1, which was simple shear. For many practical rheometers, this flow is only achieved approximately in the limit of narrow gaps, small angles, and relatively slow flows (see Chapter 10).[1] In processing equipment shear flow occurs near walls (Figure 4.3). Sketches of actual experimental devices used to produce or to approximate shear flow are shown in Figure 4.4.

The velocity profile for simple shear flow is defined in Cartesian coordinates as follows [26]:

Definition
of shear flow

$$\underline{v} = \begin{pmatrix} v_1 \\ v_2 \\ v_3 \end{pmatrix}_{123} = \begin{pmatrix} \dot\varsigma(t)x_2 \\ 0 \\ 0 \end{pmatrix}_{123} \tag{4.1}$$

Shear flow may be produced in many ways, and it is standard practice to call the flow direction of shear flow in Cartesian coordinates the 1-direction, with the 2-direction reserved for the direction in which the velocity changes (the gradient direction), and the 3-direction called the neutral direction, since flow neither occurs in this direction nor changes in this direction (Figure 4.5). The function $\dot\varsigma(t)$,[2] which equals the derivative $\partial v_1/\partial x_2$, is often denoted by the symbol $\dot\gamma_{21}(t)$ because it is equal to the 21-component of the shear-rate tensor $\dot{\underline{\underline{\gamma}}}$ for this flow [Equation (4.2)].

---

[1] See Bird et al. [26] Section 3.7 for a discussion of the classification of shear flows that are not *simple* shear flow. As discussed there, a more general class of shear flows can also be used for measuring the material functions we will define in Chapter 5 for simple shear flow.

[2] The symbol ς is a variation of σ (sigma), the Greek letter s. This version appears at the ends of words in written Greek. The symbol $\dot\varsigma$ is referred to as *sigma dot*.

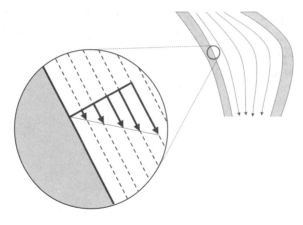

Figure 4.3 Even in complex mixed flows such as that shown schematically on the right, the flow near the walls is shear flow.

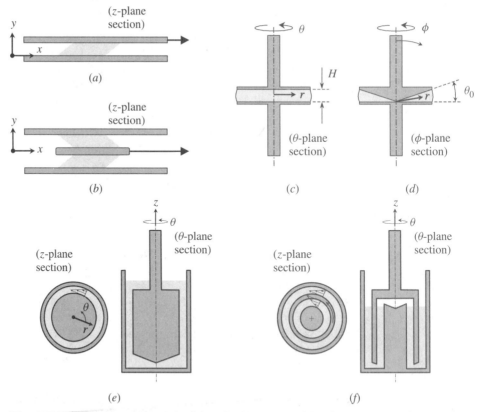

Figure 4.4 Geometries used to produce shear flow in commercial and research rheometers. (*a*) Rectilinear parallel plate. (*b*) Rectilinear double parallel plate. (*c*) Torsional parallel plate or parallel disk. (*d*) Torsional cone and plate. (*e*) Couette or cup and bob. (*f*) Double-walled Couette.

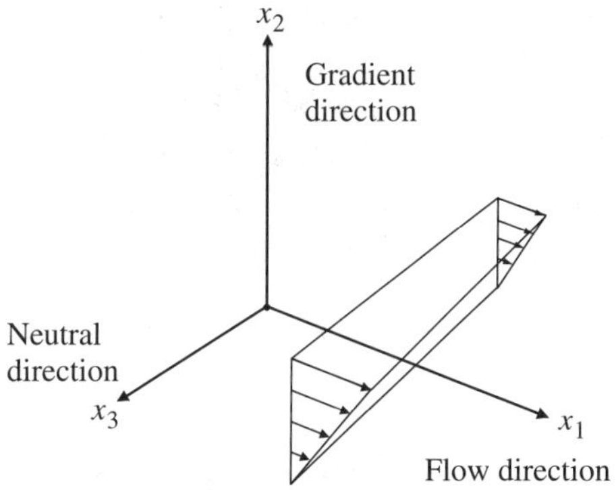

Figure 4.5 Standard coordinate system used for describing shear flow.

$$\underline{\underline{\dot{\gamma}}} \equiv \nabla \underline{v} + (\nabla \underline{v})^T \tag{4.2}$$

$$= \begin{pmatrix} \frac{\partial v_1}{\partial x_1} & \frac{\partial v_2}{\partial x_1} & \frac{\partial v_3}{\partial x_1} \\ \frac{\partial v_1}{\partial x_2} & \frac{\partial v_2}{\partial x_2} & \frac{\partial v_3}{\partial x_2} \\ \frac{\partial v_1}{\partial x_3} & \frac{\partial v_2}{\partial x_3} & \frac{\partial v_3}{\partial x_3} \end{pmatrix}_{123} + \begin{pmatrix} \frac{\partial v_1}{\partial x_1} & \frac{\partial v_1}{\partial x_2} & \frac{\partial v_1}{\partial x_3} \\ \frac{\partial v_2}{\partial x_1} & \frac{\partial v_2}{\partial x_2} & \frac{\partial v_2}{\partial x_3} \\ \frac{\partial v_3}{\partial x_1} & \frac{\partial v_3}{\partial x_2} & \frac{\partial v_3}{\partial x_3} \end{pmatrix}_{123} \tag{4.3}$$

$$= \begin{pmatrix} 0 & 0 & 0 \\ \dot{\varsigma}(t) & 0 & 0 \\ 0 & 0 & 0 \end{pmatrix}_{123} + \begin{pmatrix} 0 & \dot{\varsigma}(t) & 0 \\ 0 & 0 & 0 \\ 0 & 0 & 0 \end{pmatrix}_{123} \tag{4.4}$$

$$= \begin{pmatrix} 0 & \dot{\varsigma}(t) & 0 \\ \dot{\varsigma}(t) & 0 & 0 \\ 0 & 0 & 0 \end{pmatrix}_{123} \tag{4.5}$$

$$\dot{\gamma}_{21}(t) = \dot{\varsigma}(t) \tag{4.6}$$

$$\underline{v} = \begin{pmatrix} \dot{\gamma}_{21} x_2 \\ 0 \\ 0 \end{pmatrix}_{123} \tag{4.7}$$

The magnitude of $\underline{\underline{\dot{\gamma}}}$ for shear flow is called the *shear rate* or *rate of strain* and is denoted by the symbol $\dot{\gamma}(t)$. Applying the definition of tensor magnitude from Equation (2.174) we obtain

$$\dot{\gamma}(t) = \left| \underline{\underline{\dot{\gamma}}}(t) \right| = \frac{+\sqrt{\underline{\underline{\dot{\gamma}}} : \underline{\underline{\dot{\gamma}}}}}{2} \tag{4.8}$$

$$= |\dot{\varsigma}(t)| = \pm \dot{\varsigma}(t) \tag{4.9}$$

The magnitude of a tensor is always positive, and thus $\dot{\gamma}$ is always a positive number, whereas $\dot{\varsigma}(t) = \dot{\gamma}_{21}$ may be positive or negative. From this we see that for shear flow written in the Cartesian coordinate system, the rate of strain $\dot{\gamma}$ and $\dot{\gamma}_{21}$ are equal or differ only by sign. In other coordinate systems or for flows other than simple shear, however, the deformation rate $\dot{\gamma} = |\dot{\underline{\underline{\gamma}}}|$ and the 21-component of $\dot{\underline{\underline{\gamma}}}$ may be quite different. Notice also that the shear rate $[\dot{\gamma} = |\dot{\varsigma}(t)|]$ is independent of position (not a function of x_1, x_2, or x_3) for simple shear flow as defined by the velocity field in Equation (4.1). A flow for which the rate of deformation is independent of position is called a *homogeneous* flow. Homogeneous flows may be functions of time $[\dot{\gamma} = \dot{\gamma}(t)]$.

One reason shear flow is used as a standard flow is that $\dot{\underline{\underline{\gamma}}}$ for shear flow is simple,

$$\dot{\underline{\underline{\gamma}}} = \begin{pmatrix} 0 & \dot{\gamma}_{21}(t) & 0 \\ \dot{\gamma}_{21}(t) & 0 & 0 \\ 0 & 0 & 0 \end{pmatrix}_{123} \tag{4.10}$$

Recall that we first encountered $\dot{\underline{\underline{\gamma}}}$ in the Newtonian constitutive equation:

$$\underline{\underline{\tau}} = -\mu \dot{\underline{\underline{\gamma}}} \tag{4.11}$$

For shear flow as defined here, the stress tensor predicted by the Newtonian constitutive equation is also simple:

$$\underline{\underline{\tau}} = \begin{pmatrix} \tau_{11} & \tau_{12} & \tau_{13} \\ \tau_{21} & \tau_{22} & \tau_{23} \\ \tau_{31} & \tau_{32} & \tau_{33} \end{pmatrix}_{123} \tag{4.12}$$

$$= -\mu \dot{\underline{\underline{\gamma}}} = \begin{pmatrix} 0 & -\mu\dot{\gamma}_{21} & 0 \\ -\mu\dot{\gamma}_{21} & 0 & 0 \\ 0 & 0 & 0 \end{pmatrix}_{123} \tag{4.13}$$

For a Newtonian incompressible fluid in shear flow only two of the nine Cartesian components of the stress tensor are nonzero, and these two components are equal. Thus, shear flow for an incompressible Newtonian fluid reduces to one simple scalar equation:

$$\tau_{21} = \tau_{12} = -\mu\dot{\gamma}_{21} = -\mu\dot{\varsigma}(t) \tag{4.14}$$

$$= -\mu \frac{\partial v_1}{\partial x_2} \tag{4.15}$$

This equation is the familiar Newton's law of viscosity [85]. We see now that Newton's law of viscosity is a simplification of the full, tensorial Newtonian constitutive equation, $\underline{\underline{\tau}} = -\mu \dot{\underline{\underline{\gamma}}}$. Newton's law of viscosity only describes material response for shear flow.

A second reason that shear flow is a standard rheological flow is that it is a simple version of a sliding flow. Because shear is in one direction and only varies in the 2-direction, planes of constant x_2 move together in the flow direction, sliding over one another but never mixing (Figure 4.6). Two fluid particles that lie in the same $\hat{e}_2$-plane (a plane whose unit normal is $\hat{e}_2$) a distance r apart will always be a distance r apart. When two fluid particles are located at different values of x_2, however, they will get farther apart in steady shear. Sliding flows occur near most boundaries in processing flows.

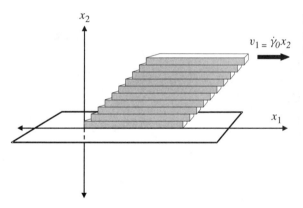

Figure 4.6 Schematic of shear planes sliding over each other during steady shear flow, $\dot{\varsigma}(t) = \dot{\gamma}_{21}(t) = \dot{\gamma}_0$.

$v_1 = \dot{\gamma}_0 x_2$

To clarify the sliding nature of shear, consider two fluid elements undergoing a steady shear flow (Figure 4.7). The two points P_1 and P_2 are initially separated by a distance l_0. For simplicity we choose these two elements to be in the same $\hat{e}_3$-plane and initially at the same value of x_1. If we choose our coordinate system such that the particles are on the x_2-axis, then at time $t = 0$ the coordinates of the particles are $P_1(0, l_1, 0)$ and $P_2(0, l_2, 0)$, with $l_2 - l_1 = l_0$. After experiencing shear flow for some time, the two fluid particles will have separated because the particle located at the larger value of x_2 will move faster than the other particle. The new separation distance l can be calculated by noting that the $\hat{e}_1$-component of velocity, given by Equation (4.7), is just the rate of change of the particle position x_1 with time,

$$v_1 = \frac{dx_1}{dt} = \dot{\varsigma}(t)x_2 \tag{4.16}$$

For steady shear flow $\dot{\varsigma}(t) = \text{constant} \equiv \dot{\gamma}_0$, and this equation is easy to evaluate for each particle. For the particle at P_1,

$$\frac{dx_1}{dt} = \dot{\gamma}_0 \, l_1 \tag{4.17}$$

$$x_1 = \dot{\gamma}_0 \, l_1 t + C_1 \tag{4.18}$$

$$x_1 = \dot{\gamma}_0 \, l_1 t \tag{4.19}$$

C_1 is an integration constant, and in arriving at the last equation we have used the initial condition, $t = 0$, $x_1 = 0$. Likewise for the second particle, $x_1 = \dot{\gamma}_0 l_2 t$.

The final separation distance can be calculated to be (see Figure 4.7)

$$l = \sqrt{l_0^2 + (\dot{\gamma}_0 \, l_0 t)^2} \tag{4.20}$$

$$= l_0 \sqrt{1 + (\dot{\gamma}_0 \, t)^2} \tag{4.21}$$

As the flow continues, l gets very large, and we can take the limit as t goes to infinity,

$$l = \lim_{t \longrightarrow \infty} \left[l_0 \sqrt{1 + (\dot{\gamma}_0 t)^2} \right] \approx l_0 \, \dot{\gamma}_0 t \tag{4.22}$$

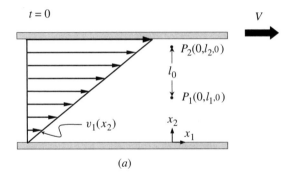

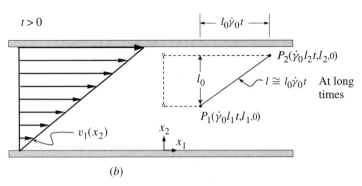

Figure 4.7 Separation by shear flow of two fluid particles that lie in the same e_3-plane but start at different values of x_2.

$$\frac{l}{l_0} = \dot{\gamma}_0 t \qquad (4.23)$$

Thus, in shear flow the relative distance between two fluid elements in different shear planes increases linearly with time. In other words, the longer you shear, the distance separating particles increases in proportion to the shearing time. Although this result was derived for two particles on the same vertical axis, it can be shown that it is equally valid for any pair of particles not in the same shear plane. This characteristic, that particles separate linearly with time, makes shearing flow a rather mild flow in terms of particle separation. The next flows we discuss, simple shear-free flows, are much more dramatic in the kind of particle separations they produce.

4.3 Simple Shear-Free (Elongational) Flows

We have divided the standard flows into shear and shear-free flows. Shear flows in general, including simple shear discussed in Section 4.2, are flows that include nonzero off-diagonal components to the rate-of-strain tensor $\underset{=}{\dot{\gamma}}$ and the stress tensor $\underset{=}{\tau}$. Elongational flow is a shear-free flow, that is, one with no nonzero, off-diagonal components in $\dot{\underline{\gamma}}$ or $\underline{\underline{\tau}}$, as we shall see. The diagonal components of $\underline{\underline{\tau}}$ are called the *normal stresses* since they represent stresses acting perpendicularly to a surface. The off-diagonal components of $\underline{\underline{\tau}}$ are called the *shear components*.

There are three types of simple, shear-free flows that are commonly discussed in rheology, and, like shear flow, they are defined in terms of their velocity profiles. In describing these flows we follow the nomenclature of Bird, Armstrong, and Hassager [26].

4.3.1 UNIAXIAL ELONGATIONAL FLOW

The choice of uniaxial elongational flow as a standard flow is based on the importance of this type of flow in polymer-processing operations such as fiber spinning and injection molding [236, 179]. Near the centerline of the flow in fiber spinning, for example, fluid particles are stretched uniformly (Figure 4.8). The idealized version of this stretching flow is called *uniaxial extensional flow* or *uniaxial elongational flow* and is defined by the following velocity profile:

$$
\begin{matrix}
\text{Definition of} \\
\text{uniaxial} \\
\text{elongational flow}
\end{matrix}
\qquad
\underline{v} = \begin{pmatrix} v_1 \\ v_2 \\ v_3 \end{pmatrix}_{123} = \begin{pmatrix} -\frac{\dot{\epsilon}(t)}{2} x_1 \\ -\frac{\dot{\epsilon}(t)}{2} x_2 \\ \dot{\epsilon}(t) x_3 \end{pmatrix}_{123} , \quad \dot{\epsilon}(t) > 0
\qquad (4.24)
$$

The function $\dot{\epsilon}(t)$ is called the *elongation rate*, and for uniaxial elongational flow $\dot{\epsilon}(t)$ is positive. The flow pattern that this velocity profile describes is three-dimensional, with a strong stretch occurring in the x_3-direction and contraction occurring equally in the x_1- and x_2-directions. Representing the velocity field of uniaxial elongational flow graphically is more difficult than representing shear, as can be seen in Figure 4.9. The particle path lines for uniaxial elongational flow are shown in Figure 4.10. Uniaxial elongational flow is a more

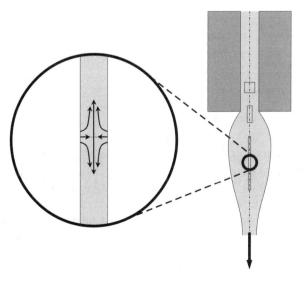

Figure 4.8 In fiber-spinning, a polymeric fluid is forced through a circular channel and emerges into the atmosphere. The fluid fiber is pulled by an external take-up reel, which stretches the fluid elements in the fiber. Following one fluid element as it passes out into the atmosphere, we see that the particles are stretched uniaxially as they are convected along with the flow. The stretching flow is uniaxial extension.

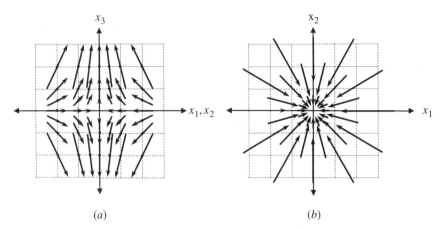

(a)   (b)

Figure 4.9 Two-dimensional representations of the velocity field in uniaxial elongational flow. At each point the vector indicates the direction of the flow at the point at which the vector is centered, and the length of the vector is proportional to the velocity at that point. (*a*) The velocity field is the same in the x_3x_1- and x_3x_2-planes as indicated. In fact, it is the same in every plane that includes the x_3-axis. (*b*) Velocity field in the x_1x_2-plane, which is perpendicular to the main flow direction and includes the stagnation point at the origin.

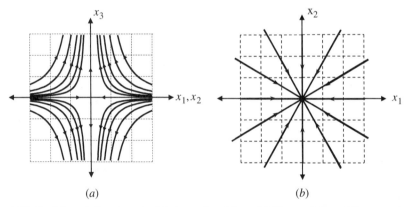

(a)   (b)

Figure 4.10 Particle path lines in uniaxial elongational flow. (*a*) The particle path lines are the same in the x_3x_1- and x_3x_2-planes. (*b*) Particle path lines in the x_1x_2-plane, which is perpendicular to the flow direction and includes the stagnation point at the origin.

complicated flow than shear flow since the velocity components are nonzero in all three directions for most points. Only on the coordinate axes is one of the velocity components zero. The only point not moving at all in elongational flow is the particle located at (0, 0, 0); this is called the *stagnation point*. By contrast, in shear flow the flow is unidirectional, $v_2 = v_3 = 0$ for all points, and points at $x_2 = 0$ are not moving. In the lab, elongational flow is approximated by such devices as shown schematically in Figure 4.11*a–c*. This important flow is very difficult to realize, and researchers are actively pursuing the design of better instruments for producing elongational flow.

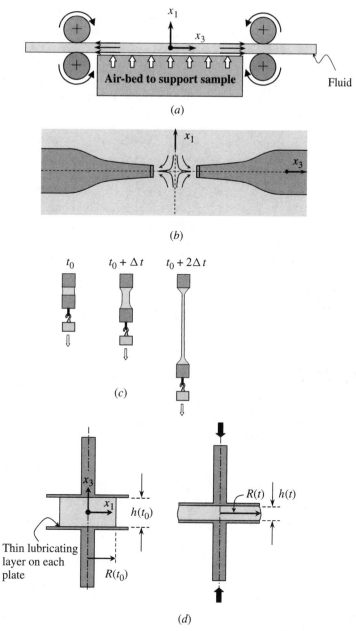

Figure 4.11 Geometries used to produce shear-free flows in commercial and research rheometers. (*a*) Uniaxial stretching by a pulling device. (*b*) Uniaxial extension in an opposed-nozzle suction device. (*c*) Uniaxial extension by filament stretching of polymer solutions. (*d*) Biaxial extension through lubricated squeezing [162].

We can consider the characteristics of this flow more closely by examining Equation (4.24). All three components of the velocity for elongational flow are a function of position. Further, the x_1-direction velocity v_1 is a function of x_1, the x_2-direction velocity is a function of x_2, and the x_3-direction velocity is a function of x_3. We can see then that a fluid particle originally at position x_{1_0} will move in the x_1-direction with velocity $v_1(x_{1_0})$, and this will move the fluid to a new x_1-position, x_{1_1}. At this new position, however, the velocity of the fluid will be different; it will be $v_1(x_{1_1})$. Thus, in elongational flow the velocities of fluid particles are continuously changing as the flow progresses. This is in stark contrast to shear flow, $\underline{v} = \dot{\varsigma}(t)x_2\hat{e}_1$. In shear flow, the only nonzero velocity is in the x_1-direction; it does not vary with x_1 but rather with x_2. Fluid particles in shear flow move along in planes of constant x_2. Thus the velocities of fluid particles are constant in time when $\dot{\varsigma}(t) = \dot{\gamma}_{21}(t)$ is constant. For elongational flow, even when $\dot{\epsilon}(t) = \dot{\epsilon}_0$ is constant, the velocity of a fluid particle is not constant since $v_1 = -(\dot{\epsilon}_0/2) x_1$, $v_2 = -(\dot{\epsilon}_0/2) x_2$, $v_3 = \dot{\epsilon}_0 x_3$, and x_1, x_2, and x_3, the coordinate positions of the fluid particle, are continuously changing as the fluid flows.

Shear and elongation also differ qualitatively in the form of the rate-of-strain tensor $\dot{\underline{\underline{\gamma}}}$, as we see by constructing $\dot{\underline{\underline{\gamma}}}$ for this flow:

$$\dot{\underline{\underline{\gamma}}} = \nabla\underline{v} + (\nabla\underline{v})^T \tag{4.25}$$

$$= \begin{pmatrix} -\frac{\dot{\epsilon}(t)}{2} & 0 & 0 \\ 0 & -\frac{\dot{\epsilon}(t)}{2} & 0 \\ 0 & 0 & \dot{\epsilon}(t) \end{pmatrix}_{123} + \begin{pmatrix} -\frac{\dot{\epsilon}(t)}{2} & 0 & 0 \\ 0 & -\frac{\dot{\epsilon}(t)}{2} & 0 \\ 0 & 0 & \dot{\epsilon}(t) \end{pmatrix}_{123} \tag{4.26}$$

$$= \begin{pmatrix} -\dot{\epsilon}(t) & 0 & 0 \\ 0 & -\dot{\epsilon}(t) & 0 \\ 0 & 0 & 2\dot{\epsilon}(t) \end{pmatrix}_{123} \tag{4.27}$$

This tensor is diagonal when written in this Cartesian coordinate system and thus has no shear components. Looking at their rate-of-strain tensors, we see that elongational flow is qualitatively different from shear flow in a mathematical sense. The magnitude of $\dot{\underline{\underline{\gamma}}}$ can be calculated in a straightforward manner:

$$\dot{\gamma} = \left| \dot{\underline{\underline{\gamma}}} \right| \tag{4.28}$$

$$= +\sqrt{\frac{1}{2} \begin{pmatrix} -\dot{\epsilon}(t) & 0 & 0 \\ 0 & -\dot{\epsilon}(t) & 0 \\ 0 & 0 & 2\dot{\epsilon}(t) \end{pmatrix}_{123} : \begin{pmatrix} -\dot{\epsilon}(t) & 0 & 0 \\ 0 & -\dot{\epsilon}(t) & 0 \\ 0 & 0 & 2\dot{\epsilon}(t) \end{pmatrix}_{123}} \tag{4.29}$$

$$= |\dot{\epsilon}(t)| \sqrt{3} \tag{4.30}$$

In the previous section we used the Newtonian constitutive equation to calculate the stress tensor generated by a Newtonian fluid subjected to steady shear flow. This calculation showed us that Newton's law of viscosity is predicted by the Newtonian constitutive equation in simple shear flow. Now we can examine what the Newtonian constitutive equation predicts for the stress tensor in steady uniaxial elongational flow. We begin with the Newtonian constitutive equation for incompressible fluids,

$$\underline{\underline{\tau}} = -\mu \underline{\underline{\dot{\gamma}}} \tag{4.31}$$

For steady elongational flow, $\dot{\epsilon}(t) = \dot{\epsilon}_0 = $ constant. Substituting this into $\underline{\underline{\dot{\gamma}}}$ for elongational flow, we obtain for the stress tensor for a Newtonian fluid in steady elongational flow,

$$\underline{\underline{\tau}} = -\mu \begin{pmatrix} -\dot{\epsilon}_0 & 0 & 0 \\ 0 & -\dot{\epsilon}_0 & 0 \\ 0 & 0 & 2\dot{\epsilon}_0 \end{pmatrix}_{123} \tag{4.32}$$

$$= \begin{pmatrix} \mu\dot{\epsilon}_0 & 0 & 0 \\ 0 & \mu\dot{\epsilon}_0 & 0 \\ 0 & 0 & -2\mu\dot{\epsilon}_0 \end{pmatrix}_{123} \tag{4.33}$$

There is little about this equation for stress to remind us of shear flow. There are three nonzero terms out of nine, two of which are equal. There are no off-diagonal elements, and thus Newton's law of viscosity, $\tau_{21} = -\mu \, \partial v_1/\partial x_2$, is not at all relevant in this flow. The limitations of thinking about even simple Newtonian fluid response purely in terms of that scalar law are now apparent.

Equation (4.33) predicts that if we subject a Newtonian fluid to steady, uniaxial elongational flow, stresses will be generated such that $\tau_{11} = \tau_{22} = -\frac{1}{2}\tau_{33} = \mu\dot{\epsilon}_0$. We can define a different "law of viscosity" for this flow, define a new material function, perhaps an elongational viscosity, and set about studying material response in this type of flow. We would expect that we would observe some materials that will follow the Newtonian predictions (Newtonian fluids) and others that do not (non-Newtonian fluids). This is precisely the case. The definitions of various types of material functions for both shear and shear-free flows will be addressed in Chapter 5.

In addition to being different from shear flows in the predictions of the Newtonian constitutive equation, elongation flow is also qualitatively different from shear flow in how the fluid particles move apart. In the discussion on shear flow in the previous section, we found that at long times two fluid particles in different shear planes move apart linearly in time. Now we can calculate how two material particles move apart in elongational flow.

We choose to look at two elements originally on the x_3-axis, separated by a distance l_0 (Figure 4.12). The origin of the Cartesian coordinate system is midway between the two fluid elements, and thus the coordinates of the two fluid elements under consideration are $(0, 0, l_0/2)$ and $(0, 0, -l_0/2)$. Since our particles are on the x_3-axis, the velocity that is experienced by these two fluid elements is

$$\underline{v} = \begin{pmatrix} -\frac{\dot{\epsilon}_0}{2}x_1 \\ -\frac{\dot{\epsilon}_0}{2}x_2 \\ \dot{\epsilon}_0 x_3 \end{pmatrix}_{123} = \begin{pmatrix} 0 \\ 0 \\ \dot{\epsilon}_0 x_3 \end{pmatrix}_{123} \tag{4.34}$$

Recalling that $v_3 = dx_3/dt$, we can solve for the positions of the particles at some time t in the future,

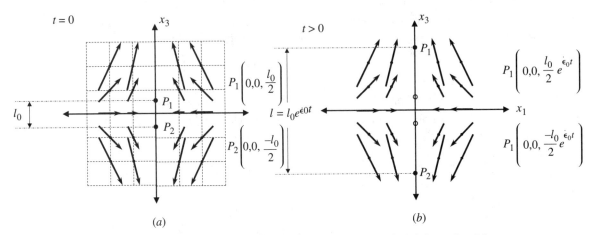

Figure 4.12 Separation of two fluid particles in simple uniaxial elongational flow.

$$v_3 = \frac{dx_3}{dt} \tag{4.35}$$

$$\frac{dx_3}{dt} = \dot{\epsilon}_0 x_3 \tag{4.36}$$

$$\frac{dx_3}{x_3} = \dot{\epsilon}_0 dt \tag{4.37}$$

$$\ln x_3 = \dot{\epsilon}_0 t + C_1 \tag{4.38}$$

Applying the initial conditions for each particle to solve for C_1 we obtain

$$x_3 \text{ (particle 1)} = \frac{l_0}{2} e^{\dot{\epsilon}_0 t} \tag{4.39}$$

$$x_3 \text{ (particle 2)} = -\frac{l_0}{2} e^{\dot{\epsilon}_0 t} \tag{4.40}$$

The distance l separating the two particles is the difference between these two coordinate positions,

$$\frac{l}{l_0} = e^{\dot{\epsilon}_0 t} \tag{4.41}$$

Thus, as was the case for steady shear flow, as time increases, the separation of the two particles increases, but unlike steady shear flow, the separation is not linear in time, but rather exponential. This is quite a bit more rapid than in steady shear flow (see Problem 4.14), and this difference shows up in the stresses that are generated in elongational flow. This rapid particle separation results from the fact that as particles on the x_3-axis move in the x_3-direction, they accelerate. Remember that acceleration $\underline{a}$ is $\partial \underline{v}/\partial t$, and for steady elongational flow,

$$\underline{a}\;\underset{\text{elongation}}{\text{steady}} = \begin{pmatrix} -\frac{1}{2}\dot{\epsilon}_0\frac{\partial x_1}{\partial t} \\ -\frac{1}{2}\dot{\epsilon}_0\frac{\partial x_2}{\partial t} \\ \dot{\epsilon}_0\frac{\partial x_3}{\partial t} \end{pmatrix}_{123} = \begin{pmatrix} -\frac{1}{2}\dot{\epsilon}_0 v_1 \\ -\frac{1}{2}\dot{\epsilon}_0 v_2 \\ \dot{\epsilon}_0 v_3 \end{pmatrix}_{123} = \begin{pmatrix} \frac{1}{4}\dot{\epsilon}_0^2 x_1 \\ \frac{1}{4}\dot{\epsilon}_0^2 x_2 \\ \dot{\epsilon}_0^2 x_3 \end{pmatrix}_{123} \tag{4.42}$$

and thus the particles will accelerate in a linearly increasing fashion in all three coordinate directions. Again we can contrast this flow behavior with steady shear flow where the acceleration is zero:

$$\underline{a}\;\underset{\text{shear}}{\text{steady}} = \begin{pmatrix} \frac{\partial(\dot{\gamma}_0 x_2)}{\partial t} \\ 0 \\ 0 \end{pmatrix}_{123} = \begin{pmatrix} \dot{\gamma}_0 v_2 \\ 0 \\ 0 \end{pmatrix}_{123} = \begin{pmatrix} 0 \\ 0 \\ 0 \end{pmatrix}_{123} \tag{4.43}$$

Because of the linear acceleration of particles in elongational flow, an acceleration that results in particle separation that increases exponentially in time, elongational flow is considered a strong flow in terms of deformation.

4.3.2 Biaxial Stretching Flow

The second type of shear-free flow that we study is biaxial stretching. This flow has the same form of velocity profile as uniaxial elongational flow [Equation (4.24)], but $\dot{\epsilon}(t)$ is always negative for this flow. Biaxial stretching may be produced by squeezing a sample between two lubricated surfaces (see, for example, Figure 4.11d), or by inflating a film (Figure 4.13):

Definition of biaxial elongational flow

$$\underline{v} = \begin{pmatrix} v_1 \\ v_2 \\ v_3 \end{pmatrix}_{123} = \begin{pmatrix} -\frac{\dot{\epsilon}(t)}{2}x_1 \\ -\frac{\dot{\epsilon}(t)}{2}x_2 \\ \dot{\epsilon}(t)x_3 \end{pmatrix}_{123} \;,\quad \dot{\epsilon}(t) < 0 \tag{4.44}$$

Since the flows are the same except for flow direction, the comments of the previous section on the severity of deformation in uniaxial elongation apply to biaxial stretching

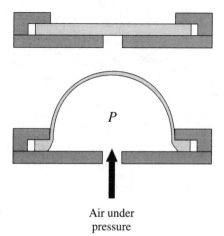

P

Air under
pressure

Figure 4.13 Testing configuration that produces biaxial flow. A sheet of material to be tested is clamped over a hole. A gas is then pumped through the hole, inflating the film. The pressure of the gas can be related to the stress in the film. The deformation can be measured by marking a grid on the film in the rest state and noting the shape of the grid in the inflated state.

as well. In practice biaxial flows are carried out to lesser extensional strains than uniaxial elongational flows. The conceptual differences between uniaxial elongational flow and biaxial stretching may be described by looking at the deformation experienced by fluid elements in the two flows. In uniaxial elongational flow, in one direction ($\hat{e}_3$, the primary flow direction) the fluid elements are stretched, while contraction occurs in the other two directions. Thus, if we look at a cube of incompressible fluid undergoing simple, uniaxial elongational flow, after some time the cube would be distorted to a rectangular solid with one side being, for example, twice as long as it was initially, while the other two sides would have compressed by a factor of $1/\sqrt{2}$ (Figure 4.14a). For uniaxial elongational flow of an incompressible fluid then,

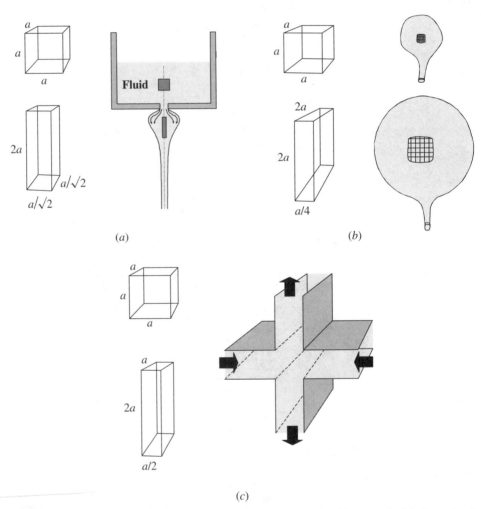

Figure 4.14 Schematics of the deformation (shape change) produced by (a) uniaxial elongational flow, (b) biaxial stretching flow, and (c) planar elongational flow.

$$\text{volume before} = a^3 = \text{volume after}$$

$$a^3 = (2a)\left(\frac{a}{\sqrt{2}}\right)\left(\frac{a}{\sqrt{2}}\right) \tag{4.45}$$

By comparison, in biaxial stretching (or biaxial extension as it is sometimes called) the flow is considered to occur in the 1- and 2-directions at the same rates, while contraction occurs in the 3-direction. Thus, a cube of fluid experiencing biaxial stretching to twice its original length would be distorted as shown in Figure 4.14b and would contract by a factor of 4 in the 3-direction:

$$\text{volume before} = a^3 = \text{volume after}$$

$$a^3 = (2a)(2a)\left(\frac{a}{4}\right) \tag{4.46}$$

Two flows in which biaxial stretching occurs are film blowing, a process used to manufacture plastic bags, and blow molding, a process that is used to create plastic bottles and other hollow parts [236, 179].

4.3.3 PLANAR ELONGATIONAL FLOW

The final type of shear-free flow that we will discuss is planar elongational flow, defined by the following velocity profile:

$$
\begin{array}{c}
\text{Definition of} \\
\text{planar} \\
\text{elongational flow}
\end{array}
\quad
\boxed{
\underline{v} = \begin{pmatrix} -\dot{\epsilon}(t)x_1 \\ 0 \\ \dot{\epsilon}(t)x_3 \end{pmatrix}_{123} , \quad \dot{\epsilon}(t) > 0
}
\tag{4.47}
$$

In planar elongational flow, no deformation is allowed in the 2-direction ($v_2 = 0$). The deformation experienced by a cube of incompressible fluid in planar elongation is shown in Figure 4.14c. If the side of the cube is stretched to twice its length in the flow direction (3-direction), then the cube must contract by a factor of 2 along the 1-direction to satisfy conservation of mass:

$$\text{volume} = a^3 = (2a)(a)\left(\frac{a}{2}\right)$$

This flow has gained the attention of experimentalists since constraining one direction of the flow makes the experiment a bit easier.[3] Planar elongation occurs in cross-channel dies like the one shown in Figure 4.14c.

    All three shear-free flows mentioned, along with many others, can be described by a single expression for the velocity profile [26]. The different flows are produced by different values for the two parameters $\dot{\epsilon}(t)$ and b. In the Cartesian coordinate system these shear-free flows can be written as follows:

---

[3] This flow is also called pure shear [155]. It can be shown by using the methods of Appendix C.6 that this flow is equivalent to a shear flow without rotation of the principal axes. For more discussion of principal axes see Appendix C.6 and [155].

TABLE 4.1
Summary of Parameters Defining Standard Shear-Free Flows

Elongational flow	$b = 0, \dot{\epsilon}(t) > 0$
Biaxial stretching flow	$b = 0, \dot{\epsilon}(t) < 0$
Planar elongational flow	$b = 1, \dot{\epsilon}(t) > 0$

$$\underline{v} = \begin{pmatrix} v_1 \\ v_2 \\ v_3 \end{pmatrix}_{123} = \begin{pmatrix} -\frac{1}{2}\dot{\epsilon}(t)(1+b)x_1 \\ -\frac{1}{2}\dot{\epsilon}(t)(1-b)x_2 \\ \dot{\epsilon}(t)x_3 \end{pmatrix}_{123} \tag{4.48}$$

For the flows we have discussed, the values and ranges of the parameters b and $\dot{\epsilon}(t)$ are listed in Table 4.1. The parameter b in the velocity profile affects the way that the streamlines of the flow change with rotations around the flow direction [26].

4.4 Forms of the Stress Tensor in Standard Flows

We have presented several flows that are accepted as standard flows for the study of the rheology of fluids. The velocity fields that define the flows are simple and lead to simple rate-of-strain tensors $\underline{\underline{\dot{\gamma}}}$. This is one argument for adopting these flows as standard flows.

There is another compelling reason to concentrate on these flows. The aim of much of the study of rheology is to determine the stress tensor $\underline{\underline{\tau}}$:

$$\underline{\underline{\tau}} = \begin{pmatrix} \tau_{11} & \tau_{12} & \tau_{13} \\ \tau_{21} & \tau_{22} & \tau_{23} \\ \tau_{31} & \tau_{32} & \tau_{33} \end{pmatrix}_{123} \tag{4.49}$$

Since this tensor has, in general, six unknown coefficients (for a symmetric stress tensor), this is a rather daunting task. Both shear flow and elongational flow are highly symmetric, however. The symmetry of a velocity field places simplifying constraints on the stress tensor, as we shall see.

4.4.1 SIMPLE SHEAR FLOW

For simple shear flow, the mathematical description of the velocity field is completely unchanged by a 180° rotation of the coordinate system around the neutral direction, as we will now show (Figure 4.15). Since the stress tensor depends on the velocity field, the symmetry of the velocity field implies something about $\underline{\underline{\tau}}$. To demonstrate the implications of symmetry in shear flow, we can write the flow in two different coordinate systems, one rotated 180° around $\hat{e}_3$ from the other. As we will see, in order for $\underline{\underline{\tau}}$ to be invariant to this rotation, several of its components must be zero.

Consider a shear flow described in our usual Cartesian coordinate system, with $\hat{e}_1$ being the flow direction, $\hat{e}_2$ the gradient direction, and $\hat{e}_3$ the neutral direction (Figure 4.15). The three space variables are x_1, x_2, and x_3. The velocity field is shown in Equation (4.50)

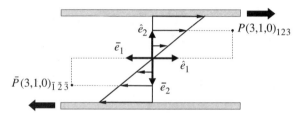

Figure 4.15 The symmetry axis for shear flow. The velocity and stress fields are unaffected by a 180° rotation around $\hat{e}_3$.

By symmetry:

$$\underline{v}\Big|_P = \underline{v}\Big|_{\bar{P}}$$

$$\underline{\underline{\tau}}\Big|_P = \underline{\underline{\tau}}\Big|_{\bar{P}}$$

$$\underline{v} = \begin{pmatrix} v_1 \\ v_2 \\ v_3 \end{pmatrix}_{123} = \begin{pmatrix} \dot{\varsigma}(t)x_2 \\ 0 \\ 0 \end{pmatrix}_{123} \tag{4.50}$$

$$\underline{v} = \dot{\varsigma}(t)x_2\hat{e}_1 = \dot{\gamma}_{21}(t)x_2\hat{e}_1 \tag{4.51}$$

by definition of shear flow. Now we can also write the velocity field in a different coordinate system, one we will call the $\bar{e}$ system, defined by

$$\bar{e}_1 = -\hat{e}_1 \tag{4.52}$$

$$\bar{e}_2 = -\hat{e}_2 \tag{4.53}$$

$$\bar{e}_3 = \hat{e}_3 \tag{4.54}$$

This is a Cartesian coordinate system arrived at by rotating the original coordinate system by 180° around $\hat{e}_3$. The spatial variables in this coordinate system are $\bar{x}_1 = -x_1, \bar{x}_2 = -x_2$, and $\bar{x}_3 = x_3$. The coefficients of $\underline{v}$ in this coordinate system we will write as

$$\underline{v} = \bar{v}_1\bar{e}_1 + \bar{v}_2\bar{e}_2 + \bar{v}_3\bar{e}_3 \tag{4.55}$$

$$= \begin{pmatrix} \bar{v}_1 \\ \bar{v}_2 \\ \bar{v}_3 \end{pmatrix}_{\bar{1}\bar{2}\bar{3}} \tag{4.56}$$

To distinguish between the expressions written in the two different coordinate systems, we will indicate the coordinate system being used by writing 123 or $\bar{1}\bar{2}\bar{3}$ at the lower right of a matrix of coefficients, as appropriate.

We can also write $\underline{v}$ in the $\bar{e}$ coordinate system by substituting the appropriate expressions for $\hat{e}_1$ and x_2 into Equation (4.51),

$$\underline{v} = \dot{\varsigma}(t)x_2\hat{e}_1 \tag{4.57}$$

$$= \dot{\varsigma}(t)(-\bar{x}_2)(-\bar{e}_1) = \dot{\varsigma}(t)\bar{x}_2\bar{e}_1 \tag{4.58}$$

$$= \begin{pmatrix} \dot{\varsigma}(t)\bar{x}_2 \\ 0 \\ 0 \end{pmatrix}_{\bar{1}2\bar{3}} \tag{4.59}$$

Thus, the velocity vector written with respect to the two different coordinate systems is

$$\underline{v} = \begin{pmatrix} \dot{\varsigma}(t)x_2 \\ 0 \\ 0 \end{pmatrix}_{123} = \begin{pmatrix} \dot{\varsigma}(t)\bar{x}_2 \\ 0 \\ 0 \end{pmatrix}_{\bar{1}2\bar{3}} \tag{4.60}$$

We see from the foregoing that simple shear flow written in these two different coordinate systems has the exact same form: a scalar function $\dot{\varsigma}(t)$, multiplied by the coordinate variable in the 2-direction. Consider two points that have the same numerical coefficients in the two different systems, for example, $(3, 1, 0)_{123}$ and $(3, 1, 0)_{\bar{1}2\bar{3}}$ (see Figure 4.15). From Equation (4.60) fluid particles located at these two points will have the same velocities since the values of x_2 and $\bar{x}_2$ are equal. This is true for all pairs of points with common coefficients expressed in the two coordinate systems. Further, because the velocities are equal, any function of $\underline{v}$ that we could write (stress, for example) must have the same numerical values of stress coefficients in the two coordinate systems. Again, at points that look the same in the two coordinate systems, for example, $(3, 1, 0)_{123}$ and $(3, 1, 0)_{\bar{1}2\bar{3}}$, the stress coefficients are equal: τ_{pk} at point $(3,1,0)$ in the $\hat{e}$-system equals $\bar{\tau}_{pk}$ at point $(3, 1, 0)$ in the $\bar{e}$-system, where τ_{pk} are the coordinates of the stress tensor written in the 123 coordinate system, and $\bar{\tau}_{pk}$ are the coordinates of the stress tensor written in the $\bar{1}2\bar{3}$ coordinate system.

This has important implications. The extra stress tensor $\underline{\underline{\tau}}$, written in both coordinate systems, is

$$\underline{\underline{\tau}} = \tau_{pk}\hat{e}_p\hat{e}_k \tag{4.61}$$

$$= \tau_{11}\hat{e}_1\hat{e}_1 + \tau_{12}\hat{e}_1\hat{e}_2 + \tau_{13}\hat{e}_1\hat{e}_3$$
$$+ \tau_{21}\hat{e}_2\hat{e}_1 + \tau_{22}\hat{e}_2\hat{e}_2 + \tau_{23}\hat{e}_2\hat{e}_3$$
$$+ \tau_{31}\hat{e}_3\hat{e}_1 + \tau_{32}\hat{e}_3\hat{e}_2 + \tau_{33}\hat{e}_3\hat{e}_3 \tag{4.62}$$

$$\underline{\underline{\tau}} = \bar{\tau}_{pk}\bar{e}_p\bar{e}_k \tag{4.63}$$

$$= \bar{\tau}_{11}\bar{e}_1\bar{e}_1 + \bar{\tau}_{12}\bar{e}_1\bar{e}_2 + \bar{\tau}_{13}\bar{e}_1\bar{e}_3$$
$$+ \bar{\tau}_{21}\bar{e}_2\bar{e}_1 + \bar{\tau}_{22}\bar{e}_2\bar{e}_2 + \bar{\tau}_{23}\bar{e}_2\bar{e}_3$$
$$+ \bar{\tau}_{31}\bar{e}_3\bar{e}_1 + \bar{\tau}_{32}\bar{e}_3\bar{e}_2 + \bar{\tau}_{33}\bar{e}_3\bar{e}_3 \tag{4.64}$$

By the symmetry arguments made, $\tau_{pk} = \bar{\tau}_{pk}$ for all p and k,

$$\begin{matrix} \tau_{11} = \bar{\tau}_{11} & \tau_{21} = \bar{\tau}_{21} & \tau_{31} = \bar{\tau}_{31} \\ \tau_{12} = \bar{\tau}_{12} & \tau_{22} = \bar{\tau}_{22} & \tau_{32} = \bar{\tau}_{32} \\ \tau_{13} = \bar{\tau}_{13} & \tau_{23} = \bar{\tau}_{23} & \tau_{33} = \bar{\tau}_{33} \end{matrix} \tag{4.65}$$

We can now substitute the relationships between the $\hat{e}$ and the $\bar{e}$ vectors, that is, $\hat{e}_1 = -\bar{e}_1$, and so on, into the expression for $\underline{\underline{\tau}}$ in Equation (4.62) to obtain another relationship between τ_{pk} and $\bar{\tau}_{pk}$:

$$\underline{\underline{\tau}} = \tau_{pk}\hat{e}_p\hat{e}_k \tag{4.66}$$

$$= \tau_{11}\hat{e}_1\hat{e}_1 + \tau_{12}\hat{e}_1\hat{e}_2 + \tau_{13}\hat{e}_1\hat{e}_3$$
$$+ \tau_{21}\hat{e}_2\hat{e}_1 + \tau_{22}\hat{e}_2\hat{e}_2 + \tau_{23}\hat{e}_2\hat{e}_3$$
$$+ \tau_{31}\hat{e}_3\hat{e}_1 + \tau_{32}\hat{e}_3\hat{e}_2 + \tau_{33}\hat{e}_3\hat{e}_3 \tag{4.67}$$

$$= \tau_{11}\bar{e}_1\bar{e}_1 + \tau_{12}\bar{e}_1\bar{e}_2 - \tau_{13}\bar{e}_1\bar{e}_3$$
$$+ \tau_{21}\bar{e}_2\bar{e}_1 + \tau_{22}\bar{e}_2\bar{e}_2 - \tau_{23}\bar{e}_2\bar{e}_3$$
$$- \tau_{31}\bar{e}_3\bar{e}_1 - \tau_{32}\bar{e}_3\bar{e}_2 + \tau_{33}\bar{e}_3\bar{e}_3 \tag{4.68}$$

Note that in Equation (4.68) the coefficients are in terms of the 123 coefficients of $\underline{\underline{\tau}}$ (no bar) and the vectors are now the $\bar{1}\bar{2}\bar{3}$ vectors. Finally we compare Equations (4.68) and (4.64). For these two expressions to equate (as they must since they both express $\underline{\underline{\tau}}$, and like all tensors, $\underline{\underline{\tau}}$ does not depend on a coordinate system), the coefficients of like dyads must be equal. After careful comparison and using Equation (4.65), the symmetry conditions, we see contradictions for $\bar{\tau}_{13}$, $\bar{\tau}_{23}$, $\bar{\tau}_{31}$, and $\bar{\tau}_{32}$:

$$\bar{\tau}_{13} = -\bar{\tau}_{13} \tag{4.69}$$

$$\bar{\tau}_{23} = -\bar{\tau}_{23} \tag{4.70}$$

$$\bar{\tau}_{31} = -\bar{\tau}_{31} \tag{4.71}$$

$$\bar{\tau}_{32} = -\bar{\tau}_{32} \tag{4.72}$$

This contradiction is only resolved if $\bar{\tau}_{23} = \tau_{23} = \bar{\tau}_{32} = \tau_{32} = \bar{\tau}_{13} = \tau_{13} = \bar{\tau}_{31} = \tau_{31} = 0$.

This important conclusion points out a significant reason for choosing shear flow as a standard flow in rheology. Since we have determined that four of the nine stress components are zero from the start (due to the symmetry of the flow), there are only five stress components to be measured. This is a tremendous advantage from both experimental and computational points of view.

For simple shear flow then, even for the most complex fluid, we can write

Total stress tensor
in shear flow
(general fluid)

$$\underline{\underline{\Pi}} = p\underline{\underline{I}} + \underline{\underline{\tau}} = \begin{pmatrix} p + \tau_{11} & \tau_{12} & 0 \\ \tau_{21} & p + \tau_{22} & 0 \\ 0 & 0 & p + \tau_{33} \end{pmatrix}_{123} \tag{4.73}$$

The stress tensor in shear flow has five nonzero components, but two are equal ($\tau_{21} = \tau_{12}$), leaving four unknowns. Compared to six unknown coefficients in the general symmetric stress tensor, this is a worthwhile simplification.

4.4.2 ELONGATIONAL FLOW

The kinematics of elongational flow were also chosen because of their symmetry and the consequent simplicity of the stress tensor for this flow. Looking back at the sketch of elongational flow in Figure 4.9 and examining the velocity field for this flow [Equation (4.24)], we see that the velocity field is unchanged for 180° rotations around any of the three Cartesian axes ($\hat{e}_1$, $\hat{e}_2$, $\hat{e}_3$). This high degree of symmetry may be used to show that the stress tensor for simple elongational flow of a general fluid can be written as

Total stress tensor in elongational flow (general fluid)

$$\underline{\underline{\Pi}} = p\underline{\underline{I}} + \underline{\underline{\tau}} = \begin{pmatrix} p + \tau_{11} & 0 & 0 \\ 0 & p + \tau_{22} & 0 \\ 0 & 0 & p + \tau_{33} \end{pmatrix}_{123} \tag{4.74}$$

The procedure for showing this is analogous to that followed in Section 4.4.1 for shear. Elongational flow is thus simpler in one way than shear since there are only three nonzero components of $\underline{\underline{\tau}}$ for this flow.

4.5 Measuring Stresses in Standard Flows

Our final topic in this chapter concerns the relationship between pressure and flow-induced normal stresses and a complication that crops up in the study of incompressible fluids. When measuring stresses on a surface, it is the total stress, $\underline{\underline{\Pi}} = \underline{\underline{\tau}} + p\underline{\underline{I}}$, that is sensed by any measuring device:

$$\underline{\underline{\Pi}} = \begin{pmatrix} p + \tau_{11} & \tau_{12} & \tau_{13} \\ \tau_{21} & p + \tau_{22} & \tau_{23} \\ \tau_{31} & \tau_{32} & p + \tau_{33} \end{pmatrix}_{123} \tag{4.75}$$

In the case of normal stresses, there are two parts to each Π_{ii}, the pressure and the extra stress. In gases, pressure is related to other thermodynamic variables through an equation of state. For example, for ideal gases the equation of state is the ideal-gas law

$$p\hat{V} = RT \tag{4.76}$$

where $\hat{V}$ is the specific volume of the gas (volume/moles), T is the temperature, and R is the ideal-gas constant. The pressure is related to the density of the gas by

$$p = \frac{\rho RT}{M} \tag{4.77}$$

where M is the molecular weight (molar mass) of the fluid, or the average molecular weight if it is a mixture. Thus, for a gas we can calculate the two contributions to the total stress, p and τ_{ii} separately. First we measure temperature and density to deduce pressure using the equation of state; second we measure Π_{ii} and calculate the extra normal stress from $\tau_{ii} = \Pi_{ii} - p$. This procedure will work for any compressible fluid for which an equation of state is known.

For incompressible and nearly incompressible fluids, however, it is problematic to make an independent measurement of p. If we plot density ρ versus p for an ideal gas and for an incompressible fluid (Figure 4.16), we see the problem immediately. For the ideal gas, density is a known linear function of pressure given by the ideal-gas law; if the gas is not ideal, the density as a function of pressure is given by an appropriate equation of state. For the incompressible fluid, however, pressure does not affect the fluid density since, by definition, the density of an incompressible fluid does not change. For polymers, although density is a function of pressure, it is a weak function, and very little change in density is observed over the widest range of pressures encountered. In addition, if the pressure portion of Π_{ii} were sought for a polymer melt flow, an independent and highly accurate measurement of density would be required. Thus we encounter a difficulty: we cannot separate the measured quantity Π_{ii} into τ_{ii} and p for incompressible (or nearly incompressible) non-Newtonian fluids in shear flow. Note that this problem with pressure is not encountered for Newtonian fluids in shear flow (e.g., water in a pipe) because the normal stresses τ_{11}, τ_{22}, τ_{33} are zero [see Equation (4.13)]. For a Newtonian fluid in shear flow a sensor on a pipe (transducer, manometer) measures just p.

Without the ability to measure p independently, it is impossible to separate p from normal-stress measurements for general incompressible fluid flow. This is not a problem in calculating momentum flux and solving flow problems, however, because it is only the divergence of the stress ($\nabla \cdot \underline{\underline{\Pi}} = \nabla p + \nabla \cdot \underline{\underline{\tau}}$) that appears in the equation of motion; the absolute magnitude of $\underline{\underline{\tau}}$ is not needed. It is a problem when considering measurements of the stress tensor, however, since we cannot measure all five nonzero components of $\underline{\underline{\tau}}$ in shear flow nor all three nonzero components of $\underline{\underline{\tau}}$ in elongational flow.

There is a way around this problem. The approach taken by rheologists is to consider normal-stress differences instead of normal stresses. In simple shear flow, for example, the five nonzero stress components are the shear stresses, $\tau_{21} = \tau_{12}$, and the three normal stresses τ_{11}, τ_{22}, and τ_{33}. The shear stress can be measured directly, and for the normal stresses two differences of normal stresses can be measured:

First normal-
stress difference

$$\boxed{N_1 \equiv \Pi_{11} - \Pi_{22} = \tau_{11} - \tau_{22}}$$ (4.78)

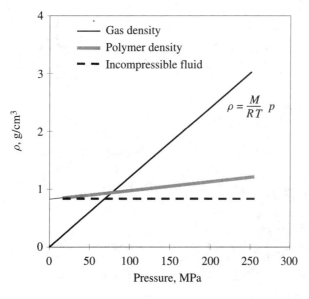

Figure 4.16 Density versus pressure for an ideal gas, an incompressible fluid, and a typical polymer melt.

Second normal-
stress difference

$$N_2 \equiv \Pi_{22} - \Pi_{33} = \tau_{22} - \tau_{33}$$ (4.79)

By considering differences in normal stresses rather than raw normal stresses, the issue of the indeterminacy of pressure for incompressible fluids is avoided. The same philosophy holds for measurements of elongational flow.

In summary, for shear flow there are three nonzero, unique stress quantities that are measured, the shear stress, the first normal-stress difference, and the second normal-stress difference. For elongational flow, there are two stress quantities to be measured for incompressible fluids, $\tau_{33} - \tau_{11}$ and $\tau_{22} - \tau_{11}$. These quantities are not usually named.

We have now completed most of the work needed to understand how rheologists classify fluids. What remains are the formal definitions of the material functions. These formal definitions are based on the stress quantities defined in this chapter and include kinematic functions from the flow fields used, that is, $\dot{\varsigma}(t)$ and $\dot{\epsilon}(t)$. Material functions will be the subject of the next chapter.

4.6 PROBLEMS

4.1 Compare and contrast shear and shear-free flows.

4.2 Do the fluid particles experience acceleration in steady shear flow? in steady uniaxial elongational flow? Explain your answers.

4.3 Sketch the flow field in simple shear and in uniaxial elongational flow.

4.4 What are the characteristics of shear and uniaxial elongational flow that makes them appealing choices for standard flows in rheology? How does the stress tensor simplify for these flows?

4.5 Show that the third invariant of $\dot{\underline{\underline{\gamma}}}$, $III_{\dot{\underline{\underline{\gamma}}}}$, is equal to zero for simple shear flow. (See Section 2.3.4 for the definition of tensor invariants.)

4.6 What is the rate-of-deformation tensors, $\dot{\underline{\underline{\gamma}}}$, for shear flow and for uniaxial elongational flow? The velocity fields for each of these flows are given.

$$shear: \quad \underline{v} = \begin{pmatrix} \dot{\varsigma}(t)x_2 \\ 0 \\ 0 \end{pmatrix}_{123}$$

$$uniaxial: \quad \underline{v} = \begin{pmatrix} \frac{-\dot{\epsilon}(t)}{2}x_1 \\ \frac{-\dot{\epsilon}(t)}{2}x_2 \\ \dot{\epsilon}(t)x_3 \end{pmatrix}_{123}$$

4.7 What are the values for the three invariants (see definitions in Section 2.3.4) of $\dot{\underline{\underline{\gamma}}}$ for steady uniaxial elongational flow?

4.8 What is the magnitude of the rate-of-deformation tensor $\dot{\gamma} \equiv |\dot{\underline{\underline{\gamma}}}|$ for biaxial extension? planar elongation? uniaxial extension?

4.9 Classify the following flows as elongational, biaxial stretching, or planar elongational flows.

(a) $\underline{v} = \begin{pmatrix} -4x_1 \\ 0 \\ 4x_3 \end{pmatrix}_{123}$

(b) $\underline{v} = \begin{pmatrix} -6x_1 \\ -6x_2 \\ 12x_3 \end{pmatrix}_{123}$

(c) $\underline{v} = \begin{pmatrix} 3x_1 \\ 3x_2 \\ -6x_3 \end{pmatrix}_{123}$

4.10 Show that simple shear and the three shear-free flows satisfy continuity for incompressible fluids.

4.11 We showed in steady shear flow ($\underline{v} = \dot{\gamma}_0 x_2 \hat{e}_1$) that as t goes to infinity, the separation l between two particles not in the same shear plane follows the following equation:

$$l = l_0 \dot{\gamma}_0 t$$

where the two particles are initially separated in the y-direction by a distance l_0 and t is time. This result was derived for two particles in the x_1x_2-plane. Show that the equation above is equally valid for particles not in the same x_1x_2-plane.

4.12 What is $\dot{\gamma} \equiv |\underline{\underline{\dot{\gamma}}}|$ for a flow with the following velocity field?

$$\underline{v} \text{ (cm/s)} = \begin{pmatrix} 3x_1 \\ 2x_1 + x_2 \\ 0 \end{pmatrix}_{123}$$

Recall that $\underline{\underline{\dot{\gamma}}} \equiv \nabla \underline{v} + (\nabla \underline{v})^T$.

4.13 For the velocity field $\underline{v}$ (cm/s) $= \begin{pmatrix} 10x_2 \\ 5x_1 + 5x_2 \\ -5x_3 \end{pmatrix}_{123}$

(a) What is $\dot{\gamma} = |\underline{\underline{\dot{\gamma}}}|$? Recall that $\underline{\underline{\dot{\gamma}}} = \nabla \underline{v} + (\nabla \underline{v})^T$.

(b) Is the fluid compressible or incompressible? Prove your answer.

4.14 Consider the points in a fluid with the Cartesian coordinates given below (Figure 4.17):

$$A(0, 0, 0) \quad B(1, 0, 0) \quad C(1, 1, 0)$$
$$D(0, 1, 0) \quad E(0, 0, 1)$$

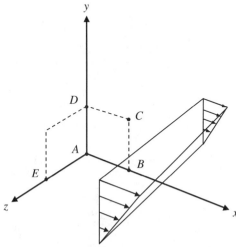

Figure 4.17 Flow domain for Problem 4.14.

(a) At $t = 0$ a shear flow given by $\underline{v}$ below is imposed on the flow domain above:

$$\underline{v} \text{ (cm/s)} = \begin{pmatrix} 3y \\ 0 \\ 0 \end{pmatrix}_{xyz}$$

Where will points A, B, C, D, and E be after 10 seconds?

(b) Where would the points be after 10 seconds if instead the following elongational flow were imposed at $t = 0$?

$$\underline{v} = \begin{pmatrix} -x \\ -y \\ 2z \end{pmatrix}_{xyz}$$

(c) Calculate the rate-of-deformation tensor $\underline{\underline{\dot{\gamma}}}$ and the magnitude of the rate-of-deformation tensor $\dot{\gamma} \equiv |\underline{\underline{\dot{\gamma}}}|$ for each of the above flows.

(d) Comment on the similarities and differences between these two flows.

4.15 By performing $180°$ rotations of the coordinate system around the basis vectors, show that the stress tensor for uniaxial elongational flow simplifies to

$$\underline{\underline{\tau}} = \begin{pmatrix} \tau_{11} & 0 & 0 \\ 0 & \tau_{22} & 0 \\ 0 & 0 & \tau_{33} \end{pmatrix}_{123}$$

4.16 Draw the particle paths $[\underline{x} = \underline{x}(x_0, y_0, z_0, t)]$ for the following flow fields using a computer program. Label axes quantitatively. Draw the flow fields in the xy, xz, and yz planes.

(a) $\underline{v}$ (cm/s) $= \begin{pmatrix} 6y \\ 0 \\ 0 \end{pmatrix}_{xyz}$

(b) $\underline{v}$ (cm/s) $= \begin{pmatrix} -3x \\ -3y \\ 6z \end{pmatrix}_{xyz}$

4.17 Consider the coordinate systems shown in Figure 4.18.

(a) Write $\underline{v} = \begin{pmatrix} 6x_2 \\ 0 \\ 0 \end{pmatrix}_{123} = 6x_2\hat{e}_1$ in the $\bar{x}_i$ coordinate system shown above for ψ in general and for $\psi = 30°$.

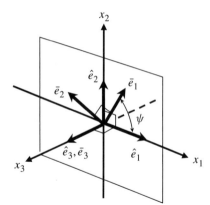

Figure 4.18 Coordinate systems for Problem 4.17.

(b) Write $\underline{\underline{\tau}} = \begin{pmatrix} 3 & 4 & 0 \\ 4 & 2 & 0 \\ 0 & 0 & 1 \end{pmatrix}_{123}$ in the $\bar{x}_i$ coordinate

system for ψ in general and for $\psi = 30°$.

(c) Write $\underline{\underline{I}} \equiv \begin{pmatrix} 1 & 0 & 0 \\ 0 & 1 & 0 \\ 0 & 0 & 1 \end{pmatrix}_{123}$ in the same two

coordinate systems.

4.18 The velocity field of a flow is given in one Cartesian coordinate system $(\bar{x}, \bar{y}, \bar{z})$ as

$$\underline{v} = (\bar{y} - \sqrt{2}\bar{x}) \begin{pmatrix} \frac{1}{\sqrt{2}} \\ 1 \\ -\sqrt{\frac{3}{2}} \end{pmatrix}_{\bar{x}\bar{y}\bar{z}}$$

You are told that the flow is steady, simple shear in some other Cartesian coordinate system.

(a) Find the unit vectors $\hat{e}_x, \hat{e}_y, \hat{e}_z$ for the coordinate system in which this flow has the classic form $[\dot{\varsigma}(t) = \text{constant} = \dot{\gamma}_0]$

$$\underline{v} = \begin{pmatrix} \dot{\gamma}_0 y \\ 0 \\ 0 \end{pmatrix}_{xyz} = \dot{\gamma}_0 y \hat{e}_x$$

Express the unit vectors $\hat{e}_x$, $\hat{e}_y$, and $\hat{e}_z$ in the $\bar{x}$, $\bar{y}$, $\bar{z}$ coordinate system.

(b) What is $\dot{\varsigma}(t) = \dot{\gamma}_0$ for this flow?

CHAPTER

ʊʊʊ 5

Material Functions

For incompressible Newtonian fluids, flow properties are governed by the continuity equation (conservation of mass), the equation of motion (conservation of momentum), and the Newtonian constitutive equation:

$$\nabla \cdot \underline{v} = 0 \tag{5.1}$$

$$\rho \left(\frac{\partial \underline{v}}{\partial t} + \underline{v} \cdot \nabla \underline{v} \right) = -\nabla p - \nabla \cdot \underline{\underline{\tau}} + \rho \underline{g} \tag{5.2}$$

$$\underline{\underline{\tau}} = -\mu [\nabla \underline{v} + (\nabla \underline{v})^T] \tag{5.3}$$

There are two material parameters that appear in these equations, the density ρ and the viscosity μ. The values of these two material constants are all the material information that is needed to predict the behavior of incompressible Newtonian fluids.

For incompressible non-Newtonian fluids, the continuity equation and the equation of motion remain the same, but a different constitutive equation is needed, $\underline{\underline{\tau}} = f(\nabla \underline{v}, \underline{v}, \text{material information, etc.})$. Material information in this case is contained in ρ, in the format of the constitutive equation, and in material-based information contained in the unknown constitutive equation. For polymers and other non-Newtonian fluids, $\underline{\underline{\tau}}$ is a much more complicated function of material properties than for Newtonian fluids, and it is an important challenge in the field of rheology to find appropriate constitutive equations for non-Newtonian fluids.

To find constitutive equations, experiments are performed on materials using standard flows. There are numerous standard flows that may be constructed from the two sets of flows (shear and elongational flows) presented in Chapter 4. The differences are arrived at by varying the functions $\dot{\varsigma}(t)$ and $\dot{\epsilon}(t)$, and in the case of shear-free flows, the parameter b. When the flow field is established on a fluid of interest, three stress-related quantities can be measured in shear flow (τ_{21}, $\tau_{11} - \tau_{22} = N_1$, and $\tau_{22} - \tau_{33} = N_2$), and two stress differences can be measured in elongational flow ($\tau_{33} - \tau_{11}$ and $\tau_{22} - \tau_{11}$). The stress responses that are observed depend on what material is studied and the type of flow imposed on the material. The responses will, in general, be functions of time, strain, or strain rate (or other kinematic parameters related to flow) and will depend on the chemical nature of the material.

We see, then, that it is quite a bit more complicated to characterize non-Newtonian fluids than Newtonian fluids. For Newtonian fluids, when we impose shear flow, for example, we measure a single stress τ_{21}, the only nonzero component of $\underline{\underline{\tau}}$ for this flow [see Equation (4.13)]. From τ_{21} and the imposed deformation rate $\dot{\varsigma} = dv_1/dx_2$, we can calculate the viscosity $\mu = -\tau_{21}/(dv_1/dx_2)$, which is all the material information we need to specify the constitutive equation. For non-Newtonian fluids, the process is much more involved. First of all, since we do not know the form of the constitutive equation, we do not know what experiments to conduct. If, following the Newtonian example, we choose to impose shear flow on a non-Newtonian fluid, we will measure the three nonzero stress quantities τ_{21}, N_1, and N_2, and we will usually find that $-\tau_{21}/(dv_1/dx_2)$ is a function of dv_1/dx_2, rather than being constant as for the Newtonian case. Thus, for fluids that are not Newtonian, we find ourselves observing a wide variety of material properties to be functions of kinematic parameters such as $\dot{\gamma} = |dv_1/dx_2|$, rather than constants. The functions of kinematic parameters that characterize the rheological behavior of fluids are called *rheological material functions*. The standardized material functions defined in this chapter provide a common language and framework for measuring and predicting the rheological behavior of non-Newtonian fluids.

If you are a bit confused about material functions so far, do not be alarmed. Until one is familiar with non-Newtonian fluid response, no amount of careful explanation ahead of time (such as was attempted here) will make the subject crystal clear. We recommend that you read this entire chapter and perhaps push on through the next two chapters, and later, when the subject material is more familiar, you can revisit the meaning and importance of material functions.

5.1 Introduction and Definitions

In this chapter we define the material functions that are most commonly used in rheological investigations of non-Newtonian fluids. They are equally valid for Newtonian fluids, as we will see. In all cases, the definitions of a material function will consist of three parts, as outlined next:

1. *Choice of flow type, i.e., shear or elongation.* We will only consider material functions based on flows of these two types.

2. *Details of the functions $\dot{\varsigma}(t)$ or $\dot{\epsilon}(t)$ and the parameter b that appear in the definitions of the flows.* Since shear flow is defined as

$$\underline{v} = \begin{pmatrix} \dot{\varsigma}(t)x_2 \\ 0 \\ 0 \end{pmatrix}_{123} \tag{5.4}$$

and elongation as

$$\underline{v} = \begin{pmatrix} -\frac{1}{2}\dot{\epsilon}(t)(1+b)x_1 \\ -\frac{1}{2}\dot{\epsilon}(t)(1-b)x_2 \\ \dot{\epsilon}(t)x_3 \end{pmatrix}_{123} \tag{5.5}$$

we must choose the actual functional forms of $\dot{\zeta}(t)$ or $\dot{\epsilon}(t)$ and b in order to fully specify the flow. These functions may be constants, in which case we are considering steady flows, or they may be time varying (unsteady flows).

3. *Material function definitions.*   The material functions themselves will be based on the measured stress quantities τ_{21}, N_1, and N_2 in shear flow, and $\tau_{33} - \tau_{11}$ or $\tau_{22} - \tau_{11}$ for elongational flows.

Material functions are either predicted or measured. If we are predicting material functions, we will use the kinematics [the type of flow and the chosen form of the functions $\dot{\zeta}(t)$ or $\dot{\epsilon}(t)$] and the constitutive equation to predict the stress components and subsequently calculate the material function. If we are measuring material functions, we will impose the kinematics on a material in a flow cell and measure the stress components.[1] When choosing a constitutive equation to describe a material, we need both to measure material functions on the fluid in question and to predict these material functions for a variety of constitutive equations. The constitutive equation that is found to predict a fluid's actual measured material functions most closely is the most appropriate constitutive equation to use when modeling flows of that material. Material functions are also used in qualitative rheological analysis, such as in quality control applications and the evaluation of new materials (Figure 5.1) [61]. When dealing with the rheological properties of

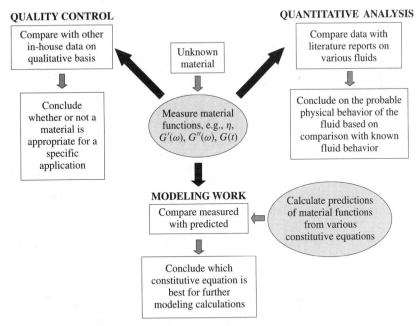

Figure 5.1 Role of material functions in rheological analysis.

---

[1] An exception to this order of operation is creep, where the stress is prescribed rather than the function $\dot{\zeta}(t)$ or $\dot{\epsilon}(t)$; see Sections 5.2.2.3 and 5.3.2.2.

non-Newtonian fluids one must have a framework for the discussion; material functions form this framework.

5.2 Shear Flow

Recall that the velocity field given here defines simple shear flow:[2]

$$
\underline{v} = \begin{pmatrix} \dot{\varsigma}(t)x_2 \\ 0 \\ 0 \end{pmatrix}_{123} \tag{5.6}
$$

The different types of shear-flow material functions are defined for different shear-rate functions $\dot{\varsigma}(t)$. We pause here to reiterate the difference between $\dot{\gamma} = |\underline{\underline{\dot{\gamma}}}|$ and $\dot{\varsigma}(t) = \dot{\gamma}_{21}(t)$. For shear flow, the quantity $\dot{\gamma}(t)$ has the same magnitude as $\dot{\varsigma}(t) = \dot{\gamma}_{21}(t)$; the quantity $\dot{\gamma}$, however, is always positive, since it is defined as the magnitude of a tensor. On the other hand, $\dot{\varsigma}(t) = \dot{\gamma}_{21}(t)$ may be positive or negative, depending on the direction of flow and the coordinate system being used. It is the quantity $\dot{\varsigma}(t)$ that appears in the definitions we are going to give here. The reason for this is that when the direction of flow changes, the sign on the shear stress also changes. For material functions (such as viscosity) to remain positive quantities independent of the direction of shear, we define them such that both the numerator and the denominator change sign when the direction of flow changes. When we plot material functions as a function of deformation rate, however, we use $|\dot{\varsigma}| = \dot{\gamma}$ as the independent variable so that we are plotting the positive material function[3] versus a positive quantity.

5.2.1 STEADY SHEAR

For steady shear flow, $\dot{\varsigma}(t) = \dot{\gamma}_0$ is a constant, that is, the flow is at steady state:

Kinematics for steady shear

$$
\boxed{\dot{\varsigma}(t) = \dot{\gamma}_0} \tag{5.7}
$$

This flow is typically produced in a rheometer where the fluid is forced through a capillary at a constant rate, and the steady pressure required to maintain the flow is measured. Another common method is to use a cone-and-plate or parallel-plate geometry and to rotate the cone or plate at a constant angular velocity while measuring the torque generated by the fluid being tested (see Chapter 10 and Figure 4.4).

---

[2] We will only discuss simple shear. See footnote 1 in Chapter 4 for more on nonsimple shear flows for which the shear material functions defined here are equally valid.

[3] Material functions are usually defined so that they are positive. One exception to this is the shear normal-force material function Ψ_2; for polymers, this is usually negative. In addition, there are materials for which some usually positive material functions become negative. For example, for most polymers the normal-force material function Ψ_1 is positive; for polymer liquid crystals, however, Ψ_1 can be positive or negative.

For the case of steady-state flow, the stress tensor is constant in time, and three constant stress quantities are measured, τ_{21}, N_1, and N_2. There are three material functions that are defined with these three stress quantities, namely:[4]

Viscosity
$$\eta(\dot{\gamma}) \equiv \frac{-\tau_{21}}{\dot{\gamma}_0} \qquad (5.8)$$

First normal-
stress coefficient
$$\Psi_1(\dot{\gamma}) \equiv \frac{-N_1}{\dot{\gamma}_0^2} = \frac{-(\tau_{11} - \tau_{22})}{\dot{\gamma}_0^2} \qquad (5.9)$$

Second normal-
stress coefficient
$$\Psi_2(\dot{\gamma}) \equiv \frac{-N_2}{\dot{\gamma}_0^2} = \frac{-(\tau_{22} - \tau_{33})}{\dot{\gamma}_0^2} \qquad (5.10)$$

where $\dot{\gamma}_0$ may be positive or negative, depending on the flow direction and the choice of coordinate system.

We now see that the viscosity η is defined for any fluid subjected to steady shear flow as the ratio of the steady-state shear stress to the constant shear rate $\dot{\gamma}_0$. For Newtonian fluids this is the same meaning as before (recall Newton's law of viscosity), and $\eta = \mu$ is independent of the shear rate, $\dot{\gamma} = |\dot{\gamma}_0|$. For non-Newtonian fluids, the quantity $-\tau_{21}/\dot{\gamma}_0 = \eta$ will be a function of the shear rate, $\eta = \eta(\dot{\gamma})$. The zero-shear viscosity η_0 is defined as

$$\lim_{\dot{\gamma} \to 0} \eta(\dot{\gamma}) \equiv \eta_0 \qquad (5.11)$$

The material functions Ψ_1 and Ψ_2 are also functions of $\dot{\gamma}$ in general; both functions are zero for Newtonian fluids (see the example that follows). For most polymers Ψ_1 is positive, and Ψ_2 is small and negative ($\Psi_2 \approx -0.1\Psi_1$ [26]). Examples of $\eta(\dot{\gamma})$, $\Psi_1(\dot{\gamma})$, and $\Psi_2(\dot{\gamma})$ for various polymeric systems are presented and discussed in more detail in Chapter 6.

EXAMPLE

Calculate $\eta(\dot{\gamma})$, $\Psi_1(\dot{\gamma})$, and $\Psi_2(\dot{\gamma})$ for an incompressible Newtonian fluid.

SOLUTION

These material functions are defined with respect to the following kinematics:

$$\underline{v} = \begin{pmatrix} \dot{\varsigma}(t)x_2 \\ 0 \\ 0 \end{pmatrix}_{123} \qquad (5.12)$$

---

[4] The Society of Rheology has official nomenclature for shear and elongational material functions [60], and, for the most part, we have adhered to this adopted nomenclature. There are some exceptions, however, such as our τ_{21} versus their σ for shear stress, where we have chosen to follow instead the widely used nomenclature of Bird, Armstrong, and Hassager [26]. There are also some instances of newer material functions where no nomenclature has officially been adopted. In these cases we have defined symbols that follow the pattern of The Society of Rheology system of nomenclature.

$$\dot{\varsigma}(t) = \dot{\gamma}_0 = \text{constant} \tag{5.13}$$

The stress for an incompressible Newtonian fluid is given by its constitutive equation:

$$\underline{\underline{\tau}} = -\mu \underline{\underline{\dot{\gamma}}} \tag{5.14}$$

From these kinematics, the rate-of-deformation tensor $\underline{\underline{\dot{\gamma}}}$ can be calculated as follows:

$$\underline{\underline{\dot{\gamma}}} = \nabla \underline{v} + (\nabla \underline{v})^T \tag{5.15}$$

$$= \begin{pmatrix} 0 & 0 & 0 \\ \dot{\varsigma}(t) & 0 & 0 \\ 0 & 0 & 0 \end{pmatrix}_{123} + \begin{pmatrix} 0 & \dot{\varsigma}(t) & 0 \\ 0 & 0 & 0 \\ 0 & 0 & 0 \end{pmatrix}_{123} \tag{5.16}$$

$$= \begin{pmatrix} 0 & \dot{\varsigma}(t) & 0 \\ \dot{\varsigma}(t) & 0 & 0 \\ 0 & 0 & 0 \end{pmatrix}_{123} \tag{5.17}$$

Since $\dot{\varsigma}(t) = \dot{\gamma}_0$ is a constant, the tensor $\underline{\underline{\dot{\gamma}}}$ is a constant, and we can calculate $\underline{\underline{\tau}}$ for all times for an incompressible Newtonian fluid in steady shear flow:

$$\underline{\underline{\tau}} = \begin{pmatrix} 0 & -\mu\dot{\gamma}_0 & 0 \\ -\mu\dot{\gamma}_0 & 0 & 0 \\ 0 & 0 & 0 \end{pmatrix}_{123} = \text{constant} \tag{5.18}$$

Turning to the definitions of the material functions η, Ψ_1, and Ψ_2 we can now calculate these quantities for an incompressible Newtonian fluid:

$$\eta \equiv \frac{-\tau_{21}}{\dot{\gamma}_0} = \mu \tag{5.19}$$

$$\Psi_1 \equiv \frac{-(\tau_{11} - \tau_{22})}{\dot{\gamma}_0^2} = 0 \tag{5.20}$$

$$\Psi_2 \equiv \frac{-(\tau_{22} - \tau_{33})}{\dot{\gamma}_0^2} = 0 \tag{5.21}$$

5.2.2 Unsteady Shear

Most polymers differ not only in their steady-state responses to shear flow but also in their unsteady-state responses. Non-steady-state measurements are made in the same geometries as steady-state measurements, that is, capillary flow, torsional cone-and-plate flow, and so on. The pressures and torques measured are functions of time when the flow is unsteady.

There are many different types of time-dependent shear flows, several of which we will discuss next. Shear-flow material functions are compared and summarized in Figure 5.2.

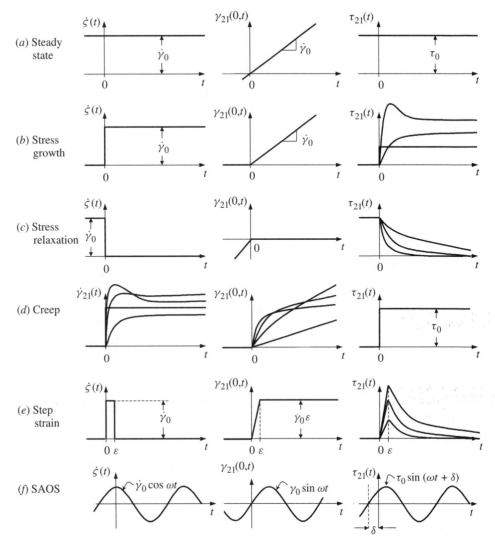

Figure 5.2 Summary of prescribed shear rate $\dot{\varsigma}(t)$, strain $\gamma_{21}(0, t)$, and shear stress τ_{21} in various shear-flow material functions. The shear strain is defined in Section 5.2.2.3.

5.2.2.1 Shear-Stress Growth

Measurements of the viscosity are the most common rheological measurements, and these are carried out in steady-state flow. Before steady state is reached, however, there is a start-up portion to the experiment in which the stress grows from its zero at-rest value to the steady-state value. This start-up experiment is one time-dependent shear-flow experiment that can be carried out easily.

The unsteady shear flows we are discussing only differ in the function $\dot{\varsigma}(t)$ used in the definition of $\underline{v}$ for shear flow, $\underline{v} = \dot{\varsigma}(t) x_2 \hat{e}_1$. For steady-state flow (Figure 5.2a), $\dot{\varsigma}(t) = \dot{\gamma}_0 =$

constant at all time. The function $\dot{\varsigma}(t)$ that is used in the stress-growth experiment is (see Figure 5.2b)

Kinematics for
shear-stress growth

$$\dot{\varsigma}(t) = \begin{cases} 0 & t < 0 \\ \dot{\gamma}_0 & t \geq 0 \end{cases} \qquad (5.22)$$

where again $\dot{\gamma}_0$ may be positive or negative. This function implies that there is no flow for time before $t = 0$; at $t = 0$ a constant shear rate $\dot{\gamma}_0$ is imposed on the fluid.

The response of a general fluid to this flow will be measured as three time-dependent stress quantities, $\tau_{21}(t, \dot{\gamma})$, $N_1(t, \dot{\gamma})$, and $N_2(t, \dot{\gamma})$. The three material functions defined are

Shear-stress
growth coefficient

$$\eta^+(t, \dot{\gamma}) \equiv \frac{-\tau_{21}}{\dot{\gamma}_0} \qquad (5.23)$$

First normal-stress
growth coefficient

$$\Psi_1^+(t, \dot{\gamma}) \equiv \frac{-(\tau_{11} - \tau_{22})}{\dot{\gamma}_0^2} \qquad (5.24)$$

Second normal-stress
growth coefficient

$$\Psi_2^+(t, \dot{\gamma}) \equiv \frac{-(\tau_{22} - \tau_{33})}{\dot{\gamma}_0^2} \qquad (5.25)$$

In general, all of these material functions depend on time and the magnitude of the applied shear rate $\dot{\gamma}$, as indicated. Note that these definitions closely resemble the material functions for steady shear flow, except that the stresses depend on time. At steady state these material functions become the steady-state functions:

$$\lim_{t \to \infty} \eta^+(t, \dot{\gamma}) = \eta(\dot{\gamma}) \qquad (5.26)$$

$$\lim_{t \to \infty} \Psi_1^+(t, \dot{\gamma}) = \Psi_1(\dot{\gamma}) \qquad (5.27)$$

$$\lim_{t \to \infty} \Psi_2^+(t, \dot{\gamma}) = \Psi_2(\dot{\gamma}) \qquad (5.28)$$

EXAMPLE

Calculate $\eta^+(t, \dot{\gamma})$, $\Psi_1^+(t, \dot{\gamma})$, and $\Psi_2^+(t, \dot{\gamma})$ for an incompressible Newtonian fluid.

SOLUTION

Whenever we seek to calculate a material function we begin with the kinematics. For the material functions $\eta^+(t, \dot{\gamma})$, $\Psi_1^+(t, \dot{\gamma})$, and $\Psi_2^+(t, \dot{\gamma})$ the kinematics are given by

$$\underline{v} = \begin{pmatrix} \dot{\varsigma}(t)x_2 \\ 0 \\ 0 \end{pmatrix}_{123} \qquad (5.29)$$

$$\dot{\varsigma}(t) = \begin{cases} 0 & t < 0 \\ \dot{\gamma}_0 & t \geq 0 \end{cases} \tag{5.30}$$

The stress for an incompressible Newtonian fluid is $\underline{\underline{\tau}} = -\mu\underline{\underline{\dot{\gamma}}}$, and in the same manner as in the last example, we can calculate $\underline{\underline{\dot{\gamma}}}$ from the given kinematics:

$$\underline{\underline{\dot{\gamma}}} = \nabla\underline{v} + (\nabla\underline{v})^T \tag{5.31}$$

$$= \begin{pmatrix} 0 & \dot{\varsigma}(t) & 0 \\ \dot{\varsigma}(t) & 0 & 0 \\ 0 & 0 & 0 \end{pmatrix}_{123} \tag{5.32}$$

For this flow $\dot{\varsigma}(t)$ is not constant, and thus $\underline{\underline{\dot{\gamma}}}$ and $\underline{\underline{\tau}}$ are not constant either but vary with time. For $t < 0$, $\dot{\varsigma}(t) = 0$, and therefore $\underline{\underline{\dot{\gamma}}}$ and the stress tensor $\underline{\underline{\tau}}$ are both the zero tensor,

$$\underline{\underline{\tau}} = \underline{\underline{0}} \qquad t < 0 \tag{5.33}$$

For time greater than or equal to zero, $\dot{\varsigma}(t) = \dot{\gamma}_0 =$ constant, and $\underline{\underline{\dot{\gamma}}}$ and $\underline{\underline{\tau}}$ are given by

$$\underline{\underline{\dot{\gamma}}} = \begin{pmatrix} 0 & \dot{\gamma}_0 & 0 \\ \dot{\gamma}_0 & 0 & 0 \\ 0 & 0 & 0 \end{pmatrix}_{123} \qquad t \geq 0 \tag{5.34}$$

$$\underline{\underline{\tau}} = \begin{pmatrix} 0 & -\mu\dot{\gamma}_0 & 0 \\ -\mu\dot{\gamma}_0 & 0 & 0 \\ 0 & 0 & 0 \end{pmatrix}_{123} \qquad t \geq 0 \tag{5.35}$$

Turning to the definitions of the material functions $\eta^+(t, \dot{\gamma})$, $\Psi_1^+(t, \dot{\gamma})$, and $\Psi_2^+(t, \dot{\gamma})$ [Equations (5.23)–(5.25)] we obtain

$$\eta^+ \equiv \frac{-\tau_{21}(t)}{\dot{\gamma}_0} = \frac{-1}{\dot{\gamma}_0}\begin{cases} 0 & t < 0 \\ -\mu\dot{\gamma}_0 & t \geq 0 \end{cases} \tag{5.36}$$

$$= \begin{cases} 0 & t < 0 \\ \mu & t \geq 0 \end{cases} \tag{5.37}$$

$$\Psi_1^+(t) \equiv \frac{-[\tau_{11}(t) - \tau_{22}(t)]}{\dot{\gamma}_0^2} = \frac{-1}{\dot{\gamma}_0^2}\begin{cases} 0 & t < 0 \\ 0 & t \geq 0 \end{cases} \tag{5.38}$$

$$= 0 \tag{5.39}$$

$$\Psi_2^+(t) \equiv \frac{-[\tau_{22}(t) - \tau_{33}(t)]}{\dot{\gamma}_0^2} = \frac{-1}{\dot{\gamma}_0^2}\begin{cases} 0 & t < 0 \\ 0 & t \geq 0 \end{cases} \tag{5.40}$$

$$= 0 \tag{5.41}$$

We see that no nonzero normal stresses, transient or otherwise, are predicted by the Newtonian constitutive equation. The shear-stress growth coefficient η^+ is nonzero,

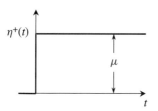

Figure 5.3 Stress growth function predicted by the Newtonian constitutive equation. $\Psi_1^+(t, \dot{\gamma})$ and $\Psi_2^+(t, \dot{\gamma})$ are both zero and η^+ is independent of the shear rate $\dot{\gamma}$.

however, and is given by a function that jumps instantaneously from zero to the steady value of viscosity μ at $t = 0$, the time at which the flow is imposed on the fluid (Figure 5.3). This instantaneous stress response is a signature of Newtonian fluids. Note also that η^+ is not a function of the shear rate $\dot{\gamma}$.

5.2.2.2 Shear-Stress Decay

Some information about the relaxation properties of non-Newtonian fluids may be obtained by observing how the steady-state stresses in shear flow relax when the flow is stopped. The cessation of steady shearing in the experiment corresponds to a shear flow with the function $\dot{\varsigma}(t, \dot{\gamma})$ defined as follows (Figure 5.2c):

$$\text{Kinematics of shear-stress decay} \qquad \dot{\varsigma}(t) = \begin{cases} \dot{\gamma}_0 & t < 0 \\ 0 & t \geq 0 \end{cases} \qquad (5.42)$$

The time-dependent material functions for shear-stress decay after cessation of steady shear are defined analogously to the stress growth material functions; in general they also vary with $\dot{\gamma}$.

$$\text{Shear-stress decay coefficient} \qquad \eta^-(t, \dot{\gamma}) \equiv \frac{-\tau_{21}}{\dot{\gamma}_0} \qquad (5.43)$$

$$\text{First normal-stress decay coefficient} \qquad \Psi_1^-(t, \dot{\gamma}) \equiv \frac{-(\tau_{11} - \tau_{22})}{\dot{\gamma}_0^2} \qquad (5.44)$$

$$\text{Second normal-stress decay coefficient} \qquad \Psi_2^-(t, \dot{\gamma}) \equiv \frac{-(\tau_{22} - \tau_{33})}{\dot{\gamma}_0^2} \qquad (5.45)$$

Calculating η^-, Ψ_1^-, and Ψ_2^- for a Newtonian fluid is straightforward.

While Newtonian fluids relax instantaneously when the flow stops (stress is proportional to the rate of deformation), for many non-Newtonian fluids relaxation takes a finite amount of time. The time that characterizes a material's stress relaxation after deformation is called the *relaxation time* λ. A dimensionless number that is used to characterize the importance of relaxation time in the analysis of a flow is the Deborah number De, which is the ratio of the material relaxation time scale to the time scale of the flow being studied:

$$\text{De} \equiv \frac{\text{material relaxation time}}{\text{flow time scale}} \tag{5.46}$$

$$= \frac{\lambda}{t_{\text{flow}}} \tag{5.47}$$

The Deborah number can help predict the response of a system to a particular deformation. For example, if De is large, material relaxation determines the response. If De is small or zero, the flow time scale determines the system response. The Deborah number is therefore important in determining whether material relaxation effects dominate in a given application [215].

5.2.2.3 Shear Creep

An alternative way of producing steady shear flow is to drive the flow at constant stress τ_0, rather than at constant shear rate $\dot{\gamma}_0$. This can be done in a capillary flow by imposing a constant-pressure driving stress or in a torsional cone-and-plate or parallel-plate rheometer by driving the plate with a constant-torque motor. Another way to produce a constant driving force is to use a weight attached through a pulley [37] to drive the flow (Figure 5.4). This method had widespread use in the early days of rheological testing [37]. Since both the stress and the shear rate arc constant at steady state, whether we drive the flow at a constant stress or at a constant shear rate, the same steady state results.

The unsteady response to shear flow when a constant stress is imposed is necessarily different from the response when a constant strain rate is imposed. In the constant-strain-rate experiment, the buildup in stress is measured in the start-up experiment (as discussed in Section 5.2.2.1), whereas in the constant-stress experiment, the time-dependent deformation of the sample is measured during the transient flow. The unsteady shear experiment where the stress is held constant is called *creep*. In creep, rather than prescribing the shear-rate

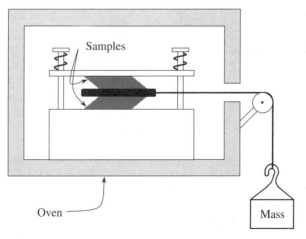

Figure 5.4 In this apparatus [37] a shear flow is created in two samples situated on either side of a moving centerpiece. The force of gravity on a weight connected to a pulley drives the flow. This flow geometry produces a constant driving stress in the shear flows in the two samples. *Source: Physical Properties of Polymers*, F. Bueche, Copyright © 1962 by Wiley Interscience. Reprinted by permission of John Wiley & Sons, Inc.

function $\dot{\zeta}(t)$, as we have been doing up until now (and will do for all material functions other than creep), we will prescribe the shear stress (Figure 5.2d):

Prescribed
stress function
for creep

$$\tau_{21}(t) = \begin{cases} 0 & t < 0 \\ \tau_0 = \text{constant} & t \geq 0 \end{cases} \qquad (5.48)$$

What is measured in creep is the deformation of a sample, that is, how the sample changes shape over some time interval as a result of the imposition of the stress τ_0. To understand the measurement of deformation we need to spend time at this point discussing in some detail the concept of strain.

To measure deformation in shear we use a quantity called the *shear strain*. Strain is a measure of the change of the shape of a fluid particle, that is, how much stretching or contracting a fluid experiences. Shear strain is denoted by $\gamma_{21}(t_{\text{ref}}, t)$, in which the subscript 21 identifies the strain that results from a shear flow in which $\hat{e}_2$-planes (planes with unit normal $\hat{e}_2$) slide over each other in the 1-direction. There are two arguments of γ_{21} because strain measures shape at a particular time with respect to the shape of the fluid particle at some other time. The expression $\gamma_{21}(t_{\text{ref}}, t)$ refers to the strain at time t relative to the configuration of the fluid at time t_{ref}. In the discussion that follows the reference time t_{ref} is often taken to be $t_{\text{ref}} = 0$. The shear strain $\gamma_{21}(t_{\text{ref}}, t)$ may be abbreviated as $\gamma_{21}(t)$, or simply $\gamma(t)$, where shear flow and $t_{\text{ref}} = 0$ are understood.

Although the concept of strain is straightforward—a measure of the deformation of fluid particles in a flow—the formal definition is somewhat involved since it is important to use a deformation measure that is applicable to all flows. For now we will discuss strain in shear flow only. For short time intervals, shear strain is defined as

Shear strain
(small deformations)

$$\gamma_{21}(t_{\text{ref}}, t) \equiv \frac{\partial u_1}{\partial x_2} \qquad (5.49)$$

where $u_1 = u_1(t_{\text{ref}}, t)$ is called the *displacement function* in the x_1-direction (Figure 5.5). The displacement function gives the position of a particle in a flow at time t relative to its position at time $t = t_{\text{ref}}$. In general, let $\underline{r}(t_{\text{ref}})$ be a vector that indicates the position of a fluid particle at time t_{ref}, and let $\underline{r}(t)$ be the same vector at time t. Then the displacement function $\underline{u}(t_{\text{ref}}, t)$ is

$$\underline{r}(t_{\text{ref}}) = \begin{pmatrix} x_1(t_{\text{ref}}) \\ x_2(t_{\text{ref}}) \\ x_3(t_{\text{ref}}) \end{pmatrix}_{123} \qquad (5.50)$$

$$\underline{r}(t) = \begin{pmatrix} x_1(t) \\ x_2(t) \\ x_3(t) \end{pmatrix}_{123} \qquad (5.51)$$

and

Displacement
function

$$\underline{u}(t_{\text{ref}}, t) \equiv \underline{r}(t) - \underline{r}(t_{\text{ref}}) \qquad (5.52)$$

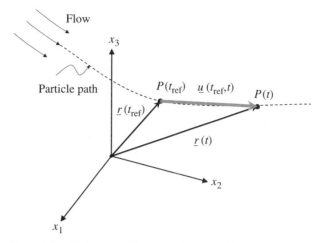

Figure 5.5 Definition of the displacement function $\underline{u}(t_{\mathrm{ref}}, t)$, which gives the position of a particle P at time t relative to its position at t_{ref}. The displacement function is a vector, which in the x_1, x_2, x_3 coordinate system has components u_1, u_2, and u_3. In shear flow, $u_2 = u_3 = 0$.

The x_1-direction displacement function u_1 is just the 1-component of $\underline{u}$; $u_1(t_{\mathrm{ref}}, t) = x_1(t) - x_1(t_{\mathrm{ref}})$.

A physical understanding of the definition of shear strain can be gained from Figure 5.6. For two points P_1 and P_2, which at time t_{ref} are on the x_2-axis in a shear flow, the displacement functions $\underline{u}(t_{\mathrm{ref}}, t)$ will be lines parallel to the x_1-axis, $\underline{u} = u_1\hat{e}_1$. Since point P_2 is located in the part of the flow with higher velocity, u_1 for point P_2 will be longer than u_1 for point P_1 for the time interval $t - t_{\mathrm{ref}}$. We see from Figure 5.6 that the shear strain $\gamma_{21}(t_{\mathrm{ref}}, t) = \partial u_1/\partial x_2 \approx \Delta u_1/\Delta x_2$ is the inverse of the slope of the side of the deformed particle. Thus strain is related to the change in shape of a fluid particle in the vicinity of points P_1 and P_2, and the formal definition of shear strain [Equation (5.49)] matches our qualitative understanding that strain is a measure of the deformation of fluid particles in a flow.

Returning to the formal definition of strain, we can now relate $\gamma_{21}(0, t)$ to $\dot{\gamma}_0$, the shear rate in steady shear flow. For steady shear flow over short time intervals, the particle position vector $\underline{r}(t)$ is

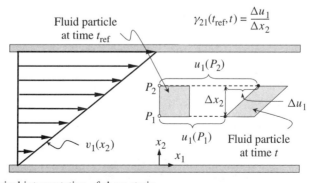

Figure 5.6 Physical interpretation of shear strain.

$$\underline{r} = \begin{pmatrix} x_1(t) \\ x_2(t) \\ x_3(t) \end{pmatrix}_{123} = \begin{pmatrix} x_1(t_{\text{ref}}) + (t - t_{\text{ref}})\dot{\gamma}_0 x_2 \\ x_2(t_{\text{ref}}) \\ x_3(t_{\text{ref}}) \end{pmatrix}_{123} \tag{5.53}$$

where $\underline{r}(t_{\text{ref}})$ is the initial particle position, and the velocity is $\underline{v} = \dot{\gamma}_0 x_2 \hat{e}_1$. With $\underline{r}$ now written for any particle in the shear flow, we can calculate the strain in the flow using Equation (5.49). For steady simple shear flow over a short time interval from 0 to t then, we calculate the displacement function and strain as

$$u_1(t_{\text{ref}}, t) = (t - t_{\text{ref}})\,\dot{\gamma}_0 x_2 \tag{5.54}$$

and

Strain in steady shear over short interval

$$\gamma_{21}(0, t) = \frac{\partial u_1}{\partial x_2} = \dot{\gamma}_0 t \tag{5.55}$$

Recall that we are interested in strain at this point in the text because we need a way to describe sample deformation in the creep experiment, since for creep, stress is imposed and deformation is measured. The deformation in the creep experiment occurs over a long time interval, and therefore Equation (5.49), which gives the strain over short intervals, is not sufficient for calculating strain in this flow. We can, however, break a large strain into a sequence of N smaller strains:

$$\gamma_{21}(0, t) = \gamma(0, t_1) + \gamma(t_1, t_2) + \cdots + \gamma(t_p, t_{p+1}) + \cdots + \gamma[(N-1)\Delta t, t] \tag{5.56}$$

where $t_p = p\Delta t$ and $\Delta t = t/N$. The total strain from 0 to t is then the sum of all the smaller strains. For each of the small strains, Equation (5.49) may be used along with the displacement function in Equation (5.54). We will try the calculation first for steady flow. In terms of the variables defined here, the steady shear-flow displacement function $u_1(t_p, t_{p+1})$ [Equation (5.54)] is given for short even intervals Δt by

$$u_1(t_p, t_{p+1}) = \Delta t\,\dot{\gamma}_0\,x_2 \tag{5.57}$$

Therefore, for each small-strain interval,

$$\gamma(t_p, t_{p+1}) = \frac{\partial u_1}{\partial x_2} = \Delta t\,\dot{\gamma}_0 \tag{5.58}$$

which is independent of time. Therefore, the total strain over the entire interval from 0 to t is given by

$$\gamma_{21}(0, t) = \sum_{p=0}^{N-1} \gamma_{21}(t_p, t_{p+1}) \tag{5.59}$$

$$= N\Delta t\,\dot{\gamma}_0 = t\dot{\gamma}_0 \tag{5.60}$$

This is the same as the result we obtained earlier for short time intervals [Equation (5.55)], and it is valid in steady shear flow.

For unsteady shear flows (creep is unsteady), the relationship between $\gamma_{21}(0, t)$ and the measured shear rate $\dot{\gamma}_{21}(t)$ is a bit more complicated since $\dot{\gamma}_{21}$ varies with time. The displacement function [Equation (5.57)] is the same, except that the constant shear rate $\dot{\gamma}_0$ is replaced by the measured time-dependent shear-rate function $\dot{\gamma}_{21}(t)$ (the only nonzero component of the rate-of-deformation tensor $\dot{\underline{\underline{\gamma}}}$). We will now consider the general case of strain between two times t_1 and t_2. First we break up the interval into N pieces of duration Δt:

$$t_p = t_1 + p\Delta t, \quad p = 0, 1, 2, \ldots, N - 1 \tag{5.61}$$

$$u_1(t_p, t_{p+1}) = \Delta t \, \dot{\gamma}_{21}(t_{p+1}) x_2 \tag{5.62}$$

The strain for each interval is calculated using Equation (5.49):

$$\gamma_{21}(t_p, t_{p+1}) = \frac{\partial u_1}{\partial x_2} = \Delta t \, \dot{\gamma}_{21}(t_{p+1}) \tag{5.63}$$

which varies with time because of the time dependence of $\dot{\gamma}_{21}(t_{p+1})$. Thus, for unsteady shear flow, a large strain between times t_1 and t_2 is given by

$$\gamma_{21}(t_1, t_2) = \sum_{p=0}^{N-1} \gamma_{21}(t_p, t_{p+1}) = \sum_{p=0}^{N-1} \Delta t \, \dot{\gamma}_{21}(t_{p+1}) \tag{5.64}$$

In the limit that Δt goes to zero, Equation (5.64) becomes the integral of $\dot{\gamma}_{21}(t')$ between t_1 and t_2:

$$\gamma_{21}(t_1, t_2) = \lim_{\Delta t \to 0} \left[\sum_{p=0}^{N-1} \Delta t \, \dot{\gamma}_{21}(t_{p+1}) \right] \tag{5.65}$$

and

Strain at t_2
with respect to
fluid configuration at t_1
in unsteady shear flow

$$\boxed{\gamma_{21}(t_1, t_2) = \int_{t_1}^{t_2} \dot{\gamma}_{21}(t') \, dt'} \tag{5.66}$$

This expression for strain is valid in unsteady shear flows such as creep.

From Equation (5.66) we see that the shear strain in the creep experiment may be obtained by measuring the instantaneous shear rate $\dot{\gamma}_{21}(t)$ as a function of time and integrating it over the time interval. The quantity $\dot{\gamma}_{21}(t)$ can be measured in a straightforward manner in the torsional cone-and-plate or parallel-plate geometries by recording the time-dependent angular velocity of the cone (or plate) (see Chapter 10).

Now that we have a measure for sample deformation in shear flow, we are ready to proceed with the definitions of material functions for shear creep. In the creep experiment, because the stress is prescribed rather than measured, the material functions relate the measured sample deformation (strain) to the prescribed (constant) stress τ_0. The material function that is defined for creep is called the *creep compliance* $J(t, \tau_0)$:

$$\text{Shear creep compliance} \qquad \boxed{J(t, \tau_0) \equiv \frac{\gamma_{21}(0, t)}{-\tau_0}} \qquad (5.67)$$

$\gamma(0, t)$ is obtained from a measurement of $\dot{\gamma}_{21}$ and Equation (5.66). The creep compliance curve $J(t, \tau_0)$ has many features, and several other material functions are defined that are related to $J(t, \tau_0)$ (see Figure 5.7). At long enough times, the strain becomes linear with time, that is, the flow reaches steady state, and the slope of $J(t, \tau_0)$ is the steady-state shear rate, $d\gamma_{21}/dt = \dot{\gamma}_\infty = $ constant, divided by the imposed stress $-\tau_0$. This ratio is just the inverse of the steady shear viscosity. The steady-state compliance J_s is defined as the difference between the compliance function at a particular time at steady state and t/η, the steady-flow contribution to the compliance function at that time:

$$\text{Steady-state compliance} \qquad \boxed{J_s(\tau_0) \equiv J(t, \tau_0)|_{\text{steady state}} - \frac{t}{\eta(\dot{\gamma}_\infty)}} \qquad (5.68)$$

$J_s(\tau_0)$ may be calculated by extrapolating the linear portion of $J(t, \tau_0)$ back to time $t = 0$ (Figure 5.7).

Also commonly measured in the creep experiment is something called the *creep recovery*. After the creep deformation has reached steady state, the shear stress is suddenly removed (set to zero) at time $t' = 0$. When the driving stress is removed, elastic and viscoelastic materials will spring back in the direction opposite to the initial flow direction, and the amount of strain that is recovered is called the steady-state *recoverable* shear strain or *recoil* strain $\gamma_r(t')$. Note that in the shear recovery experiment, the flowing sample is constrained so that no recovery may take place in the x_2-direction.

$$\gamma_r(t) \equiv \gamma_{21}(0, t_2) - \gamma_{21}(0, t) \qquad (5.69)$$

$$= \int_0^{t_2} \dot{\gamma}(t'') \, dt'' - \int_0^{t} \dot{\gamma}(t'') \, dt'' \qquad (5.70)$$

$$= \int_0^{t_2} \dot{\gamma}(t'') \, dt'' - \left[\int_0^{t_2} \dot{\gamma}(t'') \, dt'' + \int_{t_2}^{t} \dot{\gamma}(t'') \, dt'' \right] \qquad (5.71)$$

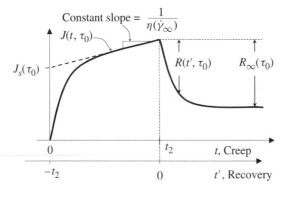

Figure 5.7 Relationships among the various material functions in the shear creep experiment. All properties are in general a function of the imposed shear stress τ_0.

$$= - \int_{t_2}^{t} \dot{\gamma}(t'') \, dt'' \tag{5.72}$$

In terms of t', the time scale that starts at the beginning of the recovery experiment, we have

$$\gamma_r(t') = - \int_{0}^{t'} \dot{\gamma}(t'') \, dt'' = -\gamma_{21}(0, t') \tag{5.73}$$

From the recoil strain, the material function *recoverable compliance* $J_r(t', \tau_0)$ is defined. The recoverable compliance is also called the *recoil function* $R(t', \tau_0)$ (see Figure 5.7) [60]:

Recoverable
creep compliance
$$\boxed{J_r(t', \tau_0) \equiv \frac{\gamma_r(t')}{-\tau_0}} \tag{5.74}$$

Recoil function
$$\boxed{R(t', \tau_0) = J_r(t', \tau_0)} \tag{5.75}$$

The *recoverable shear* γ_∞ is the ultimate strain recovered after the recoiling sample has come to rest, and it is used to define the *ultimate recoil function* $R_\infty(\tau_0)$:

Recoverable shear
$$\boxed{\gamma_\infty \equiv \lim_{t' \to \infty} \gamma_r(t')} \tag{5.76}$$

Ultimate recoil function
$$\boxed{R_\infty(\tau_0) \equiv \lim_{t' \to \infty} R(t', \tau_0) = \frac{\gamma_\infty}{-\tau_0}} \tag{5.77}$$

For τ_0 not too large (linear viscoelastic limit, see Chapter 8) the strain at all times $\gamma(t)$ is just the sum of the strain that is recoverable γ_r, as defined, and the strain that is not recoverable, that is, the strain due to steady viscous flow at $\dot{\gamma}_\infty$, calculated from Equation (5.60) (Figure 5.8):

Nonrecoverable shear strain
due to steady shear flow
$$\boxed{\dot{\gamma}_\infty t} \tag{5.78}$$

where $\dot{\gamma}_\infty$ is the shear rate attained at steady state in the creep experiment. Thus, in the linear viscoelastic limit,

$$\gamma(t) = \gamma_r(t) + t\dot{\gamma}_\infty \tag{5.79}$$

and dividing by $-\tau_0$, we obtain

$$J(t) = R(t) + \frac{t}{\eta_0} \tag{5.80}$$

where we have made the substitution $\eta_0 = -\tau_0/\dot{\gamma}_\infty$. Equation (5.80) can be used to calculate $R(t)$ in experiments in which the linear viscoelastic compliance $J(t)$ and the zero-shear viscosity η_0, are obtained. Finally, we had previously defined the steady-state compliance $J_s(\tau_0)$, which for the linear viscoelastic limit may be written as J_s^0 [see Equation (5.68)]:

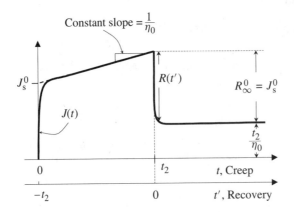

Figure 5.8 Relationships among the various material functions in the shear creep experiment in the linear viscoelastic limit. All properties are independent of the imposed shear stress τ_0, and the steady-state compliance equals the ultimate recoil function $J_s^0 = R_\infty^0$.

$$J(t)|_{\text{steady state}} = J_s^0 + \frac{t}{\eta_0} \tag{5.81}$$

By comparing this with Equation (5.80) we see that in the linear viscoelastic limit (denoted by superscript 0), the ultimate recoil is equal to the steady-state compliance:

$$R(t)|_{\text{steady state}} = R_\infty^0 \tag{5.82}$$

$$R_\infty^0 = J_s^0 \tag{5.83}$$

There are some important advantages to creep flow compared to a shear-rate-controlled flow, including a more rapid approach to steady state. In addition, the creep-recovery experiment gives important insight into elastic memory effects [211]. Another advantage of creep is that complex materials are often sensitive to the applied stress levels rather than to applied shear-rate levels. In the creep experiment, stress is maintained constant, and it is straightforward to determine any critical stresses. In a rate-controlled shear experiment [$\dot{\varsigma}(t) = \dot{\gamma}_{21}(t)$ controlled] the effects of critical stresses are often buried in complex, transient stress and rate responses that are difficult to interpret.

EXAMPLE

Calculate the shear creep compliance $J(t)$ for an incompressible Newtonian fluid.

SOLUTION

We begin, as usual, with the kinematics:

$$\tau_{21}(t) = \begin{cases} 0 & t < 0 \\ \tau_0 & t \geq 0 \end{cases} \tag{5.84}$$

$$\underline{v} = \begin{pmatrix} \dot{\gamma}_{21}(t)x_2 \\ 0 \\ 0 \end{pmatrix}_{123} \tag{5.85}$$

The case of creep differs from the other shear material functions we discuss in that $\dot{\varsigma}(t) = \dot{\gamma}_{21}(t)$ is not specified as part of the defined kinematics, rather $\tau_{21}(t)$ is prescribed. For a Newtonian fluid in shear flow, $\underline{\underline{\tau}}$ is given by

$$\underline{\underline{\tau}} = -\mu \underline{\underline{\dot{\gamma}}} \tag{5.86}$$

$$= -\mu \left[\nabla \underline{v} + (\nabla \underline{v})^T \right] \tag{5.87}$$

$$= \begin{pmatrix} 0 & -\mu\dot{\gamma}_{21}(t) & 0 \\ -\mu\dot{\gamma}_{21}(t) & 0 & 0 \\ 0 & 0 & 0 \end{pmatrix}_{123} \tag{5.88}$$

We can calculate $\dot{\gamma}_{21}(t)$ for an incompressible Newtonian fluid in creep from the 21-component of $\underline{\underline{\tau}}$:

$$\tau_{21} = -\mu\dot{\gamma}_{21}(t) = \tau_0 \tag{5.89}$$

$$\dot{\gamma}_{21}(t) = \frac{-\tau_0}{\mu} = \text{constant} \quad t \geq 0 \tag{5.90}$$

The definition of shear creep compliance $J(t)$ includes the strain $\gamma_{21}(0, t)$, which can be calculated from $\dot{\gamma}_{21}(t)$ using Equation (5.66):

$$\gamma_{21}(0, t) = \int_0^t \dot{\gamma}_{21}(t')\, dt' \tag{5.91}$$

$$= \int_0^t \frac{-\tau_0}{\mu}\, dt' = \frac{-\tau_0}{\mu} t \tag{5.92}$$

Compliance is then calculated from its definition:

$$J(t) \equiv \frac{\gamma_{21}(0, t)}{-\tau_0} = \frac{1}{\mu} t \quad t \geq 0 \tag{5.93}$$

This result is sketched in Figure 5.9.

The recoverable compliance or recoil function is calculated from Equation (5.80). Note that for a Newtonian fluid $\eta_0 = \mu$ and therefore,

$$J_r(t) = R(t) = J(t) - \frac{t}{\eta_0} = 0 \tag{5.94}$$

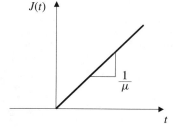

$J(t)$

$\dfrac{1}{\mu}$

t

Figure 5.9 Shear creep compliance $J(t)$ for an incompressible Newtonian fluid of viscosity μ.

We see that there is no recovery upon cessation of shearing of a Newtonian fluid. This is because Newtonian fluids generate a shear stress based on the value of the instantaneous strain rate, which is zero when flow stops.

5.2.2.4 Step Shear Strain

The three sets of unsteady shear material functions that we have described so far (startup, cessation, creep) have all been variations of basic steady shearing flow. There are also important material functions that have been defined for shear deformations that have little to do with steady shear flow. The step-strain experiment discussed in this section falls into that category.

One of the interesting properties of polymers and other viscoelastic materials is that they have partial memory, that is, the stresses generated in viscoelastic materials do not relax immediately but rather decay over time. The time for the decay to occur is a kind of memory time or relaxation time for the fluid. To investigate relaxation time, one of the most commonly employed experiments is the step-strain experiment in shear flow. In this experiment, a sample at rest between parallel plates is suddenly disturbed by the imposition of a constant, large, shear rate, but only for a small time ε (see Figure 5.2e). In a cone-and-plate or parallel-plate instrument this is accomplished by rotating the plate rapidly through a set angle. The time-dependent stress generated by this action is then recorded. This flow can also be generated in a sliding-plate rheometer by displacing one plate rapidly by a desired distance and measuring the shear stress with a flush-mounted transducer [88, 89, 61].

The shear-rate function for the step-strain experiment is

$$\text{Kinematics of step shear strain} \qquad \dot{\varsigma}(t) = \lim_{\varepsilon \to 0} \begin{cases} 0 & t < 0 \\ \dot{\gamma}_0 & 0 \leq t < \varepsilon \\ 0 & t \geq \varepsilon \end{cases} \tag{5.95}$$

$$\dot{\gamma}_0 \varepsilon = \text{constant}$$

The limit expresses that the shearing should occur as rapidly as possible. The condition $\dot{\gamma}_0 \varepsilon = \text{constant}$ relates to the magnitude of the shear strain imposed, as will be discussed next.

In the previous discussion of creep, we defined shear strain as

$$\gamma_{21}(t_{\text{ref}}, t) = \frac{\partial u_1}{\partial x_2} \tag{5.96}$$

Further we saw that for a general shear flow (steady or unsteady):

$$\gamma_{21}(t_{\text{ref}}, t) = \int_{t_{\text{ref}}}^{t} \dot{\gamma}_{21}(t') \, dt' \tag{5.97}$$

If we take the time derivative of both sides of this equation and apply the Leibnitz rule, we obtain

$$\text{Relationship between strain and strain rate in shear} \qquad \frac{d\gamma(t_{\text{ref}}, t)}{dt} = \dot{\gamma}_{21}(t) \tag{5.98}$$

Thus, the shear-rate component $\dot{\gamma}_{21}(t)$ is just the time derivative of the strain $\gamma_{21}(t_{ref}, t)$. Further, we can now see that the dot nomenclature used in writing the shear rate $\dot{\gamma}_{21}(t)$, which is notation familiar from differential equation courses to represent time derivatives, has this same meaning in the nomenclature for strain and strain rate in rheology.

For the step-strain experiment, the integral in Equation (5.97) is quite simple. The reference time is taken to be $t_{ref} = -\infty$.[5] For time t less than zero, the strain is zero since no flow has yet occurred, and the integral for strain is zero. For times greater than zero, the integral gives nonzero values:

$$\gamma_{21}(t) = \int_{-\infty}^{t} \dot{\gamma}_{21}(t') \, dt' \tag{5.99}$$

$$= \int_{-\infty}^{0} 0 \, dt' + \int_{0}^{\varepsilon} \dot{\gamma}_0 \, dt' + \int_{\varepsilon}^{t} 0 \, dt' \tag{5.100}$$

$$= \dot{\gamma}_0 \varepsilon \equiv \gamma_0 \tag{5.101}$$

We see why this is called the step-strain experiment: this flow involves a fixed strain $\gamma_0 \equiv \dot{\gamma}_0 \varepsilon$ applied rapidly to a test sample at time $t = 0$.

The prescribed shear-rate function $\dot{\varsigma}(t)$ for the step-strain experiment can be written in terms of γ_0 as follows:

$$\dot{\varsigma}(t) = \lim_{\varepsilon \to 0} \begin{cases} 0 & t < 0 \\ \dot{\gamma}_0 & 0 \leq t < \varepsilon \\ 0 & t \geq \varepsilon \end{cases} \tag{5.102}$$

$$= \gamma_0 \lim_{\varepsilon \to 0} \begin{cases} 0 & t < 0 \\ \frac{1}{\varepsilon} & 0 \leq t < \varepsilon \\ 0 & t \geq \varepsilon \end{cases} \tag{5.103}$$

The function multiplying γ_0 is a standard math function, an asymmetric impulse or delta function $\delta_+(t)$ [129]:

$$\boxed{\text{Asymmetric delta function} \qquad \delta_+(t) \equiv \lim_{\varepsilon \to 0} \begin{cases} 0 & t < 0 \\ \frac{1}{\varepsilon} & 0 \leq t < \varepsilon \\ 0 & t \geq \varepsilon \end{cases}} \tag{5.104}$$

and

$$\int_{-\infty}^{\infty} \delta_+(t) \, dt = 1 \tag{5.105}$$

Thus we can write

$$\dot{\varsigma}(t) = \gamma_0 \delta_+(t) \tag{5.106}$$

---

[5] The choice of $t_{ref} = -\infty$ versus $t_{ref} = 0$ is immaterial here since no flow occurs for $t < 0$. It is useful from a mathematical point of view to have $t_{ref} = -\infty$ for the step-strain experiment, as we will see in Chapter 8.

The strain function $\gamma_{21}(-\infty, t)$ can also be written in terms of a standard math function, the Heaviside unit step function $H(t)$:

$$\gamma_{21}(-\infty, t) = \int_{-\infty}^{t} \gamma_0 \delta_+(t') \, dt' \tag{5.107}$$

$$= \begin{cases} 0 & t < 0 \\ \gamma_0 & t \geq 0 \end{cases} \tag{5.108}$$

$$= \gamma_0 H(t) \tag{5.109}$$

and

$$\text{Heaviside step function} \qquad \boxed{H(t) \equiv \begin{cases} 0 & t < 0 \\ 1 & t \geq 0 \end{cases}} \tag{5.110}$$

The response of a non-Newtonian fluid to the imposition of a step strain is a rapid increase in shear and normal stresses (if the material produces normal stresses) followed by a relaxation of these stresses. The material functions for the step-strain experiment are based on the idea of modulus rather than viscosity. Modulus is the ratio of stress to strain and is a concept that is quite useful for elastic materials. We will discuss elastic modulus in more detail in the next section and in Chapter 8.

The material functions for the step shear strain experiment are given next. In general γ_0 could be positive or negative, depending on the coordinate system chosen. We will choose the coordinate system so that γ_0 is positive.

$$\text{Relaxation modulus} \qquad \boxed{G(t, \gamma_0) \equiv \frac{-\tau_{21}(t, \gamma_0)}{\gamma_0}} \tag{5.111}$$

$$\begin{matrix} \text{First normal-stress} \\ \text{step shear relaxation} \\ \text{modulus} \end{matrix} \qquad \boxed{G_{\Psi_1}(t, \gamma_0) \equiv \frac{-(\tau_{11} - \tau_{22})}{\gamma_0^2}} \tag{5.112}$$

$$\begin{matrix} \text{Second normal-stress} \\ \text{step shear relaxation} \\ \text{modulus} \end{matrix} \qquad \boxed{G_{\Psi_2}(t, \gamma_0) \equiv \frac{-(\tau_{22} - \tau_{33})}{\gamma_0^2}} \tag{5.113}$$

$G_{\Psi_2}(t, \gamma_0)$ is seldom measured since it is small and requires specialized equipment [199, 127]. Note that the material functions for the step-strain experiment are functions of time and of the strain amplitude of the step γ_0. For small strains, $G(t, \gamma_0)$ and $G_{\Psi_1}(t, \gamma_0)$ are found to be independent of strain; this limit is called the *linear viscoelastic regime*. In the linear viscoelastic regime $G(t, \gamma_0)$ is written as $G(t)$, and often high strain data are reported relative to $G(t)$ through the use of a material function called the *damping function* $h(\gamma_0)$:

$$h(\gamma_0) \equiv \frac{G(t, \gamma_0)}{G(t)} \tag{5.114}$$

This function is only reported when the resulting function $h(\gamma_0)$ is independent of time (see Section 6.2.2).

5.2.2.5 Small-Amplitude Oscillatory Shear

The final set of unsteady shear material functions that we wish to introduce are very widely used to characterize complex fluids by chemists, chemical engineers, and materials scientists. The flow is again shear flow, and the time-dependent shear-rate function $\dot{\varsigma}(t)$ used for this flow is periodic (a cosine function) (see Figure 5.2f). This flow is called *small-amplitude oscillatory shear (SAOS)*:

$$
\begin{array}{c}
\text{Kinematics} \\
\text{for SAOS}
\end{array}
\quad
\boxed{
\begin{aligned}
\underline{v} &= \begin{pmatrix} \dot{\varsigma}(t)x_2 \\ 0 \\ 0 \end{pmatrix}_{123} \\
\dot{\varsigma}(t) &= \dot{\gamma}_0 \cos \omega t
\end{aligned}
}
\qquad (5.115)
$$

The frequency of the cosine function is ω (rad/s), and $\dot{\gamma}_0$ is the constant amplitude of the shear-rate function. This flow is almost always carried out in a cone-and-plate or parallel-plate torsional rheometer (see Figure 4.4), although the concentric-cylinder (Couette) geometry is also used (see Chapter 10).

From the strain we can calculate the wall motion required to produce SAOS in, for example, the cone-and-plate apparatus. We saw in Section 5.2.2.3 and Figure 5.6 that small shear strains can be written as

$$
\gamma_{21} = \frac{\Delta u_1}{\Delta x_2} \qquad (5.116)
$$

If we call $b(t)$ the time-dependent displacement of the upper plate and h the gap between the plates, then for small strains

$$
\gamma_{21}(0, t) = \frac{b(t)}{h} \qquad (5.117)
$$

Thus $b(t)$ is related to the strain, which we can calculate from the strain rate using Equation (5.97):

$$
\gamma_{21}(0, t) = \int_0^t \dot{\gamma}_{21}(t')\, dt' \qquad (5.118)
$$

$$
= \int_0^t \dot{\gamma}_0 \cos \omega t'\, dt' = \frac{\dot{\gamma}_0}{\omega} \sin \omega t \qquad (5.119)
$$

$$
= \gamma_0 \sin \omega t \qquad (5.120)
$$

where $\gamma_0 = \dot{\gamma}_0/\omega$ is the strain amplitude. Note that zero was chosen as the lower limit of the integral, that is, the reference state for the strain was taken to be $t = 0$. Thus the motion of the wall, $b(t) = h\gamma_0 \sin \omega t$, is a sine function (Figure 5.10). Moving the wall of a shear cell in a sinusoidal manner does not guarantee that the shear-flow velocity profile [Equation (5.115)] will be produced, but one can show (see Problem 5.17) that a linear velocity profile will be produced for sufficiently low frequencies or high viscosities.

When a sample is strained in this way at low strain amplitudes, the shear stress that is produced will be a sine wave of the same frequency as the input strain wave. The shear stress, however, usually will not be in phase with the input strain. We can write this as follows:

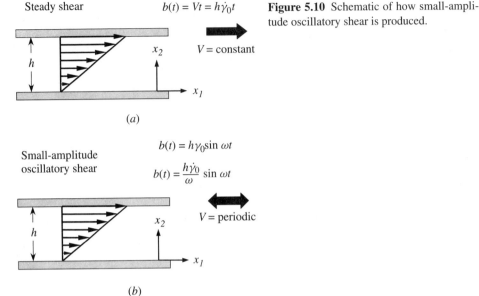

Steady shear $b(t) = Vt = h\dot{\gamma}_0 t$

$V = \text{constant}$

(a)

Figure 5.10 Schematic of how small-amplitude oscillatory shear is produced.

Small-amplitude oscillatory shear

$b(t) = h\gamma_0 \sin \omega t$

$b(t) = \dfrac{h\dot{\gamma}_0}{\omega} \sin \omega t$

$V = \text{periodic}$

(b)

$$- \tau_{21}(t) = \tau_0 \sin (\omega t + \delta) \tag{5.121}$$

where the quantity δ [not to be confused with the asymmetric impulse function $\delta_+(t)$] gives the phase difference between the strain wave and the stress response.

The material functions for SAOS are defined based on this sinusoidal shear-stress output. We can expand the preceding expression by using trigonometric identities.

$$-\tau_{21}(t) = \tau_0 \sin (\omega t + \delta) \tag{5.122}$$

$$= \tau_0(\sin \omega t \cos \delta + \sin \delta \cos \omega t) \tag{5.123}$$

$$= (\tau_0 \cos \delta) \sin \omega t + (\tau_0 \sin \delta) \cos \omega t \tag{5.124}$$

By splitting up the shear stress in this way, we see that there is a portion of the stress wave that is in phase with the imposed strain (i.e., proportional to $\sin \omega t$) and a portion of the stress wave that is in phase with the imposed strain rate (proportional to $\cos \omega t$).

To appreciate the significance of this observation, recall that for Newtonian fluids, the shear-stress response is proportional to the imposed shear rate:

$$\tau_{21} = -\mu \dot{\gamma}_{21} \tag{5.125}$$

For elastic materials, discussed in more detail in Chapter 8, shear stress is proportional to the imposed strain:

Hooke's law
(shear only) $\boxed{\tau_{21} = -G\gamma_{21}}$ (5.126)

This latter observation is called *Hooke's law* for elastic solids, and it is an empirical, scalar rule much like Newton's law of viscosity. It serves as a definition of the elastic modulus G

and describes the stress of a limited class of solids that are called elastic. For elastic materials the stress generated is directly proportional to the strain, that is, to the deformation. This is similar to the response of mechanical springs, which generate stress that is directly proportional to the change in length (deformation) of the spring (Figure 5.11). The stress response in SAOS described by Equation (5.124) contains both a part that is Newtonian-like (proportional to $\dot{\gamma}_{21}$) and an elastic part (proportional to γ_{21}). Thus the SAOS experiment is ideal for probing viscoelastic materials, defined as materials that show both viscous and elastic properties.

The material functions for SAOS are the storage modulus $G'(\omega)$ and the loss modulus $G''(\omega)$, and they are defined as follows:

SAOS
material functions

$$\frac{-\tau_{21}}{\gamma_0} = G' \sin \omega t + G'' \cos \omega t \tag{5.127}$$

Storage modulus

$$G'(\omega) \equiv \frac{\tau_0}{\gamma_0} \cos \delta \tag{5.128}$$

Loss modulus

$$G''(\omega) \equiv \frac{\tau_0}{\gamma_0} \sin \delta \tag{5.129}$$

G' is equal to the amplitude of the portion of the stress wave that is in phase with the strain wave divided by the amplitude of the strain wave. G'' is defined analogously as the amplitude of the portion of the stress wave that is out of phase with the strain wave, divided by the amplitude of the strain wave.

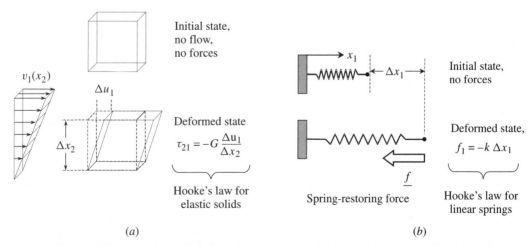

(a)   (b)

Figure 5.11 Hooke's law for elastic fluids resembles the force–deformation relationship of linear springs, which is also called Hooke's law. The spring-restoring force is $\underline{f}$, f_1 is the 1-component of $\underline{f}$, Δx_1 is the change in length of the spring, and k is the force constant of the spring.

For a Newtonian fluid in SAOS, the response is completely in phase with the strain rate, $G' = 0$, and $\eta' = G''/\omega = \mu$. This can be shown by using the Newtonian constitutive equation and specifying the SAOS kinematics. For an elastic solid that follows Hooke's law (a Hookean solid), the shear-stress response in SAOS is completely in phase with the strain, $G' = G$ and $G'' = 0$. For viscoelastic materials both $G'(\omega)$ and $G''(\omega)$ are nonzero and are generally functions of frequency. Some experimental data for G' and G'' for polymers are shown in Section 6.2.1.

Several other material functions related to G' and G'' are also used by the rheological community, although they contain no information not already present in the two dynamic moduli already defined. These other material functions are summarized in Table 5.1. G' and J' (see Problem 5.12) are called the *storage modulus* and *storage compliance*, respectively, because they are related to elastic energy storage by the material. G'' and J'' are called the *viscous loss modulus* and *viscous loss compliance*, respectively, since they are related to the viscous response of the fluid. An in-depth presentation of experimental results for all the SAOS material functions is contained in Ferry [75].

In the next section we show how complex number notation can also be used to describe SAOS. Many of the definitions of the ancillary material functions (η^*, η', η'', etc.) make more sense when complex notation is applied.

5.2.2.6 Small-Amplitude Oscillatory Shear—Complex Notation

An alternative way of expressing a periodic function is to use complex notation. When doing complex algebra recall that the conjugate $x^{(*)}$ of a complex number x

$$x = a + bi \tag{5.130}$$

is given by

$$x^{(*)} = a - bi \tag{5.131}$$

TABLE 5.1
Definitions of Material Functions for Small-Amplitude Oscillatory Shear (SAOS) in Terms of Storage Modulus G' and Loss Modulus G''

Complex modulus magnitude	$\lvert G^* \rvert = \sqrt{G'^2 + G''^2}$
Loss tangent	$\tan \delta = \frac{G''}{G'}$
Dynamic viscosity	$\eta' = \frac{G''}{\omega}$
Out-of-phase component of η^*	$\eta'' = \frac{G'}{\omega}$
Complex viscosity magnitude	$\lvert \eta^* \rvert = \sqrt{\eta'^2 + \eta''^2}$
Complex compliance magnitude	$\lvert J^* \rvert = \frac{1}{\lvert G^* \rvert}$
Storage compliance	$J' = \frac{1/G'}{1 + \tan^2 \delta}$
Loss compliance	$J'' = \frac{1/G''}{1 + (\tan^2 \delta)^{-1}}$

where[6]

$$a = \Re(x) \tag{5.134}$$

$$b = \Im(x) \tag{5.135}$$

$\Re(x)$ denotes the real part of x, and $\Im(x)$ denotes the imaginary part of x. The magnitude of x, $|x|$, is calculated from

$$|x| = +\sqrt{x\, x^{(*)}} = +\sqrt{a^2 + b^2} \tag{5.136}$$

If we write e^{ix} in series form and compare the result to the series for $\sin x$ and $\cos x$, we can see that the following relationships, called *Euler's formulas* [239], hold:

Euler's formulas
$$\boxed{\begin{aligned} e^{i\alpha} &= \cos \alpha + i \sin \alpha \\ e^{-i\alpha} &= \cos \alpha - i \sin \alpha \end{aligned}} \tag{5.137}$$

From Euler's formulas we can deduce that

$$\cos \alpha = \frac{e^{i\alpha} + e^{-i\alpha}}{2} \tag{5.138}$$

$$\sin \alpha = \frac{e^{i\alpha} - e^{-i\alpha}}{2i} \tag{5.139}$$

Using Euler's formulas we see that the strain rate, strain, and stress for SAOS can be written as

$$\dot{\gamma}_{21}(t) = \dot{\gamma}_0 \cos \omega t = \Re\left(\dot{\gamma}_0 e^{i\omega t}\right) \tag{5.140}$$

$$\gamma_{21}(t) = \int_{-\infty}^{t} \dot{\gamma}_{21}(t')\, dt' = \Re\left(\int_{-\infty}^{t} \dot{\gamma}_0 e^{i\omega t'}\, dt'\right) \tag{5.141}$$

$$= \Re\left(\frac{\dot{\gamma}_0}{i\omega} e^{i\omega t}\right) = \Re\left(-i\gamma_0 e^{i\omega t}\right) \tag{5.142}$$

$$= \gamma_0 \sin \omega t \tag{5.143}$$

Recall that $\gamma_0 = \dot{\gamma}_0/\omega$. For stress we write,

$$-\tau_{21}(t) = \tau_0 \sin(\omega t + \delta) = \Re\left(-i\tau_0 e^{i(\omega t + \delta)}\right) \tag{5.144}$$

---

[6] When taking the real part of an algebraically complex expression, one must always rearrange the expression into the form $a+bi$, and then a is the real part and b is the imaginary part. For example [232],

$$\frac{a + bi}{c + di} = \frac{(a + bi)\,(c - di)}{(c + di)\,(c - di)} = \left(\frac{ac + bd}{c^2 + d^2}\right) + \left(\frac{bc - ad}{c^2 + d^2}\right)i \tag{5.132}$$

$$\Re\left(\frac{a + bi}{c + di}\right) = \frac{ac + bd}{c^2 + d^2} \tag{5.133}$$

$$= \Re\left(-i\tau_0 e^{i\delta} e^{i\omega t}\right) \tag{5.145}$$

$$= \Re\left(-\tilde{\tau}_0 e^{i\omega t}\right) \tag{5.146}$$

where $\tilde{\tau}_0 = \tilde{\tau}_0(\omega) = i\tau_0 e^{i\delta}$ is the complex coefficient of the stress wave. The phase difference between τ_{12} and $\dot{\gamma}_{21}$ is represented by the complex nature of the prefactor $\tilde{\tau}_0$. From here on we will drop the $\Re$ notation except where we explicitly wish to call the reader's attention to the action of taking the real part.

The material function for SAOS is the complex modulus G^*, defined in complex notation as follows:

$$G^*(\omega) \equiv \frac{-\tau_{21}(t)}{\gamma_{21}(t)} = \frac{-\tilde{\tau}_0 e^{i\omega t}}{-i\gamma_0 e^{i\omega t}} \tag{5.147}$$

$$= \frac{\tilde{\tau}_0(\omega)}{i\gamma_0} \tag{5.148}$$

$$= \frac{\tau_o e^{i\delta}}{\gamma_0} = \frac{\tau_0}{\gamma_0}(\cos\delta + i\sin\delta) \tag{5.149}$$

$$= G' + iG'' \tag{5.150}$$

where G' and G'' have the same definitions as before [compare with Equations (5.128) and (5.129)]. The complex viscosity $\eta^*(\omega)$ and the complex compliance $J^*(\omega)$ are defined as follows:

$$\eta^*(\omega) \equiv \frac{-\tau_{21}(t)}{\dot{\gamma}_{21}(t)} \tag{5.151}$$

$$= \frac{-\tilde{\tau}_0 e^{i\omega t}}{\dot{\gamma}_0 e^{i\omega t}} = \frac{-\tilde{\tau}_0}{\dot{\gamma}_0} = \frac{-i\tau_0 e^{i\delta}}{\gamma_0 \omega} \tag{5.152}$$

$$= \frac{G^*}{i\omega} = \eta' - i\eta'' \tag{5.153}$$

$$J^*(\omega) \equiv \frac{\gamma_{21}(t)}{-\tau_{21}(t)} \tag{5.154}$$

$$= \frac{-i\gamma_0 e^{i\omega t}}{-\tilde{\tau}_0 e^{i\omega t}} = \frac{\gamma_0 e^{-i\delta}}{\tau_0} = \frac{1}{G^*} \tag{5.155}$$

$$= \frac{\gamma_0}{\tau_0}(\cos\delta - i\sin\delta) \tag{5.156}$$

$$= J' - iJ'' \tag{5.157}$$

Using complex notation we see that the material functions in SAOS are defined analogously to other shear material functions: complex viscosity is the ratio of a shear stress to a shear rate, complex modulus is the ratio of a shear stress to a shear strain, and complex compliance is the ratio of a shear strain to a shear stress. Because these functions are complex, they have both real parts [denoted by a prime, i.e., G', η', J'] and imaginary parts [denoted by a double prime, i.e., G'', η'', J''). The magnitudes of complex quantities are found by multiplying a complex number by its complex conjugate and taking the square root:

$$|G^*| = \sqrt{(G' + iG'')(G' - iG'')} \qquad (5.158)$$

$$= \sqrt{G'^2 + G''^2} \qquad (5.159)$$

$$|\eta^*| = \sqrt{(\eta' + i\eta'')(\eta' - i\eta'')} \qquad (5.160)$$

$$= \sqrt{\eta'^2 + \eta''^2} \qquad (5.161)$$

$$|J^*| = \sqrt{(J' - iJ'')(J' + iJ'')} \qquad (5.162)$$

$$= \sqrt{J'^2 + J''^2} \qquad (5.163)$$

These expressions correspond to the definitions given in the previous section.

SAOS is the last of the shear-flow kinematics that we will discuss. Next is a discussion of the elongational-flow material functions.

5.3 Elongational Flow

All elongational-flow material functions are based on the velocity field shown here and discussed in Chapter 4:

$$\underline{v} = \begin{pmatrix} -\frac{1}{2}\dot{\epsilon}(t)(1 + b)x_1 \\ -\frac{1}{2}\dot{\epsilon}(t)(1 - b)x_2 \\ \dot{\epsilon}(t)x_3 \end{pmatrix}_{123} \qquad (5.164)$$

Differences among the flows are achieved by varying the function $\dot{\epsilon}(t)$ and the parameter b.

As pointed out earlier, only two stress-related quantities can be measured for elongational flow of incompressible fluids, $\tau_{33} - \tau_{11}$ and $\tau_{22} - \tau_{11}$. Elongational flow is very difficult to produce, and, in addition, stress measurements are very challenging to make in the elongational geometries. In some geometries one can measure directly the stress to drive the flow, for example, the force on the end of a stretching sample. In the opposing jets device (see Figure 4.11b) the force is measured by the deflection of the nozzle creating the jet. In many elongational measurements flow birefringence is used. Flow birefringence is an optical property that is proportional to stress, and it is exhibited by many polymers; see Chapter 10 and Appendix E. Measurements of elongational strain are sometimes made by videotaping marker particles in the flow and analyzing the images using computer software.

5.3.1 STEADY ELONGATION

Steady-state elongational flows [uniaxial, biaxial, planar; see Equation (4.48) and Table 4.1] are produced by choosing the following kinematics:

$$\text{Kinematics of steady elongation} \quad \boxed{\dot{\epsilon}(t) = \dot{\epsilon}_0 = \text{constant}} \quad (5.165)$$

For these flows, constant stress differences are measured. The material functions defined are two elongational viscosities based on the measured normal-stress differences. For both uniaxial and biaxial extension, the elongational viscosity based on $\tau_{22} - \tau_{11}$ is zero for all fluids; planar elongational flow has two nonzero elongational viscosities.

Uniaxial elongation ($b = 0$, $\dot{\epsilon}_0 > 0$):

$$\underline{v} = \begin{pmatrix} -\frac{1}{2}\dot{\epsilon}_0 x_1 \\ -\frac{1}{2}\dot{\epsilon}_0 x_2 \\ \dot{\epsilon}_0 x_3 \end{pmatrix}_{123} \quad \dot{\epsilon}_0 > 0 \quad (5.166)$$

$$\text{Uniaxial elongational viscosity} \quad \boxed{\bar{\eta}(\dot{\epsilon}_0) \equiv \frac{-(\tau_{33} - \tau_{11})}{\dot{\epsilon}_0}} \quad (5.167)$$

Biaxial elongation ($b = 0$, $\dot{\epsilon}_0 < 0$):[7]

$$\underline{v} = \begin{pmatrix} -\frac{1}{2}\dot{\epsilon}_0 x_1 \\ -\frac{1}{2}\dot{\epsilon}_0 x_2 \\ \dot{\epsilon}_0 x_3 \end{pmatrix}_{123} \quad \dot{\epsilon}_0 < 0 \quad (5.168)$$

$$\text{Biaxial elongational viscosity} \quad \boxed{\bar{\eta}_B(\dot{\epsilon}_0) \equiv \frac{-(\tau_{33} - \tau_{11})}{\dot{\epsilon}_0}} \quad (5.169)$$

Planar elongation ($b = 1$, $\dot{\epsilon}_0 > 0$):

$$\underline{v} = \begin{pmatrix} -\dot{\epsilon}_0 x_1 \\ 0 \\ \dot{\epsilon}_0 x_3 \end{pmatrix}_{123} \quad \dot{\epsilon}_0 > 0 \quad (5.170)$$

$$\text{First planar elongational viscosity} \quad \boxed{\bar{\eta}_{P_1}(\dot{\epsilon}_0) \equiv \frac{-(\tau_{33} - \tau_{11})}{\dot{\epsilon}_0} = \bar{\eta}_P(\dot{\epsilon}_0)} \quad (5.171)$$

$$\text{Second planar elongational viscosity} \quad \boxed{\bar{\eta}_{P_2}(\dot{\epsilon}_0) \equiv \frac{-(\tau_{22} - \tau_{11})}{\dot{\epsilon}_0}} \quad (5.172)$$

---

[7] We have followed Bird et al. [26] in our notation here. Note that biaxial and uniaxial elongational flows are identical, except for the sign of the stretch rate. This convention draws on the fact that both flows are axisymmetric, and axisymmetric flows are usually analyzed in the cylindrical coordinate system with the flow in the z- or x_3-direction. If uniaxial and biaxial extension are defined so that the elongation rate in each case is positive, of magnitudes $\dot{\epsilon}$ and $\dot{\epsilon}_B$, respectively, and in the x_1-direction, then $\dot{\epsilon} = -2\dot{\epsilon}_B$, and the definition of extensional viscosity looks a little different in both cases (see Problem 5.18).

Steady elongational flow is difficult to achieve because of the rapid rate of particle deformation that is required (see Section 4.3.1). Very few reliable data are available for this important flow.

The strain ϵ in elongational flow is defined analogously to the shear strain as the integral of the deformation rate between two times:

$$\text{Elongational strain} \qquad \boxed{\epsilon(t_{\text{ref}}, t) \equiv \int_{t_{\text{ref}}}^{t} \dot{\epsilon}(t')\, dt'} \qquad (5.173)$$

For steady elongational flow where $\dot{\epsilon}(t') = \dot{\epsilon}_0$ we can carry out the integral from time $t_{\text{ref}} = 0$ to the current time t, obtaining

$$\text{Hencky strain} \qquad \boxed{\epsilon = \dot{\epsilon}_0 t = \ln \frac{l}{l_0}} \qquad (5.174)$$

where we have used Equation (4.41) in obtaining the last result. The strain as defined here is called the *Hencky strain* to distinguish it from the extension ratio l/l_0, which is used to measure strain in studies of metals and other solid materials [78].

EXAMPLE

Calculate the planar elongational viscosities $\bar{\eta}_{P_1}$ and $\bar{\eta}_{P_2}$ for an incompressible Newtonian fluid.

SOLUTION

The kinematics for steady, planar elongation are given by

$$\underline{v} = \begin{pmatrix} -\dot{\epsilon}_0 x_1 \\ 0 \\ \dot{\epsilon}_0 x_3 \end{pmatrix}_{123} \qquad \dot{\epsilon}_0 > 0 \qquad (5.175)$$

where $\dot{\epsilon}_0$ is a constant. The planar elongational viscosities are defined as

$$\bar{\eta}_{P_1}(\dot{\epsilon}_0) \equiv \frac{-(\tau_{33} - \tau_{11})}{\dot{\epsilon}_0} \qquad (5.176)$$

$$\bar{\eta}_{P_2}(\dot{\epsilon}_0) \equiv \frac{-(\tau_{22} - \tau_{11})}{\dot{\epsilon}_0} \qquad (5.177)$$

We need to calculate the stress tensor $\underline{\underline{\tau}}$ for a Newtonian fluid subjected to the kinematics given:

$$\underline{\underline{\tau}} = -\mu[\nabla \underline{v} + (\nabla \underline{v})^T] \qquad (5.178)$$

$$= -\mu \left[\begin{pmatrix} -\dot{\epsilon}_0 & 0 & 0 \\ 0 & 0 & 0 \\ 0 & 0 & \dot{\epsilon}_0 \end{pmatrix}_{123} + \begin{pmatrix} -\dot{\epsilon}_0 & 0 & 0 \\ 0 & 0 & 0 \\ 0 & 0 & \dot{\epsilon}_0 \end{pmatrix}_{123} \right] \qquad (5.179)$$

$$= \begin{pmatrix} 2\mu\dot{\epsilon}_0 & 0 & 0 \\ 0 & 0 & 0 \\ 0 & 0 & -2\mu\dot{\epsilon}_0 \end{pmatrix}_{123} \tag{5.180}$$

Now we can calculate $\bar{\eta}_{P_1}$ and $\bar{\eta}_{P_2}$.

$$\bar{\eta}_{P_1}(\dot{\epsilon}_0) \equiv \frac{-(\tau_{33} - \tau_{11})}{\dot{\epsilon}_0} = 4\mu \tag{5.181}$$

$$\bar{\eta}_{P_2}(\dot{\epsilon}_0) \equiv \frac{-(\tau_{22} - \tau_{11})}{\dot{\epsilon}_0} = 2\mu \tag{5.182}$$

5.3.2 UNSTEADY ELONGATION

5.3.2.1 Elongational Stress Growth

Startup of steady elongational flow has the same experimental difficulties as steady elongational flow, but some start-up curves have been reported (see Section 6.4.2). For the startup of steady elongational flow the kinematics are

Kinematics of
startup of steady
uniaxial elongation

$$\begin{array}{|c} \underline{v} = \begin{pmatrix} -\frac{1}{2}\dot{\epsilon}(t)(1+b)x_1 \\ -\frac{1}{2}\dot{\epsilon}(t)(1-b)x_2 \\ \dot{\epsilon}(t)x_3 \end{pmatrix}_{123} \\ \dot{\epsilon}(t) = \begin{cases} 0 & t < 0 \\ \dot{\epsilon}_0 & t \geq 0 \end{cases} \end{array} \tag{5.183}$$

The material functions for the startup of steady elongation are defined analogously to those for startup of steady shearing (see Table 5.2).

Material functions for stress decay after steady elongation could be defined, but in practice steady state is almost never reached in an elongational experiment. Thus, it is not very useful to define such material functions.

5.3.2.2 Elongational Creep

If instead of a constant elongational rate $\dot{\epsilon}_0$ a constant elongational stress σ_0 is applied to drive

TABLE 5.2
Definitions of Material Functions for Startup of Steady Elongation

Uniaxial elongational stress growth coefficient $(b = 0, \dot{\epsilon}_0 > 0)$	$\bar{\eta}^+(t, \dot{\epsilon}_0) \equiv \frac{-(\tau_{33}-\tau_{11})}{\dot{\epsilon}_0}$
Biaxial elongational stress growth coefficient $(b = 0, \dot{\epsilon}_0 < 0)$	$\bar{\eta}_B^+(t, \dot{\epsilon}_0) \equiv \frac{-(\tau_{33}-\tau_{11})}{\dot{\epsilon}_0}$
Planar elongational stress growth coefficients $(b = 1, \dot{\epsilon}_0 > 0)$ }	$\bar{\eta}_{P_1}^+(t, \dot{\epsilon}_0) = \bar{\eta}_P^+(t, \dot{\epsilon}_0) \equiv \frac{-(\tau_{33}-\tau_{11})}{\dot{\epsilon}_0}$
	$\bar{\eta}_{P_2}^+(t, \dot{\epsilon}_0) \equiv \frac{-(\tau_{22}-\tau_{11})}{\dot{\epsilon}_0}$

a flow, the flow is called *elongational creep*. Elongational creep can be produced simply by hanging a weight on a cylindrical sample. As was seen in shear creep, the measured quantity then becomes the deformation (length change) of the sample, expressed as a strain. The kinematics for elongational creep are

Kinematics of
elongational creep
$$\tau_{33} - \tau_{11} = \begin{cases} 0 & t < 0 \\ \sigma_0 = \text{constant} & t \geq 0 \end{cases}$$

(5.184)

and the material function that is defined is the elongational compliance $D(t, \sigma_0)$:

Elongational
creep compliance
$$D(t, \sigma_0) \equiv \frac{\epsilon(0, t)}{-\sigma_0}$$

(5.185)

where the elongational strain ϵ is defined in Equation (5.173) and is calculated from measurements of the length change as a function of time.

An experiment that gives some information about relaxation after elongational deformation is the unconstrained or free recoil experiment. In this experiment, at some time in the creep elongation experiment the sample is cut free of the driving mechanism and allowed to relax. The material is able to relax in all three directions. The amount of contraction that occurs can be expressed as an amount of recoil strain and is an indication of the amount of elasticity in the material:

Ultimate recoverable
elongational strain
$$\epsilon_r = \ln \left[\frac{l(t_\infty)}{l(0)} \right]$$

(5.186)

where $l(0)$ is the length of the sample at the time at which the sample is cut free of the driving mechanism ($t = 0$), and $l(t_\infty)$ is the length of the sample after it has had a chance to relax completely.

5.3.2.3 Step Elongational Strain

The step-strain experiment can be performed in elongational flows. The lubricated squeezing experiment in particular (see Section 10.2.2.2) provides reasonable measurements of step biaxial extensions. The kinematics of step extension are analogous to the kinematics in the step-shear experiment:

Kinematics of
step elongational
strain

$$\underline{v} = \begin{pmatrix} -\frac{1}{2}\dot{\epsilon}(t)(1 + b)x_1 \\ -\frac{1}{2}\dot{\epsilon}(t)(1 - b)x_2 \\ \dot{\epsilon}(t)x_3 \end{pmatrix}_{123}$$

$$\dot{\epsilon}(t) = \lim_{\varepsilon \to 0} \begin{cases} 0 & t < 0 \\ \dot{\epsilon}_0 & 0 \leq t < \varepsilon \\ 0 & t \geq \varepsilon \end{cases}$$

$$\dot{\epsilon}_0 \varepsilon = \epsilon_0 = \text{constant}$$

(5.187)

The quantity ϵ_0 is the magnitude of the elongational strain imposed on the fluid.

The material functions in these flows are elongational relaxation moduli $E(t, \epsilon_0)$. Uniaxial and biaxial elongation each have one nonzero step elongational modulus; for planar elongational flow, there are two such moduli. In step elongation, as in step shear, the relaxation moduli are defined as the ratio of a stress to a measure of strain. By convention, however, the strain measure is not the simple elongational strain ϵ_0, but rather the difference between two components of a strain tensor called the *Finger strain tensor* $\underline{\underline{C}}^{-1}$. The Finger strain tensor is defined and discussed in detail in Chapter 9. The relevant components of the Finger tensor for step elongational flow are

Uniaxial
step elongational
relaxation modulus

$$E(t, \epsilon_0) = \frac{-(\tau_{33} - \tau_{11})}{C_{33}^{-1} - C_{11}^{-1}} = \frac{-(\tau_{33} - \tau_{11})}{e^{2\epsilon_0} - e^{-\epsilon_0}}$$

(5.188)

Biaxial
step elongational
relaxation modulus

$$E_B(t, \epsilon_0) = \frac{-(\tau_{33} - \tau_{11})}{C_{33}^{-1} - C_{11}^{-1}} = \frac{-(\tau_{33} - \tau_{11})}{e^{2\epsilon_0} - e^{-\epsilon_0}}$$

(5.189)

Planar
step elongational
relaxation moduli

$$E_{P_1}(t, \epsilon_0) = \frac{-(\tau_{33} - \tau_{11})}{C_{33}^{-1} - C_{11}^{-1}} = \frac{-(\tau_{33} - \tau_{11})}{e^{2\epsilon_0} - e^{-2\epsilon_0}}$$

$$E_{P_2}(t, \epsilon_0) = \frac{-(\tau_{22} - \tau_{11})}{C_{22}^{-1} - C_{11}^{-1}} = \frac{-(\tau_{22} - \tau_{11})}{1 - e^{-2\epsilon_0}}$$

(5.190)

For biaxial extension sometimes the biaxial extensional strain $\epsilon_B = -\epsilon_0/2$ is reported instead of ϵ_0 (see Problem 5.18).

5.3.2.4 Small-Amplitude Oscillatory Elongation

A small-amplitude oscillatory deformation can be imposed in elongational flow by squeezing a sample between two small plates in an oscillatory mode. This geometry is used in instruments such as the differential mechanical analyzer (DMA) [204]. As was true with small-amplitude oscillatory shear (SAOS), if small-amplitude oscillatory elongation (SAOE) is performed at low enough amplitude, the output stresses will oscillate with the same frequency as the input deformation. Since the flow is oscillating (changing direction), the SAOE flow has aspects of both uniaxial and biaxial elongation. There is no past or current use of a planar SAOE flow.

The kinematics of this flow are

Kinematics
of SAOE

$$\underline{v}(t) = \begin{pmatrix} -\frac{1}{2}\dot{\epsilon}(t)x_1 \\ -\frac{1}{2}\dot{\epsilon}(t)x_2 \\ \dot{\epsilon}(t)x_3 \end{pmatrix}_{123}$$

$$\dot{\epsilon}(t) = \dot{\epsilon}_0 \cos \omega t$$

(5.191)

The rate of deformation tensor for this flow is therefore

$$\dot{\underline{\underline{\gamma}}}(t) = \begin{pmatrix} -\dot{\epsilon}_0 \cos \omega t & 0 & 0 \\ 0 & -\dot{\epsilon}_0 \cos \omega t & 0 \\ 0 & 0 & 2\dot{\epsilon}_0 \cos \omega t \end{pmatrix}_{123} \tag{5.192}$$

We can calculate the strain between $t_{\text{ref}} = 0$ and the current time t by integrating the deformation rate $\dot{\epsilon}(t)$:

$$\dot{\epsilon}(t) = \dot{\epsilon}_0 \cos \omega t \tag{5.193}$$

$$\epsilon(0, t) = \int_0^t \dot{\epsilon}_0 \cos \omega t' \, dt' = \frac{\dot{\epsilon}_0}{\omega} \sin \omega t \tag{5.194}$$

$$= \epsilon_0 \sin \omega t \tag{5.195}$$

where $\epsilon_0 = \dot{\epsilon}_0/\omega$.

To analyze this flow we will assume that the three nonzero components of the stress tensor for this flow are related to one another in the same way that the three nonzero components of the rate-of-deformation tensor are related; that is, we assume that $\underline{\underline{\tau}}$ can be written as

$$\underline{\underline{\tau}} = \begin{pmatrix} \tau_{11} & 0 & 0 \\ 0 & \tau_{11} & 0 \\ 0 & 0 & -2\tau_{11} \end{pmatrix}_{123} \tag{5.196}$$

This is a good assumption for materials when they are deformed at the small rates of deformation that are seen in the SAOE experiment.[8]

For small deformations and small deformation rates, the stresses generated in the SAOE flow will be oscillatory functions of time with the same frequency ω as the input deformation wave. The stress will, in general, be out of phase with respect to both deformation rate, $\dot{\epsilon}(t) = \dot{\epsilon}_0 \cos \omega t$, and deformation, $\epsilon(0, t) = \epsilon_0 \sin \omega t$. If we designate δ as the phase difference between stress and strain, we can express the 11-component of the stress as

$$\tau_{11}(t) = \tau_0 \sin (\omega t + \delta) \tag{5.197}$$

where τ_0 is the amplitude of τ_{11}, and δ is the phase difference between τ_{11} and the strain wave $\epsilon(0, t)$. The stress difference on which all the material functions of SAOE will be based is $\tau_{33} - \tau_{11}$:

$$\tau_{33} - \tau_{11} = -2\tau_{11} - \tau_{11} = -3\tau_{11} \tag{5.198}$$

Following the same methods that were used for SAOS, we can then develop material functions for this flow. Expanding $\tau_{33} - \tau_{11}$ using trigonometric identities we obtain

---

[8] At the small rates of deformation where the SAOE test is valid part of the stress tensor will be proportional to the rate-of-deformation tensor, and part of the stress tensor will be proportional to a strain tensor (see Chapter 8). Both of these tensors are diagonal tensors with the 11- and 22-components equal and the 33-component equal to -2 times the 11-component.

$$-(\tau_{33} - \tau_{11}) = 3\tau_0 \sin{(\omega t + \delta)} \tag{5.199}$$

$$= 3\tau_0(\sin{\omega t}\cos{\delta} + \cos{\omega t}\sin{\delta}) \tag{5.200}$$

The definitions of the material functions for SAOE are

SAOE
material functions

$$\boxed{\frac{-(\tau_{33} - \tau_{11})}{\epsilon_0} = E' \sin{\omega t} + E'' \cos{\omega t}} \tag{5.201}$$

Elongational
storage modulus

$$\boxed{E'(\omega) = \frac{3\tau_0}{\epsilon_0} \cos{\delta}} \tag{5.202}$$

Elongational
loss modulus

$$\boxed{E''(\omega) = \frac{3\tau_0}{\epsilon_0} \sin{\delta}} \tag{5.203}$$

Using the linear viscoelastic constitutive equation developed in Chapter 8, we can show that the dynamic moduli E' and E'' of SAOE are related to the dynamic moduli G' and G'' of SAOS as follows (see Problem 8.19):

$$E' = 3G' \tag{5.204}$$

$$E'' = 3G'' \tag{5.205}$$

The amount of rheological data taken in shear far exceeds that taken in elongation because of the experimental difficulties faced in producing elongational flows. In the next chapter we will look at some actual data on polymer systems for both shear and elongational flows. As we will see, a very wide variety of responses is observed, depending on the material studied.

5.4 PROBLEMS

5.1 What is a constitutive equation? What is a material function? What is the difference and how are they related?

5.2 Compare and contrast the quantities η, $\bar{\eta}$, η_0, and η^*.

5.3 Compare and contrast the quantities $\underline{\dot{\gamma}}$, $\dot{\gamma}$, $\underline{\underline{\varsigma}}(t)$, $\dot{\gamma}_{21}(t)$, and $\dot{\gamma}_0$.

5.4 Why are $G'(\omega)$ and $G''(\omega)$ called the *storage* and *loss* moduli, respectively?

5.5 What are η^+, Ψ_1^+, and Ψ_2^+ predicted to be for an incompressible Newtonian fluid? What are η^-, Ψ_1^-, and Ψ_2^- predicted to be for an incompressible Newtonian fluid?

5.6 What is the steady elongational viscosity $\bar{\eta}$ for an incompressible Newtonian fluid?

5.7 What is the elongational stress growth function $\bar{\eta}^+(t)$ for an incompressible Newtonian fluid?

5.8 What are $G'(\omega)$ and $G''(\omega)$ for an incompressible Newtonian fluid? Work out both in real and in complex notation.

5.9 Starting with the fact that $J^* = 1/G^*$, show that the following relations hold:

$$J' = \frac{G'}{|G^*|^2}, \quad J'' = \frac{G''}{|G^*|^2}$$

5.10 In Figure 5.8 the steady state value of $J(t)$ after recovery is shown as equal to t_2/η_0. Show why this is the case for the linear viscoelastic limit.

5.11 Show that the second extensional viscosity $\bar{\eta}_2 \equiv -(\tau_{22} - \tau_{11})/\epsilon_0$, is equal to zero for uniaxial and biaxial extension.

5.12 The small-amplitude oscillatory shear compliance material function J' is defined as the ratio of the amplitude of the strain in phase with stress to the amplitude of the stress, and J'' is defined as the ratio of the amplitude of the strain 90° out of phase with the amplitude of stress to the stress [75]. Using this definition, show that the following relationships between G', G'' and J', J'' are valid (also given in Table 5.1):

$$J' = \frac{1/G'}{1 + \tan^2 \delta}$$

$$J'' = \frac{1/G''}{1 + (\tan^2 \delta)^{-1}}$$

5.13 For this hypothetical constitutive equation,

$$\underline{\underline{\tau}} = -\zeta_0 \nabla \underline{v} \cdot (\nabla \underline{v})^T$$

where ζ_0 is a constant parameter of the model, answer the questions below:

(a) What are the units of ζ_0?

(b) Calculate the steady elongational viscosity $\bar{\eta}$ and the steady shear viscosity η predicted by this constitutive equation.

(c) From what you know about rheology so far, is this an appropriate constitutive equation? Explain.

5.14 Calculate the steady shear viscosity η and the steady elongational viscosity $\bar{\eta}$ for this constitutive equation:

$$\underline{\underline{\tau}} = -\left[\alpha \nabla \underline{v} + \beta (\nabla \underline{v})^T \right]$$

where α and β are constant parameters of the model. From what you know about rheology so far, is this an appropriate constitutive equation? Explain.

5.15 A know-it-all coworker in your department says, "Rheologists just make life difficult. It is not so hard to find a good constitutive equation. Here— this one predicts shear thinning and will work for

our elastic materials too." The proposed constitutive equation is

$$\underline{\underline{\tau}} = -\left(\frac{\zeta_0}{\dot{\gamma}_0} \right) \underline{\underline{\dot{\gamma}}}$$

where ζ_0 is a constant parameter of the model and $\dot{\gamma}_0$ is the constant shear rate in steady shear flow $[\dot{\varsigma}(t) = \dot{\gamma}_0 = \text{constant}]$.

(a) Calculate the steady shear viscosity $\eta(\dot{\gamma}_0)$ for this model. Does the model predict shear thinning? Sketch your result.

(b) Calculate the steady elongational viscosity $\bar{\eta}$ for this model. What difficulties do you encounter?

(c) From what you know about rheology so far, is this an appropriate constitutive equation? Explain.

5.16 An incompressible Newtonian fluid is subjected to the velocity profile

$$\underline{v} \text{ (m/s)} = \begin{pmatrix} \alpha x + \beta y \\ \alpha x \\ \beta z \end{pmatrix}_{xyz}$$

If $\dot{\gamma} = 10 \text{ s}^{-1}$, solve for the velocity profile (i.e., find α and β.).

5.17 (a) Under what conditions will the velocity profile in a Newtonian fluid undergoing small-amplitude oscillatory shear (SAOS) be linear as required in the defintions of the SAOS material functions? See the analysis in Problem 3.23.

(b) For typical values of density, frequency, and so on for a polymer melt, what is the minimum viscosity as a function of gap between parallel plates that satisfies this criterion? Is linearity of velocity profile a good assumption under most conditions of experimentation on polymer melts?

5.18 We have defined uniaxial and biaxial extension using the axisymmetric flow convention (see Section 5.3.1). In this convention, both uniaxial and biaxial extension have the same kinematics, but the sign of the extension rate $\dot{\epsilon}$ is different in the two cases. An alternative convention puts the flow directions always in the x_1-direction, and the two different extension rates $\dot{\epsilon}$ and $\dot{\epsilon}_B$ as always positive:

Uniaxial elongation:

$$\underline{v} = \begin{pmatrix} \dot{\epsilon} x_1 \\ -\frac{1}{2} \dot{\epsilon} x_2 \\ -\frac{1}{2} \dot{\epsilon} x_3 \end{pmatrix}_{123} \qquad \dot{\epsilon} > 0$$

$$\eta_E(\dot{\epsilon}) \equiv \frac{-(\tau_{11} - \tau_{33})}{\dot{\epsilon}}$$

Biaxial elongation:

$$\underline{v} = \begin{pmatrix} \dot{\epsilon}_B x_1 \\ \dot{\epsilon}_B x_2 \\ -2\dot{\epsilon}_B x_3 \end{pmatrix}_{123} \qquad \dot{\epsilon}_B > 0$$

$$\eta_B(\dot{\epsilon}_B) \equiv \frac{-(\tau_{11} - \tau_{33})}{\dot{\epsilon}_B}$$

What are the equivalencies between $\dot{\epsilon}$ and $\dot{\epsilon}_B$ and among $\bar{\eta}$, η_E, $\bar{\eta}_B$, and η_B? Prove your answers.

CHAPTER

ʊʊʊ **6**

Experimental Data

The division of fluids into two types, Newtonian and non-Newtonian, is straightforward: Newtonian fluids obey the Newtonian constitutive equation, $\underline{\underline{\tau}} = -\mu\underline{\dot{\gamma}}$, and non-Newtonian fluids do not. The ways in which fluids fail to follow the Newtonian constitutive equation, however, vary enormously. As was discussed in the last chapter, material functions provide a common basis on which to compare the flow responses of non-Newtonian materials. In this chapter we compare and contrast the measured material functions for several polymer systems. In order to choose or develop constitutive equations to model non-Newtonian behavior one must be familiar with observed rheological behavior. Other important uses of rheological testing are quality control and other types of qualitative analysis (see Figure 5.1), and these applications also draw heavily on our knowledge of non-Newtonian behavior. The material in this chapter is just a brief introduction to this subject. More discussion of rheological responses of non-Newtonian fluids can be found in the texts of Bird et al. [26], Macosko [162], and especially in Ferry [75], among other sources.

This chapter is the last of the background chapters of this text. In Chapters 7 through 9 we will describe constitutive equations that can model some of the non-Newtonian behavior we will describe in this chapter.

6.1 Steady Shear Flow

6.1.1 GENERAL EFFECTS—LINEAR POLYMERS

The steady shear viscosity η is one of the most widely measured material functions, and nonconstant measurements of η are usually the first sign that a fluid is non-Newtonian. Figures 6.1 and 6.2 show $\eta(\dot{\gamma})$ for concentrated solutions of polybutadiene (PB), and Figure 6.3 shows $\eta(\dot{\gamma})$ for linear and branched polydimethylsiloxane melts (PDMS). At low shear rates the viscosity is constant (η_0), but at higher rates the viscosity decreases and continues to drop as the shear rate increases further. This high-rate behavior, termed shear-thinning, is almost universally observed for high-molecular-weight polymer melts. Note that viscosity versus shear rate is plotted on a double-log plot. This is necessary since η decreases by several orders of magnitude over the usual range of shear rates ($10^{-2} < \dot{\gamma} < 10^2 \text{s}^{-1}$ or higher).

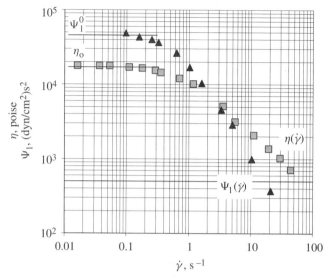

Figure 6.1 Steady shear viscosity η and first normal-stress coefficient Ψ_1 as a function of shear rate $\dot{\gamma}$ for a concentrated solution of a narrowly distributed polybutadiene; replotted from Menezes and Graessley [177]. $M_w = 350$ kg/mol; $M_w/M_n < 1.05$; concentration $= 0.0676$ g/cm^3 in Flexon 391, a hydrocarbon oil. *Source:* From "Nonlinear rheological behavior of polymer systems for several shear-flow histories," by E. V. Menezes and W. W. Graessley, *Journal of Polymer Science, Polymer Physics Edition,* Copyright © 1982 by John Wiley & Sons, Inc. Reprinted by permission of John Wiley & Sons, Inc.

The other two steady shear material functions are the normal-stress coefficients Ψ_1 and Ψ_2. The first normal-stress coefficient is fairly easy to measure in a cone-and-plate apparatus, and for the same polymers shown previously, the first normal-stress coefficients versus shear rate are shown in Figures 6.1, 6.2, and 6.4. For some of the polybutadiene solutions the first normal-stress coefficient also shows a zero-shear value Ψ_1^0, followed by shear-thinning. For the high-molecular-weight PDMS melts no zero-shear first normal-stress coefficient is observed, perhaps due to the limited range over which $\Psi_1(\dot{\gamma})$ was measured. For the $M_w = 350$ kg/mol polybutadiene solution, viscosity and first normal-stress coefficient are compared directly in Figure 6.1. The zero-shear value of the first normal-stress coefficient Ψ_1^0 is greater than η_0, and Ψ_1 shear-thins more strongly than does the viscosity.

In Figure 6.5 we show η and Ψ_1 for a dilute solution of polyisobutylene in a viscous solvent, a mixture of polybutene and kerosene [24]. This type of fluid is called a Boger fluid, and these fluids have been studied extensively because they are elastic ($\Psi_1 > 0$) but not shear-thinning [$\eta(\dot{\gamma}) = $ constant]. By studying Boger fluids researchers hope to separate the effects caused by shear-thinning and elasticity [31].

The second normal-stress coefficient is difficult to measure and is consequently more rarely encountered in the literature. By adding special pressure transducers to a cone-and-plate apparatus, however, Ψ_2 can be measured [181]. Such measurements for two polystyrene solutions are shown in Figure 6.6; note that Ψ_2 is negative and small. For these solutions none of the steady shear material functions vary much with the shear rate, whereas η and Ψ_1 were quite shear-thinning for PDMS (Figures 6.3 and 6.4) and polybutadiene solutions (Figures 6.1 and 6.2). There are literature reports that both normal-stress

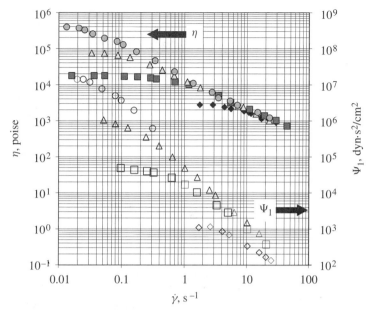

Figure 6.2 Steady shear viscosity η and first normal-stress coefficient Ψ_1 as a function of shear rate $\dot\gamma$ for concentrated solutions of four narrowly distributed polybutadienes; replotted from Menezes and Graessley [177]. Filled symbols are viscosity η; open symbols are first normal-stress coefficient Ψ_1. $\diamond$, $M_w = 200$ kg/mol; $\square$, $M_w = 350$ kg/mol; $\triangle$, $M_w = 517$ kg/mol; $\bigcirc$, $M_w = 813$ kg/mol. For all materials $M_w/M_n < 1.05$; concentration $= 0.0676$ g/cm^3 in Flexon 391, a hydrocarbon oil. Note that the Ψ_1 data are displayed three decades lower than the η data. *Source:* From "Nonlinear rheological behavior of polymer systems for several shear-flow histories," by E. V. Menezes and W. W. Graessley, *Journal of Polymer Science, Polymer Physics Edition*, Copyright © 1982 by John Wiley & Sons, Inc. Reprinted by permission of John Wiley & Sons, Inc.

coefficients are shear-thinning for aqueous solutions of polyacrylamide and polyethylene oxide [45].

6.1.2 Limits on Measurements–Instability

There are experimental limits on the range of shear rates that can be explored when measuring η and Ψ_1. At the low shear-rate end, measurements are bounded by the sensitivity with which torque or pressure drop can be measured. For the high-shear-rate limit, it is often an instability that intrudes on the measurement. For example, for the cone-and-plate viscometer, when the rate of rotation of the cone (or plate, since either can be turned) exceeds a critical value, the edges of the sample start to deform [137] (Figures 6.7 and 6.8). If shearing continues, the edge deforms more severely, and for some materials the sample splits and twists out of the gap [250]. The shear rate at which the instability appears is related to the cone angle employed, such that maximum shear rate times cone angle is a constant [203]. Because edge instabilities occur at rather modest shear rates, most high-shear-rate data are measured in a capillary rheometer. Recent results by Lee et al. [149] have shown that the onset of the edge instability is controlled by the value of the second normal-stress coefficient Ψ_2.

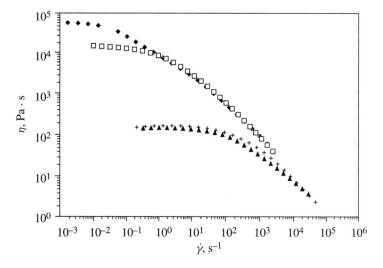

Figure 6.3 Shear viscosity η as a function of shear rate $\dot{\gamma}$ for linear and branched polydimethyl-siloxanes. $+$, M_w=131 kg/mol, M_w/M_n=1.9, linear; $\triangle$, M_w=156 kg/mol, M_w/M_n=2.8, branched; $\square$, M_w=418 kg/mol, M_w/M_n=3.2, linear; $\diamond$, M_w=428 kg/mol, M_w/M_n=2.9, branched; from Piau et al. [207]. *Source:* Reprinted from *Journal of Non-Newtonian Fluid Mechanics*, **30**, J. M. Piau, N. El Kissi, and B. Tremblay, "Low Reynolds number flow visualization of linear and branched silicones upstream of orifice dies," 197–232, Copyright © 1988, with permission from Elsevier Science.

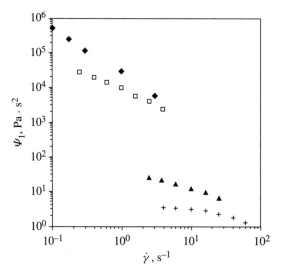

Figure 6.4 Shear first normal stress coefficient Ψ_1 as a function of shear rate $\dot{\gamma}$ for linear and branched polydimethylsiloxanes. $+$, $M_w = 131$ kg/mol, $M_w/M_n = 1.9$, linear; $\triangle$, $M_w = 156$ kg/mol, $M_w/M_n = 2.8$, branched; $\square$, $M_w = 418$ kg/mol, $M_w/M_n = 3.2$, linear; $\diamond$, $M_w = 428$ kg/mol, $M_w/M_n = 2.9$, branched; from Piau et al. [207]. *Source:* Reprinted from *Journal of Non-Newtonian Fluid Mechanics*, **30**, J. M. Piau, N. El Kissi, and B. Tremblay, "Low Reynolds number flow visualization of linear and branched silicones upstream of orifice dies," 197–232, Copyright © 1988, with permission from Elsevier Science.

While capillary rheometers are capable of measuring much higher shear rates than cone-and-plate rheometers, they are also subject to flow instabilities. At high rates it becomes impossible to produce a smooth stream of polymer at the exit of the capillary die. The extrudate becomes distorted in a way that depends on the type of polymer being examined. For some materials a mild surface distortion occurs, known as sharkskin (Figure 6.9).

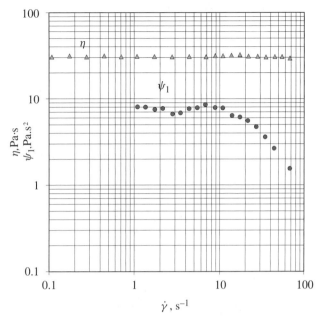

Figure 6.5 Viscosity and first normal-stress coefficient as a function of shear rate for a dilute solution of polyisobutylene in a viscous solvent, a mixture of polybutene and kerosene; from Binnington and Boger [24]. This is a Boger fluid; note that the viscosity is constant, but the fluid is clearly not Newtonian since $\Psi_1 \neq 0$. *Source:* From the *Journal of Rheology,* Copyright © 1985, The Society of Rheology. Reprinted by permission.

PDMS [208] shows a spiraling instability, whereas polystyrene (PS) and polyethylene (PE) produce a wavy [231] and a completely distorted stream [119], respectively. For some monodisperse polymers (PB, PI [253]) and high-molecular-weight linear polyethylene (high-density polyethylene, HDPE [30]), the flow rate becomes discontinuous at a critical value of the shear stress (Figure 6.10), shooting material from the exit of the capillary (spurt instability). This instability occurs if the flow is driven by a constant pressure mechanism. If the flow is driven at a constant flow rate, wildly oscillating pressure drops are observed (Figure 6.11). Because the flow rate Q, and hence the shear rate $\dot{\gamma}$, is not stable in this flow regime, the data shown in Figure 6.10 are not represented as viscosity versus shear rate but rather as raw experimental data of apparent shear rate at the wall $4Q/\pi R^3$ versus wall shear stress $\Delta PR/2L$, where ΔP is the pressure drop, R and L are the radius and length of the capillary, respectively, and Q is the volumetric flow rate. The definitions of apparent shear rate at the wall and wall shear rate and other details of capillary rheometry can be found in Chapter 10.

The capillary instabilities described are grouped under the general phenomenological name *melt fracture*, and their causes are the subject of ongoing research [206, 65]. Implicated in at least some of the manifestations of melt fracture are inlet and exit effects, melt compressibility, wall slip, and elastic properties of the material. From the point of view of capillary measurements of viscosity, melt fracture limits the highest values of shear rate for which steady flow can be achieved.

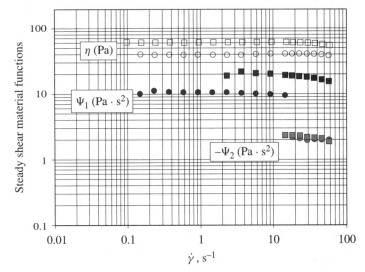

Figure 6.6 All three shear-flow material functions (η, Ψ_1, Ψ_2) as a function of shear rate for two polystyrene solutions; from Magda et al. [163]. ○, 28% polystyrene in dioctylpthalate (DOP); □, 28% polystyrene in tricresylphosphate (TCP). The polystyrene in these solutions is broadly distributed with $M_w = 47$ kg/mol. *Source:* Reprinted with permission, "Rheology, flow instabilities, and shear-induced diffusion in polystyrene solutions," J. J. Magda, C. S. Lee, S. J. Muller, and R. G. Larson, *Macromolecules*, **26**, 1696–1706 (1993). Copyright © 1993, American Chemical Society.

6.1.3 MATERIAL EFFECTS

The most significant factors that influence measured steady shear-flow properties are molecular composition and structure. A great deal of research has been conducted to correlate composition and structure with rheological properties. In fact, industrial polymeric-materials research is often focused on modifying materials chemically or physically to produce desired rheological properties. A detailed discussion of molecular differences among polymers can be found in the literature [75, 77, 97]. Beyond chemical composition, the most significant properties that affect rheology are molecular weight and molecular-weight distribution (MWD) as well as molecular architecture.

6.1.3.1 Molecular Weight and Molecular-Weight Distribution

The effect of molecular weight can best be appreciated by examining the now classic plot by Berry and Fox [21] of zero-shear viscosity η_0 versus molecular weight for monodisperse polymer melts (Figure 6.12). For a wide range of materials, η_0 increases proportionally to the first power of molecular weight below M_c, the critical molecular weight for entanglement. Above M_c the dependence of viscosity on molecular weight changes considerably, $\eta_0 \propto M^{3.4-3.5}$. The low-molecular-weight regime is sometimes called the *Rouse regime* [138] since a molecular and constitutive model for short-chain polymer dynamics by Rouse predicts the observed M^1 dependence of viscosity as well as many other aspects of short-chain behavior [138]. No purely theoretical model predicts the $M^{3.4}$ dependence above M_c, but the Doi–Edwards model comes close [70], predicting $\eta_0 \propto M^3$.

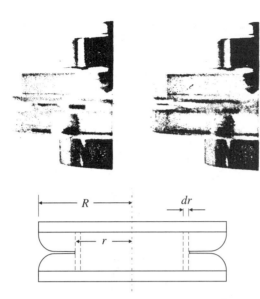

Figure 6.7 Edge fracture of a poly-dimethylsiloxane melt in a cone-and-plate rheometer; from Hutton [113]. At high rates (7.2 s^{-1}) the meniscus at the edge draws in. *Source:* From "Fracture and secondary flow of elastic liquids," J. F. Hutton, *Rheologica Acta*, **8**, 54–59 (1969), Figure 2. Copyright © 1969, Springer-Verlag.

Figure 6.8 Conical instabilities in a cone-and-plate flow ($\dot{\gamma} = 6.25$ s^{-1}) of an 11 wt % solution of polystyrene in decalin; from Kulicke et al. [134]. Conical vortices that disturb the edges of the sample are observed at high rates. *Source:* "Visual observation of flow irregularities in polymer solutions at theta-conditions," by W. M. Kulicke, H. E. Jerebiern, H. Kiss and R. S. Porter, *Rheologica Acta*, **18**, 711–716 (1979), Figure 2. Copyright © 1979, Springer-Verlag.

The existence of the two regimes of molecular-weight dependence of viscosity is one of the strongest pieces of evidence for the existence of entanglements in polymers (Figure 6.13). In the entanglement picture, below a critical molecular weight M_c, polymer chains are too short to knot up and entangle. They have high viscosities, but doubling their molecular weight only doubles the viscosity ($\eta \propto M^1$). Above a threshold molecular

Figure 6.9 Surface of extrude from a capillary rheometer exhibiting sharkskin and wavy-flow melt fracture instabilities; from Pomar et al. [213]. The sample is a solution of 25 wt % octadecane, 75 wt % linear low-density polyethylene (M_w = 114 kg/mol, M_w/M_n = 3.9). (a) Early in the sharkskin region. (b) Just prior to the transition to wavy flow. (c) Just after the transition to wavy flow. *Source:* Reprinted from *Journal of Non-Newtonian Fluid Mechanics*, **54**, G. Pomar, S. J. Muller, and M. M. Denn, "Extrudate distortions in linear low-density polyethylene solutions and melt," 143–151, Copyright © 1994, with permission from Elsevier Science.

(a)

(b)

(c)

weight, knots and entanglements can exist among the polymer chains, and the viscosity increases dramatically compared to unentangled chains. In this regime if the molecular weight is doubled, the viscosity increases by a factor of $2^{3.4} \approx 10$. Other properties, such as the diffusion coefficient, also show a change in molecular-weight dependence at M_c [189].

Molecular-weight distribution influences the shape of the viscosity function, as illustrated in Figure 6.14. The transition between the constant zero-shear portion of the curve and the shear-thinning portion is more abrupt and occurs at a higher shear rate for the narrow molecular-weight-distribution material than for the broadly distributed material. Increasing M_w/M_n also has the effect of decreasing the slope of the shear-thinning region of $\eta(\dot{\gamma})$. Narrow MWD polymers are known to show melt-flow instabilities more easily than their broad MWD counterparts. This stabilization effect is attributed to the wider distribution of relaxation times and processes present in the broadly distributed material. A material with a wide distribution of relaxation times has more options of how to respond to the imposed stresses in a flow and thus can avoid instability. The stabilizing effect of a broad molecular-weight distribution can be seen in the polybutadiene spurt data of Vinogradov (Figure 6.15) [252], where the instability disappears as polydispersity, measured by M_w/M_n, increases.

6.1.3.2 Chain Architecture

Chain architecture refers to the physical arrangement of monomeric building blocks in a polymer material. Up until now we have been concentrating on linear homopolymers—chains in which the monomers are arranged sequentially. Many industrially important polymers are branched, however. Branching itself can have several forms (Figure 6.16), including long-chain branching, short-chain branching, radial branching (star polymers), and dense

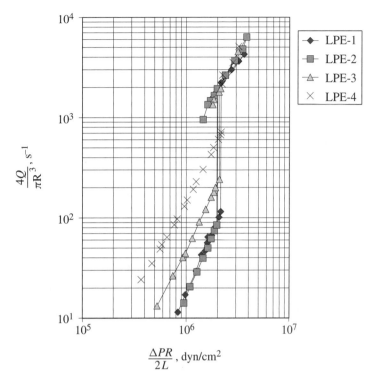

Figure 6.10 Spurt instability in a linear polyethylene at 160°C; replotted from Blyler and Hart [30]. At a critical wall shear stress $\Delta PR/2L = 2 \times 10^6$ dyn/cm², the apparent shear rate $4Q/\pi R^3$ increases abruptly, and polymer spurts from the capillary. See the figure for a definition of the symbols used. Materials are linear polyethylenes: LPE-1 melt index (MI) = 0.1, M_w unknown; LPE-2 MI = 0.2, M_w = 200 kg/mol; LPE-3 MI = 0.9, M_w = 131 kg/mol; LPE-4 MI = 5, M_w = 79 kg/mol. See Glossary for a definition of melt-flow index. *Source:* From *Polymer Engineering and Science,* Copyright © 1970, Society of Plastics Engineers. Reprinted by permission.

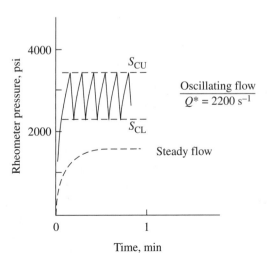

Figure 6.11 Pressure fluctuations in the capillary flow at 190°C of a high-density polyethylene (HDPE); from Lupton and Regester [160]. The physical properties of the HDPE are $\rho = 0.952$ g/cm³, melt index = 1. S_{CU} and S_{CL} are the upper and lower critical wall stresses (wall stress = $\Delta PR/2L$) between which the pressure fluctuates, and Q^* is the apparent shear rate, $Q^* = 4\times$flow rate$/\pi R^3$. *Source:* From *Polymer Engineering and Science,* Copyright © 1965, Society of Plastics Engineers. Reprinted by permission.

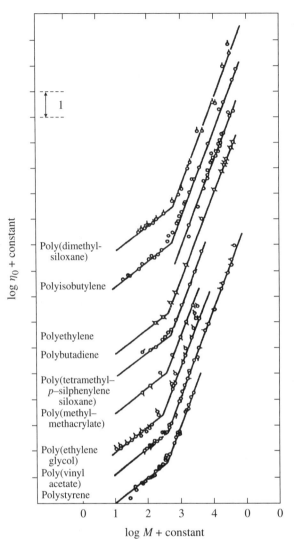

Figure 6.12 Effect of molecular weight on the measured zero-shear viscosity for a variety of polymers; from Berry and Fox [21]. Below a critical molecular weight M_c, $\eta_0 \propto M^1$; at higher molecular weights $\eta_0 \propto M^{3.4-3.5}$. *Source:* From "The viscosity of polymers and their concentrated solutions," G. C. Berry and T. G Fox, *Advances in Polymer Science*, **5**, 261–357 (1968), Figure 1. Copyright © 1968, Springer-Verlag.

branching upon branches, known as hyperbranching. A type of material that essentially is composed entirely of branches is the dendrimer. Dendrimers are a new kind of polymer whose structures resemble fractals—the molecule emanates from a multifunctional center, and two repeat units are added to each of the end-group sites. These form the next generation of sites for subsequent branching. Chemical assembly of the structure proceeds layer after layer, and a very dense molecule results [242, 71] (Figure 6.16*e*).

Branching tends to affect the shape of the viscosity versus shear-rate curve in the same way as does broadening the molecular-weight distribution. More significantly, long-chain branching has a strong effect on the zero-shear viscosity (Figure 6.17). Long-chain branching adds considerably to the viscosity of a polymer at low shear rates because of the increase in relaxation times caused by the branches. Polymers tend to relax by motion along

$M < M_c$ $M > M_c$

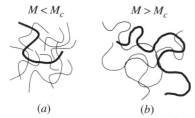

(a)   (b)

Figure 6.13 Schematic of (a) unentangled and (b) entangled polymers. Unentangled polymers are able to rapidly relax interactions with neighboring molecules. Entangled polymers form knots and are slowed down in their ability to relax interactions with their neighbors.

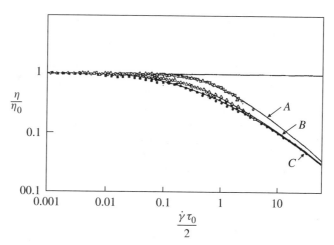

Figure 6.14 Effect of molecular weight distribution on viscosity; from Uy and Graessley [250a]. Shown are viscosity versus shear-rate master curves for poly(vinyl acetate) concentrated solutions in diethyl phthalate. Polymers of different M_w values were shifted empirically using the zero-shear viscosity η_0 and a time constant $\tau_0/2$ to produce the master curves for comparison. The different curves represent samples with different molecular weight distributions: A—$M_w/M_n = 1.09$; B—$M_w/M_n = 2.0$; C—branched. *Source:* Reprinted with permission, "Viscosity and normal stresses in poly(vinyl acetate) systems," W. C. Uy and W. W. Graessley, *Macromolecules*, **4**, 458–463 (1971). Copyright © 1971, American Chemical Society.

their backbones (Figure 6.18). This motion is hindered by branch points, and the relaxation time increases dramatically. In addition, these long relaxation times introduce a dependence on the shear history into rheological behavior as follows [96]. Once a branched polymer has been processed for some time, it adopts some chain configurations that take a very long time to randomize. This hysteresis can introduce irreproducibility into rheological testing and into performance properties.

At high shear rates the viscosity of a branched system is lower than that of a linear system of the same molecular weight (Figure 6.19). This is attributed to the more compact structure of the branched system. At high rates both linear and branched chains disentangle and flow past each other as whole units, and the compact structure of branched polymers reduces the interactions among molecules and reduces viscosity (Figure 6.18). The zero-shear first normal-stress difference Ψ_1^0 also increases dramatically with the introduction of long-chain branching [61]. Branched polymers are preferred in processes with a highly elongational character such as film blowing, where long-chain branching increases extension-thickening

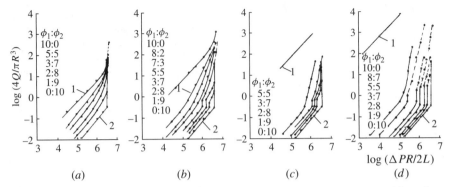

Figure 6.15 Effect of molecular-weight distribution on the spurt instability in capillary flow of polybutadiene melts; from Vinogradov [252]. The flow curves shown ($4Q/\pi R^3$ versus $\Delta PR/2L$; see Chapter 10 for details) are for blends of polybutadiene of molecular weight $M_2 = 240$ kg/mol blended with polybutadiene of molecular weights $M_1 = $ (a) 76, (b) 38, (c) 20, and (d) 8.7 kg/mol. Mass ratios $\phi_1 : \phi_2$ are indicated. *Source:* From "Critical regimes of deformation of liquid polymeric systems," by G. V. Vinogradov, *Rheologica Acta*, **12**, 357–373 (1973), Figure 13. Copyright © 1973, Springer-Verlag.

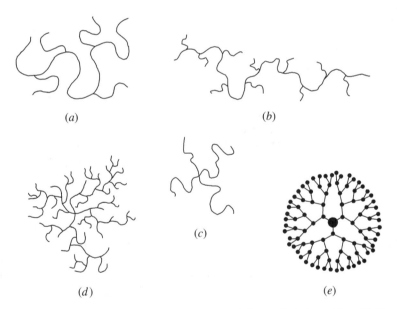

Figure 6.16 Various branched polymeric architectures. (a) Long-chain branching. (b) Short-chain branching. (c) Star polymers. (d) Hyperbranching. (e) Dendrimers.

behavior (the tendency for the extensional viscosity to increase as the rate of extension increases; see Section 6.3) and thus stabilizes the blown film.

Short-chain branches actually reduce viscosity compared to a linear polymer of the same molecular weight [61]. This is because the molecule extends less far out into the melt and therefore has fewer interactions with its neighbors. This is only true when the length

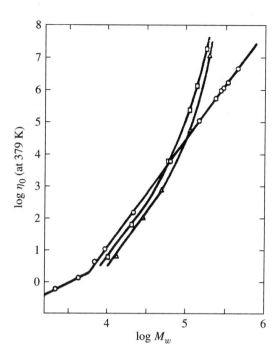

Figure 6.17 Effect of branching on zero-shear viscosity η_0, that is, at low shear rates; from Kraus and Gruver [131]. Curves are for ○ linear, □ three-armed star-branched (linear with one long side chain), and △ four-armed star-branched polybutadienes at 379 K. At lower molecular weights the branched polymers have lower η_0; at very high molecular weights, however, the viscosity of the branched polymers greatly exceeds that of the linear polymer. *Source:* From "Rheological properties of multichain polybutadienes," by G. Kraus and J. T. Gruver, *Journal of Polymer Science, Part A,* Copyright © 1965 by John Wiley & Sons, Inc. Reprinted by permission of John Wiley & Sons, Inc.

of the branches is less than the entanglement molecular weight of the polymer, however [61]. Hyperbranched materials, an outgrowth of the invention of the dendrimer, are new materials, and little is known of their rheological properties. They hold promise of having very useful chemical properties because of their high molecular weight and relatively low viscosity when compared to a linear polymer of the same molecular weight. Dendrimers are the most compact polymers of all. They exhibit much lower viscosities than linear polymers of comparable molecular weight and do not shear-thin [248] (Figure 6.20).

6.1.3.3 Mixtures and Copolymers

As mentioned at the start of this section, chemical composition is the most significant material trait that causes different polymers to exhibit distinct rheological properties. An in-depth discussion of the properties of various homopolymers is beyond our scope, but there are some general observations that we can make on the changes in rheological properties observed when chemical composition is changed by blending polymers with other polymers, with low-molecular-weight materials, or with inert fillers, or when a copolymer is formed. Many of these processes have signature effects on viscosity and other rheological functions.

It is common in the polymer processing industry to mix fillers such as carbon black or talc into polymers to produce a composite that has greater stiffness and greater resistance to flow at high temperatures than the unfilled polymer melt. In general, fillers have little effect on viscosity at high rates, but at low rates fillers can increase the viscosity by an order of magnitude. Figure 6.21 shows the effect on the viscosity of polypropylene of adding varying amounts of talc (hydrated magnesium silicate) as filler [40]. The magnitude of the

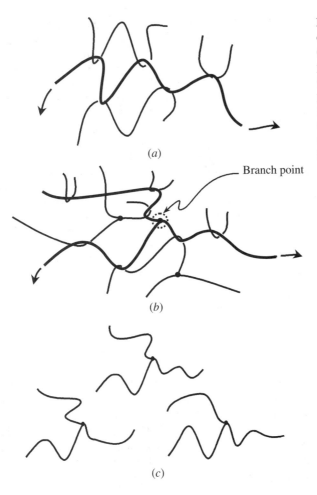

(a)

(b)

Branch point

(c)

Figure 6.18 Relaxation motions of linear and branched polymers. (*a*) Entangled linear: chains move principally along their contour. (*b*) Entangled branched: branch points retard motion along contour. (*c*) Disentangled branched: once disentangled (high $\dot{\gamma}$), branched and linear polymers flow more freely.

effect depends on the concentration of the filler. For the system shown, talc particles are believed to interact at low rates, forming a network-like structure that must be broken for flow. The extra stress it takes to break the structure is measured as an increase in viscosity at low rates. The first normal-stress difference is also increased by filler [237]. Even fillers that do not interact to form a network increase low-rate viscosity, as pointed out by Einstein [73, 162] and as observed in the data in Figure 6.22 on polypropylene mixed with glass beads.

Because the high-rate viscosity of filled systems is unaffected by the filler, processing at high rates is unchanged by the presence of the filler. In mold-filling operations such as injection molding, however, the flow into the cold mold can be quite slow. In this low-shear-rate process, the viscosity of the filled system will be appreciably higher than that of the unfilled system, requiring higher pressures and causing the process to take a longer amount of time for a filled polymer compared to an unfilled polymer. In addition, undesirable effects such as mold lines and surface structures are observed in products made with filled polymers [40]. These effects result from the slow relaxation times of filled compounds during mold filling.

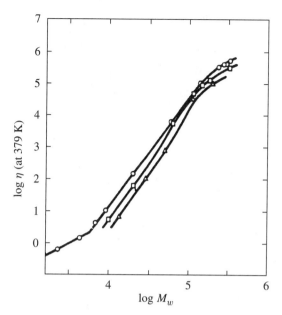

Figure 6.19 Effect of branching on viscosity at $\dot{\gamma} = 20$ s^{-1}; from Kraus and Gruver [131]. Symbols are defined in Figure 6.17. $\eta(\dot{\gamma} = 20$ s$^{-1})$ is lower for branched polybutadienes than for linear polybutadienes of the same molecular weight. *Source:* From "Rheological properties of multichain polybutadienes," by G. Kraus and J. T. Gruver, *Journal of Polymer Science, Part A*, Copyright © 1965 by John Wiley & Sons, Inc. Reprinted by permission of John Wiley & Sons, Inc.

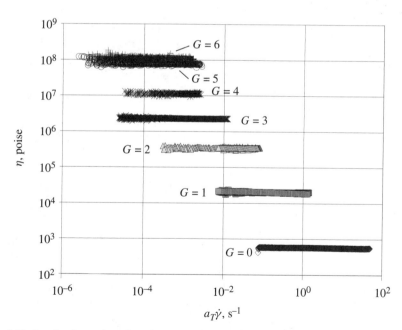

Figure 6.20 Steady shear viscosity of generations (G) 0 through 6 ethylenediamine (EDA)-core, poly(amidoamine) (PAMAM)-bulk dendrimers at 70°C; from Uppuluri [248]. Each generation represents an additional growth cycle of the dendrimer, that is, increasing molecular weight.

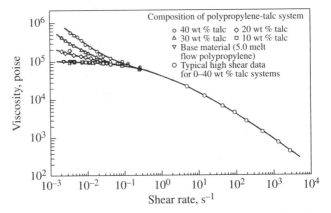

Figure 6.21 Viscosity versus shear rate at 200°C for talc-filled polypropylene; from Chapman and Lee [40]. Data at low rates were taken in a cone-and-plate rheometer; high-rate data were taken in a capillary rheometer. *Source:* From *SPE Journal,* Copyright © 1970, Society of Plastics Engineers. Reprinted by permission.

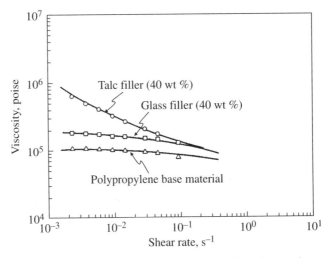

Figure 6.22 Viscosity versus shear rate for polypropylene and for polypropylene mixed with inert glass filler and interacting talc filler; from Chapman and Lee [40]. Although the absence of specific interactions reduces the viscosity of the glass-filled compared to the talc-filled polypropylene, the viscosity of the glass-filled system is still greater than that of the neat polymer. *Source:* From *SPE Journal,* Copyright © 1970, Society of Plastics Engineers. Reprinted by permission.

Polymer blending is a common technique for producing new rheological properties. Most polymers are incompatible, that is, when they are mixed together, the two constituent homopolymers phase-separate into domains. The sizes of the phase-separated domains range from 100 nm to several μm, depending on the mixing methods employed. The shear viscosity of one such incompatible blend is shown in Figure 6.23 for various temperatures. The viscosity of the homopolymer nylon is independent of the shear rate over

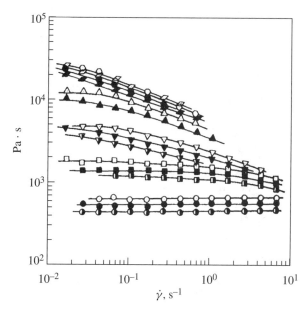

Figure 6.23 Viscosity η versus shear rate $\dot{\gamma}$ for blends of nylon and a branched polyolefin at various temperatures; from Chuang and Han [46]. Various blend compositions are shown. The nylon wt % of the blends varies as follows: ○, 100; □, 80; ▽, 60; ◔, 40; ◇, 20; △, 0. Temperatures (°C) are indicated by the shading of the symbols: left half filled, 220; open, 230; filled in, 240; right half filled, 250. *Source:* Reprinted with permission, "Rheological behavior of blends of nylon with a chemically modified polyolefin," H.-K Chuang and C. D. Han, *Advances in Chemistry: Polymer Blends and Composites in Multiphase Systems*, **206**, 171–183 (1984). Copyright © 1984, American Chemical Society.

the experimental range explored (○). This material is mixed with a branched polyolefin that shear-thins at higher rates and which has a much higher viscosity (△). The viscosity curves of the nylon-rich blends of these two materials lie inbetween the curves of the two homopolymers (□, ▽). The nylon-lean blends, however, exhibit a viscosity that is shear-thinning everywhere and that is higher than the viscosity of either blend component (◔, ◇). This viscosity enhancement is caused by the filler effect of the stiffer polyolefin domains, which are quite large at these low wt % nylon compositions. Viscosities of all of the blends depend on temperature.

The first normal-stress differences measured for the same incompatible blends are shown in Figure 6.24 as a function of shear stress. Unlike viscosity, N_1 is nearly independent of temperature for these blends. Note also that the curves for N_1 for all blend compositions appear in sequence between the curves for the two homopolymers, indicating that the filler effect does not influence the first normal-stress difference and hence the elasticity of these blends. Note that if these data were plotted as the material function $\Psi_1 = |N_1|/\dot{\gamma}_0^2$ instead of as bare N_1, the blends would be shear-thinning. Note also that the opposite-sign convention for stress was used in the reporting of these data.

As noted earlier, physical blending of two polymers usually results in a phase-separated morphology. The detailed structure of a phase-separated blend depends on the conditions under which the blend was mixed and on the thermal history of the blend. Another, more controlled way of combining the properties of two polymers is to create a block copolymer. A block copolymer is a polymer that has long sequences or blocks of one type of polymer connected to long sequences of a second type of polymer.

Block copolymers are important industrial materials since at room temperature they are tough and elastic, but at high temperatures they flow easily [4, 8]. They are also important

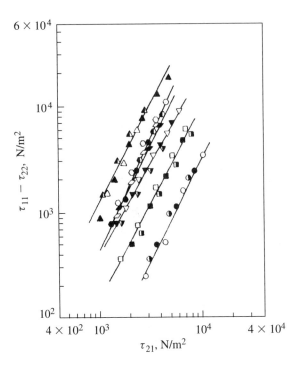

Figure 6.24 First normal-stress difference $N_1 = \tau_{11} - \tau_{22}$ versus shear stress τ_{21} for blends of nylon and a branched polyolefin at various temperatures; from Chuang and Han [46]. Various blend compositions are shown. The symbols are defined in Figure 6.23. *Source:* Reprinted with permission, "Rheological behavior of blends of nylon with a chemically modified polyolefin," H.-K Chuang and C. D. Han, *Advances in Chemistry: Polymer Blends and Composites in Multiphase Systems*, **206**,171–183 (1984). Copyright © 1984, American Chemical Society.

components in adhesives. The steady shear viscosity of a (*b*)polystyrene–(*b*)polybutadiene–(*b*)polystyrene (SBS) triblock copolymer versus shear rate is shown in Figure 6.25. Block copolymers are shear-thinning, but the shear-thinning seen for SBS occurs at much lower shear rates than was seen for linear homopolymers. The shear-thinning in block copolymers is caused by their microstructure. For industrially important block copolymers such as the SBS shown in Figure 6.25, the two constituent homopolymers, polystyrene and polybutadiene, are not thermodynamically compatible, that is, blends of these two homopolymers phase separate. In a block copolymer, however, the covalent bonds that link the blocks of styrene and butadiene together prevent phase-separation They do not produce a homogeneous polymer melt, however. Rather, SBS microphase-separates— the blocks of polystyrene find like blocks on neighboring molecules and form small microdomains of polystyrene that are present both at room temperature and at elevated temperatures, where the polystyrene is molten (Figure 6.26). The same happens for the polybutadiene blocks. The extra energy it takes to disrupt this thermodynamic microstructure accounts for the observed high viscosity at low rates for microphase-separated SBS (Figure 6.25).

Suspensions of solids and liquids are an important class of fluids since they include inks, processing fluids, and many foods. We will not give a review of the extensive literature on suspensions (see Macosko [162, chap. 10], as well as [218, 112]), except to indicate that concentrated suspensions of small particles can exhibit shear-thickening behavior in steady shearing flows [178, 98] (Figure 6.27). The data in Figure 6.27 show that at low shear rates the TiO$_2$ suspensions studied shear-thin, but at high shear rates the high-concentration suspensions shear-thicken, that is, the viscosities of the suspensions increase with the shear

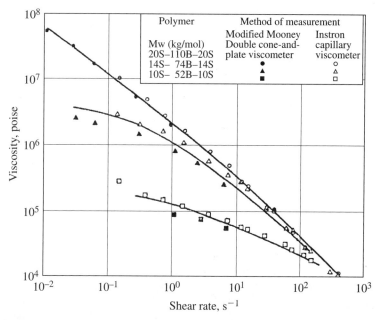

Figure 6.25 Steady shear viscosities versus shear rate of three (*b*)polystyrene–(*b*)polybutadiene–(*b*)polystyrene (SBS) triblock copolymers at 175°C; from Holden et al. [111]. Block molecular weights of each copolymer are also listed. *Source:* From "Thermoplastic elastomers," by G. Holden, E. T. Bishop and N. R. Legge, *Journal of Polymer Science, Part C*, Copyright © 1969 by John Wiley & Sons, Inc. Reprinted by permission of John Wiley & Sons, Inc.

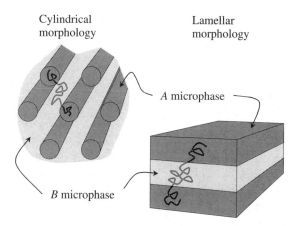

Cylindrical morphology

Lamellar morphology

A microphase

B microphase

Figure 6.26 Schematic of two morphologies adopted by microphase-separated block copolymers. *ABA* triblock copolymer chains are shown arranging their configurations so that the *A* blocks group with other *A* blocks to form a distinct microphase. The same holds for the *B* blocks. The shapes of the microdomains are primarily controlled by the volume fractions of *A* and *B* in the block copolymer.

rate. In suspensions, shear-thickening is associated with the presence of volume dilatancy, that is, the tendency of the volume of the suspension to increase with shearing. Metzner and Whitlock [178] showed that the onset of rheological dilatancy (shear-thickening) occurs after the onset of volumetric dilatancy (volume expansion). Shear-thickening has also been seen in some polymer solutions [146] where the mechanism must be different.

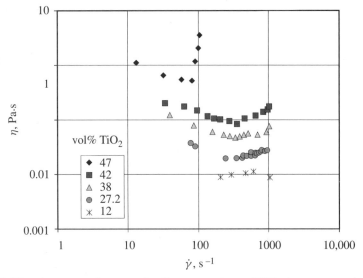

Figure 6.27 Viscosity versus shear rate for five suspensions of TiO_2 in water; recalculated and replotted from Metzner and Whitlock [178]. The diameters of the TiO_2 particles were between 0.2 and 1 μm, and the data were taken on a Couette (cup-and-bob) rheometer. *Source:* From the *Transactions of the Society of Rheology,* Copyright © 1958, The Society of Rheology. Reprinted by permission.

6.1.4 TEMPERATURE AND PRESSURE EFFECTS

Steady shear viscosity is a strong function of temperature, as seen for polybutadiene in Figure 6.28 [203]. For a temperature increase of 125°C the zero-shear viscosity of this polybutadiene drops by a factor of 25. For many liquids, including polymers, the temperature dependence of viscosity is exponential:

$$\eta_0 = A e^{B/T} \tag{6.1}$$

where η_0 is the zero-shear viscosity, T is absolute temperature in kelvins, and A and B are constants that vary from material to material [180].

The variation of viscosity with shear rate complicates the consideration of the temperature dependence of viscosity. The curves at different temperatures have similar shapes, however, and often can be represented more compactly by shifting the data vertically until the zero-shear values line up, and subsequently by shifting horizontally on the shear-rate axis until the curves superimpose. This process is called time–temperature superposition, and it has a theoretical basis that is discussed in depth in Section 6.2.1.3.

Pressure also has an effect on viscosity [172] (Figure 6.29); viscosity increases with increasing pressure due to the increased frictional forces among molecules [172, 262]. The usual form for the pressure dependence of viscosity is also exponential:

$$\eta_0 = K e^{aP} \tag{6.2}$$

where η_0 is the zero-shear viscosity, P is pressure, and K and a are constants [162, 61]. Although the effect of pressure on viscosity is not as large as the effect of temperature,

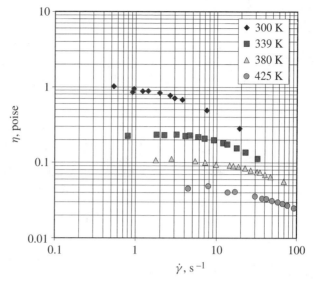

Figure 6.28 Viscosity versus shear rate at four different temperatures for a polybutadiene melt. Raw capillary data from Gruver and Kraus [102] corrected for nonparabolic velocity profile (Rabinowitsch correction; see Chapter 10) and plotted. $M_w = 145$ kg/mol, $M_w/M_n = 1.1$. From "Rheological properties of polybutadienes prepared by n-butyllithium initiation," by J. T. Gruver and G. Kraus, *Journal of Polymer Science, Part A*, Copyright © 1964 by John Wiley & Sons, Inc. Reprinted by permission of John Wiley & Sons, Inc.

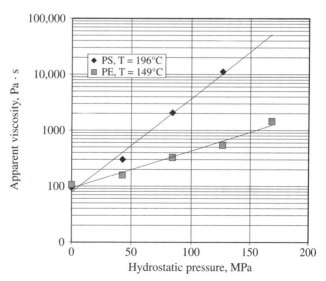

Figure 6.29 Apparent viscosity of polystyrene and branched polyethylene as a function of pressure; replotted from Maxwell and Jung [172]. Use of the term "apparent viscosity" reflects that the Rabinowitsch correction has not been applied; see Chapter 10. *Source:* Reprinted from *Modern Plastics*, March, 1957, pp. 174–182, 276, a publication of the McGraw-Hill Companies, Inc.

pressure effects can be important in practical flows because pressures in excess of 50 MPa (7000 psig; extruders) or even 100 MPa (14,000 psig; injection molders [236]) can develop in processing equipment. In capillary viscometers a pressure gradient is used to produce the flow in which viscosity is measured (see Chapter 10), and as a consequence, viscosity can vary along the length of the capillary. This effect is small at modest pressures (less than 1 kPa [145]), but it can become significant for measurements in long dies. The magnitude of the viscosity–pressure effect varies with the type of polymer under consideration. For polyethylene the addition of a hydrostatic pressure of 172 MPa (25,000 psig) causes viscosity to increase by between a factor of 3 and 10. For polystyrene, however, such a pressure increases viscosity by a factor of 100 [262] (Figure 6.29).

In summary, the steady shear rheological properties vary widely with material composition, molecular weight, and temperature. Significant but of lesser importance are the influences of molecular-weight distribution, pressure, and the presence of a filler. We see then, even when considering the simplest rheological measurements (steady shear material functions), that the situation is complex. This is one of the challenges of the practice of rheology.

We are not limited, of course, to studying steady shear flow. We now move on to discuss results from non-steady shear flows and, in the subsequent sections, elongational flows.

6.2 Unsteady Shear Flow

6.2.1 SMALL-STRAIN UNSTEADY SHEAR FLOW—SAOS AND STEP STRAIN

The steady shear data discussed in the last section give important information on large-strain behavior that is quite relevant in polymer processing flows. The small-strain region is also important, but in a different venue—as an analytical tool to quantify material relaxation characteristics. Two of the most widely used small-strain tests are the small-amplitude oscillatory shear (SAOS) and the step-strain tests; these will be discussed next.

The SAOS experiment introduced in Section 5.2.2.5 is used quite widely because of its accuracy. This experiment has the advantage that data can be taken over several cycles and averaged, increasing the reliability of the measurement. The step-strain experiment, which gives similar information when performed at small strains, depends on the measurement of rapidly changing transient stresses, making it inherently less accurate than SAOS. The step-strain experiment is more useful in the nonlinear regime (high strains) where the nonlinear property known as damping function $h(\gamma_0)$ may be obtained. Large-amplitude step strains and damping functions are discussed in detail in Section 6.2.2.

SAOS and small-strain step strains both measure rheological properties in what is called the linear viscoelastic limit. In the linear viscoelastic limit strains are additive, that is, the effect of two strains is just the sum of the effects of each strain imposed individually [see Equation (5.56) and the equations that follow]. As a result of this property of linear superposition, it is possible to formulate an accurate constitutive equation for materials in the linear viscoelastic limit. We will study this equation in Chapter 8. Because the constitutive equation is known, moreover, it is possible to measure one linear viscoelastic property, such as $G'(\omega)$ or $G''(\omega)$ in SAOS, and from the measured results we can calculate all other material functions in the linear viscoelastic limit, for example, $G(t)$ (small step strains), η_0 (low steady shear rates), and so on.

6.2.1.1 General Effects—Linear Polymers

SAOS data are sensitive to molecular properties. As an example, consider high-molecular-weight linear homopolymers [75]. These materials typically have curves of $G'(\omega)$ and $G''(\omega)$ as shown in Figure 6.30. At high frequencies, all homopolymers exhibit a glassy modulus of approximately $G' = 10^9$ Pa. At lower frequencies (the transition region) the storage modulus G' drops, reflecting the greater amount of relaxation that can occur when the polymer is deformed at a lower frequency. For entangled melts, a plateau appears at these intermediate frequencies at a modulus level of between 10^4 and 10^6 Pa, and a minimum is observed in the loss modulus G'' [75, 77]. The modulus level of the G' plateau, called the *rubbery plateau,* is known as the plateau modulus G_N^0. It is inversely proportional to the molecular weight between entanglements [75]; the breadth of the rubbery plateau is proportional to the molecular weight [196]. At the lowest frequencies, representing time scales long enough to permit the unraveling of entanglements, the moduli drop further, eventually to unmeasurable levels. At these lower frequencies, $G' \propto \omega^2$ and $G'' \propto \omega$. These powers of 2 and 1 can be predicted from the linear viscoelastic model mentioned earlier (see Chapter 8). The low-frequency region is called the *terminal zone* [75], and the frequency of the onset of the terminal zone is often associated with the inverse of the longest relaxation time of the material, $\lambda_1 = 1/\omega_x$. Again, this assignment can be justified by the use of the linear viscoelastic equation discussed in Chapter 8. The actual values of frequency that are classified as high, intermediate, and low depend on the temperature at which the moduli are measured. A more extensive discussion of the shape of linear viscoelastic rheological functions for a wide variety of polymers can be found in Ferry [75, chap. 2].

As mentioned earlier, the constitutive equation for the linear viscoelastic limit is known, and therefore rheological data in the linear viscoelastic region can be converted from one type of material function to another. As an example, the $G(t)$ curve calculated for the data of Figure 6.30 using the linear viscoelastic model of Chapter 8 is shown in Figure 6.31. In many ways $G(t)$ resembles a mirror image of $G'(\omega)$. This reciprocal relationship between t and ω is not unexpected since they are both measures of the time of relaxation.

Another reason for the importance of SAOS data is the observation [53] that for many materials, the steady-shear-viscosity versus shear-rate curve (η vs. $\dot{\gamma}$) has the same shape and values as the complex-viscosity versus frequency curve ($|\eta^*|$ vs. ω) if they are compared at $\dot{\gamma}(\text{s}^{-1}) = \omega$ (rad/s),

$$\text{Cox–Merz rule} \qquad \boxed{\eta(\dot{\gamma}) = |\eta^*(\omega)||_{\dot{\gamma}=\omega}} \qquad (6.3)$$

This empirical rule is known as the Cox–Merz relationship, and it has been seen experimentally to hold for many polymeric systems in the low- and intermediate-frequency regimes (Figure 6.32). Because of the existence of this rule, experimentalists often use the easier to measure linear viscoelastic $\eta^*(\omega)$ curves as a substitute for the true nonlinear-property viscosity $\eta(\dot{\gamma})$. This should be done with caution, since the validity of the Cox–Merz rule must be evaluated for each individual system. The Cox–Merz rule is known to hold for some

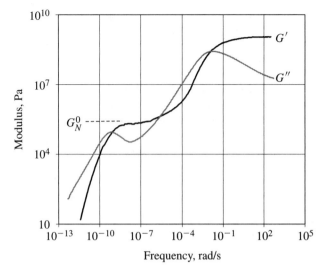

Figure 6.30 Master curve of the material functions $G'(\omega)$ and $G''(\omega)$ for an atactic linear polystyrene; data taken by D. J. Plazek and V. M. O'Rourke as plotted in Ferry [75]. $M_w = 600$ kg/mol, narrow molecular-weight distribution. The reference temperature is 100°C. *Source:* From *Viscoelastic Properties of Polymers*, J. D. Ferry, Copyright © 1980 by John Wiley & Sons. Reprinted by permission of John Wiley & Sons, Inc.

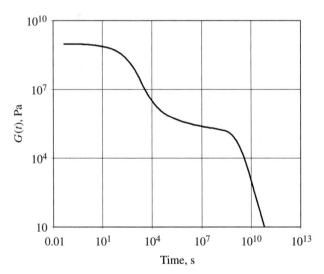

Figure 6.31 Master curve of the material function $G(t)$, the relaxation modulus, for an atactic linear polystyrene; calculated from the G' and G'' data in Figure 6.30 calculated by Ferry [75] from the data taken by D. J. Plazek and V. M. O'Rourke. The reference temperature is 100°C. *Source:* From *Viscoelastic Properties of Polymers*, J. D. Ferry, Copyright © 1980 by John Wiley & Sons. Reprinted by permission of John Wiley & Sons, Inc.

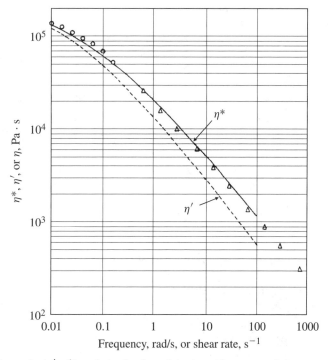

Figure 6.32 Dynamic (η', η^*) and steady shear (η) viscosities versus frequency and shear rate, respectively, for a low-density polyethylene (LDPE) at 175°C; from Venkatraman et al. [251]. ○, data taken in a cone-and-plate rheometer; △, corrected capillary data. *Source:* From *Polymer Engineering and Science,* Copyright © 1990, Society of Plastics Engineers. Reprinted by permission.

low-density (branched) polyethylenes [251] and for polystyrene [53] and to break down for some linear [249, 251] and branched [144, 119] polyethylenes, for block copolymers [87], and for rigid molecules [221].

6.2.1.2 Material Effects

As with steady shear properties, linear viscoelastic properties vary widely from material to material. For linear homopolymers, the basic shapes of the SAOS curves are as described in the last section, but the details (the value of G_N^0, the location of the curve on the frequency scale, the detailed shapes of the transitions, etc.) vary with the chemical composition [75]. The differences in linear viscoelastic (LVE) properties are such that SAOS and the related linear viscoelastic material functions can be used to help determine the chemical makeup of unknown systems by comparing curves of unknown samples to a database of known materials.

Molecular weight and molecular-weight distribution have important effects on linear viscoelastic properties. As mentioned earlier, the rubbery plateau in G' ($G' \approx 10^4 - 10^6$) only appears in entangled polymers, that is, only above a critical molecular weight, and its breadth increases with increasing molecular weight, as can be seen from the polystyrene and polybutadiene data in Figures 6.33 and 6.34. The minimum in G'' also broadens as

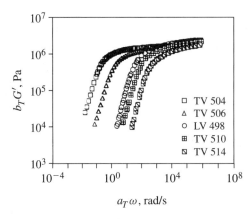

Figure 6.33 Master curves of storage modulus $b_T G'$ versus $a_T \omega$ for narrow-molecular-weight-distribution polybutadienes of various molecular weights; from Palade et al. [201]. b_T is the vertical shift factor, $b_T \equiv T_{ref} \rho_{ref} / T\rho$. Master curves were obtained from time–temperature superposition. $T_{ref} = 25°C$, M_w (kg/mol): TV504 = 464, TV506 = 229, LV498 = 85.8, TV510 = 70.6, TV514 = 51.3. *Source:* Reprinted with permission, "Time–temperature superposition and linear viscoelasticity of polybutadienes," L. I. Palade, V. Verney, and P. Attane, *Macromolecules*, **28**, 7051–7057 (1995). Copyright © 1995, American Chemical Society.

the molecular weight increases (Figures 6.35 and 6.36). Polydispersity tends to broaden the transition in G' from the plateau to the terminal region, as shown in the data for polystyrene in Figure 6.37 [168]. The minimum in G'' that appears for monodisperse polymers disappears with increasing polydispersity (Figure 6.38). Note that both G' and G'' are independent of polydispersity at high frequencies, that is, in the portion of the curve that reflects the transition to glassy behavior. The effect of long-chain branching on linear viscoelastic properties resembles the effect of broadening molecular-weight distribution. In addition, branching increases the longest relaxation time of the molecule and thus pushes the terminal zone (the range of frequency where $G' \propto \omega^2$ and $G'' \propto \omega$) to lower frequencies, often outside of the experimental window.

The effect of copolymerization can be seen clearly in the SAOS data shown in Figure 6.39b for random copolymers of styrene and butadiene. These data are in terms of log G' versus temperature rather than log G' versus $a_T \omega$, but, as discussed in the next section, log G' versus temperature exhibits the same features as log G' versus log $a_T \omega$ (in mirror image; see Section 6.2.1.3). The pure polybutadiene in this study has a glass transition temperature of about $-75°C$, as reflected by the transition of G' at this temperature from a value of 2×10^{10} dyn/cm^2 (10^9 Pa) to the plateau modulus ($< 10^7$ dyn/cm^2). As the amount of styrene monomer is increased in the copolymers, the G' transition shifts to higher temperatures, reflecting the increase in the glass transition temperature. The shifting of the temperature of the glass–rubber transition occurs progressively with the addition of styrene comonomer until the curve of homopolystyrene is obtained, reflecting the polystyrene glass-transition temperature of about 100°C. In contrast, for block copolymers that microphase-separate (see Section 6.1.3) a different effect is seen. Figure 6.39a shows the curves of storage modulus versus temperature for styrene-isoprene block copolymers. For compositions of the block copolymer with less than 50% polystyrene, glass transitions are observed at $-15°C$ and 110°C, reflecting homopolymer properties.[1] In between these two transition zones the

---

[1] The measurement of T_g in a temperature sweep is affected by the rate at which temperature is increased in the experiment. When the scan rate is high, the measured value of T_g is higher than the equilibrium value, which is associated with infinitely slow changes in temperature [75].

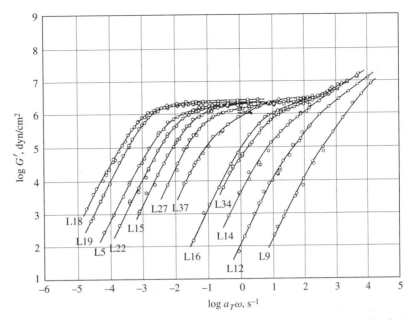

Figure 6.34 Master curves of storage modulus G' versus $a_T\omega$ for narrow-molecular-weight-distribution polystyrenes of various molecular weights; from Onogi et al. [196]. Master curves were obtained from time–temperature superposition. $T_{\text{ref}} = 160°C$, M_w (kg/mol): L18 = 581, L19 = 513, L5 = 351, L22 = 275, L15 = 215, L27 = 167, L37 = 113, L16 = 58.7, L34 = 46.9, L14 = 28.9, L12 = 14.8, L9 = 8.9. *Source:* Reprinted with permission, "Rheological properties of anionic polystyrenes. I. Dynamic viscoelasticity of narrow-distribution polystyrenes," Onogi, S., T. Masuda, and K. Kitagawa, *Macromolecules*, **3**, 109–116 (1970). Copyright © 1970, American Chemical Society.

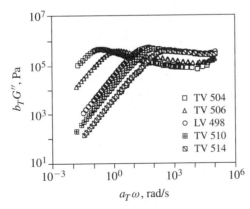

Figure 6.35 Master curves of loss modulus G'' versus $a_T\omega$ for narrow-molecular-weight-distribution polybutadienes of various molecular weights; from Palade et al. [201]. Master curves were obtained from time–temperature superposition. $T_{\text{ref}} = 25°C$; M_w listed in Figure 6.33. *Source:* Reprinted with permission, "Time–temperature superposition and linear viscoelasticity of polybutadienes," L. I. Palade, V. Verney, and P. Attane, *Macromolecules*, **28**, 7051–7057 (1995). Copyright © 1995, American Chemical Society.

value of G' is the plateau modulus G_N^0. The value of G_N^o in microphase-separated block copolymers varies with the styrene content because of the filler effect caused by polystyrene microdomains.

The linear viscoelastic properties of block copolymers are quite unique, as can be seen by looking at their SAOS moduli as a function of frequency and temperature (Figures 6.40

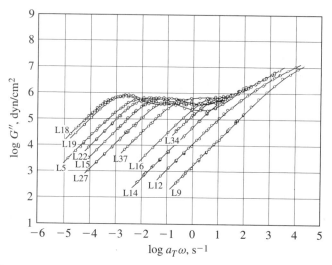

Figure 6.36 Master curves of loss modulus G'' versus $a_T\omega$ for narrow-molecular-weight-distribution polystyrenes of various molecular weights; from Onogi et al. [196]. Master curves were obtained from time–temperature superposition. $T_{\mathrm{ref}} = 160°C$; M_w listed in Figure 6.34. *Source:* Reprinted with permission, "Rheological properties of anionic polystyrenes. I. Dynamic viscoelasticity of narrow-distribution polystyrenes," Onogi, S., T. Masuda, and K. Kitagawa, *Macromolecules*, **3**, 109–116 (1970). Copyright © 1970, American Chemical Society.

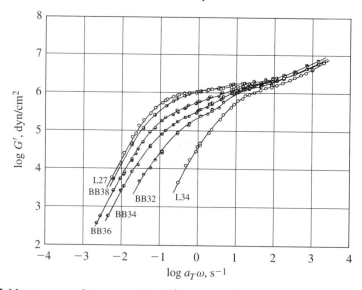

Figure 6.37 Master curves of storage modulus G' versus $a_T\omega$ for narrow-molecular-weight-distribution polystyrene melts and blends; from Masuda et al. [168]. $T_{\mathrm{ref}} = 160°C$. Mixtures of two molecular weights: L34 = 46.9 kg/mol and L27 = 167 kg/mol. Binary blends have compositions as follows: BB38 = 80 wt % L27, BB36 = 60 wt % L27, BB34 = 40 wt % L27, and BB32 = 20 wt % L27. *Source:* Reprinted with permission, "Rheological properties of anionic polystyrenes. II. Dynamic viscoelasticity of blends of narrow-distribution polystyrenes," T. Masuda, K. Kitagawa, T. Inoue, and S. Onogi, *Macromolecules*, **3**, 116–125 (1970). Copyright © 1970, American Chemical Society.

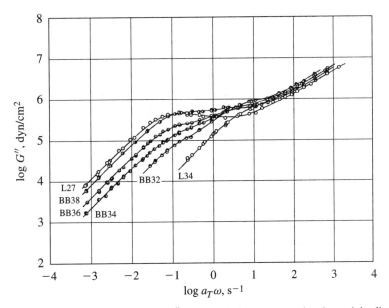

Figure 6.38 Master curves of loss modulus G'' versus $a_T \omega$ for narrow-molecular-weight-distribution polystyrene melts and blends from Masuda et al. [168]. $T_{ref} = 160°C$. Samples are identified in Figure 6.37. *Source:* Reprinted with permission, "Rheological properties of anionic polystyrenes. II. Dynamic viscoelasticity of blends of narrow-distribution polystyrenes," T. Masuda, K. Kitagawa, T. Inoue, and S. Onogi, *Macromolecules*, **3**, 116–125 (1970). Copyright © 1970, American Chemical Society.

and 6.41). In the terminal region (low frequencies) block copolymers do not show the scaling laws of $G' \propto \omega^2$ and $G'' \propto \omega$ that are characteristic of linear homopolymers, but instead show scaling laws of $G' \propto \omega^{0.5}$ and $G'' \propto \omega^{0.5}$ [47, 14]. This behavior is caused by microphase separation, the same thermodynamic effect that causes block copolymers to show high steady shear viscosities at low shear rates, as discussed in Section 6.1.3. For some lower-molecular-weight block copolymers it is possible to dissolve the microdomains at high temperature. When the microdomains dissolve, the block copolymer becomes a homogeneous material, and it exhibits a constant zero-shear viscosity in steady shear. The dynamic viscosity η' also approaches a constant value at low frequencies (Figure 6.42; $T \geq 146°C$). In the homogeneous phase the slopes of the curves of the low-frequency moduli increase from 0.5 to near 1, and then ultimately follow the homopolymer scaling rules, $G' \propto \omega^2$ and $G'' \propto \omega^1$ (see Figures 6.40 and 6.41). It is highly desirable to process block copolymers in the homogeneous state if one exists, since less stress is required to make the polymer flow in the homogeneous regime. Temperature has an important effect on the rheological properties of all systems, as discussed in the next section.

6.2.1.3 Temperature Effects—Time–Temperature Superposition

Linear viscoelastic properties, like steady shear properties, are a strong function of temperature. Figure 6.43 shows the dynamic compliance $J'(\omega)$ for a poly(n-octyl methacrylate) as a function of frequency over a wide temperature range. $J'(\omega)$ changes qualitatively in shape

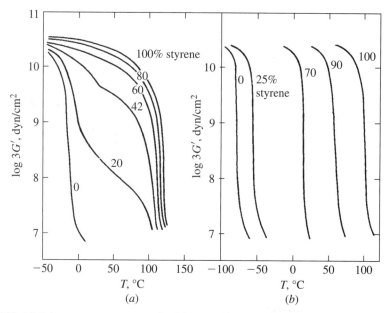

Figure 6.39 Modulus versus temperature for (*a*) styrene isoprene block copolymers and (*b*) styrene butadiene random copolymers as a function of styrene content; from Cooper and Tobolsky [52]. *Source:* From "Properties of linear elastomeric polyurethanes," by S. L. Cooper and A. V. Tobolsky, *Journal of Applied Polymer Science,* Copyright © 1966 by John Wiley & Sons, Inc. Reprinted by permission of John Wiley & Sons, Inc.

as well as quantitatively over almost four decades in magnitude as temperature changes by ≈ 145°C. The change in the rheological properties of polymers with temperature is due to several factors; the dominant effect is that relaxation times decrease strongly as the temperature increases. A second and less important effect is that the moduli associated with various relaxations in a polymer are proportional to absolute temperature. An important experimental observation that facilitates our understanding of the temperature dependence of rheological functions is that for many (but not all [76, 17, 209]) materials, all the relaxation times and moduli have the same functional dependence on temperature. This fact greatly simplifies how rheological material functions change with temperature and has led to the important time–temperature superposition principle, which we will now discuss in detail.

To understand the temperature effect on material functions, consider the SAOS moduli $G'(\omega)$ and $G''(\omega)$. These moduli are a function of the time scale of deformation through their dependence on frequency ω. They are also a function of temperature through various relaxation times λ_i and moduli parameters g_i that characterize the particular material response of the fluid being considered. In Chapter 8 we will derive the functional forms of $G'(\omega, \lambda_i, g_i)$ and $G''(\omega, \lambda_i, g_i)$ that result from a specific linear viscoelastic model known as the generalized Maxwell model. The functions that describe G' and G'' have an important property that affects the temperature dependence of the SAOS moduli: the frequency and relaxation times always appear together, that is, G' and G'' are a function of the product $\omega\lambda_i$ rather than of ω and λ_i individually.

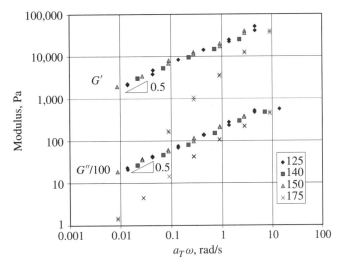

Figure 6.40 Linear viscoelastic moduli $G'(\omega)$ and $G''(\omega)$ for a polystyrene–polybutadiene–poly-styrene block copolymer (block M_w =7–43–7 kg/mol) [47]; from Chung and Gale. The data have been time–temperature shifted to a reference temperature of 125°C. The moduli at low frequency do not follow the power laws $G' \propto \omega^2$ and $G'' \propto \omega$ that are exhibited by linear polymers. Rather for microphase-separated block copolymers $G' \propto \omega^{0.5}$ and $G'' \propto \omega^{0.5}$. At 175°C the microphase structure is melted, $G' \propto \omega^{1.2}$ and $G'' \propto \omega^{0.84}$. *Source:* From "Newtonian behavior of a styrene-butadiene-styrene block copolymer," by C. I. Chung and J. C. Gale, *Journal of Polymer Science, Polymer Physics Edition*, Copyright © 1976 by John Wiley & Sons. Reprinted by permission of John Wiley & Sons, Inc.

To see how this observation impacts the general temperature dependence of rheological functions, we can define a function $a_T(T)$ which represents the temperature dependence of the relaxation times λ_i. Thus we can write $\lambda_i = \tilde{\lambda}_i a_T$, where $\tilde{\lambda}_i$ is not a function of temperature. We noted that G' and G'' are a function of the product $\omega \lambda_i = (a_T \omega) \tilde{\lambda}_i$. Thus if all of the temperature dependence of the SAOS moduli enters through the relaxation times, we can conclude that a plot of G' and G'' versus $a_T \omega$ will be independent of temperature.

This is approximately correct. There is another minor contribution to the temperature dependence of SAOS moduli that enters through the individual moduli g_i associated with each relaxation time. The temperature dependence of g_i is $g_i = \tilde{g}_i T \rho(T)$, where $\rho(T)$ is the polymer density, which is a weak function of temperature, and $\tilde{g}_i$ is not a function of temperature. The g_i enter into the functions for G' and G'' such that the product $T\rho$ can be factored out of the function (see Chapter 8), that is, $G'/\rho T$ and $G''/\rho T$ are functions of $a_T \omega$ only. (Note that T is in units of absolute temperature.)

Thus, all of the temperature dependence of G' and G'' can be suppressed if, instead of plotting these bare functions versus ω, the following reduced functions G'_r and G''_r are plotted:

$$G'_r \equiv \frac{G'(T) T_{\text{ref}} \rho_{\text{ref}}}{T \rho} \quad \text{versus} \quad a_T \omega \qquad (6.4)$$

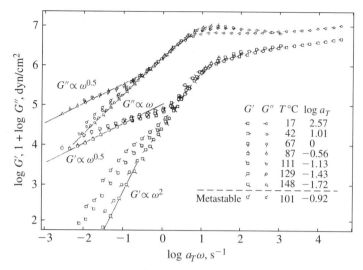

Figure 6.41 Linear viscoelastic moduli $G'(\omega)$ and $G''(\omega)$ for a 1,4-polybutadiene–1,2-polybutadiene block copolymer (block M_w =27.3–44.5 kg/mol) [14]; from Bates. The data have been time–temperature shifted to a reference temperature of 67°C. Near 100°C the microphase structure is melted, and at the highest temperature shown $G' \propto \omega^2$ and $G'' \propto \omega^1$. *Source:* Reprinted with permission, "Block copolymers near the microphase separation transition. 2. Linear dynamic mechanical properties," F. Bates, *Macromolecules*, **17**, 2607–2613 (1984). Copyright © 1984, American Chemical Society.

$$G''_r \equiv \frac{G''(T)T_{\text{ref}}\rho_{\text{ref}}}{T\rho} \quad \text{versus} \quad a_T\omega \tag{6.5}$$

where T_{ref} is a reference temperature, and ρ_{ref} is the density of the material at T_{ref}. Since J' and J'' are related to the inverses of G' and G'' (see Table 5.1), the reduced functions for compliance are given by

$$J'_r \equiv \frac{J'(T)T\rho}{T_{\text{ref}}\rho_{\text{ref}}} \quad \text{versus} \quad a_T\omega \tag{6.6}$$

$$J''_r \equiv \frac{J''(T)T\rho}{T_{\text{ref}}\rho_{\text{ref}}} \quad \text{versus} \quad a_T\omega \tag{6.7}$$

This technique is known as time–temperature superposition, or the method of reduced variables [75]. The plot in Figure 6.30 of G' and G'' versus frequency for polystyrene was obtained with the help of time–temperature superposition.

Time–temperature superposition is observed for a wide variety of materials. As an example, we can apply the time–temperature superposition principle to the data in Figure 6.43. The measurements of complex compliance $J'(\omega)$ reported in Figure 6.43 span the frequency range $20 < \omega < 3000$ Hz. We see that these curves have a definite relationship to one another, specifically that horizontal shifts will cause the curves to coincide. This is the equivalent of the data-reduction technique described before, as we can see by examining an arbitrary function $f(a_T\omega)$ plotted in the same way:

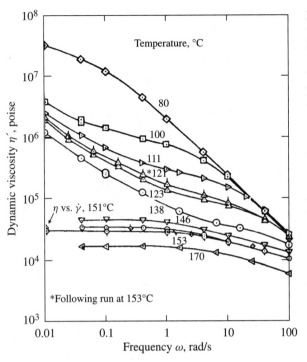

Figure 6.42 Dynamic viscosity η' versus frequency ω and steady shear viscosity η versus shear rate $\dot{\gamma}$ for a (b)polystyrene–(b)polybutadiene–(b)polystyrene triblock copolymer of block $M_w = 7(PS)$–$43(PB)$–$7(PS)$ kg/mol; from Gouinlock and Porter [94]. For η' various temperatures are shown as indicated; steady shear viscosity is given at a single temperature, 151°C. At 150°C the block copolymer goes through the microphase-separation transition, and both viscosities become independent of the rate of deformation. *Source:* From *Polymer Engineering and Science,* Copyright © 1977, Society of Plastics Engineers. Reprinted by permission.

$$\log f \text{ versus } \log (a_T\omega) = \log f \text{ versus } \log a_T + \log \omega \qquad (6.8)$$

On a log–log plot of f versus ω, the function $a_T(T)$ can be calculated by measuring the shift along the frequency scale $\log a_T$ required to bring a plot of $\log f(T)$ versus $\log \omega$ into superposition with the reference curve, $\log f(T_{\text{ref}})$ versus $\log \omega$. The results of such shifting of the data in Figure 6.43 is shown in Figure 6.44.

The function $a_T(T)$ has been widely studied for many polymers [75], and when time–temperature superposition is performed correctly (see cautions), a_T contains important information about the temperature dependence of material relaxation times. The function $a_T(T)$ usually takes on one of two functional forms, depending on the proximity of the experimental temperature range to the glass-transition temperature, T_g [75]. For temperatures within 100 K of the polymer's glass transition temperature, most polymers show Arrhenius dependence.

$$\begin{matrix} \text{Arrhenius} \\ \text{equation} \end{matrix} \qquad \boxed{a_T = \exp\left[\frac{-\Delta \bar{H}}{R}\left(\frac{1}{T} - \frac{1}{T_{\text{ref}}}\right)\right]} \qquad (6.9)$$

where $\Delta \bar{H}$ is the activation energy for flow, R is the ideal gas constant, T is the temperature in K, and T_{ref} is the reference temperature in K. For temperatures closer to T_g, a_T follows the Williams-Landel Ferry (WLF) equation [265].

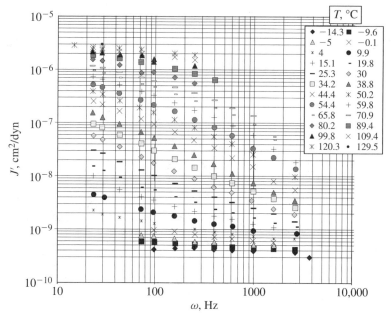

Figure 6.43 Linear viscoelastic storage compliance J' of poly(n-octyl methacrylate) as a function of frequency and temperature; replotted from Dannhauser et al. [56]. *Source:* From the *Journal of Colloid Science,* Copyright © 1958, Academic Press. Reprinted by permission.

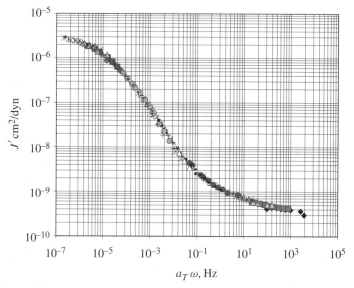

Figure 6.44 Master curve of linear viscoelastic storage compliance J' of poly(n-octyl methacrylate) at $T_{ref} = -14.3°C = 258.9$ K as a function of reduced frequency $a_T\omega$; recalculated from data in Figure 6.43 from Dannhauser et al. [56]. *Source:* From the *Journal of Colloid Science,* Copyright © 1958, Academic Press. Reprinted by permission.

$$\text{WLF equation} \qquad \boxed{\log a_T = \frac{-c_1^0 \, (T - T_{\text{ref}})}{c_2^0 + (T - T_{\text{ref}})}} \qquad (6.10)$$

where c_1^0 and c_2^0 are model parameters. Note that in the Arrhenius equation temperature must be expressed in absolute temperature units, K (kelvins).

For many materials, the constants c_1^0 and c_2^0 of the WLF equation have been found to have the universal values of $c_1^o = 17.44$ and $c_2^0 = 51.6\text{K}$ when $T_{\text{ref}} = T_g$ [75]. Alternatively c_1^0 and c_2^0 may be used as fitting parameters and can be calculated from a linear regression on a plot of $(T - T_{\text{ref}})/\log a_T$ versus $(T - T_{\text{ref}})$. The values of a_T used to produce the master curve in Figure 6.44 are shown in Figure 6.45 along with the fit of a_T versus T to the WLF equation with $c_1^0 = 13.53$ and $c_2^0 = 97.5$ K ($T_{\text{ref}} = -14.3°\text{C} = 258.9$ K).

Although time–temperature superposition (TTS) has a theoretical basis for some systems [27], it is also widely used for systems for which there is no theoretical justification. The empirical observation of successful time–temperature superposition has been found for G' and G'' as well as for other rheological functions, including steady shear viscosity, first normal-stress coefficient, dynamic compliance (as demonstrated before), and creep compliance [75, 26]. It was, in fact, the observation that such horizontal shifts could be used to collapse creep compliance data that led to the discovery of time–temperature superposition [147]. Note that since $\eta^* \equiv G^*/\omega$, $\eta' \equiv G''/\omega$, and $\eta'' \equiv G'/\omega$, the reduced complex viscosity functions are obtained by dividing the complex-viscosity-related functions by a_T in addition to applying the density–temperature factor used for moduli:

$$\eta_r^*(a_T \omega) = \frac{G^*(T) T_{\text{ref}} \rho_{\text{ref}}}{a_T \omega T \rho} = \frac{\eta^* T_{\text{ref}} \rho_{\text{ref}}}{a_T T \rho} \qquad (6.11)$$

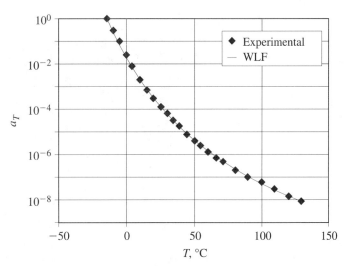

Figure 6.45 Shift parameters versus temperature for poly (*n*-octyl methacrylate) in Figure 6.44; $T_{\text{ref}} = -14.3°\text{C} = 258.9\text{K}$. Also included is the fit to the WLF equation; calculated from data in Dannhauser et al. [56]. *Source:* From the *Journal of Colloid Science,* Copyright © 1958, Academic Press. Reprinted by permission.

$$\eta_r'(a_T\omega) = \frac{G''(T)T_{\text{ref}}\rho_{\text{ref}}}{a_T\omega T\rho} = \frac{\eta'T_{\text{ref}}\rho_{\text{ref}}}{a_T T\rho} \tag{6.12}$$

$$\eta_r''(a_T\omega) = \frac{G'(T)T_{\text{ref}}\rho_{\text{ref}}}{a_T\omega T\rho} = \frac{\eta''T_{\text{ref}}\rho_{\text{ref}}}{a_T T\rho} \tag{6.13}$$

Steady shear viscosity also shifts in the same manner:

$$\eta_r(a_T\dot\gamma) = \frac{\eta(T)T_{\text{ref}}\rho_{\text{ref}}}{a_T T\rho} \tag{6.14}$$

Thus the time–temperature superposition principle predicts that one function, a_T, accounts for both the vertical shift in η [Equation (6.14)] and the horizontal shift $a_T\dot\gamma$. Figure 6.46 shows an example of time–temperature superposition applied to the steady shear viscosity data in Figure 6.28.

The time–temperature superposition concept is used principally to allow a large amount of data to be represented compactly in just two graphs, the master curve and the curve of shift factors versus temperature. Master curves are used also to infer rheological properties under conditions that are difficult or impossible to achieve experimentally, for example, to predict moduli at very low or very high frequencies. These types of extrapolations are somewhat dangerous, however, since they cannot be verified experimentally and depend on the validity of time–temperature superposition, that is, on the assumption that all relaxation

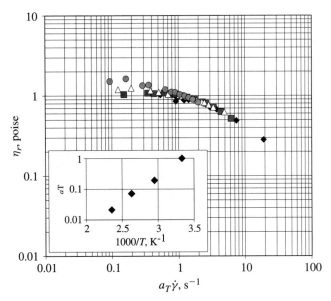

Figure 6.46 Master curve of steady shear viscosity η_r versus reduced shear rate $a_T\dot\gamma$ obtained through time–temperature superposition for the polybutadiene data shown in Figure 6.28; $T_{\text{ref}} = 300$ K. *Source:* From "Rheological properties of polybutadienes prepared by *n*-butyllithium initiation," by J. T. Gruver and G. Kraus, *Journal of Polymer Science, Part A*, Copyright © 1964 by John Wiley & Sons, Inc. Reprinted by permission of John Wiley & Sons, Inc.

times have the same temperature dependence. The allure of time–temperature superposition is often irresistible, however, since this method allows the rheological response of a fluid to be measured over a limited dynamic range (range of ω or $\dot{\gamma}$) at different temperatures, and then to be shifted to produce a master curve that can represent the rheological properties over a much wider dynamic range. An initial experimental range of four decades of frequency easily can be extended to eight or more decades [75].

When materials fail to shift smoothly as discussed, this indicates that the relaxation times associated with the material do not share the same temperature dependence. This is expected in phase-separated blends, for example, where if one component of the blend is nearing its glass-transition temperature, the relaxation times associated with that component would not be expected to exhibit the same temperature dependence as the relaxation times associated with the more fluid component. Deviations from the time–temperature superposition principle have been observed in semicrystalline polymers [75] and star-branched polymers [96].

Some simple linear homopolymers show deviations from time–temperature superposition. In polystyrene, Plazek [209] showed that shift factors a_T calculated from viscosity [Equation (6.14)], when applied to compliance, effectively shift data in the terminal zone (long times, high temperatures) while failing to reduce the data in the transition region (short times, low temperatures); see Figure 6.47. Plazek's values of recoverable compliance J_r can be shifted empirically in the transition region, but the shift factors obtained are different from those calculated from viscosity. This behavior results from differences in the temperature dependence of the relaxation processes of the terminal region and the glassy region. Penwell

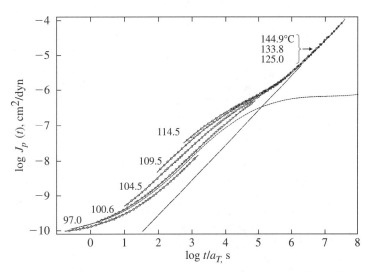

Figure 6.47 Shear creep compliance corrected for vertical shift, $J_p(t) = J(t)T\rho/(T_{\text{ref}}\rho_{\text{ref}})$, versus shifted time, t/a_T for nearly monodisperse polystyrene (46.9 kg/mol) reduced to $T_{\text{ref}} = 100°C$ by using values of a_T calculated from $a_T = \eta(T)/\eta(T_{\text{ref}})$; from Plazek [209]. Poor superposition is seen due to the proximity of T_g for the lower experimental temperatures. *Source:* Reprinted with permission from "Temperature dependence of the viscoelastic behavior of polystyrene," D. J. Plazek, *Journal of Physical Chemistry,* **69,** 3480–3487 (1965). Copyright © 1965, American Chemical Society.

et al. also saw deviations from time–temperature superposition in polystyrene steady shear viscosity below 190°C [203]. Deviations from time–temperature superposition were seen by Ferry and coworkers [76, 17] in polymethacrylates, again due to the presence of two sets of relaxation modes, each with its own temperature dependence. In the case of poly(n-butyl methacrylate), the entanglement spacing varies with temperature in the terminal region causing a breakdown in time–temperature superposition, while for polymethacrylates with longer side chains, specific interactions due to the side chains can cause some lack of superposition near the glassy zone [75]. Other pitfalls of time–temperature superposition have been discussed in the literature [210]. Materials that do not follow the time–temperature superposition principle are termed thermorheologically complex, and their behavior must be examined separately at every temperature and deformation rate of interest.

Finally, recall that the time–temperature superposition principle springs from the fact that the product $\lambda_i \omega$ appears in the linear viscoelastic functions rather than λ_i or ω individually. Thus, as we pointed out earlier, if the temperature dependence of λ_i is given by $\tilde{\lambda}_i a_T(T)$, plots of rheological functions versus $a_T \omega$ will be independent of temperature. Usually data are taken at a variety of temperatures and frequencies and are combined to create a master curve. Another possible approach is to take data at a fixed frequency and to vary the temperature. Since $a_T(T)$ is not known a priori, this approach would appear to be flawed, but the quantity $\log a_T$ is reasonably close to being a linear function of temperature (see Figure 6.45). Therefore plots of $\log G'$ and $\log G''$ versus T (not $\log T$) resemble slightly skewed versions of the mirror images (mirror image because high temperature is equivalent to low frequency) of $\log G'$ and $\log G''$ versus $\log \omega$. Runs of G' and G'' at fixed frequency versus temperature are straightforward to carry out and require a single test and no data manipulation. Comparing this procedure to the data taking required to produce a frequency master curve [runs of $G'(\omega)$ and $G''(\omega)$ at multiple temperatures, followed by time–temperature shifting], it is clear why the temperature sweep is a popular test, particularly in an industrial setting. In addition, when time–temperature superposition fails (e.g., incompatible blends), $\log G'$ and $\log G''$ versus T can still be measured, and the shapes obtained can be correlated with industrial variables. One field where this technique is used is adhesives development [219]. Figure 6.48 shows the effects on $\log G'$ versus T of adding different amounts of a plasticizer to a block copolymer. Research has shown that adhesive properties can be correlated with the height of the plateau modulus G_N^0 and to the position of the transition region on the temperature scale. Different proposed adhesive formulations could be evaluated by comparing the different shapes of temperature sweeps (such as those in Figure 6.48) conducted under the same conditions (temperature ramp rate, sample preparation conditions, etc.) [219].

In this section we have reviewed a wide range of material behavior in small-strain unsteady shearing flows. Now we move on to large-strain unsteady flows.

6.2.2 LARGE-STRAIN UNSTEADY SHEAR FLOW

Unsteady large-strain flows produce a breakdown of structure in a fluid. In the startup of steady shear of high-molecular-weight melts and concentrated solutions, the most striking feature is the appearance of an overshoot at high rates in both $\eta^+(t)$ and $\Psi_1^+(t)$. This is shown in Figures 6.49 and 6.50 for a concentrated polybutadiene solution. The highest values of η^+ and Ψ_1^+ are in the low-rate region, and these low-rate curves form linear viscoelastic

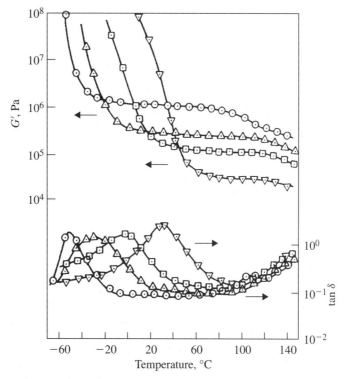

Figure 6.48 Storage modulus G' and $\tan\delta$ for a block copolymer (Kraton 1111, (*b*)polystyrene–(*b*)polyisoprene–(*b*)polystyrene block copolymer, Shell Development Company) and three mixtures with tackifier (Piccotac 95BHT, a hydrocarbon resin; Hercules, Inc.) showing that adding the tackifier reduces the plateau modulus and moves the glass transition to higher temperatures (transition zone shifts to the right): ○, Kraton 1111; △, 30 wt % tackifier; □, 50 wt % tackifier; ▽, 70 wt % tackifier; from Kim et al. [126]. *Source:* From "Viscoelastic behavior and order-disorder transition in mixtures of a block copolymer and a midblock-associating resin," by J. Kim, C. D. Han, and S. G. Chu, *Journal of Polymer Science, Polymer Physics Edition,* Copyright © 1988 by John Wiley & Sons, Inc. Reprinted by permission of John Wiley & Sons, Inc.

envelopes that contain the higher rate data. The maxima in η^+ and Ψ_1^+ occur at low strains and are attributed to structural breakdown. For complex materials (block copolymers [87]; suspensions, liquid crystalline polymers, dilute solutions [163]) other features are observed occasionally, including minima and secondary maxima. The time to reach steady state can be considerable for higher-molecular-weight materials. On modern rheological equipment an option is often provided of being able to linearly increase shear rate during a run. This allows the entire steady-state flow curve η versus $\dot{\gamma}$ to be measured at once. Care must be taken, however, that adequate time is allowed between tests for transients such as stress overshoots to die out and for steady state to be achieved.

When steady shearing stops, the stresses relax monotonically to zero. The curves for the cessation material functions $\eta^-(t)$ and $\Psi_1^-(t)$ for the same concentrated polybutadiene solution discussed before are shown in Figures 6.51 and 6.52. Both stresses relax more quickly as the initial shear rate increases, with Ψ_1^- decreasing more rapidly than η^-. Due

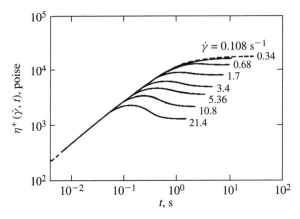

Figure 6.49 Shear-stress growth function as a function of time t at several shear rates for a concentrated solution of a narrowly distributed polybutadiene; from Menezes and Graessley [177]. $M_w = 350$ kg/mol; $M_w/M_n < 1.05$; concentration $= 0.0676$ g/cm^3 in Flexon 391, a hydrocarbon oil. *Source:* From "Nonlinear rheological behavior of polymer systems for several shear-flow histories," by E. V. Menezes and W. W. Graessley, *Journal of Polymer Science, Polymer Physics Edition,* Copyright © 1982 by John Wiley & Sons, Inc. Reprinted by permission of John Wiley & Sons, Inc.

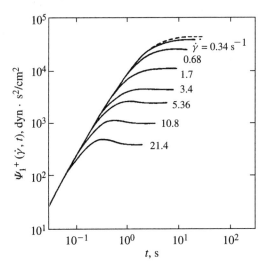

Figure 6.50 First normal-stress growth function as a function of time t at several shear rates for a concentrated solution of a narrowly distributed polybutadiene; from Menezes and Graessley [177]. $M_w = 350$ kg/mol, $M_w/M_n < 1.05$; concentration $= 0.0676$ g/cm^3 in Flexon 391, a hydrocarbon oil. *Source:* From "Nonlinear rheological behavior of polymer systems for several shear-flow histories," by E. V. Menezes and W. W. Graessley, *Journal of Polymer Science, Polymer Physics Edition,* Copyright © 1982 by John Wiley & Sons, Inc. Reprinted by permission of John Wiley & Sons, Inc.

to the difficulty involved in measuring Ψ_2 there are no reported start-up curves of this material function.

The creep compliance for polystyrene at various temperatures is shown in Figure 6.53. The data in this figure have been corrected for the vertical shift due to temperature using the relations in Section 6.2.1.3. Like viscosity and the linear viscoelastic functions, $J(t)$ is a strong function of temperature. For high temperatures or long times these data follow the time–temperature superposition principle and form a master curve with shift factors $a_T = \eta(T)T_{\text{ref}}\rho_{\text{ref}}/[\eta(T_{\text{ref}})T\rho]$ as discussed previously in Figure 6.47 (long times). The data at short times and low temperatures can be shifted when the recoverable compliance, rather than the total compliance, is considered. The recoverable compliance can be obtained by subtracting the quantity t/η_0 from the data [see Equation (5.80)]. The

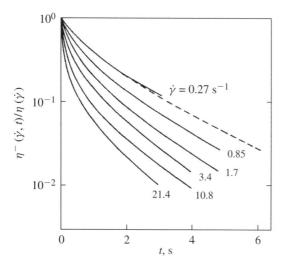

Figure 6.51 Normalized shear stress cessation function versus time since cessation of flow t at several shear rates $\dot{\gamma}$ for a concentrated solution of a narrowly distributed polybutadiene; from Menezes and Graessley [177]. $M_w = 350$ kg/mol; $M_w/M_n < 1.05$; concentration = 0.0676 g/cm^3 in Flexon 391, a hydrocarbon oil. *Source:* From "Nonlinear rheological behavior of polymer systems for several shear-flow histories," by E. V. Menezes and W. W. Graessley, *Journal of Polymer Science, Polymer Physics Edition,* Copyright © 1982 by John Wiley & Sons, Inc. Reprinted by permission of John Wiley & Sons, Inc.

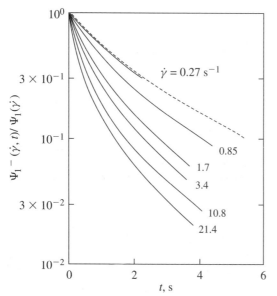

Figure 6.52 First normal-stress stress cessation function at several shear rates for a concentrated solution of a narrowly distributed polybutadiene; from Menezes and Graessley [177]. $M_w = 350$ kg/mol; $M_w/M_n < 1.05$; concentration = 0.0676 g/cm^3 in Flexon 391, a hydrocarbon oil. *Source:* From "Nonlinear rheological behavior of polymer systems for several shear-flow histories," by E. V. Menezes and W. W. Graessley, *Journal of Polymer Science, Polymer Physics Edition,* Copyright © 1982 by John Wiley & Sons, Inc. Reprinted by permission of John Wiley & Sons, Inc.

recoverable compliance curves $J_r(t)$ are shown in Figure 6.54. These curves shift quite successfully to produce a master curve (Figure 6.55) although, as mentioned earlier, the recoverable creep data give different values of a_T than those calculated from viscosity [see Equation (6.14); Figure 6.56]. The shape of the recoverable compliance master curve indicates that an increasing amount of deformation is stored (is recoverable) as the creep experiment progresses. The amount of stored deformation plateaus, however, as the flow reaches steady state. Recall that Newtonian fluids do not store energy—when the flow of a Newtonian fluid is stopped, no recovery is seen. Thus, the curve of elastic recovery for a polymeric fluid is an effective way of characterizing the elastic nature of the fluid.

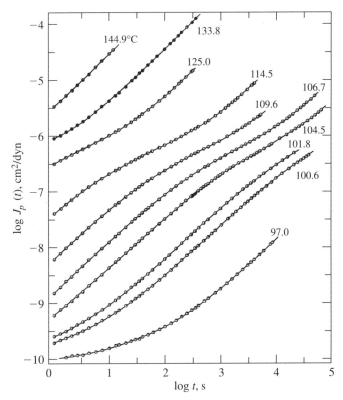

Figure 6.53 Shear creep compliance corrected for vertical shift, $J_p(t) = J(t)T\rho/(T_{ref}\rho_{ref})$, of nearly monodisperse polystyrene (46.9 kg/mol) at various temperatures; from Plazek [209]. *Source:* Reprinted with permission from "Temperature dependence of the viscoelastic behavior of polystyrene," D. J. Plazek, *Journal of Physical Chemistry*, **69**, 3480–3487 (1965). Copyright © 1965, American Chemical Society.

The large-strain step-strain experiment is perhaps the most important of the large-strain unsteady shear experiments. Data for a concentrated solution of polystyrene are shown in Figure 6.57. At low strain the data are independent of strain, and this is the linear viscoelastic limit discussed earlier. At higher strains the curves begin to be strain-dependent, but the shapes of the curves do not change much. The invariance of the shapes of $G(t, \gamma_0)$ with strain indicates that all the data have the same time dependence. If we hypothesize that time and strain can be separated as follows:

$$G(t, \gamma_0) = G(t)h(\gamma_0) \tag{6.15}$$

then on a log–log plot, curves of $\log G(t, \gamma_0)$ versus $\log t$ at different strains should be just shifted vertically from one another by an amount $\log h(\gamma_0)$,

$$\log G(t, \gamma_0) = \log G(t) + \log h(\gamma_0) \tag{6.16}$$

The composite curve resulting from such a shift is shown in Figure 6.58, and we see that it is quite successful for times greater than about 20 s. When the data for a material shift

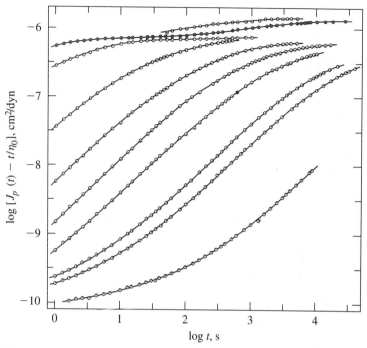

Figure 6.54 Recoverable shear creep compliance $J_p - t/\eta_0$ of nearly monodisperse polystyrene (46.9 kg/mol). Data have been corrected for vertical shift, $J_p(t) = J(t)T\rho/(T_{ref}\rho_{ref})$; from Plazek [209]. *Source:* Reprinted with permission from "Temperature dependence of the viscoelastic behavior of polystyrene," D. J. Plazek, *Journal of Physical Chemistry*, **69**, 3480–3487 (1965). Copyright © 1965, American Chemical Society.

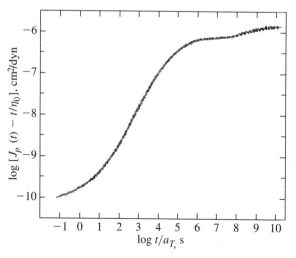

Figure 6.55 Master curve at 100°C of recoverable shear compliance versus time t of a nearly monodisperse polystyrene (46.9 kg/mol) calculated from the data in Figure 6.54; from Plazek [209]. *Source:* Reprinted with permission from "Temperature dependence of the viscoelastic behavior of polystyrene," D. J. Plazek, *Journal of Physical Chemistry*, **69**, 3480–3487 (1965). Copyright © 1965, American Chemical Society.

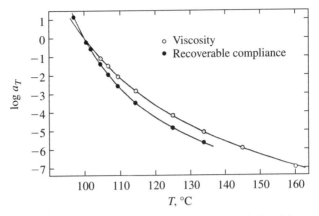

Figure 6.56 Shift factors as a function of temperature. ○, values calculated from viscosity [Equation (6.14)], ●, values used to produce the master curve of recoverable compliance in Figure 6.55 for monodisperse polystyrene (46.9 kg/mol; $T_{ref} = 100°C$); from Plazek [209]. *Source:* Reprinted with permission from "Temperature dependence of the viscoelastic behavior of polystyrene," D. J. Plazek, *Journal of Physical Chemistry,* **69**, 3480–3487 (1965). Copyright © 1965, American Chemical Society.

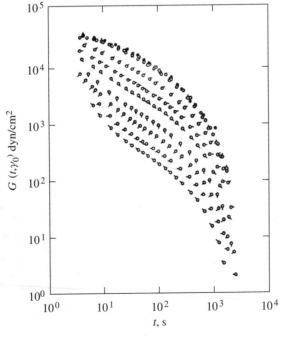

Figure 6.57 Nonlinear shear relaxation modulus $G(t, \gamma_0)$ as a function of time t at 33.5°C, measured in relaxation after step shear strain for 20% solutions of narrow molecular-weight-distribution polystyrene ($M_w = 1.8 \times 10^6$) in chlorinated diphenyl; from Einaga et al. [72]. Different curves are for different strain amplitudes γ_0 : ○, $\gamma_0 = 0.41$; pip up $\gamma_0 = 1.87$; for successive clockwise 45° rotations of pip, $\gamma_0 = 3.34, 5.22, 6.68, 10.0, 13.4, 18.7,$ and 25.4. *Source:* From *Polymer Journal,* Copyright © 1971, The Society of Polymer Science, Japan. Reprinted by permission.

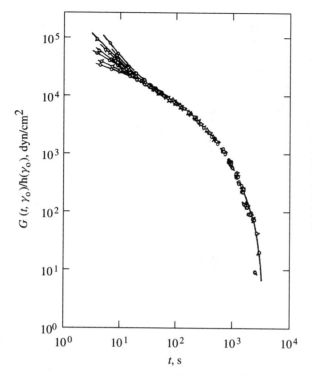

Figure 6.58 Shifted relaxation modulus $G(t, \gamma_0)/h(\gamma_0)$ at 33.5°C for the polystyrene solution data shown in Figure 6.57; from Einaga et al. [72]. $h(\gamma_0)$ is the damping function and is equal to $G(t, \gamma_0)/G(t)$, where $G(t)$ is the relaxation modulus at small strains, and $\log h$ is the vertical shift required to superpose the curves of Figure 6.57. Symbols are the same as in Figure 6.57. *Source:* From *Polymer Journal,* Copyright © 1971, The Society of Polymer Science, Japan. Reprinted by permission.

successfully such as those presented in Figure 6.58, we say that the material is time-strain factorable. This property is predicted by some constitutive equations (see Section 9.4.1.3 and [138]). The function $h(\gamma_0)$ is called the *damping function*, and it represents the strain dependence of the polymer.

Lodge and Meissner [159, 157] showed that the following relation holds for a wide class of materials[2] subjected to the step-strain deformation:

Lodge–Meissner
relationship
$$\boxed{\frac{G(t, \gamma_0)}{G_{\Psi_1}(t, \gamma_0)} = 1} \tag{6.17}$$

This relationship says that the shapes of the relaxation functions for shear stress and for the first normal-stress difference are the same. This relationship is predicted by several viscoelastic constitutive equations (see [138] and Chapter 9), and it has been found to hold for polyethylene [142], polybutadiene solutions (narrow molecular-weight distribution) [254], a slightly branched polybutadiene melt (broad molecular-weight distribution) [127], polystyrene melts [143], and concentrated polystyrene solutions when the molecular weight is not too high [140, 198, 199, 254].

---

[2] They showed this relationship to hold for isotropic liquids [7] for which the principal directions of stress and strain are always coincident. For a discussion of principal axes, see Appendix C.6.

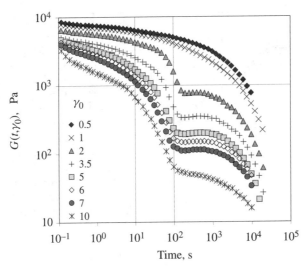

Figure 6.59 Nonlinear shear relaxation modulus $G(t, \gamma_0)$ versus time for a concentrated solution ($c = 0.23$ g/cm^3) of nearly monodisperse polystyrene (8.42×10^6 g/mol) in tricresyl phosphate; 30°C; from Morrison and Larson [184].

For high-molecular-weight materials, the large-amplitude step-shear test is subject to instabilities that cause $G(t, \gamma_0)$ to develop a kink (Figure 6.59). This effect has been seen in polystyrene solutions [83, 184, 6] and in polybutadiene and polyisoprene melts [255, 254, 185], and is believed to be related to wall slip or to internal strain inhomogeneities [184, 6]. For these materials the Lodge–Meissner relationship is violated.

6.3 Steady Elongational Flow

The experimental literature on elongational-flow behavior is a fraction of the size of the shear-flow literature, despite the importance of elongational properties in film blowing, fiber spinning, and other processing operations. This subject is an area of active research.

As with the measurement of shear viscosity, steady state must be reached for an elongational viscosity to be obtained. As noted previously, it is difficult to reach steady state in elongation. A limited amount of data for the steady uniaxial elongational viscosity of several polystyrenes is shown in Figure 6.60. The steady uniaxial elongational viscosity $\bar{\eta}$ at low rates for all four polystyrenes is equal to the three times the zero-shear steady shear viscosity η_0. This relationship is predicted for a Newtonian fluid, and $\bar{\eta} = 3\eta_0$ is also predicted by other constitutive equations (see Chapter 9). The quantity $3\eta_0$ is called the *Trouton viscosity* and $\bar{\eta}/\eta_0$ is called the *Trouton ratio*.

The variation of $\bar{\eta}$ with the shear rate is not strong in polystyrene, and weak extension-thinning as well as extensional-thickening regions (also called tension-thinning and tension-thickening) are observed (Figure 6.60). It is often important to know whether a material is tension-thinning or tension-thickening since processes such as film blowing and fiber spinning are only stable for materials that are tension-thickening; tension-thinning materials become weaker as they are stretched, and this causes breakage [207]. Unfortunately, in many cases it is not possible to reach high enough extension rates $\dot{\epsilon}_0$ to determine whether

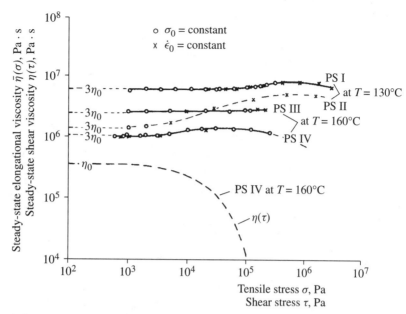

Figure 6.60 Steady uniaxial elongational viscosities $\bar{\eta}(\sigma)$ of several polystyrenes at the temperatures indicated; from Münstedt [187]. The elongational viscosities are plotted as a function of $\sigma = \tau_{33} - \tau_{11}$, while shear viscosity $\eta(\tau)$ for PS IV is plotted as a function of $\tau = \tau_{21}$ Molecular weights: PS I, $M_w = 74$ kg/mol, $M_w/M_n = 1.2$; PS II, $M_w = 39$ kg/mol, $M_w/M_n = 1.1$; PS III, $M_w = 253$ kg/mol, $M_w/M_n = 1.9$; PS IV, $M_w = 219$ kg/mol, $M_w/M_n = 2.3$. *Source:* From the *Journal of Rheology*, Copyright © 1980, The Society of Rheology. Reprinted by permission.

a material is tension-thickening or -thinning. New techniques in this area may yield new data, however (see Chapter 10).

The uniaxial elongational viscosity as a function of the extension rate for a filled polymer system is shown in Figure 6.61. The unfilled polyisobutylene (PIB) shows a constant elongational viscosity, but when the filler is added, the mixture is tension-thinning [101]. This resembles the effect of filler on shear viscosity (see Figure 6.22)). For both the unfilled polymer and the mixture of PIB and 42% α-alumina powder the Trouton ratio was found to be 3.

6.4 Unsteady Elongational Flow

6.4.1 SMALL-STRAIN UNSTEADY ELONGATIONAL FLOW

In the small-strain limit, elongational flows such as small-amplitude oscillatory elongation (SAOE) give results that are directly and very simply related to small-strain shear experiments [see Equations (5.204) and (5.205)]. Usually no distinction is made for linear viscoelastic measurements taken in shear versus extension, and G' and G'' are reported instead of $E' = 3G'$ and $E'' = 3G''$ (see Figure 6.39). The choice of whether to use shear or elongation for measuring linear viscoelastic properties is purely one of convenience.

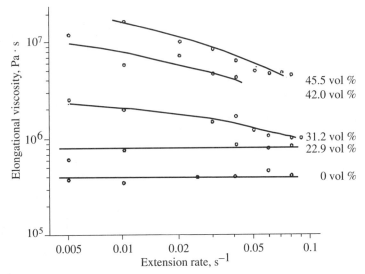

Figure 6.61 Uniaxial elongational viscosity as a function of elongation rate for polyisobutylene (PIB) and PIB filled with α-alumina powder to various volume fractions; from Greener and Evans [101]. The data are at 294 K. The filler causes the PIB to become tension-thinning. *Source:* From the *Journal of Rheology,* Copyright © 1998, The Society of Rheology. Reprinted by permission.

6.4.2 LARGE-STRAIN UNSTEADY ELONGATIONAL FLOW

Because of the difficulties involved in creating elongational flow, most of what we know about elongational behavior we have learned from elongational startup curves rather than curves of steady elongational viscosity. Data for elongational startup have been taken for uniaxial, biaxial, and planar elongational flows.

The uniaxial elongational growth curves for two different polystyrenes are shown in Figure 6.62, for low-density polyethylene (LDPE) in Figure 6.63, and for polypropylene at 180°C in Figure 6.64. Like the shear start-up curves, the low-rate data form a boundary shared by all higher-rate data; but while in shear the low-rate data were an upper bound [in $\eta^+(t)$], in elongation they are a lower bound. At high values of time and high rates, the elongational stresses become unbounded. We know little about the high-rate regime of $\bar{\eta}^+(\dot{\epsilon}_0)$ because when samples are stretched out in these experiments, the cross-sectional area of the sample becomes quite small, and the filament becomes weak. Surface tension and interactions with the surroundings have a strong effect on data collected at high elongational strains.

Recently a method has been developed for measuring elongational start-up curves for polymer solutions. The filament-stretching technique [234, 171, 42] is described in Chapter 10, and data on the Trouton ratio Tr $= \bar{\eta}^+(t, \dot{\epsilon}_0)/\eta_0$ for a Boger fluid are shown in Figure 6.65. At short times this solution shows a modest transient followed by a turnup toward very high values of Tr at a critical time. The time at which the turnup occurs decreases with increasing deformation rate. The response of dilute solutions of this type is very similar to what is observed for polymer melts. Figure 6.66 shows the Trouton ratio as a function of time for high-molecular-weight polystyrene solutions [153, 197, 105]. These data were also

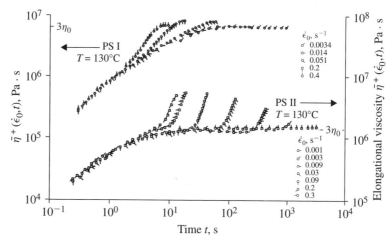

Figure 6.62 Start-up uniaxial elongational viscosity $\bar{\eta}^+(\dot{\epsilon}_0, t)$ versus time t for two polystyrenes at 130°C; from Münstedt [187]. For PS I, $M_w = 74$ kg/mol, $M_w/M_n = 1.2$; for PS II, $M_w = 39$ kg/mol, $M_w/M_n = 1.1$. *Source:* From the *Journal of Rheology,* Copyright © 1980, The Society of Rheology. Reprinted by permission.

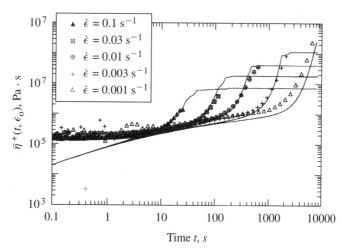

Figure 6.63 Start-up uniaxial elongational viscosity, $\bar{\eta}^+(t, \dot{\epsilon}_0)$ versus time t for low-density polyethylene at 140°C, $M_w = 250$ kg/mol, $M_w/M_n = 15$; from Inkson et al. [130a]. Solid curves represent a fit to a molecular constitutive equation called the 12 mode pompom melt. *Source:* From the *Journal of Rheology,* Copyright © 1999, The Society of Rheology. Reprinted by permission.

taken using the filament stretching technique (see Chapter 10). In Figure 6.66 the results are compared to Brownian dynamics simulations of a bead-spring model of polymer solutions (see Section 9.4.2).

Figure 6.67 shows data on two polymers for the startup of steady biaxial elongation. These data were measured using the lubricated squeezing technique (see Chapter 10). Note that the results in biaxial extension resemble the uniaxial extensional data. Soskey and Winter also measured large-amplitude biaxial step strains on the same materials

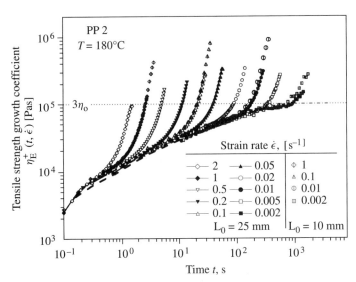

Figure 6.64 Start-up uniaxial elongational viscosity, $\bar{\eta}^+(t, \dot{\epsilon}_0)$, versus time t for polypropylene at 180°C; from Kurzbeck et al. [135]. Two initial sample lengths were used, $L_0 = 25$ mm and 10 mm as indicated on the figure. *Source:* From the *Journal of Rheology,* Copyright © 1999, The Society of Rheology. Reprinted by permission.

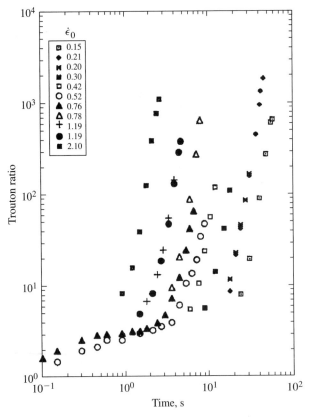

Figure 6.65 Trouton ratio Tr $= \bar{\eta}^+(t, \dot{\epsilon}_0)/\eta_0$ versus time for a Boger fluid made from 0.185% polyisobutylene in a mixture of kerosene and polybutene; from Sridhar et al. [234]. Since Boger fluids exhibit constant viscosity, Tr versus t has the same shape as the curves of $\bar{\eta}^+$ versus t. *Source:* Reprinted from *Journal of Non-Newtonian Fluid Mechanics*, **40**, T. Sridhar, V. Tirtaatmadja, D. A. Nguyen, and R. K. Gupta, "Measurement of extensional viscosity of polymer solutions," 271–280, Copyright © 1991, with permission from Elsevier Science.

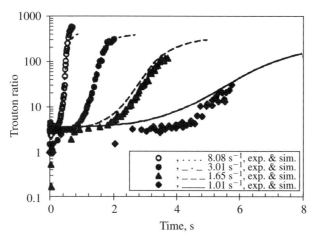

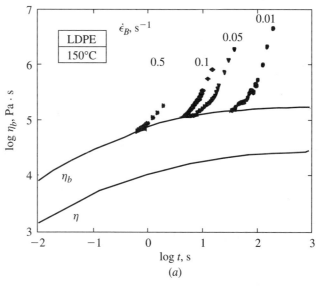

Figure 6.67 Startup of steady biaxial elongation, from Soskey and Winter [230]. (*a*) Low-density polyethylene at 150°C. (*b*) Polystyrene at 180°C. The solid curves are calculations of the shear (lower curve) and biaxial elongational start-up curves using the generalized linear viscoelastic constitutive equation (see Chapter 8). The data shown were calculated using the positive biaxial strain rate $\dot{\epsilon}_B = -\frac{1}{2}\dot{\epsilon}$, and thus η_b in the figure is equal to $2\bar{\eta}_B$ (see Problem 5.18). *Source:* From the *Journal of Rheology,* Copyright © 1985, The Society of Rheology. Reprinted by permission.

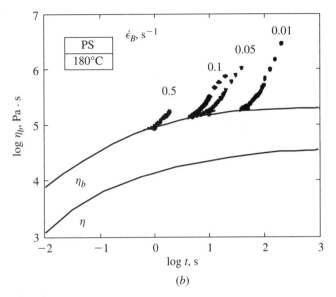

(b)

Figure 6.67 Continued

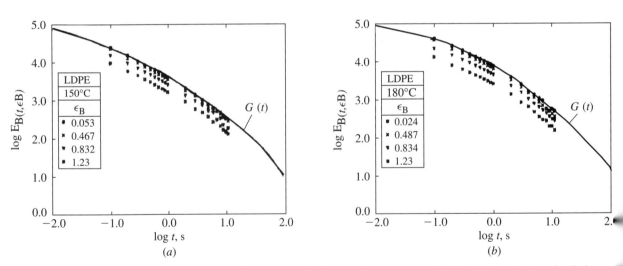

Figure 6.68 Nonlinear elongational stress-relaxation modulus $E_B(t, \epsilon_B)$, where $\epsilon_B = -\epsilon_0/2$, for (a) low-density polyethylene at 150°C and (b) polystyrene at 180°C; from Soskey and Winter [230]. At low strains the step-strain curves are independent of strain, which is also observed in shear. *Source:* From the *Journal of Rheology,* Copyright © 1985, The Society of Rheology. Reprinted by permission.

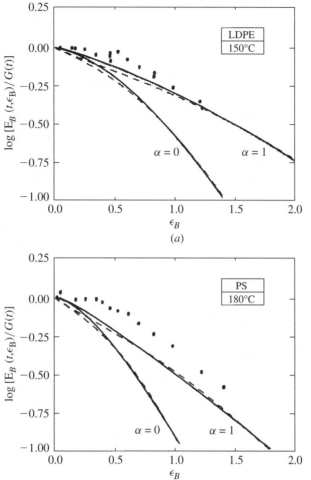

(a)

(b)

Figure 6.69 Step biaxial damping functions for (a) low-density polyethylene and (b) polystyrene calculated from the data in Figure 6.68; from Soskey and Winter [230]. The solid and dashed lines are empirical fits of the curves to the models of Soskey and Winter, where α is a parameter in their model, and $\epsilon_B = -\epsilon_0/2$. *Source: From the Journal of Rheology,* Copyright © 1985, The Society of Rheology. Reprinted by permission.

(Figure 6.68). At small strains the elongational stress-relaxation modulus $E(t)$ equals the small-strain shear stress relaxation modulus $G(t)$. At larger strains $E(t, \epsilon_0)$ drops below the linear viscoelastic limit, and as was done for shear step strains, an elongational damping function h_B can be calculated (Section 6.2.2):

$$E(t, \epsilon_0) = E(t)h_B(\epsilon_0) \qquad (6.18)$$

The damping functions for the data in Figure 6.68 are shown in Figure 6.69. The shear damping function $h(\gamma_0)$ and the biaxial damping function h_B are not usually equal.

6.5 Summary

Much more information on the rheological behavior of polymers and other systems can be found in the technical literature. A good starting place for learning more about polymer linear viscoelastic behavior is Ferry's text [75]. Recent mongraphs on rheology by Larson [139], Macosko [162], and Dealy and Wissbrun [61] also contain very useful discussions of material behavior, including systems not discussed here such as liquid crystals and suspensions.

We now seek constitutive models that can capture non-Newtonian behavior quantitatively. The variety of material behaviors exhibited by polymeric systems makes modeling these systems quite challenging. To keep our task from overwhelming us, we will initially consider models that capture only the simplest of non-Newtonian behaviors. These models are discussed in the next chapter.

6.6 PROBLEMS

6.1 For most linear polymers the steady shear viscosity at low rates η_0 is observed to follow the following proportionality:

$$\eta_0 \quad \propto \quad M^{3.4}$$

If a polymer of a molecular weight of 25 kg/mol has a zero-shear viscosity of 1.2×10^3 Pa · s, what is the zero-shear viscosity of a polymer that is three times longer?

6.2 What is the spurt phenomenon? What is the impact of this phenomenon on rheological measurements?

6.3 The steady elongational viscosity of a polymer is found to be tension-thinning, and the polymer was therefore rejected for use in a process that would have spun the polymer into a fiber. Why would tension-thinning be disadvantageous for a polymer to be spun into a fiber?

6.4 For an unknown polymer, how could you determine M_c, the critical molecular weight for entanglement?

6.5 The nonlinear shear stress-relaxation modulus $G(t, \gamma_0)$ at 33.5°C for a 20% solution of narrow-molecular-weight-distribution polystyrene ($M_w = 1.8 \times 10^6$) in chlorinated diphenyl is given in Figure 6.57. Calculate the damping function $h(\gamma_0)$ for these data. Plot your results.

6.6 The storage moduli versus time–temperature shifted frequency for two monodisperse polymers are sketched in Figure 6.70. Which of the two polymers has a higher molecular weight? Explain your answer.

6.7 The steady shear viscosities versus shear rate for three batches of the same type of polymer are given in Figure 6.71. Which of the batches has the broadest molecular-weight distribution? Explain your answer.

6.8 The zero-shear viscosity versus molecular weight for a polymer is shown in Figure 6.72. What is the entanglement molecular weight of this polymer? Explain your answer. What is the significance of M_e?

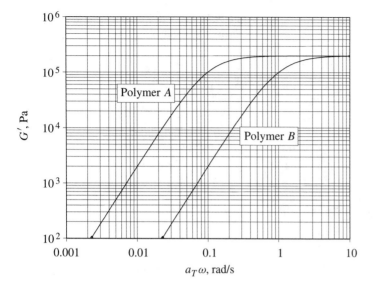

Figure 6.70 Problem 6.6: storage modulus versus time–temperature-shifted frequency for two monodisperse polymers.

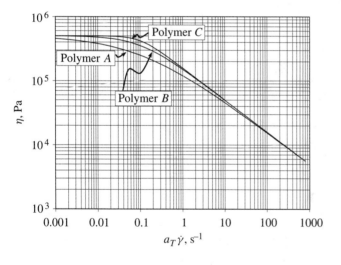

Figure 6.71 Problem 6.7: steady shear viscosities versus time–temperature-shifted shear rate for three batches of the same type of polymer.

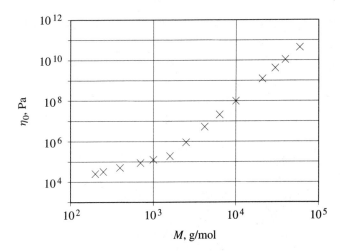

Figure 6.72 Problem 6.8: zero-shear viscosity versus molecular weight for a polymer.

CHAPTER 7

No Memory
Generalized Newtonian Fluids

We are now ready to model non-Newtonian flow behaviors such as those described in Chapter 6. The approach to developing constitutive equations that we will follow is based on continuum mechanics. We will propose empirical constitutive equations and then predict material functions. The validity of a constitutive equation will be judged on the ability of the equation to predict observed behavior, in particular, the shape of the viscosity versus shear-rate curve. The continuum approach will yield some interesting and useful non-Newtonian constitutive equations, although the results will not be totally satisfying since this trial-and-error method is not based on information about molecular structure. A more rigorous approach would be to investigate the chemical and physical structure of a system of interest, model the interactions of the particles using the laws of physics, and derive a constitutive equation from these first principles. There is a problem with the molecular modeling method, however; the interactions between particles are very complex in rheologically interesting systems such as polymers, colloids, suspensions, and other mixtures. Even with modern methods one cannot make interesting rheological calculations on relatively simple systems such as polymer solutions or polymer melts.

Besides leading to simple and useful constitutive equations, the continuum approach to constitutive modeling has established the framework for all constitutive modeling, including molecular modeling approaches and modern thermodynamic and stochastic methods. Chapter 9 gives a brief introduction to these other approaches. After studying the material in this book, the reader will be well prepared to tackle the literature dealing with these advanced methods.

In this chapter we introduce the generalized Newtonian fluid, a simple constitutive equation that captures some observed non-Newtonian behavior. We begin with a discussion of several constraints that must be met by all constitutive equations. This is followed by the introduction of the generalized Newtonian fluid (GNF) model. Finally, we work out some example problems and discuss the limitations of the GNF model.

7.1 Constitutive Constraints

Constitutive modeling is the art and science of looking for appropriate tensorial expressions for stress as a function of deformation to match observed material behavior. These relations

225

are then used with the equations of motion and continuity to solve flow problems. The ultimate measure of the validity and usefulness of a given constitutive equation is how well its predictions match observations.

There is no recipe to follow in developing useful constitutive equations. There are, however, some physical and mathematical constraints that must be met by successful constitutive equations. These constraints are the minimum criteria that ensure that the equation makes mathematical sense.

- **Observation:** *Stress is a second-order quantity.* As shown in Chapter 3, there are always two directions associated with stress—the direction of the stress and the orientation of the surface on which the stress acts.

  Implication: Constitutive equations are always equations of second order, that is, all the terms in a constitutive equation must have two directions (unit vectors) associated with them. This is the first and most easily met constraint.

- **Observation:** *Stress is independent of the coordinate system used to describe it, that is, stress is more than just of second order; stress is a second-order tensor.*

  Implication: Constitutive equations must be coordinate invariant, that is, the functions contained in constitutive equations must not include any variables whose values change when the coordinate system changes. The coefficients of vectors and tensors, for instance, differ in different coordinate systems. Therefore vector and tensor coefficients that are specific to one particular coordinate system must not appear explicitly in a constitutive function. The only scalar functions that may appear in a constitutive equation are functions of invariants. For example, only magnitudes of vectors (vectors have one invariant, the magnitude) or any of the three scalar invariants of a second-order tensor (see Section 2.3.4 and below) may appear in constitutive equations.

- **Observation:** *The stress tensor is symmetric for most materials and for all conventional polymer melts and solutions.*

  Implication: The constitutive equation must predict a symmetric stress tensor. We have seen that when a tensor is added to or multiplied by its transpose, a symmetric tensor results. This is useful to remember when constructing constitutive equations.

- **Observation:** *The response of a material to imposed stresses or to imposed deformation is the same for all observers (requirement of material objectivity).* For example, if a child inflates a balloon on a merry-go-round, the mathematical description of the deformation of the balloon must be independent of whether the equation is written with respect to an observer on the merry-go-round or with respect to an observer on the ground outside.

  Implication: This is a subtle and restrictive requirement that constrains the mathematical forms of constitutive equations. We will postpone discussion of this point to the end of Chapter 8, when we will discover how easy it is to violate the requirement of material objectivity.

Constraints other than those outlined here may be adopted if it is desired that a particular class of material behavior be predicted by the constitutive equation [247, 195, 238, 7]. For example, if the Lodge–Meissner rule is followed by the materials being studied, this forms an additional constraint on the constitutive equation: the constitutive equation must predict the Lodge–Meissner relation. Until an empirical rule is shown to be followed by all materials, however, it is the choice of the individual rheologist whether to make a particular class of material behavior a constraint.

EXAMPLE

Constitutive equations must be coordinate invariant, which implies that only vectors, tensors, and vector and tensor invariants may appear in constitutive equations. For shear flow of an incompressible fluid, what scalar quantities associated with the rate-of-deformation tensor $\underline{\underline{\dot{\gamma}}}$ may appear in constitutive equations?

SOLUTION

We begin by examining the invariants of the rate-of-deformation tensor $\underline{\underline{\dot{\gamma}}}$. To calculate the invariants, we write $\underline{\underline{\dot{\gamma}}}$ in an orthonormal coordinate system since the definitions for scalar invariants presented in Section 2.3.4 only apply to orthonormal coordinate systems:

$$\underline{\underline{\dot{\gamma}}} = \begin{pmatrix} \frac{\partial v_1}{\partial x_1} & \frac{\partial v_2}{\partial x_1} & \frac{\partial v_3}{\partial x_1} \\ \frac{\partial v_1}{\partial x_2} & \frac{\partial v_2}{\partial x_2} & \frac{\partial v_3}{\partial x_2} \\ \frac{\partial v_1}{\partial x_3} & \frac{\partial v_2}{\partial x_3} & \frac{\partial v_3}{\partial x_3} \end{pmatrix}_{123} \tag{7.1}$$

$$I_{\underline{\underline{\dot{\gamma}}}} = \text{trace}(\underline{\underline{\dot{\gamma}}}) = \dot{\gamma}_{ii} \tag{7.2}$$

$$= \frac{\partial v_1}{\partial x_1} + \frac{\partial v_2}{\partial x_2} + \frac{\partial v_3}{\partial x_3} = \nabla \cdot \underline{v} = 0 \text{ for incompressible fluids} \tag{7.3}$$

$$II_{\underline{\underline{\dot{\gamma}}}} = \text{trace}(\underline{\underline{\dot{\gamma}}} \cdot \underline{\underline{\dot{\gamma}}}) = \underline{\underline{\dot{\gamma}}} : \underline{\underline{\dot{\gamma}}} \tag{7.4}$$

$$= \sum_{p=1}^{3} \sum_{k=1}^{3} \dot{\gamma}_{pk} \dot{\gamma}_{kp} \tag{7.5}$$

$$III_{\underline{\underline{\dot{\gamma}}}} = \text{trace}(\underline{\underline{\dot{\gamma}}} \cdot \underline{\underline{\dot{\gamma}}} \cdot \underline{\underline{\dot{\gamma}}}) = \sum_{p=1}^{3} \sum_{k=1}^{3} \sum_{j=1}^{3} \dot{\gamma}_{pk} \dot{\gamma}_{kj} \dot{\gamma}_{jp} \tag{7.6}$$

We see that for incompressible fluids $I_{\underline{\underline{\dot{\gamma}}}}$ is always zero due to the continuity equation. The magnitude of $\underline{\underline{\dot{\gamma}}}$, $|\underline{\underline{\dot{\gamma}}}| = \dot{\gamma}(t)$, is related to the second invariant:

$$\left| \underline{\underline{\dot{\gamma}}} \right| = \dot{\gamma}(t) = +\sqrt{\frac{\underline{\underline{\dot{\gamma}}} : \underline{\underline{\dot{\gamma}}}}{2}} = +\sqrt{\frac{II_{\underline{\underline{\dot{\gamma}}}}}{2}} \tag{7.7}$$

Thus the shear rate $\dot{\gamma}(t)$ may appear in a constitutive equation. For simple shear flow, where $\underline{\underline{\dot{\gamma}}}$ is given by

$$\underline{\underline{\dot{\gamma}}} = \begin{pmatrix} 0 & \dot{\gamma}(t) & 0 \\ \dot{\gamma}(t) & 0 & 0 \\ 0 & 0 & 0 \end{pmatrix}_{123} \tag{7.8}$$

it is straightforward to show that the third invariant of $\underline{\underline{\dot{\gamma}}}$, $III_{\underline{\underline{\dot{\gamma}}}}$, is zero, and we leave this exercise up to the reader. Thus, in shear flow $\dot{\gamma} = |\underline{\underline{\dot{\gamma}}}|$ is the only scalar quantity associated with $\underline{\underline{\dot{\gamma}}}$ that may appear in a constitutive equation for incompressible fluids.

We begin now our discussion of empirical constitutive equations. These equations were proposed by researchers who based their hypotheses on observed material behavior while following the rules outlined earlier. The strategy when working with these and all other constitutive equations is to be aware of the origin of a particular equation and of its predictions, and to limit the use of these equations to situations where they perform well. The first group of constitutive equations we will discuss are the generalized Newtonian fluids.

7.2 GNF Constitutive Equation

The generalized Newtonian fluid (GNF) constitutive equation was developed from the Newtonian constitutive equation for incompressible fluids,

$$\text{Newtonian constitutive equation} \qquad \boxed{\underline{\underline{\tau}} = -\mu \underline{\underline{\dot{\gamma}}}} \qquad (7.9)$$

Since the Newtonian equation predicts a constant viscosity in steady shear, $\eta(\dot{\gamma}) = \mu$, it must be modified for materials for which viscosity is not a constant:

$$\text{Generalized Newtonian fluid constitutive equation} \qquad \boxed{\underline{\underline{\tau}} = -\eta(\dot{\gamma}) \underline{\underline{\dot{\gamma}}}} \qquad (7.10)$$

where $\eta(\dot{\gamma})$ is a scalar function and $\dot{\gamma} = |\underline{\underline{\dot{\gamma}}}|$.

The GNF constitutive equation, Equation (7.10), is an equation of tensor order, and thus it satisfies our first criterion for a constitutive equation. It produces a symmetric tensor $\underline{\underline{\tau}}$ since the equation consists of a scalar function $\eta(\dot{\gamma})$ multiplying a symmetric tensor, $\underline{\underline{\dot{\gamma}}} = \nabla \underline{v} + (\nabla \underline{v})^T$. Finally, it is coordinate invariant since it is only a function of the tensor $\underline{\underline{\dot{\gamma}}}$ and invariant scalars. The only scalar variable present in the GNF constitutive equation is $\dot{\gamma}$, which is related to the second invariant of $\underline{\underline{\dot{\gamma}}}$. Recall that $\dot{\gamma} = |\underline{\underline{\dot{\gamma}}}|$ is by definition a positive quantity, since the positive square root is taken in the definition of magnitude.

The use of the symbol η, which is usually associated with viscosity, for the function multiplying $\underline{\underline{\dot{\gamma}}}$ in the GNF equation is not accidental. We can see this by calculating the stress tensor predicted by the generalized Newtonian fluid in steady shear flow.

EXAMPLE

Calculate the steady shear material functions, that is, viscosity $\eta(\dot{\gamma})$ and the normal-stress coefficients $\Psi_1(\dot{\gamma})$ and $\Psi_2(\dot{\gamma})$, for the generalized Newtonian fluid.

SOLUTION

Steady shear flow is defined by

$$\underline{v} = \begin{pmatrix} \dot{\varsigma}(t)x_2 \\ 0 \\ 0 \end{pmatrix}_{123}, \qquad \dot{\varsigma}(t) = \dot{\gamma}_0 = \text{constant} \qquad (7.11)$$

and $\underline{\underline{\dot{\gamma}}}$ is therefore

$$\underline{\underline{\dot{\gamma}}} = \nabla \underline{v} + (\nabla \underline{v})^T \tag{7.12}$$

$$= \begin{pmatrix} 0 & \dot{\gamma}_0 & 0 \\ \dot{\gamma}_0 & 0 & 0 \\ 0 & 0 & 0 \end{pmatrix}_{123} \tag{7.13}$$

where $\dot{\gamma}_o$ may be positive or negative. Inserting the kinematics into the GNF equation we obtain

$$\underline{\underline{\tau}} = -\eta(\dot{\gamma})\underline{\underline{\dot{\gamma}}} \tag{7.14}$$

$$= \begin{pmatrix} 0 & -\eta(\dot{\gamma})\dot{\gamma}_0 & 0 \\ -\eta(\dot{\gamma})\dot{\gamma}_0 & 0 & 0 \\ 0 & 0 & 0 \end{pmatrix}_{123} \tag{7.15}$$

Thus the GNF model predicts that there are only two nonzero components of $\underline{\underline{\tau}}$, and these two components are equal. Using the definitions of the steady shear material functions, Equations (5.8)–(5.10), we can calculate $\eta(\dot{\gamma})$, $\Psi_1(\dot{\gamma})$, and $\Psi_2(\dot{\gamma})$ for the generalized Newtonian fluid:

$$\eta \equiv \frac{-\tau_{21}}{\dot{\gamma}_0} = \eta(\dot{\gamma}) \tag{7.16}$$

$$\Psi_1 \equiv \frac{-(\tau_{11} - \tau_{22})}{\dot{\gamma}_0^2} = 0 \tag{7.17}$$

$$\Psi_2 \equiv \frac{-(\tau_{22} - \tau_{33})}{\dot{\gamma}_0^2} = 0 \tag{7.18}$$

Thus the scalar function $\eta(\dot{\gamma})$ that multiplies $\underline{\underline{\dot{\gamma}}}$ in the GNF equation is just equal to the steady shear viscosity. We see that, like the Newtonian constitutive equation, the GNF constitutive equation predicts $\Psi_1 = \Psi_2 = 0$ in steady shear flow.

The GNF model is quite general since the functional form of $\eta(\dot{\gamma})$ has not yet been specified. It must be given or fit to data in order for flow properties to be predicted using the GNF equation. We will introduce three models for $\eta(\dot{\gamma})$. Many other functional forms for η are used with the GNF constitutive model, and these may be found in the literature [26, 227] or in flow simulation software [212, 182, 38].

7.2.1 POWER-LAW MODEL

The power-law or Ostwald–de Waele model describes viscosity with a function that is proportional to some power of the shear rate $\dot{\gamma}$:

Power-law model
for viscosity
(GNF)

$$\eta(\dot{\gamma}) = m\dot{\gamma}^{n-1}$$

(7.19)

The power-law equation has two parameters that must be fit to experimental data. One parameter is the exponent of $\dot{\gamma}$, $n-1$, which is the slope of $\log\eta$ versus $\log\dot{\gamma}$. The second parameter is m, which is called the *consistency index;* $\log m$ is the y-intercept of the $\log\eta$ versus $\log\dot{\gamma}$ plot, and m is related to the magnitude of the viscosity. The units of the two parameters of the power-law model can be deduced from Equation (7.19):

$$m \quad [=] \quad Pa \cdot s^n$$

(7.20)

$$n \quad [=] \quad \text{dimensionless}$$

(7.21)

These unusual units are the result of the model's choice to raise the dimensional quantity $\dot{\gamma}$ to a fractional exponent, $n-1$.

The power-law model can describe a Newtonian fluid. In that case $m = \mu$ and $n = 1$ (Figure 7.1). For $n > 1$, the plot of $\log\eta$ versus $\log\dot{\gamma}$ slants upward, and the material is called *dilatant* or *shear-thickening* (it becomes thicker as it is sheared). For $n < 1$, the plot of $\log\eta$ versus $\log\dot{\gamma}$ slants downward, and the behavior is called *shear-thinning* (Figure 7.1). Note that in the power-law model the slope of the $\log\eta$ versus $\log\dot{\gamma}$ plot is constant for a given material. Thus this model cannot describe the viscosity of a material that has a Newtonian plateau at small shear rates (η_0) and which then shear-thins at high rates (for example, see Figure 6.1).

The power-law GNF model has been used quite widely in calculations applied to polymer manufacturing processes [236]. In processes such as extrusion the shear rates are very high, and models capturing only the high-shear-rate region of the viscosity behavior can do a fair job of predicting flow-rate versus pressure-drop data [236, 26]. There are disadvantages to this model, however. It is strictly empirical, and thus we obtain no molecular insight from it. It is not possible, for example, to predict how a new material of higher molecular weight will perform given the power-law fit of a similar material of lower molecular weight. Also, this model does not have a material relaxation time as one of its parameters. It is useful to have a measure of a material's relaxation time in order to be able to predict time-dependent behavior such as how rapidly the fluid will relax upon cessation

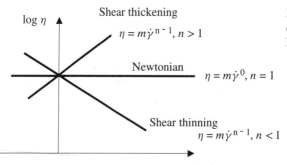

Figure 7.1 Viscosity behavior as predicted by the power-law generalized Newtonian fluid model.

of flow. The relaxation time is also used in the Deborah number to scale time-dependent processes involving the flow of a viscoelastic fluid (see Section 5.2.2.2). Finally, the power-law model, like all generalized Newtonian fluids, does not predict nonzero normal stresses in shear flow and thus misses some important nonlinear effects.

The advantages of the power-law GNF model are the ease with which calculations can be made and the success the model has in predicting flow-rate versus pressure-drop measurements. Two example flow calculations carried out with the power-law model are discussed in Section 7.4.

7.2.2 CARREAU–YASUDA MODEL

A viscosity model that captures more details of the shape of experimentally measured $\eta(\dot{\gamma})$ curves is the Carreau–Yasuda model. The Carreau–Yasuda model uses five parameters (compared with two parameters in the power-law model):

$$\text{Carreau–Yasuda model} \qquad \frac{\eta(\dot{\gamma}) - \eta_\infty}{\eta_0 - \eta_\infty} = \left[1 + (\dot{\gamma}\lambda)^a\right]^{\frac{n-1}{a}} \qquad (7.22)$$

The five parameters in the Carreau–Yasuda model have the following effects on the shape of the predicted $\eta(\dot{\gamma})$ curve (Figure 7.2):

- η_∞ The viscosity function approaches the constant value η_∞ as $\dot{\gamma}$ gets large.
- η_0 The viscosity function approaches the constant value η_0 as $\dot{\gamma}$ becomes small.
- a The exponent affects the shape of the transition region between the zero-shear-rate plateau and the rapidly decreasing (power-law-like) portion of the viscosity versus shear-rate curve. Increasing a sharpens the transition.
- λ The parameter is a time constant for the fluid. The value of λ determines the shear rate at which the transition occurs from the zero-shear-rate plateau to the power-law portion. It also governs the transition from power-law to $\eta = \eta_\infty$.
- n The exponent is a power-law-like parameter that describes the slope of the rapidly decreasing portion of the η curve.

This model can effectively fit most viscosity versus shear-rate data. A disadvantage of the Carreau–Yasuda model is that it contains five parameters that must be fit simultaneously,

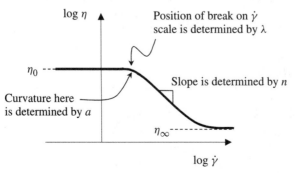

log η

Position of break on $\dot{\gamma}$ scale is determined by λ

η_0

Slope is determined by n

Curvature here is determined by a

η_∞

log $\dot{\gamma}$

Figure 7.2 Viscosity behavior as predicted by the Carreau–Yasuda generalized Newtonian fluid model.

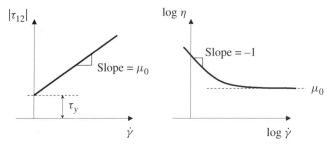

Figure 7.3 Viscosity behavior as predicted by the Bingham generalized Newtonian fluid model.

although this is now fairly easy with modern software tools. It is difficult to arrive at analytical solutions for velocity and stress fields using the Carreau–Yasuda model, but it is straightforward to use this model in numerical calculations. Finally, like the power-law model, the Carreau–Yasuda model does not give molecular insight into polymer behavior; for example, it cannot predict molecular-weight dependence of viscosity. The Carreau–Yasuda model does contain a material relaxation time λ, which can be correlated with molecular structure.

7.2.3 BINGHAM MODEL

The Bingham model represents behavior that is fundamentally different from either the power-law or the Carreau–Yasuda models. This model describes fluids that exhibit yield stresses (Figure 7.3):

$$\text{Bingham model} \qquad \eta(\dot{\gamma}) = \begin{cases} \infty & \tau \leq \tau_y \\ \mu_0 + \dfrac{\tau_y}{\dot{\gamma}} & \tau > \tau_y \end{cases} \qquad (7.23)$$

where $\tau \equiv |\underline{\underline{\tau}}|$, and τ_y, called the *yield stress,* is positive. In words, the Bingham model says that the fluid will not flow until a stress exceeding the yield stress τ_y is applied. At stresses much higher than the yield stress ($\dot{\gamma} \longrightarrow \infty$) the fluid flows with a constant viscosity. This is a two-parameter model:

- τ_y No flow occurs until the absolute value of the shear stress exceeds the value of the yield stress τ_y. This parameter is always positive.
- μ_0 The viscosity of the fluid at high shear rates. This parameter is always positive.

7.3 Material Function Predictions

We showed in a previous example problem that we could predict the shear material functions η, Ψ_1, and Ψ_2 using the GNF model. Since the latter is a constitutive equation we can, in fact, calculate any material function, even those defined for non-shear flows. It is not clear how useful these predictions will be since the GNF model was developed from the Newtonian model by fixing up the steady shear viscosity to match non-Newtonian observations. The way to find out whether the GNF model is useful in nonsteady shear flows is to make predictions of nonsteady shear-flow material functions and to compare the predictions to observations. For practice with these types of calculations we will evaluate the step-strain material functions and the steady elongational viscosity for two generalized Newtonian fluids.

EXAMPLE

Calculate the predictions of the Carreau–Yasuda generalized Newtonian fluid for the material functions for the step-strain experiment, $G(t, \gamma_0)$, $G_{\Psi_1}(t, \gamma_0)$, and $G_{\Psi_2}(t, \gamma_0)$.

SOLUTION

The step-strain experiment is defined in shear flow, and thus the velocity field is given by

$$\underline{v} = \begin{pmatrix} \dot{\varsigma}(t)x_2 \\ 0 \\ 0 \end{pmatrix}_{123} \tag{7.24}$$

The shear-rate tensor is

$$\underline{\dot{\gamma}} = \begin{pmatrix} 0 & \dot{\varsigma}(t) & 0 \\ \dot{\varsigma}(t) & 0 & 0 \\ 0 & 0 & 0 \end{pmatrix}_{123} \tag{7.25}$$

and the magnitude of the shear-rate tensor is $\dot{\gamma} = |\dot{\varsigma}(t)|$.

For a Carreau–Yasuda generalized Newtonian fluid, the stress tensor in this flow is calculated from

$$\underline{\tau}(t) = -\eta(\dot{\gamma})\underline{\dot{\gamma}} \tag{7.26}$$

$$= -\left\{\eta_\infty + (\eta_0 - \eta_\infty)\left[1 + (\dot{\gamma}\lambda)^a\right]^{\frac{n-1}{a}}\right\} \begin{pmatrix} 0 & \dot{\varsigma}(t) & 0 \\ \dot{\varsigma}(t) & 0 & 0 \\ 0 & 0 & 0 \end{pmatrix}_{123} \tag{7.27}$$

For the step-strain experiment, $\dot{\varsigma}(t)$ is given by (see Section 5.2.2.4)

$$\dot{\varsigma}(t) = \lim_{\varepsilon \to 0} \begin{cases} 0 & t < 0 \\ \dfrac{\gamma_0}{\varepsilon} & 0 \le t < \varepsilon \\ 0 & t \ge \varepsilon \end{cases} \tag{7.28}$$

$$= \gamma_0 \delta_+(t) \tag{7.29}$$

where $\delta_+(t)$ is the asymmetric delta function, and γ_0 is the magnitude of the step, which will be taken to be positive.

The material functions in step shear strain are $G(t, \gamma_0)$, $G_{\Psi_1}(t, \gamma_0)$, and $G_{\Psi_2}(t, \gamma_0)$, which are defined as

$$G(t, \gamma_0) = \frac{-\tau_{21}(t)}{\gamma_0} \tag{7.30}$$

$$G_{\Psi_1}(t, \gamma_0) = \frac{-(\tau_{11} - \tau_{22})}{\gamma_0^2} \tag{7.31}$$

$$G_{\Psi_2}(t, \gamma_0) = \frac{-(\tau_{22} - \tau_{33})}{\gamma_0^2} \tag{7.32}$$

We need $\tau_{21}(t)$ to calculate $G(t, \gamma_0)$, and we can calculate this from Equation (7.27):

$$\tau_{21}(t) = -\eta \dot{\varsigma}(t) = -\eta \gamma_0 \delta_+(t) \tag{7.33}$$

$$= -\left\{ \eta_\infty + (\eta_0 - \eta_\infty) \left(1 + [\gamma_0 \delta_+(t) \lambda]^a \right)^{\frac{n-1}{a}} \right\} \gamma_0 \delta_+(t) \tag{7.34}$$

We can now calculate $G(t, \gamma_0)$:

$$G(t, \gamma_0) = \left\{ \eta_\infty + (\eta_0 - \eta_\infty) \left(1 + [\gamma_0 \delta_+(t) \lambda]^a \right)^{\frac{n-1}{a}} \right\} \delta_+(t) \tag{7.35}$$

Note that the delta function is zero except near time $t = 0$, when it is very large. This means that $G(t, \gamma_0)$ is also zero except near $t = 0$. Near $t = 0$ the delta-function term dominates in the expression $1 + [\gamma_0 \delta_+(t) \lambda]^a$, and we can neglect the 1. Further, the delta-function term dominates the resulting expression, $\eta_\infty + (\eta_0 - \eta_\infty) \left[\gamma_0 \delta_+(t) \lambda \right]^{n-1}$, and we can neglect the effect of the first η_∞. Thus the expression for step-strain modulus becomes

$$\boxed{G(t, \gamma_0) = (\eta_0 - \eta_\infty) \gamma_0^{n-1} \lambda^{n-1} [\delta_+(t)]^n} \tag{7.36}$$

Note that the units are correct in this expression, since the delta function has units of s^{-1} and γ_0 is unitless. This is the final expression, and it tells us that the step-strain modulus predicted by the Carreau–Yasuda generalized Newtonian fluid is a modified impulse function at $t = 0$. This is not a very realistic prediction, and it shows that this GNF model is not very useful in the step-strain flow.

We can see from Equation (7.27) that the other two material functions in step shear strain are equal to zero for this model:

$$\boxed{\begin{aligned} G_{\Psi_1}(t, \gamma_0) &= \frac{-(\tau_{11} - \tau_{22})}{\gamma_0^2} = 0 \\ G_{\Psi_2}(t, \gamma_0) &= \frac{-(\tau_{22} - \tau_{33})}{\gamma_0^2} = 0 \end{aligned}} \tag{7.37}$$

The preceding example shows one of the problems with GNF models: they were derived by fixing up the steady shear predictions, but there is no guarantee that their predictions in nonsteady shear flows will be meaningful. We check how a GNF model behaves in uniaxial elongational flow in the next example.

EXAMPLE

Calculate the predictions of the power-law generalized Newtonian fluid model in steady uniaxial elongation.

SOLUTION

The kinematics for steady uniaxial elongation are given by

$$\underline{v} = \begin{pmatrix} -\frac{1}{2}\dot{\epsilon}_0 x_1 \\ -\frac{1}{2}\dot{\epsilon}_0 x_2 \\ \dot{\epsilon}_0 x_3 \end{pmatrix}_{123} \qquad \dot{\epsilon}_0 > 0 \tag{7.38}$$

where $\dot{\epsilon}_0$ is a positive constant. The rate-of-deformation tensor is then

$$\underline{\dot{\gamma}} = \begin{pmatrix} -\dot{\epsilon}_0 & 0 & 0 \\ 0 & -\dot{\epsilon}_0 & 0 \\ 0 & 0 & 2\dot{\epsilon}_0 \end{pmatrix}_{123} \tag{7.39}$$

and the magnitude of the rate-of-deformation tensor is $\dot{\epsilon}_0\sqrt{3}$. The uniaxial elongational viscosity is defined as

$$\bar{\eta}(\dot{\epsilon}_0) \equiv \frac{-(\tau_{33} - \tau_{11})}{\dot{\epsilon}_0} \tag{7.40}$$

Thus we must calculate the stress tensor $\underline{\tau}$ for a power-law generalized Newtonian fluid subjected to the kinematics given by Equation (7.38):

$$\underline{\tau} = -\eta(\dot{\gamma})\underline{\dot{\gamma}} \tag{7.41}$$

$$= -m3^{\frac{n-1}{2}}\dot{\epsilon}_0^{n-1} \begin{pmatrix} -\dot{\epsilon}_0 & 0 & 0 \\ 0 & -\dot{\epsilon}_0 & 0 \\ 0 & 0 & 2\dot{\epsilon}_0 \end{pmatrix}_{123} \tag{7.42}$$

$$= \begin{pmatrix} m3^{\frac{n-1}{2}}\dot{\epsilon}_0^n & 0 & 0 \\ 0 & m3^{\frac{n-1}{2}}\dot{\epsilon}_0^n & 0 \\ 0 & 0 & -2m(3^{\frac{n-1}{2}})\dot{\epsilon}_0^n \end{pmatrix}_{123} \tag{7.43}$$

Now we can calculate $\bar{\eta}$,

$$\bar{\eta}(\dot{\epsilon}_0) \equiv \frac{-(\tau_{33} - \tau_{11})}{\dot{\epsilon}_0} = 3^{\frac{n+1}{2}} m\dot{\epsilon}_0^{n-1} \tag{7.44}$$

The power-law GNF model predicts that the elongational viscosity will parallel the shear viscosity, $\eta = m\dot{\gamma}^{n-1}$. The Trouton ratio is defined as the ratio of elongational to shear viscosity at the same value of deformation rate $\dot{\gamma}$, the magnitude of the rate-of-deformation tensor for each flow. For Newtonian fluids the Trouton ratio can be shown to be equal to 3. The Trouton ratio for the power-law GNF model is calculated as

Trouton ratio
power-law GNF

$$\frac{\bar{\eta}}{\eta} = \frac{3m(\dot{\epsilon}_0\sqrt{3})^{n-1}}{m(\dot{\gamma})^{n-1}} = 3 \tag{7.45}$$

which is the same result as is obtained for Newtonian fluids.

The predictions of the power-law GNF model in steady uniaxial elongation seem reasonable. The only way to check if they are correct is to compare them with experimental data. The result of these two examples shows us that generalized Newtonian fluids are a mixed bag: sometimes they make reasonable predictions, as we saw for uniaxial elongation, but sometimes their predictions do not capture what is observed, as was seen in the step-strain example.

We turn now to two examples to show how the GNF constitutive equation may be used to calculate velocity profiles and flow rates in some simple flows. For these examples we employ the power-law model for the viscosity function.

7.4 Flow Problems: Power-Law Generalized Newtonian Fluid

7.4.1 PRESSURE-DRIVEN FLOW IN A TUBE

Calculate the velocity profile, pressure profile, and stress tensor $\underline{\underline{\tau}}$ for pressure-driven flow of an incompressible power-law liquid in a tube of circular cross section. The pressure at an upstream point is P_0, and at a point a distance L downstream the pressure is P_L. Assume that the flow between these two points is fully developed and at steady state.

This problem is worked in cylindrical coordinates, as was the Newtonian case. The difference in how the two problems are worked is that for the power-law generalized Newtonian fluid the general equation of motion is used instead of the Navier–Stokes equation. The correct cylindrical components of the equations of motion and of continuity can be found in Table C.7 of Appendix C.2.

The equation of mass conservation is

$$0 = \nabla \cdot \underline{v} \tag{7.46}$$

$$= \frac{1}{r}\frac{\partial(rv_r)}{\partial r} + \frac{1}{r}\frac{\partial v_\theta}{\partial \theta} + \frac{\partial v_z}{\partial z} \tag{7.47}$$

Since the flow is only in the z-direction, the r- and θ-components of $\underline{v}$ are zero,

$$\underline{v} = \begin{pmatrix} v_r \\ v_\theta \\ v_z \end{pmatrix}_{r\theta z} = \begin{pmatrix} 0 \\ 0 \\ v_z \end{pmatrix}_{r\theta z} \tag{7.48}$$

The continuity equation thus gives us the result

$$\frac{\partial v_z}{\partial z} = 0 \tag{7.49}$$

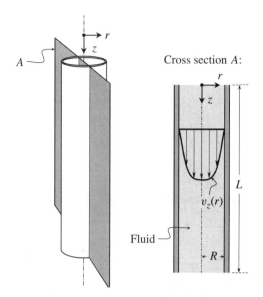

Figure 7.4 Flow problem solved in Section 7.4.1: Poiseuille flow of a power-law generalized Newtonian fluid in a tube.

The equation of motion for an incompressible non-Newtonian fluid is

$$\rho \left(\frac{\partial \underline{v}}{\partial t} + \underline{v} \cdot \nabla \underline{v} \right) = -\nabla p - \nabla \cdot \underline{\underline{\tau}} + \rho \underline{g} \tag{7.50}$$

In cylindrical coordinates the terms become (see Appendix C.2)

$$\rho \frac{\partial \underline{v}}{\partial t} = \begin{pmatrix} \rho \frac{\partial v_r}{\partial t} \\ \rho \frac{\partial v_\theta}{\partial t} \\ \rho \frac{\partial v_z}{\partial t} \end{pmatrix}_{r\theta z} \tag{7.51}$$

$$\rho \underline{v} \cdot \nabla \underline{v} = \rho \begin{pmatrix} v_r \frac{\partial v_r}{\partial r} + v_\theta \left(\frac{1}{r} \frac{\partial v_r}{\partial \theta} - \frac{v_\theta}{r} \right) + v_z \frac{\partial v_r}{\partial z} \\ v_r \frac{\partial v_\theta}{\partial r} + v_\theta \left(\frac{1}{r} \frac{\partial v_\theta}{\partial \theta} + \frac{v_r}{r} \right) + v_z \frac{\partial v_\theta}{\partial z} \\ v_r \frac{\partial v_z}{\partial r} + v_\theta \left(\frac{1}{r} \frac{\partial v_z}{\partial \theta} \right) + v_z \frac{\partial v_z}{\partial z} \end{pmatrix}_{r\theta z} \tag{7.52}$$

$$\nabla p = \begin{pmatrix} \frac{\partial p}{\partial r} \\ \frac{1}{r} \frac{\partial p}{\partial \theta} \\ \frac{\partial p}{\partial z} \end{pmatrix}_{r\theta z} \tag{7.53}$$

$$\nabla \cdot \underline{\underline{\tau}} = \begin{pmatrix} \frac{1}{r} \frac{\partial}{\partial r}(r\tau_{rr}) + \frac{1}{r} \frac{\partial \tau_{\theta r}}{\partial \theta} + \frac{\partial \tau_{zr}}{\partial z} - \frac{\tau_{\theta\theta}}{r} \\ \frac{1}{r^2} \frac{\partial}{\partial r}(r^2 \tau_{r\theta}) + \frac{1}{r} \frac{\partial \tau_{\theta\theta}}{\partial \theta} + \frac{\partial \tau_{z\theta}}{\partial z} + \frac{\tau_{\theta r} - \tau_{r\theta}}{r} \\ \frac{1}{r} \frac{\partial}{\partial r}(r\tau_{rz}) + \frac{1}{r} \frac{\partial \tau_{\theta z}}{\partial \theta} + \frac{\partial \tau_{zz}}{\partial z} \end{pmatrix}_{r\theta z} \tag{7.54}$$

$$\rho \underline{g} = \begin{pmatrix} \rho g_r \\ \rho g_\theta \\ \rho g_z \end{pmatrix}_{r\theta z} \tag{7.55}$$

Substituting in what we know already about $\underline{v}$ and assuming symmetry of $\underline{\underline{\tau}}$ and steady state, we obtain

$$\rho\frac{\partial \underline{v}}{\partial t} = \begin{pmatrix} 0 \\ 0 \\ 0 \end{pmatrix}_{r\theta z} \tag{7.56}$$

$$\rho\underline{v}\cdot\nabla\underline{v} = \begin{pmatrix} 0 \\ 0 \\ 0 \end{pmatrix}_{r\theta z} \tag{7.57}$$

$$\nabla p = \begin{pmatrix} \frac{\partial p}{\partial r} \\ \frac{1}{r}\frac{\partial p}{\partial \theta} \\ \frac{\partial p}{\partial z} \end{pmatrix}_{r\theta z} \tag{7.58}$$

$$\nabla\cdot\underline{\underline{\tau}} = \begin{pmatrix} \frac{1}{r}\frac{\partial}{\partial r}(r\tau_{rr}) + \frac{1}{r}\frac{\partial \tau_{\theta r}}{\partial \theta} + \frac{\partial \tau_{zr}}{\partial z} - \frac{\tau_{\theta\theta}}{r} \\ \frac{1}{r^2}\frac{\partial}{\partial r}(r^2\tau_{r\theta}) + \frac{1}{r}\frac{\partial \tau_{\theta\theta}}{\partial \theta} + \frac{\partial \tau_{z\theta}}{\partial z} \\ \frac{1}{r}\frac{\partial}{\partial r}(r\tau_{rz}) + \frac{1}{r}\frac{\partial \tau_{\theta z}}{\partial \theta} + \frac{\partial \tau_{zz}}{\partial z} \end{pmatrix}_{r\theta z} \tag{7.59}$$

$$\rho\underline{g} = \begin{pmatrix} 0 \\ 0 \\ \rho g \end{pmatrix}_{r\theta z} \tag{7.60}$$

Note that gravity is taken to be in the flow direction.

Putting these together we obtain (all terms written in the r, θ, z coordinate system)

$$\begin{pmatrix} 0 \\ 0 \\ 0 \end{pmatrix} = \begin{pmatrix} -\frac{\partial p}{\partial r} \\ -\frac{1}{r}\frac{\partial p}{\partial \theta} \\ -\frac{\partial p}{\partial z} \end{pmatrix} - \begin{pmatrix} \frac{1}{r}\frac{\partial}{\partial r}(r\tau_{rr}) + \frac{1}{r}\frac{\partial \tau_{\theta r}}{\partial \theta} + \frac{\partial \tau_{zr}}{\partial z} - \frac{\tau_{\theta\theta}}{r} \\ \frac{1}{r^2}\frac{\partial}{\partial r}(r^2\tau_{r\theta}) + \frac{1}{r}\frac{\partial \tau_{\theta\theta}}{\partial \theta} + \frac{\partial \tau_{z\theta}}{\partial z} \\ \frac{1}{r}\frac{\partial}{\partial r}(r\tau_{rz}) + \frac{1}{r}\frac{\partial \tau_{\theta z}}{\partial \theta} + \frac{\partial \tau_{zz}}{\partial z} \end{pmatrix} + \begin{pmatrix} 0 \\ 0 \\ \rho g \end{pmatrix} \tag{7.61}$$

To proceed further we need a relationship between $\underline{\underline{\tau}}$ and the velocity components. This is provided by the power-law GNF constitutive equation:

$$\underline{\underline{\tau}} = -\eta\underline{\underline{\dot{\gamma}}} \tag{7.62}$$

$$= -\eta\left[\nabla\underline{v} + (\nabla\underline{v})^T\right] \tag{7.63}$$

$$\nabla\underline{v} = \begin{pmatrix} \frac{\partial v_r}{\partial r} & \frac{\partial v_\theta}{\partial r} & \frac{\partial v_z}{\partial r} \\ \frac{1}{r}\frac{\partial v_r}{\partial \theta} - \frac{v_\theta}{r} & \frac{1}{r}\frac{\partial v_\theta}{\partial \theta} + \frac{v_r}{r} & \frac{1}{r}\frac{\partial v_z}{\partial \theta} \\ \frac{\partial v_r}{\partial z} & \frac{\partial v_\theta}{\partial z} & \frac{\partial v_z}{\partial z} \end{pmatrix}_{r\theta z} \tag{7.64}$$

$$= \begin{pmatrix} 0 & 0 & \frac{\partial v_z}{\partial r} \\ 0 & 0 & \frac{1}{r}\frac{\partial v_z}{\partial \theta} \\ 0 & 0 & \frac{\partial v_z}{\partial z} \end{pmatrix}_{r\theta z} \tag{7.65}$$

where $v_r = v_\theta = 0$ has been used to simplify $\nabla \underline{v}$ in the last step. Because of θ-symmetry and the continuity equation, two of the three remaining components of $\nabla \underline{v}$ are also zero, and thus $\dot{\underline{\underline{\gamma}}}$ is given by

$$\dot{\underline{\underline{\gamma}}} = \nabla \underline{v} + (\nabla \underline{v})^T \tag{7.66}$$

$$= \begin{pmatrix} 0 & 0 & \frac{\partial v_z}{\partial r} \\ 0 & 0 & 0 \\ 0 & 0 & 0 \end{pmatrix}_{r\theta z} + \begin{pmatrix} 0 & 0 & 0 \\ 0 & 0 & 0 \\ \frac{\partial v_z}{\partial r} & 0 & 0 \end{pmatrix}_{r\theta z} \tag{7.67}$$

$$= \begin{pmatrix} 0 & 0 & \frac{\partial v_z}{\partial r} \\ 0 & 0 & 0 \\ \frac{\partial v_z}{\partial r} & 0 & 0 \end{pmatrix}_{r\theta z} \tag{7.68}$$

Then $\underline{\underline{\tau}}$ becomes

$$\underline{\underline{\tau}} = -\eta \dot{\underline{\underline{\gamma}}} = \begin{pmatrix} 0 & 0 & -\eta \frac{\partial v_z}{\partial r} \\ 0 & 0 & 0 \\ -\eta \frac{\partial v_z}{\partial r} & 0 & 0 \end{pmatrix}_{r\theta z} \tag{7.69}$$

Now that we know more about the stress tensor we can simplify Equation (7.61) (all terms written in the r, θ, z coordinate system):

$$\begin{pmatrix} 0 \\ 0 \\ 0 \end{pmatrix} = \begin{pmatrix} -\frac{\partial p}{\partial r} \\ -\frac{1}{r}\frac{\partial p}{\partial \theta} \\ -\frac{\partial p}{\partial z} \end{pmatrix} - \begin{pmatrix} \frac{\partial \tau_{zr}}{\partial z} \\ 0 \\ \frac{1}{r}\frac{\partial}{\partial r}(r\tau_{rz}) \end{pmatrix} + \begin{pmatrix} 0 \\ 0 \\ \rho g \end{pmatrix} \tag{7.70}$$

$$= \begin{pmatrix} -\frac{\partial p}{\partial r} \\ -\frac{1}{r}\frac{\partial p}{\partial \theta} \\ -\frac{\partial p}{\partial z} \end{pmatrix} - \begin{pmatrix} \frac{\partial}{\partial z}\left(-\eta \frac{\partial v_z}{\partial r}\right) \\ 0 \\ \frac{1}{r}\frac{\partial}{\partial r}\left(-r\eta \frac{\partial v_z}{\partial r}\right) \end{pmatrix} + \begin{pmatrix} 0 \\ 0 \\ \rho g \end{pmatrix} \tag{7.71}$$

The function for $\eta(\dot{\gamma})$ is given by the power-law equation

$$\eta = m\dot{\gamma}^{n-1} \tag{7.72}$$

To evaluate this we must calculate $\dot{\gamma}$ from the tensor $\dot{\underline{\underline{\gamma}}}$:

$$\dot{\gamma} = |\dot{\underline{\underline{\gamma}}}| = +\sqrt{\frac{\dot{\underline{\underline{\gamma}}} : \dot{\underline{\underline{\gamma}}}}{2}} \tag{7.73}$$

$$= +\sqrt{\left(\frac{\partial v_z}{\partial r}\right)^2} \tag{7.74}$$

$$= \pm \frac{\partial v_z}{\partial r} \tag{7.75}$$

The definition of magnitude requires that the result be a positive number. Thus our choice of the sign that precedes $\partial v_z/\partial r$ in Equation (7.75) depends on whether the derivative $\partial v_z/\partial r$ is positive or negative. In the current example, as r increases, the velocity decreases—the velocity is at its maximum at the center of the tube (Figure 7.5). The derivative $\partial v_z/\partial r$ is negative, and the correct sign in Equation (7.75), therefore, is the negative:

$$\dot{\gamma} = |\underline{\dot{\gamma}}| = -\frac{\partial v_z}{\partial r} > 0 \tag{7.76}$$

The power-law equation thus becomes

$$\eta = m\left(-\frac{\partial v_z}{\partial r}\right)^{n-1} = m\left(-\frac{dv_z}{dr}\right)^{n-1} \tag{7.77}$$

Since we have assumed that v_z is not a function of θ and the continuity equation told us that v_z is not a function of z, we know that $v_z = v_z(r)$, and we have changed the partial derivatives $\partial/\partial r$ to total derivatives d/dr in the second expression of Equation (7.77).

Substituting this back into the simplified equation of motion [Equation (7.71)] we obtain (all terms written in r, θ, z coordinates)

$$\begin{pmatrix} 0 \\ 0 \\ 0 \end{pmatrix} = \begin{pmatrix} -\frac{\partial p}{\partial r} \\ -\frac{1}{r}\frac{\partial p}{\partial \theta} \\ -\frac{\partial p}{\partial z} \end{pmatrix} - \begin{pmatrix} \frac{\partial}{\partial z}\left[-m\left(-\frac{dv_z}{dr}\right)^{n-1}\frac{dv_z}{dr}\right] \\ 0 \\ \frac{1}{r}\frac{\partial}{\partial r}\left[-r\,m\left(-\frac{dv_z}{dr}\right)^{n-1}\frac{dv_z}{dr}\right] \end{pmatrix} + \begin{pmatrix} 0 \\ 0 \\ \rho g \end{pmatrix} \tag{7.78}$$

$$= \begin{pmatrix} -\frac{\partial p}{\partial r} \\ -\frac{1}{r}\frac{\partial p}{\partial \theta} \\ -\frac{\partial p}{\partial z} \end{pmatrix} - \begin{pmatrix} 0 \\ 0 \\ \frac{m}{r}\frac{d}{dr}\left[r\left(-\frac{dv_z}{dr}\right)^{n}\right] \end{pmatrix} + \begin{pmatrix} 0 \\ 0 \\ \rho g \end{pmatrix} \tag{7.79}$$

The r-component of $\nabla \cdot \underline{\tau}$ is zero since v_z is not a function of z. The equation of motion (EOM) is now quite simple, yielding that the pressure is only a function of z, and an equation to be solved for $v_z(r)$:

$$r\text{-component of EOM:} \quad \frac{\partial p}{\partial r} = 0 \tag{7.80}$$

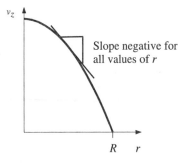

Figure 7.5 Illustration of negative $\partial v_z/\partial r$ for all values of r for Poiseuille flow in a tube.

v_z

Slope negative for
all values of r

R r

$$\theta\text{-component of EOM:}\quad \frac{1}{r}\frac{\partial p}{\partial \theta} = 0 \tag{7.81}$$

$$z\text{-component of EOM:}\quad \frac{\partial p}{\partial z} = -\frac{m}{r}\frac{d}{dr}\left[r\left(-\frac{dv_z}{dr}\right)^n\right] + \rho g \tag{7.82}$$

Note that since p is only a function of z, $\partial p/\partial z$ becomes dp/dz in Equation (7.82). Equation (7.82) is a differential equation for p and v_z and can be solved using separation of variables, as was shown earlier for Newtonian fluids (Section 3.5.2). The boundary conditions for this problem match those in the Newtonian case, and allow us to solve for the three integration constants and the constant that arises from the separation of variables technique (see Section 3.5.2).

$$
\begin{array}{lll}
z = 0 & p = P_0 \\[4pt]
z = L & p = P_L \\[4pt]
r = 0 & \dfrac{dv_z}{dr} = 0 \\[4pt]
r = R & v_z = 0
\end{array}
\tag{7.83}
$$

The final results for v_z and p are

$$\boxed{p = \frac{P_L - P_0}{L}z + P_0} \tag{7.84}$$

$$\boxed{v_z = R^{\frac{1}{n}+1}\left(\frac{P_0 - P_L + \rho g L}{2mL}\right)^{\frac{1}{n}}\left(\frac{n}{n+1}\right)\left[1 - \left(\frac{r}{R}\right)^{\frac{1}{n}+1}\right]} \tag{7.85}$$

The stress tensor is given by Equation (7.69). To calculate $\underline{\underline{\tau}}$ explicitly we must evaluate $\tau_{rz} = -\eta \partial v_z/\partial r$:

$$\tau_{rz} = -\eta\frac{\partial v_z}{\partial r} \tag{7.86}$$

$$= -m\left(-\frac{dv_z}{dr}\right)^{n-1}\left(\frac{dv_z}{dr}\right) = m\left(-\frac{dv_z}{dr}\right)^n \tag{7.87}$$

$$= \frac{(P_0 - P_L + \rho g L)r}{2L} \tag{7.88}$$

$$\boxed{\tau_{rz}(r) = \frac{(\mathcal{P}_0 - \mathcal{P}_L)r}{2L}} \tag{7.89}$$

where as before $\mathcal{P} = p - \rho g z$ is used to fold the influence of gravity into the modified pressure $\mathcal{P}$. The stress tensor is then

$$\underline{\underline{\tau}} = \begin{pmatrix} 0 & 0 & \frac{(\mathcal{P}_0 - \mathcal{P}_L)r}{2L} \\ 0 & 0 & 0 \\ \frac{(\mathcal{P}_0 - \mathcal{P}_L)r}{2L} & 0 & 0 \end{pmatrix}_{r\theta z} \tag{7.90}$$

Calculating the flow rate and average velocity from the velocity profile is straightforward:

$$Q = \int_A v_z \, dA \tag{7.91}$$

Poiseuille flow
power-law GNF
$$Q = \left[\frac{(\mathcal{P}_0 - \mathcal{P}_L)R}{2mL} \right]^{\frac{1}{n}} \left(\frac{n\pi R^3}{1 + 3n} \right) \tag{7.92}$$

The predicted velocity profile is shown in Figure 7.6. For $n = 1$ the Newtonian case is recovered, and as the power-law index n decreases from 1, the profiles flatten. Plug flow is reached for $n \longrightarrow 0$. The shear stress is found to be the same simple linear function of r that was calculated for the Newtonian case [compare Equations (3.225) and (7.90)].

7.4.2 COMBINED DRAG AND POISEUILLE FLOW THROUGH A SLIT

Calculate the velocity profile for pressure-driven flow of an incompressible power-law liquid confined between two infinite plates if the top plate is moving at a constant velocity V. The pressure at an upstream point is P_0, and at a point a distance L downstream the pressure is P_L ($P_0 > P_L$). Assume that the flow between these two

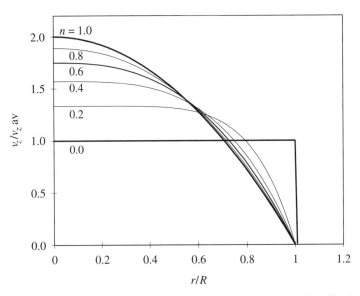

Figure 7.6 Velocity profiles for different values of the power-law index n, predicted by the power-law generalized Newtonian fluid model for steady Poiseuille flow in a tube.

points is fully developed and at steady state. The separation between the two plates is H, and the effect of gravity is negligible. This problem arises in sheet coating [236].

This problem is worked in rectangular coordinates. From mass conservation we have

$$0 = \nabla \cdot \underline{v} \tag{7.93}$$

$$= \frac{\partial v_x}{\partial x} + \frac{\partial v_y}{\partial y} + \frac{\partial v_z}{\partial z} \tag{7.94}$$

We will take the flow to be in the x-direction (Figure 7.7), and thus the y- and z-components of $\underline{v}$ are zero,

$$\underline{v} = \begin{pmatrix} v_x \\ v_y \\ v_z \end{pmatrix}_{xyz} \tag{7.95}$$

$$= \begin{pmatrix} v_x \\ 0 \\ 0 \end{pmatrix}_{xyz} \tag{7.96}$$

The continuity equation thus gives us the result

$$\frac{\partial v_x}{\partial x} = 0 \tag{7.97}$$

The equation of motion for an incompressible fluid is

$$\rho \left(\frac{\partial \underline{v}}{\partial t} + \underline{v} \cdot \nabla \underline{v} \right) = -\nabla p - \nabla \cdot \underline{\underline{\tau}} + \rho \underline{g} \tag{7.98}$$

For steady-state unidirectional flow where gravity is neglected this becomes

$$\underline{0} = -\nabla p - \nabla \cdot \underline{\underline{\tau}} \tag{7.99}$$

$$\begin{pmatrix} 0 \\ 0 \\ 0 \end{pmatrix}_{xyz} = \begin{pmatrix} -\frac{\partial p}{\partial x} \\ -\frac{\partial p}{\partial y} \\ -\frac{\partial p}{\partial z} \end{pmatrix}_{xyz} - \begin{pmatrix} \frac{\partial \tau_{xx}}{\partial x} + \frac{\partial \tau_{yx}}{\partial y} + \frac{\partial \tau_{zx}}{\partial z} \\ \frac{\partial \tau_{xy}}{\partial x} + \frac{\partial \tau_{yy}}{\partial y} + \frac{\partial \tau_{zy}}{\partial z} \\ \frac{\partial \tau_{xz}}{\partial x} + \frac{\partial \tau_{yz}}{\partial y} + \frac{\partial \tau_{zz}}{\partial z} \end{pmatrix}_{xyz} \tag{7.100}$$

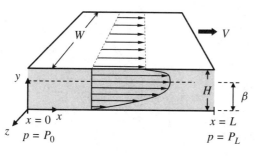

Figure 7.7 Flow example: combined drag and Poiseuille flow of a power-law generalized Newtonian fluid in a slit.

To proceed further we must relate $\underline{\underline{\tau}}$ to the velocity components. As in the previous example, we use the power-law, GNF equation:

$$\underline{\underline{\tau}} = -\eta \underline{\underline{\dot{\gamma}}} = -\eta[\nabla \underline{v} + (\nabla \underline{v})^T)] \tag{7.101}$$

$$\nabla \underline{v} = \begin{pmatrix} \dfrac{\partial v_x}{\partial x} & \dfrac{\partial v_y}{\partial x} & \dfrac{\partial v_z}{\partial x} \\[2mm] \dfrac{\partial v_x}{\partial y} & \dfrac{\partial v_y}{\partial y} & \dfrac{\partial v_z}{\partial y} \\[2mm] \dfrac{\partial v_x}{\partial z} & \dfrac{\partial v_y}{\partial z} & \dfrac{\partial v_z}{\partial z} \end{pmatrix}_{xyz} \tag{7.102}$$

$$= \begin{pmatrix} \dfrac{\partial v_x}{\partial x} & 0 & 0 \\[2mm] \dfrac{\partial v_x}{\partial y} & 0 & 0 \\[2mm] \dfrac{\partial v_x}{\partial z} & 0 & 0 \end{pmatrix}_{xyz} \tag{7.103}$$

where $v_y = v_z = 0$ has been used to simplify $\nabla \underline{v}$ in the last step. Because the plates are infinite in the z-direction, there is no variation of any quantity in the z-direction, and the velocity cannot depend on z. Also, the continuity equation tells us that $\partial v_x / \partial x = 0$, and thus $v_x = v_x(y)$ only, and the partial derivative of v_x with respect to y becomes the total derivative of v_x with respect to y $(\partial v_x / \partial y = dv_x / dy)$. The rate-of-deformation tensor is therefore

$$\underline{\underline{\dot{\gamma}}} = \nabla \underline{v} + (\nabla \underline{v})^T \tag{7.104}$$

$$= \begin{pmatrix} 0 & 0 & 0 \\[1mm] \dfrac{dv_x}{dy} & 0 & 0 \\[2mm] 0 & 0 & 0 \end{pmatrix}_{xyz} + \begin{pmatrix} 0 & \dfrac{dv_x}{dy} & 0 \\[2mm] 0 & 0 & 0 \\[1mm] 0 & 0 & 0 \end{pmatrix}_{xyz} \tag{7.105}$$

$$= \begin{pmatrix} 0 & \dfrac{dv_x}{dy} & 0 \\[2mm] \dfrac{dv_x}{dy} & 0 & 0 \\[2mm] 0 & 0 & 0 \end{pmatrix}_{xyz} \tag{7.106}$$

Then $\underline{\underline{\tau}}$ becomes

$$\underline{\underline{\tau}} = \begin{pmatrix} 0 & -\eta\dfrac{dv_x}{dy} & 0 \\[2mm] -\eta\dfrac{dv_x}{dy} & 0 & 0 \\[2mm] 0 & 0 & 0 \end{pmatrix}_{xyz} \tag{7.107}$$

Returning to the simplified momentum balance [Equation (7.100)] and using the stress tensor derived in Equation (7.107), we now obtain

$$\begin{pmatrix} 0 \\ 0 \\ 0 \end{pmatrix}_{xyz} = \begin{pmatrix} -\dfrac{\partial p}{\partial x} \\[2mm] -\dfrac{\partial p}{\partial y} \\[2mm] -\dfrac{\partial p}{\partial z} \end{pmatrix}_{xyz} - \begin{pmatrix} \dfrac{\partial \tau_{yx}}{\partial y} \\[2mm] \dfrac{\partial \tau_{xy}}{\partial x} \\[2mm] 0 \end{pmatrix}_{xyz} \tag{7.108}$$

$$= \begin{pmatrix} -\frac{\partial p}{\partial x} \\ -\frac{\partial p}{\partial y} \\ -\frac{\partial p}{\partial z} \end{pmatrix}_{xyz} - \begin{pmatrix} \frac{\partial}{\partial y}\left(-\eta\frac{dv_x}{dy}\right) \\ \frac{\partial}{\partial x}\left(-\eta\frac{dv_x}{dy}\right) \\ 0 \end{pmatrix}_{xyz} \qquad (7.109)$$

The function for $\eta(\dot{\gamma})$ is given by the power-law equation

$$\eta = m\dot{\gamma}^{n-1} \qquad (7.110)$$

We need to calculate $\dot{\gamma}$ from the tensor $\underline{\underline{\dot{\gamma}}}$,

$$\dot{\gamma} = |\underline{\underline{\dot{\gamma}}}| = +\sqrt{\frac{\underline{\underline{\dot{\gamma}}}:\underline{\underline{\dot{\gamma}}}}{2}} = +\sqrt{\left(\frac{dv_x}{dy}\right)^2} \qquad (7.111)$$

$$= \pm\frac{dv_x}{dy} \qquad (7.112)$$

As we discussed in the last example, the definition of tensor magnitude requires that the quantity obtained for the magnitude be positive. Thus our choice for the sign that precedes dv_x/dy in Equation (7.112) depends on whether that derivative is positive or negative. In the current problem, if we choose y to be zero at the stationary plate, the sign of dv_x/dy depends on the magnitude of the pressure gradient relative to the plate velocity V (Figure 7.8). We must consider two cases.

Case 1:   There is no maximum in $v_x(y)$. In this case $dv_x/dy > 0$ always, and therefore $\dot{\gamma} = +dv_x/dy$. We will use $v_{x,1}$ for the velocity profile in case 1. The power-law equation thus becomes

$$\eta = m\left(\frac{dv_{x,1}}{dy}\right)^{n-1} \qquad (7.113)$$

and the equation of motion is

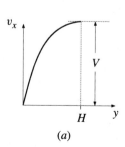

(a)

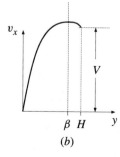

(b)

Figure 7.8 Two possible types of velocity profile generated in combined Poiseuille and drag flow. (*a*) The pressure gradient is not sufficient to produce a velocity maximum between the two confining walls at $y = 0$ and $y = H$ (case 1). (*b*) A maximum in velocity is seen at $y = \beta$ (case 2).

$$\begin{pmatrix} 0 \\ 0 \\ 0 \end{pmatrix}_{xyz} = \begin{pmatrix} -\frac{\partial p}{\partial x} \\ -\frac{\partial p}{\partial y} \\ -\frac{\partial p}{\partial z} \end{pmatrix}_{xyz} - \begin{pmatrix} \frac{\partial}{\partial y}\left[-m\left(\frac{dv_{x,1}}{dy}\right)^{n-1}\frac{dv_{x,1}}{dy}\right] \\ \frac{\partial}{\partial x}\left[-m\left(\frac{dv_{x,1}}{dy}\right)^{n-1}\frac{dv_{x,1}}{dy}\right] \\ 0 \end{pmatrix}_{xyz} \tag{7.114}$$

Since $v_{x,1}$ is only a function of y, the y-component of $\nabla \cdot \underline{\underline{\tau}}$ is zero. Thus the y- and z-components of the equation of motion (EOM) give

$$y\text{-component of EOM:} \quad \frac{\partial p}{\partial y} = 0 \tag{7.115}$$

$$z\text{-component of EOM:} \quad \frac{\partial p}{\partial z} = 0 \tag{7.116}$$

and we can conclude that $p = p(x)$ only. The x-component of the equation of motion gives

$$x\text{-component of EOM:} \quad \frac{dp}{dx} = m\frac{d}{dy}\left[\left(\frac{dv_{x,1}}{dy}\right)^n\right] \tag{7.117}$$

which is a separable differential equation. The boundary conditions are

$$\begin{aligned} x &= 0 & p &= P_0 \\ x &= L & p &= P_L \\ y &= 0 & v_{x,1} &= 0 \\ y &= H & v_{x,1} &= V \end{aligned} \tag{7.118}$$

The solution for $v_{x,1}$ subject to these boundary conditions is

Velocity
profile,
case 1
$$v_{x,1} = \frac{mL}{P_L - P_0}\left(\frac{n}{n+1}\right)\left[\left(\frac{P_L - P_0}{mL}y + C_1\right)^{\frac{n+1}{n}} - C_1^{\frac{n+1}{n}}\right] \tag{7.119}$$

where C_1 is found from the (numerical) solution of

$$V = \frac{mL}{P_L - P_0}\left(\frac{n}{n+1}\right)\left[\left(\frac{P_L - P_0}{mL}H + C_1\right)^{\frac{n+1}{n}} - C_1^{\frac{n+1}{n}}\right] \tag{7.120}$$

Case 2:   There is a maximum in $v_x(y)$ at $y = \beta$ (Figure 7.8). We will call the velocity in this case $v_{x,2}$, and then $\dot{\gamma}$ is given by

$$\dot{\gamma} = \begin{cases} +\frac{dv_{x,2}}{dy} & 0 \le y < \beta \\ -\frac{dv_{x,2}}{dy} & \beta \le y \le H \end{cases} \tag{7.121}$$

The power-law equation thus becomes

$$\eta = m \left(\frac{dv_{x,2}}{dy}\right)^{n-1} \quad 0 \le y < \beta \tag{7.122}$$

$$\eta = m \left(-\frac{dv_{x,2}}{dy}\right)^{n-1} \quad \beta \le y \le H \tag{7.123}$$

For the case where $\dot{\gamma} = +dv_{x,2}/dy$, the equation for $v_{x,2}(y)$ is the same as in the first case, except for different boundary conditions on y. The boundary conditions are

$$\begin{aligned} y = 0 \quad & v_{x,2} = 0 \\ y = \beta \quad & v_{x,2} = v_{x,1} \end{aligned} \tag{7.124}$$

where $v_{x,1}$ and $v_{x,2}$ are the solutions for v_x in cases 1 and 2, respectively. For the case where $\dot{\gamma} = -dv_{x,2}/dy$, the equation is different:

$$-\frac{dp}{dx} = m \frac{\partial}{\partial y}\left[\left(-\frac{dv_{x,2}}{dy}\right)^n\right] \tag{7.125}$$

and the solution may be obtained by separation of variables, as in case 1. The boundary conditions are

$$\begin{aligned} y = \beta \quad & v_{x,2} = v_{x,1} \\ y = H \quad & v_{x,2} = V \end{aligned} \tag{7.126}$$

In addition, recall that $dv_x/dy = 0$ at $y = \beta$ for both $v_{x,1}$ and $v_{x,2}$. The predicted velocity profiles are shown in Figures 7.9 and 7.10 for selected values of the power-law parameters. The complete solution can be found in Flumerfelt et al. [80] and is discussed in some detail in Problem 7.26.

7.5 Limitations on GNF Models

As pointed out in the introduction to this chapter, the GNF models are popular because of the relative ease with which flow calculations can be made. They also enjoy success in predicting pressure-drop versus flow curves for polymer processes [236]. There are some significant limitations to the usefulness of these models, however:

1. Some GNF models (the power-law model, for instance) do not accurately model the zero-shear region of the viscosity curve. The power-law model is popular for its simplicity, but it cannot be relied upon if the shear rate becomes small in the flow of interest.

2. Since the GNF models rely on the modeling shear viscosity $\eta(\dot{\gamma})$ to incorporate non-Newtonian effects, it is not clear whether these models will be useful in nonshearing flows. We can calculate nonshear material functions (e.g., $\bar{\eta}$, $\bar{\eta}^+$, etc.) using generalized Newtonian fluids, but in many cases they do not match observations.

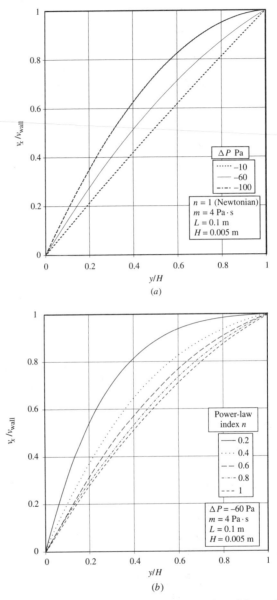

Figure 7.9 Velocity profiles for different values of (*a*) pressure drop (Newtonian case, $n = 1$) and (*b*) power-law index n predicted by the power-law generalized Newtonian fluid model for combined drag and pressure-driven flow in a slit. Case 1, no velocity maximum. Values of the parameters used in the calculations are given in the figure.

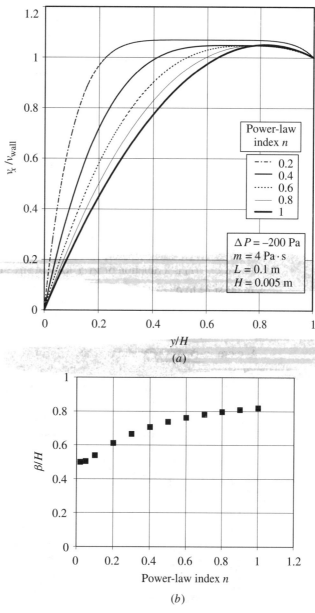

Figure 7.10 (*a*) Velocity profiles for different values of the power-law index n predicted by the power-law generalized Newtonian fluid model for combined drag and pressure-driven flow in a slit. Case 2, velocity shows a maximum. (*b*) Variation in the location of the velocity maximum as a function of power-law index. Values of the parameters used in the calculations are given in the figure.

3. The GNF models do not predict shear normal stresses N_1 and N_2, which are elastic effects. In fact, all constitutive equations that are proportional to $\dot{\gamma}$ fail to predict normal stresses in shear flow because of the form of $\dot{\underline{\underline{\gamma}}}$ for shear flow:

$$\text{shear flow:} \quad \dot{\underline{\underline{\gamma}}} = \begin{pmatrix} 0 & \dot{\gamma}_{21}(t) & 0 \\ \dot{\gamma}_{21}(t) & 0 & 0 \\ 0 & 0 & 0 \end{pmatrix}_{123} \tag{7.127}$$

$$\text{GNF:} \quad \underline{\underline{\tau}} = -\eta \dot{\underline{\underline{\gamma}}} \tag{7.128}$$

$$= \begin{pmatrix} 0 & -\eta \dot{\gamma}_{21}(t) & 0 \\ -\eta \dot{\gamma}_{21}(t) & 0 & 0 \\ 0 & 0 & 0 \end{pmatrix}_{123} \tag{7.129}$$

4. Since the GNF model is a direct empirical extension of the Newtonian fluid model, there is no guarantee that elastic effects are properly accounted for in any of the GNF models. In fact, since the GNF models are only a function of the instantaneous rate-of-deformation tensor [i.e., $\underline{\underline{\tau}}(t)$ is a function of $\dot{\underline{\underline{\gamma}}}(t)$ not $\dot{\underline{\underline{\gamma}}}$ at any times other than the present time t], it is impossible to predict observed material behavior such as strain recoil after creep and gradual stress growth, as these effects depend on the history of the rate-of-deformation tensor.

A more in-depth evaluation of the GNF models may be found in Bird et al. [26]. Several books on polymer processing contain solutions to problems employing the power-law generalized Newtonian fluid as well as problems employing other GNF models [236, 179]. There are also several practice problems using GNF models at the end of this chapter.

To move beyond the GNF models toward more realistic constitutive equations, we must consider memory effects. That is the subject of the next chapter.

7.6 PROBLEMS

7.1 How are tensor invariants important in constitutive modeling?

7.2 Why is $\dot{\gamma} = |\dot{\underline{\underline{\gamma}}}|$ the only kinematic parameter that appears in the GNF models?

7.3 What are the units of m and n in the power-law GNF constitutive equation? Explain these units.

7.4 The viscosity function for a given material is measured and found to be

$$\eta = 5450 \dot{\gamma}^{-0.34}$$

where shear rate $\dot{\gamma}$ is in units of s^{-1} and viscosity η is in Pa · s. What is the power-law index n? What are the units of 5450?

7.5 For the viscosity versus shear-rate data given in Table 7.1, calculate the consistency index m and the power-law index n.

7.6 From the solution for $v_z(r)$ in Equation (7.85), calculate the flow rate and the average velocity for Poiseuille flow in a tube (pressure-driven unidirectional flow) of a power-law generalized Newtonian fluid.

7.7 Solve Equation (7.82) for the pressure profile and velocity profile for a power-law generalized Newtonian fluid in Poiseuille flow in a tube.

7.8 Will a fluid that follows the power-law GNF constitutive equation exhibit rod climbing? Why or why not?

7.9 Can the power-law GNF constitutive equation predict shear-thickening?

7.10 When we say that a model is empirical, what does that mean?

TABLE 7.1
Viscosity η versus Shear Rate $\dot{\gamma}$ for Problem 7.5

$\dot{\gamma}$ (s^{-1})	η (Pa·s)
0.020	7.5×10^5
0.050	4.5×10^5
0.10	3.5×10^5
0.20	2.0×10^5
0.50	1.3×10^5
1.0	1.0×10^5
2.0	6.0×10^4
5.0	3.5×10^4
10	2.8×10^4
20	1.7×10^4
50	1.0×10^4
100	8.0×10^3

7.11 Calculate the viscosities in steady planar elongation, $\bar{\eta}_{P_1}$ and $\bar{\eta}_{P_2}$, for a Carreau–Yasuda, generalized Newtonian fluid.

7.12 Calculate the shear stress growth coefficients $\eta^+(t)$, $\Psi_1^+(t)$, and $\Psi_2^+(t)$ for a Bingham generalized Newtonian fluid. Sketch your results.

7.13 Calculate the viscosities $\bar{\eta}_{P_1}$ and $\bar{\eta}_{P_2}$ in steady planar elongation for a power-law generalized Newtonian fluid.

7.14 Calculate the uniaxial elongational stress growth function $\bar{\eta}^+(t)$ for a power-law generalized Newtonian fluid. Sketch your answer.

7.15 What is the stress tensor $\underline{\underline{\tau}}$ for all incompressible generalized Newtonian fluids in Poiseuille flow (pressure-driven, laminar flow) in a pipe? What is $\tau_{rz}(r)$? List all of your assumptions. Do not assume power-law or any other specific model.

7.16 The friction factor f for fully developed laminar flow in a circular pipe is given by [28]

$$f = \frac{1}{4} \left(\frac{D}{L} \right) \frac{\Delta P}{\frac{1}{2} \rho v_{z,av}^2}$$

where D is the pipe diameter, L is the pipe length, ΔP is the pressure drop, ρ is the density, and $v_{z,av}$ is the average velocity in the pipe. The Reynolds number for Newtonian fluids is $\rho v_{z,av} D/\mu$, where μ is the Newtonian viscosity. A generalized Reynolds number for power-law fluids, Re_{gen}, can be defined by requiring that $f = 16/\text{Re}_{\text{gen}}$, in analogy with

the Newtonian case. What is Re_{gen} for a power-law generalized Newtonian fluid? Note that your expression should not contain ΔP.

7.17 I propose the following constitutive equation:

$$-\underline{\underline{\tau}} = \alpha \left[\nabla \underline{v} + (\nabla \underline{v})^T \right] + \beta (\nabla \underline{v})^T \cdot \nabla \underline{v}$$

where α and β are constant parameters associated with the material described by the model. A student says, "No way." Based on what we know so far about constitutive equations, is she right? Justify your answer.

7.18 Massa et al. [167] reported the complex viscosity master curve $\eta'(25°\text{C}) = \eta'(T)/a_T(T)$ of a solution of narrow-polydispersity polystyrene ($M = 860$ kg/mol) in chlorinated diphenyl (concentration $= 0.0154$ g/cm^3) as a function of $a_T \omega$ (rad/s), where a_T is the time–temperature shift factor (see Section 6.2.1.3), and ω is the oscillation frequency in the small-amplitude oscillatory shear experiment. The data are given in Table 7.2. The solution viscosity η'_s has been subtracted from the measured η' in order to isolate the polymeric contribution to the complex viscosity, $\eta'_p = \eta' - \eta'_s$. This subtraction is done in order to be able to compare experimental results to molecular models that predict the polymeric contributions to viscosity [27].

Recall that often the Cox–Merz rule applies to polymer data (see Section 6.2.1.1). Assuming that the Cox–Merz rule applies, $\eta'_p(a_T \omega) = \eta(\dot{\gamma}_0)$, fit the data in Table 7.2 to the Carreau–Yasuda GNF model. Plot the data and the best-fit prediction of the Carreau–Yasuda GNF model. What are the values of the model parameters (η_0, η_∞, n, a, λ) that give the best fit? (*Hint:* You can obtain an acceptable fit through trial and error; another way would be to use the add-in Solver in Microsoft Excel or any other nonlinear fitting algorithm.)

7.19 Your labmate proposes the following constitutive equation:

$$\underline{\underline{\tau}} = -f(v_1, v_2) \left[\nabla \underline{v} + (\nabla \underline{v})^T \right]$$

$$f(v_1, v_2) = \frac{\alpha v_2^2 + \beta v_1}{2}$$

where α and β are scalars associated with the material, and $\underline{v} = \begin{pmatrix} v_1 \\ v_2 \\ v_3 \end{pmatrix}_{123}$. Discuss why it is or is not acceptable as a constitutive equation.

TABLE 7.2
Data for Problem 7.18*

$a_T \omega \approx \dot{\gamma}_0$ (rad/s)	$\eta'_p (25°C) \approx \eta$ (poise)
9.97E−01	1.72E+01
3.89E+00	1.69E+01
9.89E+00	1.62E+01
2.47E+01	1.40E+01
6.26E+01	9.86E+00
2.49E+02	5.40E+00
1.01E+03	3.12E+00
2.56E+03	2.28E+00
6.30E+03	1.67E+00
1.57E+04	1.30E+00
4.05E+04	1.09E+00
1.27E+05	9.20E−01
5.04E+05	7.97E−01
1.27E+06	7.27E−01
3.17E+06	7.30E−01

*A more complete data set may be found in Table F.1 in Appendix F.

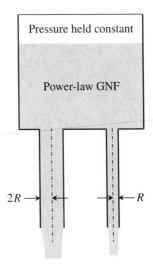

Figure 7.11 Problem 7.22: tank containing a power-law generalized Newtonian fluid.

7.20 I have decided that there is a need for a new material function for simple shear flow. I propose the following kinematics:

$$\underline{v} = \begin{pmatrix} \dot{\varsigma}(t)y \\ 0 \\ 0 \end{pmatrix}_{xyz}$$

$$\dot{\varsigma}(t) = \begin{cases} 0 & t < 0 \\ e^{at} & t \geq 0; a > 0 \end{cases}$$

(a) Sketch $\dot{\varsigma}(t) = \dot{\gamma}_{21}(t)$ and $\gamma_{21}(0, t)$.

(b) The material function I define is

$$\eta^{\exp} \equiv \frac{-\tau_{yx}(t)}{a}$$

What is $\eta^{\exp}$ for a power-law generalized Newtonian fluid?

7.21 The solution presented in the text for Poiseuille flow in a tube for a power-law fluid assumes power-law behavior at all shear rates. The shear rate at the tube center is zero, however, and power-law behavior is not expected at low shear rates. Estimate the error in flow rates calculated from the power-law generalized Newtonian fluid based on this effect [235]. For simplicity, consider flow between infinite parallel plates.

7.22 A power-law generalized Newtonian fluid is contained in a tank in which the pressure is maintained constant (Figure 7.11). The tank has two exit pipes, one of radius R and the second of radius $2R$. Both exit pipes are of length L, and the fluid exits the pipes at atmospheric pressure. The flow rate of the fluid in the larger pipe is 32 times the flow rate in the smaller pipe. What is the fluid's power-law index? The effect of gravity can be neglected.

7.23 (a) For a power-law generalized Newtonian fluid with the parameters listed below, plot the viscosity η as a function of shear rate $\dot{\gamma}$ (log–log plot).

$$m = 50,000 \text{ Pa}$$
$$n = 0.4$$

(b) For both m and n, calculate the viscosity function for twice and one-half the value of the parameter (e.g., for $2m$ and $m/2$), keeping the other parameter the same as the in base case. Plot and discuss your results.

7.24 (a) For a Carreau–Yasuda generalized Newtonian fluid with the parameters listed below, plot the viscosity as a function of shear rate (log–log plot). Plot for $10^{-3} \leq \omega \leq 10^3$.

$$\eta_\infty = 500 \text{ Pa}$$
$$\eta_0 = 50,000 \text{ Pa}$$
$$a = 1$$
$$n = 0.5$$
$$\lambda = 10 \text{ s}$$

(b) For each of the five parameters above calculate the viscosity function for twice and one-half the value of the parameter (e.g., for $2\eta_\infty$ and $\eta_\infty/2$), keeping all other parameters the same as in the base case. Plot and discuss your results for all five parameters.

7.25 (a) The generalized Newtonian constitutive equation $\underline{\underline{\tau}} = -\eta\dot{\underline{\underline{\gamma}}}$ can use any of a number of empirical relationships for $\eta(\dot{\gamma})$. One that we have used extensively is the power-law equation, but others are equally valid, for example, the Ellis equation, which is a three-parameter model (η_0, τ_0, α):

$$\frac{\eta}{\eta_0} = \frac{1}{1 + (\tau/\tau_0)^{\alpha-1}}$$

where τ is the magnitude of the stress tensor $\underline{\underline{\tau}}$. For an Ellis generalized Newtonian fluid with the parameters listed below, plot the viscosity as a function of shear rate (log–log plot).

$$\eta_0 = 50{,}000 \text{ Pa}$$
$$\tau_0 = 25{,}000 \text{ Pa}$$
$$\alpha = 3.0$$

(b) For each of the three parameters above calculate the viscosity function for twice and one-half the value of the parameter (e.g., for $2\eta_0$ and $\eta_0/2$), keeping all other parameters the same as in the base case. Plot and discuss your results.

7.26 In the text, a partial solution for combined pressure-driven/drag flow of a power-law generalized Newtonian fluid in a slit is given, and results for particular values of the material and flow parameters are plotted. The complete solution to this problem is given in Flumerfelt et al. [80]. Their solution was calculated for the coordinate system x, y, z, which differs from that used in this text (shown as $\bar{x}, \bar{y}, \bar{z}$); see Figure 7.12. For their Case I, no maximum in velocity, their result is

$$\frac{v_z}{V} = \frac{1}{\Lambda(s+1)}\left\{\left[\Lambda\left(\lambda + \frac{1}{2}\right)\right]^{s+1}\right.$$
$$\left. - \left[\Lambda\left(\lambda - \xi\right)\right]^{s+1}\right\}$$

where

$$\Lambda = \frac{(P_0 - P_L)B}{mL}\left(\frac{B}{V}\right)^n$$

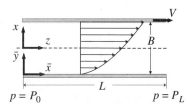

Figure 7.12 Coordinate system used by Flumerfelt et al. [80] in solving generalized Couette flow discussed in Problem 7.26.

$\xi = x/B$, $s = 1/n$, and the constant λ is found through numerical solution of the equation

$$1 = \frac{1}{\Lambda(s+1)}\left\{\left[\Lambda\left(\lambda + \frac{1}{2}\right)\right]^{s+1}\right.$$
$$\left. - \left[\Lambda\left(\lambda - \frac{1}{2}\right)\right]^{s+1}\right\}$$

(a) Show that the solution in this text and the one given here from reference [80] match.

(b) Plot Flumerfelt et al.'s results for the case when a maximum is present (their Case II); the solution is given below. Plot (i) for $\Lambda = \pm3, n = 0.2, 0.4, 0.5, 0.8, 1.0$, and (ii) for $n = 5, \Lambda = \pm1.8, \pm3.0, \pm5.0, \pm7.0, \pm10$. This flow is important in the analysis of extrusion.

Solution for Case II [80]:

$$\frac{v_z^<}{V} = \frac{\Lambda|\Lambda|^{s-1}}{s+1}\left[\left(\lambda + \frac{1}{2}\right)^{s+1} - (\lambda - \xi)^{s+1}\right]$$

$$\frac{v_z^>}{V} = \frac{\Lambda|\Lambda|^{s-1}}{s+1}\left[\left(\frac{1}{2} - \lambda\right)^{s+1} - (\xi - \lambda)^{s+1}\right] + 1$$

where λ is given by the solution to

$$1 = \frac{\Lambda|\Lambda|^{s-1}}{s+1}\left[\left(\frac{1}{2} + \lambda\right)^{s+1} - \left(\frac{1}{2} - \lambda\right)^{s+1}\right]$$

The maximum is located at $\xi = \lambda$; $v_z^<$ is the solution for the velocity profile when $\xi \leq \lambda$, and $v_z^>$ is the solution when $\xi \geq \lambda$.

(c) Flumerfelt et al. qualify their solution as follows: "There are two important assumptions in the numerical results: It is assumed that there are no appreciable viscous heating effects that would cause the rheological properties to be position-dependent. It is assumed that the power law is appropriate; in Case II, if an appreciable portion of the slit cross section is in the neighborhood of zero velocity gradient, the inadequacies of the power law may become apparent." Explain these comments.

7.27 Flow Problem: Pressure-driven flow of a power-law generalized Newtonian fluid between infinite parallel plates. Calculate the velocity profile and flow rate for pressure-driven flow of an incompressible power-law generalized Newtonian liquid confined between two infinitely wide parallel plates, separated by a gap of $2H$. (The geometry is the same as in the Newtonian case, treated in Section 3.5.2.) The pressure at an upstream point is P_0, and at a point a distance L downstream the pressure is P_L. Assume that the flow between these two points is fully developed and at steady state. Plot the velocity profile as $v_1/v_{1,av}$ versus x_2/H for $n = 0.1, 0.5, 0.99$. Comment on your results.

7.28 Flow Problem: Flow of a power-law generalized Newtonian fluid down an inclined plane.

(a) Find the velocity profile at steady state for an incompressible power-law generalized Newtonian fluid flowing down an inclined plane. The upper surface of the fluid is exposed to air. The fluid film has a constant height of H at steady state. The plane makes an angle of ψ with the vertical.

(b) Calculate the maximum velocity, the average velocity, and the flow rate of the fluid.

(c) Plot the velocity profile as $v_x/v_{x,av}$ versus h/H for $n = 0.1, 0.5, 0.99$. Comment on your results.

7.29 Flow Problem: Axial annular flow of an incompressible power-law generalized Newtonian fluid. The flow shown in Figure 3.17 takes place in the coating of electrical wires [26]. A cylinder (the wire) of radius κR is drawn through a tubular bath of fluid at a velocity V. The tube has a radius R. The upstream fluid is at pressure P_0, and the downstream fluid is also at pressure P_0. Using the constitutive equation for an incompressible power-law GNF, carry out the following calculations:

(a) Calculate the steady-state velocity profile.

(b) Show how your result above reduces to the Newtonian result in the appropriate limit (see Problem 3.19).

(c) Calculate the flow rate in the central channel.

(d) Calculate the force needed to pull the wire.

7.30 Flow Problem: Tangential annular flow of a power-law generalized Newtonian fluid. Calculate the velocity and pressure profiles for steady tangential annular flow of an incompressible power-law generalized Newtonian fluid between long concentric cylinders (length $= L$) when the inner cylinder is turning (see Figure 3.16). Also calculate the torque required to turn the cylinder. The outer cylinder has a radius R, and the inner cylinder has a radius κR and is rotating with an angular velocity Ω. For the boundary condition on pressure, assume that $P = P_R$ at the outer radius. Plot the velocity field for $n = 0.99, 0.3, 0.15, 0.1$. (See Appendix C.1, Math Hints, for help with this problem.)

7.31 Flow Problem: Poiseuille flow of an Ellis generalized Newtonian fluid. The GNF constitutive equation $\underline{\underline{\tau}} = -\eta\dot{\underline{\underline{\gamma}}}$ can use any of a number of empirical relationships for $\eta(\dot{\gamma})$. One that we have used extensively is the power-law equation, but others are equally valid, for example, the Ellis equation, which is a three-parameter model (η_0, τ_0, α):

$$\frac{\eta}{\eta_0} = \frac{1}{1 + (\tau/\tau_0)^{\alpha-1}}$$

where τ is the magnitude of the stress tensor $\underline{\underline{\tau}}$.

(a) Show that the velocity profile for steady laminar flow of an Ellis fluid in a horizontal pipe of circular cross section under a constant pressure gradient ($z = 0, P = P_0, z = L, P = P_L$) is

$$v_z = \frac{\tau_R R}{2\eta_0}\left\{\left[1 - \left(\frac{r}{R}\right)^2\right] + \left(\frac{\tau_R}{\tau_0}\right)^{\alpha-1} \times \frac{2}{\alpha+1}\left[1 - \left(\frac{r}{R}\right)^{\alpha+1}\right]\right\}$$

$$\tau_R = \frac{R\Delta P}{2L}$$

$$\Delta P = P_0 - P_L$$

(b) Plot the final velocity field as $v_z/v_{z,av}$ versus r/R for $\alpha = 0.1, 0.5, 0.99$ and for $\tau_R/\tau_0 = 1, 10, 1000$.

7.32 Flow Problem: Draining a power-law generalized Newtonian fluid from a tank. The tank and tube assembly shown in Figure 7.13 is initially filled with an incompressible power-law generalized Newtonian fluid [28, 26]. During the draining process it is assumed that the flow in the tube is laminar.

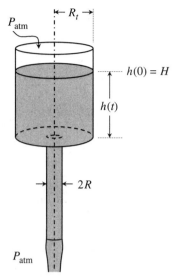

Figure 7.13 Problem 7.32. *Source: From Dynamics of Polymeric Liquids, Volume 1: Fluid Mechanics*, R. B. Bird, R. C. Armstrong, and O. Hassager, Copyright © 1987 by John Wiley & Sons, Inc. Reprinted by permission of John Wiley & Sons, Inc.

(a) What is the flow rate in the pipe as a function of the fluid height in the tank?

(b) Using a quasi-steady-state approach, set $h = h(t)$ and solve for the time to drain just the tank (not the pipe).

7.33 Flow Problem: Laminar drag flow of two layers of power-law generalized Newtonian fluids. Two immiscible power-law generalized Newtonian fluids are sandwiched between infinite plates, the top one of which is moving at a constant speed V (Figure 7.14). Assuming that the interface between the two fluids remains planar, calculate the steady-state velocity distribution.

7.34 *Flow Problem: Laminar combined drag and pressure flow of two layers of power-law generalized Newtonian fluids. Repeat Problem

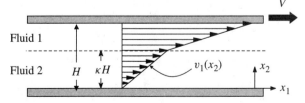

Figure 7.14 Problem 7.33.

7.33 with an additional imposed pressure gradient in the flow direction,

$$\frac{\partial P}{\partial x_1} = \frac{P_L - P_0}{L} = \text{constant}$$

(The solution is quite involved.)

7.35 Flow Problem: Flow in a conical die [179]. Solve for the pressure drop for steady, isothermal flow of an incompressible power-law fluid through a die whose boundaries are cones with a common apex (Figure 7.15). Follow the solution steps outlined below.

(a) The solution was worked out by Parnaby and Worth [202]. In their solution they use the following expression for the power law equation:

$$\eta = \frac{\tilde{m}}{\dot{\gamma}_0} \left[\frac{\dot{\gamma}}{\dot{\gamma}_0} \right]^{n-1}$$

where n is the power-law index, $\tilde{m}$ is a parameter with units of stress, and $\dot{\gamma}_0 = 1\,\text{s}^{-1}$ is a reference shear rate. Show how this expression relates to our equation for power law viscosity. What is the advantage of their expression?

(b) Assuming α and β are small (lubrication approximation) solve for the pressure drop as a function of flow rate, Q [202]. The solution is

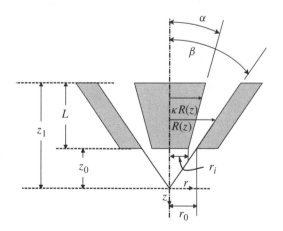

Figure 7.15 Problem 7.35 [202]. *Source:* This material has been reproduced from the Proceedings of the Institution of Mechanical Engineers, Vol. 188, 357–364, 1974, Figure 2, by J. Parnaby and R. A. Worth by permission of the Council of the Institution of Mechanical Engineers.

$$\Delta P = \left(\frac{Q}{\Omega \pi \dot{\gamma}_0}\right)^n \left(\frac{2\bar{m}}{3nr_0^{3n} \tan \beta}\right)\left(1 - Y^{3n}\right)$$

where ΔP is pressure drop, Q is flow rate, $r_0 = z_0 \tan \beta$, $Y = z_0/z_1$, and Ω is given by

$$\Omega(s, \kappa) = \int_\kappa^1 \left|\lambda^2 - \xi^2\right|^{s+1} \xi^{-s} \, d\xi$$

with $s = 1/n$, $\kappa = \kappa R/R = \tan \alpha / \tan \beta$, $\xi = r/R$, and λ is the value of the dimensionless coordinate ξ for which shear stress is zero, which can be solved from

$$\int_\kappa^\lambda \left(\frac{\lambda^2}{\xi} - \xi\right)^s \, d\xi = \int_\lambda^1 \left(\xi - \frac{\lambda^2}{\xi}\right)^s \, d\xi$$

ᘒᘒ **8**

Memory Effects
Generalized Linear Viscoelastic Fluids

Power-law generalized Newtonian fluids, the main topic of the last chapter, are most useful and accurate when applied at high shear rates to steady flows for which elastic effects are not important. At the other end of the flow spectrum are slow, time-dependent flows where elastic effects play an important role. Our current topic is to explore constitutive equations that function well in the slow, elastically dominated limit.

The constitutive equations discussed in this chapter are the first constitutive equations we will discuss that have memory effects built in. We begin with a general discussion of memory effects, then we introduce the empirical Maxwell model, which incorporates deformation history into the calculation of the stress at the current time. We then discuss two generalizations of the Maxwell model and show what these models predict. We continue with flow problems that employ the generalized Maxwell model constitutive equation, and we end with a section describing the limitations of these models. All along we will practice using the new constitutive equations by predicting material functions.

8.1 Memory Effects

Missing from the GNF approach is elasticity, or fluid memory. Recall that both the Newtonian constitutive equation and the generalized Newtonian constitutive equation are simple proportionalities between the stress tensor and the *instantaneous* rate-of-strain tensor.

$$\text{Newtonian:} \qquad \underline{\underline{\tau}}(t) = -\mu\, \underline{\underline{\dot{\gamma}}}(t) \tag{8.1}$$

$$\text{generalized Newtonian:} \qquad \underline{\underline{\tau}}(t) = -\eta(\dot{\gamma})\, \underline{\underline{\dot{\gamma}}}(t) \tag{8.2}$$

Because $\underline{\underline{\dot{\gamma}}}(t)$ represents only the instantaneous deformation, there can be no effect of the history of the deformation on the stress in these models. Yet we know that for polymers rheological behavior is strongly affected by flow history [165]: when we stretch rubber bands, they snap back; when we shear polymeric fluids, the stress does not grow instantaneously when the flow starts, but rather it takes some time to accumulate (see Figure 6.49); likewise when we stop the shearing of a polymeric fluid, the stress does not instantaneously decrease to zero as is required by the Newtonian and generalized Newtonian models, but rather there is a slow decay of stress that takes finite time (see Figure 6.51).

To account for memory effects we need a constitutive equation that depends on what happened to a material sometime in the past. The models we have examined so far, the Newtonian and generalized Newtonian models, are functions of $\dot{\underline{\underline{\gamma}}}(t)$, the rate-of-deformation tensor at the current time only. To construct a constitutive equation with memory, we must include terms that involve expressions such as $\dot{\underline{\underline{\gamma}}}(t - t_0)$, the value of $\dot{\underline{\underline{\gamma}}}$ at a time t_0 seconds (or minutes, or hours) in the past. We would expect, however, that the current and recent deformation rate would have a more important effect on the current stress than the deformation that took place several seconds or minutes ago. A constitutive equation that incorporates both of these characteristics might look like this:

$$\underbrace{\underline{\underline{\tau}}(t)}_{\substack{\text{stress at} \\ \text{current time}}} = -\tilde{\eta}\left[\underbrace{\dot{\underline{\underline{\gamma}}}(t)}_{\substack{\text{current} \\ \text{strain rate}}} + 0.8\ \underbrace{\dot{\underline{\underline{\gamma}}}(t - t_0)}_{\substack{\text{strain rate} \\ t_0 \text{ seconds ago}}} \right] \tag{8.3}$$

In this equation the current deformation rate contributes a term $-\tilde{\eta}\,\dot{\underline{\underline{\gamma}}}(t)$, which looks like a Newtonian-type contribution since it is based on the current rate of deformation $\dot{\underline{\underline{\gamma}}}(t)$. The equation contains a second term, $-0.8\tilde{\eta}\,\dot{\underline{\underline{\gamma}}}(t - t_0)$, which also contributes a portion of the stress at the current time t. This term is related not to the current deformation, but rather to the rate of deformation t_0 seconds ago. Since fluids are observed to have fading memory of what they experienced in the past, the memory term can be expected to be less important than the stress contribution due to the current rate of deformation of the fluid. Thus the factor 0.8 appears in front of the memory term, indicating that the fluid has forgotten 20% of what happened to it t_0 seconds ago.

To see how a fluid described by a memory equation such as Equation (8.3) behaves, we can calculate various material functions for this constitutive equation. We begin by calculating the steady shear viscosity.

EXAMPLE

Calculate the steady shear material functions η, Ψ_1, and Ψ_2 for the constitutive equation

$$\underline{\underline{\tau}}(t) = -\tilde{\eta}\left[\dot{\underline{\underline{\gamma}}}(t) + 0.8\dot{\underline{\underline{\gamma}}}(t - t_0)\right] \tag{8.4}$$

where $\tilde{\eta}$ and t_0 are constant parameters of the model, and both parameters are positive. $\tilde{\eta}$ has units of viscosity (e.g., Pa · s), and t_0 has units of time.

SOLUTION

Steady shear material functions are defined with respect to the following kinematics:

$$\underline{v} = \begin{pmatrix} \dot{\varsigma}(t)x_2 \\ 0 \\ 0 \end{pmatrix}_{123} \tag{8.5}$$

$$\dot{\zeta}(t) = \dot{\gamma}_0 = \text{constant} \tag{8.6}$$

The definitions of the steady shear material functions are given in Chapter 5 [Equations (5.8)–(5.10)], where the stress components are evaluated for the preceding kinematics.

We must calculate the tensor $\underline{\underline{\tau}}$, which we obtain from the constitutive equation [Equation (8.4)]. First we need $\dot{\underline{\underline{\gamma}}}$,

$$\dot{\underline{\underline{\gamma}}} = \nabla \underline{v} + (\nabla \underline{v})^T \tag{8.7}$$

$$= \begin{pmatrix} 0 & \dot{\gamma}_0 & 0 \\ \dot{\gamma}_0 & 0 & 0 \\ 0 & 0 & 0 \end{pmatrix}_{123} \tag{8.8}$$

Since $\dot{\gamma}_0$ is a constant, the tensor $\dot{\underline{\underline{\gamma}}}$ is also a constant, and therefore both $\dot{\underline{\underline{\gamma}}}(t)$ and $\dot{\underline{\underline{\gamma}}}(t - t_0)$ are given by Equation (8.8). It is thus straightforward to calculate $\underline{\underline{\tau}}$,

$$\underline{\underline{\tau}}(t) = -\tilde{\eta} \left[\dot{\underline{\underline{\gamma}}}(t) + 0.8 \dot{\underline{\underline{\gamma}}}(t - t_0) \right] \tag{8.9}$$

$$= -\tilde{\eta} \left[\begin{pmatrix} 0 & \dot{\gamma}_0 & 0 \\ \dot{\gamma}_0 & 0 & 0 \\ 0 & 0 & 0 \end{pmatrix}_{123} + 0.8 \begin{pmatrix} 0 & \dot{\gamma}_0 & 0 \\ \dot{\gamma}_0 & 0 & 0 \\ 0 & 0 & 0 \end{pmatrix}_{123} \right] \tag{8.10}$$

$$= -1.8\tilde{\eta} \begin{pmatrix} 0 & \dot{\gamma}_0 & 0 \\ \dot{\gamma}_0 & 0 & 0 \\ 0 & 0 & 0 \end{pmatrix}_{123} \tag{8.11}$$

We can now calculate the steady shear material functions for this model:

$$\eta = \frac{-\tau_{21}}{\dot{\gamma}_0} = 1.8\tilde{\eta} \tag{8.12}$$

$$\Psi_1 = 0 \tag{8.13}$$

$$\Psi_2 = 0 \tag{8.14}$$

How can we interpret this? Consider that the portion of this constitutive equation that depends on the instantaneous rate-of-deformation tensor is $-\tilde{\eta}\dot{\underline{\underline{\gamma}}}$. A fluid characterized by this instantaneous portion alone, that is, one for which $\underline{\underline{\tau}} = -\tilde{\eta}\dot{\underline{\underline{\gamma}}}$, is a Newtonian fluid with viscosity equal to $\tilde{\eta}$. From the results of this practice example we see that for the memory fluid described by Equation (8.4) the viscosity is 1.8 times higher in steady shear than it would have been if it had had no memory.

Normal-stress coefficients are both zero for this constitutive equation since the stress is proportional to the rate-of-strain tensor, and $\dot{\underline{\underline{\gamma}}}$ has zeros on the diagonal for shear flow.

When a constitutive equation includes terms that depend on the past deformation experienced by the fluid (memory), this has a quantitative effect on the steady shear viscosity, as shown in the preceding example. In unsteady flows there is also a qualitative effect on the predicted material functions. To see how memory affects unsteady shear flow, we will calculate the shear start-up material functions η^+, Ψ_1^+, and Ψ_2^+ for the simple memory fluid defined previously.

EXAMPLE

Calculate $\eta^+(t)$, $\Psi_1^+(t)$, and $\Psi_2^+(t)$ for the simple memory fluid

$$\underline{\underline{\tau}}(t) = -\tilde{\eta}\left[\underline{\underline{\dot{\gamma}}}(t) + 0.8\underline{\underline{\dot{\gamma}}}(t - t_0)\right] \tag{8.15}$$

where $\tilde{\eta}$ and t_0 are constant parameters of the model, and both parameters are positive. $\tilde{\eta}$ has units of viscosity (poise or Pa·s), and t_0 has units of time.

SOLUTION

The startup of the steady shearing experiment is defined with respect to the following kinematics:

$$\underline{v}(t') = \begin{pmatrix} \dot{\varsigma}(t')x_2 \\ 0 \\ 0 \end{pmatrix}_{123} \tag{8.16}$$

$$\dot{\varsigma}(t') = \begin{cases} 0 & t' < 0 \\ \dot{\gamma}_0 & t' \geq 0 \end{cases} \tag{8.17}$$

We have used t' when writing these kinematics since we will need to consider these equations at times other than the current time t. The variable t' is a dummy variable, which just shows us what the function is.

The startup steady shearing material functions are defined for the kinematics given as

$$\eta^+(t, \dot{\gamma}_0) \equiv -\frac{\tau_{21}}{\dot{\gamma}_0} \tag{8.18}$$

$$\Psi_1^+(t, \dot{\gamma}_0) \equiv -\frac{(\tau_{11} - \tau_{22})}{\dot{\gamma}_0^2} \tag{8.19}$$

$$\Psi_2^+(t, \dot{\gamma}_0) \equiv -\frac{(\tau_{22} - \tau_{33})}{\dot{\gamma}_0^2} \tag{8.20}$$

To evaluate these functions we must calculate the stress tensor $\underline{\underline{\tau}}(t)$ predicted by the constitutive equation. First we calculate the rate-of-deformation tensor:

$$\underline{\underline{\dot{\gamma}}}(t') = \nabla\underline{v} + (\nabla\underline{v})^T \tag{8.21}$$

$$= \begin{pmatrix} 0 & \dot{\varsigma}(t') & 0 \\ \dot{\varsigma}(t') & 0 & 0 \\ 0 & 0 & 0 \end{pmatrix}_{123} \tag{8.22}$$

Since $\dot{\underline{\underline{\varsigma}}}(t')$ is a function of time, $\dot{\underline{\underline{\gamma}}}(t')$ is also a function of time. Looking at the constitutive equation, we see that to calculate $\underline{\underline{\tau}}(t)$ we need $\dot{\underline{\underline{\gamma}}}(t)$ and $\dot{\underline{\underline{\gamma}}}(t - t_0)$. To obtain $\dot{\underline{\underline{\gamma}}}(t)$, we substitute $t' = t$ into Equations (8.17) and (8.22):

$$\dot{\underline{\underline{\gamma}}}(t) = \begin{cases} \underline{\underline{0}} & t < 0 \\ \begin{pmatrix} 0 & \dot{\gamma}_0 & 0 \\ \dot{\gamma}_0 & 0 & 0 \\ 0 & 0 & 0 \end{pmatrix}_{123} & t \geq 0 \end{cases} \tag{8.23}$$

To calculate $\dot{\underline{\underline{\gamma}}}(t - t_0)$ we substitute $t' = t - t_0$ into Equations (8.22) and (8.17),[1]

$$\dot{\underline{\underline{\gamma}}}(t - t_0) = \begin{cases} \underline{\underline{0}} & t - t_0 < 0 \\ \begin{pmatrix} 0 & \dot{\gamma}_0 & 0 \\ \dot{\gamma}_0 & 0 & 0 \\ 0 & 0 & 0 \end{pmatrix}_{123} & t - t_0 \geq 0 \end{cases} \tag{8.24}$$

$$= \begin{cases} \underline{\underline{0}} & t < t_0 \\ \begin{pmatrix} 0 & \dot{\gamma}_0 & 0 \\ \dot{\gamma}_0 & 0 & 0 \\ 0 & 0 & 0 \end{pmatrix}_{123} & t \geq t_0 \end{cases} \tag{8.25}$$

Now we can calculate $\underline{\underline{\tau}}(t)$ from the constitutive equation:

$$\underline{\underline{\tau}}(t) = -\tilde{\eta} \left[\dot{\underline{\underline{\gamma}}}(t) + \dot{\underline{\underline{\gamma}}}(t - t_0) \right] \tag{8.26}$$

To add the two rate-of-deformation tensors (the one evaluated at t and the other evaluated at $t - t_0$), we must consider three different time intervals:

$$t < 0 \qquad \underline{\underline{\tau}} = -\tilde{\eta} \left[\underline{\underline{0}} + \underline{\underline{0}} \right] \tag{8.27}$$

$$0 \leq t < t_0 \qquad \underline{\underline{\tau}} = -\tilde{\eta} \left[\begin{pmatrix} 0 & \dot{\gamma}_0 & 0 \\ \dot{\gamma}_0 & 0 & 0 \\ 0 & 0 & 0 \end{pmatrix}_{123} + \underline{\underline{0}} \right] \tag{8.28}$$

---

[1] Reminder on functions: Given a function $f(x)$ such as $f(x) = x^2 + 3x + 2$ we know that when $x = a$ we can evaluate $f(a)$ by substituting a for x in the general formula, $f(a) = a^2 + 3a + 2$. When functions appear in less familiar form, however, the same rule applies. For $f(x)$ given by

$$f(x) = \begin{cases} 6 & x < 0 \\ 10 & x \geq 0 \end{cases}$$

we can evaluate $f(x - 3)$ as follows:

$$f(x - 3) = \begin{cases} 6 & x - 3 < 0 \\ 10 & x - 3 \geq 0 \end{cases} = \begin{cases} 6 & x < 3 \\ 10 & x \geq 3 \end{cases}$$

$$t \geq t_0 \qquad \underline{\underline{\tau}} = -\tilde{\eta}\left[\begin{pmatrix} 0 & \dot{\gamma}_0 & 0 \\ \dot{\gamma}_0 & 0 & 0 \\ 0 & 0 & 0 \end{pmatrix}_{123} + 0.8\begin{pmatrix} 0 & \dot{\gamma}_0 & 0 \\ \dot{\gamma}_0 & 0 & 0 \\ 0 & 0 & 0 \end{pmatrix}_{123}\right] \qquad (8.29)$$

Thus the stress tensor for our simple memory constitutive equation in the startup of steady shearing is given by

$$\underline{\underline{\tau}}(t) = \begin{cases} \underline{\underline{0}} & t < 0 \\\\ -\tilde{\eta}\begin{pmatrix} 0 & \dot{\gamma}_0 & 0 \\ \dot{\gamma}_0 & 0 & 0 \\ 0 & 0 & 0 \end{pmatrix}_{123} & 0 \leq t < t_0 \\\\ -1.8\tilde{\eta}\begin{pmatrix} 0 & \dot{\gamma}_0 & 0 \\ \dot{\gamma}_0 & 0 & 0 \\ 0 & 0 & 0 \end{pmatrix}_{123} & t \geq t_0 \end{cases} \qquad (8.30)$$

We can now calculate the material functions η^+, Ψ_1^+, and Ψ_2^+ from the definitions given at the start of the problem [Equations (8.18)–(8.20)]:

$$\eta^+(t) = \begin{cases} 0 & t < 0 \\ \tilde{\eta} & 0 \leq t < t_0 \\ 1.8\tilde{\eta} & t \geq t_0 \end{cases} \qquad (8.31)$$

$$\Psi_1^+(t) = 0 \qquad (8.32)$$

$$\Psi_2^+(t) = 0 \qquad (8.33)$$

The shear stress growth function $\eta^+(t)$ calculated here is sketched in Figure 8.1.

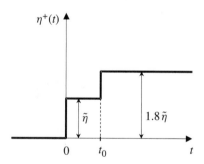

Figure 8.1 Shear stress growth function η^+ as a function of time t for the simple memory fluid discussed in text.

The simple memory fluid shows a two-step buildup of stress in the start-up experiment. In contrast, the startup of steady shearing of a purely Newtonian fluid is an instantaneous step up at $t = 0$ to the steady shear viscosity value (see Section 5.2.2.1). The gradual buildup in η^+ of the simple memory model is reminiscent of the actual response of polymeric fluids shown earlier in Figures 6.49 and 6.50, although our simple model does not, of course, closely predict the smooth rise in η^+ that is actually observed. By considering constitutive equations that depend in a continuous manner on the history of $\dot{\underline{\underline{\gamma}}}$, however, we can find models that predict the smooth buildup of η^+ correctly. Thus, by including a dependence of $\underline{\underline{\tau}}(t)$, the stress at the current time, on the history of the rate-of-deformation tensor [$\dot{\underline{\underline{\gamma}}}(t)$ evaluated at times other than the current time t], we can correctly model many memory effects, such as the gradual rise of η^+ as well as the gradual fall of η^-, the shape of $G(t)$ (step-strain experiment), and other time-dependent effects.

8.2 The Maxwell Models

In the last section we demonstrated that for a constitutive equation to include memory effects, and hence elasticity, it must contain some information about the deformation history that a fluid has experienced. We discussed a simple memory fluid constitutive equation that contained a limited amount of information about deformation history. We designed a simple example to reflect our belief that a fluid's memory of past deformations should be imperfect, that is, a fading memory. Now we would like to present another memory fluid model that improves on the simple (and rather arbitrary) memory fluid we just discussed. This is called the *Maxwell model*.

8.2.1 SIMPLE MAXWELL MODEL

The development of the Maxwell model follows a different path than that considered above. In the development of the simple memory fluid we added (integrated) the contributions of past events to the stress at the current time. The development of the Maxwell model, as we shall see in this section, follows a *differential* approach based on ideas of elasticity. At the end of the section, however, we will find that these two approaches are equivalent.

The Maxwell model results from the direct combination of viscous behavior and elastic behavior in one constitutive equation. From studying Newtonian fluids in Chapter 3, we understand a thing or two about viscous behavior. To understand elastic behavior we will now examine the elastic solid that was mentioned briefly in Section 5.2.2.5. Perfectly elastic solids respond to shear deformation according to Hooke's law (shear only):

$$\text{Hooke's law} \atop \text{(shear only)} \qquad \boxed{\tau_{21}(t) = -G\frac{\partial u_1}{\partial x_2}} \qquad (8.34)$$

$$= -G\gamma_{21}(t_{\text{ref}}, t) \qquad (8.35)$$

where G is a scalar constant, called the *elastic modulus*. In words, Hooke's law says that the shear stress is proportional to the shear strain. Strain is the deformation of the sample at the current time t with respect to the shape at some reference time t_{ref} or some other time of interest (see Section 5.2.2.3). In Hooke's law the reference time is a time when the stress is zero, $\tau_{21} = 0$. We see then that the stress generated by a Hookean material depends only on the current state of the system and the reference state of the system, but not at all on the instantaneous rate of deformation.

Hooke's law can be applied to flows other than shear if we write the strain as a tensor:

Infinitesimal
strain tensor

$$\underline{\underline{\gamma}}(t_{\text{ref}}, t) \equiv \nabla \underline{u} + (\nabla \underline{u})^T \qquad (8.36)$$

Hooke's law
(slow rates)

$$\underline{\underline{\tau}}(t) = -G\underline{\underline{\gamma}}(t_{\text{ref}}, t) \qquad (8.37)$$

The vector $\underline{u}(t_{\text{ref}}, t) = \underline{r}(t) - \underline{r}(t_{\text{ref}})$ gives the displacement of a fluid particle between times t_{ref} and the current time t, as was discussed in Section 5.2.2.3. The tensor $\nabla \underline{u}$ is called the *displacement gradient tensor*, and the strain tensor $\underline{\underline{\gamma}}(t_{\text{ref}}, t)$ is called the *infinitesimal strain tensor*. Recall that the shear strain and the shear rate are related by (see Section 5.2.2.4)

$$\gamma_{21}(t_{\text{ref}}, t) = \int_{t_{\text{ref}}}^{t} \dot{\gamma}_{21}(t') \, dt' \qquad (8.38)$$

The variable t' is a dummy variable of integration. For small displacement gradients this relationship holds for each coefficient of the tensors $\underline{\underline{\dot{\gamma}}}$ and $\underline{\underline{\gamma}}$,

$$\gamma_{pk}(t_{\text{ref}}, t) = \int_{t_{\text{ref}}}^{t} \dot{\gamma}_{pk}(t') \, dt' \qquad (8.39)$$

This expression for the components of the infinitesimal strain tensor in terms of integrals over the strain-rate components results from the assumption that strains are additive in the limit of small strain rates.

The original Hooke's law [Equation (8.34)] is just the 21-component of the tensor version of Hooke's law [Equation (8.37)] written for the specific case of shear flow. In our discussion of more advanced constitutive equations in Chapter 9 we will discuss $\underline{\underline{\gamma}}(t_{\text{ref}}, t)$ and other strain tensors in more detail. Hooke's law is empirical and is found to hold at only small displacement gradients.

The Hookean solid constitutive equation is a good model for rubbers (rubber bands) and crosslinked gels that do not flow without breaking, or for metals at very small strains. The challenge for polymeric melts, however, is to find a constitutive equation that can incorporate some elastic effects while still predicting viscous effects and flow. Such an equation was proposed for shear flow by James Clerk Maxwell in 1867:

Maxwell equation
(scalar version)

$$\tau_{21} + \frac{\mu}{G} \frac{\partial \tau_{21}}{\partial t} = -\mu \dot{\gamma}_{21} \qquad (8.40)$$

At steady state ($\partial \tau_{21}/\partial t \longrightarrow 0$) this equation becomes Newton's law of viscosity. For rapid motions at short times ($t \longrightarrow 0$), the time-derivative term is much larger than the stress term, and if we therefore neglect τ_{21} in Equation (8.40), Hooke's law for elastic solids is recovered:

$$\frac{\partial \tau_{21}}{\partial t} \gg \tau_{21} \tag{8.41}$$

$$\frac{\mu}{G} \frac{\partial \tau_{21}}{\partial t} = -\mu \dot{\gamma}_{21} \tag{8.42}$$

$$\frac{\partial \tau_{21}}{\partial t} = -G \dot{\gamma}_{21} \tag{8.43}$$

$$\tau_{21}(t) = -G \int_{t_{\text{ref}}}^{t} \dot{\gamma}_{21}(t') \, dt' \tag{8.44}$$

$$= -G \gamma_{21} (t_{\text{ref}}, t) \tag{8.45}$$

In integrating the left side of Equation (8.44), we have taken $\tau_{21}(t_{\text{ref}}) = 0$. The Maxwell equation is strictly empirical, and its validity depends only on how well it predicts observed shear behavior. The Maxwell equation is limited to small strains by its use of the relationship between γ_{21} and $\dot{\gamma}_{21}$, and it is also limited to shear flow since it is a scalar rather than a tensor equation.

Before we address some of these concerns, let us first consider the physics behind the Maxwell equation. As we have discussed, the Maxwell equation combines viscous and elastic expressions. One could, of course, combine elastic and viscous effects in a variety of ways; why did Maxwell combine them as he did?

The answer to this question can be found by considering two elementary mechanical elements that produce force, the spring and the dashpot. If a spring is made to undergo a displacement D_{spring}, it produces a restoring force in the direction opposite to the direction of the displacement:

$$f = -G_{\text{sp}} D_{\text{spring}} \tag{8.46}$$

where f is the magnitude of the resisting force and G_{sp} is the spring force constant. This force law resembles Hooke's law for elastic solids. A dashpot is a device like the shock absorbers on an automobile: a piston is made to move through a viscous Newtonian fluid. The magnitude of the resisting force f caused by the drag on the piston as it moves through the liquid at a speed dD_{dash}/dt is related to the viscosity of the fluid:

$$f = -\mu \frac{dD_{\text{dash}}}{dt} \tag{8.47}$$

where μ is the fluid viscosity and D_{dash} represents the piston displacement. If we connect a spring and a dashpot in series (Figure 8.2) and displace the combined unit, the spring will deform, and the piston will move through the liquid in the dashpot. Both units will experience the same applied force, but each element will undergo a different displacement. The total displacement of the two units in series D_{total} will just be the sum of the displacements of the individual units:

$$D_{\text{total}} = D_{\text{spring}} + D_{\text{dash}} \tag{8.48}$$

Taking the time derivative of Equation (8.48) and substituting the expressions for D_{spring}

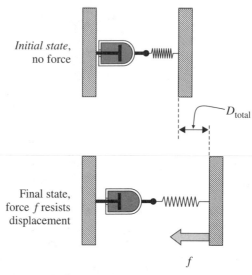

Initial state, no force

D_{total}

Final state, force f resists displacement

f

Figure 8.2 Displacement of a spring and dashpot in series. Initially the combined unit exerts no force. After displacement by an amount D_{total} the unit experiences a force of magnitude f in the direction opposite to the displacement.

and D_{dash} [Equations (8.46) and (8.47)], we obtain an equation for the total force f that is analogous to the Maxwell equation:

$$f + \frac{\mu}{G_{\text{sp}}} \frac{df}{dt} = -\mu \frac{dD_{\text{total}}}{dt} \qquad (8.49)$$

We see then that the Maxwell equation can be interpreted as the series combination of viscous and elastic effects in a fluid. There are, of course, other ways of combining the viscous and elastic contributions to stress in viscoelastic fluids (e.g., combining them in parallel, or combining an elastic contribution in parallel with the combined elastic/viscous unit, etc.), and many of these ways of combining springs and dashpots have been investigated over the years as possible rheological models [25]. Again, all of these models are empirical, and the only justification for choosing one over another is if the model correctly predicts experimental data.

Thus the Maxwell equation combines the Newtonian and Hookean equations in a way that makes some sense, both in terms of the limits at steady state and at short times and in terms of the physical analogy to combining springs and dashpots in series. It is not a constitutive equation, however, because it is not of tensor order, and we have yet to see whether the predictions of the Maxwell equation are useful.

We can remove the first objection by generalizing the Maxwell equation to a tensor form in what seems to be the most obvious way, that is, replace the appropriate kinematic and stress scalars with tensors. This is an arbitrary[2] choice, but since the entire development

---

[2] This is also dangerous, as we will show in Section 8.5 and discuss in Chapter 9.

of this equation was empirical, no obvious rules have been broken. Thus the full Maxwell fluid constitutive model is

Maxwell model
(differential form)

$$\underline{\underline{\tau}} + \lambda \frac{\partial \underline{\underline{\tau}}}{\partial t} = -\eta_0 \underline{\underline{\dot{\gamma}}} \tag{8.50}$$

We have replaced μ with η_0 since we are limited to small rates and the zero-shear viscosity is the relevant viscosity. We have also replaced μ/G with λ, which has the units of time and which we introduced earlier as the relaxation time of the material. This tensor equation has the same limits as the scalar Maxwell equation, that is, at steady state it becomes the Newtonian constitutive equation $\underline{\underline{\tau}} = -\eta_0 \underline{\underline{\dot{\gamma}}}$, and at short times it becomes the Hooke's law constitutive equation $\underline{\underline{\tau}} = -(\eta_0/\lambda) \underline{\underline{\gamma}}(t_{\text{ref}}, t)$ with $G = \eta_0/\lambda$.

Notice that the Maxwell model [Equation (8.50)] is a differential equation for $\underline{\underline{\tau}}$. To calculate stresses, we must solve the Maxwell model for the stress tensor. We can find the solution for $\underline{\underline{\tau}}$ by integrating with the help of an integrating factor, in this case $(1/\lambda)e^{\frac{t}{\lambda}}$ (see [34] and Appendix C.1). The fact that it is a tensor equation is not an impediment. We can solve the equations represented by all nine tensor components at once. The Maxwell differential equation is

$$\underline{\underline{\tau}} + \lambda \frac{\partial \underline{\underline{\tau}}}{\partial t} = -\eta_0 \underline{\underline{\dot{\gamma}}} \tag{8.51}$$

Multiplying through by $(1/\lambda)e^{\frac{t}{\lambda}}$ we obtain

$$\left(\frac{1}{\lambda}\right) e^{\frac{t}{\lambda}} \underline{\underline{\tau}} + e^{\frac{t}{\lambda}} \frac{\partial \underline{\underline{\tau}}}{\partial t} = -\eta_0 \left(\frac{1}{\lambda}\right) e^{\frac{t}{\lambda}} \underline{\underline{\dot{\gamma}}} \tag{8.52}$$

We can factor the left side to become

$$\frac{\partial}{\partial t} \left(e^{\frac{t}{\lambda}} \underline{\underline{\tau}} \right) = \frac{-\eta_0}{\lambda} e^{\frac{t}{\lambda}} \underline{\underline{\dot{\gamma}}} \tag{8.53}$$

Integration over all past times up to the time of interest t yields

$$\int_{-\infty}^{t} d\left[e^{\frac{t'}{\lambda}} \underline{\underline{\tau}}(t') \right] = \int_{-\infty}^{t} \frac{-\eta_0}{\lambda} e^{\frac{t'}{\lambda}} \underline{\underline{\dot{\gamma}}}(t') \, dt' \tag{8.54}$$

$$\left. e^{\frac{t'}{\lambda}} \underline{\underline{\tau}}(t') \right|_{-\infty}^{t} = \int_{-\infty}^{t} \frac{-\eta_0}{\lambda} e^{\frac{t'}{\lambda}} \underline{\underline{\dot{\gamma}}}(t') \, dt' \tag{8.55}$$

If we require that at $t = -\infty$ the stress be finite, we can simplify the left side, and we obtain

$$e^{\frac{t}{\lambda}} \underline{\underline{\tau}}(t) = \int_{-\infty}^{t} \frac{-\eta_0}{\lambda} e^{\frac{t'}{\lambda}} \underline{\underline{\dot{\gamma}}}(t') \, dt' \tag{8.56}$$

$$\underline{\underline{\tau}}(t) = e^{-\frac{t}{\lambda}} \int_{-\infty}^{t} \frac{-\eta_0}{\lambda} e^{\frac{t'}{\lambda}} \underline{\underline{\dot{\gamma}}}(t') \, dt' \tag{8.57}$$

Maxwell model
(integral form)

$$\underline{\underline{\tau}}(t) = -\int_{-\infty}^{t} \left[\frac{\eta_0}{\lambda} e^{\frac{-(t-t')}{\lambda}} \right] \underline{\underline{\dot{\gamma}}}(t') \, dt' \qquad (8.58)$$

Note again the different meanings of the times t' (dummy variable of integration) and t (current time), which is the time at which we are calculating $\underline{\underline{\tau}}$. The term $e^{\frac{-t}{\lambda}}$ is a constant with respect to the integral over t', and we can move it in and out of the integral with impunity.

Equation (8.58) is the integral form of the Maxwell model, a two-parameter (η_0 and λ) constitutive equation. Written in this form we see that the stress in the Maxwell model is proportional to the integral, over all past times, of the shear-rate tensor at some past time t', $\underline{\underline{\dot{\gamma}}}(t')$, multiplied by an exponentially decaying function of the interval between the current time t and all the past times being integrated over, t',

$$\underline{\underline{\tau}}(t) = -\int_{-\infty}^{t} \overbrace{\left[\frac{\eta_0}{\lambda} e^{\frac{-(t-t')}{\lambda}} \right]}^{\substack{\text{variable} \\ \text{forgetting} \\ \text{function}}} \overbrace{\underline{\underline{\dot{\gamma}}}(t')}^{\substack{\underline{\underline{\dot{\gamma}}} \text{ at} \\ \text{past} \\ \text{times}}} \, dt' \qquad (8.59)$$

Since the Maxwell model calculates the stress at the time of interest t as the integral over events at past times t', it is not only a function of the *instantaneous* shear-rate tensor, but it is also a function of the history of the shear-rate tensor. Thus the Maxwell model is a refinement of the crude memory model discussed at the beginning of this chapter. The function $(\eta_0/\lambda)e^{\frac{-(t-t')}{\lambda}}$ in the Maxwell model serves as a weighting function that indicates how much of the past deformation is remembered by the fluid at time t. It is a decreasing function of the interval between t and t', that is, the farther back the deformation occurred, the less impact that deformation has on the stress at the current time t. This is exactly the kind of continuously forgetting function that we thought would be appropriate for polymers in our discussion in Section 8.1.

Thus we have identified a new constitutive equation that satisfies many of our requirements for a polymeric constitutive equation: it is of tensor order, it predicts a symmetric stress tensor [since it is proportional to a symmetric tensor $\underline{\underline{\dot{\gamma}}}(t')$], and it contains no obvious violations of material objectivity.[3] In addition, it is a function of the history of the rate-of-deformation tensor, a requirement for a constitutive equation that will predict memory effects. Finally, it contains a continuously varying forgetting factor that meets our expectations for the needed weighting of the importance of $\underline{\underline{\dot{\gamma}}}(t')$ as t' recedes farther and farther into the past.

What next? The ultimate test of any constitutive equation is whether the predictions of the equation match experimental observations. To answer this we must calculate material functions using the Maxwell model and compare the predictions to measurements. We begin by calculating steady shear properties.

---

[3] There is a problem here, however, that is less obvious. We will discuss it at the end of the chapter.

EXAMPLE

Calculate the steady shear material functions $\eta(\dot{\gamma})$, $\Psi_1(\dot{\gamma})$, and $\Psi_2(\dot{\gamma})$, predicted by the Maxwell constitutive equation.

SOLUTION

As always, when calculating material functions, we begin with the kinematics. For steady shear flow the kinematics are given by

$$\underline{v} = \begin{pmatrix} \dot{\varsigma}(t)x_2 \\ 0 \\ 0 \end{pmatrix}_{123} \tag{8.60}$$

$$\dot{\varsigma}(t) = \dot{\gamma}_0 = \text{constant} \tag{8.61}$$

Steady shear viscosity is a material function defined for steady shear flow.

$$\eta \equiv \frac{-\tau_{21}}{\dot{\gamma}_0} \tag{8.62}$$

The stress τ_{21} for the Maxwell model is calculated as follows:

$$\underline{\underline{\tau}} = -\int_{-\infty}^{t} \frac{\eta_0}{\lambda} e^{\frac{-(t-t')}{\lambda}} \, \dot{\underline{\underline{\gamma}}}(t') \, dt' \tag{8.63}$$

$$= -\int_{-\infty}^{t} \frac{\eta_0}{\lambda} e^{\frac{-(t-t')}{\lambda}} \begin{pmatrix} 0 & \dot{\gamma}_0 & 0 \\ \dot{\gamma}_0 & 0 & 0 \\ 0 & 0 & 0 \end{pmatrix}_{123} dt' \tag{8.64}$$

$$\tau_{21}(t) = -\int_{-\infty}^{t} \frac{\eta_0}{\lambda} e^{\frac{-(t-t')}{\lambda}} \, \dot{\gamma}_0 \, dt' \tag{8.65}$$

$$= -\eta_0 \dot{\gamma}_0 \, e^{\frac{-(t-t')}{\lambda}} \Big|_{-\infty}^{t} \tag{8.66}$$

$$= -\eta_0 \dot{\gamma}_0 \tag{8.67}$$

From the definition of viscosity we calculate

$$\eta = \frac{-\tau_{21}}{\dot{\gamma}_0} = \eta_0 \tag{8.68}$$

Thus we reach the satisfying solution that, at steady state, the viscosity predicted by the Maxwell model equals η_0, the parameter we originally inserted into the Maxwell model to account for viscous effects. The normal-stress coefficients are both zero, since the Maxwell constitutive equation is proportional to $\dot{\underline{\underline{\gamma}}}$, which has zeros on the diagonal for shear flow,

$$\Psi_1 = \frac{-(\tau_{11} - \tau_{22})}{\dot{\gamma}_0^2} = 0 \tag{8.69}$$

$$\Psi_2 = \frac{-(\tau_{22} - \tau_{33})}{\dot{\gamma}_0^2} = 0 \tag{8.70}$$

Now we will examine how the Maxwell model performs when subjected to a time-dependent shear flow.

EXAMPLE

Calculate the step shear strain material functions $G(t, \gamma_0)$, $G_{\Psi_1}(t, \gamma_0)$, and $G_{\Psi_2}(t, \gamma_0)$ for the Maxwell constitutive equation.

SOLUTION

The step-strain experiment is defined in shear flow, and thus the velocity field is (see Section 5.2.2.4)

$$\underline{v}(t') = \begin{pmatrix} \dot{\varsigma}(t')x_2 \\ 0 \\ 0 \end{pmatrix}_{123} \tag{8.71}$$

$$\dot{\varsigma}(t') = \lim_{\varepsilon \to 0} \dot{\varsigma}(t', \varepsilon) \tag{8.72}$$

$$\dot{\varsigma}(t', \varepsilon) = \begin{cases} 0 & t' < 0 \\ \dfrac{\gamma_0}{\varepsilon} & 0 \le t' < \varepsilon \\ 0 & t' \ge \varepsilon \end{cases} \tag{8.73}$$

and the shear-rate tensor is

$$\underline{\underline{\dot{\gamma}}}(t') = \begin{pmatrix} 0 & \dot{\varsigma}(t') & 0 \\ \dot{\varsigma}(t') & 0 & 0 \\ 0 & 0 & 0 \end{pmatrix}_{123} \tag{8.74}$$

For a Maxwell fluid, the stress tensor in shear flow is given by

$$\underline{\underline{\tau}}(t) = -\int_{-\infty}^{t} \left(\frac{\eta_0}{\lambda}\right) e^{\frac{-(t-t')}{\lambda}} \begin{pmatrix} 0 & \dot{\varsigma}(t') & 0 \\ \dot{\varsigma}(t') & 0 & 0 \\ 0 & 0 & 0 \end{pmatrix}_{123} dt' \tag{8.75}$$

and the material functions $G(t, \gamma_0)$, $G_{\Psi_1}(t, \gamma_0)$, and $G_{\Psi_2}(t, \gamma_0)$ are defined as

$$G(t, \gamma_0) = \frac{-\tau_{21}(t)}{\gamma_0} \tag{8.76}$$

$$G_{\Psi_1}(t, \gamma_0) = \frac{-(\tau_{11} - \tau_{22})}{\gamma_0^2} \tag{8.77}$$

$$G_{\Psi_2}(t, \gamma_0) = \frac{-(\tau_{22} - \tau_{33})}{\gamma_0^2} \tag{8.78}$$

where γ_0 is the magnitude of the step.

Taking $\tau_{21}(t)$ from Equation (8.75), we can calculate the material function $G(t, \gamma_0)$ predicted by the Maxwell model:

$$\tau_{21}(t) = -\int_{-\infty}^{t} \left(\frac{\eta_0}{\lambda} \right) e^{\frac{-(t-t')}{\lambda}} \dot{\varsigma}(t') \, dt' \tag{8.79}$$

$$-\tau_{21}(t, \varepsilon) = \int_{-\infty}^{0} 0 \, dt' + \int_{0}^{\varepsilon} \left(\frac{\eta_0}{\lambda} \right) e^{\frac{-(t-t')}{\lambda}} \frac{\gamma_0}{\varepsilon} \, dt' + \int_{\varepsilon}^{t} 0 \, dt' \tag{8.80}$$

$$= \frac{\gamma_0 \eta_0}{\lambda \varepsilon} \int_{0}^{\varepsilon} e^{\frac{-(t-t')}{\lambda}} \, dt' \tag{8.81}$$

$$G(t, \gamma_0) = \lim_{\varepsilon \to 0} \frac{-\tau_{21}(t, \varepsilon)}{\gamma_0} \tag{8.82}$$

$$= \lim_{\varepsilon \to 0} \frac{\frac{\eta_0}{\lambda} \int_{0}^{\varepsilon} e^{\frac{-(t-t')}{\lambda}} \, dt'}{\varepsilon} \tag{8.83}$$

which is independent of γ_0. Since in this last expression both the numerator and the denominator vanish when we take the limit, we can use l'Hôpital's rule to complete the calculation of $G(t)$. The Leibnitz rule will allow us to take the derivative of the numerator with respect to ε,

$$G(t) = \lim_{\varepsilon \to 0} \frac{\frac{\eta_0}{\lambda} \int_{0}^{\varepsilon} e^{\frac{-(t-t')}{\lambda}} \, dt'}{\varepsilon} \tag{8.84}$$

$$= \lim_{\varepsilon \to 0} \frac{\frac{\eta_0}{\lambda} \frac{d}{d\varepsilon} \int_{0}^{\varepsilon} e^{\frac{-(t-t')}{\lambda}} \, dt'}{\frac{d}{d\varepsilon} \varepsilon} \tag{8.85}$$

$$= \lim_{\varepsilon \to 0} \frac{\eta_0}{\lambda} e^{\frac{-(t-\varepsilon)}{\lambda}} \tag{8.86}$$

Relaxation modulus for
Maxwell model
$$\boxed{G(t) = \frac{\eta_0}{\lambda} e^{\frac{-t}{\lambda}}} \tag{8.87}$$

Thus the relaxation modulus for the Maxwell model is an exponential decay with respect to time (Figure 8.3).

Because the diagonal stress components are zero for a Maxwell fluid in shear flow, the other two material functions in step shear strain are zero,

$$G_{\Psi_1}(t, \gamma_0) = \frac{-(\tau_{11} - \tau_{22})}{\gamma_0^2} = 0 \tag{8.88}$$

$$G_{\Psi_2}(t, \gamma_0) = \frac{-(\tau_{22} - \tau_{33})}{\gamma_0^2} = 0 \tag{8.89}$$

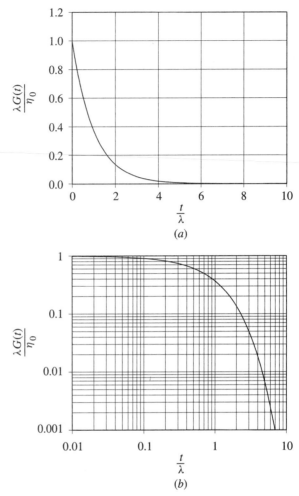

Figure 8.3 Plot of the scaled shear relaxation modulus for a single-relaxation-time Maxwell fluid. (*a*) Linear scale. (*b*) Log scale.

We see from the preceding example that the Maxwell model reasonably predicts the relaxation modulus $G(t)$. Compared with the shape of the experimental $G(t)$ for a concentrated solution of polystyrene shown in Figure 6.57, the Maxwell model correctly predicts a smoothly decreasing function of time for $G(t)$. With only two parameters, η_0 and λ, it is not possible to fit the Maxwell model to most experimental curves of relaxation modulus, but compared to the predictions of the Newtonian and generalized Newtonian fluid models (see Section 7.3), the Maxwell model represents a significant advance toward capturing real polymer shear behavior in this type of transient experiment. One remaining important qualitative test is to see what the Maxwell model predicts in elongation. We examine this in the final practice example.

EXAMPLE

Calculate the steady uniaxial elongational viscosity $\bar{\eta}$ predicted by the Maxwell constitutive equation.

SOLUTION

The kinematics for steady uniaxial elongation are given by

$$\underline{v} = \begin{pmatrix} -\frac{1}{2}\dot{\epsilon}(t)x_1 \\ -\frac{1}{2}\dot{\epsilon}(t)x_2 \\ \dot{\epsilon}(t)x_3 \end{pmatrix}_{123} \qquad \dot{\epsilon}_0 > 0 \tag{8.90}$$

where $\dot{\epsilon}(t) = \dot{\epsilon}_0$, a positive constant. The rate-of-deformation tensor is then

$$\underline{\underline{\dot{\gamma}}} = \begin{pmatrix} -\dot{\epsilon}_0 & 0 & 0 \\ 0 & -\dot{\epsilon}_0 & 0 \\ 0 & 0 & 2\dot{\epsilon}_0 \end{pmatrix}_{123} \tag{8.91}$$

and the magnitude of the rate-of-deformation tensor is $\dot{\epsilon}_0\sqrt{3}$. The uniaxial elongational viscosity is defined as

$$\bar{\eta}(\dot{\epsilon}_0) \equiv \frac{-(\tau_{33} - \tau_{11})}{\dot{\epsilon}_0} \tag{8.92}$$

Thus we need to calculate the stress tensor $\underline{\underline{\tau}}$ for a Maxwell fluid subjected to the kinematics given,

$$\underline{\underline{\tau}}(t) = -\int_{-\infty}^{t} \left(\frac{\eta_0}{\lambda}\right) e^{\frac{-(t-t')}{\lambda}} \underline{\underline{\dot{\gamma}}}\, dt' \tag{8.93}$$

$$= -\int_{-\infty}^{t} \left(\frac{\eta_0}{\lambda}\right) e^{\frac{-(t-t')}{\lambda}} \begin{pmatrix} -\dot{\epsilon}_0 & 0 & 0 \\ 0 & -\dot{\epsilon}_0 & 0 \\ 0 & 0 & 2\dot{\epsilon}_0 \end{pmatrix}_{123} dt' \tag{8.94}$$

$$= \overbrace{\left[-\int_{-\infty}^{t} \left(\frac{\eta_0}{\lambda}\right) e^{\frac{-(t-t')}{\lambda}}\, dt'\right]}^{-\eta_0} \begin{pmatrix} -\dot{\epsilon}_0 & 0 & 0 \\ 0 & -\dot{\epsilon}_0 & 0 \\ 0 & 0 & 2\dot{\epsilon}_0 \end{pmatrix}_{123} \tag{8.95}$$

$$= \begin{pmatrix} \eta_0\dot{\epsilon}_0 & 0 & 0 \\ 0 & \eta_0\dot{\epsilon}_0 & 0 \\ 0 & 0 & -2\eta_0\dot{\epsilon}_0 \end{pmatrix}_{123} \tag{8.96}$$

Now we can calculate $\bar{\eta}$:

$$\bar{\eta}(\dot{\epsilon}_0) \equiv \frac{-(\tau_{33} - \tau_{11})}{\dot{\epsilon}_0} = 3\eta_0 \tag{8.97}$$

We see that the Maxwell model predicts the same elongation viscosity as the Newtonian model. The Trouton ratio $\bar{\eta}/\eta_0$ is just 3, the same result as was obtained for Newtonian and power-law generalized Newtonian fluids, a reasonable result.

8.2.2 Generalized Maxwell Model

As was noted in the second practice example, the shear stress-relaxation modulus for a fluid that follows the Maxwell model is an exponential decay with time, and this response is independent of the magnitude of the strain γ_0. Although the single exponential decay function does describe qualitatively the relaxation of stress of many fluids, the fit is not quantitative, as we can see in Figure 8.4 for a narrow-molecular-weight-distribution polystyrene solution. Most fluids are not characterized by a single relaxation time λ, but rather have multiple relaxation times. We can obtain a good fit to the data in Figure 8.4 if we add up four Maxwell models (Figure 8.5). The Maxwell constitutive equation can be modified to describe materials with more than one relaxation time if we assume additivity of stresses, that is, that the total stress exhibited by a material is just the sum of the individual stresses due to each relaxation time.

To express mathematically what we have just described in words, we postulate that a fluid has many relaxation times λ_k, and we write a Maxwell model for each relaxation time λ_k. For each λ_k and η_k there is a contribution to the extra stress tensor $\underline{\underline{\tau}}_{(k)}$ generated:

$$\underline{\underline{\tau}}_{(k)} + \lambda_k \frac{\partial \underline{\underline{\tau}}_{(k)}}{\partial t} = -\eta_k \underline{\underline{\dot{\gamma}}} \tag{8.98}$$

These individual stress contributions now sum to give the complete extra stress tensor $\underline{\underline{\tau}}$. For a system with N relaxation times,

$$\underline{\underline{\tau}} = \sum_{k=1}^{N} \underline{\underline{\tau}}_{(k)} \tag{8.99}$$

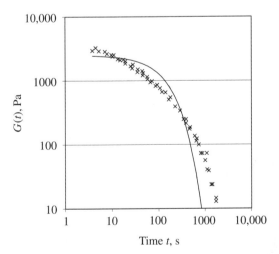

Figure 8.4 Relaxation-modulus data at 33.5°C for a polystyrene of narrow molecular-weight distribution, $M_w = 1.8 \times 10^6$, 20% solution in chlorinated diphenyl; from Einaga et al. [72]. The data are for small strains ($\gamma_0 = 0.41$, 1.87) and are independent of strain. Also shown is a predicted $G(t)$ using the Maxwell model with $\lambda = 150$ s and $g = \eta_0/\lambda = 2500$ Pa.

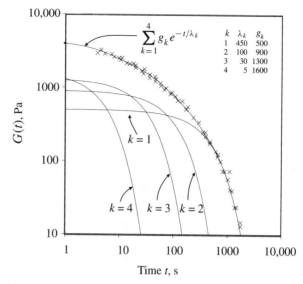

Figure 8.5 Relaxation-modulus data from Figure 8.4 fit to the sum of four $G(t)$ contributions calculated using the Maxwell model with parameters λ_k and $g_k = \eta_k/\lambda_k$ as indicated. The fit can be made arbitrarily good by choosing to use more λ_k and g_k.

$$= \sum_{k=1}^{N} \left[-\int_{-\infty}^{t} \left(\frac{\eta_k}{\lambda_k} \right) e^{\frac{-(t-t')}{\lambda_k}} \underline{\dot{\gamma}}(t')\, dt' \right] \qquad (8.100)$$

Generalized
Maxwell model
$$\underline{\underline{\tau}}(t) = -\int_{-\infty}^{t} \left[\sum_{k=1}^{N} \left(\frac{\eta_k}{\lambda_k} \right) e^{\frac{-(t-t')}{\lambda_k}} \right] \underline{\underline{\dot{\gamma}}}(t')\, dt' \qquad (8.101)$$

This last expression is called the *generalized Maxwell model*. This model has $2N$ parameters, λ_k and η_k for $k = 1$ to N, and therefore sufficient flexibility to fit any experimental stress-relaxation curve. One question remains: does this model predict the summation of exponentials that we were hoping for? We can answer this question by calculating $G(t)$ for the generalized Maxwell model.

EXAMPLE

Calculate the step shear strain material functions $G(t, \gamma_0)$, $G_{\Psi_1}(t, \gamma_0)$, and $G_{\Psi_2}(t, \gamma_0)$ for the generalized Maxwell constitutive equation.

SOLUTION

The step-strain experiment is defined in shear flow, and thus the velocity field is given by

$$\underline{v} = \begin{pmatrix} \dot{\varsigma}(t)x_2 \\ 0 \\ 0 \end{pmatrix}_{123} \tag{8.102}$$

$$\dot{\varsigma}(t') = \lim_{\varepsilon \to 0} \dot{\varsigma}(t', \varepsilon) \tag{8.103}$$

$$\dot{\varsigma}(t', \varepsilon) = \begin{cases} 0 & t' < 0 \\ \dot{\gamma}_0 = \dfrac{\gamma_0}{\varepsilon} & 0 \le t' < \varepsilon \\ 0 & t' \ge \varepsilon \end{cases} \tag{8.104}$$

and the material functions are $G(t, \gamma_0)$, $G_{\Psi_1}(t, \gamma_0)$, and $G_{\Psi_2}(t, \gamma_0)$, which were defined in the last example. The shear-rate tensor is

$$\underline{\underline{\dot{\gamma}}} = \begin{pmatrix} 0 & \dot{\varsigma}(t) & 0 \\ \dot{\varsigma}(t) & 0 & 0 \\ 0 & 0 & 0 \end{pmatrix}_{123} \tag{8.105}$$

For a generalized Maxwell fluid, the stress tensor in this flow is given by

$$\underline{\underline{\tau}}(t) = -\int_{-\infty}^{t} \left[\sum_{k=1}^{N} \left(\frac{\eta_k}{\lambda_k} \right) e^{\frac{-(t-t')}{\lambda_k}} \right] \begin{pmatrix} 0 & \dot{\varsigma}(t') & 0 \\ \dot{\varsigma}(t') & 0 & 0 \\ 0 & 0 & 0 \end{pmatrix}_{123} dt' \tag{8.106}$$

Taking $\tau_{21}(t)$ from Equation (8.106), we can calculate the material function $G(t, \gamma_0)$ predicted by the Maxwell model:

$$\tau_{21}(t) = -\int_{-\infty}^{t} \left[\sum_{k=1}^{N} \left(\frac{\eta_k}{\lambda_k} \right) e^{\frac{-(t-t')}{\lambda_k}} \right] \dot{\varsigma}(t') \, dt' \tag{8.107}$$

$$-\tau_{21}(t, \varepsilon) = \int_{-\infty}^{0} 0 \, dt' + \int_{0}^{\varepsilon} \left[\sum_{k=1}^{N} \left(\frac{\eta_k}{\lambda_k} \right) e^{\frac{-(t-t')}{\lambda_k}} \right] \frac{\gamma_0}{\varepsilon} \, dt' + \int_{\varepsilon}^{t} 0 \, dt' \tag{8.108}$$

$$= \sum_{k=1}^{N} \frac{\gamma_0 \eta_k}{\lambda_k \varepsilon} \int_{0}^{\varepsilon} e^{\frac{-(t-t')}{\lambda_k}} \, dt' \tag{8.109}$$

$$G(t, \gamma_0) = \lim_{\varepsilon \to 0} \frac{-\tau_{21}(t, \varepsilon)}{\gamma_0} \tag{8.110}$$

$$= \lim_{\varepsilon \to 0} \frac{\sum_{k=1}^{N} \frac{\eta_k}{\lambda_k} \int_{0}^{\varepsilon} e^{\frac{-(t-t')}{\lambda_k}} \, dt'}{\varepsilon} \tag{8.111}$$

As we did with the single-relaxation-time Maxwell model, we use l'Hôpital's rule to complete the calculation of $G(t, \gamma_0)$. The result is independent of γ_0:

Relaxation modulus for the
generalized Maxwell model

$$G(t) = \sum_{k=1}^{N} \frac{\eta_k}{\lambda_k} e^{\frac{-t}{\lambda_k}}$$

(8.112)

Thus the relaxation modulus for the generalized Maxwell model is a summation of exponentials. Any observed relaxation response can be fit to Equation (8.112) if enough relaxation times are chosen, that is, for N large. As was true for the single-relaxation-time Maxwell model, the diagonal stress components are zero for a generalized Maxwell fluid in shear flow, and $G_{\Psi_1}(t, \gamma_0) = G_{\Psi_2}(t, \gamma_0) = 0$.

8.3 GLVE Constitutive Equation

Both the Maxwell model [Equation (8.58)] and the generalized Maxwell model [Equation (8.101)] have the same structure, only differing in the form of the function in square brackets. We can further generalize both models and define a generalized linear viscoelastic (GLVE) model as follows:

Generalized linear
viscoelastic (GLVE) model

$$\underline{\underline{\tau}}(t) = -\int_{-\infty}^{t} g(t - t')\underline{\underline{\dot{\gamma}}}(t')\,dt'$$

(8.113)

This constitutive equation has as its parametric quantity one function, $g(t - t')$. Once this function is known, the stress can be calculated in any flow situation. The Maxwell model and the generalized Maxwell model are both GLVE constitutive equations with the GLVE function $g(t - t')$ given by

Maxwell model: $g(t - t') = \dfrac{\eta_0}{\lambda} e^{\frac{-(t-t')}{\lambda}}$ (8.114)

generalized Maxwell model: $g(t - t') = \displaystyle\sum_{k=1}^{N} \dfrac{\eta_k}{\lambda_k} e^{\frac{-(t-t')}{\lambda_k}}$ (8.115)

These two expressions are just the same functions that we obtained for the step shear relaxation modulus, only evaluated for the argument $t - t'$ instead of for t. [Compare Equations (8.114) and (8.115) with Equations (8.87) and (8.112)]. Thus, for any GLVE fluid, the function that multiplies $\underline{\underline{\dot{\gamma}}}(t')$ is the shear-relaxation-modulus function obtained in step strain,

$$g(t - t') = G(t - t')$$

(8.116)

We can verify that the response of a GLVE fluid to a step-strain experiment is just $g(t)$ by carrying out the usual step-strain calculations on the GLVE model. To remind us of the link between the function $g(t - t')$ and the step-strain material function $G(t)$, the GLVE model is usually written with the nomenclature $G(t - t')$ replacing $g(t - t')$:

$$\text{generalized linear}$$
$$\text{viscoelastic (GLVE) model:} \qquad \underline{\underline{\tau}}(t) = -\int_{-\infty}^{t} G(t - t')\underline{\underline{\dot{\gamma}}}(t') \, dt' \qquad (8.117)$$

Within the integral in the GLVE equation, note also that two separate expressions appear, one related to the kinematics, $\dot{\gamma}(t')$, and another separate function that describes the material response to the flow, $g(t - \overline{\overline{t'}}) = G(t - t')$. The GLVE equation is one of a class of constitutive equations called *separable* [138]. Separability of material and kinematic contributions to the stress is a general property exhibited by some materials at all strains and by all linear viscoelastic fluids at small strain rates. Many, but not all, constitutive equations are separable; see Larson [138] for a more in-depth discussion of the implications of separability.

We see from the GNF, Maxwell, and GLVE fluids that it is common practice for constitutive equations to be written with material functions as part of their definitions. So far we have seen the viscosity η appear in the generalized Newtonian equation, the zero-shear viscosity η_0 in the Maxwell model, and the step-strain relaxation modulus $G(t)$ in the GLVE constitutive equation. Although this is standard practice, it is somewhat unfortunate since this nomenclature causes confusion to beginners who are trying to keep constitutive equations and material functions straight. The advantage of writing the constitutive equations this way is that the user is reminded up front of the predictions of the constitutive equation with regard to some material functions.

We devote the rest of this section to some example problems for calculating material functions from the GLVE constitutive equation, followed by some flow problems using the GLVE equation. Several of the examples are related to small-amplitude oscillatory shear flow, which is perhaps the most widely used experimental application of the GLVE equation. The final section of the chapter discusses some of the limitations of the GLVE model.

EXAMPLE

Calculate the steady shear material functions η, Ψ_1, and Ψ_2 for the generalized linear viscoelastic constitutive equation.

SOLUTION

Steady shear material functions are defined with respect to the following kinematics:

$$\underline{v} = \begin{pmatrix} \dot{\varsigma}(t)x_2 \\ 0 \\ 0 \end{pmatrix}_{123} \qquad (8.118)$$

$$\dot{\varsigma}(t) = \dot{\gamma}_0 = \text{constant} \qquad (8.119)$$

The definitions of the steady shear material functions are given in Chapter 5 [Equations (5.8)–(5.10)], where the stress components are evaluated for these kinematics.

We must calculate the tensor $\underline{\underline{\tau}}$, which we obtain from the constitutive equation:

$$\underline{\underline{\tau}}(t) = -\int_{-\infty}^{t} G(t - t') \begin{pmatrix} 0 & \dot{\gamma}_0 & 0 \\ \dot{\gamma}_0 & 0 & 0 \\ 0 & 0 & 0 \end{pmatrix}_{123} dt' \qquad (8.120)$$

We can now calculate the steady shear material functions for this model:

$$\eta = \frac{-\tau_{21}}{\dot{\gamma}_0} = \int_{-\infty}^{t} G(t - t') \, dt' \tag{8.121}$$

$$\Psi_1 = \frac{-(\tau_{11} - \tau_{22})}{\dot{\gamma}_0^2} = 0 \tag{8.122}$$

$$\Psi_2 = \frac{-(\tau_{22} - \tau_{33})}{\dot{\gamma}_0^2} = 0 \tag{8.123}$$

We can simplify the expression for viscosity by introducing the variable $s = t - t'$. With this substitution, the final result for viscosity becomes

Viscosity for
GLVE model
$$\boxed{\eta = \int_{0}^{\infty} G(s) \, ds} \tag{8.124}$$

EXAMPLE

Calculate the material functions $G'(\omega)$ (storage modulus) and $G''(\omega)$ (loss modulus) for the generalized linear viscoelastic model and for the generalized Maxwell model.

SOLUTION

One of the most common uses of the generalized Maxwell model is to perform calculations on small-amplitude oscillatory shear (SAOS) data. We can find the predictions of the GLVE and generalized Maxwell models in SAOS by the same method we have been following, that is, we substitute the definition of velocity in shear flow $\underline{v} = \dot{\varsigma}(t)x_2\hat{e}_1$, with $\dot{\varsigma}(t)$ specified, into the constitutive equation and calculate the stresses and subsequently the material functions.

For SAOS, the usual shear-flow shear-rate tensor $\dot{\underline{\underline{\gamma}}}$ is obtained as

$$\dot{\underline{\underline{\gamma}}} = \begin{pmatrix} 0 & \dot{\varsigma}(t) & 0 \\ \dot{\varsigma}(t) & 0 & 0 \\ 0 & 0 & 0 \end{pmatrix}_{123} \tag{8.125}$$

with the strain rate given by

$$\dot{\varsigma}(t) = \dot{\gamma}_{21}(t) = \dot{\gamma}_0 \cos \omega t \tag{8.126}$$

The definitions of the material functions $G'(\omega)$ and $G''(\omega)$ are related to the stress wave $\tau_{21}(t)$ (see Section 5.2.2.5):

$$-\tau_{21}(t) = G'\dot{\gamma}_0 \sin \omega t + G''\dot{\gamma}_0 \cos \omega t \tag{8.127}$$

To calculate $\tau_{21}(t)$, we use the GLVE constitutive equation:

$$\underline{\underline{\tau}} = -\int_{-\infty}^{t} G(t - t') \begin{pmatrix} 0 & \dot{\varsigma}(t') & 0 \\ \dot{\varsigma}(t') & 0 & 0 \\ 0 & 0 & 0 \end{pmatrix}_{123} dt' \tag{8.128}$$

$$\tau_{21}(t) = -\int_{-\infty}^{t} G(t - t') \, \dot{\varsigma}(t') \, dt' \tag{8.129}$$

$$= -\int_{-\infty}^{t} G(t - t') \, \dot{\gamma}_0 \cos \omega t' \, dt' \tag{8.130}$$

This integration is a bit easier to follow if we substitute $s = t - t'$. This change alters the limits of the integral as well:

$$\tau_{21}(t) = \int_{\infty}^{0} G(s) \, \dot{\gamma}_0 \cos(\omega t - \omega s) \, ds \tag{8.131}$$

Expanding the cosine term allows us to identify G' and G'' (recall that $\dot{\gamma}_0 = \gamma_0 \omega$):

$$-\tau_{21}(t) = \int_{0}^{\infty} G(s) \, \dot{\gamma}_0 \cos \omega t \cos \omega s \, ds$$

$$+ \int_{0}^{\infty} G(s) \, \dot{\gamma}_0 \sin \omega t \sin \omega s \, ds \tag{8.132}$$

$$= \left[\int_{0}^{\infty} G(s) \cos \omega s \, ds \right] \gamma_0 \omega \cos \omega t$$

$$+ \left[\int_{0}^{\infty} G(s) \sin \omega s \, ds \right] \gamma_0 \omega \sin \omega t \tag{8.133}$$

Comparing this expression with the definitions of G' and G'' [Equation (8.127)] we find:

$$G''(\omega) = \omega \int_{0}^{\infty} G(s) \cos \omega s \, ds \tag{8.134}$$

$$G'(\omega) = \omega \int_{0}^{\infty} G(s) \sin \omega s \, ds \tag{8.135}$$

For the generalized Maxwell model these integrals may be carried out by using the complex notation for cosine and sine, yielding

SAOS material
functions for
generalized
Maxwell model

$$\boxed{\begin{aligned} G''(\omega) &= \sum_{k=1}^{N} \frac{g_k \lambda_k \omega}{1 + \lambda_k^2 \omega^2} \\ G'(\omega) &= \sum_{k=1}^{N} \frac{g_k \lambda_k^2 \omega^2}{1 + \lambda_k^2 \omega^2} \end{aligned}} \tag{8.136}$$

where we have made the substitution $g_k = \eta_k / \lambda_k$. $G'(\omega)$ and $G''(\omega)$ for the Maxwell model are plotted in Figure 8.6 in nondimensional form for $N = 1$. G' and G'' cross at

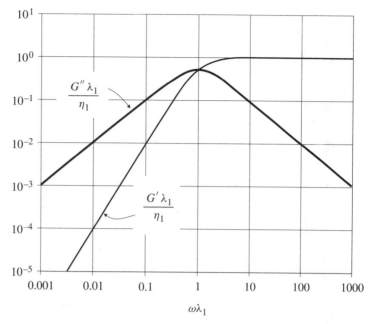

Figure 8.6 Plot of scaled dynamic moduli $G'(\omega)$ and $G''(\omega)$ for a single-relaxation-time Maxwell fluid. λ_1 is the relaxation time, η_1 is the viscosity parameter, and ω is radian frequency.

$\omega\lambda_1 = 1(G' = G'')$, or at $\omega = 1/\lambda_1$. This relationship can be used to estimate a relaxation time from experimental data of $G'(\omega)$ and $G''(\omega)$.

In the preceding example the expressions for $G'(\omega)$ and $G''(\omega)$ for the generalized Maxwell model capture some important features of actual melt-flow data for these material functions. Notice that for small ω,

$$\lim_{\omega \to 0} G' \propto \omega^2 \tag{8.137}$$

$$\lim_{\omega \to 0} G'' \propto \omega \tag{8.138}$$

These relationships are observed for straight-chain polymers at low frequencies (for example, see Figures 6.34 and 6.36). In the following example problem we see how this scaling at low frequency determines the longest relaxation time that characterizes a fluid.

EXAMPLE   Fit the relaxation spectrum of the generalized Maxwell model to the SAOS data for $G'(\omega)$ and $G''(\omega)$ listed in Table 8.1. The data are for a narrowly distributed polystyrene melt of $M_w = 59$ kg/mol, $T = 190°C$ [252].

TABLE 8.1
Data for Example Problem

ω (rad/s)	G' (Pa)	ω (rad/s)	G'' (Pa)
2.45E+01	9.27E+01	5.56E+00	4.53E+02
4.98E+01	3.33E+02	1.59E+01	1.29E+03
1.01E+02	1.16E+03	4.55E+01	3.82E+03
2.05E+02	3.76E+03	1.30E+02	1.09E+04
4.16E+02	1.10E+04	3.72E+02	3.10E+04
8.44E+02	2.79E+04	1.06E+03	6.23E+04
1.71E+03	5.19E+04	2.13E+03	8.52E+04
3.44E+03	8.40E+04	4.29E+03	8.81E+04
6.95E+03	1.18E+05	8.62E+03	1.01E+05
1.99E+04	1.84E+05	1.73E+04	1.24E+05
5.69E+04	2.50E+05	3.49E+04	1.76+05
1.63E+05	3.51E+05	9.96E+04	3.18E+05
4.66E+05	5.67E+05	2.84E+05	5.76E+05
1.34E+06	1.17E+06	5.72E+05	8.44E+05

*A more complete data set may be found in Table F.2 in Appendix F.

SOLUTION

There is commercial software available that will perform this calculation [216], but a reasonable solution can be found by trial and error using a spreadsheet program or other mathematical software.

The data are plotted in Figure 8.7. Also shown are G' and G'' calculated for a single-relaxation-time Maxwell element:

SAOS material functions
for a single-relaxation-time
Maxwell fluid

$$G'(\omega) = \frac{g_1 \omega^2 \lambda_1^2}{1 + \omega^2 \lambda_1^2}$$

$$G''(\omega) = \frac{g_1 \omega \lambda_1}{1 + \omega^2 \lambda_1^2}$$

(8.139)

The data we were given show the expected terminal behavior at low frequencies, that is, $G' \propto \omega^2$ and $G'' \propto \omega^1$, and thus the longest relaxation-time contribution to the relaxation spectrum needs to fit this portion of the curve. By trial and error with values for g_1 and λ_1 we found that $g_1 = 4 \times 10^4$ Pa and $\lambda_1 = 0.0016$ s provide a reasonable fit at low frequencies, as shown in Figure 8.7. For the single-relaxation-time Maxwell fluid, g_1 changes the height of the plateau seen in the predicted G', and λ_1 moves the curves left and right on the frequency

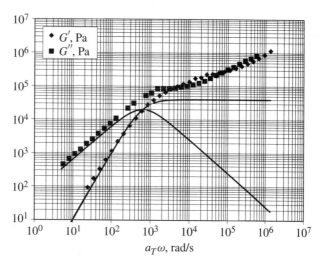

Figure 8.7 Linear viscoelastic moduli G' and G'' for a polystyrene melt at 190°C; from Vinogradov [252]. The polymer has $M_w = 59$ kg/mol, and its molecular weight is narrowly distributed. The model curves shown are for a single-relaxation-time Maxwell fluid with $g_1 = 4 \times 10^4$ Pa and $\lambda_1 = 0.0016$ s. *Source:* From "Critical regimes of deformation of liquid polymeric systems," by G. V. Vinogradov, *Rheologica Acta*, **12**, 357–373 (1973), Figure 13. Copyright © 1973, Springer-Verlag.

scale. (Increasing λ_1 shifts the crossover of G' and G'' for the predicted curves to the left, toward lower frequencies.)

The number and spacing of the relaxation times used in this fit are arbitrary, depending only on the quality of the fit desired. Our strategy is to use the fewest number of relaxation elements necessary for a reasonable fit. It is clear from the shapes of $G'(\omega)$ and $G''(\omega)$ for the single Maxwell element that more than one relaxation time is needed. To add a second element we arbitrarily choose $\lambda_2 = \lambda_1/5$ and guess values of g_2 until the combined G' and G'' curves of the two relaxation modes fits the next higher frequency portion of the experimental curves. If the G' guessed curve seems to grow too quickly when a new relaxation mode is added, we may need to go back and decrease the modulus parameter of an earlier relaxation time. Relaxation times are added and moduli guessed and adjusted until the entire curve is fit.

This method is rough, but a reasonable fit can be obtained, and if desired we can obtain a more refined fit using a nonlinear fitting program.[4] The rough fit calculated using the method described can be used as an initial guess in a nonlinear fitting program to speed up the convergence. A good objective function to minimize when using a nonlinear fitting program to fit G' and G'' data is

---

[4] Microsoft Solver in Excel works well for this purpose. A more detailed discussion of fitting relaxation spectra is given in Macosko [162].

$$O = \sum_{i=1}^{N} \left\{ \frac{[G'(\omega_i) - G'_{\text{model}}(\omega_i)]^2}{[G'(\omega_i)]^2} + \frac{[G''(\omega_i) - G''_{\text{model}}(\omega_i)]^2}{[G''(\omega_i)]^2} \right\} \qquad (8.140)$$

where O is the objective function. Equation (8.140) is preferred to a simple sum of squared differences between the model and the data, which would be dominated by the errors at large values of G' and G''. Other possible objective functions are discussed and compared in Secor [162, Appendix 3A]. The final fit is given in Figure 8.8.

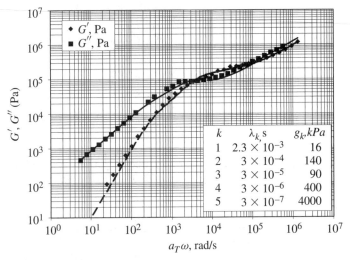

Figure 8.8 Linear viscoelastic moduli G' and G'' for a polystyrene melt at 190°C; from Vinogradov [252]. The polymer has $M_w = 59$ kg/mol, and its molecular weight is narrowly distributed. The model curves shown are for a 5-relaxation-time Maxwell fluid ($N = 5$ in Equation 8.136) with parameters given in the figure. *Source:* From "Critical regimes of deformation of liquid polymeric systems," by G. V. Vinogradov, *Rheologica Acta*, **12**, 357–373 (1973), Figure 13. Copyright © 1973, Springer-Verlag.

With an appropriate number N of relaxation times, nearly any set of G' and G'' curves can be fit to Equations (8.136) (see Figure 8.9). Keep in mind, however, that if the data to be fit do not show terminal behavior, the value of the longest relaxation time λ_1 obtained from the fit will only reflect what was the lowest frequency available in the experimental data, $\omega_{\min}$. We can obtain a fairly good fit to the generalized Maxwell model for the polyethylene data shown in Figure 8.9. If we plot the fit curves outside the experimental frequency range, however, we can see that our fit assumes $1/\omega_{\min} \approx \lambda_1$ and that the value of G' at the largest experimental frequency $\omega_{\max}$ is the plateau modulus (Figure 8.10). The prediction for the longest relaxation time is fine for these data because they are nearly terminal, but it is unclear whether G_N^o is equal to the highest measured value of G', as is assumed in the fitting.

Thus far we have used the Maxwell and the GLVE models to predict material functions. We can also solve flow problems with these constitutive equations, as we demonstrate in the next section.

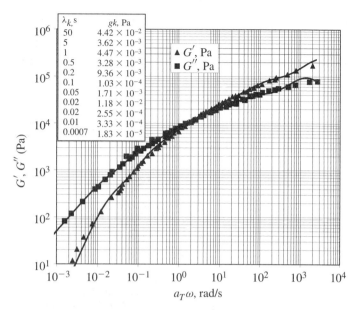

Figure 8.9 Dynamic shear moduli for polyethylene and best fit to Equation (8.136); from from Laun [142]. Note that the data are borderline terminal (they are just beginning to approach slopes of 2 and 1 at low frequencies). *Source:* From "Description of the non-linear shear behavior of a low density polyethylene melt by means of an experimentally determined strain dependent memory function," by H. M. Laun, *Rheologica Acta*, **17**, 1–5 (1978), Figure 1. Copyright © 1978, Springer-Verlag.

8.4 Flow Problems: GLVE Fluid

8.4.1 POISEUILLE FLOW OF A GLVE FLUID

Calculate the velocity profile and the stress tensor for pressure-driven flow of an incompressible generalized linear viscoelastic liquid flowing in a tube of circular cross section. The pressure at an upstream point is P_0, and at a point a distance L downstream the pressure is P_L. Assume that the flow between these two points is fully developed and at steady state.

We have seen this classic problem twice before (see Figure 7.6), and the solution follows the same pattern as in both the Newtonian case and the generalized Newtonian case. The problem is worked in cylindrical coordinates.

We assume that the velocity is in the z-direction only, and this allows us to simplify the continuity equation:

$$\underline{v} = \begin{pmatrix} v_r \\ v_\theta \\ v_z \end{pmatrix}_{r\theta z} = \begin{pmatrix} 0 \\ 0 \\ v_z \end{pmatrix}_{r\theta z} \tag{8.141}$$

$$0 = \nabla \cdot \underline{v} = \frac{\partial v_r}{\partial r} + \frac{1}{r}\frac{\partial v_\theta}{\partial \theta} + \frac{\partial v_z}{\partial z} \tag{8.142}$$

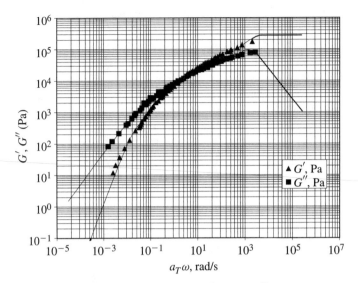

Figure 8.10 Plot of a generalized Maxwell model fit to $G'(\omega)$ and $G''(\omega)$ data of polyethylene [142]. The fit curve will take on Maxwell-like shapes outside the experimental range. Accurate values of the longest relaxation time and of the plateau modulus cannot be obtained unless the data are clearly terminal (have slopes of 2 and 1 for G' and G'' at low frequencies) and unless they reach a plateau at high frequencies. *Source:* From " Description of the non-linear shear behavior of a low density polyethylene melt by means of an experimentally determined strain dependent memory function," by H. M. Laun, *Rheologica Acta*, **17**, 1–5 (1978), Figure 1. Copyright © 1978, Springer-Verlag.

$$0 = \frac{\partial v_z}{\partial z} \tag{8.143}$$

The equation of motion for a steady ($\partial \underline{v}/\partial t = 0$) unidirectional ($\underline{v} \cdot \nabla \underline{v} = 0$) flow simplifies to

$$\underline{0} = -\nabla p - \nabla \cdot \underline{\underline{\tau}} + \rho \underline{g} \tag{8.144}$$

$$\begin{pmatrix} 0 \\ 0 \\ 0 \end{pmatrix}_{r\theta z} = -\begin{pmatrix} \frac{\partial p}{\partial r} \\ \frac{1}{r}\frac{\partial p}{\partial \theta} \\ \frac{\partial p}{\partial z} \end{pmatrix}_{r\theta z} - \begin{pmatrix} \frac{1}{r}\frac{\partial}{\partial r}(r\tau_{rr}) + \frac{1}{r}\frac{\partial \tau_{\theta r}}{\partial \theta} + \frac{\partial \tau_{zr}}{\partial z} - \frac{\tau_{\theta\theta}}{r} \\ \frac{1}{r^2}\frac{\partial}{\partial r}(r^2\tau_{r\theta}) + \frac{1}{r}\frac{\partial \tau_{\theta\theta}}{\partial \theta} + \frac{\partial \tau_{z\theta}}{\partial z} \\ \frac{1}{r}\frac{\partial}{\partial r}(r\tau_{rz}) + \frac{1}{r}\frac{\partial \tau_{\theta z}}{\partial \theta} + \frac{\partial \tau_{zz}}{\partial z} \end{pmatrix}_{r\theta z} + \begin{pmatrix} 0 \\ 0 \\ \rho g \end{pmatrix}_{r\theta z} \tag{8.145}$$

Note that gravity is taken to be in the flow direction. To simplify the equation of motion further, we must calculate $\underline{\underline{\tau}}$ from the constitutive equation, which is the GLVE constitutive equation,

$$\underline{\underline{\tau}}(t) = -\int_{-\infty}^{t} G(t - t')\, \underline{\underline{\dot{\gamma}}}(t')\, dt' \tag{8.146}$$

$$= -\int_{-\infty}^{t} G(t - t') \begin{pmatrix} 0 & 0 & \frac{\partial v_z}{\partial r} \\ 0 & 0 & 0 \\ \frac{\partial v_z}{\partial r} & 0 & 0 \end{pmatrix}_{r\theta z} dt' \tag{8.147}$$

We see that many of the stress coefficients are zero, and Equation (8.145) then simplifies to

$$
\begin{pmatrix} 0 \\ 0 \\ 0 \end{pmatrix}_{r\theta z} = -\begin{pmatrix} \frac{\partial p}{\partial r} \\ \frac{1}{r}\frac{\partial p}{\partial \theta} \\ \frac{\partial p}{\partial z} \end{pmatrix}_{r\theta z} - \begin{pmatrix} \frac{\partial \tau_{zr}}{\partial z} \\ 0 \\ \frac{1}{r}\frac{\partial}{\partial r}(r\tau_{rz}) \end{pmatrix}_{r\theta z} + \begin{pmatrix} 0 \\ 0 \\ \rho g \end{pmatrix}_{r\theta z}
\tag{8.148}
$$

The continuity equation told us that v_z is not a function of z, and we will assume that v_z is not a function of θ due to symmetry; therefore v_z is only a function of r, and $\partial \tau_{zr}/\partial z = 0$. As usual with Poiseuille flow, the r- and θ-components of the equation of motion tell us that pressure is neither a function of r nor of θ, and we will solve the z-component of the equation of motion for the velocity profile.

The z-component of the simplified equation of motion can be written as

$$
0 = -\frac{d\mathcal{P}}{dz} - \frac{1}{r}\frac{\partial}{\partial r}(r\tau_{rz})
\tag{8.149}
$$

where we have written $\mathcal{P} = p - \rho g z$. The boundary conditions on pressure, velocity, and stress are the usual ones for Poiseuille flow:

$$
\begin{aligned}
z = 0 && \mathcal{P} = \mathcal{P}_0 = P_0 \\
z = L && \mathcal{P} = P_L - \rho g L = \mathcal{P}_L \\
r = 0 && \tau_{rz} = 0 \\
r = R && v_z = 0
\end{aligned}
\tag{8.150}
$$

Previously when addressing the Poiseuille flow of a generalized Newtonian fluid, we solved Equation (8.149) with the same boundary conditions. The solutions for $\mathcal{P}$ and τ_{rz} are

$$
\boxed{\mathcal{P} = \left(\frac{\mathcal{P}_L - \mathcal{P}_0}{L}\right)z + \mathcal{P}_0}
\tag{8.151}
$$

$$
\boxed{\tau_{rz}(r) = -\left(\frac{\mathcal{P}_L - \mathcal{P}_0}{2L}\right)r}
\tag{8.152}
$$

To solve for the velocity field, we can equate this expression for τ_{rz} with the same quantity obtained from Equation (8.147):

$$
\left(\frac{\mathcal{P}_L - \mathcal{P}_0}{2L}\right)r = \int_{-\infty}^{t} G(t - t')\left(\frac{\partial v_z}{\partial r}\right)dt'
\tag{8.153}
$$

Since we are at steady state, $\partial v_z/\partial r$ is independent of time, and this derivative can be removed from the integral,

$$
\left(\frac{\mathcal{P}_L - \mathcal{P}_0}{2L}\right)r = \left(\frac{\partial v_z}{\partial r}\right)\left[\int_{-\infty}^{t} G(t - t')\,dt'\right]
\tag{8.154}
$$

As we showed in a previous example problem, the expression in square brackets is just the viscosity η_0; therefore we can solve for $v_z(r)$ in a straightforward manner. We have used the symbol η_0 for the viscosity because the GLVE equation is limited to low shear rates. After solving for $v_z(r)$ and applying the boundary conditions, the result is

Velocity profile of
a GLVE fluid in
Poiseuille flow in a tube

$$v_z(r) = \frac{(\mathcal{P}_0 - \mathcal{P}_L)R^2}{4L\eta_0}\left[1 - \left(\frac{r}{R}\right)^2\right] \quad (8.155)$$

This is the same relationship as that in the Newtonian case with the Newtonian viscosity μ replaced by the zero-shear viscosity η_0 of the GLVE fluid. The final result for the stress tensor is

$$\underline{\underline{\tau}}(t) = \begin{pmatrix} 0 & 0 & \frac{\mathcal{P}_0 - \mathcal{P}_L}{2L} \\ 0 & 0 & 0 \\ \frac{\mathcal{P}_0 - \mathcal{P}_L}{2L} & 0 & 0 \end{pmatrix} \quad (8.156)$$

In the next sections we demonstrate another flow problem, which is more complicated. It is also quite realistic as it results in the relationships that govern real commercial instruments for measuring SAOS properties.

8.4.2 Torsional Oscillatory Viscometer, Part I: Calculation of the Velocity Profile

For a generalized linear-viscoelastic fluid occupying the annular gap between two concentric cylinders (Couette geometry), calculate the velocity profile for the flow produced by tangentially driving the outer cylinder at an oscillatory shear rate. The inner cylinder is attached to a torque bar and can therefore also oscillate at a small amplitude as a result of the shear stress imposed by the fluid.

One of the most common experiments in rheology is the measurement of the small-amplitude oscillatory shear (SAOS) properties $\eta'(\omega)$ and $\eta''(\omega)$ or $G'(\omega)$ and $G''(\omega)$ [216, 32]. As we will see more extensively in Chapter 10, to make rheological measurements we must relate actual measurable quantities such as torque, pressure, flow rate, and rotational speed to the quantities that appear in material functions such as stress, shear rate, and strain. In this example and the follow-up example discussed next we see how this works for the tangential-annular-flow geometry. The first step is to calculate the velocity profile.

Our solution to this problem parallels that given by Bird et al. [26], and we have generally used the same notation as that text. Eventually we will assume that the gap between the inner and outer cylinders is narrow, and we will neglect curvature effects. We begin the problem, however, without making this assumption. We will use complex notation for the oscillating quantities (see Section 5.2.2.6).

The flow geometry and some of the geometric variables are defined in Figure 8.11. In the type of rheometer pictured, the outer cylinder, called the cup, is rotated in an oscillatory manner by a motor. The inner cylinder, called the bob, is attached to a fairly stiff torque bar. The torque exerted by the fluid on the bob rotates the bob through a small angle that is measured as a function of time. The torque bar acts like a linear spring, and thus the torque is proportional to the tiny angular displacement of the torque bar. To calculate the motion of the torque bar, we must first calculate the velocity profile of the fluid in the gap.

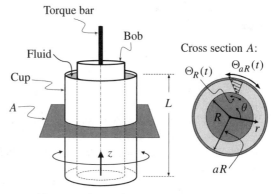

Figure 8.11 Geometry used in example: small-amplitude oscillatory tangential annular flow of a generalized linear viscoelastic fluid.

The flow is driven by the motion of the outer wall, which is moved following a cosine function of amplitude Θ_{aR}^0. The motion of the outer wall is described by the time-dependent angle $\Theta_{aR}(t)$, from which the velocity of the outer wall, $v_\theta|_{r=aR}$, can be calculated:

$$\Theta_{aR}(t) = \Theta_{aR}^0 \cos \omega t = \Re\left(\Theta_{aR}^0 e^{i\omega t}\right) \tag{8.157}$$

$$v_\theta|_{r=aR} = aR \frac{d\Theta_{aR}}{dt} = \Re\left(aRi\omega\, \Theta_{aR}^0 e^{i\omega t}\right) \tag{8.158}$$

Since the function for $\Theta_{aR}(t)$ is a cosine function, the amplitude Θ_{aR}^0 is a real number.

The stress transmits through the fluid, and the stress field in the gap is assumed to oscillate at the driving frequency ω,

$$\tau_{r\theta}(t) = \Re\left(\tilde{\tau}_0 e^{i\omega t}\right) \tag{8.159}$$

where $\tilde{\tau}_0$ is complex since $\tau_{r\theta}(t)$ will not, in general, be in phase with the driving wall motion. The shear stress causes the inner wall (the bob) to move; we define an angle $\Theta_R(t)$ to describe this motion. We assume that $\Theta_R(t)$ will also oscillate at the driving frequency ω,

$$\Theta_R(t) = \Re\left(\Theta_R^0 e^{i\omega t}\right) \tag{8.160}$$

$$v_\theta|_{r=R} = R\frac{d\Theta_R}{dt} = Ri\omega\, \Theta_R^0 e^{i\omega t} \tag{8.161}$$

The phases of Θ_R and Θ_{aR} will usually be different. For this reason, although Θ_{aR}^0 is a real number, Θ_R^0 is in general a complex number.

We now set out to solve for the velocity field in the gap. We proceed in the same manner as with the flow problems addressed in Chapters 3 and 7. We begin by assuming that v_r and v_z are zero,

$$\underline{v} = \begin{pmatrix} v_r \\ v_\theta \\ v_z \end{pmatrix}_{r\theta z} = \begin{pmatrix} 0 \\ v_\theta \\ 0 \end{pmatrix}_{r\theta z} \tag{8.162}$$

The continuity equation therefore gives $\partial v_\theta / \partial \theta = 0$, and using the tables for cylindrical coordinates we can construct $\underline{\underline{\dot{\gamma}}}$:

$$\underline{\underline{\dot{\gamma}}} = \begin{pmatrix} 0 & \frac{\partial v_\theta}{\partial r} - \frac{v_\theta}{r} & 0 \\ \frac{\partial v_\theta}{\partial r} - \frac{v_\theta}{r} & 0 & 0 \\ 0 & 0 & 0 \end{pmatrix}_{r\theta z} \tag{8.163}$$

We have assumed that the cylinders are long, and therefore $\partial v_\theta / \partial z = 0$, and the velocity is only a function of r and t. As we did with the velocity at the boundary, we assume that the velocity in the gap is a time-dependent oscillating function of frequency ω,

$$v_\theta(r) = \Re\left(\tilde{v}_0 e^{i\omega t}\right) \tag{8.164}$$

where $\tilde{v}_0$ is a complex number.

We use the GLVE constitutive equation to construct $\underline{\underline{\tau}}$:

$$\underline{\underline{\tau}} = -\int_{-\infty}^{t} G(t - t') \begin{pmatrix} 0 & \frac{\partial v_\theta(t')}{\partial r} - \frac{v_\theta(t')}{r} & 0 \\ \frac{\partial v_\theta(t')}{\partial r} - \frac{v_\theta(t')}{r} & 0 & 0 \\ 0 & 0 & 0 \end{pmatrix}_{r\theta z} dt' \tag{8.165}$$

From the preceding we can see that the only nonzero terms in $\underline{\underline{\tau}}$ are $\tau_{r\theta}$ and $\tau_{\theta r}$. This allows us to simplify the equation of motion (neglecting gravity) to (all terms written in the r, θ, z coordinate system):

$$\begin{pmatrix} 0 \\ \rho\frac{\partial v_\theta}{\partial t} \\ 0 \end{pmatrix} + \begin{pmatrix} \rho\frac{v_\theta^2}{r} \\ 0 \\ 0 \end{pmatrix} = \begin{pmatrix} -\frac{\partial p}{\partial r} \\ -\frac{1}{r}\frac{\partial p}{\partial \theta} \\ -\frac{\partial p}{\partial z} \end{pmatrix} - \begin{pmatrix} 0 \\ \frac{1}{r^2}\frac{\partial}{\partial r}\left(r^2 \tau_{r\theta}\right) \\ 0 \end{pmatrix} \tag{8.166}$$

The z-component of the equation of motion tells us that $\partial p / \partial z = 0$ (we have neglected gravity). The r-component of the equation of motion (EOM) tells us that there is a pressure gradient in the r-direction due to centrifugal force,

$$z\text{-component of EOM:} \qquad \frac{\partial p}{\partial z} = 0 \tag{8.167}$$

$$r\text{-component of EOM:} \qquad \rho\frac{v_\theta^2}{r} = -\frac{\partial p}{\partial r} \tag{8.168}$$

We can solve for the radial pressure gradient at the end of the problem if we wish, when we have an expression for v_θ. The θ-component of the equation of motion gives an equation for the shear stress:

$$\theta\text{-component of EOM:} \qquad \rho\frac{\partial v_\theta}{\partial t} = -\frac{1}{r}\frac{\partial p}{\partial \theta} - \frac{1}{r^2}\frac{\partial}{\partial r}\left(r^2 \tau_{r\theta}\right) \tag{8.169}$$

Due to symmetry in the θ-direction, we can assume that $\partial p / \partial \theta = 0$. Thus the equation to solve becomes

$$-\rho\frac{\partial v_\theta}{\partial t} = \frac{1}{r^2}\frac{\partial}{\partial r}\left(\tau_{r\theta} r^2\right) \tag{8.170}$$

Unlike the expressions we have encountered previously, this is not, in general, a separable differential equation since both v_θ and $\tau_{r\theta}$ are functions of t and r. We have assumed, however, that v_θ and $\tau_{r\theta}$ are particular time-oscillatory functions with frequency ω, and therefore we can separate the time and radial variations. Substituting the complex expressions in Equations (8.159) and (8.164) into Equation (8.170), we obtain

$$-i\omega\rho\,\tilde{v}_0 e^{i\omega t} = \frac{e^{i\omega t}}{r^2}\frac{\partial}{\partial r}\left(\tilde{\tau}_0 r^2\right) \tag{8.171}$$

$$-i\omega\rho\,\tilde{v}_0 r^2 = \frac{d}{dr}\left(\tilde{\tau}_0 r^2\right) \tag{8.172}$$

We have thus eliminated the time dependence and can concentrate on the r-dependence. We can write the r-dependence of $\tilde{\tau}_0$ in terms of $\tilde{v}_0$ and r by considering the relationship between $\tilde{\tau}_0$ and the complex material function η^*. From the definition of η^* found in Section 5.2.2.6,[5] we know that $\tilde{\tau}_0 = -\eta^*\dot{\gamma}_0$, where $\dot{\gamma}_0$ is the amplitude of $\dot{\gamma}_{21}(t)$. If we write that $\dot{\gamma}_{21}(t)$ must be an oscillatory function of ω,

$$\dot{\gamma}_{21}(t) = \dot{\gamma}_0 e^{i\omega t} \tag{8.173}$$

where $\dot{\gamma}_0$ is complex, we can calculate $\dot{\gamma}_0$ from Equations (8.163) and (8.164),

$$\frac{d\tilde{v}_0}{dr} - \frac{\tilde{v}_0}{r} = \dot{\gamma}_0 \tag{8.174}$$

and Equation (8.172) becomes

$$i\omega\rho\,\tilde{v}_0 r^2 = \eta^*\frac{d}{dr}\left[r^2\left(\frac{d\tilde{v}_0}{dr} - \frac{\tilde{v}_0}{r}\right)\right] \tag{8.175}$$

$$\frac{i\omega\rho}{\eta^*}\tilde{v}_0 r^2 = \frac{d}{dr}\left[r^3\frac{d}{dr}\left(\frac{\tilde{v}_0}{r}\right)\right] \tag{8.176}$$

The simplification on the right side of Equation (8.176) can be verified by the reader.

Substituting $\tilde{v}_0 = r\Omega(r)$, where $\Omega(r)$ is a function related to the r-dependence of the angular velocity, Equation (8.176) can be rearranged to become

$$\Omega'' + \frac{3}{r}\Omega' + \tilde{a}^2\Omega = 0 \tag{8.177}$$

where the prime indicates differentiation by r. The solution to this equation is [164]

$$\Omega = \frac{1}{r}\left[C_1 J_1(\tilde{a}r) + C_2 Y_1(\tilde{a}r)\right] \tag{8.178}$$

$$\tilde{a} \equiv \sqrt{\frac{-i\omega\rho}{\eta^*}} \tag{8.179}$$

---

[5] The derivation of the equation for η^* in Section 5.2.2.6 assumed simple parallel shear flow. It can be shown [26] that the material functions defined in simple shear flow also apply to nonsimple shear flows such as tangential annular flow. For more details on nonsimple shear flows see [26, 155, 156].

Here J_1 and Y_1 are first-order Bessel functions of the first and second kinds, respectively [232], and C_1 and C_2 are integration constants. The complete solution incorporating the appropriate boundary conditions for torsional oscillatory flow of a GLVE fluid has been worked out by Markovitz [164].

We can solve a simplified version of Equation (8.176) by making the assumption that the gap between the cylinders is narrow. This allows us to neglect the effects of curvature.[6] The equations become those written in the Cartesian (123) shear coordinate system; we will continue to use r, θ and z to maintain contact with our original coordinate system. For a narrow-gap limit,

$$\dot{\underline{\underline{\gamma}}} = \begin{pmatrix} 0 & \frac{\partial v_\theta}{\partial r} & 0 \\ \frac{\partial v_\theta}{\partial r} & 0 & 0 \\ 0 & 0 & 0 \end{pmatrix}_{r\theta z} \tag{8.180}$$

and

$$\begin{pmatrix} 0 \\ \rho \frac{\partial v_\theta}{\partial t} \\ 0 \end{pmatrix}_{r\theta z} + \begin{pmatrix} 0 \\ 0 \\ 0 \end{pmatrix}_{r\theta z} = \begin{pmatrix} -\frac{\partial p}{\partial r} \\ 0 \\ -\frac{\partial p}{\partial z} \end{pmatrix}_{r\theta z} - \begin{pmatrix} 0 \\ \frac{\partial \tau_{r\theta}}{\partial r} \\ 0 \end{pmatrix}_{r\theta z} \tag{8.181}$$

In this limit, Equation (8.176) becomes

$$\left(\frac{i\omega\rho}{\eta^*} \right) \tilde{v}_0 = \frac{d^2\tilde{v}_0}{dr^2} \tag{8.182}$$

This is a straightforward second-order ordinary differential equation with constant coefficients.

When solving an equation such as Equation (8.182), it is convenient to redefine the variables to simplify the boundary conditions. We will change Equation (8.182) from being a function of r to being a function of a dimensionless variable X, where

$$X = \frac{r - R}{aR - R} \tag{8.183}$$

The equation for $\tilde{v}_0$ becomes

$$\frac{d^2\tilde{v}_0}{dX^2} = \left(\frac{i\omega}{M_1} \right) \tilde{v}_0 \tag{8.184}$$

where

$$M_1 = \frac{\eta^*}{\rho R^2 (a - 1)^2} \tag{8.185}$$

subject to the no-slip boundary conditions at each wall, that is, at $r = aR$ and $r = R$. In terms of our new coordinates the no-slip boundary conditions become

---

[6] Note that when we neglect curvature effects, the flow becomes simple shear flow again, that is, $\dot{\underline{\underline{\gamma}}}(t) = \dot{\gamma}_{21}(t)(\hat{e}_1\hat{e}_2 + \hat{e}_2\hat{e}_1)$.

$$X = 0 \quad v_\theta = R\frac{d\Theta_R}{dt} \implies \tilde{v}_0(0) = i\omega R\Theta_R^0 \tag{8.186}$$

$$X = 1 \quad v_\theta = aR\frac{d\Theta_{aR}}{dt} \implies \tilde{v}_0(1) = i\omega aR\Theta_{aR}^0 \tag{8.187}$$

The solution for $\tilde{v}_0$ is [232]

$$\tilde{v}_0 = C_1 e^{\frac{i+1}{\sqrt{2}}\sqrt{\frac{\omega}{M_1}}X} + C_2 e^{\frac{-(i+1)}{\sqrt{2}}\sqrt{\frac{\omega}{M_1}}X} \tag{8.188}$$

Applying two boundary conditions allows us to calculate the two integration constants C_1 and C_2 and to obtain the solution. This final exercise is left up to the reader.

8.4.3 TORSIONAL OSCILLATORY VISCOMETER, PART II: MEASUREMENT OF SAOS MATERIAL FUNCTIONS

Using the velocity profile calculated in the previous example, show how the small-amplitude oscillatory shear material functions $\eta'(\omega)$ and $\eta''(\omega)$ can be calculated from measurements of the angle ratio $\Theta_R^0/\Theta_{aR}^0 = \mathcal{A}e^{i\delta}$, where $\mathcal{A}$ is the amplitude of the ratio, and δ is the phase difference between the motions of the inner and outer cylinders.

To relate the calculated velocity profile to η' and η'' we must find the connection between the shear stress imposed on the inner wall by the fluid and the restoring torque generated by the resisting torque bar. To do this, we must perform an angular momentum balance on the bob.

The shear stress in the fluid exerts a total torque on the bob that may be written as[7]

$$\begin{pmatrix} \text{torque on} \\ \text{inner surface} \end{pmatrix} = (2\pi RL)\,(-\tau_{r\theta})|_R\ R \tag{8.189}$$

The other force acting on the bob is the restoring force of the torque bar:

$$\begin{pmatrix} \text{restoring force} \\ \text{of torque bar} \end{pmatrix} = -\mathcal{K}\Theta_R \tag{8.190}$$

where $\mathcal{K}$ is the force constant of the torque bar. The resulting motion of the bob, $\Theta_R(t)$, is determined by the balance of angular momentum,

$$I\frac{d^2\Theta_R}{dt^2} = -\mathcal{K}\Theta_R + 2\pi R^2 L\,(-\tau_{r\theta})|_R \tag{8.191}$$

where I is the moment of inertia of the bob.

The shear stress at the inner wall, $(-\tau_{r\theta})|_{r=R}$, depends on material properties through the complex viscosity. For the cylindrical geometry, the complex viscosity is written as

---

[7] Note that our sign convention for stress requires that stress be positive when it is transporting momentum from high-velocity regions to low-velocity regions. Because the moving cup is causing a flux of momentum in the negative r-direction, the stress $\tau_{r\theta}$ is negative. To obtain a positive torque, a negative sign is included.

$$\eta^* \equiv \frac{-\tau_{r\theta}(t)}{\dot{\gamma}_{r\theta}(t)} \tag{8.192}$$

We can therefore calculate the stress at the inner wall in terms of η^*,

$$(-\tau_{r\theta})|_{r=R} = \eta^* \dot{\gamma}_{r\theta}(t)|_{r=R} \tag{8.193}$$

In the narrow-gap case we obtain the shear rate at the wall $\dot{\gamma}_{r\theta}$ from Equation (8.180) combined with the solution for the velocity profile from the previous exercise,

$$-\tau_{r\theta}|_{r=R} = \eta^* e^{i\omega t} \left.\frac{d\tilde{v}_0}{dr}\right|_{r=R} \tag{8.194}$$

We can now substitute the expressions for Θ_R and $(-\tau_{r\theta})_R$ into the balance of angular momentum and simplfy:

$$I\frac{d^2}{dt^2}\left(\Theta_R^0 e^{i\omega t}\right) = -\mathcal{K}\Theta_R^0 e^{i\omega t} + 2\pi R^2 L\eta^* e^{i\omega t} \left.\frac{d\tilde{v}_0}{dr}\right|_{r=R} \tag{8.195}$$

$$\Theta_R^0 = \frac{2\pi R^2 L\eta^*}{\mathcal{K} - I\omega^2} \left.\frac{d\tilde{v}_0}{dr}\right|_{r=R} \tag{8.196}$$

The derivative $d\tilde{v}_0/dr|_{r=R}$ can be obtained from Equation (8.188).

Our goal is to solve for η^* in terms of the amplitude ratio $\mathcal{A}$ and phase lag δ of the ratio of inner and outer wall motions:

$$\frac{\Theta_R^0}{\Theta_{aR}^0} = \mathcal{A}e^{i\delta} \tag{8.197}$$

To obtain the final result involves considerable algebra, but it is straightforward. First the boundary conditions [Equations (8.186) and (8.187)] are substituted into the solution for $\tilde{v}_0$ [Equation (8.188)], and a result for $\tilde{v}_0$ is obtained. This is used to calculate the derivative $d\tilde{v}_0/dr|_{r=R}$ in Equation (8.196), and that equation is solved for Θ_{aR}^0/Θ_R^0. As shown in Bird et al. [26], the result for this ratio is

$$\frac{\Theta_{aR}^0}{\Theta_R^0} = 1 + \frac{i}{\mathcal{M}}\left(\frac{\tilde{\omega}^2 - 1}{\tilde{A}\tilde{\omega}} + \frac{\tilde{\omega}}{2}\right) - \frac{1}{\mathcal{M}^2}\left(\frac{\tilde{\omega}^2 - 1}{6\tilde{A}} + \frac{\tilde{\omega}^2}{24}\right) + \text{terms of order}\left(\frac{1}{\mathcal{M}^3}\right) \tag{8.198}$$

where

$$\tilde{\omega} = \omega\sqrt{\frac{I}{\mathcal{K}}} \tag{8.199}$$

$$\mathcal{M} = M_1\sqrt{\frac{I}{\mathcal{K}}} = \frac{\eta^*}{\rho R^2(a-1)^2}\sqrt{\frac{I}{\mathcal{K}}} \tag{8.200}$$

$$\tilde{A} = 2\pi R^4\frac{L\rho(a-1)}{I} \tag{8.201}$$

Keeping only the first two terms and solving for $\mathcal{M}$, we obtain

$$\mathcal{M} = \frac{i\,\mathcal{W}}{\mathcal{A}^{-1}(\cos\delta - i\sin\delta) - 1} \tag{8.202}$$

where

$$\mathcal{W} = \frac{\tilde{\omega}^2 - 1}{\mathcal{A}\tilde{\omega}} + \frac{\tilde{\omega}}{2} \tag{8.203}$$

Separating this expression into the form $a + bi$ and incorporating the definition for $\mathcal{M}$ allows us to solve for η^* and its real and imaginary parts, η' and η'':

$$\boxed{\begin{aligned}
\eta'(\omega) &= \frac{G''}{\omega} = \frac{-\rho\mathcal{B}\mathcal{W}(\omega)\mathcal{A}\sin\delta}{1 + \mathcal{A}^2 - 2\mathcal{A}\cos\delta} \\
\eta''(\omega) &= \frac{G'}{\omega} = \frac{\rho\mathcal{B}\mathcal{W}(\omega)\mathcal{A}(\mathcal{A} - \cos\delta)}{1 + \mathcal{A}^2 - 2\mathcal{A}\cos\delta}
\end{aligned}} \tag{8.204}$$

in which $\mathcal{B}$ is given by

$$\mathcal{B} = (a - 1)^2 R^2 \sqrt{\frac{\mathcal{K}}{I}} \tag{8.205}$$

Thus measurements of η' and η'' can be made from the knowledge of instrument constants $(\mathcal{K}, I)$, geometry (a, R), material properties (ρ), and the actual dynamic measurements of amplitude $(\mathcal{A})$ and phase lag (δ) for a particular material. Equations (8.204) are the actual relations used in commercial rheometers of this type [32].

8.5 Limitations of the GLVE Model

The GLVE model succeeds in capturing the small-strain-rate behavior of most materials. The generalized Maxwell model, given that it has an essentially unlimited number of parameters η_k, λ_k, can be used to curve-fit experimental linear viscoelastic data for ease of calculation. There are important limitations to the GLVE model, however, which are enumerated as follows:

1. The GLVE model predicts a constant viscosity (no shear-thinning), which is inconsistent with experimental observations. This limits the usefulness of the GLVE model to small shear rates where $\eta = \eta_0$.
2. The definition of strain used in the derivation of the generalized Maxwell model assumes strain is additive. This limits the GLVE models to small strain rates.
3. Like the generalized Newtonian fluid models discussed in Chapter 7, the GLVE models are proportional to $\underline{\dot{\gamma}}$ and cannot predict normal stresses in shear flow.
4. The GLVE model cannot describe flows with a superimposed rigid rotation, that is, it is *not* objective.

The last item is extremely important since objectivity, that is, the requirement that predictions be independent of the choice of observer (see Section 7.1), is a requirement of all constitutive equations. It is particularly surprising to learn that the GLVE constitutive equation violates this important tenet since throughout the development of the model we seem to have followed all the rules of vector/tensor mathematics. Where, then, did we go wrong in the development of this model?

We misstepped when we generalized the scalar Maxwell equation to the tensorial Maxwell model:

$$\text{scalar Maxwell equation:} \quad \tau_{21} + \frac{\mu}{G}\frac{\partial \tau_{21}}{\partial t} = -\mu\dot{\gamma}_{21} \qquad (8.206)$$

$$\text{tensor Maxwell equation:} \quad \underline{\underline{\tau}} + \frac{\eta_0}{G}\frac{\partial \underline{\underline{\tau}}}{\partial t} = -\eta_0\underline{\underline{\dot{\gamma}}} \qquad (8.207)$$

At the time that this step was introduced, we said that since the model we were developing was strictly empirical, we could do as we liked in proposing the tensor version of the Maxwell model. This is true, as long as we follow all rules of frame invariance. As it happens, the partial time derivative used in the tensorial Maxwell model does not transform invariantly when the frame of reference is changed from a stationary to a rotating frame, as we will now show. Thus, by writing the simple partial derivative of $\underline{\underline{\tau}}$ in the tensor version of the Maxwell model, we proposed a frame-variant equation.

The violation of frame invariance in the Maxwell model and in the GLVE model can be demonstrated by finding a coordinate system in which the stress tensor predicted by the GLVE constitutive equation is incorrect. The breakdown occurs when the stress in a rotating frame of reference is written in a stationary reference frame. To show the failure of the GLVE model, we will perform a change-of-basis calculation for a shear flow occurring in a rotating frame of reference; this example was developed by Bird et al. [26].

Consider a shear flow produced by, for example, a bath of fluid and two counterrotating belts. The entire apparatus is on a rotating turntable (Figure 8.12). A Cartesian coordinate system ($\bar{x}$, $\bar{y}$, $\bar{z}$ with basis vectors $\hat{e}_{\bar{x}}$, $\hat{e}_{\bar{y}}$, $\hat{e}_{\bar{z}}$) rotating around the z- or $\bar{z}$-axis with the speed of the turntable is defined as shown. The stationary coordinate system will be x, y, and z with basis vectors $\hat{e}_x$, $\hat{e}_y$, and $\hat{e}_z$. The $\bar{z}$- and z-directions are the same ($\hat{e}_z = \hat{e}_{\bar{z}}$) and point upward. With respect to the x, y, z system the $\bar{x}$, $\bar{y}$, $\bar{z}$ system is located at position x_0, y_0, 0.

The flow that occurs on the turntable is a steady shear flow in the $\bar{x}$, $\bar{y}$, $\bar{z}$ coordinate system with a constant (positive) shear rate equal to $\dot{\gamma}_0$. Thus

$$\underline{v} = \begin{pmatrix} \dot{\gamma}_0\bar{y} \\ 0 \\ 0 \end{pmatrix}_{\bar{x}\bar{y}\bar{z}} = \dot{\gamma}_0\bar{y}\,\hat{e}_{\bar{x}} \qquad (8.208)$$

$$\underline{\underline{\dot{\gamma}}} = \begin{pmatrix} 0 & \dot{\gamma}_0 & 0 \\ \dot{\gamma}_0 & 0 & 0 \\ 0 & 0 & 0 \end{pmatrix}_{\bar{x}\bar{y}\bar{z}} \qquad (8.209)$$

The GLVE equation gives (in the $\bar{x}$, $\bar{y}$, $\bar{z}$ coordinate system)

$$\underline{\underline{\tau}} = -\int_{-\infty}^{t} G(t - t')\,\underline{\underline{\dot{\gamma}}}(t')\,dt' \qquad (8.210)$$

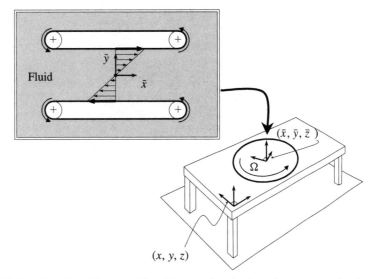

Figure 8.12 Rotating shear flow considered in text. Steady shear flow is occurring in a frame of reference that is rotating due to the motion of the turntable.

$$\tau_{\bar{y}\bar{x}} = -\int_{-\infty}^{t} G(t - t') \, \dot{\gamma}_0 \, dt' \tag{8.211}$$

$$= -\int_{0}^{\infty} G(s)\dot{\gamma}_0 \, ds \tag{8.212}$$

where $s = t - t'$. We then get the expected result for the viscosity [see Equation (8.124)]:

$$\eta = \frac{-\tau_{\bar{y}\bar{x}}}{\dot{\gamma}_0} \tag{8.213}$$

$$= \int_{0}^{\infty} G(s) \, ds \tag{8.214}$$

Now we will carry out the same calculation in the stationary x, y, z coordinate system. Our first step is to eliminate the differences in the origins of the two coordinate systems by considering the coordinate system $(x - x_0, y - y_0, z)$. This coordinate system has the same origin as the $\bar{x}, \bar{y}, \bar{z}$ system, and thus the only remaining difference between $(x - x_0, y - y_0, z)$ and $\bar{x}, \bar{y}, \bar{z}$ is the rotation of the $\bar{x}, \bar{y}, \bar{z}$ system.

When written in the $\bar{x}, \bar{y}, \bar{z}$ system, the velocity of the flowing fluid is

$$\underset{\substack{\text{with respect} \\ \text{to rotating} \\ \text{frame}}}{\underline{v}} = \dot{\gamma}_0 \bar{y} \, \hat{e}_{\bar{x}} \tag{8.215}$$

When the fluid velocity is written with respect to the stationary $(x - x_0, y - y_0, z)$ system, however, the velocity of the frame of reference $\underline{v}_{\text{frame}}$ will also appear:

$$\underline{v}_{\substack{\text{with respect} \\ \text{to stationary} \\ \text{frame}}} = \dot{\gamma}_0 \bar{y}\, \hat{e}_{\bar{x}} + \underline{v}_{\text{frame}} \tag{8.216}$$

To complete the change of frame we need to write expressions for $\bar{y}$, $\hat{e}_{\bar{x}}$, and $\underline{v}_{\text{frame}}$ in the stationary coordinate system $(x - x_0, y - y_0, z)$. We do this with the aide of Figure 8.13. The speed of the frame of reference (magnitude of the velocity $|\underline{v}_{\text{frame}}|$) is just the angular velocity Ω of the turntable times the radial position of a given point,

$$|\underline{v}_{\text{frame}}| = r\Omega \tag{8.217}$$

From the geometry in Figure 8.13 we see that the frame velocity at point P is

$$\underline{v}_{\text{frame}} = -\Omega(y - y_0)\hat{e}_x + \Omega(x - x_0)\hat{e}_y \tag{8.218}$$

To relate $(x - x_0)$ and $(y - y_0)$ with $\bar{x}$ and $\bar{y}$ we again refer to Figure 8.13:

$$\bar{x} = (y - y_0)\sin \Omega t + (x - x_0)\cos \Omega t \tag{8.219}$$

$$\bar{y} = (y - y_0)\cos \Omega t - (x - x_0)\sin \Omega t \tag{8.220}$$

$$\bar{z} = z \tag{8.221}$$

Similarly we can see that the unit vectors are related by

$$\hat{e}_{\bar{x}} = \cos \Omega t\, \hat{e}_x + \sin \Omega t\, \hat{e}_y \tag{8.222}$$

$$\hat{e}_{\bar{y}} = -\sin \Omega t\, \hat{e}_x + \cos \Omega t\, \hat{e}_y \tag{8.223}$$

Putting it all together we obtain

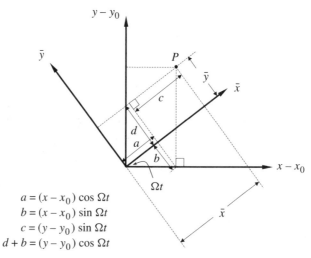

$$a = (x - x_0)\cos \Omega t$$
$$b = (x - x_0)\sin \Omega t$$
$$c = (y - y_0)\sin \Omega t$$
$$d + b = (y - y_0)\cos \Omega t$$

Figure 8.13 Relationship between velocities with respect to the rotating and stationary reference frames.

$$\underset{\substack{\text{with respect} \\ \text{to stationary} \\ \text{frame}}}{\underline{v}} = \dot{\gamma}_0 \bar{y} \hat{e}_{\bar{x}} + \underline{v}_{\text{frame}} \tag{8.224}$$

$$= \dot{\gamma}_0[(y - y_0)\cos \Omega t - (x - x_0)\sin \Omega t](\cos \Omega t \, \hat{e}_x + \sin \Omega t \, \hat{e}_y) \tag{8.225}$$

$$- \Omega(y - y_0)\hat{e}_x + \Omega(x - x_0)\hat{e}_y \tag{8.226}$$

$$= \begin{pmatrix} \dot{\gamma}_0 \cos \Omega t[(y - y_0)\cos \Omega t - (x - x_0)\sin \Omega t] - \Omega(y - y_0) \\ \dot{\gamma}_0 \sin \Omega t[(y - y_0)\cos \Omega t - (x - x_0)\sin \Omega t] + \Omega(x - x_0) \\ 0 \end{pmatrix}_{xyz} \tag{8.227}$$

Now we are ready to carry out the calculation for $\underline{\underline{\tau}}$ using the GLVE equation, but expressed in the stationary frame of reference. We have $\underline{v}$; now we calculate $\dot{\underline{\underline{\gamma}}}$:

$$\dot{\underline{\underline{\gamma}}} = \nabla \underline{v} + (\nabla \underline{v})^T \tag{8.228}$$

$$\nabla \underline{v} = \begin{pmatrix} \frac{\partial v_x}{\partial x} & \frac{\partial v_y}{\partial x} & \frac{\partial v_z}{\partial x} \\ \frac{\partial v_x}{\partial y} & \frac{\partial v_y}{\partial y} & \frac{\partial v_z}{\partial y} \\ \frac{\partial v_x}{\partial z} & \frac{\partial v_y}{\partial z} & \frac{\partial v_z}{\partial z} \end{pmatrix}_{xyz} \tag{8.229}$$

$$= \begin{pmatrix} \frac{\partial v_x}{\partial x} & \frac{\partial v_y}{\partial x} & 0 \\ \frac{\partial v_x}{\partial y} & \frac{\partial v_y}{\partial y} & 0 \\ 0 & 0 & 0 \end{pmatrix}_{xyz} \tag{8.230}$$

$$= \begin{pmatrix} -\dot{\gamma}_0 \cos \Omega t \sin \Omega t & -\dot{\gamma}_0 \sin^2 \Omega t + \Omega & 0 \\ \dot{\gamma}_0 \cos^2 \Omega t - \Omega & \dot{\gamma}_0 \sin \Omega t \cos \Omega t & 0 \\ 0 & 0 & 0 \end{pmatrix}_{xyz} \tag{8.231}$$

Therefore $\dot{\underline{\underline{\gamma}}}(t)$ is given by

$$\dot{\underline{\underline{\gamma}}} = \begin{pmatrix} -\dot{\gamma}_0 \sin 2\Omega t & \dot{\gamma}_0 \cos 2\Omega t & 0 \\ \dot{\gamma}_0 \cos 2\Omega t & \dot{\gamma}_0 \sin 2\Omega t & 0 \\ 0 & 0 & 0 \end{pmatrix}_{xyz} \tag{8.232}$$

We now use Equation (8.232) for $\dot{\underline{\underline{\gamma}}}$ in the GLVE equation:

$$\underline{\underline{\tau}}(t) = -\int_{-\infty}^{t} G(t - t')\dot{\underline{\underline{\gamma}}}(t') \, dt' \tag{8.233}$$

$$= -\int_{-\infty}^{t} G(t - t') \begin{pmatrix} -\dot{\gamma}_0 \sin 2\Omega t' & \dot{\gamma}_0 \cos 2\Omega t' & 0 \\ \dot{\gamma}_0 \cos 2\Omega t' & \dot{\gamma}_0 \sin 2\Omega t' & 0 \\ 0 & 0 & 0 \end{pmatrix}_{xyz} dt' \tag{8.234}$$

Thus for τ_{yx} in the x, y, z coordinate system we arrive at

$$\tau_{yx}(t) = -\int_{-\infty}^{t} G(t - t')\cos 2\Omega t' \dot{\gamma}_0 \, dt' \tag{8.235}$$

$$= -\int_0^\infty G(s) \cos [2\Omega(t-s)]\dot{\gamma}_0 \, ds \qquad (8.236)$$

where $s = t - t'$. To calculate the viscosity we must consider a system such that the rate-of-deformation tensor has the following form:

$$\underline{\underline{\dot{\gamma}}} = \begin{pmatrix} 0 & \dot{\gamma}_0 & 0 \\ \dot{\gamma}_0 & 0 & 0 \\ 0 & 0 & 0 \end{pmatrix}_{123} \qquad (8.237)$$

Looking at Equation (8.232) we see that this is true at $t = 0$. Using τ_{yx} from Equation (8.236) evaluated at $t = 0$ we obtain the following result for the viscosity:

$$\eta = \frac{-\tau_{yx}}{\dot{\gamma}_o} \qquad (8.238)$$

$$= \int_0^\infty G(s) \cos (2\Omega s) \, ds \qquad (8.239)$$

This result says that the viscosity of the fluid, which in the rotating coordinate system was a material constant as it should be, now depends on the rate of rotation Ω of the original coordinate system. This is clearly incorrect. We therefore conclude that the GLVE constitutive equation is not properly formulated for a change of basis between stationary and rotating frames of reference.

Does this mean that the GLVE model is invalid as a constitutive equation? As we see from Equation (8.232), if the frequency of rotation of the rotating frame of reference is sufficiently low ($\Omega \longrightarrow 0$), $\underline{\underline{\dot{\gamma}}}$ becomes the usual steady shear-rate tensor, and the GLVE equation holds and is invariant to the change of reference frame. This observation reinforces the need to limit the application of the GLVE constitutive equation to slow flows.

This near mishap with frame invariance sets the stage for our discussion in Chapter 9 of nonlinear viscoelastic constitutive equations where we will be showing how to fix the Maxwell model so that it is frame invariant and therefore valid in all reference frames. The discussion in Chapter 9 also gives the reader many tools needed to study advanced rheological topics. If your interests lie more directly in taking rheological data and learning to determine material functions for various fluids experimentally, you may wish to skip the next chapter.

8.6 PROBLEMS

8.1 What constitutive equations have we studied thus far? What are their pros and cons?

8.2 What physical significance is attached to each of the two parameters of the Maxwell model?

8.3 Does the generalized linear viscoelastic model predict rod climbing? Briefly justify your answer.

8.4 All constitutive equations we have studied so far predict a stress tensor $\underline{\underline{\tau}}$ proportional to the shear-rate tensor $\underline{\underline{\dot{\gamma}}}$. For an incompressible fluid that follows such a constitutive equation, what is $\tau_{rz}(r)$ in steady Poiseuille flow in a tube? Do not assume any particular constitutive equation in arriving at your result.

8.5 Is the following constitutive equation capable of predicting memory effects? Why or why not?

$$\underline{\underline{\tau}}(t) = -\eta_1\, \underline{\underline{\dot{\gamma}}}(t)$$

where η_1 is a function of $\dot{\gamma}$

8.6 Calculate the material functions for cessation of steady shearing, $\eta^-(t)$, Ψ_1^-, and Ψ_2^-, for the simple memory fluid discussed in the text [Equation (8.4)].

8.7 A colleague suggests the following constitutive equation for a material you are studying:

$$\underline{\underline{\tau}}(t) = -\tilde{\zeta}\left[\underline{\underline{\dot{\gamma}}}(t) + 0.8\underline{\underline{\dot{\gamma}}}(t-t_0) + 0.6\underline{\underline{\dot{\gamma}}}(t-2t_0)\right]$$

where $\tilde{\zeta}$ and t_0 are the parameters of the model.

(a) Calculate the steady shear material functions η, Ψ_1, and Ψ_2 predicted by this model.

(b) Sketch the shear stress $\tau_{21}(t)$ as a function of time that would be predicted by this model in the startup of steady shear experiment. What are $\eta^+(t)$, Ψ_1^+, and Ψ_2^+ predicted by this model?

(c) Sketch the shear stress as a function of time that would be predicted by this model in the cessation of steady shear experiment. What are $\eta^-(t)$, Ψ_1^-, and Ψ_2^- predicted by this model?

(d) What do the parameters $\tilde{\zeta}$ and t_0 represent physically?

8.8 For slow deformations, the Hookean constitutive equation can be written as

$$\underline{\underline{\tau}} = -G\underline{\underline{\gamma}}(t_{\text{ref}}, t)$$

where G is the modulus, a constant parameter of the model. Calculate the shear step-strain relaxation modulus $G(t)$ for this model. Recall that for small strains [Equation (8.39)],

$$\underline{\underline{\gamma}}(t_{\text{ref}}, t) = \int_{t_{\text{ref}}}^{t} \underline{\underline{\dot{\gamma}}}(t')\, dt'$$

8.9 What are $G'(\omega)$ and $G''(\omega)$ for a Hookean solid? [*Hint:* See Equation (8.39).]

8.10 Calculate the viscosity η for the generalized linear viscoelastic model, the Maxwell model, and the generalized Maxwell model.

8.11 What is the shear-stress response $\tau_{21}(t)$ to the step-strain experiment predicted for the generalized linear viscoelastic model?

8.12 Calculate the step shear-strain material functions $G(t, \gamma_0)$, $G_{\Psi_1}(t, \gamma_0)$, and $G_{\Psi_2}(t, \gamma_0)$ for the generalized linear viscoelastic constitutive equation.

Express your result in terms of an integral over $s = t - t'$.

8.13 Calculate the shear creep compliance $J(t)$ and the recoverable creep compliance $J_r(t') = R(t')$ for the Maxwell model. Sketch your solutions.

8.14 What is the magnitude of the complex viscosity $|\eta^*(\omega)|$ for a single-relaxation-time Maxwell fluid? Plot this material function versus frequency over a wide range of frequencies (log–log plot). What other material function does $|\eta^*(\omega)|$ resemble?

8.15 Calculate the material functions for cessation of steady shearing, $\eta^-(t)$, Ψ_1^-, and Ψ_2^-, for the generalized linear viscoelastic model. Express your result in terms of an integral over $s = t - t'$. Calculate these material functions for the Maxwell model and the generalized Maxwell model.

8.16 Calculate the material functions for startup of steady shearing, $\eta^+(t)$, Ψ_1^+, and Ψ_2^+, for the generalized linear viscoelastic model. Express your result in terms of an integral over $s = t-t'$. Calculate these material functions for the generalized Maxwell model.

8.17 Show that $G'(\omega)$ and $G''(\omega)$ are given by Equations (8.136) for the generalized Maxwell model.

8.18 The value of the relaxation modulus $G(t)$ at zero time is called the instantaneous modulus, $G(0) \equiv G^0 = \lim_{t\to 0} G(t)$ [95]. Show for a generalized linear viscoelastic fluid that the instantaneous modulus can be calculated from G'' as follows:

$$G^0 = \frac{2}{\pi}\int_{-\infty}^{\infty} G''(\omega)\, d\ln\omega$$

8.19 Show that the dynamic moduli E', E'' of small-amplitude oscillatory elongation (SAOE) are related to the dynamic moduli G', G'' of small-amplitude oscillatory shear (SAOS) as follows: $E' = 3G'$ and $E'' = 3G''$.

8.20 Consider a shear deformation where the strain rate history is given by

$$\dot{\varsigma}(t) = \begin{cases} 0 & t \le 0 \\ \dot{\gamma}_0 & 0 < t \le t_1 \\ -\dot{\gamma}_0 & t_1 < t \le t_2 \\ 0 & t > t_2 \end{cases}$$

We want to define a new material function η_Δ based on this deformation:

$$\eta_\Delta \equiv -\frac{\tau_{yx}(t)}{\dot{\gamma}_0} \quad t > t_2$$

where $\tau_{yx}(t)$ is the stress generated by the deformation history given above.

(a) Sketch $\dot{\varsigma}(t)$.

(b) Calculate and sketch the strain $\gamma(0, t)$.

(c) What is η_Δ for the generalized Maxwell fluid?

8.21 For a particular fluid, the relaxation modulus was measured and was found to fit the following equation:

$$G \text{ (Pa)} = 240 e^{-\frac{t}{25}}$$

where t is in seconds. For slow flows, calculate $\eta^-(t)$ for this fluid.

8.22 For the generalized Maxwell model parameters shown in Table 8.2:

TABLE 8.2
Parameters for
Problem 8.22

λ_k (s)	η_k (Pa · s)
0.01	2×10^4
0.1	6×10^4
1	2×10^5
10	2×10^4
100	2×10^4

(a) Plot $\log G'(\omega)$ and $\log G''(\omega)$ versus $\log \omega$ over the range $0.01 \leq \omega \leq 100$ rad/s.

(b) Plot $G(t)$ versus t and $\log G$ versus $\log t$.

(c) Calculate η_0.

8.23 For the relaxation spectrum given in Table 8.3:

TABLE 8.3
Parameters for Problem 8.23

k	λ_k (s)	g_k (Pa)
1	5×10^1	5×10^5
2	1×10^1	6×10^5
3	5×10^0	8×10^5
4	1×10^0	3×10^6
5	5×10^{-1}	3×10^6
6	2×10^{-1}	6×10^6
7	1×10^{-1}	8×10^6

(a) Plot $G'(\omega)$ and $G''(\omega)$ on a log–log plot. Use $10^{-3} \leq \omega \leq 10^2$ rad/s.

(b) What is the value of ω at the crossover of G' and G''?

(c) We saw that for the single-relaxation-time Maxwell fluid the crossover of G' and G'' occurred at $\omega = 1/\lambda_1$. Is the value of ω at the crossover equal to $1/\lambda_1$ in this case? Discuss your answer.

8.24 The Jeffreys model is as follows:

$$\underline{\underline{\tau}} + \lambda \frac{\partial \underline{\underline{\tau}}}{\partial t} = -\eta_0 \left(\underline{\dot{\gamma}} + \Lambda \frac{\partial \underline{\dot{\gamma}}}{\partial t} \right)$$

where λ is the relaxation time, η_0 is a viscosity parameter, and Λ is called a retardation time.

(a) Show that the Jeffreys model may be written as the sum of a Maxwell contribution to the stress and a Newtonian stress:

$$\underline{\underline{\tau}} = \underline{\underline{\tau}}_{\text{Maxwell}} + \underline{\underline{\tau}}_{\text{Newtonian}}$$

Note that $\underline{\underline{\tau}}_{\text{Maxwell}}$ satisfies the Maxwell equation and $\underline{\underline{\tau}}_{\text{Newtonian}}$ satisfies the Newtonian equation.

(b) What is the relationship between parameters η_0 and Λ of the Jeffreys model and the parameters of the Maxwell and Newtonian equations?

8.25 For a certain material we measure $G'(\omega)$ and $G''(\omega)$ and we find

$$G' \text{ (Pa)} = \frac{5000\omega^2}{1 + 100\omega^2}$$

$$G'' \text{ (Pa)} = \frac{500\omega}{1 + 100\omega^2}$$

where ω is in units of rad/s.

(a) Sketch $\log G'(\omega)$ and $\log G''(\omega)$ versus $\log \omega$ at low frequencies.

(b) What is $G(t)$ for this material?

(c) What is η_0?

(d) What is the shear stress as a function of time after this material is subjected to a sudden step strain of $\gamma_0 = 5$ at time $t = 0$?

8.26 The linear viscoelastic data for a generation-4 poly (amidoamine) (PAMAM) dendrimer are given in Table 8.4 [248]. What is $1/\omega_x$ (ω_x is the frequency at which G' and G'' cross)? Determine the linear viscoelastic spectrum; use the minimum number of relaxation times needed to get a good fit. What is the longest relaxation time?

TABLE 8.4
Data for Problem 8.26*

$a_T\omega$ (rad/s)	G^* (Pa)	$\tan\delta$
8.29E+00	8.06E+05	1.54E−01
3.92E+00	7.46E+05	2.82E−01
2.33E+00	6.66E+05	4.06E−01
1.47E+00	6.10E+05	5.83E−01
8.44E−01	5.94E+05	8.79E−01
5.63E−01	3.94E+05	1.28E+00
3.52E−01	2.74E+05	1.96E+00
2.18E−01	1.91E+05	2.77E+00
1.33E−01	1.22E+05	4.30E+00
7.60E−02	7.67E+04	6.64E+00
4.98E−02	5.12E+04	9.74E+00
3.29E−02	3.44E+04	1.40E+01
2.17E−02	2.25E+04	2.16E+01
1.43E−02	1.49E+04	2.93E+01
9.40E−03	9.81E+03	4.72E+01

*A more complete data set may be
found in Table F.3 in Appendix F.

TABLE 8.5
Data for Problem 8.27*

ω (rad/s)	G' (Pa)	ω (rad/s)	G'' (Pa)
6.35E+00	9.81E−01	1.02E+00	1.70E+00
1.55E+01	5.49E+00	2.54E+00	4.23E+00
3.99E+01	2.13E+01	6.22E+00	1.01E+01
9.62E+01	4.81E+01	1.56E+01	2.43E+01
2.38E+02	8.96E+01	3.91E+01	4.70E+01
5.99E+02	1.58E+02	1.54E+02	1.05E+02
1.48E+03	2.72E+02	5.93E+02	2.32E+02
3.73E+03	4.69E+02	1.90E+03	4.88E+02
7.70E+03	6.38E+02	6.08E+03	1.03E+03
1.91E+04	1.10E+03	1.53E+04	2.03E+03
4.81E+04	1.83E+03	3.76E+04	4.08E+03
1.22E+05	2.75E+03	1.20E+05	1.10E+04
3.02E+05	4.23E+03	3.05E+05	2.54E+04
		7.63E+05	6.21E+04
		1.94E+06	1.49E+05
		3.01E+06	2.44E+05

*A more complete data set may be found in Table F.4
in Appendix F.

8.27 The linear viscoelastic data for a narrowly distributed solution of polystyrene (M_w = 860 kg/mol, c = 0.015 g/cm^3 in Aroclor 1248 [167]) are given in Table 8.5. What is $1/\omega_x$ (ω_x is the frequency at which G' and G'' cross)? Determine the linear-viscoelastic spectrum; use the minimum number of relaxation times needed to get a good fit. What is the longest relaxation time?

8.28 The linear viscoelastic data for a broadly distributed high-density polyethylene are given in Table 8.6 [142].

(a) Determine the linear viscoelastic spectrum.

(b) How does the meaning of the longest relaxation time differ for the fit to these data compared to the fit performed in the example problem in Section 8.3?

8.29 For the high-density polyethylene discussed in Problem 8.28 (G', G'' given) calculate the step shear relaxation modulus $G(t)$. Plot your results.

8.30 For the high-density polyethylene discussed in Problem 8.28 (G', G'' given) calculate the dynamic compliance material functions J' and J''. Plot your results.

8.31 For the material whose $G'(\omega)$ and $G'(\omega)$ are shown in Figure 8.14, what is approximately the value of the

longest relaxation time? Please explain how you justify your answer. It may help to make some sketches on the figure.

8.32 You need a relaxation time value λ to scale up a mixer for a non-Newtonian fluid. Table 8.7 gives the data for the linear viscoelastic step-strain modulus $G(t)$ for your material. Calculate λ.

8.33 For the generation-4 PAMAM dendrimer discussed in Problem 8.26 (G^* and $\tan\delta$ given) calculate the dynamic modulus material functions G' and G'' and also the step shear relaxation modulus $G(t)$. Plot your results.

8.34 Many molecular models have been developed to describe the behaviors of polymer melts and solutions. One model that has achieved some success is the Rouse model [138]. The relaxation spectrum predicted by the Rouse model is

$$G_i = G = \text{constant for all relaxation modes}$$

$$\lambda_i = \frac{\lambda_1}{i^2} \quad \text{for } i = 1, 2, 3, \ldots$$

where λ_1 is the longest relaxation time.

(a) Plot $G'(\omega)$ and $G''(\omega)$ for a material that follows the Rouse model using only:

TABLE 8.6
Data for Problem 8.28*

$a_T\omega$ (rad/s)	G' (Pa)	$a_T\omega$ (rad/s)	G'' (Pa)
2.46E−03	1.18E+01	1.56E−03	8.24E+01
4.59E−03	3.70E+01	4.30E−03	2.12E+02
7.46E−03	7.13E+01	1.06E−02	4.38E+02
1.23E−02	1.30E+02	2.27E−02	9.17E+02
2.20E−02	2.61E+02	5.07E−02	1.97E+03
3.83E−02	4.19E+02	9.58E−02	2.91E+03
7.76E−02	9.14E+02	2.17E−01	4.48E+03
1.33E−01	1.55E+03	5.40E−01	6.26E+03
2.78E−01	2.91E+03	1.11E+00	8.73E+03
6.74E−01	5.36E+03	2.55E+00	1.18E+04
1.44E+00	8.60E+03	5.92E+00	1.70E+04
3.40E+00	1.40E+04	1.48E+01	2.24E+04
8.14E+00	2.28E+04	2.60E+01	3.00E+04
1.72E+01	3.32E+04	6.57E+01	3.85E+04
3.94E+01	4.69E+04	1.66E+02	4.94E+04
8.68E+01	6.46E+04	4.95E+02	5.99E+04
2.55E+02	9.52E+04	7.40E+02	6.60E+04
7.62E+02	1.35E+05	1.67E+03	7.68E+04
2.03E+03	1.73E+05	2.71E+03	7.89E+04

*A more complete data set may be found in Table F.5 in Appendix F.

(i) The first mode
(ii) The first two modes
(iii) The first 15 modes
(iv) 150 modes.

(b) At high frequencies, what is the slope of G' and G'' for the full Rouse model?

8.35 Calculate the final solution for the velocity v_θ, from Equation (8.188) for the torsional oscillatory annular flow exercise in Section 8.4.2.

TABLE 8.7
Data for Problem 8.31*

t,s	G(t), Pa
1.00E−03	2.70E+07
2.00E−03	2.69E+07
5.01E−03	2.65E+07
1.00E−02	2.59E+07
2.00E−02	2.48E+07
5.01E−02	2.20E+07
1.00E−01	1.84E+07
2.00E−01	1.41E+07
5.01E−01	8.75E+06
1.00E+00	5.68E+06
2.00E+00	3.29E+06
5.01E+00	1.53E+06
1.00E+01	7.95E+05
2.00E+01	3.03E+05
5.01E+01	4.49E+04
7.08E+01	1.50E+04
1.00E+02	3.40E+03
1.58E+02	1.81E+02
2.00E+02	2.32E+01

*A more complete data set may be found in Table F.5 in Appendix F.

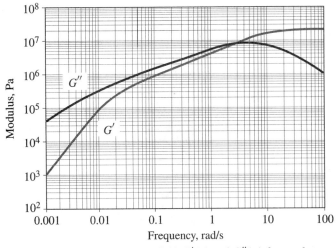

Figure 8.14 Figure for Problem 8.31 $G'(\omega)$ and $G''(\omega)$ for a polymer.

CHAPTER
ᗡᗡᖆ **9**

Introduction to More Advanced Constitutive Modeling

We have shown that the two non-Newtonian constitutive equations we have encountered thus far are useful within limited ranges. The generalized Newtonian fluid (GNF) model is accurate in calculations of pressure-drop and flow-rate information and in situations where shear-thinning of viscosity is the dominant behavior. The generalized linear viscoelastic (GLVE) constitutive equation describes the small-deformation-rate limit and is widely used by chemists and engineers to characterize material properties. Both equations have serious limitations, however: neither predicts normal stresses in shear flow, nor do they capture other nonlinear elastic effects involving fluid memory. Also, most seriously, at the end of Chapter 8 we showed that the GLVE model is not properly formulated to be invariant to a change of reference frame for a rotating frame of reference, and the GLVE equation is therefore inappropriate for use at high deformation rates.

Our goal in rheological modeling of polymers is to formulate constitutive equations that can be used to predict accurately the rheological behavior of polymers, particularly in the large-strain, high-rate regime that is used in polymer processing. The models we have studied were developed as extensions of the Newtonian constitutive model and thus were formulated in terms of the shear-rate or rate-of-deformation tensor $\dot{\underline{\underline{\gamma}}}$. For larger strain flows, however, it is the strain and the strain history that must be considered in order to develop better constitutive equations, as we will see in this chapter. Properly formulating the strain measure fixes the frame-invariance problem encountered in the GLVE model. The question of strain measure is also closely linked to the types of time derivatives that are permitted in constitutive equations, and in this chapter we will introduce the concept of using convected coordinate systems in which to take time derivatives. More comprehensive descriptions of advanced constitutive approaches may be found in the two volumes of Bird et al. [26, 27] as well as in Larson [138].

If you are anxious to move on to practical subjects related to measurements of rheological functions, please skip to Chapter 10 and return to this chapter later.

9.1 Finite Strain Measures

As mentioned before, experiments demonstrate that the deformation history has a profound effect on stresses and strains generated in polymeric systems. The Maxwell model,

$$\underset{=}{\tau} + \lambda \frac{\partial \underset{=}{\tau}}{\partial t} = -\eta_0 \underset{=}{\dot{\gamma}} \tag{9.1}$$

was introduced in Chapter 8 as an empirical attempt to add some dependence on deformation history to the Newtonian constitutive equation. The strain does not appear explicitly in this form of the Maxwell model, however.

We can reveal the strain tensor associated with the Maxwell model through a manipulation of the integral version. We will use the GLVE equation in this derivation. The GLVE model becomes the Maxwell or generalized Maxwell model with the appropriate choice of the relaxation-modulus function $G(t - t')$.

The GLVE model is,

$$\underset{=}{\tau}(t) = - \int_{-\infty}^{t} G(t - t') \underset{=}{\dot{\gamma}}(t') \, dt' \tag{9.2}$$

We now integrate to rewrite this expression in terms of $\underset{=}{\gamma}$, the infinitesimal strain tensor, which was defined in Section 5.2.2.3 as

$$\underline{u}\left(t_{\text{ref}}, t'\right) = \underline{r}(t') - \underline{r}(t_{\text{ref}}) \tag{9.3}$$

$$\underset{=}{\gamma}\left(t_{\text{ref}}, t'\right) = \nabla \underline{u}\left(t_{\text{ref}}, t'\right) + \left[\nabla \underline{u}\left(t_{\text{ref}}, t'\right)\right]^{T} \tag{9.4}$$

$$= \int_{t_{\text{ref}}}^{t'} \underset{=}{\dot{\gamma}}(t'') \, dt'' \tag{9.5}$$

where t_{ref} is the reference time for strain, which we will now take as the current time t, t' is the time of interest, and $\underline{u}(t, t') = \underline{r}(t') - \underline{r}(t) = \underline{r}' - \underline{r}$ is the vector that follows a particle's displacement between t and t'. To carry out the integration, we use integration by parts as follows:[1]

$$-\underset{=}{\tau}(t) = \int_{-\infty}^{t} G(t - t') \underset{=}{\dot{\gamma}}(t') \, dt' \tag{9.9}$$

$$= G(t - t') \underset{=}{\gamma}(t, t') \Big|_{t'=-\infty}^{t'=t} - \int_{-\infty}^{t} \frac{\partial G(t - t')}{\partial t'} \underset{=}{\gamma}(t, t') \, dt' \tag{9.10}$$

---

[1] The details of this calculation are as follows. Integration by parts,

$$\int_{a}^{b} u \, dv = u \, v \big|_{a}^{b} - \int_{a}^{b} v \, du \tag{9.6}$$

$$u = G(t - t') \qquad\qquad dv = \underset{=}{\dot{\gamma}}(t') \, dt' \tag{9.7}$$

$$du = \frac{\partial G(t - t')}{\partial t'} dt' \qquad v = \int_{t}^{t'} \underset{=}{\dot{\gamma}}(t'') \, dt'' = \underset{=}{\gamma}(t, t') \tag{9.8}$$

The choice of $t_{\text{ref}} = t$ as the lower limit of the strain integral is arbitrary. This choice results in the simplification of the $u \, v \big|_{a}^{b}$ term in this integration.

If we require that $G(\infty) = 0$ and $\underline{\underline{\gamma}}(t, -\infty)$ be finite, the first term goes to zero, and we obtain a version of the GLVE fluid model that contains the first derivative of the relaxation modulus $M(t - t')$, called the memory function, and the infinitesimal strain tensor $\underline{\underline{\gamma}}(t, t')$:

GLVE model (strain version)

$$\underline{\underline{\tau}}(t) = + \int_{-\infty}^{t} M(t - t')\, \underline{\underline{\gamma}}(t, t')\, dt' \qquad (9.11)$$

where

$$M(t - t') \equiv \frac{\partial G(t - t')}{\partial t'} \qquad (9.12)$$

In defining $\underline{\underline{\gamma}}(t, t')$ we chose the current time t as our reference state. Strain is a measure of relative deformation, and two states must always be specified, the reference state and the state of interest. There is no unique reference state.

We see that the GLVE fluid model (and by extension the Maxwell models) uses the infinitesimal strain tensor as its strain measure [26]. The use of the infinitesimal strain tensor in the GLVE model is the cause of the frame variance observed for this model, as we will now show. Consider $\underline{\underline{\gamma}}(t, t')$ written for the rotation of a rigid body around the z-axis (Figure 9.1). In Cartesian coordinates the position vector $\underline{r}$ of an arbitrary particle can be written as

$$\underline{r} = \begin{pmatrix} x \\ y \\ z \end{pmatrix}_{xyz} = \bar{\underline{r}} + z\hat{e}_z \qquad (9.13)$$

where $\bar{\underline{r}}$ is the vector projection of $\underline{r}$ into the xy-plane. After rotation of the body around the z-axis, the z-coordinates of particles do not change, and therefore the vector $\underline{r}'$, which gives the position of the particle after the rotation, can be written as

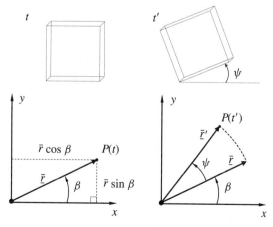

Figure 9.1 Relationship between particle position vectors in a rigid body that is rotating. $P(t)$ is a point in the body at time t, and $P(t')$ is the same point at a later time t'.

$$\underline{r}' = \bar{\underline{r}}' + z\hat{e}_z \tag{9.14}$$

where $\bar{\underline{r}}'$ is the xy-projection of $\underline{r}'$, and z is unchanged. To write $\underline{\underline{\gamma}}(t, t')$ for this rotation we must calculate the displacement vector $\underline{u} = \underline{r}' - \underline{r} = \bar{\underline{r}}' - \bar{\underline{r}}$, which we can write in Cartesian coordinates using the geometric relations shown in Figure 9.1:

$$\bar{\underline{r}} = \begin{pmatrix} x \\ y \\ 0 \end{pmatrix}_{xyz} \tag{9.15}$$

$$\bar{\underline{r}}' = \begin{pmatrix} \bar{r} \cos{(\psi + \beta)} \\ \bar{r} \sin{(\psi + \beta)} \\ 0 \end{pmatrix}_{xyz} = \begin{pmatrix} x \cos{\psi} - y \sin{\psi} \\ x \sin{\psi} + y \cos{\psi} \\ 0 \end{pmatrix}_{xyz} \tag{9.16}$$

To arrive at the last expression we have expanded the trigonometric terms and used $x = \bar{r} \cos{\beta}$, $y = \bar{r} \sin{\beta}$. The displacement vector $\underline{u}$ is therefore,

$$\underline{u} = \bar{\underline{r}}' - \bar{\underline{r}} = \begin{pmatrix} x(\cos{\psi} - 1) - y \sin{\psi} \\ x \sin{\psi} + y(\cos{\psi} - 1) \\ 0 \end{pmatrix}_{xyz} \tag{9.17}$$

and we can calculate $\nabla \underline{u}$:

$$\nabla \underline{u}(t, t') = \frac{\partial u_p}{\partial x_i} \hat{e}_i \hat{e}_p \tag{9.18}$$

$$= \begin{pmatrix} \cos{\psi} - 1 & \sin{\psi} & 0 \\ -\sin{\psi} & \cos{\psi} - 1 & 0 \\ 0 & 0 & 0 \end{pmatrix}_{xyz} \tag{9.19}$$

Finally, we calculate $\underline{\underline{\gamma}}(t, t')$:

$$\underline{\underline{\gamma}}(t, t') = \nabla \underline{u}(t, t') + \left[\nabla \underline{u}(t, t') \right]^T \tag{9.20}$$

$$= \begin{pmatrix} 2(\cos{\psi} - 1) & 0 & 0 \\ 0 & 2(\cos{\psi} - 1) & 0 \\ 0 & 0 & 0 \end{pmatrix}_{xyz} \tag{9.21}$$

Equation (9.21) shows that for a solid-body rotation, the infinitesimal strain tensor $\underline{\underline{\gamma}}(t, t')$ is a function of the rotation angle ψ. Substituting $\underline{\underline{\gamma}}(t, t')$ into the GLVE model we obtain the stress tensor predicted by this model for solid-body rotation around z:

$$\underline{\underline{\tau}} = + \int_{-\infty}^{t} M(t - t') \begin{pmatrix} 2(\cos{\psi} - 1) & 0 & 0 \\ 0 & 2(\cos{\psi} - 1) & 0 \\ 0 & 0 & 0 \end{pmatrix}_{xyz} dt' \tag{9.22}$$

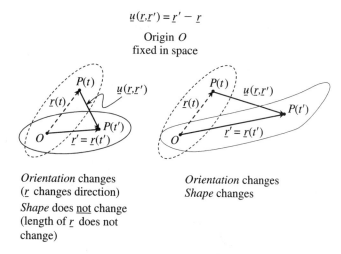

$$\underline{u}(\underline{r},\underline{r}') = \underline{r}' - \underline{r}$$

Origin O
fixed in space

Orientation changes
($\underline{r}$ changes direction)

Shape does <u>not</u> change
(length of $\underline{r}$ does not
change)

Orientation changes
Shape changes

Figure 9.2 Displacement vector $\underline{u}(\underline{r}, \underline{r}') \equiv \underline{r}' - \underline{r}$ tracks orientation changes as well as shape changes.

Thus the stress $\underline{\underline{\tau}}$ in a rigid body rotating around an axis is found to depend on the angle of rotation of the body, a completely false prediction.

The problem with $\underline{\underline{\gamma}}$ is that it measures deformation using the vector $\underline{u} = \underline{r}' - \underline{r}$ directly. This vector indicates the relative displacement of particles in a fluid, but it also conveys information about the orientation of the motion (Figure 9.2). The relative orientation of particles is irrelevant to the state of stress, and using $\underline{u}$ directly can introduce artificially large deformation rates into the mathematics describing the problem. The GLVE equation is only valid for small deformation rates, and therefore unphysical predictions like that in Equation (9.22) result when $\underline{\underline{\gamma}}$ is used in that constitutive equation. We must find a strain measure that retains information about change in shape while dropping orientation information. We pursue this next. For slow deformations even the rotation-induced artificial deformation rates are low, and the GLVE model is valid: in Equation (9.22) if we allow ψ to approach zero, the infinitesimal strain tensor vanishes, as desired for a solid-body rotation.

9.1.1 DEFORMATION GRADIENT TENSOR

We must develop a strain tensor that will accurately capture large-strain deformations without being affected by irrelevant rigid-body rotation. We begin by describing an arbitrary deformation that includes both change in shape and rotation, and then we will address how to eliminate the rotation from the description.

Consider a particle of fluid labeled P, as depicted in Figure 9.3. Two times are shown: t, which is the current time and the reference time, and t', which is a time in the past and the time of interest. At time t the location of each fluid particle is described by a vector $\underline{r}$ that points from the origin of a fixed coordinate system to the fluid particle. In the discussion that follows, the value of the vector $\underline{r}$ at time t serves as an index that identifies which particle is being discussed.

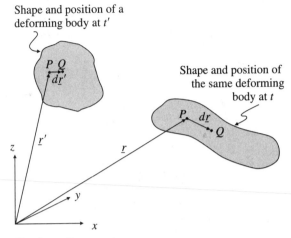

Shape and position of a deforming body at t'

Shape and position of the same deforming body at t

Fixed coordinate system

Figure 9.3 Arbitrary deformation of a particle of fluid.

At previous times t', the particle positions are designated by the vectors $\underline{r}'$. Thus, for each particle at all times there is a vector function $\underline{r}'(t')$, which describes that particle's path. The position of every particle at all times can be written as $\underline{r}'(t', \underline{r})$, which indicates the position at time t' of the particle that is now at $\underline{r}$.

To describe deformation or change in shape, consider the distance between P and the nearby particle labeled Q, shown in Figure 9.3. The distance between P and Q at time t' is $d\underline{r}'$, and at time t it is $d\underline{r}$. The deformation that these nearby particles experience between the current (reference) time and time t' can be described by writing the function that transforms $d\underline{r}$ to $d\underline{r}'$. To express this function mathematically, we begin by writing $d\underline{r}$ and $d\underline{r}'$ in Cartesian coordinates:

$$d\underline{r} = \begin{pmatrix} dx \\ dy \\ dz \end{pmatrix}_{xyz} \qquad d\underline{r}' = \begin{pmatrix} dx' \\ dy' \\ dz' \end{pmatrix}_{xyz} \qquad (9.23)$$

The particle position at t' depends only on what particle is being considered, that is, on the particle label $\underline{r}$. Thus $\underline{r}' = \underline{r}'(\underline{r})$, and we can relate the differential dx' to x, y, and z as follows:

$$dx' = \frac{\partial x'}{\partial x} dx + \frac{\partial x'}{\partial y} dy + \frac{\partial x'}{\partial z} dz \qquad (9.24)$$

$$= d\underline{r} \cdot \frac{\partial x'}{\partial \underline{r}} \qquad (9.25)$$

Analogously, for dy' and dz', we find

$$dy' = d\underline{r} \cdot \frac{\partial y'}{\partial \underline{r}} \qquad (9.26)$$

$$dz' = d\underline{r} \cdot \frac{\partial z'}{\partial \underline{r}} \tag{9.27}$$

or, equivalently,

$$d\underline{r}' = d\underline{r} \cdot \frac{\partial \underline{r}'}{\partial \underline{r}} \tag{9.28}$$

The last expression gives the transformation of the relative position vector $d\underline{r}$ at t to $d\underline{r}'$, its direction and magnitude at time t'. The tensor $\partial \underline{r}'/\partial \underline{r}$, which accomplishes this transformation, is called the *deformation-gradient tensor* denoted by $\underline{\underline{F}}(t, t')$.

<table>
<tr><td>Deformation-gradient
tensor</td><td>$$\boxed{\underline{\underline{F}}(t, t') \equiv \frac{d\underline{r}'}{d\underline{r}}}$$</td><td>(9.29)</td></tr>
</table>

and

$$d\underline{r}' = d\underline{r} \cdot \underline{\underline{F}} \tag{9.30}$$

Following the same procedure, we can write dx, dy, and dz in terms of x', y', and z' as follows:

$$d\underline{r} = d\underline{r}' \cdot \frac{d\underline{r}}{d\underline{r}'} \tag{9.31}$$

We define the inverse deformation-gradient tensor $\underline{\underline{F}}^{-1}$, such that $\underline{\underline{F}} \cdot \underline{\underline{F}}^{-1} = \underline{\underline{I}}$. By dot multiplying $\underline{\underline{F}}^{-1}$ on the right-hand side of Equation (9.30) and comparing the result with Equation 9.31) we can see that

$$d\underline{r}' \cdot \underline{\underline{F}}^{-1} = d\underline{r} \cdot \underline{\underline{F}} \cdot \underline{\underline{F}}^{-1} \tag{9.32}$$

$$d\underline{r}' \cdot \underline{\underline{F}}^{-1} = d\underline{r} \tag{9.33}$$

and

<table>
<tr><td>Inverse deformation-
gradient tensor</td><td>$$\boxed{\underline{\underline{F}}^{-1}(t', t) \equiv \frac{d\underline{r}}{d\underline{r}'}}$$</td><td>(9.34)</td></tr>
</table>

The deformation-gradient tensor and the inverse deformation-gradient tensor represent the linear function of time and position that describes the entire deformation history of each fluid particle. $\underline{\underline{F}}$ and $\underline{\underline{F}}^{-1}$ are inverses, that is, the motion described by $\underline{\underline{F}}$ is the reverse of that captured by $\underline{\underline{F}}^{-1}$. These two tensors can be used to keep track of the relative positions of particles in a flowing fluid and therefore of the changes in shape and orientation caused by flow. These tensors will be essential tools in our development of large-strain constitutive equations.

To familiarize ourselves with $\underline{\underline{F}}$ and $\underline{\underline{F}}^{-1}$, we will now calculate the inverse deformation-gradient tensor for shear flow.

EXAMPLE

Calculate $\underline{\underline{F}}^{-1}$ for steady simple shear flow at a shear rate $\dot{\gamma}_0$ ($\dot{\gamma}_0 > 0$), with $\underline{v}$ written in the usual Cartesian frame (xyz).

SOLUTION

The velocity field for steady shear flow is

$$
\underline{v} = \begin{pmatrix} \dot{\gamma}_0 y \\ 0 \\ 0 \end{pmatrix}_{xyz}
\tag{9.35}
$$

At time t' the location of a particle with respect to a fixed origin will be given by $\underline{r}'$:

$$
\underline{r}' = \begin{pmatrix} x' \\ y' \\ z' \end{pmatrix}_{xyz}
\tag{9.36}
$$

At a later time t the position will be given by the vector $\underline{r}$. The y' and z' coordinates of $\underline{r}'$ do not change in a shear deformation since the velocity components are zero in those directions. In the x'-direction, however, the coordinate changes as a result of the flow:

$$
\underline{r} = \begin{pmatrix} x \\ y \\ z \end{pmatrix}_{xyz} = \begin{pmatrix} x' + (t - t')\dot{\gamma}_0 y' \\ y' \\ z' \end{pmatrix}_{xyz}
\tag{9.37}
$$

The quantity $(t - t')\dot{\gamma}_0$ is the strain, $\gamma(t', t) = \int_{t'}^{t} \dot{\gamma}_0 \, dt''$, at time t relative to the fluid configuration at time t'. The reference time for strain in this case is $t_{\text{ref}} = t'$. From this expression for $\underline{r}$ and the definition of $\underline{\underline{F}}^{-1}$ it is straightforward to calculate $\underline{\underline{F}}^{-1}$:

$$
\underline{\underline{F}}^{-1}(t', t) = \frac{\partial \underline{r}}{\partial \underline{r}'} = \begin{pmatrix} \frac{\partial x}{\partial x'} & \frac{\partial y}{\partial x'} & \frac{\partial z}{\partial x'} \\ \frac{\partial x}{\partial y'} & \frac{\partial y}{\partial y'} & \frac{\partial z}{\partial y'} \\ \frac{\partial x}{\partial z'} & \frac{\partial y}{\partial z'} & \frac{\partial z}{\partial z'} \end{pmatrix}_{xyz}
\tag{9.38}
$$

Inverse deformation-gradient tensor for shear flow

$$
\underline{\underline{F}}^{-1}(t', t) = \begin{pmatrix} 1 & 0 & 0 \\ \gamma & 1 & 0 \\ 0 & 0 & 1 \end{pmatrix}_{xyz}
\tag{9.39}
$$

where $\gamma \equiv \dot{\gamma}_0(t - t')$ and is a positive number.

Derivatives of $\underline{\underline{F}}^{-1}$ with respect to t and t' will appear in some of the quantities we will use, and it is helpful to relate these to the velocity gradient $\nabla \underline{v}$. To calculate, for example, $\partial \underline{\underline{F}}^{-1}/\partial t$, we begin with the definition of $\underline{\underline{F}}^{-1}$ and calculate the time derivative using Einstein notation ($x = x_1, y = x_2, z = x_3$):

$$\underline{\underline{F}}^{-1} = \frac{\partial \underline{r}}{\partial \underline{r}'} = \frac{\partial x_k}{\partial x'_p} \hat{e}_p \hat{e}_k \tag{9.40}$$

$$\frac{\partial \underline{\underline{F}}^{-1}}{\partial t} = \frac{\partial}{\partial t} \frac{\partial x_k}{\partial x'_p} \hat{e}_p \hat{e}_k = \frac{\partial}{\partial x'_p} \frac{\partial x_k}{\partial t} \hat{e}_p \hat{e}_k \tag{9.41}$$

$$= \frac{\partial v_k}{\partial x'_p} \hat{e}_p \hat{e}_k \tag{9.42}$$

We can rewrite this in more familiar terms if we recognize that v_k, the kth component of velocity at time t, is a function of the particle being considered, that is, of $\underline{r}$. We can therefore write the differential components dv_k as follows:

$$dv_k = \frac{\partial v_k}{\partial x_1} dx_1 + \frac{\partial v_k}{\partial x_2} dx_2 + \frac{\partial v_k}{\partial x_3} dx_3 \tag{9.43}$$

$$\frac{\partial v_k}{\partial x'_p} = \frac{\partial v_k}{\partial x_1} \frac{\partial x_1}{\partial x'_p} + \frac{\partial v_k}{\partial x_2} \frac{\partial x_2}{\partial x'_p} + \frac{\partial v_k}{\partial x_3} \frac{\partial x_3}{\partial x'_p} \tag{9.44}$$

$$= \frac{\partial v_k}{\partial x_m} \frac{\partial x_m}{\partial x'_p} \tag{9.45}$$

Finally we can write

$$\frac{\partial \underline{\underline{F}}^{-1}}{\partial t} = \frac{\partial v_k}{\partial x'_p} \hat{e}_p \hat{e}_k = \frac{\partial v_k}{\partial x_m} \frac{\partial x_m}{\partial x'_p} \hat{e}_p \hat{e}_k \tag{9.46}$$

$$= \underline{\underline{F}}^{-1} \cdot \nabla \underline{v} \tag{9.47}$$

The time derivative of the deformation-gradient tensor $\underline{\underline{F}}$ is found similarly (see Problem 9.4),

$$\frac{\partial \underline{\underline{F}}}{\partial t} = -\nabla \underline{v} \cdot \underline{\underline{F}} \tag{9.48}$$

A summary of various derivatives of $\underline{\underline{F}}$ and $\underline{\underline{F}}^{-1}$ and of related tensors is given in Table 9.1.

9.1.2 FINGER AND CAUCHY TENSORS

We are seeking a strain tensor to capture deformation without rotation. We have defined two linked strain measures, $\underline{\underline{F}}(t, t')$ and $\underline{\underline{F}}^{-1}(t', t)$, which describe changes in the relative locations of nearby material particles and which may be written for every pair of points in the body. These tensors contain complete information on the motion of the body, including simple rotations. To describe large-strain deformation correctly we must separate rotation from stress-producing change of shape. This problem can be solved by decomposing $\underline{\underline{F}}$ or $\underline{\underline{F}}^{-1}$ into its rotation and deformation parts by a technique called *polar decomposition*.

The polar decomposition theorem [9] states that any tensor $\underline{A}$ for which an inverse $\underline{A}^{-1}$ exists ($\underline{\underline{A}} \cdot \underline{\underline{A}}^{-1} = \underline{\underline{I}}$) has two unique decompositions:

TABLE 9.1
Strain Tensors and Their Derivatives[*]

Name	$\underline{\underline{A}}$	Ref. Eq.	Definition	$\dfrac{\partial \underline{\underline{A}}}{\partial t}$	$\dfrac{\partial \underline{\underline{A}}}{\partial t'}$
Deformation gradient, $t \longrightarrow t'$	$\underline{\underline{F}}$	(9.29)	$\dfrac{\partial \underline{r}'}{\partial \underline{r}}$	$-\nabla \underline{v} \cdot \underline{\underline{F}}$	$\underline{\underline{F}} \cdot \nabla' \underline{v}'$
Inverse deformation gradient, $t' \longrightarrow t$	$\underline{\underline{F}}^{-1}$	(9.34)	$\dfrac{\partial \underline{r}}{\partial \underline{r}'}$	$\underline{\underline{F}}^{-1} \cdot \nabla \underline{v}$	$-\nabla' \underline{v}' \cdot \underline{\underline{F}}^{-1}$
Cauchy, $t \longrightarrow t'$	$\underline{\underline{C}}$	(9.108)	$\underline{\underline{F}} \cdot \underline{\underline{F}}^T$	$-\left[\underline{\underline{C}} \cdot (\nabla \underline{v})^T + \nabla \underline{v} \cdot \underline{\underline{C}}\right]$	$\underline{\underline{F}} \cdot \underline{\underline{\dot{\gamma}}}' \cdot \underline{\underline{F}}^T$
Finger, $t' \longrightarrow t$	$\underline{\underline{C}}^{-1}$	(9.109)	$(\underline{\underline{F}}^{-1})^T \cdot \underline{\underline{F}}^{-1}$	$(\nabla \underline{v})^T \cdot \underline{\underline{C}}^{-1} + \underline{\underline{C}}^{-1} \cdot \nabla \underline{v}$	$-(\underline{\underline{F}}^{-1})^T \cdot \underline{\underline{\dot{\gamma}}}' \cdot \underline{\underline{F}}^{-1}$
Finite strain, $t \longrightarrow t'$	$\underline{\underline{\gamma}}^{[0]}$	(9.128)	$\underline{\underline{C}} - \underline{\underline{I}}$	$\dfrac{\partial \underline{\underline{C}}}{\partial t}$	$\dfrac{\partial \underline{\underline{C}}}{\partial t'}$
Finite strain, $t \longrightarrow t'$	$\underline{\underline{\gamma}}_{[0]}$	(9.132)	$\underline{\underline{I}} - \underline{\underline{C}}^{-1}$	$-\dfrac{\partial \underline{\underline{C}}^{-1}}{\partial t}$	$-\dfrac{\partial \underline{\underline{C}}^{-1}}{\partial t'}$
Displacement gradient, $t \longrightarrow t'$	$\nabla \underline{u}$	(5.52)	$\underline{r}' - \underline{r}$	$\dfrac{\partial \underline{\underline{F}}}{\partial t}$	$\dfrac{\partial \underline{\underline{F}}}{\partial t'}$
Infinitesimal strain, $t \longrightarrow t'$	$\underline{\underline{\gamma}}$	(8.36)	$\nabla \underline{u} + (\nabla \underline{u})^T$	$\dfrac{\partial \underline{\underline{\gamma}}(t, t')}{\partial t} = -\underline{\underline{\dot{\gamma}}}$	$\dfrac{\partial \underline{\underline{\gamma}}(t, t')}{\partial t'} = \underline{\underline{\dot{\gamma}}}'$

[*] The symbol ∇' denotes the del operator with respect to $\underline{x}'$; $\underline{\underline{\dot{\gamma}}}'$ denotes $\underline{\underline{\dot{\gamma}}}(t')$. Note that in the entries for the infinitesimal strain tensor $\underline{\underline{\gamma}}$, the deformation indicated is from reference time t to time of interest t'. The negative sign in the result for derivative of $\underline{\underline{\gamma}}$ with respect to time t is present because this is the derivative of the infinitesimal strain tensor with respect to its reference time.

$$\underline{\underline{A}} = \underline{\underline{R}} \cdot \underline{\underline{U}} \qquad (9.49)$$

$$= \underline{\underline{V}} \cdot \underline{\underline{R}} \qquad (9.50)$$

where $\underline{\underline{U}}$, $\underline{\underline{V}}$, and $\underline{\underline{R}}$ are given by

$$\underline{\underline{U}} \equiv (\underline{\underline{A}}^T \cdot \underline{\underline{A}})^{\frac{1}{2}} \qquad (9.51)$$

$$\underline{\underline{V}} \equiv (\underline{\underline{A}} \cdot \underline{\underline{A}}^T)^{\frac{1}{2}} \qquad (9.52)$$

$$\underline{\underline{R}} \equiv \underline{\underline{A}} \cdot (\underline{\underline{A}}^T \cdot \underline{\underline{A}})^{-\frac{1}{2}} = \underline{\underline{A}} \cdot \underline{\underline{U}}^{-1} \tag{9.53}$$

$\underline{\underline{U}}$ and $\underline{\underline{V}}$ are symmetric nonsingular tensors,[2] and the tensor $\underline{\underline{R}}$ is orthogonal. An orthogonal tensor is one for which its transpose is also its inverse:

$$\begin{array}{c} \text{Orthogonal} \\ \text{tensor} \end{array} \qquad \boxed{\underline{\underline{R}}^{-1} = \underline{\underline{R}}^T} \tag{9.54}$$

$$\underline{\underline{R}}^T \cdot \underline{\underline{R}} = \underline{\underline{R}} \cdot \underline{\underline{R}}^T = \underline{\underline{I}} \tag{9.55}$$

Orthogonal tensors have an important property, which we can demonstrate by causing the orthogonal tensor $\underline{\underline{R}}$ to operate on an arbitrary vector $\underline{u}$. We begin by noting that for any vector $\underline{u}$ and tensor $\underline{\underline{R}}$, we can write:

$$\underline{v} \equiv \underline{\underline{R}} \cdot \underline{u} \tag{9.56}$$

$$= \underline{u} \cdot \underline{\underline{R}}^T \tag{9.57}$$

The equivalency of Equations (9.56) and (9.57) can be verified by carrying out the dot product using Einstein notation. We now calculate the magnitude of $\underline{v}$:

$$v = +\sqrt{\underline{v} \cdot \underline{v}} \tag{9.58}$$

$$= \sqrt{(\underline{u} \cdot \underline{\underline{R}}^T) \cdot (\underline{\underline{R}} \cdot \underline{u})} \tag{9.59}$$

$$= \sqrt{\underline{u} \cdot (\underline{\underline{R}}^T \cdot \underline{\underline{R}}) \cdot \underline{u}} \tag{9.60}$$

Since $\underline{\underline{R}}$ is orthogonal, $\underline{\underline{R}}^T \cdot \underline{\underline{R}} = \underline{\underline{I}}$, and we obtain

$$v = \sqrt{\underline{u} \cdot \underline{u}} \tag{9.61}$$

$$= u \tag{9.62}$$

or, the magnitude of a vector is unchanged after being operated on by the orthogonal tensor $\underline{\underline{R}}$. We see that $\underline{\underline{R}}$ is a pure rotation tensor—it changes the direction of the vector on which it operates, but it does not change its magnitude.

The symmetric tensors $\underline{\underline{U}}$ and $\underline{\underline{V}}$ that result from the polar decomposition of a tensor $\underline{\underline{A}}$ are called the right and left Cauchy–Green stretch tensors of $\underline{\underline{A}}$, respectively. Since the original tensor $\underline{\underline{A}}$ would, in general, change both the vector length and direction, and $\underline{\underline{R}}$ is a pure rotation tensor, $\underline{\underline{U}}$ and $\underline{\underline{V}}$ must contain all of the stretch information of the tensor $\underline{\underline{A}}$. $\underline{\underline{U}}$ and $\underline{\underline{V}}$ do contain some rotation information, however, as we will demonstrate in the example that follows.

---

[2] A nonsingular tensor is one whose determinant is nonzero. This is a sufficient condition for an inverse to the tensor to exist. $\underline{\underline{U}}$ and $\underline{\underline{V}}$ are also positive definite [9]. A tensor $\underline{\underline{U}}$ is positive definite when for any vector $\underline{v}$, the scalar $\underline{v} \cdot \underline{\underline{U}} \cdot \underline{v}$ is positive.

For the tensor $\underset{=}{A}$

$$\underset{=}{A} = \begin{pmatrix} 1 & 0 & 2 \\ 0 & 3 & 1 \\ 1 & 0 & 0 \end{pmatrix}_{xyz} \tag{9.63}$$

calculate the right stretch tensor $\underset{=}{U}$ and the rotation tensor $\underset{=}{R}$. Calculate the angle through which $\underset{=}{R}$ rotates the vector $\underline{u}$,

$$\underline{u} = \begin{pmatrix} 1 \\ 2 \\ 1 \end{pmatrix}_{xyz} \tag{9.64}$$

the angle through which $\underset{=}{U}$ rotates $\underline{u}$, and the length changes brought about by $\underset{=}{U}$ and $\underset{=}{A}$.

SOLUTION

The vectors of interest are shown schematically in Figure 9.4. We can calculate the right stretch and rotation tensors directly from their definitions [Equations (9.51) and (9.53)]. These calculations are simplified by the use of computer software.[3]

First we form the tensor $\underset{=}{A}^T \cdot \underset{=}{A}$,

$$\underset{=}{A}^T \cdot \underset{=}{A} = \begin{pmatrix} 2 & 0 & 2 \\ 0 & 9 & 3 \\ 2 & 3 & 5 \end{pmatrix}_{xyz} \tag{9.65}$$

$$= (\underset{=}{R} \cdot \underset{=}{U})^T \cdot \underset{=}{R} \cdot \underset{=}{U} \tag{9.66}$$

$$= \underset{=}{U}^T \cdot \underset{=}{R}^T \cdot \underset{=}{R} \cdot \underset{=}{U} \tag{9.67}$$

$$= \underset{=}{U}^2 \tag{9.68}$$

where the last expression follows from the symmetry of $\underset{=}{U}$ ($\underset{=}{U}^T = \underset{=}{U}$) and the orthogonality of $\underset{=}{R}$.

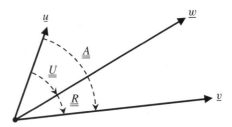

Figure 9.4 Calculation of the effects of the right stretch and rotation tensors.

[3] Programs capable of these calculations include Mathcad [170], Matlab [169], Mathematica [267], and Maple [260].

To calculate the square root of $\underline{\underline{U}}^2$ we must first express this tensor in diagonal form, that is, we must find the coordinate system ξ_1, ξ_2, ξ_3 in which $\underline{\underline{U}}$'s only nonzero elements are along the diagonal. Once in diagonal form, the square root can be calculated by taking the square root of the diagonal elements[4]:

$$\begin{pmatrix} a & 0 & 0 \\ 0 & b & 0 \\ 0 & 0 & c \end{pmatrix}_{\xi_1\xi_2\xi_3}^{\frac{1}{2}} = \begin{pmatrix} \sqrt{a} & 0 & 0 \\ 0 & \sqrt{b} & 0 \\ 0 & 0 & \sqrt{c} \end{pmatrix}_{\xi_1\xi_2\xi_3} \tag{9.69}$$

The result can then be expressed in the original coordinates by another coordinate transformation.

To carry out the change in basis needed for the square-root calculation, we follow the procedure outlined in Appendix C.6. These operations are matrix operations, not tensor operations, that is, we are calculating a representation of $\underline{\underline{U}}$, the tensor, in two different coordinate systems; $\underline{\underline{U}}$ is unchanged. To distinguish matrix from tensor operations, we will represent matrices with square brackets and tensors with parentheses with a subscript indicating the coordinate system used.

Let $[U^2]_{xyz}$ be the matrix of coefficients of $\underline{\underline{U}}^2$ in the original coordinate system; $[U^2]_{\xi_1\xi_2\xi_3}$ will be the matrix of coefficients of $\underline{\underline{U}}^2$ in the system in which $\underline{\underline{U}}^2$ is diagonal. Following Appendix C.6,

$$[U^2]_{\xi_1\xi_2\xi_3} = L^T [U^2]_{xyz} L \tag{9.70}$$

where the transformation matrix L is constructed from the eigenvectors of $\underline{\underline{U}}$ squared.[5]

The calculation of eigenvalues and eigenvectors is straightforward and is covered in advanced undergraduate math classes; this calculation is also preprogrammed into many computer packages. The eigenvalues λ_i of $\underline{\underline{U}}^2$ in this example are $\lambda_1 = 0.739$, $\lambda_2 = 4.548$, and $\lambda_3 = 10.713$, and the eigenvectors form the columns of the transformation matrix L:

$$L = \begin{bmatrix} -0.83 & 0.546 & 0.113 \\ -0.19 & -0.468 & 0.863 \\ 0.524 & 0.695 & 0.493 \end{bmatrix} \tag{9.71}$$

Carrying out the matrix calculation in Equation (9.70) we obtain

$$[U^2]_{\xi_1\xi_2\xi_3} = L^T [U^2]_{xyz} L = \begin{bmatrix} 0.739 & 0 & 0 \\ 0 & 4.548 & 0 \\ 0 & 0 & 10.713 \end{bmatrix} \tag{9.72}$$

$$\underline{\underline{U}}^2 = \begin{pmatrix} 0.739 & 0 & 0 \\ 0 & 4.548 & 0 \\ 0 & 0 & 10.713 \end{pmatrix}_{\xi_1\xi_2\xi_3} \tag{9.73}$$

---

[4] Since the stretch tensors are positive definite, the positive square root is appropriate.

[5] The eigenvalues and eigenvectors of a tensor are the same in all coordinate systems; see Appendix C.6.

The square root of the tensor $\underline{\underline{U}}^2$ can now be calculated by taking the square root of each diagonal element in Equation (9.73):

$$[U]_{\xi_1\xi_2\xi_3} = \begin{bmatrix} 0.860 & 0 & 0 \\ 0 & 2.133 & 0 \\ 0 & 0 & 3.273 \end{bmatrix} \tag{9.74}$$

$$\underline{\underline{U}} = \begin{pmatrix} 0.860 & 0 & 0 \\ 0 & 2.133 & 0 \\ 0 & 0 & 3.273 \end{pmatrix}_{\xi_1\xi_2\xi_3} \tag{9.75}$$

To convert this square root back to our original coordinate system we employ the inverse transformation (Appendix C.6):

$$[U]_{xyz} = L[U]_{\xi_1\xi_2\xi_3} L^T \tag{9.76}$$

$$= \begin{bmatrix} 1.269 & -0.09 & 0.617 \\ -0.09 & 2.936 & 0.612 \\ 0.617 & 0.612 & 2.06 \end{bmatrix} \tag{9.77}$$

$$\underline{\underline{U}} = \begin{pmatrix} 1.269 & -0.09 & 0.617 \\ -0.09 & 2.936 & 0.612 \\ 0.617 & 0.612 & 2.06 \end{pmatrix}_{xyz} \tag{9.78}$$

To calculate $\underline{\underline{R}} = \underline{\underline{A}} \cdot \underline{\underline{U}}^{-1}$, we need $\underline{\underline{U}}^{-1}$. The inverse of a matrix is also straightforward to calculate, and once again we use a mathematical software package. We obtain

$$\underline{\underline{U}}^{-1} = \begin{pmatrix} 0.946 & 0.094 & -0.311 \\ 0.094 & 0.372 & -0.139 \\ -0.311 & -0.139 & 0.620 \end{pmatrix}_{xyz} \tag{9.79}$$

$$\underline{\underline{R}} = \underline{\underline{A}} \cdot \underline{\underline{U}}^{-1} = \begin{pmatrix} 0.324 & -0.183 & 0.928 \\ -0.03 & 0.979 & 0.204 \\ 0.946 & 0.094 & -0.311 \end{pmatrix}_{xyz} \tag{9.80}$$

The reader can verify that the tensor $\underline{\underline{R}}$ is an orthogonal tensor. We now calculate the requested vectors and angles:

$$\underline{v} = \underline{\underline{A}} \cdot \underline{u} = \begin{pmatrix} 3 \\ 7 \\ 1 \end{pmatrix}_{xyz} \tag{9.81}$$

$$\underline{w} = \underline{\underline{U}} \cdot \underline{u} = \begin{pmatrix} 1.707 \\ 6.393 \\ 3.901 \end{pmatrix}_{xyz} \tag{9.82}$$

To calculate the angles between $\underline{u}$ and $\underline{w}$ and $\underline{w}$ and $\underline{v}$ we return to the definition of the dot product:

$$\underline{u} \cdot \underline{w} = uw \cos \psi_{uw} \tag{9.83}$$

$$\psi_{uw} = \cos^{-1}\left(\frac{\underline{u} \cdot \underline{w}}{uw}\right) = 0.212 \text{ rad} \tag{9.84}$$

$$\psi_{wv} = \cos^{-1}\left(\frac{\underline{w} \cdot \underline{v}}{wv}\right) = 0.424 \text{ rad} \tag{9.85}$$

The length changes brought about by $\underline{\underline{U}}$ and $\underline{\underline{A}}$ can be measured by the relative changes in length among $\underline{u}$, $\underline{v}$, and $\underline{w}$:

$$|\underline{u}| = 2.449 \tag{9.86}$$

$$|\underline{v}| = 7.681 \tag{9.87}$$

$$|\underline{w}| = 7.681 \tag{9.88}$$

$$\frac{v}{u} = \sqrt{\frac{(\underline{v} \cdot \underline{v})}{(\underline{u} \cdot \underline{u})}} = 3.136 \tag{9.89}$$

Note that the magnitudes of $\underline{w}$ and $\underline{v}$ are equal as expected, that is, the length changes brought about by $\underline{\underline{A}}$ and $\underline{\underline{U}}$ are the same. The angles ψ_{uw} and ψ_{wv} and the length change we have calculated will differ for other vectors operated on by $\underline{\underline{A}}$.

In the preceding example, which demonstrated the effect of polar decomposition on a tensor $\underline{\underline{A}}$, we see that in considering the right stretch tensor $\underline{\underline{U}}$, rather than the original tensor $\underline{\underline{A}}$, we have not totally isolated length changes from rotations. We have taken an important step toward our goal, however, since we have removed the rotation accounted for by $\underline{\underline{R}}$, and we have produced a symmetric, nonsingular tensor, $\underline{\underline{U}}$ that contains the change-in-shape information of $\underline{\underline{A}}$ as well as some rotation information. The fact that $\underline{\underline{U}}$ is symmetric is significant because this property of $\underline{\underline{U}}$ will allow us to completely isolate the shape changes from rotations, as we will now show.

There are three families of parallel vectors that are not rotated by a tensor. When a tensor operates on one of these vectors, the vector's length changes, but its orientation in space is unchanged. For the tensor $\underline{\underline{U}}$ these vectors, which we will write as $\hat{\xi}_1$, $\hat{\xi}_2$, and $\hat{\xi}_3$, satisfy the following equations:

$$\underbrace{\underline{\underline{U}} \cdot \hat{\xi}_1}_{\substack{\underline{\underline{U}} \text{ operates} \\ \text{on } \hat{\xi}_1}} = \underbrace{\lambda_1 \hat{\xi}_1}_{\substack{\text{returns a} \\ \text{stretched} \\ \text{version of } \hat{\xi}_1}} \tag{9.90}$$

$$\underline{\underline{U}} \cdot \hat{\xi}_2 = \lambda_2 \hat{\xi}_2 \tag{9.91}$$

$$\underline{\underline{U}} \cdot \hat{\xi}_3 = \lambda_3 \hat{\xi}_3 \tag{9.92}$$

where λ_i is the amount by which $\underline{\underline{U}}$ stretches $\hat{\xi}_i$. These are the familiar unit eigenvectors of $\underline{\underline{U}}$, and λ_1, λ_2, and λ_3 are the eigenvalues of $\underline{\underline{U}}$ (see Appendix C.6). Such vectors can be found for all nonsingular tensors [see footnote after Equation (9.53)]. For symmetric tensors, such as the left and right stretch tensors, it can be shown that the eigenvalues λ_i are all distinct, real numbers, and the eigenvectors $\hat{\xi}_i$ are mutually perpendicular [7] (see Problem 9.44).

To understand how the left and right stretch tensors of a deformation tensor can be used to develop a rotation-free strain measure, consider the situation in Figure 9.5. Let $\hat{\xi}_1$, $\hat{\xi}_2$, $\hat{\xi}_3$ and λ_1, λ_2, λ_3 be the eigenvectors and eigenvalues of $\underline{\underline{U}}$ and $\hat{\zeta}_1$, $\hat{\zeta}_2$, $\hat{\zeta}_3$ and v_1, v_2, v_3 be the eigenvectors and eigenvalues of $\underline{\underline{V}}$. These quantities are related. Recall the definitions of $\underline{\underline{U}}$ and $\underline{\underline{V}}$:

$$\underline{\underline{A}} = \underline{\underline{R}} \cdot \underline{\underline{U}} = \underline{\underline{V}} \cdot \underline{\underline{R}} \tag{9.93}$$

If we right-dot-multiply this equation by $\hat{\xi}_1$ we obtain

$$\underline{\underline{R}} \cdot \left(\underline{\underline{U}} \cdot \hat{\xi}_1 \right) = \underline{\underline{V}} \cdot (\underline{\underline{R}} \cdot \hat{\xi}_1) \tag{9.94}$$

$$\lambda_1 \left(\underline{\underline{R}} \cdot \hat{\xi}_1 \right) = \underline{\underline{V}} \cdot \left(\underline{\underline{R}} \cdot \hat{\xi}_1 \right) \tag{9.95}$$

Equation (9.95) is just another eigenvalue equation for $\underline{\underline{V}}$, as we can see by comparing it to the eigenvalue equation for $\hat{\zeta}_1$:

$$v_1 \hat{\zeta}_1 = \underline{\underline{V}} \cdot \hat{\zeta}_1 \tag{9.96}$$

Thus we can conclude that $\underline{\underline{R}} \cdot \hat{\xi}_1$ and λ_1 are an eigenvector and an eigenvalue of $\underline{\underline{V}}$, respectively, and, without loss of generality, we can choose $\hat{\zeta}_1 = \underline{\underline{R}} \cdot \hat{\xi}_1$ and $\lambda_1 = v_1$. The same argument can be repeated for the other eigenvalues, yielding

$$\hat{\zeta}_i = \underline{\underline{R}} \cdot \hat{\xi}_i = \hat{\xi}_i \cdot \underline{\underline{R}}^T \tag{9.97}$$

$$\lambda_i = v_i \tag{9.98}$$

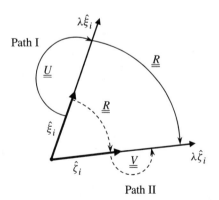

Figure 9.5 Effects of the left and right stretch tensors on their eigenvectors.

$$\hat{\xi}_i = \hat{\zeta}_i \cdot \underline{R} \tag{9.99}$$

where we have made use of the orthogonality property of $\underline{R}$.

We can interpret these relations graphically by considering Figure 9.5 and the following. When the tensor $\underline{A}$ operates on the vector $\hat{\xi}_i$, which is a unit eigenvector of $\underline{U}$, the vector can be considered to be first stretched in place by an amount λ_i, and then rotated to become $\lambda_i \hat{\zeta}_i$, where $\hat{\zeta}_i$ is a unit eigenvector of $\underline{V}$ (path I, in Figure 9.5).

$$\text{path I:} \quad \underline{\underline{A}} \cdot \hat{\xi}_i = \underline{R} \cdot \underline{U} \cdot \hat{\xi}_i \tag{9.100}$$

$$= \underline{R} \cdot \left(\lambda_i \hat{\xi}_i \right) = \lambda_i \left(\underline{R} \cdot \hat{\xi}_i \right) \tag{9.101}$$

$$= \lambda_i \hat{\zeta} \tag{9.102}$$

Alternatively, the effect of $\underline{A}$ on $\hat{\xi}_i$ can be considered to follow dashed path II: the vector $\hat{\xi}_i$ is first rotated with no length change to become $\hat{\zeta}_i$, and this new vector is then stretched in place by an amount λ_i, yielding the same net result:

$$\text{path II:} \quad \underline{A} \cdot \hat{\xi}_i = \underline{V} \cdot \left(\underline{R} \cdot \hat{\xi}_i \right) \tag{9.103}$$

$$= \underline{V} \cdot \hat{\zeta}_i \tag{9.104}$$

$$= \lambda_i \hat{\zeta}_i \tag{9.105}$$

An analogous description can be made for the transformation $\hat{\zeta}_i \cdot \underline{A}$, where a unit eigenvector of $\underline{V}$ left-multiplies $\underline{A}$.

This graphical interpretation shows us that the left and right stretch tensors do isolate pure stretch associated with a deformation tensor when the strain of an eigenvector $\hat{\xi}_i$ or $\hat{\zeta}_i$ is considered. The eigenvectors of $\underline{U}$ and $\underline{V}$ indicate the directions along which only pure stretch is occurring in the original deformation described by $\underline{A}$. The eigenvalues of these tensors tell us the amounts of stretch taking place in each of the three mutually orthogonal pure stretching directions. [Since $\underline{U}$ and $\underline{V}$ are symmetric tensors, see the definitions and discussion following Equation (9.53).] Thus we will choose to develop rotation-free large-strain measures by working with the stretch tensors of the deformation rather than with the deformation tensors $\underline{F}$ and $\underline{F}^{-1}$ themselves.

It is now time to apply these concepts to the actual deformation-gradient tensors $\underline{F}$ and $\underline{F}^{-1}$. The transformations we are concerned with are $d\underline{r}' \longrightarrow d\underline{r}$ and $d\underline{r} \longrightarrow d\underline{r}'$, which are

$$d\underline{r} = d\underline{r}' \cdot \underline{F}^{-1} = (\underline{F}^{-1})^T \cdot d\underline{r}' \tag{9.106}$$

$$d\underline{r}' = d\underline{r} \cdot \underline{F} = \underline{F}^T \cdot d\underline{r} \tag{9.107}$$

These transformations involve four deformation tensors, $\underline{F}$, $\underline{F}^T$, $\underline{F}^{-1}$, and $(\underline{F}^{-1})^T$. To remove the rotation from these tensors we apply polar decomposition, that is, we calculate $\underline{U}$ and $\underline{V}$ for each deformation tensor. The squares of the left and right stretch tensors of these deformation-gradient tensors are given in Table 9.2.

TABLE 9.2
Summary of the Squares of the Left and
Right Stretch Tensors of Various
Deformation Tensors

$\underline{\underline{A}}$	$\underline{\underline{V}}^2$	$\underline{\underline{U}}^2$
$\underline{\underline{F}}$	$\underline{\underline{F}} \cdot \underline{\underline{F}}^T$	$\underline{\underline{F}}^T \cdot \underline{\underline{F}}$
$\underline{\underline{F}}^T$	$\underline{\underline{F}}^T \cdot \underline{\underline{F}}$	$\boxed{\underline{\underline{F}} \cdot \underline{\underline{F}}^T}$
$\underline{\underline{F}}^{-1}$	$\underline{\underline{F}}^{-1} \cdot (\underline{\underline{F}}^{-1})^T$	$(\underline{\underline{F}}^{-1})^T \cdot \underline{\underline{F}}^{-1}$
$(\underline{\underline{F}}^{-1})^T$	$\boxed{(\underline{\underline{F}}^{-1})^T \cdot \underline{\underline{F}}^{-1}}$	$\underline{\underline{F}}^{-1} \cdot (\underline{\underline{F}}^{-1})^T$

Although it seems like there would be many choices for strain tensors from the entries in Table 9.2, there are several factors that reduce the number of possibilities. First, the choice of right versus left stretch tensor is arbitrary since it corresponds to the choice of eigenvectors, those of $\underline{\underline{U}}$ or $\underline{\underline{V}}$, to use to describe stretching, and both are equivalent. Referring to Table 9.2, this tells us that using $\underline{\underline{F}}^T \cdot \underline{\underline{F}}$ is equivalent to $\underline{\underline{F}} \cdot \underline{\underline{F}}^T$, and $(\underline{\underline{F}}^{-1})^T \cdot \underline{\underline{F}}^{-1}$ is equivalent to $\underline{\underline{F}}^{-1} \cdot (\underline{\underline{F}}^{-1})^T$. Thus just two unique tensors result from the decomposition of the deformation tensors, and these are $\underline{\underline{F}} \cdot \underline{\underline{F}}^T$, called the *Cauchy tensor*, and $(\underline{\underline{F}}^{-1})^T \cdot \underline{\underline{F}}^{-1}$, called the *Finger tensor*. These two tensors are inverses to each other and are denoted by the symbols $\underline{\underline{C}}$ and $\underline{\underline{C}}^{-1}$, respectively[6]:

Cauchy strain tensor $\qquad \boxed{\underline{\underline{C}} \equiv \underline{\underline{F}} \cdot \underline{\underline{F}}^T}$ $\qquad\qquad$ (9.108)

Finger strain tensor $\qquad \boxed{\underline{\underline{C}}^{-1} \equiv (\underline{\underline{F}}^{-1})^T \cdot \underline{\underline{F}}^{-1}}$ $\qquad$ (9.109)

In summary, we have used polar decomposition on the deformation-gradient and related tensors to remove unwanted rotation from the description of fluid motion. The result of the decomposition was the identification of two finite-strain tensors, $\underline{\underline{C}}$ and $\underline{\underline{C}}^{-1}$, which can now be used in place of $\underline{\underline{\gamma}}$ in constitutive equations. The tensors $\underline{\underline{C}}$ and $\underline{\underline{C}}^{-1}$ are related to the vector transformations

$$d\underline{r}' = \underline{\underline{F}}^T \cdot d\underline{r} \qquad\qquad \underline{\underline{C}} = \underline{\underline{U}}^2 \text{ of } \underline{\underline{F}}^T \qquad\qquad (9.110)$$

$$d\underline{r} = (\underline{\underline{F}}^{-1})^T \cdot d\underline{r}' \qquad \underline{\underline{C}}^{-1} = \underline{\underline{V}}^2 \text{ of } (\underline{\underline{F}}^{-1})^T \qquad (9.111)$$

Use of the squared tensors $\underline{\underline{U}}^2$ or $\underline{\underline{V}}^2$ as strain measures instead of $\underline{\underline{U}}$ or $\underline{\underline{V}}$ is justified for an elastic solid by comparing stresses predicted in shear to experimental observations, as shown in the second example that follows. In the first example, using $\underline{\underline{C}}^{-1}$ as the strain

---

[6] See Tables D. 1 and D.2 in Appendix D for other symbols used in the literature for these tensors.

measure, we calculate $\underline{\underline{\tau}}$ for an elastic body undergoing solid-body rotation. Recall that the GLVE model failed this test of objectivity.

EXAMPLE

For solid-body rotation, calculate the stress predicted by a finite-strain Hooke's law constitutive equation with $\underline{\underline{\gamma}}(t_{\text{ref}}, t)$ replaced by the negative of the Finger strain tensor, $-\underline{\underline{C}}^{-1} = -(\underline{\underline{F}}^{-1})^T \cdot \underline{\underline{F}}^{-1}$.

SOLUTION

In Chapter 8 we presented the following tensor version of Hooke's law:

$$\underline{\underline{\tau}}(t) = -G\underline{\underline{\gamma}}(t_{\text{ref}}, t) \tag{9.112}$$

where G is a scalar called the modulus, and the strain measure is the infinitesimal strain tensor. The reference time t_{ref} is taken to be at some isotropic stress state, that is, at a time when there are only isotropic stresses on the material. We will take $t_{\text{ref}} = 0$. To adapt the Hookean constitutive equation to large strains we replace $\underline{\underline{\gamma}}(0, t)$ with $-\underline{\underline{C}}^{-1}(t, 0) = -(\underline{\underline{F}}^{-1})^T \cdot \underline{\underline{F}}^{-1}$. The Finger tensor describes the particle shape at time 0 relative to the material's configuration at time t, which is the negative of the strain we desire; hence we include a negative sign:

$$\underline{\underline{\tau}}(t) = G\underline{\underline{C}}^{-1} = G \left(\underline{\underline{F}}^{-1}\right)^T \cdot \underline{\underline{F}}^{-1} \tag{9.113}$$

(See the end of the chapter for further discussion of this negative sign.)

The inverse deformation-gradient tensor for counterclockwise rotation of a material around the z-axis can be shown to be [see Equation (9.16)]

$$\underline{\underline{F}}^{-1} = \begin{pmatrix} \cos\psi & \sin\psi & 0 \\ -\sin\psi & \cos\psi & 0 \\ 0 & 0 & 1 \end{pmatrix}_{xyz} \tag{9.114}$$

where ψ is the angle between $\underline{r}(0)$ and $\underline{r}(t)$. Using this tensor we can calculate $(\underline{\underline{F}}^{-1})^T \cdot \underline{\underline{F}}^{-1}$:

$$\left(\underline{\underline{F}}^{-1}\right)^T \cdot \underline{\underline{F}}^{-1} = \begin{pmatrix} 1 & 0 & 0 \\ 0 & 1 & 0 \\ 0 & 0 & 1 \end{pmatrix}_{xyz} \tag{9.115}$$

Substituting this result into the modified Hookean constitutive equation gives

$$\underline{\underline{\tau}}(t) = G \begin{pmatrix} 1 & 0 & 0 \\ 0 & 1 & 0 \\ 0 & 0 & 1 \end{pmatrix}_{xyz} \tag{9.116}$$

Thus we arrive at the correct result for solid-body rotation, $\underline{\underline{\tau}} = G\underline{\underline{I}}$, that is, the stress of an elastic body is independent of the rotation angle in solid-body rotation and is equal to its stress at rest. The finite-strain Hooke's law that uses $\underline{\underline{C}}^{-1}$ as its strain measure passes this test of objectivity.

Using the finite-strain Hooke's law of the previous practice example, calculate the shear and normal stresses produced in an elastic body subjected to the onset of steady shear (constant shear rate $\dot{\gamma}_0$) at time $t = 0$.

SOLUTION

The finite-strain Hooke's law is

$$\underline{\underline{\tau}} = G\underline{\underline{C}}^{-1}(t, 0) \tag{9.117}$$

$$= G\left(\underline{\underline{F}}^{-1}\right)^{T} \cdot \underline{\underline{F}}^{-1} \tag{9.118}$$

The Finger tensor measures the deformation at time zero, the time at which the deformation was imposed on the sample, with respect to the fluid configuration at the current time t. For shear flow we calculated $\underline{\underline{F}}^{-1}$ in Equation (9.39). Substituting this into the constitutive equation we arrive at

$$\underline{\underline{\tau}} = G\begin{pmatrix} 1 & \gamma & 0 \\ 0 & 1 & 0 \\ 0 & 0 & 1 \end{pmatrix} \cdot \begin{pmatrix} 1 & 0 & 0 \\ \gamma & 1 & 0 \\ 0 & 0 & 1 \end{pmatrix} \tag{9.119}$$

$$= G\begin{pmatrix} 1+\gamma^2 & \gamma & 0 \\ \gamma & 1 & 0 \\ 0 & 0 & 1 \end{pmatrix} \tag{9.120}$$

where $\gamma = \gamma(0, t) = \int_0^t \dot{\gamma}_0 \, dt'' = \dot{\gamma}_0 t$. Thus the shear stress and the two normal-stress differences predicted to be generated when deforming an elastic solid are

$$\tau_{21} = G\gamma = G\dot{\gamma}_0 t \tag{9.121}$$

$$\tau_{11} - \tau_{22} = G\gamma^2 = G\dot{\gamma}_0^2 t^2 \tag{9.122}$$

$$\tau_{22} - \tau_{33} = 0 \tag{9.123}$$

For the material described by the Hooke's law constitutive equation the shear stress $\tau_{21} = G\dot{\gamma}_0 t$ does not approach steady state as time approaches infinity. This reflects the solid character of Hookean materials.

To see whether these predictions are valid, we compare them with experimental data for an elastic solid undergoing simple shear. Data for a silicon rubber subjected to torsional shear (see Chapter 10 for more on torsional shear) is shown in Figure 9.6 [162]. The best fit for the modulus G from these data is 160 kPa. The fit is good on average, although some deviations are seen at large values of strain γ. Both the model and the data show that while the shear stress can be positive or negative, the normal-stress difference is always positive.

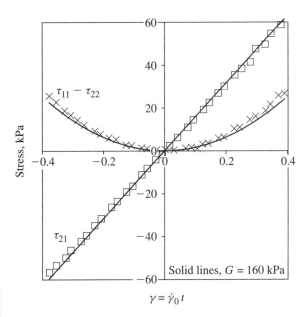

Figure 9.6 Data for shear and first normal-stress difference for a silicon rubber solid under torsional shear; from DeGroot [63] as cited in Macosko [162]. Solid lines are the fit to the finite-strain Hooke model with $G = 160$ kPa. *Source:* From *Rheology: Principles, Measurements and Applications,*" by C. W. Macosko, Copyright © 1994 by VCH Publishers, Inc. Reprinted by permission of John Wiley & Sons, Inc.

The normal-stress difference $\tau_{11} - \tau_{22}$ represents the amount of force that must be applied to the top and bottom of the deforming cylinder to maintain a constant cylinder height.

The finite-strain Hooke's law based on the Finger tensor passes an important test: it matches experimental data. This equation also makes valid predictions in uniaxial extension. This comparison with experiment confirms that the Finger tensor, the square of the left stretch tensor of $(\underline{\underline{F}}^{-1})^T$, is a useful measure of strain in an elastic body.

We have arrived at new large-strain measures to use in constitutive modeling, the Finger strain tensor $\underline{\underline{C}}^{-1}(t', t)$, and the Cauchy strain tensor $\underline{\underline{C}}(t, t')$. These strain tensors are objective,[7] that is, they are independent of the frame of reference of the observer, and thus we have avoided the problem of spurious rotation effects encountered with the Maxwell and GLVE models. In the next section we will replace $\underline{\underline{\gamma}}(t, t')$ in the GLVE model with $-\underline{\underline{C}}^{-1}(t', t)$ and examine the resulting viscoelastic constitutive equation. The ultimate test of a strain measure is how well $\underline{\tau}$ is predicted when the strain tensor is used in a constitutive equation.

Before closing this section we wish to consider two more strain tensors that are associated with $\underline{\underline{C}}, \underline{\underline{C}}^{-1}$, and $\underline{\underline{\gamma}}$. Consider the displacement vector $\underline{u}(t, t') = \underline{r}' - \underline{r}$, discussed at the beginning of the chapter. This can be related to $\underline{\underline{F}}$ as follows:

$$\nabla \underline{u} = \frac{\partial}{\partial \underline{r}} \left(\underline{r}' - \underline{r} \right) \tag{9.124}$$

---

[7] We have not proven it here, but this is discussed in detail in Lodge [155], Leal [148], and Oldroyd [192, 193, 194, 195].

$$= \underline{\underline{F}} - \underline{\underline{I}} \tag{9.125}$$

The definition of the infinitesimal strain tensor $\underline{\underline{\gamma}}$ is

$$\underline{\underline{\gamma}}(t, t') \equiv \nabla \underline{u} + (\nabla \underline{u})^T \tag{9.126}$$

We can now show that $\underline{\underline{\gamma}}$ is the small-displacement limit of a more general finite-strain tensor $\underline{\underline{\gamma}}^{[0]}$ defined as follows [26]:

$$\underline{\underline{\gamma}}^{[0]} = \underline{\underline{F}} \cdot \underline{\underline{F}}^T - \underline{\underline{I}} \tag{9.127}$$

$$\equiv \underline{\underline{C}} - \underline{\underline{I}} \tag{9.128}$$

where $\underline{\underline{C}}$ is the Cauchy tensor. Using $\underline{\underline{F}} = \nabla \underline{u} + \underline{\underline{I}}$ [Equation (9.125)] we see that at small displacement gradients $\underline{\underline{\gamma}}^{[0]}$ becomes

$$\underline{\underline{\gamma}}^{[0]} = \left(\nabla \underline{u} + \underline{\underline{I}} \right) \left(\nabla \underline{u} + \underline{\underline{I}} \right)^T - \underline{\underline{I}} \tag{9.129}$$

$$= \nabla \underline{u} \cdot (\nabla \underline{u})^T + \nabla \underline{u} + (\nabla \underline{u})^T \tag{9.130}$$

$$\lim_{\nabla \underline{u} \to 0} \left(\underline{\underline{\gamma}}^{[0]} \right) = \nabla \underline{u} + (\nabla \underline{u})^T = \underline{\underline{\gamma}}(t, t') \tag{9.131}$$

Another finite-strain tensor, $\underline{\underline{\gamma}}_{[0]}$, also becomes $\underline{\underline{\gamma}}$ at small displacement gradients (see Problem 9.5):

$$\underline{\underline{\gamma}}_{[0]} \equiv \underline{\underline{I}} - \underline{\underline{C}}^{-1} \tag{9.132}$$

The principal difference between the pairs $\underline{\underline{C}}, \underline{\underline{C}}^{-1}$, and $\underline{\underline{\gamma}}^{[0]}, \underline{\underline{\gamma}}_{[0]}$ is in their values in solid-body rotation. When no deformation takes place, $\underline{\underline{C}} = \underline{\underline{C}}^{-1} = \underline{\underline{I}}$, whereas $\underline{\underline{\gamma}}^{[0]} = \underline{\underline{\gamma}}_{[0]} = \underline{\underline{0}}$. Also, at small strains $\underline{\underline{\gamma}}^{[0]}$ and $\underline{\underline{\gamma}}_{[0]}$ reduce to $\underline{\underline{\gamma}}$, whereas in that limit $\underline{\underline{C}}^{-1}$ and $\underline{\underline{C}}$ differ from $\underline{\underline{\gamma}}$ by an isotropic constant $\underline{\underline{I}}$. Thus $\underline{\underline{\gamma}}^{[0]}$ and $\underline{\underline{\gamma}}_{[0]}$ represent relative change in shape whereas $\underline{\underline{C}}$ and $\underline{\underline{C}}^{-1}$ represent shape directly. As we will see in the discussions that follow, this difference leads to irrelevant differences in predicted stress tensors. The time derivatives of these strain measures can be worked out in the usual way, and some quantities of interest are listed in Table 9.1.

A comment on notation. The inverse deformation-gradient tensor $\underline{\underline{F}}^{-1}$ gives the change from $\underline{r}' \longrightarrow \underline{r}$, and can therefore be written as $\underline{\underline{F}}^{-1}(t', t)$. The Finger tensor $\underline{\underline{C}}^{-1} = (\underline{\underline{F}}^{-1})^T \cdot \underline{\underline{F}}^{-1}$ is likewise describing the change from $t' \longrightarrow t$ and is therefore written as $\underline{\underline{C}}^{-1}(t', t)$. The deformation-gradient tensor $\underline{\underline{F}}$ describes the reverse transformation from $\underline{r}$ to $\underline{r}'$, and $\underline{\underline{F}}$ and the related Cauchy tensor $\underline{\underline{C}}$ are written as $\underline{\underline{F}}(t, t')$ and $\underline{\underline{C}}(t, t')$, respectively. The infinitesimal strain tensor $\underline{\underline{\gamma}}$ can be written either way, but the meaning is different:

$$\underline{\underline{\gamma}}(t', t) = \int_{t'}^{t} \underline{\underline{\dot{\gamma}}}(t'') \, dt'' \tag{9.133}$$

$$\underline{\underline{\gamma}}(t, t') = \int_{t}^{t'} \underline{\underline{\dot{\gamma}}}(t'') \, dt'' \tag{9.134}$$

$$\underline{\underline{\gamma}}(t', t) = -\underline{\underline{\gamma}}(t, t') \tag{9.135}$$

This difference shows up in the small-strain limits of $\underline{\underline{C}}^{-1}$ and $\underline{\underline{C}}$, which can be derived from their definitions and Equation (9.125):

$$\lim_{(\underline{r}'-\underline{r}) \to 0} \underline{\underline{C}}^{-1}(t', t) = \underline{\underline{I}} - \underline{\underline{\gamma}}(t, t') \tag{9.136}$$

$$\lim_{(\underline{r}'-\underline{r}) \to 0} \underline{\underline{C}}(t, t') = \underline{\underline{\gamma}}(t, t') - \underline{\underline{I}} \tag{9.137}$$

At small strains the Cauchy tensor is proportional to the infinitesimal strain tensor $\underline{\underline{\gamma}}(t, t')$, whereas the Finger tensor is proportional to the negative of $\underline{\underline{\gamma}}(t, t')$. So that our finite-strain constitutive equations reduce to the experimentally valid small-strain equations, $\underline{\underline{\gamma}}(t, t')$ in the small-strain constitutive equations (Hooke's law, GLVE, Maxwell model) is replaced with $\underline{\underline{C}}(t, t')$ or $-\underline{\underline{C}}^{-1}(t', t)$, as was seen in several practice examples earlier in this chapter.

With possible finite-strain tensors now in hand, we can proceed with developing large-strain, frame-independent constitutive equations to describe real flows of polymer melts and solutions.

9.2 Lodge Equation

In this section we develop the Lodge equation by replacing the strain measure of the Maxwell equation with the Finger tensor. In Chapter 8 we saw that the Maxwell equation could be written either as a differential [Equation (8.50)] or as an integral equation [Equation 8.58]. The Lodge equation can also be written in both integral and differential forms, as we will see. The choice of which form of the Lodge equation to use is arbitrary and is dictated by convenience.

9.2.1 INTEGRAL FORM OF THE LODGE EQUATION

We seek to develop a finite-strain constitutive equation based on the Maxwell model. Using the integral version of the GLVE model in Equation (9.11) and the Maxwell relaxation function $G(t - t') = (\eta_0/\lambda)e^{-\frac{(t-t')}{\lambda}}$, we can write the Maxwell model in terms of its strain tensor:

$$\underline{\underline{\tau}}(t) = + \int_{-\infty}^{t} \frac{\eta_0}{\lambda^2} e^{-\frac{(t-t')}{\lambda}} \underline{\underline{\gamma}}(t, t') \, dt' \tag{9.138}$$

We know the strain measure $\underline{\underline{\gamma}}(t, t')$ has an unwanted dependence on solid-body rotation. We can develop a new constitutive equation from the Maxwell model by replacing $\underline{\underline{\gamma}}(t, t')$ with one of the new rotation-invariant strain tensors discussed before. Lodge [154, 158, 155] replaced $\underline{\underline{\gamma}}(t, t')$ with $-\underline{\underline{C}}^{-1}(t', t)$. The choice of which finite-strain measure to use, $\underline{\underline{C}}^{-1}$, $\underline{\underline{C}}$, or some combination of these, is arbitrary and can only be justified by examining the predictions of the resulting constitutive equation. Lodge's constitutive equation is

Lodge equation
$$\boxed{\underline{\underline{\tau}} = -\int_{-\infty}^{t} \left[\frac{\eta_0}{\lambda^2} e^{-\frac{(t-t')}{\lambda}} \right] \underline{\underline{C}}^{-1}(t', t)\, dt'} \tag{9.139}$$

This equation [154, 158, 155] is properly frame invariant because the Finger tensor correctly describes solid-body rotation as a deformation-free transformation.

As we stated before, to determine whether the Lodge equation is useful we must compare predictions of the Lodge model with known fluid behavior. We calculate material functions using the Lodge equation by substituting the Finger tensor for the flow under consideration into Equation (9.139) and calculating the predicted stress components (see Table 9.3). We will now practice carrying out constitutive calculations with the Lodge equation by considering the material function $\bar{\eta}$, the uniaxial elongational viscosity.

EXAMPLE

Calculate the steady uniaxial extensional viscosity $\bar{\eta}$ predicted by the Lodge equation.

SOLUTION

The Lodge model contains the Finger tensor $\underline{\underline{C}}^{-1}$, which is given in Table 9.3 for three common flows. For practice, however, we will calculate $\underline{\underline{C}}^{-1}$ for this flow from scratch. We begin with the kinematics:

$$\underline{v} = \begin{pmatrix} -\frac{\dot{\epsilon}(t)}{2} x_1 \\ -\frac{\dot{\epsilon}(t)}{2} x_2 \\ \dot{\epsilon}(t) x_3 \end{pmatrix}_{123} \tag{9.140}$$

$$\dot{\epsilon}(t) = \dot{\epsilon}_0 = \text{constant} \tag{9.141}$$

The definition of $\underline{\underline{C}}^{-1}$ is

$$\underline{\underline{C}}^{-1} = (\underline{\underline{F}}^{-1})^T \cdot \underline{\underline{F}}^{-1} \tag{9.142}$$

where

$$\underline{\underline{F}}^{-1} = \frac{\partial \underline{r}}{\partial \underline{r}'} \tag{9.143}$$

We must calculate the displacement function $\underline{r}(\underline{r}')$, where $\underline{r}$ is the location of a particle at the current time, and $\underline{r}'$ is the same particle's location at some time t' in the past,

TABLE 9.3
Strain Tensors for Shear and Extension in Cartesian Coordinates[*]

Tensor	Shear in 1-Direction with Gradient in 2-Direction	Uniaxial Elongation in 3-Direction	Counterclockwise Rotation around $\hat{e}_3$
$\underline{\underline{F}}(t, t')$	$\begin{pmatrix} 1 & 0 & 0 \\ -\gamma & 1 & 0 \\ 0 & 0 & 1 \end{pmatrix}_{123}$	$\begin{pmatrix} e^{\frac{\epsilon}{2}} & 0 & 0 \\ 0 & e^{\frac{\epsilon}{2}} & 0 \\ 0 & 0 & e^{-\epsilon} \end{pmatrix}_{123}$	$\begin{pmatrix} \cos\psi & -\sin\psi & 0 \\ \sin\psi & \cos\psi & 0 \\ 0 & 0 & 1 \end{pmatrix}_{123}$
$\underline{\underline{F}}^{-1}(t', t)$	$\begin{pmatrix} 1 & 0 & 0 \\ \gamma & 1 & 0 \\ 0 & 0 & 1 \end{pmatrix}_{123}$	$\begin{pmatrix} e^{-\frac{\epsilon}{2}} & 0 & 0 \\ 0 & e^{-\frac{\epsilon}{2}} & 0 \\ 0 & 0 & e^{\epsilon} \end{pmatrix}_{123}$	$\begin{pmatrix} \cos\psi & \sin\psi & 0 \\ -\sin\psi & \cos\psi & 0 \\ 0 & 0 & 1 \end{pmatrix}_{123}$
$\underline{\underline{C}}(t, t')$	$\begin{pmatrix} 1 & -\gamma & 0 \\ -\gamma & 1+\gamma^2 & 0 \\ 0 & 0 & 1 \end{pmatrix}_{123}$	$\begin{pmatrix} e^{\epsilon} & 0 & 0 \\ 0 & e^{\epsilon} & 0 \\ 0 & 0 & e^{-2\epsilon} \end{pmatrix}_{123}$	$\underline{\underline{I}}$
$\underline{\underline{C}}^{-1}(t', t)$	$\begin{pmatrix} 1+\gamma^2 & \gamma & 0 \\ \gamma & 1 & 0 \\ 0 & 0 & 1 \end{pmatrix}_{123}$	$\begin{pmatrix} e^{-\epsilon} & 0 & 0 \\ 0 & e^{-\epsilon} & 0 \\ 0 & 0 & e^{2\epsilon} \end{pmatrix}_{123}$	$\underline{\underline{I}}$
$\underline{\underline{\gamma}}^{[0]}(t, t')$	$\begin{pmatrix} 0 & -\gamma & 0 \\ -\gamma & \gamma^2 & 0 \\ 0 & 0 & 1 \end{pmatrix}_{123}$	$\begin{pmatrix} e^{\epsilon}-1 & 0 & 0 \\ 0 & e^{\epsilon}-1 & 0 \\ 0 & 0 & e^{-2\epsilon}-1 \end{pmatrix}_{123}$	$\underline{\underline{0}}$
$\underline{\underline{\gamma}}_{[0]}(t, t')$	$\begin{pmatrix} -\gamma^2 & \gamma & 0 \\ \gamma & 0 & 0 \\ 0 & 0 & 0 \end{pmatrix}_{123}$	$\begin{pmatrix} e^{-\epsilon}-1 & 0 & 0 \\ 0 & e^{-\epsilon}-1 & 0 \\ 0 & 0 & e^{2\epsilon}-1 \end{pmatrix}_{123}$	$\underline{\underline{0}}$

[*] For shear flows $\gamma = \gamma(t', t) = \int_{t'}^{t} \dot{\varsigma}(t'') \, dt'' = \int_{t'}^{t} \dot{\gamma}_{21}(t'') \, dt''$ and for elongational flows $\epsilon = \epsilon(t', t) = \int_{t'}^{t} \dot{\epsilon}(t'') \, dt''$. The angle ψ is the angle from $\underline{r}(t) = \underline{r}$ to $\underline{r}(t') = \underline{r}'$ in counterclockwise rotation around the $\hat{e}_3$-axis.

$$\underline{r}(t) = \begin{pmatrix} x_1 \\ x_2 \\ x_3 \end{pmatrix}_{123} \qquad \underline{r}'(t') = \begin{pmatrix} x_1' \\ x_2' \\ x_3' \end{pmatrix}_{123} \tag{9.144}$$

The particle positions are determined by their initial positions ($\underline{r}' = x_1'\hat{e}_1 + x_2'\hat{e}_2 + x_3'\hat{e}_3$) and their velocities [Equation (9.140)]. For the 1-direction we can calculate the relation between x_1 and x_1' as follows:

$$v_1 = \frac{dx_1}{dt} = -\frac{\dot{\epsilon}_0}{2} x_1 \tag{9.145}$$

$$\frac{dx_1}{x_1} = -\frac{\dot{\epsilon}_0}{2} dt \tag{9.146}$$

$$\ln x_1 = -\frac{\dot{\epsilon}_0}{2} t + C_1 \tag{9.147}$$

where C_1 is an arbitrary constant of integration. The initial conditions are that at $t = t'$, $x_1 = x_1'$, and thus we arrive at the following equation for the displacement in the x_1-direction:

$$x_1 = x_1' e^{-\frac{\dot\epsilon_0(t-t')}{2}} \tag{9.148}$$

$$= x_1' e^{-\frac{\epsilon}{2}} \tag{9.149}$$

where $\epsilon = \dot\epsilon_0(t - t')$ is the elongational strain in the flow. Following the same procedure for the 2- and 3-directions we obtain

$$x_2 = x_2' e^{-\frac{\epsilon}{2}} \tag{9.150}$$

$$x_3 = x_3' e^{\epsilon} \tag{9.151}$$

and therefore the displacement vector $\underline{r}$ for steady uniaxial elongational flow is

$$\underline{r} = \begin{pmatrix} x_1' e^{-\frac{\epsilon}{2}} \\ x_2' e^{-\frac{\epsilon}{2}} \\ x_3' e^{\epsilon} \end{pmatrix}_{123} \tag{9.152}$$

We can now calculate $\underline{\underline{F}}^{-1}$ and $\underline{\underline{C}}^{-1}$ from their definitions:

$$\underline{\underline{F}}^{-1} \equiv \frac{\partial \underline{r}}{\partial \underline{r}'} = \begin{pmatrix} \frac{\partial x_1}{\partial x_1'} & \frac{\partial x_2}{\partial x_1'} & \frac{\partial x_3}{\partial x_1'} \\ \frac{\partial x_1}{\partial x_2'} & \frac{\partial x_2}{\partial x_2'} & \frac{\partial x_3}{\partial x_2'} \\ \frac{\partial x_1}{\partial x_3'} & \frac{\partial x_2}{\partial x_3'} & \frac{\partial x_3}{\partial x_3'} \end{pmatrix}_{123} \tag{9.153}$$

$$= \begin{pmatrix} e^{-\frac{\epsilon}{2}} & 0 & 0 \\ 0 & e^{-\frac{\epsilon}{2}} & 0 \\ 0 & 0 & e^{\epsilon} \end{pmatrix}_{123} \tag{9.154}$$

$$\underline{\underline{C}}^{-1} \equiv (\underline{\underline{F}}^{-1})^T \cdot \underline{\underline{F}}^{-1} = \begin{pmatrix} e^{-\epsilon} & 0 & 0 \\ 0 & e^{-\epsilon} & 0 \\ 0 & 0 & e^{2\epsilon} \end{pmatrix}_{123} \tag{9.155}$$

Substituting the Finger tensor for uniaxial elongation into the Lodge model we can now calculate the stress tensor $\underline{\underline{\tau}}(t)$ for this flow:

$$\underline{\underline{\tau}}(t) = -\int_{-\infty}^{t} \frac{\eta_0}{\lambda^2} e^{-\frac{(t-t')}{\lambda}} \begin{pmatrix} e^{-\epsilon} & 0 & 0 \\ 0 & e^{-\epsilon} & 0 \\ 0 & 0 & e^{2\epsilon} \end{pmatrix}_{123} dt' \tag{9.156}$$

The extensional viscosity $\bar\eta$ is defined as

$$\bar\eta \equiv \frac{-(\tau_{33} - \tau_{11})}{\dot\epsilon_0} \tag{9.157}$$

Thus we must calculate τ_{33} and τ_{11}. Remember that $\epsilon = \dot\epsilon_0(t - t')$.

$$\tau_{33} = -\int_{-\infty}^{t} \frac{\eta_0}{\lambda^2} e^{-\frac{(t-t')}{\lambda}} e^{2\epsilon} dt' \tag{9.158}$$

$$= -\int_{0}^{\infty} \frac{\eta_0}{\lambda^2} e^{(2\dot\epsilon_0 - \frac{1}{\lambda})s} ds, \quad \text{where } s = t - t' \tag{9.159}$$

$$= \frac{\eta_0}{\lambda^2} \frac{1}{2\dot{\epsilon}_0 - \frac{1}{\lambda}} \quad 2\dot{\epsilon}_0\lambda < 1 \tag{9.160}$$

Similarly:

$$\tau_{11} = \frac{\eta_0}{\lambda} \left(\frac{-1}{1 + \lambda\dot{\epsilon}_0} \right) \tag{9.161}$$

Now we can calculate the extensional viscosity:

$$\bar{\eta} = \frac{-(\tau_{33} - \tau_{11})}{\dot{\epsilon}_0} \tag{9.162}$$

$$= \frac{3\lambda G_0}{(1 - 2\lambda\dot{\epsilon}_0)(1 + \lambda\dot{\epsilon}_0)} \tag{9.163}$$

where $G_0 \equiv \eta_0/\lambda$. Both the axial stress τ_{33} and the elongational viscosity become unbounded for $\dot{\epsilon}_0 = 1/2\lambda$.

It is straightforward to show that in steady shear the Lodge model predicts a constant viscosity, $\eta = \lambda G_0$. Thus the Trouton ratio for the Lodge model is neither 3, nor is it constant, but rather it varies with the elongation rate $\dot{\epsilon}_0$:

Trouton ratio
(Lodge model)
$$\boxed{\frac{\bar{\eta}(\dot{\epsilon}_0\sqrt{3} = \dot{\gamma})}{\eta(\dot{\gamma})} = \frac{3}{(1 - 2\lambda\dot{\gamma}/\sqrt{3})(1 + \lambda\dot{\gamma}/\sqrt{3})}} \tag{9.164}$$

Selected predictions of the Lodge model are given in Appendix D, Tables D.3 and D.4, and the steady shear and steady uniaxial elongational flow properties are illustrated in Figure 9.7. We saw before that the Lodge equation predicts an elongational viscosity that becomes unbounded at a high extension rate. In shear this model predicts a constant viscosity, like the GLVE model, but the Lodge model also predicts a nonzero first normal-stress coefficient. Thus by fixing the frame-invariance problem, we have also discovered a way of predicting first normal-stress differences in shear. The Lodge model fails to predict a nonzero second normal-stress coefficient, however, and the viscosity and first normal-stress differences predicted by the Lodge model are both constant. Thus for nonlinear polymer rheological modeling, although the Lodge model represents a significant improvement over the generalized Newtonian and the GLVE fluids, it still fails to predict some important observed behavior such as shear-thinning and nonzero second normal-stress coefficients.

We can illustrate the objectivity of the Lodge model if we return to the turntable example discussed in Chapter 8. In that example we calculated, in two different coordinate systems, the stresses predicted by the GLVE fluid in a shear flow: one coordinate system rotated slowly with the flow cell on a turntable, and the other coordinate system was a fixed laboratory reference frame. As you may recall, the GLVE predictions based on the fixed laboratory reference frame erroneously indicated that the stresses in the rotating shear flow should depend on the angular velocity Ω of the turntable.

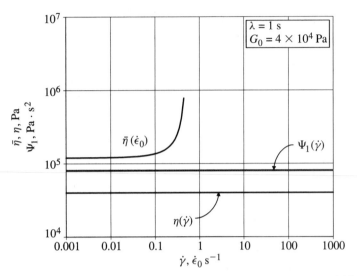

Figure 9.7 Predictions of the Lodge equation or the upper convected Maxwell model in shear and extension. Both η and Ψ_1 are independent of shear rate, whereas $\bar{\eta}$ varies with the elongation rate and becomes unbounded at $\dot{\epsilon}_0 = \frac{1}{2}\lambda$.

In this chapter we have developed new strain measures to fix the rotational problems associated with $\underline{\underline{\gamma}}(t, t')$, and the Lodge equation is the result. We now might ask, will the Lodge equation pass the rotating turntable objectivity test that the GLVE model failed? Our development of $\underline{\underline{C}}^{-1}$ was predicated on fixing this problem; we will now carry out this calculation and subject the Lodge model to the rotating reference-frame test.

Consider a shear flow carried out in a flow cell on a turntable that is slowly rotating at angular velocity Ω (see Figure 8.12). The reference frame $\bar{x}\bar{y}\bar{z}$ rotates at the speed of the turntable, and this coordinate system is a shear coordinate system for the flow, $\underline{v} = \dot{\gamma}_0 \bar{y} \hat{e}_{\bar{x}}$. The x, y, z coordinate system is stationary. To calculate the stress in the fluid with respect to the x, y, z and $\bar{x}$, $\bar{y}$, $\bar{z}$ coordinate systems using the Lodge equation, we must evaluate $\underline{\underline{C}}^{-1} = (\underline{\underline{F}}^{-1})^T \cdot \underline{\underline{F}}^{-1}$ in each coordinate system. To evaluate the Finger tensor $\underline{\underline{C}}^{-1}$, we first need the inverse deformation tensor $\underline{\underline{F}}^{-1}$. We can calculate $\underline{\underline{F}}^{-1}$ if we know the displacement function that relates the particle positions at time t, given by the vector $\underline{r}$, to their initial positions at time t', given by $\underline{r}'$,

$$\underline{\underline{F}}^{-1} = \frac{\partial \underline{r}}{\partial \underline{r}'} = \begin{pmatrix} \frac{\partial x}{\partial x'} & \frac{\partial y}{\partial x'} & \frac{\partial z}{\partial x'} \\ \frac{\partial x}{\partial y'} & \frac{\partial y}{\partial y'} & \frac{\partial z}{\partial y'} \\ \frac{\partial x}{\partial z'} & \frac{\partial y}{\partial z'} & \frac{\partial z}{\partial z'} \end{pmatrix}_{xyz} \tag{9.165}$$

We begin by calculating $\underline{r}$ and $\underline{r}'$, and hence $\underline{\underline{F}}^{-1}$ and $\underline{\underline{C}}^{-1}$, in the rotating $\bar{x}$, $\bar{y}$, $\bar{z}$ coordinate system.

In the $\bar{x}$, $\bar{y}$, $\bar{z}$ coordinate system shear flow is occurring, and we can write particle positions at time t with respect to particle positions at time t' in a straightforward manner:

$$\bar{x} = \bar{x}' + \dot{\gamma}_0 \bar{y}(t - t') \tag{9.166}$$

$$\bar{y} = \bar{y}' \tag{9.167}$$

$$\bar{z} = \bar{z}' \tag{9.168}$$

In this rotating frame we can now calculate $\underline{\underline{C}}^{-1}$ from its definition, and subsequently we obtain the Lodge equation:

$$\underline{\underline{C}}^{-1} = \begin{pmatrix} 1 + \gamma^2 & \gamma & 0 \\ \gamma & 1 & 0 \\ 0 & 0 & 1 \end{pmatrix}_{\bar{x}\bar{y}\bar{z}} \tag{9.169}$$

Lodge stress tensor, turntable example, $\bar{x}, \bar{y}, \bar{z}$ coordinate system

$$\underline{\underline{\tau}}(t) = -\int_{-\infty}^{t} \frac{\eta_0}{\lambda^2} e^{-\frac{(t-t')}{\lambda}} \begin{pmatrix} 1 + \gamma^2 & \gamma & 0 \\ \gamma & 1 & 0 \\ 0 & 0 & 1 \end{pmatrix}_{\bar{x}\bar{y}\bar{z}} dt' \tag{9.170}$$

To calculate $\underline{\underline{\tau}}$ in the stationary x, y, z coordinate system we substitute the relationships between coordinates $\bar{x}, \bar{y}, \bar{z}$ and x, y, z [(Equations (8.219)–(8.221)] into Equations (9.166)–(9.168) and solve for the displacement functions written in the x, y, z coordinate system:

$$(y - y_0) = (y' - y_0)\left[C'C + S'S + SC'\gamma\right] + (x' - x_0)\left[-CS' + SC' - SS'\gamma\right] \tag{9.171}$$

$$(x - x_0) = (y' - y_0)\left[-SC' + CS' + CC'\gamma\right] + (x' - x_0)\left[SS' + CC' - CS'\gamma\right] \tag{9.172}$$

$$z = z' \tag{9.173}$$

where $S = \sin \Omega t$, $S' = \sin \Omega t'$, $C = \cos \Omega t$, $C' = \cos \Omega t'$, and $\gamma = \dot{\gamma}_0(t - t')$. With the functions $\underline{r}(\underline{r}')$ now known, we can evaluate $\underline{\underline{F}}^{-1}$ and then $\underline{\underline{C}}^{-1}$. Although the algebra is involved, the result is quite simple:

$$\underline{\underline{C}}^{-1}(t', t) = \begin{pmatrix} 1 - 2CS\gamma + C^2\gamma^2 & (C^2 - S^2)\gamma + SC\gamma^2 & 0 \\ (C^2 - S^2)\gamma + SC\gamma^2 & 1 + 2CS\gamma + S^2\gamma^2 & 0 \\ 0 & 0 & 1 \end{pmatrix}_{xyz} \tag{9.174}$$

We can then obtain an expression for $\underline{\underline{\tau}}$ written with respect to the stationary frame by inserting this version of the Finger tensor into the Lodge equation:

Lodge stress tensor, turntable example, x, y, z coordinate system

$$\underline{\underline{\tau}}(t) = -\int_{-\infty}^{t} \frac{\eta_0}{\lambda^2} e^{-\frac{(t-t')}{\lambda}} \times$$
$$\begin{pmatrix} 1 - 2CS\gamma + C^2\gamma^2 & (C^2 - S^2)\gamma + SC\gamma^2 & 0 \\ (C^2 - S^2)\gamma + SC\gamma^2 & 1 + 2CS\gamma + S^2\gamma^2 & 0 \\ 0 & 0 & 1 \end{pmatrix}_{xyz} dt' \tag{9.175}$$

Now, to check for objectivity, we wish to calculate the viscosity using the stationary-frame expression for stress [Equation (9.175)]. We only know how to calculate viscosity in steady shear when stress is written in a shear coordinate system, however. We must use a coordinate system in which

$$
\underline{\dot{\gamma}} = \begin{pmatrix} 0 & \dot{\gamma}_0 & 0 \\ \dot{\gamma}_0 & 0 & 0 \\ 0 & 0 & 0 \end{pmatrix}_{123}
\tag{9.176}
$$

The stationary frame becomes a shear frame when $t = 0$. Setting $t = 0$, we then calculate the stress to be

$$
\underline{\tau}(0) = - \int_{-\infty}^{0} \frac{\eta_0}{\lambda^2} e^{\frac{t'}{\lambda}} \begin{pmatrix} 1 + \gamma^2 & \gamma & 0 \\ \gamma & 1 & 0 \\ 0 & 0 & 1 \end{pmatrix}_{xyz} dt'
\tag{9.177}
$$

which is independent of Ω, as it should be, and which is identical to the stress at $t = 0$ calculated from the Lodge equation expressed in the rotating shear system, $\bar{x}, \bar{y}, \bar{z}$ [Equation (9.170)]. Thus the Lodge equation passes this test of material objectivity.

We close this section with another example involving the Lodge equation. This time we illustrate how to calculate some unsteady-state material functions.

EXAMPLE

Calculate the material functions $\bar{\eta}_{P_1}^+ (t, \dot{\epsilon}_0)$ and $\bar{\eta}_{P_2}^+ (t, \dot{\epsilon}_0)$ predicted by the Lodge equation for startup of steady planar extensional flow.

SOLUTION

As usual when calculating material functions, we begin with the kinematics for the flow in question, startup of planar elongation.

$$
\underline{v} = \begin{pmatrix} -\dot{\epsilon}(t)x_1 \\ 0 \\ \dot{\epsilon}(t)x_3 \end{pmatrix}_{123}
\tag{9.178}
$$

$$
\dot{\epsilon}(t) = \begin{cases} 0 & t < 0 \\ \dot{\epsilon}_0 & t \geq 0 \end{cases}
\tag{9.179}
$$

where $\dot{\epsilon}_0$ is positive. To evaluate the Lodge constitutive equation we need the Finger tensor, which for planar elongation can be calculated to be

$$
\underline{C}^{-1}(t', t) = \begin{pmatrix} e^{-2\epsilon} & 0 & 0 \\ 0 & 1 & 0 \\ 0 & 0 & e^{2\epsilon} \end{pmatrix}_{123}
\tag{9.180}
$$

where

$$\epsilon(t', t) = \int_{t'}^{t} \dot{\epsilon}(t'') \, dt'' \tag{9.181}$$

We can now write the Lodge equation for this flow:

$$\underline{\underline{\tau}} = -\int_{-\infty}^{t} \frac{\eta_0}{\lambda^2} e^{\frac{-(t-t')}{\lambda}} \begin{pmatrix} e^{-2\epsilon} & 0 & 0 \\ 0 & 1 & 0 \\ 0 & 0 & e^{2\epsilon} \end{pmatrix}_{123} dt' \tag{9.182}$$

To calculate ϵ we must evaluate the integral in Equation (9.181) using the function $\dot{\epsilon}$ in Equation (9.179), which is sketched as a function of the dummy variable t'' in Figure 9.8. The time t in the limit of the integral for ϵ is the current time, or the time at which we seek to calculate $\underline{\underline{\tau}}$, and it is therefore greater than zero. The time t' in the lower limit is some time in the past, which can be positive or negative. We see then that we must write the integral for calculating ϵ differently, depending on the value of t'.

$$\epsilon(t', t) = \begin{cases} \int_{0}^{t} \dot{\epsilon}_0 \, dt'' = \dot{\epsilon}_0 t & t' < 0 \\ \int_{t'}^{t} \dot{\epsilon}_0 \, dt'' = \dot{\epsilon}_0(t - t') & t' \geq 0 \end{cases} \tag{9.183}$$

This expression for ϵ is then used in Equation (9.182) to calculate the stress components. The results are

$$\tau_{11} = -\frac{\eta_0}{C\lambda} \left(2\dot{\epsilon}_0 \lambda e^{-\frac{tC}{\lambda}} + 1 \right) \tag{9.184}$$

$$\tau_{22} = -\frac{\eta_0}{\lambda} \tag{9.185}$$

$$\tau_{33} = \frac{\eta_0}{A\lambda} \left(2\dot{\epsilon}_0 \lambda e^{-\frac{tA}{\lambda}} - 1 \right) \tag{9.186}$$

where $A \equiv 1 - 2\dot{\epsilon}_0\lambda$ and $C \equiv 1 + 2\dot{\epsilon}_0\lambda$. Now we calculate the material functions $\bar{\eta}_{P_1}^+(t, \dot{\epsilon}_0)$ and $\bar{\eta}_{P_2}^+(t, \dot{\epsilon}_0)$ from their definitions:

$$\bar{\eta}_{P_1}^+ = \frac{-(\tau_{33} - \tau_{11})}{\dot{\epsilon}_0} = \frac{2\eta_0}{AC} \left(2 - Ae^{-\frac{Ct}{\lambda}} - Ce^{-\frac{At}{\lambda}} \right) \tag{9.187}$$

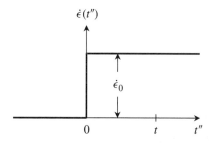

Figure 9.8 Deformation-rate function $\dot{\epsilon}(t'')$ for startup of the planar elongation experiment.

$$\bar{\eta}_{P_2}^+ = \frac{-(\tau_{22} - \tau_{11})}{\dot{\epsilon}_0} = \frac{2\eta_0}{C}\left(1 - e^{-\frac{Ct}{\lambda}}\right) \tag{9.188}$$

Note that $\bar{\eta}_{P_1}^+$ becomes unbounded for $\dot{\epsilon}_0 = 1/2\lambda$.

9.2.2 DIFFERENTIAL LODGE EQUATION—UPPER CONVECTED MAXWELL MODEL

The Lodge equation is better known in its differential form. To convert the integral equation to a differential equation, we calculate $d\underline{\underline{\tau}}/dt$ by differentiating Equation (9.139), applying the Leibnitz rule to the derivative of the integral. Table 9.1 is used to evaluate derivatives of $\underline{\underline{C}}^{-1}$.

$$-\frac{d\underline{\underline{\tau}}}{dt} = \frac{d}{dt}\int_{-\infty}^{t}\frac{\eta_0}{\lambda^2}e^{-\frac{(t-t')}{\lambda}}\underline{\underline{C}}^{-1}(t',t)\,dt' \tag{9.189}$$

$$= \int_{-\infty}^{t}\frac{\partial}{\partial t}\left[\frac{\eta_0}{\lambda^2}e^{-\frac{(t-t')}{\lambda}}\underline{\underline{C}}^{-1}(t',t)\right]dt' + \left[\frac{\eta_0}{\lambda^2}e^{-\frac{(t-t')}{\lambda}}\underline{\underline{C}}^{-1}(t',t)\right]\Big|_{t'=t} \tag{9.190}$$

Since $\underline{\underline{C}}^{-1}(t,t) = \underline{\underline{I}}$, the second term simplifies, and we can now expand the derivative of the product inside the integral to obtain

$$-\frac{d\underline{\underline{\tau}}}{dt} = \frac{\eta_0}{\lambda^2}\underline{\underline{I}} + \int_{-\infty}^{t}\frac{\eta_0}{\lambda^2}e^{-\frac{(t-t')}{\lambda}}\frac{\partial\underline{\underline{C}}^{-1}(t,t')}{\partial t}\,dt' + \frac{1}{\lambda}\underline{\underline{\tau}} \tag{9.191}$$

where we have used Equation (9.139) in writing the last term. Substituting for $\partial\underline{\underline{C}}^{-1}/\partial t$ from Table 9.1 results in [again using Equation (9.139) where appropriate]

$$-\frac{d\underline{\underline{\tau}}}{dt} = \frac{\eta_0}{\lambda^2}\underline{\underline{I}} - (\nabla\underline{v})^T\cdot\underline{\underline{\tau}} - \underline{\underline{\tau}}\cdot\nabla\underline{v} + \frac{1}{\lambda}\underline{\underline{\tau}} \tag{9.192}$$

$$-\frac{\eta_0}{\lambda^2}\underline{\underline{I}} = \left[\frac{d\underline{\underline{\tau}}}{dt} - (\nabla\underline{v})^T\cdot\underline{\underline{\tau}} - \underline{\underline{\tau}}\cdot\nabla\underline{v}\right] + \frac{1}{\lambda}\underline{\underline{\tau}} \tag{9.193}$$

The terms in square brackets have a special meaning [138] (discussed in Section 9.3) and are denoted by $\overset{\triangledown}{\underline{\underline{\tau}}}$, where the symbol $^\triangledown$ indicates the following operations on an arbitrary tensor $\underline{\underline{A}}$:

$$\begin{array}{cc}\text{Upper convected} \\ \text{derivative of } \underline{\underline{A}}\end{array}\qquad\boxed{\overset{\triangledown}{\underline{\underline{A}}} \equiv \frac{D\underline{\underline{A}}}{DT} - (\nabla\underline{v})^T\cdot\underline{\underline{A}} - \underline{\underline{A}}\cdot\nabla\underline{v}} \tag{9.194}$$

where $D\underline{\underline{A}}/Dt = d\underline{\underline{A}}/dt$ (see Section 2.6.3). The differential Lodge equation then becomes

$$\underline{\underline{\tau}} + \lambda\overset{\triangledown}{\underline{\underline{\tau}}} = -\frac{\eta_0}{\lambda}\underline{\underline{I}} \tag{9.195}$$

This equation usually appears in a slightly different form that we can derive by redefining $\underline{\underline{\tau}}$ with a constant offset in the diagonal components [138]. We can make this adjustment without any change to the significance of $\underline{\underline{\tau}}$ since for an incompressible fluid $\underline{\underline{\tau}}$ is only known to within an isotropic constant (see Section 4.5). If the new stress tensor is defined as $\underline{\underline{\varsigma}} \equiv \underline{\underline{\tau}} + (\eta_0/\lambda)\underline{\underline{I}}$, the differential Lodge equation becomes

$$\underline{\underline{\varsigma}} - \frac{\eta_0}{\lambda}\underline{\underline{I}} + \lambda\overset{\triangledown}{\underline{\underline{\varsigma}}} - \eta_0\overset{\triangledown}{\underline{\underline{I}}} = -\frac{\eta_0}{\lambda}\underline{\underline{I}} \tag{9.196}$$

$$\underline{\underline{\varsigma}} + \lambda\overset{\triangledown}{\underline{\underline{\varsigma}}} = -\eta_0\dot{\underline{\underline{\gamma}}} \tag{9.197}$$

where we have used $\overset{\triangledown}{\underline{\underline{I}}} = -\dot{\underline{\underline{\gamma}}}$ [see Equation (9.194)].

We arrive at the usual form of the differential Lodge model, which is more commonly known as the upper convected Maxwell model since it resembles the Maxwell equation:

Upper convected
Maxwell model

$$\boxed{\underline{\underline{\tau}} + \lambda\overset{\triangledown}{\underline{\underline{\tau}}} = -\eta_0\dot{\underline{\underline{\gamma}}}} \tag{9.198}$$

We have returned to using $\underline{\underline{\tau}}$ for the stress tensor. The derivative denoted by $^\triangledown$ and defined by Equation (9.194) is called the *upper convected derivative*, and the meaning of this quantity will be discussed in Section 9.3. The predictions of the upper convected Maxwell model match those for the integral Lodge equation.

9.2.3 OTHER LODGE-LIKE EQUATIONS

In developing his constitutive equation, Lodge chose the Finger tensor as the strain measure, but a Maxwell model using the Cauchy tensor could also be defined:

Cauchy–Maxwell
equation

$$\boxed{\underline{\underline{\tau}}(t) = + \int_{-\infty}^{t} \frac{\eta_0}{\lambda^2} e^{-\frac{(t-t')}{\lambda}} \underline{\underline{C}}(t, t')\, dt'} \tag{9.199}$$

The choice of strain measure hinges entirely on the accuracy of the predictions of the resulting constitutive equation. The use of the Cauchy tensor in a constitutive equation results in the prediction of nonzero second normal-stress differences, a key shortcoming of the Lodge equation and of other constitutive equations based on the Finger tensor alone. The Cauchy–Maxwell equation [Equation (9.199)] overpredicts the magnitude of second normal-stress-difference effects, however, and therefore this model is not generally used. A table of the predictions of the Cauchy–Maxwell equation is given in Appendix D.

We can create a more general version of the Lodge equation by fixing the frame variance of the GLVE model rather than of the simple Maxwell model. If we replace $\underline{\underline{\gamma}}(t, t')$ in Equation (9.11) with $-\underline{\underline{C}}^{-1}(t, t')$, we obtain the Lodge rubberlike liquid equation.

Lodge rubberlike
liquid model

$$\boxed{\underline{\underline{\tau}}(t) = - \int_{-\infty}^{t} M(t - t')\, \underline{\underline{C}}^{-1}(t', t)\, dt'} \tag{9.200}$$

The memory function $M(t - t')$ that appears in the Lodge rubberlike liquid equation was originally introduced in Equation 9.12. By replacing the single-relaxation-time memory function in the Lodge equation [function in square brackets in Equation (9.139)] with a general function $M(t - t')$, we improve the Lodge model's ability to predict linear viscoelastic behavior of systems. One common choice for $M(t - t')$ is the memory function for the generalized Maxwell model. The definition of the memory function is

$$M(t - t') \equiv \frac{\partial G(t - t')}{\partial t'} \tag{9.201}$$

For the generalized Maxwell model,

$$G(t - t') = \sum_{k=1}^{N} \left(\frac{\eta_k}{\lambda_k} \right) e^{\frac{-(t-t')}{\lambda_k}} \tag{9.202}$$

and therefore

Memory function for generalized Maxwell model

$$M(t - t') = \sum_{k=1}^{N} \left(\frac{\eta_k}{\lambda_k^2} \right) e^{\frac{-(t-t')}{\lambda_k}} \tag{9.203}$$

With this choice for memory function, the Lodge rubberlike liquid model can accurately predict linear viscoelastic behavior.

The Lodge-like models we have discussed up to now are quasi-linear models, that is, they are fixups of constitutive models based on equations linear in τ_{21} and $\dot{\gamma}_{21}$, but they are nonlinear because the strain measures $\underline{\underline{C}}^{-1}$ and $\underline{\underline{C}}$ are nonlinear in $\dot{\gamma}_{21}$. Other empirical modifications of the Lodge model are possible, particularly those producing nonlinear models. For example, we could replace $M(t - t')$ with a function of both time and invariants of the rate-of-deformation tensor $\underline{\underline{\dot{\gamma}}}$. Alternatively we could propose an integral equation that would involve both the Finger and the Cauchy tensors. These types of nonlinear integral equations will be discussed in Section 9.4.1.3.

9.3 Convected Derivatives

We saw in the last section that by changing the strain measure used in the Maxwell equation, we arrive at a modified Maxwell equation that fixes the problem of frame variance encountered in the original model and that predicts nonzero first normal-stress differences in steady shear. This represents an important first step into the domain of finite-strain constitutive models.

Also in the last section we showed that using the Finger tensor in the integral Lodge model corresponded to replacing the partial time derivative in the differential Maxwell model with a new derivative, the upper convected derivative. In the history of the development of constitutive equations based on continuum mechanics, researchers have followed both the path of developing new integral equations for $\underline{\underline{\tau}}$ based on adopting different strain measures and the path of developing new differential equations for $\underline{\underline{\tau}}$ based on proposing new time derivatives (Figure 9.9). To understand this latter path, we need some knowledge of the physical basis of convected derivatives; this is the subject of this section. We first

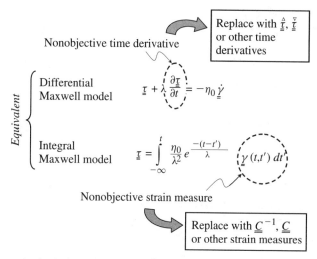

Figure 9.9 Two paths for improving constitutive equations based on continuum mechanics: new time derivatives yield new differential constitutive equations; new strain measures yield new integral constitutive equations.

begin with a discussion of nonorthonormal bases. Next we write the coefficients of the stress tensor in a nonorthonormal basis that moves and deforms with the flowing continuum. The upper convected derivative will be shown to be related to the time derivative of convected stress components. We will also see that other types of convected derivatives are admissible and lead to different, objective, differential constitutive equations, such as the differential version of the Cauchy–Maxwell equation. This section is highly technical and can be omitted on first reading. For readers more interested in advanced constitutive modeling than in the justification of the use of convected derivatives, we recommend you skip ahead to Section 9.4.

9.3.1 NONORTHONORMAL BASES

To delve into the meaning of convected derivatives, we must familiarize ourselves with the mechanics of nonorthonormal bases. Up until now the basis vectors we have used to express vector and tensor coefficients have been mutually orthogonal and of unit length. In the simplest orthonormal system, the Cartesian coordinate system, the basis vectors are also independent of position. All of the coordinate systems discussed have been fixed in the lab frame, that is, stationary (except in the example at the end of Chapter 8).

As pointed out in Chapter 2, however, basis vectors need not be orthonormal or stationary. The only requirements are that the basis vectors be linearly independent and nonzero. The convected derivative that arose in the discussion of the upper convected Maxwell model in the previous section is a special derivative, which has meaning in a coordinate system that is embedded in a flowing medium and that moves and deforms with the continuum. In such convected coordinate systems, even if the basis vectors start out as orthonormal basis vectors, after some flow, the material-embedded basis vectors will no longer be mutually orthogonal, nor will they be of unit length. The fact that material basis vectors change with time does not invalidate them as basis vectors as long as they remain linearly independent and nonzero. The first criterion of linear independence is easily satisfied by convected coordinates, since vectors that are embedded in a deforming fluid cannot become linearly dependent, as this would require that two particles of mass occupy the same space at the same time. The second criterion, that the basis vectors be nonzero, is also easily fulfilled by embedded vectors since mass is conserved—an embedded basis vector originally chosen to be nonzero cannot flow to be a zero vector if mass is conserved. Vectors embedded in a flowing material are thus well suited to serve as time-varying basis vectors for flowing media.

We begin then with the basics of working with nonorthonormal bases. When using nonorthonormal basis vectors, we must take care when calculating products between vectors. When we wish to express the dot product between two vectors $\underline{v}$ and $\underline{u}$ in terms of their Cartesian coordinates, the procedure is familiar:

$$\underline{v} \cdot \underline{u} = v_i \hat{e}_i \cdot u_k \hat{e}_k \tag{9.204}$$

$$= v_i u_k \, \delta_{ik} \tag{9.205}$$

$$= v_k u_k \tag{9.206}$$

The dot products formed between the Cartesian basis vectors give the Kronecker delta function since they are mutually orthogonal unit vectors. We can equally well write $\underline{v}$ and $\underline{u}$ in terms of nonorthonormal basis vectors $\underline{b}_{(1)}$, $\underline{b}_{(2)}$, and $\underline{b}_{(3)}$ and calculate their dot product:

$$\underline{v} \cdot \underline{u} = v_m \underline{b}_{(m)} \cdot u_p \underline{b}_{(p)} \tag{9.207}$$

$$= v_m u_p \underline{b}_{(m)} \cdot \underline{b}_{(p)} \tag{9.208}$$

Since the basis vectors $\underline{b}_{(k)}$ are not orthonormal, however, we cannot simplify the product any further without first knowing the specifics of the basis vectors $\underline{b}_{(k)}$. For the product $\underline{b}_{(m)} \cdot \underline{b}_{(p)}$ we can use the short-hand notation b_{mp}; the nine scalars b_{mp} are called the *metric coefficients* of the basis. Note that $b_{mp} = b_{pm}$ since the dot product is commutative.

There do exist special vectors that when dotted with nonorthonormal basis vectors result in the Kronecker delta [7]. These vectors are called the *reciprocal basis vectors* $\underline{b}^{(1)}$, $\underline{b}^{(2)}$, and $\underline{b}^{(3)}$, where we have placed the indices identifying the vectors in the superscript position to indicate their inverse relationship to the original subscripted vectors:

$$\underline{b}_{(p)} \cdot \underline{b}^{(j)} = \delta_{pj} \tag{9.209}$$

The reciprocal vectors are formed as follows:

$$\underline{b}^{(1)} \equiv \frac{\underline{b}_{(2)} \times \underline{b}_{(3)}}{B} \tag{9.210}$$

$$\underline{b}^{(2)} \equiv \frac{\underline{b}_{(3)} \times \underline{b}_{(1)}}{B} \tag{9.211}$$

$$\underline{b}^{(3)} \equiv \frac{\underline{b}_{(1)} \times \underline{b}_{(2)}}{B} \tag{9.212}$$

$$B \equiv \underline{b}_{(1)} \cdot \underline{b}_{(2)} \times \underline{b}_{(3)} \neq 0 \tag{9.213}$$

Notice that the indices associated with the vectors rotate through the three positions 123, 231, 312 in the definitions. The quantity $\underline{a} \cdot \underline{b} \times \underline{c}$ is called the *triple product* of the three vectors $\underline{a}, \underline{b}$, and $\underline{c}$. It can be shown that the triple product of any three vectors is equal to the volume of the parallelepiped formed by the vectors. Thus the triple product in Equation (9.213) will be nonzero as long as the basis vectors themselves are nonzero and noncoplanar.

From the use of the cross product and the triple product in the definitions of the reciprocal basis vectors, one can see that $\underline{b}_{(i)} \cdot \underline{b}^{(j)} = \underline{b}^{(i)} \cdot \underline{b}_{(j)} = \delta_{ij}$. We can also show that

$$\underline{b}^{(1)} \cdot \underline{b}^{(2)} \times \underline{b}^{(3)} = \frac{1}{B} \neq 0 \tag{9.214}$$

and thus the reciprocal basis vectors are not coplanar and may also be used as basis vectors. The reciprocal basis vectors are not, in general, mutually orthogonal, and they are not of unit length, and hence there is a set of reciprocal metric coefficients associated with the reciprocal basis vectors $\underline{b}^{(i)} \cdot \underline{b}^{(j)} \equiv b^{ij}$. For the Cartesian basis $\hat{e}_1, \hat{e}_2, \hat{e}_3$, the reciprocal vectors are equal to the basis vectors themselves, and $b_{ij} = b^{ij} = \delta_{ij}$, as can be seen from the preceding definitions.

9.3.2 CONVECTED COORDINATES

The usual way of describing fluid particle motion in a flow is to reference the motion to a fixed laboratory coordinate system and to keep track of how the coefficients of fluid particles change with time. The convected coordinate approach is in some senses opposite to this. In this approach we will hold the coordinates of the fluid particles constant and have the coordinate system (i.e., the basis vectors) change with time. In the convected approach, we wish to use coordinates that are permanently associated with particular particles of fluid (Figure 9.10). For convenience we choose the convected coordinate system to be a mutually orthogonal, equally spaced three-dimensional grid at time t, which is our reference time.

Because of flow, fluid particles separate at various rates at different times t'. In the convected coordinate approach we describe the positions of each particle at all times t' by using the same convected coordinates $\hat{x}^1, \hat{x}^2$, and $\hat{x}^3$. In writing the convected coordinates we place the indices in the superscript position following an accepted convention [7, 26]. We will have more to say about this notation a bit later. At times other than the reference time t the lines of constant $\hat{x}^1, \hat{x}^2$, and $\hat{x}^3$ will not be straight and will not form a uniform grid. Rather, the deformation of the coordinate grid will map the deformation of the fluid. The basis vectors that are associated with the convected coordinates will not be mutually orthogonal, nor of unit length. To define these basis vectors in a consistent fashion, we return briefly to a discussion of Cartesian and curvilinear coordinates.

The position of a particle with respect to an origin can be identified by the vector $\underline{r}$, which in Cartesian coordinates is given by

$$\underline{r} = x\hat{e}_x + y\hat{e}_y + z\hat{e}_z \tag{9.215}$$

The vector position $\underline{r}$ is a function of coordinates x, y, and z, and its differential $d\underline{r}$ can be written as

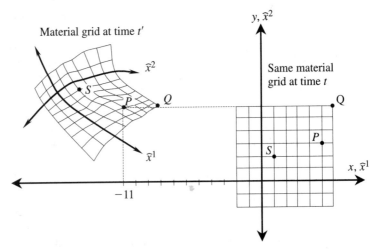

Figure 9.10 Two-dimensional schematic of the proposed convected coordinate system at two different times. The x, y, z coordinate system is fixed in space, whereas the $\hat{x}^1, \hat{x}^2, \hat{x}^3$ coordinate system moves with the flow. At time t', point P has convected coordinates $\hat{x}^1 = 5, \hat{x}^2 = 3$ and Cartesian coordinates $x = -11$, $y = 6$. At time t, convected coordinates are the same since they are attached to the material point, but Cartesian coordinates have changed to $x = 5$, $y = 3$. The convected coordinate system is chosen so that it coincides with the Cartesian system at $t' = t$.

$$dr = \frac{\partial r}{\partial x}dx + \frac{\partial r}{\partial y}dy + \frac{\partial r}{\partial z}dz \tag{9.216}$$

$$= \hat{e}_x\,dx + \hat{e}_y\,dy + \hat{e}_z\,dz \tag{9.217}$$

In cylindrical coordinates[8] $\bar{r}, \theta, z$, the position vector r may be written as

$$r = \bar{r}\hat{e}_{\bar{r}} + z\hat{e}_z \tag{9.218}$$

with its differential written as

$$dr = \frac{\partial r}{\partial \bar{r}}d\bar{r} + \frac{\partial r}{\partial \theta}d\theta + \frac{\partial r}{\partial z}dz \tag{9.219}$$

Recall that the cylindrical basis vectors are a function of position. Therefore to evaluate the partial derivatives in the expression for the differential [Equation (9.219)] we write the cylindrical basis vectors in terms of spatially independent (e.g., Cartesian) coordinates:

$$\hat{e}_{\bar{r}} = \cos\theta\,\hat{e}_x + \sin\theta\,\hat{e}_y \tag{9.220}$$

$$\hat{e}_\theta = -\sin\theta\,\hat{e}_x + \cos\theta\,\hat{e}_y \tag{9.221}$$

$$\hat{e}_z = \hat{e}_z \tag{9.222}$$

---

[8] In this section, to distinguish the magnitude of the position vector r from the cylindrical coordinate $\bar{r}$, we follow Bird et al. [26] and call the cylindrical coordinate $\bar{r}$.

We then arrive at a more familiar result for $d\underline{r}$:

$$d\underline{r} = d\bar{r}\,\hat{e}_{\bar{r}} + \bar{r}\,d\theta(-\sin\theta\,\hat{e}_x + \cos\theta\,\hat{e}_y) + dz\,\hat{e}_z \tag{9.223}$$

$$= d\bar{r}\,\hat{e}_{\bar{r}} + \bar{r}\,d\theta\,\hat{e}_\theta + dz\,\hat{e}_z \tag{9.224}$$

In spherical coordinates, the analogous procedure results in the expected differential element $d\underline{r}$ in that system (see Problem 9.41).

To define the convected basis vectors we will carry out a procedure analogous to that described for Cartesian and cylindrical coordinate systems. The vector $\underline{r}$ can be expressed as a function of the convected coordinates $\hat{x}^1$, $\hat{x}^2$, and $\hat{x}^3$, and the differential $d\underline{r}$ may be written as

$$d\underline{r} = \frac{\partial\underline{r}}{\partial\hat{x}^1}d\hat{x}^1 + \frac{\partial\underline{r}}{\partial\hat{x}^2}d\hat{x}^2 + \frac{\partial\underline{r}}{\partial\hat{x}^3}d\hat{x}^3 \tag{9.225}$$

We choose therefore to define the basis vectors $\underline{g}_{(i)}$ for our convected coordinate system as

$$\underline{g}_{(1)} \equiv \frac{\partial\underline{r}}{\partial\hat{x}^1} \tag{9.226}$$

$$\underline{g}_{(2)} \equiv \frac{\partial\underline{r}}{\partial\hat{x}^2} \tag{9.227}$$

$$\underline{g}_{(3)} \equiv \frac{\partial\underline{r}}{\partial\hat{x}^3} \tag{9.228}$$

Thus Equation (9.225) becomes

$$d\underline{r} = \sum_{i=1}^{3}\underline{g}_{(i)}\,d\hat{x}^i \tag{9.229}$$

The basis vectors $\underline{g}_{(i)}$ depend on convected coordinate position $\hat{x}^i$ as well as on time t' and are not, in general, mutually perpendicular, nor are they of unit length. The dot product between pairs of these base vectors will be denoted by g_{ij}, called the *metric* of this basis:

$$\underline{g}_{(i)} \cdot \underline{g}_{(j)} \equiv g_{ij} \tag{9.230}$$

The nonorthogonal convected basis vectors $\underline{g}_{(i)}$ have corresponding reciprocal vectors $\underline{g}^{(i)}$, defined in the usual way[9]:

$$\underline{g}^{(1)} \equiv \frac{\underline{g}_{(2)} \times \underline{g}_{(3)}}{\sqrt{g}} \tag{9.231}$$

$$\underline{g}^{(2)} \equiv \frac{\underline{g}_{(3)} \times \underline{g}_{(1)}}{\sqrt{g}} \tag{9.232}$$

---

[9] The choice of writing $\sqrt{g}$ for the value of the triple product in these definitions is arbitrary but standard [26]. The scalar g is equal to the determinant of the 3×3 matrix with elements g_{ij}.

$$\underline{g}^{(3)} \equiv \frac{\underline{g}_{(1)} \times \underline{g}_{(2)}}{\sqrt{g}} \tag{9.233}$$

$$\sqrt{g} \equiv \underline{g}_{(1)} \cdot \underline{g}_{(2)} \times \underline{g}_{(3)} \tag{9.234}$$

Any vector may be expressed in the coordinate systems formed by the convected basis vectors or the convected reciprocal basis vectors:

$$\underline{v} = v_i \hat{e}_i \tag{9.235}$$

$$= \widehat{v}^i \underline{g}_{(i)} \tag{9.236}$$

$$= \widehat{v}_i \underline{g}^{(i)} \tag{9.237}$$

The notation convention for nonorthonormal basis vectors is to match superscripted coefficients with subscripted basis vectors and vice versa [7, 26]. Thus in Equation (9.236) we write $\widehat{v}^j$ for the convected coordinates of $\underline{v}$ that are written with respect to the basis vectors $\underline{g}_{(j)}$. Vector coefficients of $\underline{v}$ written with respect to the reciprocal basis vectors $\underline{g}^{(k)}$ are written as $\widehat{v}_k$. The coefficients of $\underline{v}$ in the Cartesian system (or in stationary curvilinear systems), v_i, are called the *physical* components of $\underline{v}$. The coefficients in the convected basis system, $\widehat{v}^i$, are called the *contravariant* components, and those in the reciprocal basis, $\widehat{v}_i$, are called the *covariant* coefficients of $\underline{v}$. The dot products of the pairs of basis vectors, g_{ij}, are called the *covariant metric* coefficients, and the analogous quantities for the reciprocal base vectors, g^{ij}, are called the *contravariant metric* coefficients.[10] Also, the contravariant and covariant coefficients do not, in general, have the same units as $\underline{v}$ because the $\underline{g}^{(i)}$ and $\underline{g}_{(j)}$ can have units.

Tensor covariant and contravariant coefficients may also be written:

$$\underline{\underline{A}} = A_{ij} \hat{e}_i \hat{e}_j \tag{9.238}$$

$$= \widehat{A}_{ij} \underline{g}^{(i)} \underline{g}^{(j)} \tag{9.239}$$

$$= \widehat{A}^{ij} \underline{g}_{(i)} \underline{g}_{(j)} \tag{9.240}$$

To calculate, for example, the contravariant coefficients of a tensor $\underline{\underline{A}}$, we can pre- and postmultiply $\underline{\underline{A}}$ by the reciprocal basis vectors:

$$\underline{g}^{(p)} \cdot \underline{\underline{A}} \cdot \underline{g}^{(k)} = \underline{g}^{(p)} \cdot \widehat{A}^{jn} \underline{g}_{(j)} \underline{g}_{(n)} \cdot \underline{g}^{(k)} \tag{9.241}$$

$$= \widehat{A}^{jn} (\underline{g}^{(p)} \cdot \underline{g}_{(j)})(\underline{g}_{(n)} \cdot \underline{g}^{(k)}) \tag{9.242}$$

$$= \widehat{A}^{jn} \delta_{pj} \delta_{nk} \tag{9.243}$$

---

[10] The terms contravariant and covariant refer to the methods by which the coordinate coefficients transform from one coordinate system to another. The basis vectors and reciprocal basis vectors transform differently, because while the basis vectors are always associated with the same material particles (they are embedded vectors), the reciprocal basis vectors are not. To learn more about these concepts, see Appendix C.7 and Lodge [155].

$$= \widehat{A}^{pk} \tag{9.244}$$

For practice in using nonorthonormal bases, we will prove an identity involving the metric coefficients $g_{ij} \equiv g_{(i)} \cdot g_{(j)}$ and $g^{ij} \equiv g^{(i)} \cdot g^{(j)}$.

EXAMPLE

Prove that $\sum_{i=1}^{3} g_{ji} g^{ip} = \delta_{jp}$.

SOLUTION

We begin by writing an arbitrary vector $\underline{v}$ with respect to the two different bases. To make clear which indices are summed, we will place the summation signs explicitly:

$$\underline{v} = \sum_{i=1}^{3} \underline{g}_{(i)} \widehat{v}^i = \sum_{k=1}^{3} \underline{g}^{(k)} \widehat{v}_k \tag{9.245}$$

We now dot-multiply both sides of Equation (9.245) with one of the basis vectors $\underline{g}_{(p)}$:

$$\sum_{i=1}^{3} \underline{g}_{(p)} \cdot \underline{g}_{(i)} \widehat{v}^i = \sum_{k=1}^{3} \underline{g}_{(p)} \cdot \underline{g}^{(k)} \widehat{v}_k \tag{9.246}$$

$$\sum_{i=1}^{3} g_{pi} \widehat{v}^i = \sum_{k=1}^{3} \delta_{pk} \widehat{v}_k = \widehat{v}_p \tag{9.247}$$

Now we dot-multiply the same expression [Equation (9.245)] with $\underline{g}^{(p)}$:

$$\sum_{i=1}^{3} \underline{g}^{(p)} \cdot \underline{g}_{(i)} \widehat{v}^i = \sum_{k=1}^{3} \underline{g}^{(p)} \cdot \underline{g}^{(k)} \widehat{v}_k \tag{9.248}$$

$$\sum_{i=1}^{3} \delta_{pi} \widehat{v}^i = \widehat{v}^p = \sum_{k=1}^{3} g^{pk} \widehat{v}_k \tag{9.249}$$

We can combine Equations (9.247) and (9.249) to arrive at

$$\widehat{v}_p = \sum_{i=1}^{3} g_{pi} \widehat{v}^i \tag{9.250}$$

$$= \sum_{i=1}^{3} g_{pi} \sum_{k=1}^{3} g^{ik} \widehat{v}_k \tag{9.251}$$

$$= \sum_{k=1}^{3} \sum_{i=1}^{3} g_{pi} g^{ik} \widehat{v}_k \tag{9.252}$$

For the last equation to be true, the following relation must hold:

$$\delta_{pk} = \sum_{i=1}^{3} g_{pi} g^{ik} \tag{9.253}$$

Having defined convected coordinates and having practiced a bit with using these nonorthogonal basis vectors, we can now use convected coordinates to describe deformation and flow.

9.3.2.1 Deformation

Our goal in constitutive modeling is to develop mathematical expressions that relate the deformation of a material with the stress that is generated by the deformation [192, 193, 194, 195, 155, 25].[11] Deformation is the change in shape of a body. To describe change in shape, or strain, it is sufficient to know the change in the distances between every pair of particles in the body. When solving flow problems in fixed orthonormal coordinates, we have kept track of the location of every particle as a function of time. If we know the location of every particle at every time, we know the shape, at every instant, of the body made up of the particles. We also have additional information that does not pertain to shape, however; namely, we know the orientation of the body as a function of time.

To just describe shape, we must focus on the separations of particles as a function of time. Consider two particles S and W whose positions relative to the origin O of the convected coordinate system (another material point) are given by vectors $\underline{w}$ and $\underline{s}$ (Figure 9.11). We can write these vectors in the convected coordinate system $\underline{g}_{(i)}$ as follows:

$$\underline{w} = \widehat{w}^p \underline{g}_{(p)} \tag{9.254}$$

$$\underline{s} = \widehat{s}^m \underline{g}_{(m)} \tag{9.255}$$

The vector between S and W is just $\underline{w} - \underline{s}$, and the distance between these two points is the magnitude of that vector,

$$|\underline{w} - \underline{s}|^2 = (\underline{w} - \underline{s}) \cdot (\underline{w} - \underline{s}) \tag{9.256}$$

$$= (\widehat{w}^p - \widehat{s}^p) \underline{g}_{(p)} \cdot (\widehat{w}^k - \widehat{s}^k) \underline{g}_{(k)} \tag{9.257}$$

$$= (\widehat{w}^p - \widehat{s}^p)(\widehat{w}^k - \widehat{s}^k) \, \underline{g}_{(p)} \cdot \underline{g}_{(k)} \tag{9.258}$$

$$= (\widehat{w}^p - \widehat{s}^p)(\widehat{w}^k - \widehat{s}^k) g_{pk} \tag{9.259}$$

Recall that in our convected coordinates the coefficients of the vectors never change. The length and orientation of the basis vectors $\underline{g}_{(i)}$ do change. Since $(\widehat{w}^p - \widehat{s}^p)(\widehat{w}^k - \widehat{s}^k)$ is

---

[11] Our presentation here has been heavily influenced by the discussions of Oldroyd's work [192, 193, 194, 195] by Lodge [155] and Bird et al. [26].

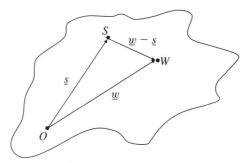

Figure 9.11 Schematic showing two material vectors. Points S, W, and the origin O are all material points. As the material deforms, the three vectors shown always connect the same pairs of material points.

constant, all of the shape information of the body must be contained in the time- and particle-varying metric of the basis, g_{pk}.

Since $g_{pk} \equiv \underline{g}_{(p)} \cdot \underline{g}_{(k)}$ and vector dot products are commutative, then $g_{pk} = g_{kp}$. There are nine coefficients g_{pk}, but due to symmetry, only six of these are independent. The original vectors indicating the positions of particles W and S contained three independent components each, which in combination yield nine independent quantities. The reduction from nine independent quantities in the combination of $\underline{w}$ and $\underline{s}$ to the six in g_{pk} reflects that the g_{pk} do not contain orientation information. The missing three independent quantities are the equivalent of the information contained in the coefficients of a unit vector pointing from S to W. We can see that if we construct a constitutive equation including only the g_{pk} to account for change in shape (strain), the resulting expression will be free from spurious effects of particle orientation. This is the great advantage of using convected coordinate systems.

9.3.2.2 Stress

We now turn to examine the properties of the coefficients of the total stress tensor $\underline{\underline{\Pi}}$, written in the convected coordinate system. Recall that the force per unit area on a surface is equal to the unit normal to that surface dotted with the stress tensor [Equation (3.58)]. Therefore the force $\underline{f}_{(i)}$,[12] on a surface perpendicular to $\underline{g}_{(i)}$ is given by (no summation implied)

$$\underline{f}_{(i)} = \frac{\underline{g}_{(i)}}{|\underline{g}_{(i)}|} \cdot \underline{\underline{\Pi}} \tag{9.260}$$

where the normalization is needed since $\underline{g}_{(i)}$ is not a unit vector. We can postmultiply this equation by $\underline{g}_{(j)}$ to obtain

---

[12] This notation $\underline{f}_{(i)}$ for the force on a surface with unit normal $\underline{g}_{(i)}/|\underline{g}_{(i)}|$ is unfortunate, but hard to avoid. Normally the parenthesized subscript is reserved for nonnormal basis vectors; this is the one exception.

$$\underline{f}_{(i)} \cdot \underline{g}_{(j)} = \frac{\underline{g}_{(i)}}{|\underline{g}_{(i)}|} \cdot \underline{\underline{\Pi}} \cdot \underline{g}_{(j)} \tag{9.261}$$

If we choose now to write $\underline{\underline{\Pi}}$ in the basis formed by the reciprocal basis vectors, expression (9.261) simplifies considerably. To avoid confusion, all summations will be written explicitly in this sequence:

$$\underline{f}_{(i)} \cdot \underline{g}_{(j)} = \frac{\underline{g}_{(i)}}{|\underline{g}_{(i)}|} \cdot \underline{\underline{\Pi}} \cdot \underline{g}_{(j)} \tag{9.262}$$

$$= \sum_{p=1}^{3}\sum_{k=1}^{3} \frac{1}{|\underline{g}_{(i)}|} \underline{g}_{(i)} \cdot \widehat{\Pi}_{pk} \, \underline{g}^{(p)} \underline{g}^{(k)} \cdot \underline{g}_{(j)} \tag{9.263}$$

$$= \sum_{p=1}^{3}\sum_{k=1}^{3} \frac{1}{\sqrt{\underline{g}_{(i)} \cdot \underline{g}_{(i)}}} \widehat{\Pi}_{pk} \, (\underline{g}_{(i)} \cdot \underline{g}^{(p)})(\underline{g}^{(k)} \cdot \underline{g}_{(j)}) \tag{9.264}$$

$$= \sum_{p=1}^{3}\sum_{k=1}^{3} \frac{1}{\sqrt{\underline{g}_{(i)} \cdot \underline{g}_{(i)}}} \widehat{\Pi}_{pk} \, \delta_{ip}\delta_{kj} \tag{9.265}$$

$$= \frac{1}{\sqrt{\underline{g}_{(i)} \cdot \underline{g}_{(i)}}} \widehat{\Pi}_{ij} \tag{9.266}$$

Solving now for $\widehat{\Pi}_{ij}$, the covariant coefficients of $\underline{\underline{\Pi}}$, we obtain

$$\widehat{\Pi}_{ij} = (\underline{f}_{(i)} \cdot \underline{g}_{(j)})\sqrt{\underline{g}_{(i)} \cdot \underline{g}_{(i)}} \tag{9.267}$$

From this relation we see that the covariant coefficients of $\underline{\underline{\Pi}}$ only depend on the dot products between vectors and not on the vectors themselves. These particular coefficients of $\underline{\underline{\Pi}}$ are therefore independent of instantaneous particle position and orientation. They depend only on the shape through scalar products among the basis vectors $\underline{g}_{(i)}$, and on forces through the scalar product of the force vector $\underline{f}_{(i)}$ with the $\underline{g}_{(j)}$. Since the $\widehat{\Pi}_{ij}$ are independent of particle orientation, we can safely avoid introducing into the constitutive equation unwanted dependence on solid-body rotation when taking the time derivative of this coefficient.

Oldroyd [192, 193, 194] formulated rules for constructing admissible constitutive equations based on the arguments presented [26]. The complete deformation history of a material is contained in the metric coefficients $g_{pk}(\widehat{x}^1, \widehat{x}^2, \widehat{x}^3, t')$, and the complete stress history is contained in the stress coefficients $\widehat{\Pi}_{pk}(\widehat{x}^1, \widehat{x}^2, \widehat{x}^3, t')$. Admissible constitutive equations must be able to be written in terms of only these two functions plus material constants. For nonisothermal problems the temperature field must be included as well. Similar arguments can be used to show that the *contravariant* coefficients of $\underline{\underline{\Pi}}$, $\widehat{\Pi}^{pk}$, and the contravariant metric coefficients g^{pk} may be used analogously to construct admissible constitutive equations [155, 192, 193, 194, 195].

9.3.3 RELATION TO CARTESIAN COORDINATES

In the last section we showed that time derivatives of contravariant and covariant coefficients of the stress tensor could be used to construct objective constitutive equations. Working directly in the convected coordinate system, however, is awkward and unfamiliar. We turn now to the task of relating quantities expressed in the convected coordinate systems to the more familiar Cartesian laboratory frame. We will see in this section that when the time derivatives of the contravariant coefficients of $\underline{\underline{\Pi}}$, $\partial \widehat{\Pi}^{ij}/\partial t$, are expressed in stationary coordinates, the upper convected derivative $\overset{\triangledown}{\underline{\underline{\Pi}}}$ results. An additional convected derivative (the lower convected time derivative, associated with the Cauchy tensor) results when the time derivatives of the covariant coefficients are used.

A tensor $\underline{\underline{\Pi}}$ may be written with respect to any basis, such as the stationary Cartesian basis or the basis formed by the convected basis vectors $\underline{g}_{(i)}$:

$$\underline{\underline{\Pi}} = \Pi_{ij}\hat{e}_i\hat{e}_j \tag{9.268}$$

$$= \widehat{\Pi}^{ij}\underline{g}_{(i)}\underline{g}_{(j)} \tag{9.269}$$

We would like to relate the physical coefficients Π_{ij} with the contravariant convective coordinates $\widehat{\Pi}^{ij}$. We have not specified much about the basis $\underline{g}_{(i)}$. We now choose that at time t (the reference time) the convected coordinate system $\underline{g}_{(i)}$ exactly coincides with the fixed Cartesian system $\hat{e}_i$. With this choice we know that the convected coordinates of each particle, $\widehat{x}^1$, $\widehat{x}^2$, and $\widehat{x}^3$, at all times t' will be numerically equal to their Cartesian coordinates at time t. This is the fundamental property of the convected basis: all of the change-of-orientation and -shape information for a particle is encoded in the changes in the basis vectors $\underline{g}_{(i)}$, allowing the convected coefficients $\widehat{x}^i$ that identify the particle to remain forever constant.

We can now relate the convected basis vectors to the familiar deformation tensors $\underline{\underline{F}}$ and $\underline{\underline{F}}^{-1}$. The basis vectors $\underline{g}_{(i)}$ at time t' are defined as

$$\underline{g}_{(i)} \equiv \frac{\partial \underline{r}'}{\partial \widehat{x}^i} \tag{9.270}$$

By our choice in the previous paragraph, the convected coordinate $\widehat{x}^i$ is equal to the ith coefficient of the vector $\underline{r}$ that identifies where the particle is at time t. If we denote this by x_i and write $\underline{r}'$ in the Cartesian basis $\hat{e}_1, \hat{e}_2, \hat{e}_3$, we obtain

$$\underline{g}_{(i)} = \frac{\partial x'_p}{\partial x_i}\hat{e}_p \tag{9.271}$$

Recalling the definition of the deformation-gradient tensor $\underline{\underline{F}}$, we can write

$$\underline{\underline{F}} \equiv \frac{\partial \underline{r}'}{\partial \underline{r}} \tag{9.272}$$

$$\underline{g}_{(i)} = \hat{e}_i \cdot \underline{\underline{F}} \tag{9.273}$$

$$= \underline{\underline{F}}^T \cdot \hat{e}_i \tag{9.274}$$

where the last two expressions are equivalent. We can derive analogous expressions[13] for the reciprocal base vectors $\underline{g}^{(i)}$:

$$\underline{g}^{(i)} = \frac{\partial \widehat{x}^i}{\partial \underline{r}'} \tag{9.280}$$

$$= \frac{\partial x_i}{\partial x_p'} \hat{e}_p \tag{9.281}$$

$$= \underline{\underline{F}}^{-1} \cdot \hat{e}_i \tag{9.282}$$

$$= \hat{e}_i \cdot (\underline{\underline{F}}^{-1})^T \tag{9.283}$$

Note that in these expressions the reciprocal relationship between $\underline{g}_{(i)}$ and $\underline{g}^{(j)}$ is preserved:

$$\underline{g}_{(i)} \cdot \underline{g}^{(j)} = \hat{e}_i \cdot \underline{\underline{F}} \cdot \underline{\underline{F}}^{-1} \cdot \hat{e}_j$$

$$= \delta_{ij} \tag{9.284}$$

We can now write down a general tensor expression for $\widehat{\Pi}^{pk}$:

$$\underline{\underline{\Pi}} = \sum_{i=1}^{3} \sum_{j=1}^{3} \widehat{\Pi}^{ij} \underline{g}_{(i)} \underline{g}_{(j)} \tag{9.285}$$

---

[13] The starting point for this second derivation, Equation (9.280), can be obtained if we write $\widehat{x}^i$ as a function of $\underline{r}'$. For the differentials $d\widehat{x}^i$ we can then write

$$d\widehat{x}^i = \frac{d\widehat{x}^i}{d\underline{r}'} \cdot d\underline{r}' \tag{9.275}$$

From the definition of $\underline{g}_{(i)}$,

$$d\underline{r}' = \underline{g}_{(i)} d\widehat{x}^i \tag{9.276}$$

Substituting this into expression (9.275), we arrive at

$$d\widehat{x}^i = \frac{d\widehat{x}^i}{d\underline{r}'} \cdot \underline{g}_{(k)} d\widehat{x}^k \tag{9.277}$$

For this to be true, the following must hold:

$$\frac{d\widehat{x}^i}{d\underline{r}'} \cdot \underline{g}_{(k)} = \delta_{ik} \tag{9.278}$$

which gives us that $d\widehat{x}^i/d\underline{r}'$ must be the vector reciprocal to $\underline{g}_{(k)}$, or

$$\frac{d\widehat{x}^i}{d\underline{r}'} = \underline{g}^{(i)} \tag{9.279}$$

$$\underline{g}^{(p)} \cdot \underline{\underline{\Pi}} \cdot \underline{g}^{(k)} = \sum_{i=1}^{3} \sum_{j=1}^{3} \underline{g}^{(p)} \cdot \widehat{\Pi}^{ij} \underline{g}_{(i)} \underline{g}_{(j)} \cdot \underline{g}^{(k)} \tag{9.286}$$

$$= \sum_{i=1}^{3} \sum_{j=1}^{3} \delta_{pi} \widehat{\Pi}^{ij} \delta_{jk} \tag{9.287}$$

$$= \widehat{\Pi}^{pk} \tag{9.288}$$

$$\widehat{\Pi}^{pk} = \underline{g}^{(p)} \cdot \underline{\underline{\Pi}} \cdot \underline{g}^{(k)} \tag{9.289}$$

$$= \hat{e}_p \cdot \left(\underline{\underline{F}}^{-1}\right)^T \cdot \underline{\underline{\Pi}} \cdot \underline{\underline{F}}^{-1} \cdot \hat{e}_k \tag{9.290}$$

We can relate $\widehat{\Pi}^{pk}$ to the Cartesian coordinates Π_{ms} by substituting $\underline{\underline{\Pi}} = \sum_{m=1}^{3} \sum_{s=1}^{3} \Pi_{ms} \hat{e}_m \hat{e}_s$ into Equation (9.290):

$$\widehat{\Pi}^{pk} = \sum_{m=1}^{3} \sum_{s=1}^{3} \hat{e}_p \cdot \left(\underline{\underline{F}}^{-1}\right)^T \cdot \Pi_{ms} \hat{e}_m \hat{e}_s \cdot \underline{\underline{F}}^{-1} \cdot \hat{e}_k \tag{9.291}$$

$$= \sum_{m=1}^{3} \sum_{s=1}^{3} \left[\hat{e}_p \cdot \left(\underline{\underline{F}}^{-1}\right)^T \cdot \hat{e}_m\right] \left(\hat{e}_s \cdot \underline{\underline{F}}^{-1} \cdot \hat{e}_k\right) \Pi_{ms} \tag{9.292}$$

We wish to examine the time derivative at any time t' of the contravariant convected coefficients of $\underline{\underline{\Pi}}$. The stress is a function of time t' and particle label $\underline{r}$ or, equivalently, of particle label $\widehat{x}^i$:

$$\underline{\underline{\Pi}}(t', \underline{r}) = \underline{\underline{\Pi}}(t', \widehat{x}^1, \widehat{x}^2, \widehat{x}^3) \tag{9.293}$$

From the chain rule then,

$$\frac{d\widehat{\Pi}^{pk}}{dt'} = \left(\frac{\partial \widehat{\Pi}^{pk}}{\partial t'}\right)_{\widehat{x}} + \sum_{s=1}^{3} \left(\frac{\partial \widehat{\Pi}^{pk}}{\partial \widehat{x}^s}\right)_{t', \widehat{x}^{j, j \neq s}} \frac{\partial \widehat{x}^s}{\partial t'} \tag{9.294}$$

$$= \left(\frac{\partial \widehat{\Pi}^{pk}}{\partial t'}\right)_{\widehat{x}} \tag{9.295}$$

where we have used the fact that the $\widehat{x}^s$ are always constant. Thus we will take the partial derivative at constant convected coefficients (i.e., following the same material particle), which is equivalent to the total time derivative of $\widehat{\Pi}^{pk}$ since the convected coefficients do not change with time. To evaluate this derivative in fixed coordinates, we will use Equation (9.290) and expand using the product rule of differentiation. The following results are written in Gibbs (vector–tensor) notation, but the details of the calculation at every step can be verified using Einstein notation:

$$\frac{d\widehat{\Pi}^{pk}}{dt'} = \frac{\partial}{\partial t'} \left[\hat{e}_p \cdot \left(\underline{\underline{F}}^{-1}\right)^T \cdot \underline{\underline{\Pi}} \cdot \underline{\underline{F}}^{-1} \cdot \hat{e}_k\right] \tag{9.296}$$

$$
= \hat{e}_p \cdot \left\{ \left(\underline{\underline{F}}^{-1} \right)^T \cdot \underline{\underline{\Pi}} \cdot \left(\frac{\partial \underline{\underline{F}}^{-1}}{\partial t'} \right)_{\hat{x}} + \left(\underline{\underline{F}}^{-1} \right)^T \cdot \left(\frac{\partial \underline{\underline{\Pi}}}{\partial t'} \right)_{\hat{x}} \cdot \underline{\underline{F}}^{-1} \right.
$$

$$
+ \left. \left[\frac{\partial (\underline{\underline{F}}^{-1})^T}{\partial t'} \right]_{\hat{x}} \cdot \underline{\underline{\Pi}} \cdot \underline{\underline{F}}^{-1} \right\} \cdot \hat{e}_k \tag{9.297}
$$

$$
= \hat{e}_p \cdot \left(\underline{\underline{F}}^{-1} \right)^T \cdot \left\{ \left(\frac{\partial \underline{\underline{\Pi}}}{\partial t'} \right)_{\hat{x}} - \left[\underline{\underline{\Pi}} \cdot \nabla' \underline{v}' + (\nabla' \underline{v}')^T \cdot \underline{\underline{\Pi}} \right] \right\} \cdot \underline{\underline{F}}^{-1} \cdot \hat{e}_k \tag{9.298}
$$

We have used Table 9.1 to arrive at Equation (9.298).

To interpret this expression, we will now let t' go to t. For most of the terms, this limit is straightforward:

$$
\underline{\underline{F}}^{-1} \longrightarrow \underline{\underline{I}} \tag{9.299}
$$

$$
\left(\underline{\underline{F}}^{-1} \right)^T \longrightarrow \underline{\underline{I}} \tag{9.300}
$$

$$
\nabla' \underline{v}' \longrightarrow \nabla \underline{v} \tag{9.301}
$$

To write $(\partial \underline{\underline{\Pi}} / \partial t')_{\hat{x}}$ as $t \longrightarrow t'$, we must consider two ways of writing the total stress tensor $\underline{\underline{\Pi}}$. While we have been writing $\underline{\underline{\Pi}}$ as a function of t' and particle label $\underline{r}$, it also may be conceived as a straightforward function of position $\underline{r}'$ and time t'. If we write $d\underline{\underline{\Pi}}/dt'$ using both representations and invoke the chain rule, it becomes clear what $(\partial \underline{\underline{\Pi}} / \partial t')_{\hat{x}}$ becomes as $t \longrightarrow t'$. We had previously calculated $d\underline{\underline{\Pi}}(t', \hat{x})/dt'$ [Equation (9.295)] and arrived at the following result:

$$
\frac{d\underline{\underline{\Pi}}(t', \hat{x})}{dt'} = \left(\frac{\partial \underline{\underline{\Pi}}}{\partial t'} \right)_{\hat{x}} \tag{9.302}
$$

Calculating the time derivative of $\underline{\underline{\Pi}}$ using the representation $\underline{\underline{\Pi}} = \underline{\underline{\Pi}}(t', \underline{r}')$ we obtain

$$
\frac{d\underline{\underline{\Pi}}(t', \underline{r}')}{dt'} = \left(\frac{\partial \underline{\underline{\Pi}}}{\partial t'} \right)_{\underline{r}'} + \sum_{s=1}^{3} \left(\frac{\partial \underline{\underline{\Pi}}}{\partial r'_s} \right)_{t', r'_{j, j \neq s}} \frac{\partial r'_s}{\partial t'} \tag{9.303}
$$

$$
= \left(\frac{\partial \underline{\underline{\Pi}}}{\partial t'} \right)_{\underline{r}'} + \sum_{s=1}^{3} \left(\frac{\partial \underline{\underline{\Pi}}}{\partial r'_s} \right)_{t', r'_{j, j \neq s}} v'_s \tag{9.304}
$$

$$
= \left(\frac{\partial \underline{\underline{\Pi}}}{\partial t'} \right)_{\underline{r}'} + \underline{v}' \cdot \nabla' \underline{\underline{\Pi}} \tag{9.305}
$$

In the stationary laboratory coordinates we see that the total derivative of $\underline{\underline{\Pi}}$ with respect to t' is also just the substantial derivative at t' [Equation (9.305)] [7]. Thus as $t \longrightarrow t'$,

$$
\left(\frac{\partial \underline{\underline{\Pi}}}{\partial t'} \right)_{\hat{x}} = \frac{d\underline{\underline{\Pi}}}{dt'} \longrightarrow \frac{D\underline{\underline{\Pi}}}{Dt} \tag{9.306}
$$

Combining all expressions for the limit $t \longrightarrow t'$, the expression for the contravariant convected derivative now becomes

$$\frac{d\widehat{\Pi}^{pk}}{dt} = \hat{e}_p \cdot \left\{ \frac{D\underline{\underline{\Pi}}}{Dt} - \left[\underline{\underline{\Pi}} \cdot \nabla\underline{v} + (\nabla\underline{v})^T \cdot \underline{\underline{\Pi}} \right] \right\} \cdot \hat{e}_k \qquad (9.307)$$

Relationship between
convected coefficients and
upper convected derivative

$$\boxed{\frac{d\widehat{\Pi}^{pk}}{dt} = \hat{e}_p \cdot \overset{\triangledown}{\underline{\underline{\Pi}}} \cdot \hat{e}_k} \qquad (9.308)$$

where we have written the final expression using the symbol for upper convected derivative that was defined in Section 9.2.2. Recall that pre- and postmultiplying a tensor with orthonormal coordinates, as is done in Equation (9.308), just yields the coefficients of the tensor (in this case $\overset{\triangledown}{\underline{\underline{\Pi}}}$) in that orthonormal basis (i.e., $\hat{e}_p \cdot \underline{\underline{A}} \cdot \hat{e}_k = A_{pk}$).

We started with the goal of relating the convected coefficients $\widehat{\Pi}^{pk}$ to a fixed Cartesian coordinate system. The result of our manipulations is shown in Equation (9.308). Equation (9.308) tells us that the Cartesian coordinates of the tensor $\overset{\triangledown}{\underline{\underline{\Pi}}} \equiv D\underline{\underline{\Pi}}/Dt - \left[\underline{\underline{\Pi}} \cdot \nabla\underline{v} + (\nabla\underline{v})^T \cdot \underline{\underline{\Pi}} \right]$, the upper convected derivative of the total stress tensor, are equal to the time derivatives of the convected coordinates $\widehat{\Pi}^{pk}$. We had earlier shown that taking the time derivative of the convected coordinates $\widehat{\Pi}^{pk}$ is an objective way of calculating the time changes in stress in a flow. Thus we have found a way to calculate, in fixed Cartesian coordinates, the frame-invariant time derivative of a quantity such as stress—the answer is to calculate the upper convected derivative.

This rather lengthy exercise serves to show the physical meaning of the upper convected derivative: the upper convected derivative represents the rate of change of a quantity as seen from a coordinate system that is convected along with the flow, that is, a coordinate system that translates, rotates, and deforms with the fluids. The upper convected derivative is an objective, that is, a frame-invariant method of taking into account the deformation that fluid particles experience in flow. The frame invariance results from using the convected coordinates, which contain information about particle separations without including information about particle orientations. Using the upper convected derivative to formulate a constitutive equation is just one choice of many possible convected derivatives, as we will see next. It is a choice, however, that yields a properly objective constitutive equation.

9.3.4 OTHER CONVECTED DERIVATIVES

In the preceding derivation we sought a frame-invariant time derivative by using time derivatives of convected components. We started with Equation (9.296) by taking the time derivative of the contravariant coefficients of $\underline{\underline{\Pi}}$. We could have equally well chosen to use the covariant coefficients, however, and we would have arrived at a different but equally valid time derivative of stress called the *lower convected derivative*, which is denoted by the symbol $\triangle$:

Lower convected
derivative of $\underline{\underline{A}}$

$$\boxed{\overset{\triangle}{\underline{\underline{A}}} \equiv \frac{D\underline{\underline{A}}}{Dt} + \nabla\underline{v} \cdot \underline{\underline{A}} + \underline{\underline{A}} \cdot (\nabla\underline{v})^T} \qquad (9.309)$$

The lower convected time derivative appears in the differential version of the Maxwell model when it is corrected for frame variance by using the Cauchy tensor $\underline{\underline{C}}$, rather than the negative of the Finger tensor $\underline{\underline{C}}^{-1}$. The result is the lower convected Maxwell model:

$$\text{Lower convected Maxwell model} \qquad \boxed{\underline{\underline{\tau}} + \lambda \overset{\triangle}{\underline{\underline{\tau}}} = -\eta_0 \dot{\underline{\underline{\gamma}}}} \qquad (9.310)$$

The approach for showing the equivalence of the Cauchy–Maxwell equation [Equation (9.199)] and the lower convected Maxwell model is the same as was followed in Section 9.2, where we manipulated the Lodge equation to get the upper convected Maxwell model.

Our use of convected coordinate systems was motivated by our need to find frame-invariant ways of expressing time-changing stresses and deformations. We have discussed two convected coordinate systems, $\underline{g}_{(i)}$ and $\underline{g}^{(i)}$, that produce frame-invariant time derivatives, but there are many other coordinate systems that may be used to produce properly invariant time derivatives. One such coordinate system is the corotational frame, in which the time derivative following a fluid particle is taken with respect to a coordinate system that rotates with the instantaneous fluid angular velocity [27]. The fixed-coordinates expression for the derivative that results, called the *corotational* or *Jaumann* derivative $\overset{\circ}{\underline{\underline{A}}}$, is shown in Table 9.4 along with the definitions of the derivatives we have seen thus far. The vorticity tensor, $\underline{\underline{\omega}} = \nabla v - (\nabla v)^T$, appears in the definition of the corotational derivative.

In addition, empirical modifications may be made to the convected derivative expressions to develop different constitutive equations. For example, Gordon and Schowalter [93] used the scalar parameter ξ to introduce nonaffine motion into their picture of how dilute

TABLE 9.4
Time Derivatives and Corresponding Strain Measures of Continuum Mechanics*

Name of Derivative	Derivative in Gibbs Notation	Strain Measure
Substantial or material	$\dfrac{D\underline{\underline{A}}}{Dt} = \dfrac{d\underline{\underline{A}}}{dt} = \dfrac{\partial \underline{\underline{A}}}{\partial t} + \underline{v} \cdot \nabla \underline{\underline{A}}$	None
Covariant or lower convected	$\overset{\triangle}{\underline{\underline{A}}} \equiv \dfrac{D\underline{\underline{A}}}{Dt} + \nabla \underline{v} \cdot \underline{\underline{A}} + \underline{\underline{A}} \cdot (\nabla \underline{v})^T$	$\underline{\underline{C}}$
Contravariant or upper convected	$\overset{\triangledown}{\underline{\underline{A}}} \equiv \dfrac{D\underline{\underline{A}}}{Dt} - \left[(\nabla \underline{v})^T \cdot \underline{\underline{A}} + \underline{\underline{A}} \cdot \nabla \underline{v} \right]$	$\underline{\underline{C}}^{-1}$
Corotational or Jaumann	$\overset{\circ}{\underline{\underline{A}}} \equiv \dfrac{D\underline{\underline{A}}}{Dt} + \dfrac{1}{2} \left(\underline{\underline{\omega}} \cdot \underline{\underline{A}} - \underline{\underline{A}} \cdot \underline{\underline{\omega}} \right)$	
Gordon–Schowalter	$\overset{\square}{\underline{\underline{A}}} \equiv \dfrac{D\underline{\underline{A}}}{Dt} - (\nabla \underline{v})^T \cdot \underline{\underline{A}} - \underline{\underline{A}} \cdot \nabla \underline{v} + \dfrac{\xi}{2}(\underline{\underline{A}} \cdot \dot{\underline{\underline{\gamma}}} + \dot{\underline{\underline{\gamma}}} \cdot \underline{\underline{A}})$	$\underline{\underline{E}}$

*$\underline{\underline{\omega}} = \nabla v - (\nabla v)^T$, and $\underline{\underline{E}}$ is the solution of the differential equation $\overset{\square}{\underline{\underline{E}}} = 0$. The parameter ξ in the Gordon–Schowalter derivative is related to the amount of nonaffine motion allowed in a deformation [93].

solutions flow [138]. In the resulting convected derivative, ξ determines the character of the derivative (see Table 9.4). For $\xi = 0$ the Gordon–Schowalter derivative $\underset{=}{\overset{\square}{A}}$ becomes the upper convected derivative, for $\xi = 1$ it is the corotational derivative, and for $\xi = 2$, the lower convected derivative. The only constraints on proposing new convected derivatives are that the derivative be objective and that the constitutive equation that results make meaningful predictions. These are simple but stringent conditions.

Often an equivalent strain measure can be deduced for a proposed invariant time derivative (see Table 9.4). Thus the two ways we have discussed of arriving at better constitutive equations—proposing better strain measures or proposing better time derivatives—are in many respects equivalent. Again, the proof of any constitutive equation is how well its predictions match reality. For a comparison of how well several constitutive equations predict actual polymer behavior, see Larson [138] and other references [124, 125].

9.4 Other Constitutive Approaches

To do meaningful simulations of stress and deformation of non-Newtonian fluids, we must have accurate constitutive models that work in the high-strain or high strain rate (nonlinear viscoelastic) regimes. In this text we have discussed several constitutive models for non-Newtonian fluids: the generalized Newtonian model, the Maxwell model, the generalized linear viscoelastic model, and, in this chapter, the Lodge model and various convected versions of the Maxwell model (see Table 9.5). Of these, the GLVE model is inappropriate for high-rate flows, and the Maxwell model is inappropriate for general flows. In this chapter we have been studying models valid at high deformation rates, but the usefulness of the constitutive equations studied so far depends on the fluid being modeled. We have still not identified any constitutive equations that are appropriate for fluids exhibiting both normal-stress effects and shear-thinning, and these are among the most common types of non-Newtonian fluids.

It is beyond the scope of this text to explore nonlinear viscoelastic constitutive models in great detail, but it will be helpful to discuss some of the major relations here. We will discuss the two methods that have dominated the search for better constitutive models, the continuum approach and the molecular modeling approach.

9.4.1 CONTINUUM APPROACH

The approach we have followed thus far has been based on continuum mechanics. We began with the Newtonian equation, which works for many fluids, and we modified it empirically, hoping for good non-Newtonian constitutive equations. The only other advance we have discussed is substituting the convected derivatives for simple time derivatives in order to preserve material objectivity.

The high-strain constitutive equations developed in Sections 9.2 and 9.3 were all based on the Maxwell model. The original Maxwell model is a simple equation that is linear in both strain rate $\dot{\gamma}_{21}$ and stress τ_{21}. When modeling any process, it is always prudent to begin with the simplest models (i.e., linear models) and to move to more complex, nonlinear equations only if the linear equations are inadequate. In the next section we discuss other

TABLE 9.5
Summary of Non-Newtonian Constitutive Equations Studied Thus Far

Name	Equation	Advantages/Limitations
Generalized Newtonian model	$\underline{\underline{\tau}} = -\eta(\dot{\gamma})\underline{\underline{\dot{\gamma}}}$	Easy to use/does not predict normal stresses in shear; no memory effects
Maxwell model	$\underline{\underline{\tau}} + \lambda\dfrac{\partial \underline{\underline{\tau}}}{\partial t} = -\eta_0 \underline{\underline{\dot{\gamma}}}$	Includes some memory effects/invalid in rotating frame of reference
Generalized linear viscoelastic model	$\underline{\underline{\tau}} = -\displaystyle\int_{-\infty}^{t} G(t-t')\underline{\underline{\dot{\gamma}}}(t')\ dt'$	Valid for all materials in linear viscoelastic limit/only valid at low strain rates; does not predict normal stresses in shear
Upper convected Maxwell model	$\underline{\underline{\tau}} + \lambda\overset{\triangledown}{\underline{\underline{\tau}}} = -\eta_0\underline{\underline{\dot{\gamma}}}$	Includes memory effects and shear 1st normal stresses/does not predict shear-thinning or nonzero shear 2nd normal-stress difference
Lodge model	$\underline{\underline{\tau}} = -\displaystyle\int_{-\infty}^{t} \dfrac{\eta_0}{\lambda^2} e^{-\frac{(t-t')}{\lambda}}\underline{\underline{C}}^{-1}(t',t)\ dt'$	Same as upper convected Maxwell
Lower convected Maxwell model	$\underline{\underline{\tau}} + \lambda\overset{\triangle}{\underline{\underline{\tau}}} = -\eta_0\underline{\underline{\dot{\gamma}}}$	Includes memory effects and shear normal stresses/does not predict shear-thinning and overpredicts shear 2nd normal-stress difference
Cauchy–Maxwell model	$\underline{\underline{\tau}}(t) = +\displaystyle\int_{-\infty}^{t} \dfrac{\eta_0}{\lambda^2} e^{-\frac{(t-t')}{\lambda}}\underline{\underline{C}}(t,t')\ dt'$	Same as lower convected Maxwell

linear scalar models besides the Maxwell model, and the quasi-linear tensorial models we can develop from them by including convected derivatives and nonlinear strain tensors. In a subsequent section we will introduce explicitly nonlinear models constructed by other methods, such as including scalar nonlinear functions and adding higher-order terms to the Maxwell model.

9.4.1.1 Linear and Quasi-Linear Models

The Maxwell model contains linear terms containing stress τ_{21}, rate of deformation $\dot{\gamma}_{21}$, and also a linear term involving the time derivative of stress, $\partial\tau_{21}/\partial t$. One linear term that is missing from the Maxwell model is the time derivative of the rate of strain. The Jeffreys model [26] is the model formed by adding this term to the Maxwell model:

$$\text{Jeffreys model (scalar version)} \qquad \boxed{\tau_{21} + \lambda_1\frac{\partial\tau_{21}}{\partial t} = -\eta_0\left(\dot{\gamma}_{21} + \lambda_2\frac{\partial\dot{\gamma}_{21}}{\partial t}\right)} \qquad (9.311)$$

The Jeffreys model contains two time constants, λ_1 and λ_2, called the *relaxation* and *retardation times*, respectively, and a viscosity parameter η_0. Recall that for the Maxwell model we could rationalize the form of the equation by invoking the image of a spring and a dashpot acting in series (see Figure 8.2). The mechanical system whose force equation is analogous to the Jeffreys model is shown in Figure 9.12.

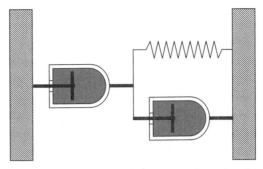

Figure 9.12 Mechanical system composed of two dashpots and a spring, for which the differential equation relating force and displacement is analogous to the Jeffreys model.

When converted to a tensor form, the Jeffreys model suffers from the same frame-variance problem as the Maxwell and the generalized linear viscoelastic models—stresses calculated with the Jeffreys model will depend on superposed coordinate rotations of the system (see the discussion in Section 8.5). As shown at the beginning of this chapter, we can fix this frame-variance problem by replacing partial differentiations with respect to time with any of the convected time derivatives in Table 9.4. Converting the Jeffreys model to a tensor form and using the upper convected and the lower convected derivatives, we obtain the following constitutive equations:

Upper convected Jeffreys
or Oldroyd B fluid [192]
$$\underline{\underline{\tau}} + \lambda_1 \overset{\triangledown}{\underline{\underline{\tau}}} = -\eta_0 \left(\underline{\underline{\dot{\gamma}}} + \lambda_2 \overset{\triangledown}{\underline{\underline{\dot{\gamma}}}} \right) \qquad (9.312)$$

Lower convected Jeffreys
or Oldroyd A fluid
$$\underline{\underline{\tau}} + \lambda_1 \overset{\triangle}{\underline{\underline{\tau}}} = -\eta_0 \left(\underline{\underline{\dot{\gamma}}} + \lambda_2 \overset{\triangle}{\underline{\underline{\dot{\gamma}}}} \right) \qquad (9.313)$$

These models are no longer linear because the convected derivative terms introduce nonlinear terms in the velocity gradient $\nabla \underline{v}$. Bird et al. [26] call the convected Maxwell and Jeffreys models quasi-linear.

The Oldroyd A or lower convected Jeffreys model turns out not to be a useful model, as it suffers from many of the drawbacks of the lower convected Maxwell model due to its use of the lower convected derivative (it overpredicts Ψ_2; see Appendix D). The Oldroyd B model resembles the Lodge model in some ways—it predicts constant viscosity and first normal-stress coefficient, as well as zero second normal-stress coefficient. The predictions of these two models differ slightly in elongation, however: the Lodge model predicts a tension-thickening elongational viscosity, whereas the Oldroyd B model can predict either tension-thickening ($\lambda_1 > \lambda_2$) or constant elongational viscosity ($\lambda_1 = \lambda_2$) (see Appendix D). For $\lambda_1 < \lambda_2$ the Oldroyd B model predicts negative first normal stresses. The Oldroyd B or upper convected Jeffreys model is used widely for Boger fluids, fluids that exhibit constant viscosity but also pronounced normal-stress effects [24, 31].

We must still find models that will predict shear thinning of viscosity and the first normal-stress coefficient. To search out new models, we can think of more linear models by adding more terms to the Jeffreys model, using as our inspiration different combinations of springs and dashpots. We can then hope to produce shear-thinning nonlinear models by fixing up the linear models with convected derivatives, as we did the Maxwell and Jeffreys models. Unfortunately this strategy will not introduce qualitatively different constitutive equations. We can see why by examining the integral version of the Jeffreys model. Consider the tensor version of the Jeffreys model:

$$\underline{\underline{\tau}} + \lambda_1 \frac{\partial \underline{\underline{\tau}}}{\partial t} = -\eta_0 \left(\underline{\underline{\dot{\gamma}}} + \lambda_2 \frac{\partial \underline{\underline{\dot{\gamma}}}}{\partial t} \right) \tag{9.314}$$

Although this constitutive equation is inadmissible due to its use of the partial time derivative, this equation is the starting point for calculating the integral version of the Jeffreys model, which we can make objective by replacing its nonobjective strain measure with the Finger or Cauchy tensors (Section 9.1). We can solve the Jeffreys equation, Equation (9.314), using the same integrating-factor technique that we used in Section 8.2.1 to solve for stress in the Maxwell equation. After multiplying by the integrating factor $(1/\lambda_1)e^{t/\lambda_1}$ and factoring the left side, we obtain

$$\frac{\partial}{\partial t} \left(e^{t/\lambda_1} \underline{\underline{\tau}} \right) = -\frac{\eta_0}{\lambda_1} \left(e^{t/\lambda_1} \underline{\underline{\dot{\gamma}}} + \lambda_2 e^{\frac{t}{\lambda_1}} \frac{\partial \underline{\underline{\dot{\gamma}}}}{\partial t} \right) \tag{9.315}$$

Then taking $\underline{\underline{\tau}}$ to be finite at $-\infty$, we can integrate this equation:

$$e^{t/\lambda_1} \underline{\underline{\tau}}(t) = -\frac{\eta_0}{\lambda_1} \int_{-\infty}^{t} \left[e^{t'/\lambda_1} \underline{\underline{\dot{\gamma}}}(t') + \lambda_2 e^{t'/\lambda_1} \frac{\partial \underline{\underline{\dot{\gamma}}}(t')}{\partial t'} \right] dt' \tag{9.316}$$

$$\underline{\underline{\tau}}(t) = -\int_{-\infty}^{t} \frac{\eta_0}{\lambda_1} e^{\frac{-(t-t')}{\lambda_1}} \underline{\underline{\dot{\gamma}}}(t')\, dt' - e^{-t/\lambda_1} \int_{-\infty}^{t} \frac{\lambda_2 \eta_0}{\lambda_1} e^{t'/\lambda_1} \frac{\partial \underline{\underline{\dot{\gamma}}}(t')}{\partial t'}\, dt' \tag{9.317}$$

The second term on the right side can be integrated by parts if we take $\underline{\underline{\dot{\gamma}}}$ to be finite at $-\infty$, and we obtain the final result:

Integral Jeffreys model
$$\boxed{\underline{\underline{\tau}}(t) = -\int_{-\infty}^{t} \left[\frac{\eta_0}{\lambda_1} \left(1 - \frac{\lambda_2}{\lambda_1} \right) e^{\frac{-(t-t')}{\lambda_1}} \right] \underline{\underline{\dot{\gamma}}}(t')\, dt' - \frac{\lambda_2 \eta_0}{\lambda_1} \underline{\underline{\dot{\gamma}}}(t)} \tag{9.318}$$

We can see that the integral Jeffreys model is the equivalent of the sum of two models: a Newtonian model with viscosity $\lambda_2 \eta_0/\lambda_1$ and a generalized linear viscoelastic model with the relaxation modulus function $G(t - t')$ given by the terms in square brackets. If the two terms on the right side are combined, the Jeffreys model can also be cast into the form of a generalized linear viscoelastic model with the relaxation modulus function $G(t - t')$ given as [26]

Relaxation modulus function for Jeffreys model
$$\boxed{G(t - t') = \left[\frac{\eta_0}{\lambda_1} \left(1 - \frac{\lambda_2}{\lambda_1} \right) e^{\frac{-(t-t')}{\lambda_1}} \right] + 2 \frac{\eta_0 \lambda_2}{\lambda_1} \delta(t - t')} \tag{9.319}$$

where $\delta(t - t')$ is the symmetric or Dirac delta function [129, 27]. The Dirac delta function is defined in terms of the following relations:

$$\int_{-a}^{+b} f(x)\delta(x)\, dx = f(0) \tag{9.320}$$

$$\int_{-a}^{+b} f(x)\frac{d\delta(x)}{dt}\, dx = -\frac{df}{dx}(0) \tag{9.321}$$

where $f(x)$ is an arbitrary function, both a and b are positive, and $a < b$. The Dirac delta function can be represented by the continuous function

$$\delta(x) = \lim_{n \to \infty} \sqrt{\frac{n}{\pi}} e^{-nx^2} \tag{9.322}$$

Recall that the function $G(t - t')$ is just the response of the fluid to a step strain imposed at $t = t'$. Thus the Jeffreys model responds to a step strain with an exponential stress decay and an added infinite contribution to the shear stress at the time of the step. This additional contribution is generated by the added Newtonian term—Newtonian fluids generate stresses proportional to instantaneous shear rate, and during the step $\dot{\gamma}_{21}$ is infinite.

We see then that the Jeffreys model is a GLVE model, that is, it can be written as an integral over all past times over a time-dependent material-based function times the time-dependent rate-of-deformation tensor $\underline{\underline{\dot{\gamma}}}$:

$$\text{GLVE model:} \quad \underline{\underline{\tau}}(t) = -\int_{-\infty}^{t} G(t - t')\, \underline{\underline{\dot{\gamma}}}(t')\, dt' \tag{9.323}$$

By adding the term $-\eta_0\lambda_2\, \partial\underline{\underline{\dot{\gamma}}}/\partial t$ to the Maxwell model, we changed the relaxation modulus function $G(t - t')$, but we left the overall structure of the model intact—an integral over a time-dependent function times $\underline{\underline{\dot{\gamma}}}$. Other linear modifications of the Maxwell model motivated by other combinations of springs and dashpots in series and parallel result in a similar effect: the relaxation-modulus function changes, but the overall structure of the model remains the same [25]. Since all these linear models are encompassed by the GLVE model, we know from the discussion in Section 9.1 that all spring–dashpot models can be written explicitly in terms of their strain measure $\underline{\underline{\gamma}}(t, t')$:

$$\text{GLVE model (strain explicit):} \quad \underline{\underline{\tau}}(t) = +\int_{-\infty}^{t} M(t - t')\, \underline{\underline{\gamma}}(t, t')\, dt' \tag{9.324}$$

where

$$M(t - t') \equiv \frac{\partial G(t - t')}{\partial t'} \tag{9.325}$$

and we can make them objective by substituting an objective strain measure such as the Finger or Cauchy tensor for $\underline{\underline{\gamma}}(t, t')$. The objective versions of any spring–dashpot-generated model then will be Lodge-like (if $\underline{\underline{C}}^{-1}$ is used) or Cauchy–Maxwell-like (if $\underline{\underline{C}}$ is used) and will be quasi-linear due to the nonlinear strain measure. Thus we cannot obtain qualitatively new behavior such as shear-thinning or the proper ratio of Ψ_1 / Ψ_2 by tinkering with dashpot and spring models in this way. We need to try a different approach.

9.4.1.2 Nonlinear Differential Constitutive Models

In the last section we showed that we are unable to model both shear-thinning and nonzero normal-stress differences in shear using quasi-linear models. We now discuss nonlinear models, that is, models that are nonlinear in either $\dot{\underline{\underline{\gamma}}}$ or $\underline{\underline{\tau}}$, or that have non-linearities introduced by scalar functions of the invariants of these tensors. Nonlinear models lead to the calculation of flow fields or material functions in the same way as linear models, but due to the complexities of the equations, we must employ advanced solution methods, usually numerical methods [54], to obtain results. We will introduce the reader to these more complex constitutive models here but leave an in-depth discussion of carrying out constitutive calculations with nonlinear models to the literature [26, 238, 54, 212].

The models in this section do succeed, in many cases, in predicting shear-thinning and reasonable normal-stress effects. The job of distinguishing among the models and choosing the best one for a particular application thus comes down to examining the detailed model predictions and comparing them with the experimental data on a material of interest. Since for all of the equations in this section it is a fairly involved process to make constitutive predictions, the added accuracy of the nonlinear models does not always justify the heavy computational price. Making the decision between using a more accurate but computationally intense nonlinear model and a simpler but qualitatively incorrect model such as the power-law generalized Newtonian fluid is the most delicate decision left up to the rheologist. For some guidance on this process, see Tables 9.6 and 9.7.

A simple way of requiring a model to shear-thin is to introduce a viscosity function that varies with the shear rate $\dot{\gamma}$. This is the approach that was taken in the development of the generalized Newtonian fluid (GNF) model:

$$\underline{\underline{\tau}} = -\eta(\dot{\gamma})\dot{\underline{\underline{\gamma}}} \tag{9.326}$$

As discussed in Chapter 7, the GNF equation will not predict nonzero normal stresses in shear flow because the predicted stresses are proportional to the rate-of-deformation tensor $\dot{\underline{\underline{\gamma}}}$, and $\dot{\gamma}_{11} = \dot{\gamma}_{22} = \dot{\gamma}_{33}$ for shear flow. Also, memory effects are absent from the GNF model.

We can create a nonlinear model that contains memory effects and that predicts normal stresses in shear flow by replacing the viscosity parameter η_0 in the upper convected Maxwell model (this model contains memory effects but does not shear-thin) with a viscosity function, $\eta(\dot{\gamma})$ (which explicitly shear-thins). Such a constitutive equation is admissible because it uses the upper convected derivative and because the function we propose to introduce is only a function of an invariant of $\dot{\underline{\underline{\gamma}}}$. The resulting equation is called the *White–Metzner* equation [264] [compare to equation (8.40)]:

$$\text{White–Metzner model} \qquad \boxed{\underline{\underline{\tau}} + \frac{\eta(\dot{\gamma})}{G_0} \overset{\triangledown}{\underline{\underline{\tau}}} = -\eta(\dot{\gamma})\dot{\underline{\underline{\gamma}}}} \tag{9.327}$$

TABLE 9.6
Summary of Predictions of Several Nonlinear Constitutive Equations in Shear and Elongation

Name	Advantages/Limitations
Upper convected Jeffreys or Oldroyd B [Equation (9.312)]	Does not predict shear-thinning; does predict shear 1st normal-stress differences; shape of elongational viscosity versus elongation rate depends on value of retardation time λ_2.
Lower convected Jeffreys or Oldroyd A [Equation (9.313)]	Does not predict shear-thinning; does predict shear 1st and 2nd normal-stress differences; predictions of Ψ_2 are too large, however.
White–Metzner [Equation (9.327)]	Predicts shear-thinning; does predict shear 1st normal-stress differences; does not predict Ψ_2 and does not predict a linear viscoelastic limit for all choices of $\eta(\dot\gamma)$.
Oldroyd 8-constant [Equation (9.328)]	Predicts shear-thinning; does predict shear 1st and 2nd normal-stress differences; predicts multiple overshoots in startup of steady shearing (not observed experimentally); predicts a maximum in elongational viscosity as a function of elongation rate (often observed); contains upper convected Maxwell and Jeffreys as well as many other models as special cases [26].
Giesekus [Equation (9.329)]	Predicts shear-thinning; does predict shear 1st and 2nd normal-stress differences; predicts overshoots in shear and 1st normal stresses in start-up experiment; contains upper convected Maxwell and Jeffreys models as special cases [26].
Factorized Rivlin-Sawyers [Equation (9.330)]	Very general model; predictions of model depend on choices of functions Φ_1 and Φ_2; for $\Phi_2 \neq 0$ the Rivlin–Sawyers equation predicts $\Psi_2 \neq 0$.
Factorized K-BKZ [Equation (9.331)]	Very general model; is a subset of the factorized Rivlin–Sawyers equation; predictions of model depend on choices of potential function $U(I_1, I_2)$.

In the White–Metzner equation G_0 is a constant modulus parameter associated with the model, and $\eta(\dot\gamma)$ is a function that must be specified. The ratio $\eta(\dot\gamma)/G_0$ can be thought of as a deformation-rate-dependent relaxation time $\lambda(\dot\gamma)$.

The White–Metzner model is appealing in that it is relatively easy to make calculations of material functions and flow fields with this model. This model does not have a unique linear viscoelastic limit, however, because of the function $\eta(\dot\gamma)$—the linear viscoelastic limit ($\dot\gamma \rightarrow 0$) depends on the function $\eta(\dot\gamma)$. If the function $\eta(\dot\gamma)$ is chosen to have a low-shear-rate plateau, then at low shear rates the linear viscoelastic behavior of the upper convected Maxwell equation will be obtained. Some of the characteristics of the White-Metzner model are discussed in Larson [138]. Calculations of the journal-bearing problem (eccentric rotating cylinders) are reported by Beris et al. [18] and are discussed in Bird et al. [26]. The material functions calculated with the White–Metzner model in steady flows are the same as for the Lodge (upper convected Maxwell) model with the constant parameter η_0 replaced with a function $\eta(\dot\gamma)$ and λ replaced with $\lambda(\dot\gamma) = \eta(\dot\gamma)/G_0$. The White–Metzner model performs qualitatively differently from the upper convected Maxwell model in step strains because of the effect of rapid deformation on the relaxation times $\lambda(\dot\gamma)$ [138]. The White–Metzner model does not predict the Lodge–Meissner relationship [138].

TABLE 9.7
Summary of the Kinds of Applications That Might Be Modeled with the Constitutive Equations Discussed in This Book[*]

Application	Important Effects	Potential Constitutive Equation
Fiber spinning	Elongational flow; want fluid to be tension-thickening	Cannot use straight power-law GNF since shear and elongation thin together; a hybrid GNF could be used, i.e., one with different power laws for shear and elongation; alternatively could use UCM, UCJ-Oldroyd-B, which do not shear-thin, or a nonlinear model that shear-thins
Extrusion	Viscoelasticity plays a minor role [61]	GNF with appropriate viscosity function
Profile extrusion	Die swell important; this is an elastic effect	UCM, UCJ-Oldroyd-B (for non-shear-thinning fluids), or any of the nonlinear constitutive equations that shear-thin
Injection molding	Pressure drop and flow rate important; nonisothermal flows; elongation can be important in fountain flow inside mold [44]	GNF with appropriate viscosity function; need to include energy effects; may use nonlinear model if fountain flow is important
Contraction flows	Elongation is important; elasticity affects the character of the vortices that are formed [103, 104]	GNF models can be used, but the fixed Trouton ratio of the standard GNF (Tr=3) will limit the type of vortex behavior that is predicted; if viscoelasticity is suspected to be an important effect, use UCM, UCJ-Oldroyd-B, or nonlinear models if shear-thinning is important
Pumping	Viscoelasticity is rarely important in pumping [44]	GNF
Mixing	Highly elastic materials may strongly affect the power consumed in a mixer; also rod climbing will occur	UCM, UCJ-Oldroyd-B, or nonlinear models if shear-thinning is important
Coating	Normal stresses are important in instabilities in coating flows	UCM, UCJ-Oldroyd-B, or nonlinear models if shear-thinning is important

[*] Help is provided in identifying which constitutive equations might be appropriate for the type of simulation attempted. Models considered include GNF and upper convected Maxwell (UCM) and Jeffreys (UCJ-Oldroyd-B).

A direct way of including nonlinearity in a proposed constitutive equation is to include terms involving expressions such as $\underline{\underline{\dot\gamma}} \cdot \underline{\underline{\dot\gamma}}$ or $\underline{\underline{\dot\gamma}} : \underline{\underline{\dot\gamma}}$. Oldroyd proposed a constitutive equation that added to the upper convected Maxwell model all possible terms that are linear in either $\underline{\underline{\dot\gamma}}$ or $\underline{\underline{\tau}}$ and at most quadratic in $\underline{\underline{\dot\gamma}}$. The resulting constitutive equation is called the *Oldroyd 8-constant model*:[14]

---

[14] The terms in the Oldroyd 8-constant model were arrived at by imposing the conditions of frame invariance on an arbitrary function of $\underline{\underline{\tau}}$ and $\underline{\underline{\dot\gamma}}$ and then retaining only the terms linear in $\underline{\underline{\tau}}$ and at most quadratic in $\underline{\underline{\dot\gamma}}$. The expressions are simplified to those shown in Equation (9.328) by using the Cayley–Hamilton theorem (see glossary in Appendix B). There is a nice discussion of the Oldroyd 8-constant model in Larson [138].

Oldroyd 8-constant model

$$
\begin{aligned}
&\underline{\underline{\tau}} + \lambda_1 \overset{\triangledown}{\underline{\underline{\tau}}} + \tfrac{1}{2}(\lambda_1 - \mu_1)(\dot{\underline{\underline{\gamma}}} \cdot \underline{\underline{\tau}} + \underline{\underline{\tau}} \cdot \dot{\underline{\underline{\gamma}}}) + \tfrac{1}{2}\mu_0(\text{tr } \underline{\underline{\tau}})\dot{\underline{\underline{\gamma}}} + \tfrac{1}{2}\nu_1(\underline{\underline{\tau}} : \dot{\underline{\underline{\gamma}}})\underline{\underline{I}} \\
&\qquad = -\eta_0 \left[\dot{\underline{\underline{\gamma}}} + \lambda_2 \overset{\triangledown}{\dot{\underline{\underline{\gamma}}}} + (\lambda_2 - \mu_2)(\dot{\underline{\underline{\gamma}}} \cdot \dot{\underline{\underline{\gamma}}}) + \tfrac{1}{2}\nu_2(\dot{\underline{\underline{\gamma}}} : \dot{\underline{\underline{\gamma}}})\underline{\underline{I}} \right]
\end{aligned}
\qquad (9.328)
$$

η_0, λ_1, and λ_2 are the zero-shear viscosity, relaxation time, and retardation time that are familiar from the Maxwell and Jeffreys models. The other five constants, μ_0, μ_1, μ_2, ν_1, and ν_2, are associated with the additional nonlinear terms. Oldroyd wrote his equation in terms of the corotational derivative (see Table 9.4), and therefore some of the parameters appear in combinations rather than singly in the equation as written here, which uses the upper convected derivative.

The Oldroyd 8-constant model contains many of the constitutive equations we have already discussed as special cases (Table 9.8). The Oldroyd 8-constant model predicts shear-thinning, normal stresses in shear flow, and many other qualitative effects. Bird et al. give a discussion of the predictions of a simplified version of this model, which they call the Oldroyd 4-constant model. It is the equivalent of the Oldroyd B plus the term containing μ_0 from the Oldroyd 8-constant model. The values of the parameters in the Oldroyd 4-constant model are listed in Table 9.8. The second-order fluid, mentioned in Table 9.8, also a subset of the Oldroyd 8-constant model, is a model that results from assuming that the stress tensor may be expressed as a polynomial in the rate-of-deformation tensor $\dot{\underline{\underline{\gamma}}}$. This expansion, known as the retarded-motion expansion, has also been considered to higher orders, most notably to third order in the constitutive equation known as the third-order fluid [26].

Omitted from the Oldroyd 8-constant model are terms that are second order in stress. There is no reason a priori to avoid these terms, and in fact the Giesekus equation, which

TABLE 9.8
Values of Parameters in the Oldroyd 8-Constant Model That Produce Various Other Constitutive Equations [26]

Name	λ_1	λ_2	μ_0	μ_1	μ_2	ν_1	ν_2
Newtonian	0	0	0	0	0	0	0
Upper convected Maxwell	λ_1	0	0	λ_1	0	0	0
Lower convected Maxwell	λ_1	0	0	$-\lambda_1$	λ_2	0	0
Corotational Maxwell	λ_1	0	0	0	0	0	0
Upper convected Jeffreys (Oldroyd B)	λ_1	λ_2	0	λ_1	λ_2	0	0
Lower convected Jeffreys (Oldroyd A)	λ_1	λ_2	0	$-\lambda_1$	$-\lambda_2$	0	0
Corotational Jeffreys	λ_1	λ_2	0	0	0	0	0
Second-order fluid	0	λ_2	0	λ_1	μ_2	0	0
Oldroyd 4-constant model	λ_1	λ_2	μ_0	λ_1	λ_2	0	0

was developed from molecular arguments involving anisotropic drag [138], includes such a term.

<table>
<tr><td>Giesekus
equation</td><td>$$\underline{\underline{\tau}} + \lambda \overset{\triangledown}{\underline{\underline{\tau}}} + \frac{\alpha\lambda}{\eta_0}\underline{\underline{\tau}} \cdot \underline{\underline{\tau}} = -\eta_0\dot{\underline{\underline{\gamma}}}$$</td><td>(9.329)</td></tr>
</table>

This model has three parameters, the zero-shear viscosity η_0, the relaxation time λ, and α, which is a parameter that relates to the anisotropy of the drag encountered by flowing polymer segments. A more in-depth discussion of this model and of the other constitutive equations mentioned here can be found in Larson [138]. Larson also compares the predictions of some of these models with experimental data and with other models not mentioned here.

9.4.1.3 Integral Models with Nonlinearity

We have discussed the idea of using frame-invariant finite-strain measures to eliminate spurious rotation effects from the Maxwell model (see Figure 9.9). The resulting finite-strain models, for example, the Lodge equation [Equation (9.139)] and the Cauchy–Maxwell equation [Equation (9.199)], are quasi-linear since the strain measures used, $\underline{\underline{C}}^{-1}$ and $\underline{\underline{C}}$, are nonlinear in strain. An explicitly nonlinear model can be formulated by introducing scalar functions of strain invariants into the integral formulation. A general constitutive equation that follows this approach is the factorized Rivlin–Sawyers equation [26][15]:

<table>
<tr><td>Factorized
Rivlin–Sawyers
equation</td><td>$$\underline{\underline{\tau}}(t) = + \int_{-\infty}^{t} M(t - t')\left[\Phi_2(I_1, I_2)\underline{\underline{C}} - \Phi_1(I_1, I_2)\underline{\underline{C}}^{-1}\right]dt'$$</td><td>(9.330)</td></tr>
</table>

where $M(t - t')$ is the memory function [Equation 9.12)], I_1 and I_2 are the first and second invariants of $\underline{\underline{C}}$ or $\underline{\underline{C}}^{-1}$ as defined in Appendix C.6 (the invariants of a tensor and its inverse are related; see Problem 9.45), and Φ_1 and Φ_2 are scalar functions that must be specified. The third invariant of $\underline{\underline{C}}^{-1}$ is not needed in these functions because of the incompressibility condition ($I_3 = \det|\underline{\underline{C}}^{-1}| = 1$ for incompressible fluids). Note that both the Finger tensor and the Cauchy tensor appear in the factorized Rivlin–Sawyers equation. Constitutive equations that use only the Finger tensor predict a zero second normal-stress coefficient in shear, whereas equations that only use the Cauchy tensor overpredict Ψ_2 (for examples, see Tables D3 and D.5). Thus, with appropriate choices for the functions Φ_1 and Φ_2, the correct magnitude of Ψ_2 can be achieved with a Rivlin–Sawyers equation.

The choice of the functions Φ_1 and Φ_2 depends on what type of behavior is most important to the user. The Rivlin–Sawyers class of equations includes many successful molecular-model-based constitutive equations, such as the Rouse–Zimm, Doi–Edwards, and Curtiss–Bird models [26, 27, 138]. It is essential to bring in material-specific mechanistic information to help specify Φ_1 and Φ_2, since the range of behavior that falls within the Rivlin–Sawyers class of equations is extremely broad.

---

[15] This model is considered factorized because the memory function $M(t - t')$ is the only explicit function of time, and it is factored out of each term, leaving the pure strain functions Φ_1 and Φ_2.

A subset of the Rivlin–Sawyers class of equations is the K-BKZ class, due to Kaye [122] and Bernstein, Kearsley, and Zapas [20]:

Factorized
K-BKZ
equation

$$\underline{\underline{\tau}} = \int_{-\infty}^{t} M(t - t') \left[2\frac{\partial U}{\partial I_2} \underline{\underline{C}}(t, t') - 2\frac{\partial U}{\partial I_1} \underline{\underline{C}}^{-1}(t', t) \right] dt' \qquad (9.331)$$

In the factorized K-BKZ equation the strain-dependent functions multiplying the strain tensors are related to each other—they are the derivatives of a potential function $U(I_1, I_2)$ with respect to the first two invariants of $\underline{\underline{C}}^{-1}$. Using the K-BKZ equation instead of the Rivlin–Sawyers equation eliminates a possible violation of the second law of thermodynamics in the Rivlin–Sawyers equation [141].

The predictions of the K-BKZ equations depend on the choice of the potential function $U(I_1, I_2)$, but reasonable fits to experimental data on linear polymers can be obtained, as discussed in Larson [138]. The K-BKZ equation does not perform well in flows that include flow reversal. See Bird et al. [26] for some sample calculations involving the K-BKZ model.

9.4.2 MOLECULAR APPROACH—POLYMERIC CONSTITUTIVE MODELS

We have studied several continuum-based constitutive equations, such as the generalized Newtonian fluid and the generalized linear viscoelastic fluid, and we have discussed how they can be used for some rheological modeling purposes. In the previous several sections we discussed the kinds of nonlinear rheological models that have been developed using the continuum approach. We have learned how to make constitutive equations frame invariant in at least two ways, and we have seen that as we demand that constitutive equations model a wider variety of behavior, the equations themselves become quite complex, and the choices we have in how to specify functions in the new constitutive equations (such as potential function U in the K-BKZ or the Rivlin–Sawyers functions) become unmanageable.

Progress in constitutive modeling is unlikely to come by random searching through all the possible differential models or all the possible Rivlin–Sawyers models using trial and error. To pilot our way through what is still an enormously wide variety of equations, we must turn to techniques that derive information from fluid structure, that is, we must pursue molecular models of rheological behavior. To illustrate the molecular modeling approach we will focus on rheological modeling of polymeric systems.

The molecular approach to polymer constitutive modeling begins with structural pictures of polymers and builds up equations for macroscopic stresses based on the model used. The advantage of molecular-based modeling is that predictions may be made, not only of the form of the constitutive equations, but also of the effect on flow properties of molecular parameters such as molecular weight, monomeric friction coefficient, and molecular relaxation times. A disadvantage of the molecular approach is the complexity that is inherent in macromolecular descriptions or, alternatively, the inaccuracies inherent in simplifying assumptions.

The molecular approach to constitutive modeling is discussed in detail in Bird et al.'s second volume [27] as well as in Larson [138] and Doi and Edwards [70]. Here we will content ourselves with a description of the most important classical molecular models of polymeric systems and a few remarks on trends in constitutive modeling from a molecular perspective.

9.4.2.1 Probabilistic Approach—Configuration Distribution Function

If we review our initial discussion in Chapter 3 of the extra stress tensor $\underline{\underline{\tau}}$, we recall that this tensor is meant to model all molecular stresses acting at a point in a fluid. If we can calculate the molecular *tension* force $\underline{\tilde{f}}$ acting on an arbitrary surface of area dA with unit normal $\hat{n}$, then we can calculate $\underline{\underline{\tau}}$ from[16]

$$\underline{\tilde{f}} = -dA\,\hat{n} \cdot \underline{\underline{\tau}} \tag{9.332}$$

We then must calculate the forces generated in a deforming polymeric system from a molecular picture of that system. First we will consider the forces generated when deforming a single polymer chain. We will then take the result for a single chain and apply it first to a crosslinked polymer solid, and then to an entangled polymer melt. Our discussion on this subject follows that of Larson [138] with some input from Bird et al.'s second volume [27].

Consider a polymer chain such as that shown schematically in Figure 9.13. Research on polymer chains in a melt of other chains has shown that polymers prefer to adopt a random coil configuration at equilibrium. The coil shape can be described in terms of the end-to-end vector $\underline{R}$, shown in Figure 9.13. The forces we seek are the forces generated when the random coil is deformed. A simple way to keep track of this deformation is to consider the change in orientation and length of the end-to-end vector $\underline{R}$.

From physics we know that the work dW is equal to force dotted with displacement. For the polymer coil, the work performed in extending the end-to-end vector by an amount $d\underline{R}$ is given by Equation (9.333)

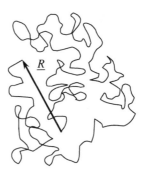

Figure 9.13 Two-dimensional schematic of the definition of the end-to-end vector $\underline{R}$ in a polymer coil. The polymer coil can be modeled as a random walk.

---

[16] Although in Equation (3.58) it is the total-stress tensor that appears in an equation of this form, we are dealing with incompressible fluids, and we are unconcerned by the addition of an isotropic force such as $p\underline{\underline{I}}$. We will only be predicting normal stress differences with the constitutive equations developed. Note also that $\underline{f} = -\underline{\tilde{f}}$.

$$dW = \tilde{\underline{f}} \cdot d\underline{R} \tag{9.333}$$

where $\tilde{\underline{f}}$ is the tension force necessary to deform the polymer coil. If $\tilde{\underline{f}}$ is a conservative force, we can calculate $\tilde{\underline{f}}$ from

$$\tilde{\underline{f}} = \frac{\partial W}{\partial \underline{R}} \tag{9.334}$$

The work W is just the coil free energy which equals $-TS$, where T is absolute temperature and S is the chain entropy. To calculate the force, then, we need an expression for the entropy.

To calculate the entropy we must turn to statistical thermodynamics. Boltzmann showed that the system entropy is related to the number of configurations that the system can adopt,

$$S = k \ln \Omega \tag{9.335}$$

where k is Boltzmann's constant, and Ω is the number of configurations of the polymer chain that result in a coil with end-to-end vector $\underline{R}$. If we model the coil as a random walk of N steps of step length a, we can use some straightforward results of studies of such random walks to determine Ω. For a random walk beginning at the origin, the probability $\psi_0 \, dR_1 dR_2 dR_3$ that the other end of the walk lies at a position between $\underline{R}$ and $\underline{R} + d\underline{R}$ is approximately given by a Gaussian function:

$$\psi_0(\underline{R}) \cong \left(\frac{\beta}{\sqrt{\pi}}\right)^3 e^{-\beta^2 \underline{R} \cdot \underline{R}} \tag{9.336}$$

where $\beta^2 = 3/(2Na^2)$. $\psi_0(\underline{R})$ is called the *configuration distribution function*, and it is associated with the equilibrium configuration of the polymer chain.

The number of configurations that have an end-to-end vector of $\underline{R}$ is proportional to ψ_0. If we write $\Omega = c\psi_0$, we can calculate the entropy from Equation (9.335) and subsequently the force needed to deform the coil:

$$S = k \ln \Omega = k(\ln c + \ln \psi_0) \tag{9.337}$$

$$= k \ln c + 3k \ln \left(\frac{\beta}{\sqrt{\pi}}\right) - k\beta^2 \underline{R} \cdot \underline{R} \tag{9.338}$$

$$\tilde{\underline{f}} = \frac{\partial W}{\partial \underline{R}} = -T \frac{\partial S}{\partial \underline{R}} \tag{9.339}$$

$$= 2kT\beta^2 \underline{R} \tag{9.340}$$

Force to deform
a Gaussian spring
$$\boxed{\tilde{\underline{f}} = \frac{3kT}{Na^2} \underline{R}} \tag{9.341}$$

where N is the number of steps of length a of a random walk used to model the polymer chain. The final result for the force necessary to elongate a polymer coil [Equation (9.341)] indicates that a polymer coil resists stretching with a force that is proportional to its length. This is the same as the force law for a linear spring [Equation (8.46)].

We can use the result for the force required to deform a single chain [Equation (9.341)] to derive an expression for the force on an arbitrary surface within the fluid [27]. Consider a plane of area dA with unit normal $\hat{n}$ located arbitrarily within a polymer melt. There are ν polymer chains per unit volume in the melt. We now construct a cube bisected by dA, as shown in Figure 9.14. If we choose the box sides to be of length $\nu^{-1/3}$, then there will be just one chain in the box on average. We will call $\underline{R}$ the end-to-end vector of the chain in the box, and we will consider an end-to-end vector to be *in* the box when its origin (tail) is in the box.

We can then write the force on $dA = \nu^{-2/3}$ due to chains of end-to-end vector $\underline{R}$ as follows:

$$\begin{pmatrix} \text{force on surface } dA \\ \text{due to chains of} \\ \text{end-to-end vector } \underline{R} \end{pmatrix} =$$

$$\begin{pmatrix} \text{probability} \\ \text{chain has} \\ \text{end-to-end} \\ \text{vector } \underline{R} \end{pmatrix} \begin{pmatrix} \text{probability a chain} \\ \text{of end-to-end} \\ \text{vector } \underline{R} \text{ crosses} \\ \text{surface } dA \end{pmatrix} \begin{pmatrix} \text{force exerted on} \\ \text{surface due to} \\ \text{chain of end-to-} \\ \text{end vector } \underline{R} \end{pmatrix} \qquad (9.342)$$

We have already calculated the expression for the force [Equation (9.341)]. The probability that a chain has end-to-end vector $\underline{R}$ is given by the conformation distribution function $\psi(\underline{R})\, dR_1\, dR_2\, dR_3$,

$$\begin{pmatrix} \text{probability a} \\ \text{nonequilibrium chain has} \\ \text{an end-to-end vector} \\ \text{between } \underline{R} \text{ and } \underline{R} + d\underline{R} \end{pmatrix} = \psi(\underline{R})\, dR_1\, dR_2\, dR_3 \qquad (9.343)$$

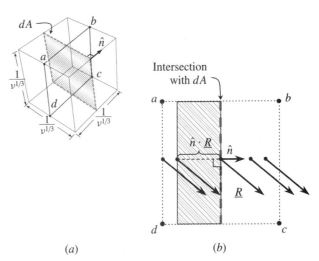

(a)   (b)

Figure 9.14 Schematic of a cube of fluid that contains one polymer chain on average. The chain is modeled by its end-to-end vector $\underline{R}$. The probability that $\underline{R}$ crosses the surface dA is just the ratio of the shaded area in (b) to the area $abcd$.

This is not the equilibrium distribution because we are considering the general case where the chain may be deformed. To calculate the probability that the vector $\underline{R}$ passes through dA, consider Figure 9.14.

The end-to-end vector $\underline{R}$ can fit into the box in a variety of places. Recall that we have defined that $\underline{R}$ is in the box if its tail (indicated by a black dot) is in the box. If we consider $\underline{R}$ located along the left edge of the section of the box shown in Figure 9.14b, we can see that it is the length $\hat{n} \cdot \underline{R}$ that determines whether $\underline{R}$ crosses the plane dA. As we move $\underline{R}$ to the right, the vector crosses dA when its tail is a distance $\hat{n} \cdot \underline{R}$ from the centerline of the cube. Moving $\underline{R}$ further to the right, the vector stops intersecting dA when its tail is at the centerline. This thought experiment is equally valid at any cube cross section and thus we can calculate the probability of $\underline{R}$ crossing dA as the ratio of the volume in which the tail of $\underline{R}$ is located when it crosses dA to the total volume of the box:

$$\begin{pmatrix} \text{probability a chain} \\ \text{of end-to-end} \\ \text{vector } \underline{R} \text{ crosses} \\ \text{surface } dA \end{pmatrix} = \frac{(\hat{n} \cdot \underline{R})\left(v^{-\frac{1}{3}}\right)^2}{v^{-1}} = (\hat{n} \cdot \underline{R})v^{\frac{1}{3}} \qquad (9.344)$$

We can now assemble the terms of Equation (9.342) to obtain the tension force on dA due to chains of end-to-end vector $\underline{R}$:

$$\begin{pmatrix} \textit{tension} \text{ force on surface} \\ dA \text{ due to chains of} \\ \text{end-to-end vector } \underline{R} \end{pmatrix} = \{\psi(\underline{R})\, dR_1\, dR_2\, dR_3\} \left\{v^{\frac{1}{3}}(\hat{n} \cdot \underline{R})\right\} \left\{\frac{3kT}{Na^2}\underline{R}\right\} \qquad (9.345)$$

Finally, to include the contributions of chains of all end-to-end vectors, we integrate the expression for force over all of end-to-end-vector space:

$$\begin{pmatrix} \textit{tension} \text{ force} \\ \text{on surface } dA \\ \text{due to all} \\ \text{polymer chains} \end{pmatrix} = \int_{-\infty}^{\infty}\int_{-\infty}^{\infty}\int_{-\infty}^{\infty} \begin{pmatrix} \text{force on surface } dA \\ \text{due to chains of} \\ \text{end-to-end vector } \underline{R} \end{pmatrix} dR_1\, dR_2\, dR_3 \qquad (9.346)$$

$$\tilde{\underline{f}} = \frac{3kT v^{\frac{1}{3}}}{Na^2} \int_{-\infty}^{\infty}\int_{-\infty}^{\infty}\int_{-\infty}^{\infty} (\hat{n} \cdot \underline{R}\,\underline{R})\,\psi(\underline{R})\, dR_1\, dR_2\, dR_3 \qquad (9.347)$$

$$= \frac{3kT v^{\frac{1}{3}}}{Na^2} (\hat{n} \cdot \langle \underline{R}\,\underline{R} \rangle) \qquad (9.348)$$

where $\langle \underline{R}\,\underline{R} \rangle$ is the integral of the tensor $\underline{R}\,\underline{R}$ over the nonequilibrium configuration distribution function $\psi(\underline{R})$,

$$\langle \,\cdot\, \rangle \equiv \int_{-\infty}^{\infty}\int_{-\infty}^{\infty}\int_{-\infty}^{\infty} \cdot\, \psi(\underline{R})\, dR_1\, dR_2\, dR_3 \qquad (9.349)$$

Comparing this result with Equation (9.332) and taking $dA = \left(v^{-1/3}\right)^2$, we obtain the result for the stress tensor predicted from this analysis:

$$\tilde{\underline{f}} = -v^{-\frac{2}{3}} \, \hat{n} \cdot \underline{\underline{\tau}} \tag{9.350}$$

Stress tensor generated
by elongating a
Gaussian chain

$$\boxed{\underline{\underline{\tau}} = -\frac{3vkT}{Na^2} \, \langle \underline{R}\,\underline{R} \rangle} \tag{9.351}$$

Note that v is the number of chains per unit volume. In probability theory the quantity $\langle \underline{R}\,\underline{R} \rangle$ is called the second moment of the function $\psi(\underline{R})$ (see Equation 9.349). Thus, the stress due to polymer chains described in this way is proportional to the second moment of the configuration distribution function.

9.4.2.2 Elastic Solids—Rubber-Elasticity Theory

Equation (9.351) gives us the stress in terms of an integral over the configuration distribution function $\psi(\underline{R})$. This function has not yet been specified, however. The configuration distribution function contains the information on how the end-to-end vectors of all the different polymers in a system deform for all possible deformations. Thus, the amount of unknown information in $\psi(\underline{R})$ is about equal to the amount of unknown information in the original stress tensor $\underline{\underline{\tau}}$. What we have gained by this analysis, however, is a change in question. Instead of asking what the stress is at all points in a fluid for all possible deformations, we are now asking, how do the end-to-end vectors of all the different polymers in a system deform for all possible deformations? To address this question, we must propose a model for how these end-to-end vectors move under all circumstances. We will discuss two models: the elastic network model for crosslinked polymers and the Green–Tobolsky temporary-network model for polymer melts.

Consider an ideal crosslinked network in which, between every two crosslinks, there is a polymer strand that follows a random walk of N steps of length a (Figure 9.15). We assume that at equilibrium, that is, at time t' before deformation, each strand is characterized by the same equilibrium configuration-distribution function $\psi_0(\underline{R}') \, dR_1' \, dR_2' \, dR_3'$.

The simplest type of deformation we can consider is an elongational deformation characterized by this inverse deformation-gradient tensor $\underline{\underline{F}}^{-1}$:

$$\underline{\underline{F}}^{-1} = \begin{pmatrix} \lambda_1 & 0 & 0 \\ 0 & \lambda_2 & 0 \\ 0 & 0 & \lambda_3 \end{pmatrix}_{123} \tag{9.352}$$

λ_1, λ_2, and λ_3 are the elongation ratios in the three coordinate directions. For example, for steady uniaxial elongation in the x_3-direction, we showed that $\lambda_3 = l/l_0 = e^{\dot{\epsilon}_0 t}$ [Equation (4.41)]. We now make the key assumption that the crosslinks follow the macroscopic deformation. This means that the deformation that characterizes the fluid at a macroscopic level can also be applied to microscopic motion, such as the motion of the crosslinks in our polymer network. This is called *affine motion*.

For affine deformation we can use the macroscopic elongation ratios λ_1, λ_2, and λ_3 to relate the microscopic end-to-end vectors before and after the deformation:

$$\lambda_1 = \frac{R_1}{R_1'} \tag{9.353}$$

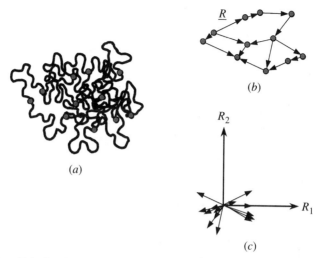

Figure 9.15 Crosslinked polymer network. (*a*) Between junction points, the polymer strands are random coils. (*b*) The stress on the network can be modeled by studying the distribution of vectors that connect the crosslink points. (*c*) If we translate all the crosslink-to-crosslink or end-to-end vectors to the origin, we can see what the configuration distribution function looks like. Here the function is random.

$$\lambda_2 = \frac{R_2}{R_2'} \tag{9.354}$$

$$\lambda_3 = \frac{R_3}{R_3'} \tag{9.355}$$

$$\underline{R}(t) = \begin{pmatrix} \lambda_1 R_1' \\ \lambda_2 R_2' \\ \lambda_3 R_3' \end{pmatrix}_{123} \tag{9.356}$$

$\underline{R}(t)$ is the end-to-end vector of the strand after deformation. The 1,2,3 coordinate system is a Cartesian system in which the deformation is elongational.

Our goal is to write the configuration distribution function $\psi(\underline{R})$ for the system after deformation. We can do this if we notice that since we deformed affinely, if we started with a certain number of polymer strands with end-to-end vector between $\underline{R}'$ and $\underline{R}' + d\underline{R}'$, after the deformation we will have the same number of strands having end-to-end vector between $\underline{R}$ and $\underline{R} + d\underline{R}$. In other words, every strand in the equilibrium state with initial end-to-end vector $\underline{R}'$ will map to a strand in the deformed state with end-to-end vector $\underline{R}$. Thus the probability that the end-to-end vector of a strand is between $\underline{R}$ and $\underline{R} + d\underline{R}$ in the deformed state is given by the equilibrium distribution function ψ_0:

$$\begin{pmatrix} \text{probability of} \\ \text{end-to-end vector} \\ \text{between } \underline{R} \\ \text{and } \underline{R} + d\underline{R} \end{pmatrix} = \psi(\underline{R}) \, dR_1 \, dR_2 \, dR_3 \tag{9.357}$$

$$= \psi_0(\underline{R}') \, dR_1' \, dR_2' \, dR_3' \tag{9.358}$$

$$\psi(\underline{R}) \, dR_1, \, dR_2, \, dR_3 = \left(\frac{\beta}{\sqrt{\pi}}\right)^3 e^{-\beta^2 \underline{R}' \cdot \underline{R}'} \, dR_1' \, dR_2' \, dR_3' \tag{9.359}$$

$$= \left(\frac{\beta}{\sqrt{\pi}}\right)^3 e^{-\beta^2 \left\{ \left(\frac{R_1}{\lambda_1}\right)^2 + \left(\frac{R_2}{\lambda_2}\right)^2 + \left(\frac{R_3}{\lambda_3}\right)^2 \right\}} \, dR_1 \, dR_2 \, dR_3 \tag{9.360}$$

Configuration distribution function after deformation

$$\boxed{\psi(\underline{R}) = \left(\frac{\beta}{\sqrt{\pi}}\right)^3 e^{-\beta^2 \left\{ \left(\frac{R_1}{\lambda_1}\right)^2 + \left(\frac{R_2}{\lambda_2}\right)^2 + \left(\frac{R_3}{\lambda_3}\right)^2 \right\}}} \tag{9.361}$$

Note that $dR_1' dR_2' dR_3' = dR_1 dR_2 dR_3$ because of incompressibility.

Now that we know $\psi(\underline{R})$, we can calculate the stress tensor from Equation (9.351),

$$\underline{\underline{\tau}} = -\frac{3kT\nu}{Na^2} \langle \underline{R} \, \underline{R} \rangle \tag{9.362}$$

$$= -\frac{3kT\nu}{Na^2} \int_{-\infty}^{\infty} \int_{-\infty}^{\infty} \int_{-\infty}^{\infty} \underline{R} \, \underline{R} \, \psi(\underline{R}) \, dR_1 \, dR_2 \, dR_3 \tag{9.363}$$

$$= -\nu kT \lambda_i^2 \, \hat{e}_i \hat{e}_i \tag{9.364}$$

Much algebra is left out between these last two steps. To write this result in a more familiar form, compare with the Finger tensor for this deformation:

$$\underline{\underline{C}}^{-1} = (\underline{\underline{F}}^{-1})^T \cdot \underline{\underline{F}}^{-1} = \begin{pmatrix} \lambda_1^2 & 0 & 0 \\ 0 & \lambda_2^2 & 0 \\ 0 & 0 & \lambda_3^2 \end{pmatrix}_{123} \tag{9.365}$$

The constitutive equation we derive for this affine-deformation model for an elastic solid is [257, 79, 115, 243]

Affine deformation of a solid rubber (rubber-elasticity theory)

$$\boxed{\underline{\underline{\tau}} = -\nu kT \underline{\underline{C}}^{-1}} \tag{9.366}$$

This is just the finite-strain Hooke's law discussed in Section 9.1.2 [Equation (9.113)] with modulus $G = \nu kt$. We derived Equation (9.366) by discussing an arbitrary elongational deformation, that is, a deformation where the Finger tensor is diagonal. Since the Finger tensor is always symmetric (by construction, since $\underline{\underline{A}}^T \cdot \underline{\underline{A}}$ is always symmetric), it can always be expressed in some coordinate system, the principal frame of $\underline{\underline{C}}^{-1}$, in which it will be diagonal (see Appendix C). Thus our result for $\underline{\underline{\tau}}$ is valid for any deformation of an incompressible elastic solid.

9.4.2.3 Polymer Melts—Temporary Network Model

Using the rubber-elasticity theory described in the last section, Green and Tobolsky [99] modeled polymer melts. Unlike elastic solids, polymer melts do not have fixed crosslinks that can be followed during a deformation. Polymer melts form a network with temporary crosslinks formed from entanglements (Figure 6.13), as discussed in Chapter 6. As will be described, the Green–Tobolsky temporary network model uses crosslink points that break and reform at equal rates to model the flowing polymer melt.

Consider a network with ν junction points per unit volume. The junctions are not permanent, but ν is constant, that is, the junctions break and reform at equal rates. Recall that in rubber-elasticity theory, we followed the vectors between junctions since these were the end-to-end vectors of the polymer strands connecting the junctions. The end-to-end vectors were given by $\underline{R}'$ at some time t' in the past, and after deformation they rotated and stretched to the vector $\underline{R}$. In the Green–Tobolsky theory, these end-to-end vectors have finite lifetimes, during which they deform affinely and after which they disappear. For the number of junctions per unit volume to remain constant, when an old junction point dies, a new end-to-end vector must be born. Green and Tobolsky proposed that when a new end-to-end vector is created, it adopts the equilibrium distribution function ψ_0.

We define the following probabilities:

$$\begin{pmatrix} \text{probablity per unit time} \\ \text{that a strand dies and is} \\ \text{reborn at equilibrium} \end{pmatrix} \equiv \frac{1}{\lambda} \tag{9.367}$$

$$\begin{pmatrix} \text{probablity that a strand} \\ \text{retains the same end-to-end} \\ \text{vector from time } t' \text{ to } t \\ \text{(survival probability)} \end{pmatrix} \equiv P_{t',t} \tag{9.368}$$

We can solve for $P_{t',t}$ in terms of λ by considering the following question: what is the probability that a strand retains the same end-to-end vector from time t' to time $t + \Delta t$?

$$P_{t',t+\Delta t} = \begin{pmatrix} \text{probability that} \\ \text{strand retains the same} \\ \text{end-to-end vector} \\ \text{from } t' \text{ to } t \end{pmatrix} \begin{pmatrix} \text{probability that} \\ \text{strand does not die} \\ \text{and reappear with the} \\ \text{equilibrium configuration} \\ \text{over the interval } \Delta t \end{pmatrix} \tag{9.369}$$

$$P_{t',t+\Delta t} = (P_{t',t}) \left(1 - \frac{1}{\lambda} \Delta t \right) \tag{9.370}$$

Rearranging and taking the limit as Δt goes to zero, we obtain the following differential equation for $P_{t',t}$, which we can then solve:

$$\lim_{\Delta t \to 0} \frac{P_{t',t+\Delta t} - P_{t',t}}{\Delta t} = \frac{d P_{t',t}}{dt} = -\frac{1}{\lambda} P_{t',t} \tag{9.371}$$

$$\ln P_{t't} = -\frac{t}{\lambda} + C_1 \tag{9.372}$$

where C_1 is an arbitrary constant of integration. We can evaluate C_1 from the boundary condition that at $t' = t$, $P_{t',t'} = 1$, that is, the probability that a strand survives from t' to t' is certain, and we obtain the final result:

Strand survival probability

$$P_{t',t} = e^{\frac{-(t-t')}{\lambda}}$$

(9.373)

Now we apply the other two assumptions of the Green–Tobolsky model: (1) if a strand does not break and reform, it deforms affinely, and (2) if a strand does escape its constraints (die), it takes on the equilibrium configuration distribution ψ_0. We can thus calculate the contribution to the stress tensor of the individual strands by following them from birth to death:

$$\begin{pmatrix} \text{stress from} \\ \text{strands born} \\ \text{between} \\ t' \text{ and } t' + dt' \end{pmatrix} = \begin{pmatrix} \text{probability} \\ \text{that a strand} \\ \text{is born between} \\ t' \text{ and } t' + dt' \end{pmatrix} \begin{pmatrix} \text{probability} \\ \text{that a strand} \\ \text{survives from} \\ t' \text{ to } t \end{pmatrix} \begin{pmatrix} \text{stress generated} \\ \text{by an affinely} \\ \text{deforming strand} \\ \text{between } t' \text{ and } t \end{pmatrix}$$

$$d\underline{\underline{\tau}} = \left[\frac{1}{\lambda} dt' \right] \left[P_{t',t} \right] \left[-G \underline{\underline{C}}^{-1}(t', t) \right]$$

(9.374)

Note that when the strand is born it is characterized by ψ_0, and the expression $-G\underline{\underline{C}}^{-1}(t', t)$ is for the stress generated in deforming a strand from the equilibrium distribution at time t' to the nonequilibrium distribution at time t. Thus the expression in Equation (9.374) incorporates the two Green–Tobolsky assumptions listed before.

We can now integrate $d\underline{\underline{\tau}}$ over all past times t' to include strands born at all possible times. This gives us the final constitutive equation:

Green–Tobolsky temporary network model (Lodge model)

$$\underline{\underline{\tau}} = -\int_{-\infty}^{t} \frac{G}{\lambda} e^{-\frac{(t-t')}{\lambda}} \underline{\underline{C}}^{-1}(t', t) \, dt'$$

(9.375)

Recall that $G = \nu k T$.

We have reached the remarkable result that the Green–Tobolsky temporary network model results in a constitutive equation that is identical to the Lodge equation [Equation (9.139)]. Recall also that the Lodge equation is equivalent to the upper convected Maxwell equation, and thus so is the Green–Tobolsky equation. It seems then that the molecular approach to constitutive modeling results in the same constitutive equations, at least in this case, as the continuum approach. Rather than a disadvantage, however, this is an encouraging result, since it reassures us that both approaches are equally valid. The molecular modeling result has the advantage that there is a physical interpretation to parameters such as G and λ, and further that there are predictions on how these parameters should vary with quantities such as temperature or the number of strands per unit volume.

We know that the Lodge model does not capture shear-thinning or second normal-stress effects. The Green–Tobolsky temporary network model implies that we must assume something other than affine motion of strands with equal birth and death rates to predict these effects. Some molecular effects that have been incorporated into models successfully include anisotropic drag [90, 91] and nonaffine motion [66, 67, 68, 69]. For more on molecular models of polymer melts, see Larson [138] and Bird et al. [27].

A qualitatively different approach to molecular modeling polymer melts is the reptation concept, which we describe next.

9.4.2.4 Polymer Melts–Reptation Theories

In 1971 de Gennes [86] proposed a qualitatively different way to look at the constraints formed by neighboring chains on a particular chain of interest. Instead of thinking of a polymer melt as a temporary network, de Gennes proposed that the neighboring chains make a test chain behave as if it were confined to a tube (Figure 9.16). The tube diameter is approximately the length of the strand end-to-end vectors in a network of the Green–Tobolsky type, but the new idea introduced by the tube picture was that polymer chains will only relax by a snake-like motion along their backbones, that is, along the tubes. De Gennes called this motion reptation, which involved the back-and-forth motion of the chain within its constraining tube. As the chain emerges from an end, it forgets part of its original tube. Even if it reverses direction, the chain will not necessarily retrace its steps; rather it will create a new tube. Eventually the chain will have forgotten all of its original tube and will be relaxed.

De Gennes' reptation idea was developed into a constitutive equation by Doi and Edwards [66, 67, 68, 69]. In their picture when a polymer melt is deformed, the chains and their tubes deform affinely with the macroscopic deformation. Doi and Edwards proposed that soon after the deformation, however, chains retract within their tubes, rapidly attaining their original length. With the retraction mechanism, the net effect of the deformation is to reorient chain segments without stretching them [138, 70].

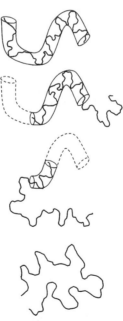

Figure 9.16 Reptating chain. Initially, the chain is associated with a tube of lateral constraints (top). The chain relaxes by moving along its contour, abandoning part of the tube. The chain can go in either direction, moving back and forth, forgetting more and more of the tube, until it finally remembers none of the original tube (bottom). When the chain has escaped from the original tube, it has relaxed.

The Doi–Edwards constitutive equation, like several other models we have discussed, may be written as an integral over all past times of a memory function times a strain tensor (see Larson [138] for the derivation):

$$\text{Doi–Edwards}\atop\text{equation} \qquad \boxed{\underline{\underline{\tau}} = -\int_{-\infty}^{t} M(t - t')\underline{\underline{Q}}(t', t)\, dt'} \qquad (9.376)$$

where

$$\underline{\underline{Q}}(t', t) = \frac{1}{4\pi} \int_0^{2\pi} \int_0^{\pi} 5 \left(\frac{\hat{u}' \cdot \underline{\underline{F}}^{-1}\, \hat{u}' \cdot \underline{\underline{F}}^{-1}}{|\hat{u}' \cdot \underline{\underline{F}}^{-1}|^2} \right) \sin\theta\, d\theta d\phi \qquad (9.377)$$

$$M(t - t') = \sum_{i\ \text{odd}} \frac{G_i}{\lambda_i} e^{-\frac{t-t'}{\lambda_i}} \qquad (9.378)$$

where $\hat{u}'$ is a unit vector that indicates the orientation of strands at time t', and $\underline{\underline{F}}^{-1}$ is the inverse deformation tensor. The integral is over the usual angles θ and ϕ of the spherical coordinate system, and G_i and λ_i are given by the expressions

$$G_i = \frac{8G_N^0}{\pi^2 i^2} \qquad (9.379)$$

$$\lambda_i = \frac{\lambda_1}{i^2} \qquad (9.380)$$

Here G_N^0 is the plateau modulus, and λ_1 is the longest relaxation time of the chain. These are the only two parameters in the Doi–Edwards model. Note that the Doi–Edwards equation is of the factorized K-BKZ type [26].

The Doi–Edwards model is quite successful in predicting some observed rheological behavior of polymers, including the ratio of Ψ_1/Ψ_2, the shape of the start-up curves, and the shape of the step shear damping function $h(\gamma_0)$. The Doi–Edwards model is also successful from a molecular point of view, predicting that zero-shear viscosity increases with the third power of molecular weight, not far from the observed exponent of $M^{3.4}$ (Figure 6.12). The Doi–Edwards model predicts shear-thinning of both viscosity and Ψ_1, and tension-thinning of elongational viscosity $\bar{\eta}$. The Doi-Edwards model performs less well in reversing flows. Refinements of the Doi–Edwards model are discussed in Larson [138].

9.4.2.5 Polymer Solutions–Elastic Dumbbell Model

A simple molecular model for a polymer in solution is the elastic dumbbell model (Figure 9.17). In this model the polymer chain end-to-end vector is represented explicitly as a spring connecting two beads. The forces that cause $\underline{R}$ to stretch are the solvent drag force on the beads and the Brownian motion force caused by collisions between solvent molecules and the polymer chain. Resisting these forces is the restoring force of the spring. This model is for dilute solutions where no two polymer chains interact.

Because the elastic dumbbell model invokes both macroscopic-type forces (drag) and molecular or statistical-type forces (Brownian motion), it is not solvable in the usual sense. Equations of this type are called *Langevin equations*. Recall from the rubber-elasticity

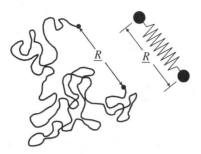

Figure 9.17 Elastic motion of the ends of a polymer coil in solution can be modeled by elastic dumbbell. The beads on the ends of the dumbbell experience solvent drag.

calculations that we can calculate the constitutive equation associated with a molecular model if we know the configuration distribution function $\psi(\underline{R})$. We can evaluate $\psi(\underline{R})$ for the dumbbell model by calculating $\underline{R}$ for a large number of elastic dumbbells (an ensemble of dumbbells) using the appropriate Langevin equation and determine the probability $\psi(\underline{R}) \, dR_1 dR_2 dR_3$ that an elastic dumbbell has an end-to-end vector between $\underline{R}$ and $\underline{R} + d\underline{R}$. We can then obtain the constitutive equation [138].[17]

When we carry out the analysis described, we obtain the following constitutive equation for the elastic dumbbell model:

$$\underline{\underline{\tau}} + \lambda \overset{\triangledown}{\underline{\underline{\tau}}} = -G\lambda \underline{\underline{\dot{\gamma}}} \tag{9.381}$$

where $G = \nu kT$, $\lambda = \zeta/8kT\beta^2$, ν is the number of dumbbells per unit volume, ζ is the bead friction coefficient, k is Boltzmann's constant, T is absolute temperature, and $\beta^2 = 3/2Na^2$ is the usual parameter associated with the random walk [Equation (9.336)]. This constitutive equation is just the upper convected Maxwell model.

More refined molecular models of the bead-spring type lead to more complex constitutive equations of the types that we have already discussed, such as K-BKZ class equations, Rivlin–Sawyers class equations, and Oldroyd 8-constant-type equations. By using a molecular picture of a system of interest, we can zero in on the type of equation that would be most helpful for that system. We must still be careful, however, not to infer that any particular molecular model is true based on its prediction of the stress tensor, since we saw before that two very different molecular models (temporary network and elastic dumbbell in solution) can lead to the same constitutive equation. [Different configuration distribution functions $\psi(\underline{R})$ can have the same second moments.]

As a final comment on advanced constitutive equations, we note that there are some new approaches that may lead to greater understanding of the rheological behavior of materials, particularly of polymers. An example of this is the mixed micro/macro stochastic method pioneered by Öttinger and coworkers [200]. Solving for the constitutive equation predicted by a particular molecular model is often mathematically difficult. At the end of

---

[17] Actually, we do not need to calculate ψ explicitly to get $\underline{\underline{\tau}}$, since we only need the second moment of ψ to evaluate $\underline{\underline{\tau}}$; see Larson [138].

the exercise we obtain an equation we can test, and if the results are not to our liking, we must revise the molecular model and solve again, if possible. In this traditional method, it may be difficult to determine whether the failure to predict observations is due to an incorrect model or whether it is due to an inappropriate assumption made in solving for the constitutive equation.

In Öttinger et al.'s method [200], dubbed CONNFFESSIT (Calculation of non-Newtonian Flow: Finite Element Stochastic Simulation Techniques), they combine finite-element calculations of the equation of motion and the continuity equation with a stochastic simulation for the stress tensor $\underline{\tau}$. This method is computationally intense, since a large number of simulations must be averaged for the evaluation of $\underline{\underline{\tau}}$ at every element in the finite-element solution, but it is free of any mathematical simplifications used to increase solvability. Also, the methods are nimble—the molecular model can be changed readily and the flow resimulated without a great deal of work by the rheologist (although with a great deal of additional work for the computer). These methods can be used to solve for material functions and also for the velocity field and stress tensor in complex flows.

Another approach to rheological modeling is to examine the flow of complex fluids from a thermodynamic point of view. Beris and Edwards [19] describe how to model flowing systems by examining the structure and interactions among molecules or other subunits within complex flowing systems. If relatively simple macroscopic parameters can be found to describe the microscopic behavior of the fluid, the simpler macroscopic model that is found can be used to calculate flow properties. By using a macroscopic model of the fluid structure, the approach of Beris and Edwards avoids some of the tremendous computational complexity of pure ab initio calculations. In addition, this approach enforces thermodynamic consistency of the resulting equations for flow properties–for some constitutive equations developed using the approaches described in this book, the second law of thermodynamics can be violated, depending on the choice of model parameters [141, 138]. We anticipate that important advances in our understanding of rheology will come from the thermodynamic approach as it develops.

In this text we have endeavored to open up to the reader the world of polymer rheology and constitutive modeling. The scope of this book is introductory, although we have explored some advanced topics in this chapter. More on molecular modeling of polymeric and complex systems can be found in the literature [138, 139, 27, 70, 19]. The reader should understand that a constitutive model is only as good as its predictions. In some cases the models we have studied will suffice, and under other circumstances more advanced models are called for. Readers now have the tools needed to explore the rheological literature and to evaluate and choose models for their own uses.

The last chapter of the text deals with the practical problem of measuring rheological properties to compare to constitutive calculations. To choose the correct constitutive equation for modeling purposes good rheological data are essential.

9.5 PROBLEMS

9.1 You work for a company that uses polymeric materials. Your boss comes in with a 50-pound bag of polymeric pellets. She says, "We want to use this in our injection molding line, but we need to run some molding simulations. Please find out what constitutive equation we should use." Outline your strategy for complying with your boss' request.

9.2 Derive the inverse deformation-gradient tensor $\underline{\underline{F}}^{-1}$ for steady uniaxial elongational flow in the z-direction.

9.3 Derive $\underline{\underline{F}}^{-1}$ for the counterclockwise rotation of a rigid body through an angle ψ around the z-axis.

9.4 The deformation-gradient tensor was defined by writing $\underline{\underline{F}} \cdot \underline{\underline{F}}^{-1} = \underline{\underline{I}}$. By taking the time derivative of $\underline{\underline{I}}$ written this way, derive the expression for $\partial \underline{\underline{F}} / \partial t$.

9.5 For small displacements we can relate the two strain tensors $\underline{\underline{\gamma}}$ and $\underline{\underline{C}}^{-1}$ as follows: when $t - t'$ is not too large, the position vector $\underline{r}'$ at time t' is related to the same quantity at the current time by the equation

$$\underline{r} = \underline{r}' + \frac{\partial \underline{r}'}{\partial t'}(t - t')$$
$$= \underline{r}' + \underline{v}'(t - t')$$

Using this relation and the definition of $\underline{\underline{\gamma}}$ in Equation (9.5), derive the relationship between $\underline{\underline{C}}^{-1}$ and $\underline{\underline{\gamma}}$ at small strains.

9.6 Show that the Finger tensor is the inverse of the Cauchy tensor.

9.7 Consider the polar decomposition of an arbitrary tensor $\underline{\underline{A}}$ into rotational tensor $\underline{\underline{R}}$ and left and right stretch tensors $\underline{\underline{V}}$ and $\underline{\underline{U}}$, respectively. Sketch the effect of $\underline{\underline{A}}$, $\underline{\underline{R}}$, and $\underline{\underline{U}}$ on $\hat{\xi}$, a unit eigenvector of $\underline{\underline{U}}$. Do the same for $\hat{\zeta}$, a unit eigenvector of $\underline{\underline{V}}$.

9.8 Consider the polar decomposition of the deformation-gradient tensor $\underline{\underline{F}}$ for counter-clockwise rotation around z, into rotational tensor $\underline{\underline{R}}$ and left and right stretch tensors $\underline{\underline{V}}$ and $\underline{\underline{U}}$, respectively. What are $\underline{\underline{R}}$, $\underline{\underline{U}}$, and $\underline{\underline{V}}$? Show that $\underline{\underline{R}}$ is orthogonal. Discuss the meaning of your solution.

9.9 Derive the Finger and Cauchy strain tensors for counterclockwise rotation of a solid body through an angle ψ around the z-axis.

9.10 Derive the Cauchy strain tensor for shear flows. Do not assume the flow is steady.

9.11 Derive the Cauchy strain tensor for uniaxial elongational flows. Do not assume that the flow is steady.

9.12 Derive the Finger and Cauchy strain tensors for planar elongational flows. Do not assume the flow is steady.

9.13 What are the Finger and Cauchy strain tensors for uniaxial and biaxial elongational flows as expressed

in Society of Rheology nomenclature, given in Problem 5.18?

9.14 Calculate the viscosity $\eta(\dot{\gamma})$ and normal-stress coefficients $\Psi_1(\dot{\gamma})$ and $\Psi_2(\dot{\gamma})$ for the Lodge model.

9.15 Calculate the step-strain shear material functions $G(t, \gamma_0)$, $G_{\Psi_1}(t, \gamma_0)$, and $G_{\Psi_2}(t, \gamma_0)$ for the Lodge model.

9.16 Calculate the startup of steady shear material functions $\eta^+(t, \dot{\gamma}_0)$, $\Psi_1^+(t, \dot{\gamma}_0)$, and $\Psi_2^+(t, \dot{\gamma}_0)$ for the Lodge model.

9.17 Calculate the cessation of steady shear material functions $\eta^-(t, \dot{\gamma}_0)$, $\Psi_1^-(t, \dot{\gamma}_0)$, and $\Psi_2^-(t, \dot{\gamma}_0)$ for the Lodge model.

9.18 Calculate the steady-state planar extensional viscosities $\bar{\eta}_{P_1}(\dot{\epsilon}_0)$ and $\bar{\eta}_{P_2}(\dot{\epsilon}_0)$ predicted by the Lodge model.

9.19 Calculate the startup of the steady uniaxial elongation material function $\bar{\eta}^+(t, \dot{\epsilon}_0)$ for the Lodge model.

9.20 Calculate the small-amplitude oscillatory shear material functions $G'(\omega)$ and $G''(\omega)$ for the Lodge model. Also calculate the time-dependent normal-stress differences $N_1(t)$ and $N_2(t)$.

9.21 Calculate the response of the Lodge model to superimposed small-amplitude oscillatory shear and steady shear [157]. For this flow, the shear-rate function is

$$\dot{\varsigma}(t) = \dot{\gamma}_0 + a\omega \cos \omega t$$

Sketch your answers for $\tau_{21}(t)$, $N_1(t)$, and $N_2(t)$.

9.22 What is $\eta^*(\omega)$ for the single-relaxation-time Maxwell model? Does the upper convected Maxwell model follow the Cox–Merz rule?

9.23 Calculate the steady shear-flow material functions $\eta(\dot{\gamma}_0)$, $\Psi_1(\dot{\gamma}_0)$, and $\Psi_2(\dot{\gamma}_0)$ for the integral finite-strain Maxwell model based on the Cauchy tensor:

$$\underline{\underline{\tau}}(t) = + \int_{-\infty}^{t} \frac{\eta_0}{\lambda^2} e^{-\frac{(t-t')}{\lambda}} \underline{\underline{C}}(t, t') \, dt'$$

9.24 Calculate the startup of steady shear material functions $\eta^+(t, \dot{\gamma}_0)$, $\Psi_1^+(t, \dot{\gamma}_0)$ and $\Psi_2^+(t, \dot{\gamma}_0)$ for the integral finite-strain Maxwell model based on the Cauchy tensor (see Problem 9.23 for the constitutive equation).

9.25 Calculate the cessation of steady shear material functions $\eta^-(t, \dot{\gamma}_0)$, $\Psi_1^-(t, \dot{\gamma}_0)$, and $\Psi_2^-(t, \dot{\gamma}_0)$ for the integral finite-strain Maxwell model based on the

Cauchy tensor (see Problem 9.23 for the constitutive equation).

9.26 Calculate the step shear strain material functions $G(t, \gamma_0)$, $G_{\Psi_1}(t, \gamma_0)$, and $G_{\Psi_2}(t, \gamma_0)$ for the integral finite-strain Maxwell model based on the Cauchy tensor (see Problem 9.23 for the constitutive equation).

9.27 Calculate the steady uniaxial elongational viscosity material function $\bar{\eta}(\dot{\epsilon}_0)$ for the integral finite-strain Maxwell model based on the Cauchy tensor (see Problem 9.23 for the constitutive equation).

9.28 Calculate the steady planar elongational viscosity material functions $\bar{\eta}_{P_1}(\dot{\epsilon}_0)$ and $\bar{\eta}_{P_2}(\dot{\epsilon}_0)$ for the integral finite-strain Maxwell model based on the Cauchy tensor (see Problem 9.23 for the constitutive equation).

9.29 Calculate the startup of steady uniaxial elongational viscosity material function $\bar{\eta}^+(t, \dot{\epsilon}_0)$ for the integral finite-strain Maxwell model based on the Cauchy tensor (see Problem 9.23 for the constitutive equation).

9.30 Calculate the startup of steady planar elongational viscosity material functions $\bar{\eta}_{P_1}^+(t, \dot{\epsilon}_0)$ and $\bar{\eta}_{P_2}^+(t, \dot{\epsilon}_0)$ for the integral finite-strain Maxwell model based on the Cauchy tensor (see Problem 9.23 for the constitutive equation).

9.31 Calculate the material functions for step shear strain $G(t, \gamma_0)$, $G_{\Psi_1}(t, \gamma_0)$, and $G_{\Psi_2}(t, \gamma_0)$ for the finite-strain Hooke's law based on the Finger tensor $\underline{\underline{C}}^{-1}$.

9.32 Calculate the predictions of the finite-strain Hooke's law (based on $\underline{\underline{C}}^{-1}$) for stress in uniaxial extension. Compare the predictions to the data given in Table 9.9. What is the best-fit value of the modulus, G?

TABLE 9.9
Data for Problem 9.32.

Elongation Ratio l/l_0	Stress (MPa)
1.61	0.69
2.08	1.39
2.34	1.50
2.91	2.79
3.51	3.87
3.95	5.10
4.34	6.45
4.73	7.76
5.19	10.46
5.73	13.38

9.33 Show that $\overset{\triangledown}{\underline{\underline{C}}^{-1}} = 0$.

9.34 Show that $\overset{\triangle}{\underline{\underline{C}}} = 0$.

9.35 If three vectors $\underline{a}$, $\underline{b}$, and $\underline{c}$ are noncoplanar, show that $\underline{a} \cdot (\underline{b} \times \underline{c}) \neq 0$.

9.36 Show that $B \equiv \underline{b}_{(1)} \cdot \underline{b}_{(2)} \times \underline{b}_{(3)}$ is the volume of the parallelepiped formed by the vectors $\underline{b}_{(1)}$, $\underline{b}_{(2)}$, and $\underline{b}_{(3)}$.

9.37 Show using Einstein notation that even permutations of the triple product are equal, that is, that for any three nonzero vectors $\underline{a}$, $\underline{b}$, and $\underline{c}$, the following is true: $\underline{a} \cdot \underline{b} \times \underline{c} = \underline{c} \cdot \underline{a} \times \underline{b} = \underline{b} \cdot \underline{c} \times \underline{a}$.

9.38 Show that $\underline{b}_{(p)} \cdot \underline{b}^{(k)} = \delta_{pk}$ for non-orthonormal basis vectors $\underline{b}_{(p)}$ and their reciprocal vectors $b^{(k)}$.

9.39 In Chapter 2 we showed that tensors are linear vector functions. In that derivation we expressed a vector in terms of orthonormal basis vectors. Repeat the derivation without referring to orthonormal basis vectors.

9.40 Using the following vector identity:

$$(\underline{a} \times \underline{b}) \cdot (\underline{c} \times \underline{d}) = \begin{vmatrix} \underline{a} \cdot \underline{c} & \underline{b} \cdot \underline{c} \\ \underline{a} \cdot \underline{d} & \underline{b} \cdot \underline{d} \end{vmatrix}$$

show that

$$\underline{b}^{(1)} \cdot \underline{b}^{(2)} \times \underline{b}^{(3)} = \frac{1}{B}$$

and thus that the reciprocal basis vectors $\underline{b}^{(1)}$, $\underline{b}^{(2)}$, and $\underline{b}^{(3)}$ form a basis.

9.41 Beginning with the expression for $d\underline{r}$, $\underline{r} = \underline{r}(r, \theta, \phi)$ in spherical coordinates, show that the differential space vector $d\underline{r}$ is given by

$$d\underline{r} = dr\,\hat{e}_r + r\,d\theta\,\hat{e}_\theta + r\sin\theta\,d\phi\,\hat{e}_\phi$$

Note the following:

$$\hat{e}_r = \sin\theta\cos\phi\,\hat{e}_x + \sin\theta\sin\phi\,\hat{e}_y + \cos\theta\,\hat{e}_z$$

$$\hat{e}_\theta = \cos\theta\cos\phi\,\hat{e}_x + \cos\theta\sin\phi\,\hat{e}_y - \sin\theta\,\hat{e}_z$$

$$\hat{e}_\phi = -\sin\phi\,\hat{e}_x + \cos\phi\,\hat{e}_y$$

9.42 Show that $\underline{\underline{I}} = g^{ij}\underline{g}_{(i)}\underline{g}_{(j)} = g_{ij}\underline{g}^{(i)}\underline{g}^{(j)}$.

9.43 Show that g defined in Equation (9.234) is the determinant of the matrix with elements g_{ij}.

9.44 For a symmetric tensor with distinct eigenvectors ($\lambda_1 \neq \lambda_2 \neq \lambda_3$), show that the eigenvectors are mutually perpendicular. For the case where the eigenvalues are not distinct (one or more eigenvalues are repeated) see Aris [7].

9.45 Show how the invariants of a tensor and its inverse are related. Note the alternative definitions of tensor invariants given in Appendix C.6.

9.46 Show that the time derivative of the covariant convected coefficients of a tensor $\underset{=}{A}$ can be written in vector–tensor notation as follows:

$$\overset{\triangle}{\underset{=}{A}} = \frac{D\underset{=}{A}}{Dt} + \nabla \underline{v} \cdot \underset{=}{A} + \underset{=}{A} \cdot (\nabla v)^T$$

where $\underline{v}$ is the velocity field.

9.47 Show that using the Cauchy tensor in the integral Maxwell model results in the appearance of the lower convected derivative when that equation is transformed to differential form.

9.48 Consider a simple shear flow occurring on a rotating turntable (see Section 8.5 and Figure 8.12). Write the stress tensor predicted by the Lodge equation using a stationary coordinate system and one that is rotating around the z-axis with the speed of the turntable. Compare the two results. What does this allow you to conclude about the Lodge equation?

9.49 The Oldroyd B model has found success in modeling polymer solutions. This model can be thought of as a straightforward sum of an upper convected Maxwell model for polymeric contributions and a Newtonian model for solvent contributions:

$$\underset{=}{\tau} = \underset{=}{\tau}^p + \underset{=}{\tau}^s$$

where the polymeric contribution $\underset{=}{\tau}^p$ follows the upper convected Maxwell equation, and the solvent contribution $\underset{=}{\tau}^s$ follows the Newtonian equation. Beginning with this formulation, show that the Oldroyd B model can also be written as

$$\underset{=}{\tau} + \lambda_1 \overset{\triangledown}{\underset{=}{\tau}} = -\eta_0 \left(\underset{=}{\dot{\gamma}} + \lambda_2 \overset{\triangledown}{\underset{=}{\dot{\gamma}}} \right)$$

where η_0, λ_1, and λ_2 are scalar parameters of the model.

9.50 In the derivation of the integral version of the Jeffreys model, show that Equation (9.317) becomes Equation (9.318).

9.51 Show that the generalized linear viscoelastic equation with relaxation modulus $G(t - t')$ given below is the equivalent of the integral Jeffreys model:

$$G(t - t') = \left[\frac{\eta_0}{\lambda_1} \left(1 - \frac{\lambda_2}{\lambda_1} \right) e^{\frac{-(t-t')}{\lambda_1}} \right] + 2 \frac{\eta_0 \lambda_2}{\lambda_1} \delta(t - t')$$

9.52 Graphically compare the predictions of the Lodge and Oldroyd B (convected Jeffreys) models in steady shear and in steady extension. Carry out your comparison using the same value of the relaxation time λ_1, but use different values for the retardation time λ_2.

9.53 Rewrite the Oldroyd 8-constant model in terms of the corotational time derivatives of stress and deformation rate, $\overset{\circ}{\underset{=}{\tau}}$ and $\overset{\circ}{\underset{=}{\dot{\gamma}}}$.

9.54 Show that the determinant of $\underset{=}{C}^{-1} = 1$ for incompressible fluids.

9.55 Show that for an encompressible fluid the two finite-strain tensors $\underset{=[0]}{\gamma}$ and $\overset{[0]}{\underset{=}{\gamma}}$ are related as follows [26]:

$$\overset{[0]}{\underset{=}{\gamma}} = (I_2 - I_1)\underset{=}{I} + (I_1 - 2)\underset{=[0]}{\gamma} + (\underset{=[0]}{\gamma} \cdot \underset{=[0]}{\gamma})$$

where I_1 and I_2 are the first two invariants of $\underset{=}{C}^{-1}$ as defined in Appendix C.6. (*Hint:* use the Cayley–Hamilton theorem, which is defined in the glossary in Appendix B.)

9.56 Show how the invariants of $\underset{=}{C}^{-1}$ and $\underset{=[0]}{\gamma}$ are related [26].

9.57 Show that the stress tensor in rubber elasticity theory is as given below. In other words, show that Equation (9.364) results from Equation (9.362):

$$\underset{=}{\tau} = -\nu k T \lambda_i^2 \, \hat{e}_i \hat{e}_i$$

9.58 Show that

$$\int_{-a}^{+b} f(x)\delta(x)dx = f(0)$$

using the following approximation for $\delta(x)$:

$$\delta(x) = \lim_{n \to \infty} \sqrt{\frac{n}{\pi}} e^{-nx^2}$$

$f(x)$ is an arbitrary function, a and b are positive, and $a < b$.

CHAPTER 10

Rheometry

In this text we have concentrated on the study of the behavior of a continuum subjected to external forces. Three equations were discussed: the equation of motion (momentum conservation), the continuity equation (mass conservation), and the constitutive equation, which is a material-specific equation that indicates how deformation and stress are related. These equations are all the relations that are needed to solve flow problems. On the practical side, we seek to apply continuum calculations to real-life problems involving complex fluids and complex geometries, and we want our results to be accurate. The decision that has perhaps the most profound impact on the accuracy of such simulations is the choice of constitutive equation.

The only way to determine whether a constitutive equation reflects the behavior of a material accurately is to measure properties of that material and to compare the measured results with predictions made by the constitutive equation. Making measurements of rheological material functions is called *rheometry*. To measure a material function we must design an experiment to produce the kinematics prescribed in the definition of the material function, and then we must measure the stress components needed and calculate the material function.[1] In this chapter we will discuss several of the more common techniques used to measure shear and elongational material functions. A more extensive discussion of flow geometries, including formulas for stresses and strain rates, may be found in Bird et al. [26]. Many experimental effects due to instrument design and operation also affect rheological measurement; Walters' classic text [258] provides a comprehensive discussion of such experimental issues.

Stress is usually measured mechanically using, for example, a strain-gauge or a force-rebalance transducer [216, 32], but this is not always the case. The last section of this chapter introduces a nonmechanical method for measuring stress, flow birefringence, which may be coupled with any of the flow geometries described. In contrast to most mechanical methods, which involve integrating a force or torque over a measurement surface, birefringence allows for the local measurement of stress.

---

[1] Creep is an exception to this. In creep experiments the stress is applied, and the deformation is measured; see Chapter 5.

10.1 Shear Flow

Most rheological measurements are performed in one of the four shear geometries discussed in this section: capillary flow, parallel-plate and cone-and-plate torsional flow, and Couette flow. This is because of ease of experimentation. Shear flow is important in situations where the viscosity is the dominant material property, such as in flows near walls and in mixing applications.

10.1.1 CAPILLARY FLOW

Flow through a capillary is a unidirectional flow in which cylindrical surfaces slide past each other as in a collapsible telescope (Figure 10.1). Near the tube walls, except for the curvature of these surfaces in the θ-direction, this flow is the same as the simple shear prescribed in the definition of viscosity.[2] To see how to calculate viscosity from measurable properties in capillary flow we must relate cylindrical coordinates, which are the natural coordinates in which to analyze flow in a tube, and shear coordinates, 1, 2, 3, in which the material functions are defined:

$$\underline{v} = \dot{\gamma}_0 x_2 \hat{e}_1 = \begin{pmatrix} \dot{\gamma}_0 x_2 \\ 0 \\ 0 \end{pmatrix}_{123} \tag{10.1}$$

We can relate the usual cylindrical coordinates of this problem, r, θ, z, with the shear coordinate system near the wall, 1, 2, 3, as follows: $\hat{e}_z$ is the flow (1) direction, $-\hat{e}_r$ is the gradient (2) direction, and $-\hat{e}_\theta$ is the neutral (3) direction. We take $-\hat{e}_r$ as the gradient direction so that the shear stress τ_{21} represents a positive flux of momentum in the negative x_2-direction, as is usual in our definition of shear flow (Figure 10.2). We must take $-\hat{e}_\theta$ as the 3-direction to maintain a right-handed coordinate system. Thus we can relate the stress and shear rate in capillary flow with these quantities in the shear coordinate system:

$$\tau_{21} = -\tau_{rz}|_{r=R} \tag{10.2}$$

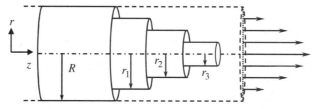

Figure 10.1 Shearing surfaces for Poiseuille flow in a tube are cylinders sliding past one another.

---

[2] Here we concentrate on simple shear flow. One can show [26] that the shear flows discussed in this chapter are all part of a more general classification of shear flows called viscometric flows. The material functions of Chapter 5 are all equally valid for viscometric flows. This implies that issues of curvature mentioned here and in subsequent sections are not significant.

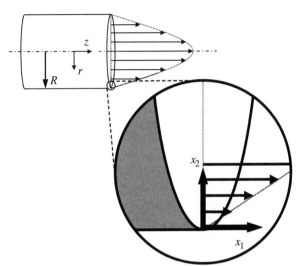

Figure 10.2 Coordinate system for shear flow, x_1, x_2, x_3, as compared to the usual cylindrical coordinate system, r, θ, z, for flow in a tube.

$$\dot{\gamma}_0 = \frac{\partial v_z}{\partial(-r)} = -\frac{\partial v_z}{\partial r} \tag{10.3}$$

The rate-of-deformation tensor for Poiseuille flow in a tube is

$$\underline{\dot{\gamma}} = \nabla \underline{v} + (\nabla \underline{v})^T \tag{10.4}$$

$$= \begin{pmatrix} 0 & 0 & \frac{\partial v_z}{\partial r} \\ 0 & 0 & 0 \\ \frac{\partial v_z}{\partial r} & 0 & 0 \end{pmatrix}_{r\theta z} \tag{10.5}$$

We see that $-\partial v_z/\partial r|_{r=R}$ is the shear rate at the wall $\dot{\gamma}_R$:

$$\dot{\gamma} = |\underline{\dot{\gamma}}| = -\frac{\partial v_z}{\partial r} \tag{10.6}$$

$$\dot{\gamma}(R) = -\left.\frac{\partial v_z}{\partial r}\right|_{r=R} \equiv \dot{\gamma}_R \tag{10.7}$$

We can now calculate viscosity in terms of variables associated with capillary flow:

$$\eta = \frac{-\tau_{21}}{\dot{\gamma}_0} = \frac{\tau_{rz}|_{r=R}}{-\left.\frac{\partial v_z}{\partial r}\right|_{r=R}} = \frac{\tau_{rz}|_{r=R}}{\dot{\gamma}_R} \tag{10.8}$$

Therefore to determine viscosity from capillary-flow experiments, we need expressions for wall shear stress $\tau_{rz}|_{r=R}$ and wall shear rate $\dot{\gamma}_R$ in terms of experimental variables.

10.1.1.1 Shear Stress in Capillary Flow

We wish to consider pressure-driven flow (Poiseuille flow) in a tube of circular cross section (see Figure 3.10) for a general fluid, that is, for a fluid for which the constitutive equation is unknown. We assume that the fluid is incompressible and the flow is unidirectional. The problem is addressed in cylindrical coordinates,

$$\underline{v} = \begin{pmatrix} v_r \\ v_\theta \\ v_z \end{pmatrix}_{r\theta z} = \begin{pmatrix} 0 \\ 0 \\ v_z \end{pmatrix}_{r\theta z} \tag{10.9}$$

$$\nabla \cdot \underline{v} = \frac{\partial v_z}{\partial z} = 0 \tag{10.10}$$

For steady-state unidirectional flow, the left side of the equation of motion (inertial contribution) is zero. By combining the pressure and gravity terms as was done in the Newtonian and power-law solutions, the equation of motion for the general fluid in Poiseuille flow simplifies to

$$\underline{0} = -\nabla \mathcal{P} - \nabla \cdot \underline{\underline{\tau}} \tag{10.11}$$

$$\begin{pmatrix} 0 \\ 0 \\ 0 \end{pmatrix}_{r\theta z} = \begin{pmatrix} -\frac{\partial \mathcal{P}}{\partial r} \\ -\frac{1}{r}\frac{\partial \mathcal{P}}{\partial \theta} \\ -\frac{\partial \mathcal{P}}{\partial z} \end{pmatrix}_{r\theta z} - \begin{pmatrix} \frac{1}{r}\frac{\partial}{\partial r}(r\tau_{rr}) + \frac{1}{r}\frac{\partial \tau_{\theta r}}{\partial \theta} + \frac{\partial \tau_{zr}}{\partial z} - \frac{\tau_{\theta\theta}}{r} \\ \frac{1}{r^2}\frac{\partial}{\partial r}(r^2\tau_{r\theta}) + \frac{1}{r}\frac{\partial \tau_{\theta\theta}}{\partial \theta} + \frac{\partial \tau_{z\theta}}{\partial z} + \frac{\tau_{\theta r}-\tau_{r\theta}}{r} \\ \frac{1}{r}\frac{\partial}{\partial r}(r\tau_{rz}) + \frac{1}{r}\frac{\partial \tau_{\theta z}}{\partial \theta} + \frac{\partial \tau_{zz}}{\partial z} \end{pmatrix}_{r\theta z} \tag{10.12}$$

where $\mathcal{P} = p - \rho g z$.

To proceed further we must make some assumptions that are compatible with the experimental realities of actually implementing this flow. The first assumption is that the stresses and pressure are θ-independent; thus each term in Equation (10.12) with a derivative with respect to θ may be eliminated. The flow field does vary with z in an actual measurement since the material enters the capillary from an upstream reservoir of larger diameter, as shown in Figure 10.3. Velocity rearrangement and elasticity due to the elongational nature of the contraction flow at the entrance affect the stresses in that area. Also, when the melt leaves the capillary, the velocity field near the exit will differ from the fully-developed flow field in the main section of the tube. If the capillary is long, however, the impact of these end effects, as they are called, is diminished. We will assume that the capillary tube is long and thus that there are no z-variations in velocity or stress components. (We will discuss how to take end effects into account in Section 10.1.1.3.) Finally, the stress tensor will be assumed to be symmetric. Thus the equation of motion becomes

$$\begin{pmatrix} 0 \\ 0 \\ 0 \end{pmatrix}_{r\theta z} = \begin{pmatrix} -\frac{\partial \mathcal{P}}{\partial r} \\ 0 \\ -\frac{\partial \mathcal{P}}{\partial z} \end{pmatrix}_{r\theta z} - \begin{pmatrix} \frac{1}{r}\frac{\partial}{\partial r}(r\tau_{rr}) - \frac{\tau_{\theta\theta}}{r} \\ \frac{1}{r^2}\frac{\partial}{\partial r}(r^2\tau_{r\theta}) \\ \frac{1}{r}\frac{\partial}{\partial r}(r\tau_{rz}) \end{pmatrix}_{r\theta z} \tag{10.13}$$

The θ-component can be solved for $\tau_{\theta r}$:

$$\tau_{\theta r} = \frac{C_1}{r^2} \tag{10.14}$$

The integration constant C_1 can be evaluated for the boundary condition that at $r = 0$ the stress is finite. Thus, $\tau_{\theta r} = 0$. The z-component of the equation of motion gives us an expression for the shear stress $\tau_{rz}(r)$:

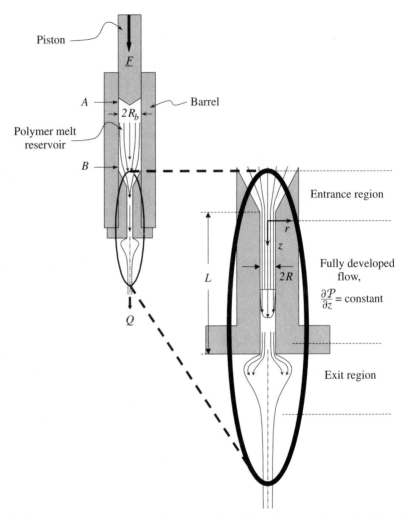

Piston

$\underline{F}$

A

Barrel

$2R_b$

Polymer melt
reservoir

B

Entrance region

L

r

z

$2R$

Fully developed
flow,

$\dfrac{\partial \mathcal{P}}{\partial z} = $ constant

Q

Exit region

Figure 10.3 Geometry used in one type of commercial capillary rheometer. The bulk of the polymer being tested resides in the upstream reservoir. A piston pushes the fluid through a small capillary of radius R, and the material exits at the bottom at a flow rate Q. Many polymers exhibit die swell, a phenomenon in which the diameter of the exiting fluid can be several times the diameter of the capillary from which it is flowing. The figure is not drawn to scale; the ratio of the capillary radius R to the barrel radius R_b is typically 10 or 12 to 1.

z-component:
$$-\frac{\partial \mathcal{P}(r, z)}{\partial z} = \frac{1}{r}\frac{\partial}{\partial r}\left[r \tau_{rz}(r)\right] \qquad (10.15)$$

So far the modified pressure field $\mathcal{P}$ is a function of r and z. To explore the r-dependence of $\mathcal{P}$, we examine the r-component of the equation of motion:

r-component:
$$-\frac{\partial \mathcal{P}}{\partial r} = \frac{1}{r}\frac{\partial}{\partial r}(r \tau_{rr}) - \frac{\tau_{\theta\theta}}{r} \qquad (10.16)$$

We can write the normal-stress components in terms of the second normal-stress difference, $N_2 = \tau_{rr} - \tau_{\theta\theta}$,

$$-\frac{\partial \mathcal{P}}{\partial r} = \frac{\partial \tau_{rr}}{\partial r} + \frac{\tau_{rr}}{r} - \frac{\tau_{\theta\theta}}{r} \tag{10.17}$$

$$= \frac{\partial N_2}{\partial r} + \frac{N_2}{r} + \frac{\partial \tau_{\theta\theta}}{\partial r} \tag{10.18}$$

From Equation (10.18) we see that for materials for which N_2 is small or zero and for which $\tau_{\theta\theta}$ is independent of r, $\mathcal{P}$ is only a function of z, and we can readily solve Equation 10.15 by separation of variables as we did for Newtonian fluids in Chapter 3 and for GNFs in Chapter 7. As we discussed in Chapter 5, $\Psi_2 = -N_2/\dot{\gamma}_0^2$ has been found to be a very small (negative) quantity for polymers. Less is specifically known about $\tau_{\theta\theta}$, but it seems reasonable to assume that this stress will be small or zero in a flow with assumed θ-symmetry. Thus the condition $N_2 = 0 = \partial \tau_{\theta\theta}/\partial r$ should be met easily by most materials.

For materials with nonzero N_2 or $\partial \tau_{\theta\theta}/\partial r$, we can still solve Equation 10.15 as long as the derivative $\partial \mathcal{P}/\partial z$ is constant in the flow domain. This condition is compatible with r-dependence of $\mathcal{P}$, as we can see by integrating $\partial \mathcal{P}/\partial z$ to obtain $\mathcal{P} = (\text{constant})z + f(r)$, where $f(r)$ is an unknown function of r. Thus, as long as the pressure distribution is of this form for the fluid in question, we may proceed with the solution of Equation 10.15 even if $N_2 \neq 0$ or $\partial \tau_{\theta\theta}/\partial r \neq 0$.

Returning to the z-component of the equation of motion, we arrive at the following:

$$-\frac{d\mathcal{P}(z)}{dz} = \frac{1}{r}\frac{\partial}{\partial r}\left[r\tau_{rz}(r)\right] \tag{10.19}$$

This is the same separable differential equation we encountered when solving this flow problem for Newtonian and power-law generalized Newtonian fluids. If the boundary conditions on pressure are $\mathcal{P}(0) = \mathcal{P}_0$, $\mathcal{P}(L) = \mathcal{P}_L$, the solution is

$$\tau_{rz} = \frac{\mathcal{P}_0 - \mathcal{P}_L}{L}\frac{r}{2} + \frac{C_1}{r} \tag{10.20}$$

where C_1 is the integration constant. For finite stress at $r = 0$, the integration constant is zero, and we obtain

Shear stress in
capillary flow

$$\boxed{\tau_{rz} = \frac{(\mathcal{P}_0 - \mathcal{P}_L)r}{2L} = \tau_R\frac{r}{R}} \tag{10.21}$$

where $\tau_R = (\mathcal{P}_0 - \mathcal{P}_L)R/2L$ is the shear stress at the wall. Our assumptions in arriving at this point are listed in Table 10.1.

We have derived an expression for shear stress at the wall τ_R, which applies for nearly all materials and is calculable from experimental measurements of $\Delta \mathcal{P}$ and a knowledge of the geometric constants R and L. To obtain the viscosity, we now need to find a way to express the wall shear rate $\dot{\gamma}_R$ in terms of experimentally measured quantities.

10.1.1.2 Shear Rate in Capillary Flow

We seek an expression for $-\partial v_z/\partial r|_{r=R}$. If the velocity field is known, it is straightforward to calculate the wall shear rate $\dot{\gamma}_R$. For Newtonian fluids, for example (see Section 3.5.2),

TABLE 10.1
Assumptions for Poiseuille Flow in a Capillary

1. Unidirectional flow
2. Incompressible fluid
3. θ-symmetry
4. Long capillary so that z-variation is negligible
5. Symmetric stress tensor
6. $\partial \mathcal{P} / \partial z = $ constant
7. Finite stress at $r = 0$

$$v_z(r) = \frac{2Q}{\pi R^2} \left[1 - \left(\frac{r}{R} \right)^2 \right] \tag{10.22}$$

$$\dot{\gamma} = -\frac{dv_z}{dr} = \frac{4Q}{\pi R^3} \frac{r}{R} \tag{10.23}$$

$$\dot{\gamma}_R = \frac{4Q}{\pi R^3} \tag{10.24}$$

and the viscosity is calculated as

$$\eta \equiv \frac{-\tau_{21}}{\dot{\gamma}_0} = \frac{\tau_R}{\dot{\gamma}_R} \tag{10.25}$$

$$= \frac{(\mathcal{P}_0 - \mathcal{P}_L)R}{2L} \left(\frac{\pi R^3}{4Q} \right) = \mu \tag{10.26}$$

where we have used the expression for Q we derived earlier for a Newtonian fluid [Hagen–Poiseuille law, Equation (3.219)] to simplify the last step. From Equation (10.26) we see that if data of pressure drop and flow rate are taken for a Newtonian fluid, an accurate calculation of the viscosity can be obtained by plotting the wall shear rate $4Q/\pi R^3$ versus the wall shear stress $(\mathcal{P}_0 - \mathcal{P}_L)R/2L$ and taking the inverse of the slope (Figure 10.4):

$$\frac{4Q}{\pi R^3} \equiv \dot{\gamma}_a = \frac{1}{\mu} \frac{(\mathcal{P}_0 - \mathcal{P}_L)R}{2L} \tag{10.27}$$

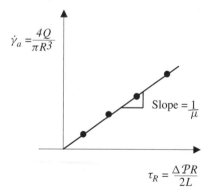

Figure 10.4 Viscosity is obtained from pressure-drop and flow-rate information on a Newtonian fluid.

Newtonian fluid
in capillary flow

$$\dot{\gamma}_a = \frac{1}{\mu}\tau_R \qquad (10.28)$$

The quantity $4Q/\pi R^3$, the shear rate at the wall for a Newtonian fluid, is also called the *apparent* shear rate when non-Newtonian fluids are studied, and it is given the symbol $\dot{\gamma}_a$. There is actually nothing apparent about this quantity; $\dot{\gamma}_a$ is simply what the shear rate at the wall would have been if the material had been Newtonian. This combination of variables appears in non-Newtonian expressions, as we shall see.

We can carry out the calculation of shear rate at the wall for a power-law generalized Newtonian fluid since, again, the velocity field [Equation (7.85)] is known. The result is

$$v_z = R^{\frac{1}{n}+1}\left(\frac{\mathcal{P}_0 - \mathcal{P}_L}{2mL}\right)^{\frac{1}{n}}\left(\frac{1}{1/n+1}\right)\left[1-\left(\frac{r}{R}\right)^{\frac{1}{n}+1}\right] \qquad (10.29)$$

$$\dot{\gamma}_R = -\left.\frac{dv_z}{dr}\right|_{r=R} \qquad (10.30)$$

$$= \left(\frac{\tau_R}{m}\right)^{\frac{1}{n}} = \left(\frac{4Q}{\pi R^3}\right)\left(\frac{1/n+3}{4}\right) \qquad (10.31)$$

where to obtain the last equation we have employed Equation (7.92) for the flow rate Q for a power-law fluid. Note that for a Newtonian fluid ($n = 1$, $m = \mu$), this reduces to the expression obtained earlier.

The equation for wall shear rate for a power-law fluid has the unknown parameter n in it. A closer look at Equation (10.31) shows that we can calculate n from a double-log plot of experimentally measured $\dot{\gamma}_a$ versus τ_R:

$$\log\left(\frac{4Q}{\pi R^3}\right) = \frac{1}{n}\log\tau_R + \log\left(\frac{4m^{-\frac{1}{n}}}{1/n+3}\right) \qquad (10.32)$$

Power-law GNF
in capillary flow

$$\log\dot{\gamma}_a = \frac{1}{n}\log\tau_R + \log\left(\frac{4m^{-\frac{1}{n}}}{1/n+3}\right) \qquad (10.33)$$

The parameters n and m of the power-law model may be calculated from the values obtained for the slope and intercept of this line (Figure 10.5). Thus, to measure the viscosity for an unknown fluid believed to be a power-law generalized Newtonian fluid, pressure-drop and flow-rate data are collected on the fluid (the pressure drop is set, and the flow rate is measured, or vice versa [114, 92]), and n and m are obtained from the log–log graph of $\dot{\gamma}_a$ versus τ_R. The raw pressure-drop and flow-rate data can then be converted to viscosity versus shear rate using the value of n calculated, the equation for the wall shear stress $\tau_R = (\mathcal{P}_0 - \mathcal{P}_L)R/2L$, and Equations (10.25) and (10.31).

For both Newtonian and power law fluids we used $v_z(r)$ to calculate $\dot{\gamma}_R$, which is needed to calculate viscosity. For a general fluid we must calculate $\dot{\gamma}_R$ without knowing $v_z(r)$. To see how to proceed, we observe that in both cases quantities related to the viscosity

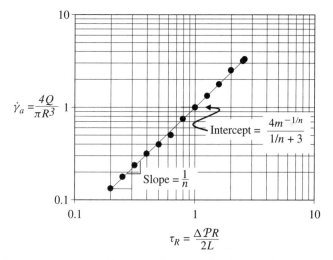

Figure 10.5 Parameters m and n of the power-law generalized Newtonian fluid model are obtained from pressure-drop and flow-rate information.

were obtained from different types of plots of $\dot{\gamma}_a$ versus τ_R. We also saw that the shear rate was related to the flow rate and geometric parameters. By pursuing expressions for $\dot{\gamma}_a$ as a function of τ_R for a general fluid, we will see that one can relate pressure drop and flow rate to viscosity without assuming a velocity field.

The general expression for viscosity from capillary data is due to Weissenberg and Rabinowitsch [261]. To get $\dot{\gamma}_R$, we begin by manipulating the general equation for flow rate in a tube:

$$Q = 2\pi \int_0^R v_z(r) r \, dr \tag{10.34}$$

The shear rate for this flow is $\dot{\gamma} = |\dot{\underline{\underline{\gamma}}}| = -dv_z/dr$. We can introduce $\dot{\gamma}$ as a variable in Equation (10.34) by integrating by parts; the result is[3]

$$Q = \pi \int_0^R \dot{\gamma} \, r^2 \, dr \tag{10.38}$$

We have assumed no slip at the wall, $v_z(R) = 0$. We have related flow rate to shear rate, but from the results for Newtonian and generalized Newtonian fluids, we suspect that

---

[3] The details of this calculation are as follows. Integration by parts,

$$\int_a^b u \, dv = u \, v|_a^b - \int_a^b v \, du \tag{10.35}$$

$$u = v_z \qquad dv = r \, dr \tag{10.36}$$

$$du = \frac{dv_z}{dr} dr \qquad v = \frac{r^2}{2} \tag{10.37}$$

we want an equation involving shear rate and shear stress. To introduce shear stress into Equation (10.38) we can use Equation (10.21) for τ_{rz} and perform a change of variable to eliminate r:

$$\tau_{rz} = \tau_R \frac{r}{R} \tag{10.39}$$

$$Q = \frac{\pi R^3}{\tau_R^3} \int_0^{\tau_R} \dot{\gamma} \, \tau_{rz}^2 \, d\tau_{rz} \tag{10.40}$$

$$\frac{4Q}{\pi R^3} \equiv \dot{\gamma}_a = \frac{4}{\tau_R^3} \int_0^{\tau_R} \dot{\gamma}(\tau_{rz}) \, \tau_{rz}^2 \, d\tau_{rz} \tag{10.41}$$

We replaced $4Q/\pi R^3$ with $\dot{\gamma}_a$ in Equation (10.41) in anticipation of obtaining an equation relating $\dot{\gamma}_a$ and τ_R. To eliminate the integral we can now differentiate Equation (10.41) with respect to τ_R. We will use the Leibnitz rule,

$$\dot{\gamma}_a \tau_R^3 = 4 \int_0^{\tau_R} \dot{\gamma}(\tau_{rz}) \, \tau_{rz}^2 \, d\tau_{rz} \tag{10.42}$$

$$\frac{d}{d\tau_R} \left(\dot{\gamma}_a \tau_R^3 \right) = 4 \int_0^{\tau_R} \frac{\partial}{\partial \tau_R} \left[\dot{\gamma}(\tau_{rz}) \tau_{rz}^2 \right] d\tau_{rz} + 4\dot{\gamma}(\tau_R) \, \tau_R^2 \tag{10.43}$$

The first term on the right side is zero, and we can expand the left side using the product rule. If we recall that $d \ln x = dx/x$, we arrive at the following compact form for $\dot{\gamma}(\tau_R) \equiv \dot{\gamma}_R$, the shear rate at the wall in capillary flow:

Wall shear rate in capillary flow of power-law GNF

$$\dot{\gamma}(\tau_R) \equiv \dot{\gamma}_R = \dot{\gamma}_a \left[\frac{1}{4} \left(3 + \frac{d \ln \dot{\gamma}_a}{d \ln \tau_R} \right) \right] \tag{10.44}$$

The quantity in square brackets is called the *Weissenberg–Rabinowitsch correction*. For Newtonian fluids the correction becomes 1, and $\dot{\gamma}_R = \dot{\gamma}_a$ as before. This correction allows us to calculate the shear rate at the wall without assuming any form for the velocity profile. The Weissenberg–Rabinowitsch correction accounts for the differences in shear rates between the Newtonian case and the general case due to the fact that the velocity profiles for non-Newtonian fluids in capillary flow are nonparabolic (Figure 10.6).

The viscosity is now obtained from Equation (10.8):

$$\eta \equiv \frac{-\tau_{21}}{\dot{\gamma}_0} \tag{10.45}$$

$$\eta(\dot{\gamma}_R) = \frac{\tau_R}{\dot{\gamma}_R} \tag{10.46}$$

Any homogeneous fluid in capillary flow

$$\eta(\dot{\gamma}_R) = \frac{4\tau_R}{\dot{\gamma}_a} \left(3 + \frac{d \ln \dot{\gamma}_a}{d \ln \tau_R} \right)^{-1} \tag{10.47}$$

The viscosity may therefore be determined in capillary flow from measurements of Q (needed to calculate $\dot{\gamma}_R$) and $\Delta P \equiv P_0 - P_L$ (needed to calculate τ_R) and the geometric

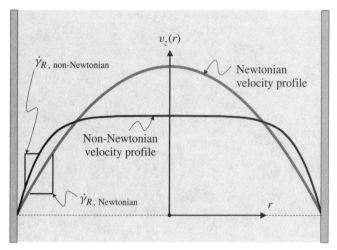

Figure 10.6 Schematic of the effect of material properties on the shear rate at the wall. For shear-thinning fluids such as shown in the figure, the shear rate at the wall is higher than in a Newtonian fluid with the same average velocity.

constants R and L. The slope of the double-logarithmic plot of $\dot{\gamma}_a$ versus τ_R is used with each data pair, $(\tau_R, \dot{\gamma}_a)$ to calculate $\eta(\dot{\gamma}_R)$ (Figure 10.7).

The expression derived for the viscosity is based on properties of the fluid near the wall. If the properties of the fluid at the wall are representative of the properties of the fluid in general (which they are believed to be for polymer melts and other similar systems), the viscosity measured in capillary flow may be relied upon. Care must be taken, however, with suspensions and other complex systems that have been shown to exhibit unusual behavior at the walls [162].

The next two sections describe corrections that may be needed in capillary rheometry to account for end effects and wall slip.

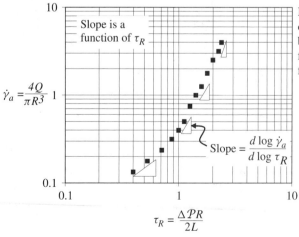

Figure 10.7 Schematic of how the derivative in the Weissenberg–Rabinowitsch correction is obtained from pressure-drop and flow-rate information for any type of fluid.

10.1.1.3 Entrance and Exit Effects—Bagley Correction

One of our assumptions in the derivation of τ_R in a capillary was that the capillary is long and that therefore velocity variations in the z-direction may be neglected. We described a flow where the modified pressure difference was $\mathcal{P}_0 - \mathcal{P}_L$ over a length L.

In an actual capillary rheometer, however, the flow takes some time to develop at the inlet, and for polymers, the flow at the exit is disturbed by die swell. These two effects introduce some z-variation in the velocity. This z-variation is not a problem as long as we only calculate $\mathcal{P}_0 - \mathcal{P}_L$ and L over the portion of the capillary where steady fully developed unidirectional flow occurs (see Figure 10.3). The problem is that the pressure drop over the portion of the flow that is steady and fully developed is unknown. We also do not know the true length of the region that is in steady fully developed flow.

Measurements of pressure drop across a capillary are typically made in an apparatus such as that shown in Figure 10.3. Gravity effects are neglected, $\mathcal{P} = p$. The pressure at the top of the capillary p_0 (point B in Figure 10.3) is related to the force per unit area $F/\pi R_b^2$ that is required to move the piston at a steady rate. The pressure drop across the wide barrel is neglected. The pressure at the bottom of the capillary p_L is atmospheric. The pressure drop over the entire capillary is then just the difference between these two pressures:

$$p_0 = P_{\text{atm}} + \frac{F}{\pi R_b^2} \tag{10.48}$$

$$p_L = P_{\text{atm}} \tag{10.49}$$

$$p_0 - p_L = \frac{F}{\pi R_b^2} \tag{10.50}$$

where R_b is the radius of the barrel and P_{atm} is atmospheric pressure. The total length of the flow is taken to be the entire capillary length L. Ignored is the fact that the polymer flow must rearrange at the capillary entrance and exit.

The effect of the pressure drop in the barrel can be eliminated by measuring the pressure at the inlet of the capillary independently, that is, by placing a transducer at point B in Figure 10.3. This feature is present on some rheometers [92]. The end effects can be accounted for by observing the effect of changing the ratio of capillary length to diameter at constant shear rate, as will now be explained.

Runs performed on a capillary rheometer at constant wall shear rate (constant flow rate Q) should always generate the same wall shear stresses, that is, $\tau_R = \Delta p R/2L = $ constant for a given material at a fixed temperature. Runs performed at constant flow rate but with different capillaries (changing R or L) should therefore result in measured values of Δp as follows:

$$\Delta p = 2\tau_R \frac{L}{R} \tag{10.51}$$

Thus a plot of Δp versus L/R at constant shear rate (i.e., constant Q) is a line through the origin of slope $2\tau_R$ when no end effects are present. When end effects are present, however, this line will not go through the origin.

Experimental results on short capillaries for some highly elastic materials show that while plots of Δp versus L/R do form straight lines, the y-intercept is not zero ([10], see

Figure 10.8). The value of the y-intercept on a plot of Δp versus L/R is the combined entrance and exit pressure loss due to rearrangements of the velocity profile at the entrance and exit. Capillary data may be corrected for end effects by subtracting the pressure-axis intercept of the Δp versus L/R plot from the values of pressure used to calculate τ_R. Equivalently, the end effects can be corrected by adding e to the value of L/R used in calculating τ_R ([61], see Figure 10.8). This is called the Bagley correction [10].

10.1.1.4 Wall Slip

In the derivation of the Weissenberg–Rabinowitsch equation we assumed no slip at the wall of the capillary. This is standard for tube flow, but there is evidence that this condition may occasionally be violated.

To determine what observations would accompany wall slip, we can examine a slipping system such as that shown in Figure 10.9. The effect of slip is to reduce the deformation experienced by the fluid. In the case where slip is occurring, as compared to the no-slip case, the shear rate is reduced throughout, but especially near the wall. The shear stress at the wall, $\tau_R = \Delta \mathcal{P} R/2L$, is unaffected by the change in boundary condition (see the beginning of Section 10.1.1).

To calculate a viscosity in a situation where slip is occurring, one must calculate the true shear rate near the wall; this analysis is due to Mooney [183]. The first step is to correct

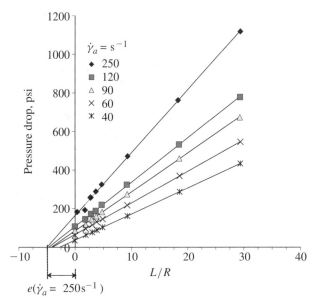

Figure 10.8 Pressure drop versus L/R for polyethylene; melt index $= 2.9$, $T = 190°C$ [10]. Each curve is taken at a constant value of apparent shear rate $\dot{\gamma}_a = 4Q/\pi R^3$. The correction for end effects at each apparent shear rate may be calculated either from the y-intercept or from the extrapolated x-intercept e. Reprinted with permission from E. B. Bagley, "End Corrections in the Capillary Flow of Polyethylene," *Journal of Applied Physics*, **28**, 624–627, (1957). Copyright © 1957, American Institute of Physics.

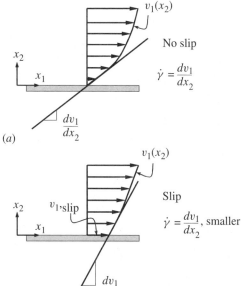

Figure 10.9 System undergoing shear flow. (a) No slip. (b) Slip at the wall. For flow in a capillary, $v_1(x_2) = v_z(-r)$.

the apparent shear rate $\dot{\gamma}_a$ for slip. The corrected $\dot{\gamma}_a$ may then be used in the Weissenberg–Rabinowitsch calculation [Equation (10.44)] to account for nonparabolic velocity profile. The apparent shear rate is normally given by (no slip)

$$\dot{\gamma}_a \equiv \frac{4Q}{\pi R^3} = \frac{4v_{z,av}}{R} \tag{10.52}$$

where $v_{z,av} = Q/\pi R^2$ is the average fluid velocity in the tube. When slip occurs, this calculation of $\dot{\gamma}_a$ is too large since much of $v_{z,av}$ goes into slip at the wall. A corrected value for $\dot{\gamma}_a$ may be obtained by substituting $v_{z,av} - v_{z,slip}$ for $v_{z,av}$ in this expression, where $v_{z,slip}$ is the wall slip velocity,

$$\dot{\gamma}_{a,\text{slip-corrected}} = \frac{4v_{z,av}}{R} - \frac{4v_{z,slip}}{R} \tag{10.53}$$

If we postulate that the slip velocity $v_{z,slip}$ is only a function of wall shear stress τ_R then plots of $4v_{z,av}/R = 4Q_{\text{measured}}/\pi R^3$ versus $1/R$ at constant τ_R would give straight lines with a slope of $4v_{z,slip}$ and an intercept of $\dot{\gamma}_{a,\text{slip-corrected}}$. Ramamurthy's data on a linear low-density polyethylene melt [214] show just this trend (Figure 10.10). Conversely, if the plots of $4Q/\pi R^3$ versus $1/R$ at constant τ_R have a slope of zero, no slip has been achieved in the experiments. However, the Mooney technique is only an indirect measurement of $v_{z,slip}$, based on a postulate that slip is occurring. Other violations of our assumptions could be responsible for the nonzero slope measured, for instance, the possible contributions of entrance losses [208], instability, compressibility, or normal stresses.

An alternative method for reporting on slip effects is based on the quantity called the extrapolation length b (Figure 10.11), which is the distance in the negative x_2-direction at

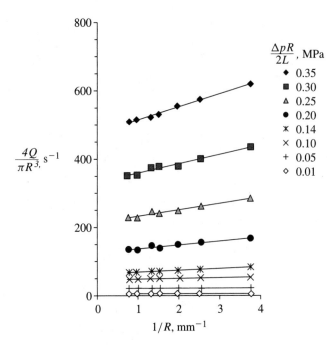

Figure 10.10 Apparent shear rate, uncorrected for slip, versus inverse capillary radius for a linear low-density polyethylene; melt index = 1, $T = 220°C$; from Ramamurthy [214]. Each curve represents data at different constant values of wall shear stress, $\Delta p R/2L$. For the lines that are horizontal, no wall slip is inferred. For the lines with positive slopes, the wall slip velocity may be calculated as slope/4. From the *Journal of Rheology*, Copyright © 1986, The Society of Rheology. Reprinted by permission.

$\dfrac{\Delta p R}{2L}$, MPa

- ◆ 0.35
- ■ 0.30
- ▲ 0.25
- ● 0.20
- ✶ 0.14
- ✕ 0.10
- + 0.05
- ◇ 0.01

Figure 10.11 Definition of extrapolation length b in a flow in which the fluid slips along the wall.

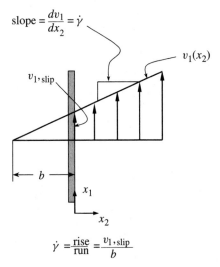

$$\text{slope} = \frac{dv_1}{dx_2} = \dot{\gamma}$$

$$\dot{\gamma} = \frac{\text{rise}}{\text{run}} = \frac{v_{1,\text{slip}}}{b}$$

which the velocity profile extrapolates to zero [35]. This way of looking at slip is favored by those investigating the molecular causes of slip because molecular models can predict the variation of b with parameters such as molecular weight and monomer properties [35]. The parameter b and the slip velocity are related as follows:

$$\dot{\gamma}_{R,\text{slip-corrected}} = \frac{v_{1,\text{slip}}}{b} \tag{10.54}$$

where $\dot{\gamma}_{R,\text{slip-corrected}}$ is the true shear rate near the wall. Thus b is calculated from the slip velocities obtained in the Mooney analysis and from the doubly corrected shear rates, that is, those obtained after the Mooney and Weissenberg–Rabinowitsch corrections have been applied.

Capillary rheometry is in wide use, particularly for obtaining viscosities at the high shear rates used in polymer processing. With appropriate attention paid to end effects and slip effects, the measured viscosities can be very accurate.

10.1.2 DRAG FLOW—PARALLEL DISKS

Measuring viscosity in a capillary typically requires 40 grams of material. In contrast, measurements may be made on less than 1 gram of sample in a parallel-disk torsional rheometer, as depicted schematically in Figure 10.12. The parallel-disk apparatus is thus preferred for the study of small quantities of materials or for substances that would be adversely affected by the severe contraction at the inlet of capillary flow. Edge fracture occurs at high rates in the parallel-disk apparatus, however (see Section 6.1.2), and thus the maximum shear rate obtainable in parallel-disk flow is less than in capillary flow.

For a Newtonian fluid we solved the problem of torsional flow between parallel disks in Section 3.5.4. There we used the equation of motion, the continuity equation, and the Newtonian constitutive equation to solve for $v_\theta(r, z)$. As we discussed in the last section, we do not want to assume a constitutive equation here, since we want to measure unknown fluid properties with the parallel-disk apparatus. Instead of making an assumption about the constitutive relation of the fluid, we will make some assumptions about the velocity profile, much as we did in the case of capillary flow.

In the parallel-disk rheometer, when the upper disk is rotated at a constant angular velocity Ω, the only nonzero component of $\underline{v}$ is v_θ:

$$\underline{v} = \begin{pmatrix} 0 \\ v_\theta \\ 0 \end{pmatrix}_{r\theta z} \tag{10.55}$$

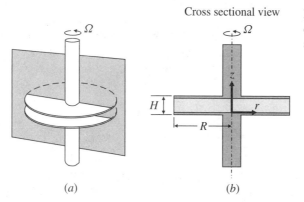

Cross sectional view

Figure 10.12 Torsional parallel-disk rheometer for viscosity measurement.

(a)   (b)

With this velocity field, and assuming incompressible flow, the continuity equation tells us that $\partial v_\theta / \partial \theta = 0$. If we assume that simple shear flow takes place in the θ-direction with the gradient in the z-direction (i.e., the velocity profile is linear in z) we can write

$$v_\theta = A(r)z + B(r) \tag{10.56}$$

where $A(r)$ and $B(r)$ are (so far) unknown functions of r. For the cylindrical coordinate system shown in Figure 10.12, the velocity boundary conditions are that $v_\theta = 0$ at $z = 0$, and $v_\theta = r\Omega$ at $z = H$, where H is the gap between the parallel disks. Applying these boundary conditions to the equation for v_θ yields

$$v_\theta = \frac{r\Omega z}{H} \tag{10.57}$$

The rate-of-deformation tensor is then

$$\dot{\underline{\underline{\gamma}}} = \nabla \underline{v} + (\nabla \underline{v})^T \tag{10.58}$$

$$= \begin{pmatrix} 0 & \frac{\partial v_\theta}{\partial r} - \frac{v_\theta}{r} & 0 \\ \frac{\partial v_\theta}{\partial r} - \frac{v_\theta}{r} & 0 & \frac{\partial v_\theta}{\partial z} \\ 0 & \frac{\partial v_\theta}{\partial z} & 0 \end{pmatrix}_{r\theta z} = \begin{pmatrix} 0 & 0 & 0 \\ 0 & 0 & \frac{r\Omega}{H} \\ 0 & \frac{r\Omega}{H} & 0 \end{pmatrix}_{r\theta z} \tag{10.59}$$

$$\dot{\gamma} = |\dot{\underline{\underline{\gamma}}}| = \frac{r\Omega}{H} \tag{10.60}$$

At the outer edge of the parallel disks we can write $\dot{\gamma} = \dot{\gamma}_R$, and thus

$$\dot{\gamma} = \dot{\gamma}_R \frac{r}{R} \tag{10.61}$$

where $\dot{\gamma}_R = R\Omega / H$. The strain $\gamma(0, t)$ also depends on the radial position:

$$\gamma(0, t) = \int_0^t \dot{\gamma}(t') \, dt' \tag{10.62}$$

$$= \int_0^t \frac{r\Omega}{H} \, dt' = \frac{r\Omega t}{H} \tag{10.63}$$

Examining Equations (10.57) and (10.59) we see that if the θ-curvature is neglected (see footnote 2 this chapter) and if we consider flow at a particular value of r, parallel-disk flow resembles shear flow, $\underline{v} = \dot{\gamma}_0 x_2 \hat{e}_1$. We can make the following assignments: θ is the flow (1) direction, z is the gradient (2) direction, and r is the neutral (3) direction. The assumption of nearly unidirectional flow is best at the rim, $r = R$, and we therefore can calculate viscosity from

$$\tau_{21} = \tau_{z\theta}|_{r=R} \tag{10.64}$$

$$\dot{\gamma}_0 = \frac{R\Omega}{H} = \dot{\gamma}_R \tag{10.65}$$

$$\eta = \frac{-\tau_{21}}{\dot{\gamma}_0} = \frac{-\tau_{z\theta}|_{r=R}}{\dot{\gamma}_R} \tag{10.66}$$

We already have an expression for $\dot{\gamma}_R$ in terms of experimentally accessible variables. Now we seek such an expression for $\tau_{z\theta}|_{r=R}$.

For parallel-disk flow with the assumed velocity profile the stress tensor takes on a simple form due to symmetry (see Section 4.4). The stress tensor written in the r, θ, z coordinate system looks a bit different from when it is written in the 1, 2, 3 coordinate system,

$$\underline{\underline{\tau}} = \begin{pmatrix} \tau_{rr} & 0 & 0 \\ 0 & \tau_{\theta\theta} & \tau_{z\theta} \\ 0 & \tau_{z\theta} & \tau_{zz} \end{pmatrix}_{r\theta z} \tag{10.67}$$

Using this form of the stress tensor and the assumed velocity profile, the equation of motion simplifies to

$$\begin{pmatrix} -\frac{\rho v_\theta^2}{r} \\ 0 \\ 0 \end{pmatrix}_{r\theta z} = \begin{pmatrix} -\frac{\partial p}{\partial r} \\ -\frac{1}{r}\frac{\partial p}{\partial \theta} \\ -\frac{\partial p}{\partial z} \end{pmatrix}_{r\theta z} - \begin{pmatrix} \frac{1}{r}\frac{\partial}{\partial r}(r\tau_{rr}) - \frac{\tau_{\theta\theta}}{r} \\ \frac{\partial \tau_{z\theta}}{\partial z} \\ \frac{\partial \tau_{zz}}{\partial z} \end{pmatrix}_{r\theta z} \tag{10.68}$$

Note that θ-derivatives have been canceled since v_θ is independent of θ and $\underline{\underline{\tau}}$ depends only on v_θ. If we further assume that pressure does not vary with θ, we can integrate the θ-component of the equation of motion to obtain

$$\frac{\partial \tau_{z\theta}(r, z)}{\partial z} = 0 \tag{10.69}$$

$$\tau_{z\theta} = C(r) \tag{10.70}$$

where $C(r)$ is an unknown function of r. Thus, to measure shear stress (at the top plate, for example), we must take measurements at specific values of r and evaluate viscosity at each position.

Although it is possible to measure $\tau_{z\theta}$ as a function of radial position [59], it is much easier to measure the total torque $\mathcal{T}$ required to turn the upper disk (or to maintain the lower disk immobile). We can relate $\mathcal{T}$ to the viscosity at the rim $\eta(\dot{\gamma}_R)$ by following a derivation resembling that used for the Weissenberg–Rabinowitsch equation for capillary flow [229], as we will now show. The torque on the top disk is given by[4]

$$\mathcal{T} = \int_A (\text{stress})(\text{lever arm})\, dA \tag{10.71}$$

$$= \int_0^R (-\tau_{z\theta}|_{z=H})\, (r)\, (2\pi r\, dr) \tag{10.72}$$

Viscosity at any value of r can be written as

$$\eta \equiv \frac{-\tau_{21}}{\dot{\gamma}_0} = \frac{-\tau_{z\theta}(r)}{\dot{\gamma}(r)} = \eta(r) \tag{10.73}$$

---

[4] The stress is negative due to the sign convention for stress; see the footnote before Equation (8.189).

We can replace $\tau_{z\theta}$ in Equation (10.72) with viscosity, which, in general, is a function of the radial position r,

$$\mathcal{T} = 2\pi \int_0^R \eta \dot{\gamma} \, r^2 \, dr \qquad (10.74)$$

We must convert the equation containing torque, which is currently an integral over viscosity and shear rate, into an expression for viscosity obtainable from torque. Following the same steps that gave us the Weissenberg–Rabinowitsch equation in capillary flow (Section 10.1.1.2), we will take the derivative of Equation (10.74) to remove the integral. We can foresee that we will obtain an equation that will require plots of torque versus rim shear rate, and therefore, before differentiating, we will replace the variable r in the integral with $\dot{\gamma}$, using $\dot{\gamma} = \dot{\gamma}_R r / R$,

$$\mathcal{T} = \frac{2\pi R^3}{\dot{\gamma}_R^3} \int_0^{\dot{\gamma}_R} \eta \dot{\gamma}^3 \, d\dot{\gamma} \qquad (10.75)$$

Now, to eliminate the integral, we differentiate both sides by $\dot{\gamma}_R$ using the Leibnitz rule:

$$\left(\frac{\mathcal{T}}{2\pi R^3} \right) \dot{\gamma}_R^3 = \int_0^{\dot{\gamma}_R} \eta \dot{\gamma}^3 \, d\dot{\gamma} \qquad (10.76)$$

$$\frac{d}{d\dot{\gamma}_R} \left[\left(\frac{\mathcal{T}}{2\pi R^3} \right) \dot{\gamma}_R^3 \right] = \int_0^{\dot{\gamma}_R} \frac{\partial}{\partial \dot{\gamma}_R} \left(\eta \dot{\gamma}^3 \right) d\dot{\gamma} + \eta(\dot{\gamma}_R) \dot{\gamma}_R^3 \qquad (10.77)$$

The first term on the right-hand side is zero, and after rearrangement we arrive at an equation for steady shear viscosity measured in a torsional parallel-disk viscometer:

Viscosity in
parallel-disk flow
$$\boxed{\eta(\dot{\gamma}_R) = \frac{\mathcal{T}/2\pi R^3}{\dot{\gamma}_R} \left[3 + \frac{d \ln (\mathcal{T}/2\pi R^3)}{d \ln \dot{\gamma}_R} \right]} \qquad (10.78)$$

Thus, to measure the viscosity of a fluid in a parallel-disk rheometer at the rim shear rate $\dot{\gamma}_R$, data at a variety of rim shear rates $\dot{\gamma}_R$, that is, rotational speeds Ω, must be taken, the torque differentiated in the manner described by Equation (10.78) (Figure 10.13), and a correction applied to each $(\mathcal{T}, \dot{\gamma}_R)$ data pair.

Recall that in the capillary viscometer the viscosity was calculated at the wall, $\eta = \eta(\dot{\gamma}_R)$. In the parallel-disk viscometer the viscosity is calculated at the rim, $\eta = \eta(\dot{\gamma}_R)$. In the case of the capillary viscometer we warned that structured fluids such as suspensions might have material properties that are affected by the presence of the wall and that this should be considered when interpreting capillary viscosities. For the parallel-disk viscometer we offer a similar warning. Since the strain varies with the radius [Equation (10.63)], not all material elements experience the same strain, $\gamma(0, t) = r\Omega t / H$. The torque, however, is a quantity measured with contributions from fluid elements at all values of r. For materials that are strain sensitive (e.g., phase-separated blends, liquid crystals), the parallel-disk viscometer gives results that represent a blurring of the material properties exhibited at each radius, that is, at a variety of shear strains.

The parallel-disk geometry is popular in the small-amplitude oscillatory shear (SAOS) mode [216, 32]. When the amplitude is small enough to give a linear profile (see Problem

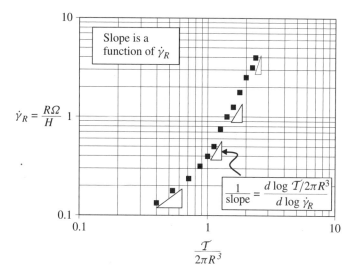

Figure 10.13 Derivative in the viscosity expression [Equation (10.78)] is obtained from torque and angular-velocity information for any type of fluid.

3.23), the preceding analysis remains valid, and $\dot{\gamma}$ is given by Equation (10.60) with $\Omega(t) = d\theta/dt$, where $\theta(t) = \theta_0 \Re\{e^{i\omega t}\}$ is the angular displacement of the plate. The SAOS material functions $\eta'(\omega)$ and $\eta''(\omega)$ can be related to the measured torque amplitude $\mathcal{T}_0$ and phase lag δ through a calculation similar to that given in Section 8.4.2 for the Couette geometry. Details of this calculation may be found in Bird et al. [26]. The results are

$$
\begin{aligned}
\text{SAOS material functions for} && \eta'(\omega) &= \frac{2H\mathcal{T}_0 \sin\delta}{\pi R^4 \omega \theta_0} \\
\text{parallel-disk apparatus} && \eta''(\omega) &= \frac{2H\mathcal{T}_0 \cos\delta}{\pi R^4 \omega \theta_0}
\end{aligned}
\qquad (10.79)
$$

10.1.3 DRAG FLOW—CONE AND PLATE

The problem of the radial dependence of the shear rate (and shear strain) in the torsional parallel-disk experiment can be eliminated if the cone-and-plate geometry is used. Although this may not be an intuitive improvement to the parallel-disk system, we will see that a homogeneous flow (no radial dependence) is produced in the limit of small angles Θ_0. Loading highly viscous materials can be difficult in a cone-and-plate viscometer, however, since the cone must be pressed into the sample; also, the cone-and-plate geometry suffers from the same edge distortions at high shear rates that were discussed in the previous section and in Section 6.1.2.

The cone-and-plate geometry is shown in Figure 10.14. In this experiment, if the curvature of the flow lines can be neglected, simple shear flow in the ϕ-direction is produced when the cone is rotated at a constant angular velocity Ω. This flow is analyzed in spherical coordinates,

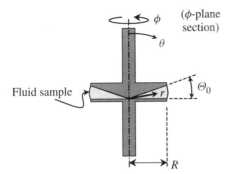

(ϕ-plane section) **Figure 10.14** Torsional cone-and-plate rheometer for viscosity measurement.

Fluid sample

$$\underline{v} = \begin{pmatrix} 0 \\ 0 \\ v_\phi \end{pmatrix}_{r\theta\phi} \tag{10.80}$$

Note that the shear surfaces for this flow are surfaces of constant θ, where θ is the usual spherical-coordinate-system angle, measured down from the vertical. These conical surfaces are approximately planes when the cone angle is small.

Our analysis will resemble that used for the parallel-disk geometry. For a shallow cone, the region of space of interest is near the bottom plate, and in this region $-r\theta$ is nearly the same as z. If we assume that simple shear flow takes place in the ϕ-direction with the gradient in the $(-r\theta)$-direction (and neglect curvature in the ϕ-direction), then the continuity equation tells us that $\partial v_\phi / \partial \phi = 0$, and we can write

$$v_\phi = C_1(-r\theta) + C_2 \tag{10.81}$$

where C_1 and C_2 are constants. For the coordinate system shown in Figure 10.14, the boundary conditions are that $v_\phi = 0$ at $\theta = \pi/2$, and $v_\phi = r\Omega$ at $\theta = \pi/2 - \Theta_0$, where Θ_0 is the (small) cone angle. Applying these boundary conditions to the equation for v_ϕ yields

$$v_\phi = \frac{r\Omega}{\Theta_0}\left(\frac{\pi}{2} - \theta\right) \tag{10.82}$$

The rate-of-deformation tensor $\dot{\underline{\underline{\gamma}}}$ in this flow (for $v_r = v_z = 0$) is

$$\dot{\underline{\underline{\gamma}}} = \begin{pmatrix} 0 & 0 & r\frac{\partial}{\partial r}\left(\frac{v_\phi}{r}\right) \\ 0 & 0 & \frac{\sin\theta}{r}\frac{\partial}{\partial \theta}\left(\frac{v_\phi}{\sin\theta}\right) \\ r\frac{\partial}{\partial r}\left(\frac{v_\phi}{r}\right) & \frac{\sin\theta}{r}\frac{\partial}{\partial \theta}\left(\frac{v_\phi}{\sin\theta}\right) & 0 \end{pmatrix}_{r\theta\phi} \tag{10.83}$$

$$= \begin{pmatrix} 0 & 0 & 0 \\ 0 & 0 & \dot{\gamma}_{\theta\phi} \\ 0 & \dot{\gamma}_{\theta\phi} & 0 \end{pmatrix}_{r\theta\phi} \tag{10.84}$$

Since θ is close to $\pi/2$ (Θ_0 is small), $\sin\theta \approx 1$, and we can simplify $\dot{\gamma}_{\theta\phi}$ as follows:

$$\dot{\gamma}_{\theta\phi} = \frac{\sin\theta}{r}\frac{\partial}{\partial\theta}\left(\frac{v_\phi}{\sin\theta}\right) \tag{10.85}$$

$$= \frac{1}{r}\frac{\partial v_\phi}{\partial\theta} = -\frac{\Omega}{\Theta_0} \tag{10.86}$$

Thus,

$$\underline{\dot{\gamma}} = \begin{pmatrix} 0 & 0 & 0 \\ 0 & 0 & -\frac{\Omega}{\Theta_0} \\ 0 & -\frac{\Omega}{\Theta_0} & 0 \end{pmatrix}_{r\theta\phi} \tag{10.87}$$

$$\dot{\gamma} = |\underline{\dot{\gamma}}| = +\frac{\Omega}{\Theta_0} \tag{10.88}$$

The strain is calculated for the cone-and-plate geometry as follows:

$$\gamma(0,t) = \int_0^t \dot{\gamma}(t')\,dt' \tag{10.89}$$

$$= \int_0^t \frac{\Omega}{\Theta_0}\,dt' = \frac{\Omega t}{\Theta_0} \tag{10.90}$$

Comparing Equations (10.82) and (10.87) with the velocity and rate-of-deformation tensor for the shear flow in the definition of viscosity, we see that

$$\dot{\gamma}_0 = \frac{1}{r}\frac{\partial v_\phi}{\partial(-\theta)} = \frac{\Omega}{\Theta_0} = \dot{\gamma} \tag{10.91}$$

$$\tau_{21} = -\tau_{\theta\phi} \tag{10.92}$$

$$\eta = \frac{-\tau_{21}}{\dot{\gamma}_0} = \frac{\tau_{\theta\phi}}{\dot{\gamma}} \tag{10.93}$$

Both the shear rate and the shear strain are independent of position in the cone-and-plate geometry, which makes it straightforward to calculate the viscosity from a total torque measurement. The torque $\mathcal{T}$ on the bottom plate may be calculated as[5]

$$\mathcal{T} = \int_A (\text{stress})(\text{lever arm})\,dA \tag{10.94}$$

$$= \int_0^{2\pi}\int_0^R \left(\tau_{\theta\phi}\big|_{\theta=\frac{\pi}{2}}\right)(r)(r\,d\phi dr) \tag{10.95}$$

Since the shear rate is constant throughout the flow domain, the viscosity and the shear stress are constant too, and $\tau_{\theta\phi}$ may be removed from the integral. For the boundary condition $\tau_{\theta\phi} = 0$, $r = 0$, we obtain for the torque on the plate,

---

[5] In the chosen spherical coordinate system, as θ increases, v_ϕ decreases, and thus $\tau_{\theta\phi}$ is positive; see the footnote before Equation (8.189).

$$\mathcal{T} = \frac{2}{3}\pi R^3 \tau_{\theta\phi}\Big|_{\theta=\frac{\pi}{2}} \tag{10.96}$$

Thus the viscosity may be calculated directly:

Viscosity in
cone-and-plate flow
$$\boxed{\eta \equiv \frac{-\tau_{21}}{\dot{\gamma}_0} = \frac{\tau_{\theta\phi}}{\dot{\gamma}} = \frac{3\mathcal{T}\,\Theta_0}{2\pi R^3 \Omega}} \tag{10.97}$$

We see that in the limit of a small cone angle, the cone-and-plate geometry produces constant shear rate, constant shear stress, and homogeneous strain throughout the sample. This makes the calculations of viscosity [Equation (10.97)] quite simple in the cone-and-plate geometry. The uniformity of the flow in the cone-and-plate geometry is also a great advantage when working with structure-forming materials such as liquid crystals, incompatible blends, and suspensions that are strain or rate sensitive. In addition, the cone-and-plate geometry has the advantage that the first normal-stress difference can be calculated from measurement of the axial thrust on the cone, as we will now show, following the derivation of Bird et al. [26].

The total thrust on the plate due to the fluid flow $\mathcal{F}$ is just the integral over the area of the plate of the normal stress on the plate minus the atmospheric thrust $\pi R^2 P_{\text{atm}}$ [26]:

$$\mathcal{F} = \left[2\pi \int_0^R \Pi_{\theta\theta}|_{\theta=\frac{\pi}{2}}\, r\,dr \right] - \pi R^2 P_{\text{atm}} \tag{10.98}$$

To calculate $\Pi_{\theta\theta}|_{\theta=\frac{\pi}{2}}$, we turn to the r-component of the equation of motion for this problem. Taking $v_r = v_\theta = 0$ and assuming incompressible fluid, the r-component of the equation of motion becomes

$$-\frac{\rho v_\phi^2}{r} = -\frac{\partial p}{\partial r} - \frac{1}{r^2}\frac{\partial}{\partial r}(r^2 \tau_{rr}) - \frac{1}{r\sin\theta}\frac{\partial}{\partial\theta}(\tau_{\theta r}\sin\theta)$$
$$- \frac{1}{r\sin\theta}\frac{\partial \tau_{\phi r}}{\partial\phi} + \frac{\tau_{\theta\theta} + \tau_{\phi\phi}}{r} \tag{10.99}$$

The flows we are considering are limited to relatively slow rates to minimize inertial effects and edge instabilities, and we can therefore neglect the centrifugal-force term $\rho v_\phi^2/r$. Since the shear rate is constant throughout the flow field in this experiment, all components of the stress tensor $\underline{\underline{\tau}}$ are also constant. This allows us to simplify this equation further by eliminating stress derivatives. Note also that $\partial \Pi_{\theta\theta}/\partial r = \partial p/\partial r$. Finally, we are interested in one particular surface, that for which $\theta = \pi/2$, and therefore $\sin\theta = 1$. Expanding the derivatives in Equation (10.99) and simplifying as described, we obtain:

$$0 = -\frac{\partial \Pi_{\theta\theta}}{\partial r} - \frac{2\tau_{rr}}{r} + \frac{\tau_{\theta\theta} + \tau_{\phi\phi}}{r} \tag{10.100}$$

We can write Equation (10.100) in terms of the first and second normal-stress coefficients for steady shear flow, Ψ_1 and Ψ_2, defined previously as:

$$\Psi_1 = -\frac{\tau_{11} - \tau_{22}}{\dot{\gamma}_0^2} = -\frac{\tau_{\phi\phi} - \tau_{\theta\theta}}{\dot{\gamma}_0^2} \tag{10.101}$$

$$\Psi_2 = -\frac{\tau_{11} - \tau_{22}}{\dot{\gamma}_0^2} = -\frac{\tau_{\theta\theta} - \tau_{rr}}{\dot{\gamma}_0^2} \qquad (10.102)$$

The final result is

$$\frac{\partial \Pi_{\theta\theta}}{\partial \ln r} = -\dot{\gamma}_0^2(\Psi_1 + 2\Psi_2) \qquad (10.103)$$

We can integrate Equation (10.103), but we need a boundary condition. At the rim, the radial normal stress is the atmospheric pressure, $\Pi_{rr}(R) = P_{atm}$, and thus $\tau_{rr}(R) = 0$. From this and from Equation (10.102) we see that $\tau_{\theta\theta}(R) = -\Psi_2 \dot{\gamma}_0^2$. At the rim, therefore, $\Pi_{\theta\theta}(R) = P_{atm} - \Psi_2 \dot{\gamma}_0^2$. Carrying out the integration of Equation (10.103) with this boundary condition, we obtain

$$\Pi_{\theta\theta}|_{\theta=\frac{\pi}{2}} = \left[-\dot{\gamma}^2(\Psi_1 + 2\Psi_2)\right] \ln\left(\frac{r}{R}\right) + \left(P_{atm} - \Psi_2 \dot{\gamma}_0^2\right) \qquad (10.104)$$

With this expression and no further assumptions, Equation (10.98) may be evaluated, and after some straightforward algebra, the following simple result is obtained:

First normal-stress
coefficient in
cone-and-plate flow

$$\boxed{\Psi_1 = \frac{2\mathcal{F}\Theta_0^2}{\pi R^2 \Omega^2}} \qquad (10.105)$$

Thus, in the cone-and-plate geometry, values of the viscosity in steady shear are obtained from measurements of the total torque required to hold the plate immobile and from knowledge of the rate of angular rotation Ω and geometric factors [Equation (10.97)]. We obtain the first normal-stress coefficient from the additional measurement of the total thrust on the plate. The simplicity of these measurements accounts for the popularity of this system.

Like the parallel-disk geometry, the cone-and-plate geometry is also widely used in linear viscoelastic, small-amplitude oscillatory shear (SAOS) measurements. For the SAOS experiment, the material functions are calculated for the cone-and-plate geometry as follows [26]:

SAOS material functions
for cone-and-plate apparatus

$$\boxed{\begin{aligned} \phi &= \phi_0 \Re\{e^{i\omega t}\} \\[4pt] \eta' &= \frac{3\Theta_0 \mathcal{T}_0 \sin\delta}{2\pi R^3 \omega \phi_0} \\[4pt] \eta'' &= \frac{3\Theta_0 \mathcal{T}_0 \cos\delta}{2\pi R^3 \omega \phi_0} \end{aligned}} \qquad (10.106)$$

where ϕ is the torsional angle through which the cone oscillates, $\mathcal{T}_0$ is the amplitude of the torque, and δ is the phase difference between torque and torsional angle.

10.1.4 DRAG FLOW—COUETTE

In Chapter 8 we discussed SAOS in the cup-and-bob or Couette geometry in depth (see Figure 8.11). In that problem the generalized linear viscoelastic (GLVE) constitutive equation was used along with the continuity equation and the equation of motion to obtain a solution relating unknown linear viscoelastic properties η' and η'' to measurements of the stress amplitude ratio $\mathcal{A}$ and phase angle δ in forced SAOS. The use of a particular constitutive equation, the GLVE equation, in this derivation does not limit the usefulness of the analysis since the small-amplitude experiment is already limited to small deformation rates. Steady shear viscosity can also be measured conveniently in the Couette flow geometry, as we will now see.

In the concentric-cylinder or Couette geometry, the fluid to be tested is confined to the narrow space between two cylinders. When the inner cylinder (the bob) is rotated at a constant angular velocity Ω, the only nonzero component of $\underline{v}$ is v_θ:

$$\underline{v} = \begin{pmatrix} 0 \\ v_\theta \\ 0 \end{pmatrix}_{r\theta z} \tag{10.107}$$

With this velocity field, the continuity equation tells us that for an incompressible fluid $\partial v_\theta / \partial \theta = 0$. For a long bob, z-variations can also be neglected. If we follow the procedure used for the parallel-disk geometry and assume that simple shear flow takes place in the θ-direction with the gradient in the r-direction (i.e., the velocity profile is linear in r), we can write

$$v_\theta = C_1 r + C_2 \tag{10.108}$$

where C_1 and C_2 are as yet unknown constants. For the cylindrical coordinate system shown in Figure 8.11, the boundary conditions are that $v_\theta = 0$ at $r = R$, and $v_\theta = \kappa R \Omega$ at $r = \kappa R$, where R is the radius of the outer cylinder (the cup), and κR is the radius of the bob. Applying these boundary conditions to the equation for v_θ yields

$$v_\theta = \frac{\kappa \Omega (r - R)}{\kappa - 1} \tag{10.109}$$

The rate-of-deformation tensor $\dot{\underline{\underline{\gamma}}}$ in this flow is calculated to be

$$\dot{\underline{\underline{\gamma}}} = \begin{pmatrix} 0 & \frac{\partial v_\theta}{\partial r} & 0 \\ \frac{\partial v_\theta}{\partial r} & 0 & 0 \\ 0 & 0 & 0 \end{pmatrix}_{r\theta z} \tag{10.110}$$

$$\dot{\gamma} = |\dot{\underline{\underline{\gamma}}}| = +\frac{\partial v_\theta}{\partial r} = \frac{\kappa \Omega}{\kappa - 1} \tag{10.111}$$

Following the same arguments used in the previous sections, we can make the assignments that θ is the flow (1) direction, $-r$ is the gradient (2) direction, and z is the neutral (3) direction and conclude that

$$\tau_{21} = -\tau_{r\theta} \tag{10.112}$$

$$\dot{\gamma}_0 = \frac{-\partial v_\theta}{\partial r} = \frac{-\kappa \Omega}{\kappa - 1} = -\dot{\gamma} \tag{10.113}$$

$$\eta = \frac{-\tau_{21}}{\dot{\gamma}_0} = -\frac{\tau_{r\theta}}{\dot{\gamma}} \tag{10.114}$$

We can measure the total torque $\mathcal{T}$ required to turn the bob (or to maintain the cup immobile), and $\mathcal{T}$ can be related to the shear stress and hence to the viscosity through the following manipulations:

$$\mathcal{T} = (\text{stress})(\text{lever arm})(\text{area}) \tag{10.115}$$

$$= (\tau_{r\theta}|_{r=\kappa R})(\kappa R)(2\pi \kappa R L) \tag{10.116}$$

Viscosity in
Couette flow
(bob turning)

$$\eta = \frac{\mathcal{T}(\kappa - 1)}{2\pi R^2 L \kappa^3 \Omega} \tag{10.117}$$

The case where the cup turns and the bob is stationary gives equivalent results (see Table 10.2).

Low-viscosity fluids generate little stress in torsional experiments. In Couette flow, however, the large fluid contact area $2\pi \kappa R L$ boosts the torque, improving the measurement. The gap must be small enough so that the assumption of a linear velocity profile holds. A separate analysis may be performed to calculate viscosity using a wide-gap Couette [26], or for rotating a bob in an infinite fluid [152, 227, 132]. The Couette geometry is limited to modest rotational speeds by the appearance of high-shear-rate instabilities leading to three-dimensional flows. In some cases these instabilities are due to inertia (Taylor cells [246]), and in other cases they are due to elasticity (see [186] and Figure 1.8).

One final comment on the Couette geometry is in order. In general, the majority of the stress transmitted to the torque transducer in the Couette geometry is due to the fluid in the narrow gap, as intended. The fluid in the bottom of the cup, however, does transfer a small stress. Two approaches are used to account for effects due to the fluid at the bottom. The first approach is to make the bottom of the bob in the shape of a cone (Figure 10.15a); thus the shear-stress contribution of the bottom can be accounted for by using the equations from the cone-and-plate analysis. A second approach is to shape the bob as shown in Figure 10.15b. With this shape, an air bubble is trapped at the bottom of the bob, and the low stresses associated with the air can be neglected. Both shapes give acceptable results in most situations, although regardless of shape the variation in shear rate at the bottom of the cup can lead to particle segregation in suspensions, particularly in viscoelastic suspensions examined at high shear rates [118].

The four shear geometries discussed—capillary, parallel-disk, cone-and-plate, and Couette—all lead in a fairly straightforward way to measurements of steady shear viscosity. In the case of cone-and-plate torsional flow we can also measure $\Psi_1(\dot{\gamma})$. The choice of which geometry to use is dictated by convenience, that is, which rheometer you have available, and by experimental factors such as limits on torque measurement capability, sample size, sample loading issues, desired shear rates, and so on. A comparison of these four shear geometries is given in Table 10.3.

All four geometries can be used for unsteady shear experiments such as startup, cessation, and creep, and the torsional geometries (parallel disk, cone and plate, Couette) are well designed for producing the SAOS, step-strain, and creep-recovery kinematics. For unsteady flows there will be a limit to how rapidly an instrument can respond due to inertia of the geometry and due to the speed with which the driving motor or driving pressure can respond. These issues vary from instrument to instrument. When considering nonsteady-state experiments, it is important to consult the manufacturer's literature on a rheometer to determine what these limitations are.

TABLE 10.2
Summary of the Expressions for Steady Shear Rheological Quantities for Common Geometries[*]

Geometry	Magnitude of Shear Stress $|\tau_{21}|$	Shear Rate $\dot{\gamma}$	Measured Material Function
Capillary flow (wall conditions)			
$\mathcal{P}_0, \mathcal{P}_L$ = modified pressure at $z = 0, L$	$\dfrac{(\mathcal{P}_0 - \mathcal{P}_L)R}{2L}$	$\dfrac{4Q}{\pi R^3}\mathcal{R}$	$\eta = \dfrac{\tau_R}{4Q/\pi R^3}\mathcal{R}^{-1}$
Q = flow rate			
L = capillary length			
$\mathcal{R} = \frac{1}{4}\left[3 + \frac{d\ln(4Q/\pi R^3)}{d\ln\tau_R}\right]$			
$\tau_R = \tau_{rz}\|_{r=R}$			
Parallel disk (at rim)			
$\mathcal{T}$ = torque on top plate	$\dfrac{2\mathcal{T}}{\pi R^3}\mathcal{R}$	$\dfrac{r\Omega}{H}$	$\eta = \dfrac{2\mathcal{T}}{\pi R^3 \dot{\gamma}_R}\mathcal{R}$
Ω = angular velocity of top plate, > 0			
H = gap			
$\mathcal{R} = \frac{1}{4}\left[3 + \frac{d\ln(\mathcal{T}/2\pi R^3)}{d\ln\dot{\gamma}_R}\right]$			
$\dot{\gamma}_R = \dot{\gamma}(R)$			
Cone and plate			
$\mathcal{T}$ = torque on plate	$\dfrac{3\mathcal{T}}{2\pi R^3}$	$\dfrac{\Omega}{\Theta_0}$	$\eta = \dfrac{3\mathcal{T}\Theta_0}{2\pi R^3 \Omega}$
$\mathcal{F}$ = thrust on plate			
Ω = angular velocity of cone, > 0			$\Psi_1 = \dfrac{2\mathcal{F}\Theta_0^2}{\pi R^2 \Omega^2}$
Θ_0 = cone angle			
Couette (bob turning)			
$\mathcal{T}$ = torque on inner cylinder, < 0	$\dfrac{-\mathcal{T}}{2\pi R^2 L\kappa^2}$	$\dfrac{\kappa\Omega}{1-\kappa}$	$\eta = \dfrac{\mathcal{T}(\kappa - 1)}{2\pi R^2 L\kappa^3\Omega}$
Ω = angular velocity of bob, > 0			
R = outer radius			
κR = inner radius			
L = length of bob			
Couette (cup turning)			
$\mathcal{T}$ = torque on inner cylinder, > 0	$\dfrac{\mathcal{T}}{2\pi R^2 L\kappa^2}$	$\dfrac{\kappa\Omega}{1-\kappa}$	$\eta = \dfrac{\mathcal{T}(1 - \kappa)}{2\pi R^2 L\kappa^3\Omega}$
Ω = angular velocity of cup, > 0			
R = outer radius			
κR = inner radius			
L = length of bob			

[*] R is radius of fixture. To calculate strain in each case, multiply shear rate by time t. Note that $\eta \equiv -\tau_{21}/\dot{\gamma}_0 = |\tau_{21}|/\dot{\gamma}$.

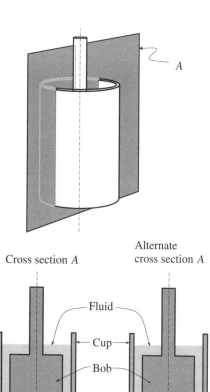

Figure 10.15 Two possible shapes for the bottom of the bob in the Couette geometry. (*a*) The stress is accounted for by using the cone-and-plate equations. (*b*) The stress is minimized by trapping an air bubble.

Cross section *A*

Alternate cross section *A*

Fluid

Cup

Bob

Air pocket

(*a*)   (*b*)

10.2 Elongational Flows

Although most rheological measurements are made in shear flow, this is not because elongational flows are unimportant. On the contrary, all industrial flows have an elongational component (e.g., flow through a contraction, jet impingement), and often the elongation flow dominates a process (e.g., fiber spinning, bubble inflation, coating [179, 236]). Shear measurements dominate the rheological literature because they are easy to perform, and in some applications information on the viscosity function or on the linear viscoelastic behavior is all that is needed, such as in narrow-gap molding flows or for quality control.

Elongational flow properties, although important, are very difficult to measure. As was discussed in Section 4.3.1, one of the important characteristics of elongational flows is the very rapid and large deformation of fluid elements. This very feature, however, makes the

TABLE 10.3
Comparison of Experimental Features of Four Common Shear Geometries

Feature	Parallel Disk	Cone and Plate	Capillary	Couette (Cup and Bob)
Stress range	Good for high viscosity	Good for high viscosity	Good for high viscosities	Good for low viscosities
Flow stability	Edge fracture at modest rates	Edge fracture at modest rates	Melt fracture at very high rates, i.e., distorted extrudates and pressure fluctuations are observed	Taylor cells are observed at high Re due to inertia; elastic cells are observed at high De
Sample size and sample loading	< 1 g; easy to load	< 1 g; highly viscous materials can be difficult to load	40 g minimum; easy to load	10–20 g; highly viscous materials can be difficult to load
Data handling	Correction on shear rate needs to be applied; this correction is ignored in most commercial software packages	Straightforward	Multiple corrections need to be applied	Straightforward
Homogeneous?	No; shear rate and shear stress vary with radius	Yes (small core angles)	No; shear rate and shear stress vary with radius	Yes (narrow gap)
Pressure effects	None	None	High pressures in reservoir cause problems with compressibility of melt	None
Shear rates	Maximum shear rate is limited by edge fracture; usually cannot obtain shear-thinning data	Maximum shear rate is limited by edge fracture; usually cannot obtain shear-thinning data	Very high rates accessible	Maximum shear rate is limited by sample leaving cup due to either inertia or elastic effects; also 3-D secondary flows develop (instability)
Special features	Good for stiff samples, even gels; wide range of temperatures possible	Ψ_1 measurable; wide range of temperatures possible	Constant-Q or constant-ΔP modes available; wide range of temperatures possible	Narrow gap required; usually limited to modest temperatures (e.g., $0 < T < 60°C$)

generation of elongational flow a challenge. In this section we will discuss some current methods for determining elongational-flow properties of polymer melts and solutions. The calculations for stress and deformation are fairly straightforward in elongational flow; the principal difficulty is in producing the flow reliably. For the most part, these measurements are not routine but rather are the focus of current research efforts.

10.2.1 UNIAXIAL EXTENSION

10.2.1.1 Melt Stretching

Uniaxial elongational flow may be produced by stretching a polymer melt sample between rapidly separating crossheads of, for example, a tensile-testing machine (Figure 4.1*a*). The motion of the ends is programmed so that the length of the melt sample increases in a manner so as to give the desired elongation-rate function $\dot{\epsilon}(t)$. Alternatively, the flow may be designed to impart a desired tensile-stress function, $-[\tau_{33}(t) - \tau_{11}(t)] = -(\tau_{zz} - \tau_{rr})$.

The time-dependent total force needed to deform a sample can be measured by a load cell (pressure transducer) positioned at one end of the sample (Figure 10.16); this can be related to Π_{zz} for the flow by

$$\Pi_{zz}(t) = P_{atm} - \frac{f(t)}{A(t)} \tag{10.118}$$

where P_{atm} is atmospheric pressure and is equal to Π_{rr}, $f(t)$ is the magnitude of the tensile force, and $A(t)$ is the (changing) cross-sectional area of the sample. The normal-stress difference is thus

$$\tau_{zz} - \tau_{rr} = \Pi_{zz} - \Pi_{rr} = -\frac{f(t)}{A(t)} \tag{10.119}$$

from which the desired material functions may be calculated. If the flow is such that elongational flow is produced everywhere throughout the sample, that is, if the flow is

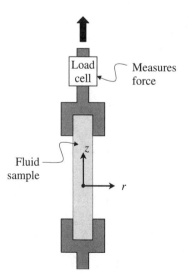

Figure 10.16 Geometry for producing a uniaxial extensional flow on a cylindrical sample. The force to bring about the deformation is measured by the load cell positioned at the top.

Load cell — Measures force

Fluid sample

z

r

homogeneous, then for the startup of steady elongation the area varies as $A(t) = A_0 e^{-\dot{\epsilon}_0 t}$ [can be calculated from Equation (5.174)], and the elongational viscosity growth function can be calculated from a measurement of $f(t)$ alone:

$$\bar{\eta}^+ = \frac{f(t)e^{\dot{\epsilon}_0 t}}{A_0 \dot{\epsilon}_0} \tag{10.120}$$

The steady elongational viscosity $\bar{\eta}$ can be calculated from the steady-state value of $f(t)$, denoted by f_∞:

$$\bar{\eta} = \frac{f_\infty e^{\dot{\epsilon}_0 t_\infty}}{A_0 \dot{\epsilon}_0} \tag{10.121}$$

Note that although the term $e^{\dot{\epsilon}_0 t_\infty}$ increases exponentially with time, f_∞ decreases exponentially with time, and the product should be constant at steady state, that is, at one particular value of t_∞, where t_∞ is large.

This comparatively simple experiment is, unfortunately, fraught with difficulties. The condition of homogeneous flow is difficult to achieve in practice since any defects in the initial sample will cause significant heterogeneities in the flow. Heterogeneity is also introduced into the flow near the clamped ends since the clamps that hold the ends of the sample do not usually adjust in width as the sample narrows during the deformation (Figure 10.17). Also, the tensile-testing equipment typically used for these experiments cannot reach sample elongations l/l_0 much above 50. This is equivalent to strains $\epsilon = \ln(l/l_0) < 4$, which are insufficient to achieve steady state. If attempted in air, small temperature gradients and drafts will strongly distort the measured forces at even modest elongations (see Figure 10.17). Only with the utmost care, and using chemically inert fluids to thermally insulate and to support the vertically or horizontally stretching sample, can credible elongational results at modest strains be obtained in this configuration [26]. When using an inert fluid, care is needed to eliminate any chemical interactions between the sample and the supporting fluid, and the highest temperatures available are limited by the stability of the supporting fluid.

The problem of the limitation on strain imposed by the size of the tensile tester can be solved by drawing a sample of fixed length through counterrotating rollers or metal belts [26, 173, 175, 176] (see Figures 4.11a and 10.18). This type of instrument for elongational viscosity measurement was invented by J. Meissner and collaborators. In the Meissner instrument the tensile force is measured by mounting one set of clamps or rollers at the lower end of a strain-gauge transducer. The motion of the strain gauge is proportional to the force experienced by the metal-belt clamp or roller. In the commercial version of this instrument [216] the samples are suspended over a porous table on a cushion of inert gas, eliminating chemical interactions present in designs that incorporated a supporting liquid. Small glass beads are sprinkled on the sample to mark the deformation (Figure 10.19), and the entire experiment is monitored from the top and side by video equipment. The true strain rate in the flow is calculated from the video images, and very impressive reproducibility is being obtained [176, 175].

The Meissner instrument is also capable of measuring unconstrained recoil after elongation at a constant rate. This is accomplished by cutting the sample at a desired time and then monitoring the sample shrinkage with the video equipment. It is also possible

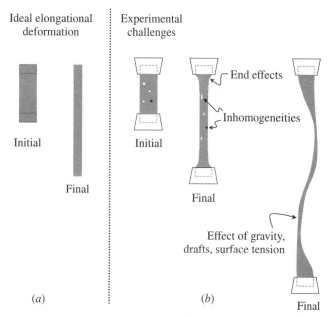

Ideal elongational   Experimental
deformation   challenges

Initial   Initial

Final

Final

Effect of gravity,
drafts, surface tension

(*a*)   (*b*)

Final

Figure 10.17 (*a*) Idealized uniaxial elongation experiment. (*b*) Some of the experimental difficulties that may be encountered. It is difficult to make a sample that is free of bubbles and dirt, and these inhomogeneities may affect the deformation in significant ways. The ends also introduce strain inhomogeneities. Finally, gravity, surface tension, and drafts can affect the flow at long times.

to measure stress relaxation after the cessation of constant elongational deformation by stopping the flow at a desired time and monitoring the decay in tensile stress. In all cases the stress is measured through the transducer mounted on the drawing clamp or belt, and strain is measured by visually observing markers on the sample. The material functions are then calculated in a straightforward manner using Equation (10.119) and the material-function definitions.

The Meissner rheometer is the best instrument available for measurements on viscous polymer melts. Since the sample must float on a bed of air, there is a minimum viscosity that can be measured, however. The maximum strain rate is limited by the speed of the clamps and the stability of the flow. The maximum viscosity is limited by the range of the strain-gauge transducer.

10.2.1.2 Filament Stretching—Sridhar's Apparatus

Polymer solutions cannot be grabbed at each end of a specimen and stretched, but a variation on this geometry was introduced by Sridhar et al. [234] to measure the elongational viscosity of polymer solutions (Figure 10.20). Based on ideas explored by Matta and Tytus [171] and Chen et al. [42], this technique stretches a small quantity of polymer solution placed between two round plates. The plates are then separated at a rapid rate, approximating a constant deformation rate. The force on the filament is measured by a load cell attached to the stationary end, and the deformation rate is monitored by video or other optical methods.

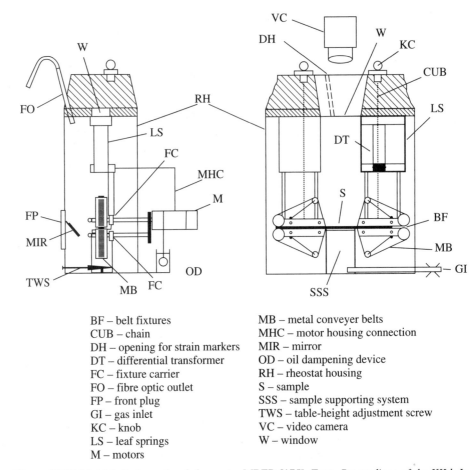

BF – belt fixtures
CUB – chain
DH – opening for strain markers
DT – differential transformer
FC – fixture carrier
FO – fibre optic outlet
FP – front plug
GI – gas inlet
KC – knob
LS – leaf springs
M – motors

MB – metal conveyer belts
MHC – motor housing connection
MIR – mirror
OD – oil dampening device
RH – rheostat housing
S – sample
SSS – sample supporting system
TWS – table-height adjustment screw
VC – video camera
W – window

Figure 10.18 Metal-belt elongational rheometer MBER [175]. From *Proceedings of the XIIth International Congress on Rheology,* Copyright © 1996, Canadian Rheology Group. Reprinted by permission.

Variations of this device have been built and refined by several groups [228, 233, 241, 16, 133]. The equations to analyze the flow, if homogeneous uniaxial extension is achieved, are the same as discussed at the beginning of this section. This technique has the advantage that the sample starts from a well-defined initial rest state. Second, except near the ends, the strain of each material element in the sample is the same in this geometry.

Experiments and calculations on the filament-stretching apparatus show, however, that due to gravity, surface tension, and the no-slip condition at the plates, the deformation near the ends is not homogeneous uniaxial extension. At short times there is an induction period during which a secondary flow occurs near the plates due to the interaction of gravitational and surface tension forces [227], and this delays the development of a uniform cylindrical column during the flow. Furthermore, elongational rates calculated on length changes versus those based on radius changes give substantially different results [228]. The latter problem can be addressed through a two-step experiment in which a constant elongational rate based

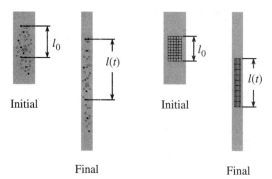

Figure 10.19 Strain measurement in an elongational flow experiment. Marker particles or a grid are placed on the surface of the sample. If the flow is homogeneous, we can assume that the deformation marked on the surface is characteristic of the sample throughout. Greater reliability is obtained by taking measurements of the elongation based on multiple pairs of points.

on the filament length is first imposed and the mid-filament diameter is measured [5, 197]. A calibration curve of Hencky strain based on length vs. Hencky strain based on mid-filament diameter is produced, and this curve is then used in a second experiment to program the plate separation that will result in an exponentially decreasing mid-filament diameter. Even with this fix, however, inhomogeneities near the plates complicate the interpretation of the measured stress. Thus, although this is a promising technique that gives reproducible results, at present there are still some important difficulties in interpreting the data. Researchers are attempting to eliminate gravity effects in this test geometry by conducting tests in zero-gravity aircraft [268]. If successful, effects due to gravity can be separated from other causes, although zero-gravity rheological testing is not likely to become routine any time soon.

10.2.2 BIAXIAL AND PLANAR EXTENSION

10.2.2.1 Meissner's Apparatus

Meissner and collaborators have applied their rotary clamp design [174] and the newer metal-belt design [175] to produce biaxial extension and other shear-free flows, including planar elongation [174, 26, 175] (Figure 10.21). Rotary clamps or belts and automatic scissors are used to pull apart a carefully prepared polymer sheet. The force $f(t)$ required to stretch the film is related to the biaxial-flow material function $\bar{\eta}_b^+$:

$$\tau_{33} - \tau_{11} = -\frac{f(t)}{A(t)} \tag{10.122}$$

$$\bar{\eta}^+(\dot{\epsilon}_0, t) = \frac{-(\tau_{33} - \tau_{11})}{\dot{\epsilon}_0} = \frac{f(t)}{A(t)\dot{\epsilon}_0} \tag{10.123}$$

where $A(t)$ is the time-dependent cross-sectional area of the thin sheet.

The original apparatus that utilized rotary clamps required an enormous sample (350-mm diameter, 5 mm thick, about 450 g) [174]. The new design based on the metal-belt system has the potential of allowing these measurements to be carried out on more modest quantities of polymer sample (2–4 g) [175]. The same issues as those discussed with the uniaxial version of this instrument apply to the biaxial flow. There is no commercial version of the biaxial instrument at this time.

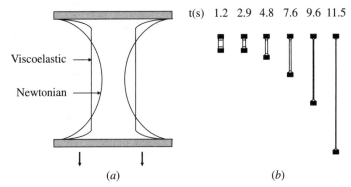

t(s) 1.2 2.9 4.8 7.6 9.6 11.5

Viscoelastic

Newtonian

(a)   (b)

Figure 10.20 Flow produced in filament-stretching experiment. (a) Change of shape of the boundary [226]. Reprinted from *Journal of Non-Newtonian Fluid Mechanics*, **40**, R. W. G. Shipman, M. M. Denn, and R. Keunings, "Mechanics of the 'falling plate' extensional rheometer," 281–288, Copyright © 1991, with permission from Elsevier Science. (b) Time sequence for an actual experiment [234]. Reprinted by permission. Reprinted from *Journal of Non-Newtonian Fluid Mechanics*, **40**, T. Sridhar, V. Tirtaatmadja, D. A. Nguyen, and R. K. Gupta, "Measurement of extensional viscosity of polymer solutions," 271–280, Copyright © 1991, with permission from Elsevier Science.

10.2.2.2 Lubricated Squeezing

One elongational flow technique that is accessible to most rheologists is lubricated squeezing [41] (see Figure 4.11d). In this technique a small disk of the sample (1–5 g) is placed between two lubricated plates, and the plates are brought together at a programmed rate to produce steady or transient biaxial stretching flows. The lubrication of the plates allows the melt to slip along the surfaces, eliminating the wall shear flow that would normally occur [162]. Experiments [41] and analysis [224] show that in order to achieve the postulated flow, the viscosity of the lubricating fluid must be chosen to be between 500 and 1000 times smaller than η_0 for the fluid being tested.

For steady biaxial squeezing the plate separation $h(t)$ (see Figure 4.11d) must follow the following relation:

$$\dot{\epsilon}_0 = \text{constant} = \frac{1}{h}\frac{dh}{dt} < 0 \tag{10.124}$$

The normal-stress difference can be calculated from the magnitude of the force $f(t)$ on the bottom plate,

$$\tau_{33} - \tau_{11} = \tau_{zz} - \tau_{rr} = \frac{f(t)}{\pi R^2} \tag{10.125}$$

where $R = R(t)$ is the time-dependent radius of the deforming disk. $R(t)$ may be measured using transparent plates and a video camera or, if homogeneous flow is produced, $R(t)$ may be deduced from conservation of mass:

$$\pi R^2(0)h(0) = \pi R^2(t)h(t) \tag{10.126}$$

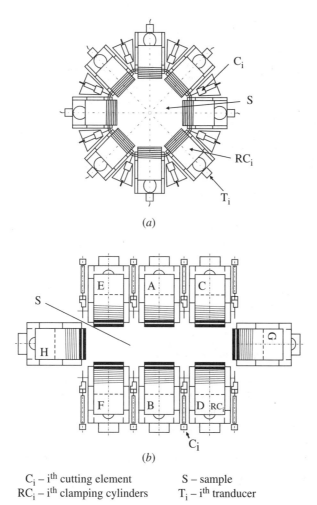

Figure 10.21 Meissner's rotary clamp rheometer for producing (*a*) equibiaxial and (*b*) planar elongational flow; from Meissner [174]. For both, the sample is a polymer film in the center (S) and rotary clamps pull the film outward while cutting the growing film in several places. In (*a*) the force is measured by 8 transducers (T). In (*b*) only clamps A–F rotate; clamps G and H measure the force required to prevent the lateral contraction. From *Chemical Engineering Communications,* Copyright © 1985, Gordon and Breach Publishers. Reprinted by permission.

C_i – i^{th} cutting element   S – sample
RC_i – i^{th} clamping cylinders T_i – i^{th} tranducer

The biaxial elongation material functions can be calculated from their definitions in the usual way.

Reaching strains higher than $\epsilon = 1$ in lubricated squeezing is prevented by the thinning of the lubricant layer and subsequent buildup of pressure in the thin film [162]. Secor [224] used a double-syringe pump to replenish the lubricant and extended his experimental regime to $\epsilon = 2.5$. Venerus and coworkers [128, 250] used porous plates to continuously supply lubricant. They were able to achieve a more uniform pressure, but were still limited to $\epsilon_{max} = 1$. The lubricated squeezing technique has also been used to obtain nonlinear step-strain data in biaxial elongation ([230] and Figure 6.68). Although this technique is limited to small strains, it is easy to implement and requires small samples. Lubricated squeezing is thus a worthwhile technique to pursue in order to obtain a limited amount of extensional flow information.

10.2.2.3 Stagnation Flows

A technique used to produce biaxial elongation in low-viscosity solutions and melts is the opposed-jets geometry (Figure 4.11*b*). In this flow two streams are made to impinge upon one another, producing a stagnation point at their intersection. Alternatively the flow can be run backward with two opposing nozzles sucking fluid from a bath, producing uniaxial elongation. The normal stresses are determined by mounting one nozzle on a stiff spring and measuring its small deflection during flow.

A simpler version of this technique directs two liquid streams at one another through crossed slots (Figure 4.14*c*). At the very center, planar elongational flow is hoped to be produced. The stresses at the stagnation point can be measured by directing a laser along the line of symmetry where the greatest extension occurs and recording the flow birefringence (see Section 10.3). The opposing-jets technique and other similar methods employing crossed slots and lubricated dies (Figure 10.22) are not very accurate because of the uncertainty about the flow field produced and also the presence of the upstream flow, which introduces an unknown flow history into the experiment. Simulations with the Carreau–Yasuda generalized Newtonian fluid model [225] show that the region of pure elongational flow at the stagnation line in these dies is vanishingly small. A method that is simpler than producing a stagnation flow, although fraught with some of the same drawbacks, is flow through a sudden contraction, which is discussed in the next section.

10.2.2.4 Contraction Flow—Cogswell and Binding Analyses

One experiment that gives some indication of the elongational behavior of a fluid is the measurement of pressure drop across flow into an abrupt contraction (Figure 10.23). This is not a homogeneous elongational flow and therefore no material functions can be measured directly, but techniques developed by Cogswell [48] and Binding [22] do permit the calculation of an elongational viscosity for some materials. These methods involve many approximations and are therefore of limited use, but the flow into an abrupt contraction can be produced on a standard laboratory capillary rheometer [11], and thus these techniques provide an easy way to estimate trends in elongational viscosity as a function of the elongation rate.

The flow into an abrupt contraction has elements of shear flow, introduced by the no-slip conditions at the walls, and elements of elongational flow, caused by the pronounced stretching of material elements near the centerline of the flow (Figure 10.23). Even when the

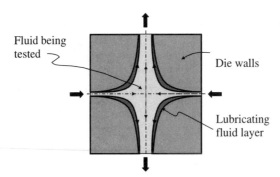

Fluid being tested

Die walls

Lubricating fluid layer

Figure 10.22 Type of lubricated die that may be used to produce a highly extensional flow. A lower viscosity fluid is introduced near the walls to eliminate the no-slip boundary condition there.

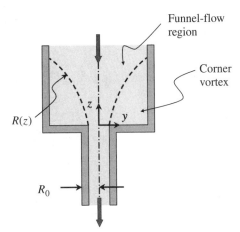

Funnel-flow region

Corner vortex

$R(z)$

z

y

R_0

Figure 10.23 Flow geometry addressed by Cogswell [48] and Binding [22]. The flow into a contraction has an extensional character along the centerline. For many fluids recirculating vortices form in the corners, and the main flow is through a funnel-shaped section near the center of the flow field. Contraction flow is not pure elongation, however, since there is shear near the walls. Both Cogswell and Binding made several assumptions about the flow field and the importance of elongational, shear, and elastic effects in order to analyze the flow.

constitutive equation is known, the analysis of this flow requires a numerical solution [23] and is quite challenging to carry out due to high stresses generated at the entrance to the smaller channel. To use contraction flow in measuring rheological material functions such as viscosity and elongational viscosity, we follow the same procedure as was followed in the analyses of, for example, capillary flow and cone-and-plate flow, that is, we will make some reasonable assumptions about the flow field.

When polymeric liquids flow into an abrupt contraction, the fluid usually channels toward the centerline and forms regions of recirculation in the corners of the contraction. On the boundaries of the central, funnel-shaped region [defined by the lines $R(z)$ shown in Figure 10.23], the velocity is not zero, but small compared to the velocity at the centerline [48], and we take it to be zero. The recirculating regions or vortices dissipate energy, and this loss in energy is reflected in the measurement of large entrance-pressure losses, Δp_{ent} (see Section 10.1.1.3). The simplest analysis of entrance flow is due to Cogswell [48, 49], who assumed that Δp_{ent} can be written as the direct summation of two pressure drops, one due to shear viscosity and the other due to elongational viscosity. Cogswell then solves for each of these two pressure drops individually, applying a force balance on a differential section of the funnel-shaped entry-flow region and integrating over the entire entry section. Details of the analysis are found in the literature [48]. In order to arrive at a tractable solution, Cogswell makes several assumptions in addition to those already mentioned. All of the assumptions invoked in Cogswell's analysis are listed in Table 10.4.

Employing these assumptions, the average elongational rate in contraction flow is calculated to be

$$\begin{array}{c} \text{Average elongation rate,} \\ \text{Cogswell analysis} \end{array} \qquad \boxed{\dot{\epsilon}_0 = \frac{\tau_R \dot{\gamma}_a}{2(\tau_{11} - \tau_{22})}} \qquad (10.127)$$

where $\tau_R = m\dot{\gamma}_a^n$ and $\dot{\gamma}_a = 4Q/\pi R^3$, associated with the calculation of the entrance pressure loss (see Figure 10.8). The average stress difference $\tau_{11} - \tau_{22}$ is obtained from the entry pressure loss:

TABLE 10.4
Assumptions for the Cogswell Analysis of Contraction Flow

1. Incompressible fluid
2. Funnel-shaped flow; no slip ($v_z = 0$) on funnel surface
3. Unidirectional flow in funnel region
4. Fully developed flow upstream and downstream
5. θ-symmetry
6. Pressure drops due to shear and elongation may be calculated separately and summed to give total entrance pressure loss
7. Neglect Weissenberg–Rabinowitsch correction, $\dot{\gamma}_R = \dot{\gamma}_a = 4Q/\pi R^3$
8. Shear stress is related to shear rate through a power law, $\tau_R = m\dot{\gamma}_a^n$
9. Elongational viscosity is constant
10. Shape of funnel is determined by the minimum generated pressure drop
11. No effect of elasticity (shear normal stresses neglected)
12. Neglect inertia

Average normal-stress
difference,
Cogswell analysis

$$\tau_{11} - \tau_{22} = -\frac{3}{8}\Delta p_{\text{ent}}(n + 1) \qquad (10.128)$$

where n is the power-law parameter for shear viscosity. The elongational viscosity is then calculated from the elongational stress and the elongational rate:

Elongational viscosity,
Cogswell analysis

$$\bar{\eta} \approx \frac{-(\tau_{11} - \tau_{22})}{\dot{\epsilon}_0} = \frac{\frac{9}{32}(n + 1)^2 \Delta p_{\text{ent}}^2}{\tau_R \dot{\gamma}_a} \qquad (10.129)$$

Thus if the shear viscosity function $\eta = m\dot{\gamma}_a^{n-1}$ is known, the elongational viscosity can be calculated from the measurement of Δp_{ent} and the flow rate Q. Similar equations can be derived for flow into a planar contraction [48, 162]. When compared to other techniques for calculating $\bar{\eta}$ for polymers such as low-density (branched) and high-density (linear) polyethylene, the Cogswell analysis is fairly accurate at high rates, but fails at lower rates [145].

A more complicated but more accurate analysis of contraction flow is due to Binding [22]. Binding makes many of the same assumptions as Cogswell, but he allows the elongational viscosity to vary with the deformation rate, and he does not neglect the Weissenberg–Rabinowitsch correction. To make the problem solvable, however, Binding resorts to some assumptions about the shape of the funnel, and, like Cogswell, he neglects any influence of elasticity on the normal stresses calculated. In his analysis Binding minimizes the overall energy required to drive the flow (Table 10.5).

In the Binding analysis the entrance pressure loss Δp_{ent} is related to the shear and elongational viscosity functions through the following equation:

$$\Delta p_{\text{ent}} = \frac{2m(1 + t)^2}{3t^2(1 + n)^2} \left[\frac{lt(3n + 1)n^t I_{nt}}{m} \right]^{\frac{1}{1+t}} \dot{\gamma}_{R_0}^{\frac{t(n+1)}{1+t}} \left[1 - \alpha^{\frac{3t(n+1)}{1+t}} \right] \qquad (10.130)$$

TABLE 10.5
Assumptions for the Binding Analysis of Contraction Flow

1. Incompressible fluid
2. Funnel-shaped flow; no slip ($v_z = 0$) on funnel surface
3. Unidirectional flow in funnel region (see assumption 10)
4. Fully developed flow upstream and downstream
5. θ-symmetry
6. Shear viscosity is related to shear rate through a power law, $\eta = m\dot{\gamma}^{n-1}$
7. Elongational viscosity is given by power law, $\bar{\eta} = l\dot{\epsilon}_0^{t-1}$
8. Shape of funnel is determined by the minimum work to drive flow
9. No effect of elasticity (shear normal stresses neglected)
10. Quantities $(dR/dz)^2$ and d^2R/dz^2 related to funnel shape are neglected; this has the implication that the radial velocity is neglected when calculating the rate of deformation
11. Neglect energy required to maintain corner circulation
12. $\tau_{\theta\theta} - \tau_{rr} = 0$
13. Neglect inertia

where m and n are the parameters associated with the shear viscosity power law ($\eta = m\dot{\gamma}^{n-1}$), l and t are the parameters associated with the elongational viscosity power law ($\bar{\eta} = l\dot{\epsilon}_0^{t-1}$), $\alpha = R_0/R_1$, R_0 is the downstream radius, R_1 is the upstream radius, Q is the flow rate, and I_{nt} and $\dot{\gamma}_{R_0}$ are defined as

$$I_{nt} = \int_0^1 \left|2 - \left(\frac{3n+1}{n}\right)\phi^{1+\frac{1}{n}}\right|^{t+1} \phi \, d\phi \tag{10.131}$$

$$\dot{\gamma}_{R_0} = \frac{(3n+1)Q}{n\pi R_0^3} \tag{10.132}$$

The variable ϕ is a dummy variable of integration. To calculate Δp_{ent} we first need to know t, the power-law index for elongational viscosity. From Equations (10.130) and (10.132) we see that a plot of $\log \Delta p_{\text{ent}}$ versus $\log Q$ will give a straight line of slope $t(n+1)/(1+t)$ from which we can calculate t if the shear viscosity power-law is known. Once t is known, the integral I_{nt} may be calculated, and finally the elongational viscosity coefficient l may be calculated from Equation (10.30). The final outcome of the analysis is the function $\bar{\eta} = l\dot{\epsilon}_0^{t-1}$.

The elongational viscosities calculated using the Binding analysis are quite reasonable for shear-thinning fluids, but for elastic fluids with constant viscosity (called Boger fluids) the analysis is not accurate. The Binding analysis is more accurate than the Cogswell analysis for commodity polymers such as linear low-density polyethylene (LLDPE) [245].

We see from the discussions in this section and in previous sections that elongational flow properties are quite challenging to measure. Many more geometries have been studied, and these are compared quantitatively and contrasted in Macosko [162] and elsewhere [205, 51]. A brief comparison of four of the most reliable methods of measuring elongational flow properties are given in Table 10.6. It is an ongoing challenge for rheologists to develop reliable, flexible techniques to evaluate elongational properties.

TABLE 10.6
A Comparison of Experimental Features of Four Elongational Geometries

Feature	Melt Stretching	MBER	Filament Stretching	Binding/ Cogswell
Stress Range	Good for high viscosity	Good for high viscosity	Good for low viscosity at room temperature	Good for high and low viscosities
Flow stability	Subject to gravity, surface tension and air currents	Can be unstable at high rates	Subject to gravity, surface tension and air currents	Unstable at very high rates
Sample size and sample loading	10 g; care must be taken to minimize end effects	<2 g; requires careful preparation and loading	<1 g; easy to load	40 g minimum; easy to load
Data handling	Straightforward, but does not result in any elongational material functions	Straightforward; more involved if strain is measured	Two tests are required to account for strain inhomogeneities	Cogswell— straightforward Binding—more complicated but not difficult
Homogeneous?	No, not at ends	Could be with care	No, not at ends	No—mixed shear and elongational flow
Pressure effects	No	No	No	Yes— compressibility of melt reservoir could cause difficulties
Elongation rates	Maximum rates depend on clamp speeds	Maximum elongation rate is limited by ability to maintain the sample in steady flow	Maximum rates depend on plate speeds; minimum rates depend on the ratio of gravity and viscous effects	High and low rates possible
Special features	Cannot reach high strains or steady state; wide range of temperatures is possible; the instrument is commercially available	Often strain is not measured but is calculated from the imposed strain rate; a wide range of temperatures is possible; the instrument is commercially available	Currently limited to room temperature liquids	Is based on a presumed funnel-shaped flow—this may not take place; wide range of temperatures possible

10.3 Flow Birefringence

The rheometric techniques we have described thus far have focused on directly measuring forces or torques and then calculating stresses and material functions from these. An alternative approach is to measure other material properties that can be related to stress rather than measuring stress directly. For transparent polymers, the optical property called *birefringence* has been found to be directly proportional to stresses [116], and thus birefringence provides a noninvasive way of measuring stress fields in some systems. Flow birefringence is applicable to both shear and elongational flows.

In this section we introduce the concept of birefringence and derive the expressions that allow shear and elongational stresses to be calculated in the most commonly used flow geometries. Fuller's text on optical rheometry [84] provides a more complete discussion of the use of polarized light in rheometry. In our discussion we assume that the reader has a level of familiarity both with light and with the optical properties of matter. A more thorough background in optics is given in Appendix E or may be obtained by reading Fowles [81].

10.3.1 INTRODUCTION

When light travels in a vacuum, its speed is the well-known result, $c = 2.997925 \times 10^8$ m/s. When light propagates through a transparent material, its speed u is less than c, and the refractive index $n = c/u$ is used to describe the speed of light in the medium. The refractive index is defined with respect to the susceptibility χ, a material property that relates the polarization state of the material (characterized by a vector $\underline{P}$) and the electric-field vector $\underline{E}$ of the incident light:

$$
\begin{array}{c}
\text{Polarization of an} \\
\text{isotropic medium}
\end{array}
\qquad
\boxed{
\begin{aligned}
\underline{P} &= \varepsilon_0 \chi \underline{E} \\
\chi &= n^2 - 1
\end{aligned}
}
\qquad (10.133)
$$

The scalar ε_0 is called the *permittivity* of free space, and it is a universal constant; $\varepsilon_0 = 8.854 \times 10^{-12}$ F/m. The interrelations between $\underline{P}$, $\underline{E}$, and the magnetic-field vector $\underline{H}$ are governed by Maxwell's equations (see Appendix E).

The preceding discussion supposes that the speed of light in a material is the same in all directions, that is, the material is isotropic. Some materials, however, are anisotropic with respect to the passage of light, and more complicated expressions are needed. The refractive index is a tensor $\underline{\underline{n}}$ in an anisotropic material, as is the susceptibility $\underline{\underline{\chi}}$:

$$
\begin{array}{c}
\text{Polarization of an} \\
\text{anisotropic medium}
\end{array}
\qquad
\boxed{
\begin{aligned}
\underline{P} &= \varepsilon_0 \underline{\underline{\chi}} \cdot \underline{E} \\
\underline{\underline{\chi}} &= \underline{\underline{n}} \cdot \underline{\underline{n}} - \underline{\underline{I}}
\end{aligned}
}
\qquad (10.134)
$$

Both $\underline{\underline{\chi}}$ and $\underline{\underline{n}}$ are symmetric tensors [81].

Having to consider the refractive index as a tensor is a considerable complication in the study of anisotropic materials. The situation can be simplified somewhat, however, by considering $\underline{\underline{n}}$ in a coordinate system $\hat{\xi}_x, \hat{\xi}_y, \hat{\xi}_z$, in which it takes on a diagonal form, that is, one in which all off-diagonal elements are zero:

$$
\underline{\underline{n}} = \begin{pmatrix} n_{xx} & 0 & 0 \\ 0 & n_{yy} & 0 \\ 0 & 0 & n_{zz} \end{pmatrix}_{\xi_x \xi_y \xi_z}
\qquad (10.135)
$$

where the subscript $\xi_x\xi_y\xi_z$ serves as a reminder that $\underline{\underline{n}}$ only takes on this simple form in a particular coordinate system denoted by ξ_x, ξ_y, ξ_z. For symmetric tensors, it is always possible to find such an orthonormal coordinate system [5a]. The vectors that form the coordinate system in which $\underline{\underline{n}}$ is diagonal are called the *principal axes* of the tensor. Details of how to find the principal axes of a tensor and of how to calculate the coefficients of a symmetric tensor in the principal frame can be found in Appendix C.6.

When $\underline{\underline{n}}$ is in its diagonal form, we need only consider three values of refractive index, $n_{xx} \equiv n_1, n_{yy} \equiv n_2$, and $n_{zz} \equiv n_3$. There are three cases to consider: $n_1 = n_2 < n_3$, called uniaxial positively birefringent, $n_1 = n_2 > n_3$, uniaxial negatively birefringent, and n_1, n_2, n_3 all distinct, characteristic of biaxial materials. Where $n_1 = n_2 = n_3$, the material is isotropic, and Equations (10.134) reduce to Equations (10.133). For uniaxially birefringent materials, the two equal refractive indices are called *ordinary* indices, $n_1 = n_2 \equiv n_0$, and the different index is called *extraordinary* index, $n_3 \equiv n_e$. The birefringence Δn is defined as the difference between the principal refractive indices, that is, $\Delta n \equiv n_e - n_0$ for uniaxial materials, and there are two independent birefringences, $\Delta n_{21} \equiv n_2 - n_1$ and $\Delta n_{32} \equiv n_3 - n_2$, for biaxial materials.

$$
\begin{array}{ll}
\text{Definition of} & \left.
\begin{array}{ll}
\Delta n = n_e - n_0 & \text{uniaxial} \\
\Delta n_{21} = n_2 - n_1 \\
\Delta n_{32} = n_3 - n_2
\end{array}
\right\} \text{biaxial}
\end{array} \tag{10.136}
$$

To measure birefringence, linear polarized light is sent through an anisotropic material along a principal axis. The electric-field vector $\underline{E}$ has two nonzero components (Figure 10.24), both orthogonal to the direction of beam travel. For example, if $\hat{\xi}_z$ is taken to be the direction of light travel, then $\underline{E}$ for the incident light may be written as

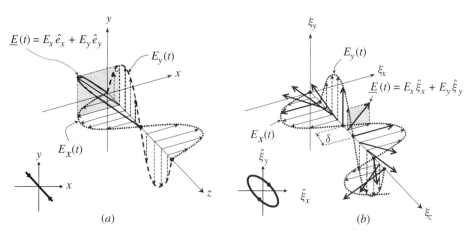

Figure 10.24 Components of the electric-field vector of linearly polarized light traveling in two media: (*a*) In an isotropic medium (e.g., air)—the light remains linearly polarized and the tip of $\underline{E}$, when projected in the xy-plane, traces back and forth on a line. (*b*) In the ξ_z-direction of an anisotropic solid, where ξ_x, ξ_y, ξ_z are the principal directions of $\underline{\underline{n}}$ for that solid—in this case the light becomes elliptically polarized, and the projected tip of $\underline{E}$ traces an ellipse in the $\xi_x\xi_y$ plane.

$$\underline{E}_{\text{incident}} = e^{i(kz-\omega t)} \begin{pmatrix} E_{0x} \\ E_{0y} \\ 0 \end{pmatrix}_{\xi_x\xi_y\xi_z} = e^{i\omega(\frac{z}{u}-t)} \begin{pmatrix} E_{0x} \\ E_{0y} \\ 0 \end{pmatrix}_{\xi_x\xi_y\xi_z} \tag{10.137}$$

where ω is the frequency of the light, $k = \omega/u$ is the magnitude of the wave vector, u is the speed of light in the medium, and E_{0x} and E_{0y} are the amplitudes of the electric-field vector in the ξ_x- and ξ_y-directions; E_{0x} and E_{0y} are constant and real for linearly polarized light. The subscript $\xi_x\xi_y\xi_z$ on the vector reminds us that we have written $\underline{E}$ in the principal frame of $\underline{\underline{n}}$, the refractive index tensor of the medium through which the light is traveling.

When light is traveling along a principal direction of an anisotropic material, it is easy to calculate the effect of the medium on the polarization state of the beam: the component of the incident-light electric-field vector in the ξ_x principal direction will travel at velocity $u_1 = c/n_1$ associated with the ξ_x principal direction; the component of the incident-light electric-field vector in the ξ_y principal direction will travel with the ξ_y-direction velocity, $u_2 = c/n_2$. The electric-field vector of the exiting light will be just the vector sum of these two components after their travel. The net effect on linearly polarized incident light of traversing an anisotropic sample of thickness h will be that it becomes elliptically polarized, that is, the ξ_x- and ξ_y-components of the exiting-light electric-field vector are out of phase by an amount δ, as shown:

$$\underline{E}_{\text{exiting}} = \begin{pmatrix} E_{0x}e^{i\omega(\frac{h}{u_1}-t)} \\ E_{0y}e^{i\omega(\frac{h}{u_2}-t)} \\ 0 \end{pmatrix}_{\xi_x\xi_y\xi_z} \tag{10.138}$$

$$= e^{i\omega\left(\frac{n_1 h}{c}\right)} e^{-i\omega t} \begin{pmatrix} E_{0x} \\ E_{0y}e^{i\delta} \\ 0 \end{pmatrix}_{\xi_x\xi_y\xi_z} \tag{10.139}$$

where

Retardance $$\boxed{\delta = \frac{\omega h \Delta n_{21}}{c} = \frac{2\pi h \Delta n_{21}}{\lambda_v}} \tag{10.140}$$

Here $\omega = 2\pi c/\lambda_v$ is the frequency of the light, n_1 and n_2 are the principal refractive indices in the ξ_x- and ξ_y-directions, $\Delta n_{21} = n_2 - n_1$ is the birefringence, h is the sample thickness, λ_v is the wavelength of light in a vacuum, and c is the speed of light in vacuum. Equation (10.139) is the equation for the electric-field vector of the elliptically polarized beam that exits the sample at $z = h$. If the beam passes next into an isotropic medium, the polarization state remains that calculated in Equation 10.139, and the wave propagates with phase factor $e^{-i\omega t}$ replaced by $e^{i(kz-\omega t)}$, where k is characteristic of the isotropic medium and z represents distance traveled in the new medium.

The phase difference or retardance δ of the birefringent sample can be measured by employing optical elements that subtract a phase difference from the light in the amount needed to return the exiting-beam polarization to its original, linearly polarized state [116] (Figure 10.25). Alternatively, a birefringent sample may be placed between crossed polarizers with light incident down a principal direction. The polarizers' transmission axes are then positioned such that they are oriented at 45° with respect to the sample's other two principal axes (Figure 10.26). Using matrix methods [84] to account for the effect of the medium on the incoming light, it can be shown that the intensity I measured in this configuration is related to δ [84, 116]:

$$I = I_0 \sin^2 \frac{\delta}{2} \tag{10.141}$$

where I_0 is the intensity of the polarized light inpinging on the sample. Measuring δ with light incident in the ξ_z-direction yields the birefringence $\Delta n_{21} = n_2 - n_1$ through Equation (10.140). If Δn_{32} or $\Delta n_{31} = \Delta n_{32} + \Delta n_{21}$ is desired, light is sent in along the ξ_x or ξ_y principal directions, respectively.

In rheology, stress causes normally isotropic polymer melts and solutions to become, in general, biaxially birefringent. Also, both positively flow-birefringent (e.g., polybutadiene) and negatively flow-birefringent (e.g., polystyrene) polymers exist [116]. The stresses in the flow are found to be directly proportional to the birefringence through a relationship, called the *stress-optical law*, described next.

10.3.2 STRESS-OPTICAL LAW

The stresses generated in the flow of polymer melts are caused by the displacement of polymer chains from their equilibrium random configurations toward elongated configurations

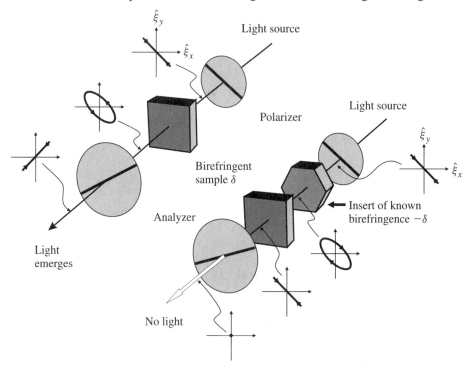

Figure 10.25 One method of measuring birefringence. Various samples of known birefringence are inserted into the light path until extinction occurs. The retardance δ of the unknown sample is then the negative of the retardance of the insert.

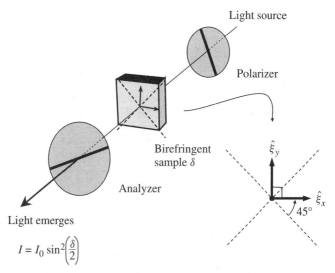

Figure 10.26 A second method of measuring birefringence. Light is sent down a principal direction, and the other two principal directions are set at an angle of 45° to the polarizer. The intensity of the exiting beam is measured and equals $I_0 \sin^2(\delta/2)$.

(see Section 9.4.2). These elongated configurations are not isotropic, and the stressed polymer displays anisotropic optical properties or birefringence (Figure 10.27). The simplest possible relationship between stress and birefringence is a linear proportionality.

For there to be a linear relationship between the stress tensor and the refractive-index tensor, these two tensors must have the same set of principal axes. This condition is called the *condition of coaxiality*. We can see that this must hold by considering a linear relationship between $\underline{\underline{n}}$ and $\underline{\underline{\tau}}$, written in the principal frame for $\underline{\underline{n}}$:

$$\underline{\underline{n}} = C\underline{\underline{\tau}} + \underline{\underline{B}} \tag{10.142}$$

No net force, isotropic chain, isotropic polarization

Figure 10.27 Elongation of a polymer chain through a tension on the chain results in anisotropic optical properties.

Force applied, anisotropic chain, anisotropic polarization = birefringent

$$
\begin{pmatrix} n_1 & 0 & 0 \\ 0 & n_2 & 0 \\ 0 & 0 & n_3 \end{pmatrix}_{\xi_x \xi_y \xi_z} = C \begin{pmatrix} \tau_{11} & \tau_{12} & \tau_{13} \\ \tau_{21} & \tau_{22} & \tau_{23} \\ \tau_{31} & \tau_{32} & \tau_{33} \end{pmatrix}_{\xi_x \xi_y \xi_z} + \begin{pmatrix} B_{11} & B_{12} & B_{13} \\ B_{21} & B_{22} & B_{23} \\ B_{31} & B_{32} & B_{33} \end{pmatrix}_{\xi_x \xi_y \xi_z} \tag{10.143}
$$

where $\underline{\underline{B}}$ is an unspecified constant tensor, and C is a scalar constant. If there is no flow, the only stresses present would be isotropic, $\underline{\underline{\tau}} = \tau_0 \underline{\underline{I}}$, and the material would be optically isotropic, $\underline{\underline{n}} = n \underline{\underline{I}}$. In this case, for no flow:

$$
\begin{pmatrix} n & 0 & 0 \\ 0 & n & 0 \\ 0 & 0 & n \end{pmatrix}_{\xi_x \xi_y \xi_z} = C \begin{pmatrix} \tau_0 & 0 & 0 \\ 0 & \tau_0 & 0 \\ 0 & 0 & \tau_0 \end{pmatrix}_{\xi_x \xi_y \xi_z} + \begin{pmatrix} B_{11} & B_{12} & B_{13} \\ B_{21} & B_{22} & B_{23} \\ B_{31} & B_{32} & B_{33} \end{pmatrix}_{\xi_x \xi_y \xi_z} \tag{10.144}
$$

and thus $B_{ii} = n - C\tau_0 \equiv B$, and all other components of $\underline{\underline{B}}$ are zero. Since $\underline{\underline{B}}$ is constant, it must be the same in the flow and no-flow cases. Thus substituting $\underline{\underline{B}} = \overline{B} \underline{\underline{I}}$ back into Equation (10.143) we see that all off-diagonal terms of $\underline{\underline{\tau}}$ are zero in this frame. Thus $\underline{\underline{n}}$ and $\underline{\underline{\tau}}$ are diagonal in the same frame, that is, they have the same principal axes. Note that $\underline{\underline{B}} = B \underline{\underline{I}}$ holds in any coordinate system since isotropic tensors are invariant to a transformation of the coordinate system (see Appendix C.6 and [7]).

The linear stress-optical law is, therefore,

$$
\text{Stress-optical law} \qquad \boxed{\underline{\underline{n}} = C \underline{\underline{\tau}} + B \underline{\underline{I}}} \tag{10.145}
$$

and it has been found to be valid for a wide variety of polymers and polymer solutions [116, 256]. The constant C is called the stress-optical coefficient, and it has units of Pa^{-1}. This relationship is also predicted by many molecularly based constitutive equations [138].

The stress-optical law is not valid for all flow conditions. To understand the limitations on the stress-optical law, consider once again Figure 10.27. When a polymer chain is deformed by flow, the chain segments orient, producing an anisotropic optical environment. The stress imposed on the polymer is proportional to the number of oriented segments, which is, in turn, proportional to birefringence. There is an upper limit to the amount of alignment that a chain can experience, however—once the chain is completely extended in the flow, applying more stress cannot produce more alignment. There is no upper limit to the amount of stress that can be imposed on the fluid, however, and thus the proportionality between stress and birefringence must fail at some point. When using flow birefringence we must always know the range of validity of the stress-optical law for the material being studied.

The stress-optical law is used to measure stresses in the same flow geometries as were discussed earlier in the chapter. One hitch is that light must be sent down a principal direction of stress.[6] As we will see, this is straightforward for simple elongational flow and a bit more complicated for shear flows.

---

[6] If the direction of the light is not a principal direction, the calculations of birefringence are more complex; see Appendix E.

10.3.3 ELONGATIONAL FLOW

In elongational flow we established in Section 4.3.1 that the symmetry of the flow implies that the stress tensor has the following form:

$$\underline{\underline{\tau}} = \begin{pmatrix} \tau_{11} & 0 & 0 \\ 0 & \tau_{22} & 0 \\ 0 & 0 & \tau_{33} \end{pmatrix}_{123} \tag{10.146}$$

The basis for these coefficients of $\underline{\underline{\tau}}$ are a Cartesian coordinate system where $\hat{e}_3$ is the direction of stretch, and the flow contracts equally along the 2- and 1-directions. The diagonal form of $\underline{\underline{\tau}}$ tells us that these conventional laboratory-frame axes are the principal axes of the stress tensor for elongational flow, and the measurement of stress birefringence is straightforward. A light beam is sent through the 2-direction, which is a principal direction, and $\Delta n_{31} = n_3 - n_1$ is measured. Knowledge of the stress-optical coefficient C allows $\tau_{33} - \tau_{11}$ to be calculated and from that, the elongational viscosity $\bar{\eta}$. If $\bar{\eta}_2$ is desired (during planar elongation), light must be sent in the 3-direction, and Δn_{21} is measured. To obtain the stress-optical coefficient initially, one must measure the flow birefringence and independently measure the stresses using conventional mechanical methods.

10.3.4 SHEAR FLOW

As we discussed in Section 4.4.1, the symmetry of shear flow allows us to conclude that the stress tensor can be written as

$$\underline{\underline{\tau}} = \begin{pmatrix} \tau_{11} & \tau_{21} & 0 \\ \tau_{21} & \tau_{22} & 0 \\ 0 & 0 & \tau_{33} \end{pmatrix}_{123} \tag{10.147}$$

This is the stress tensor written in the conventional shear coordinate system, in which $\hat{e}_1$ designates the flow direction, $\hat{e}_2$ is the gradient direction, and $\hat{e}_3$ is the neutral direction.

Unlike the stress tensor for elongational flow, this tensor is not diagonal in the conventional laboratory frame. To use birefringence for shear stress measurement we must relate the stress components above to the components of $\underline{\underline{\tau}}$ in the principal frame of reference. As discussed in Appendix C.6, to calculate the coefficients of $\underline{\underline{\tau}}$ in the principal coordinate system, we must solve the following matrix characteristic equation for the eigenvalues λ_i:

$$[\tau]_{123} \equiv \begin{bmatrix} \tau_{11} & \tau_{21} & 0 \\ \tau_{21} & \tau_{22} & 0 \\ 0 & 0 & \tau_{33} \end{bmatrix} \tag{10.148}$$

$$\det |[\tau]_{123} - \lambda I| = 0 \tag{10.149}$$

where $[\tau]_{123}$ is the matrix of coefficients of $\underline{\underline{\tau}}$ in the laboratory shear (123) frame. The results for the eigenvalues or principal stresses of shear flow are [161]

$$\sigma_1 = \lambda_1 = \frac{1}{2}(\tau_{11} + \tau_{22}) + \frac{1}{2}\sqrt{(\tau_{11} - \tau_{22})^2 + 4\tau_{21}^2} \tag{10.150}$$

$$\sigma_2 = \lambda_2 = \frac{1}{2}(\tau_{11} + \tau_{22}) - \frac{1}{2}\sqrt{(\tau_{11} - \tau_{22})^2 + 4\tau_{21}^2} \qquad (10.151)$$

$$\sigma_3 = \lambda_3 = \tau_{33} \qquad (10.152)$$

$$[\tau]_{\xi_x \xi_y \xi_z} = \begin{bmatrix} \sigma_1 & 0 & 0 \\ 0 & \sigma_2 & 0 \\ 0 & 0 & \sigma_3 \end{bmatrix} \qquad (10.153)$$

$$\underline{\underline{\tau}} = \begin{pmatrix} \sigma_1 & 0 & 0 \\ 0 & \sigma_2 & 0 \\ 0 & 0 & \sigma_3 \end{pmatrix}_{\xi_x \xi_y \xi_z} \qquad (10.154)$$

where again the subscript $\xi_x \xi_y \xi_z$ indicates that the tensor $\underline{\underline{\tau}}$ is written in its principal frame.

To interpret these results more physically, we note first that $\lambda_3 = \tau_{33}$. Solving for the eigenvector $\hat{\xi}_z$ that corresponds to λ_3, we obtain [for more on this calculation see Equation (C.72)]

$$\hat{\xi}_z = \begin{pmatrix} 0 \\ 0 \\ 1 \end{pmatrix}_{123} \qquad (10.155)$$

Note that we require the eigenvectors to be unit vectors. This can be done without loss of generality.

Thus one of the vectors is unchanged ($\hat{e}_3 = \hat{\xi}_z$) for the coordinate transformation that diagonalizes $\underline{\underline{\tau}}$. This tells us that the transformation between the shear coordinate system and the principal frame (composed of the eigenvectors $\hat{\xi}_x, \hat{\xi}_y, \hat{\xi}_z$) is a simple rotation of $\hat{e}_1$ and $\hat{e}_2$ around $\hat{e}_3$ (Figure 10.28). If the angle between $\hat{e}_1$ and $\hat{\xi}_x$ is called χ, then by the methods in Appendix C.6, the rotation matrix can be seen to be[7]

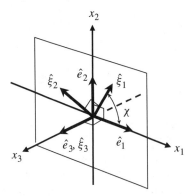

Figure 10.28 Relationship between the shear coordinate system x_1, x_2, x_3 and the principal coordinate system of the refractive index tensor, $\hat{\xi}_1, \hat{\xi}_2, \hat{\xi}_3$.

[7] The use of χ for the angle between the shear and principal frames is standard. This is not to be confused with the isotropic susceptibility, which is also called χ.

$$L = \begin{bmatrix} \cos \chi & -\sin \chi & 0 \\ \sin \chi & \cos \chi & 0 \\ 0 & 0 & 1 \end{bmatrix} \qquad (10.156)$$

The components of $\underline{\underline{\tau}}$ in the rotated coordinate system $\hat{\xi}_x, \hat{\xi}_y, \hat{\xi}_z$ can now be calculated from the matrix expression $[\tau]_{\xi_x \xi_y \xi_z} = L^T [\tau]_{123} L$ (for details see Appendixes C.5 and C.6), which yields expressions for the diagonal components σ_i, but now in terms of the angle χ,

$$\sigma_1 = (\tau_{11} - \tau_{22}) \cos^2 \chi + \tau_{22} + \tau_{21} \sin 2\chi \qquad (10.157)$$

$$\sigma_2 = (\tau_{11} - \tau_{22}) \sin^2 \chi + \tau_{22} - \tau_{21} \sin 2\chi \qquad (10.158)$$

$$\sigma_3 = \tau_{33} \qquad (10.159)$$

with

$$\tan 2\chi = \frac{2\tau_{21}}{\tau_{11} - \tau_{22}} \qquad (10.160)$$

Writing the principal stress differences $\sigma_1 - \sigma_2$ and $\sigma_2 - \sigma_3$ as $\Delta\sigma_1$ and $\Delta\sigma_2$, respectively, and rearranging, we obtain the following relationships between τ_{21}, N_1, and N_2 in the shear coordinate system, and $\Delta\sigma_1$, $\Delta\sigma_2$, and χ in the principal coordinate system:

Relation between principal stresses and shear coordinate stresses

$$\boxed{\begin{aligned} \tau_{21} &= \frac{\Delta\sigma_1 \sin 2\chi}{2} \\ N_1 &= \tau_{11} - \tau_{22} = \Delta\sigma_1 \cos 2\chi \\ N_2 &= \tau_{22} - \tau_{33} = \Delta\sigma_2 + \Delta\sigma_1 \sin^2 \chi \end{aligned}} \qquad (10.161)$$

Here χ is called the *orientation* or *extinction angle*.

The problem of measuring τ_{21}, N_1, and N_2 has now been replaced by the need to measure $\Delta\sigma_1$, $\Delta\sigma_2$, and χ. The stress-optical law is

$$\begin{pmatrix} n_1 & 0 & 0 \\ 0 & n_2 & 0 \\ 0 & 0 & n_3 \end{pmatrix}_{\xi_x \xi_y \xi_z} = C \begin{pmatrix} \sigma_1 & 0 & 0 \\ 0 & \sigma_2 & 0 \\ 0 & 0 & \sigma_3 \end{pmatrix}_{\xi_x \xi_y \xi_z} + \begin{pmatrix} B & 0 & 0 \\ 0 & B & 0 \\ 0 & 0 & B \end{pmatrix}_{\xi_x \xi_y \xi_z} \qquad (10.162)$$

Measurements of the birefringences $n_1 - n_2$ and $n_2 - n_3$, plus knowledge of the stress-optical coefficient C, yield the principal stress differences $\Delta\sigma_1$ and $\Delta\sigma_2$:

$$n_1 - n_2 = C(\sigma_1 - \sigma_2) = C\Delta\sigma_1 \qquad (10.163)$$

$$n_2 - n_3 = C(\sigma_2 - \sigma_3) = C\Delta\sigma_2 \qquad (10.164)$$

The angle χ must be measured separately. In practice, $\Delta\sigma_2 = (n_2 - n_3)/C$ is not often measured in shear flow since this would require sending light along the principal direction $\hat{\xi}_y$, and this direction is unknown a priori and varies with the material studied and the flow conditions. It is possible to measure $\Delta\sigma_1 = (n_1 - n_2)/C$, however, by sending light in along the $\hat{\xi}_z$-direction, which is a principal axis.

The simplest arrangement for measuring $n_1 - n_2$ and χ in steady flow is to bracket a Couette flow with crossed polarizers [116] (Figure 10.29). These are then turned together such that the extinction axes of the polarizers rotate, and the angle at which extinction occurs with the sample present is χ. We can see why this is true by looking at Equation (10.138). If light polarized along one of the two remaining principal directions (either $\hat{\xi}_x$ or $\hat{\xi}_y$ in this case) is sent through a birefringent medium, no change in phase is effected. For example, if the incoming light is polarized in the ξ_x-direction, then

$$\underline{E}_{\text{incident}} = e^{i\omega(\frac{z}{u}-t)} \begin{pmatrix} E_{0x} \\ 0 \\ 0 \end{pmatrix}_{\xi_x \xi_y \xi_z} \tag{10.165}$$

While passing through the birefringent sample, the ξ_x-component of $\underline{E}$ travels at speed u_1, and the ξ_y-component travels at speed u_2; hence,

$$\underline{E}_{\text{exiting}} = e^{i\omega(\frac{h}{u_1}-t)} \begin{pmatrix} E_{0x} \\ 0 \\ 0 \end{pmatrix}_{\xi_x \xi_y \xi_z} \tag{10.166}$$

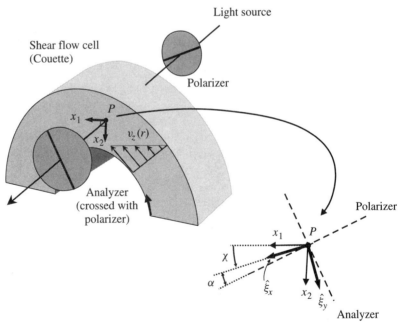

Figure 10.29 Measurement of orientation angle χ and the retardance δ for shear flow. In the configuration shown, light emerges unless the axes of the polarizer and analyzer correspond to shear principal directions ($\alpha = 0$). χ is determined by rotating the crossed polars until the light is extinguished. The polaroids are then rotated so that $\alpha = 45°$, and δ is measured as $2\sin^{-1}\sqrt{I/I_0}$ [see Equation (10.141)].

which is linearly polarized in the ξ_x-direction, that is, no change in polarization was induced upon passing through the birefringent medium in this direction. Thus a polaroid oriented to extinguish the light in its incident state will still extinguish the light with the birefringent medium present. If the incoming light is polarized in any direction but the ξ_x- or ξ_y-directions, however, the electric-field vector will have two nonzero components when written in the principal frame, each of which will travel in the birefringent medium at its respective velocity (u_1 or u_2), and the resulting phase difference will not be zero. In this case a polaroid set to extinguish the light in its original state (no sample present) will no longer extinguish the light that has passed through the birefringent medium, which is now elliptically polarized. Thus, when rotating the crossed polaroids across the steady flow field, extinction will occur when the principal birefringence (or equivalently stress) directions line up with the extinction axes of the crossed polaroids. Once χ is located, the crossed polarizers can then be positioned with their extinction axes oriented at 45° with respect to χ, the intensity measured, and the retardance δ (which is related to $n_1 - n_2$) obtained from Equation (10.141).

In unsteady shear flows the problem of measuring χ is a considerable challenge, which can be overcome by modulating the incoming polarization in some way [82, 117, 149]. One method [82] is to insert a photoelastic modulator in the laser path before the sample. The entire system is bracketed by crossed polarizers. A photoelastic modulator is a quartz crystal whose optical properties change with the strain on the crystal. The crystal is deformed sinusoidally, resulting in the light that passes through it experiencing a time-variable retardance δ_{mod}:

$$\delta_{\text{mod}} = A \sin \omega_{\text{mod}} t \tag{10.167}$$

where A is an amplitude that depends on experimental conditions, and ω_{mod} is the resonant frequency of the quartz crystal. Using matrix methods [84] to account for the optical impact of each element in the optical train, it can be shown [82] that measurements of the intensities corresponding to the first and second harmonics of ω_{mod} allow the sample retardance δ and orientation angle χ to be calculated simultaneously as a function of time. The two intensity harmonics can be measured in a straightforward manner using lock-in amplifiers.

Our preceding discussion shows how τ_{21} and N_1 can be measured with light sent in along the 3-direction of shear flow. It is sometimes easier, however, to send light along $\hat{e}_2$ or $\hat{e}_1$, such as through the plates of a parallel-disk rheometer or along the axis of a capillary tube [116]. Since $\hat{e}_2$ and $\hat{e}_1$ are not principal directions, the optics are more complex. The case of light propagating in the 2-direction of shear flow is described in detail in Appendix E. Luckily, for moderate values of the birefringence, the results of sending light down a nonprincipal axis and down a principal axis are the same, that is, a retardance δ is induced, which is related to the birefringence in the plane orthogonal to the direction of travel of light. In the shear-flow coordinate system, the refractive index tensor is

$$\underline{\underline{n}} = \begin{pmatrix} n_{11} & n_{21} & 0 \\ n_{21} & n_{22} & 0 \\ 0 & 0 & n_{33} \end{pmatrix}_{123} \tag{10.168}$$

For small values of birefringence, we show in Appendix E that sending light in the 2-direction (not a principal direction) measures the difference $n_{11} - n_{33}$, and for light in the 1-direction, $n_{22} - n_{33}$ is measured. It is important to note that these are not principal refractive index differences, $\Delta n_{21} = n_2 - n_1$ and $\Delta n_{32} = n_3 - n_2$, but rather refractive index differences associated with the shear coordinate system. They are related to the shear normal-stress differences in a straightforward manner:

$$n_{11} - n_{22} = C(\tau_{11} - \tau_{22}) \qquad (10.169)$$

$$n_{22} - n_{33} = C(\tau_{22} - \tau_{33}) \qquad (10.170)$$

10.4 Summary

We have presented an outline of only some of the rheometric techniques that are used to characterize polymeric and other non-Newtonian materials. Walters' book [258] contains important discussions on many effects that enter into the actual measurements of material functions. Recent reviews of rheometric techniques and equipment may be found in Macosko [162] and Collyer and Clegg [51]. Other representative texts that discuss rheological measurements are Dealy and Wissbrun [61], Dealy [58], and for the industrial approach, Walters' other text [259] and Barnes et al. [11]. For more on optics, see Appendix E and Fowles [81], and for more on optical rheometry, see Fuller [84].

10.5 PROBLEMS

10.1 What is the Weissenberg–Rabinowitsch correction? Why is it needed?

10.2 How can we correct capillary data for entrance and exit effects?

10.3 How can we correct capillary data for slip effects?

10.4 For a Newtonian fluid in Poiseuille flow in a capillary, show that near the wall the velocity takes on the form $\underline{v} = \dot{\gamma}_0 x_2 \hat{e}_1$.

10.5 For a power-law generalized Newtonian fluid in Poiseuille flow in a capillary, show that near the wall the velocity takes on the form $\underline{v} = \dot{\gamma}_0 x_2 \hat{e}_1$.

10.6 We showed that we can measure steady shear viscosity η in a capillary rheometer using the following equations:

$$\eta(\dot{\gamma}_R) = \frac{\tau_R}{\dot{\gamma}_R} = \frac{\tau_R}{Q/\pi R^3} \left[3 + \frac{d\ln(4Q/\pi R^3)}{d\ln \tau_R} \right]^{-1}$$

$$\dot{\gamma}_R = \frac{Q}{\pi R^3} \left[3 + \frac{d\ln(4Q/\pi R^3)}{d\ln \tau_R} \right]$$

where τ_R is the wall shear stress, R is the radius of the capillary, Q is the flow rate in the capillary, and $\dot{\gamma}_R$ is the shear rate at the wall.

(a) What are the assumptions that went into deriving this relationship?

(b) The term in brackets on the right-hand side of the equation is known as the Rabinowitsch correction. What is the Rabinowitsch correction correcting for?

(c) Describe what data you would take and how you would manipulate them to calculate the viscosity of an unknown material using a capillary rheometer.

10.7 Show that Equation (10.44) for the general case of the shear rate in capillary flow and Equation (10.31) for the capillary shear rate of a power-law generalized Newtonian fluid are consistent.

10.8 In experiments on a Newtonian fluid in a capillary rheometer, the measured pressure drop across the capillary was 200 psi for a constant flow rate Q. The capillary was 1 mm in diameter and 20 mm in length. For experiments run at the *same* flow rate, what would the pressure drop across the capillary have been if the capillary had been 1 mm in diameter but 40 mm in length? What would the pressure drop have been if the capillary had been 20 mm in length, but with a diameter of 0.5 mm? Comment on your answers.

10.9 For a power-law generalized Newtonian fluid of interest the viscosity is given by

$$\eta(\text{Pa} \cdot \text{s}) = 3.62 \times 10^6 \, \dot{\gamma}^{-0.8}$$

where $\dot{\gamma}$ is given in units of s^{-1}. In experiments on this material in a capillary rheometer, the measured pressure drop across the capillary was 200 psi for a constant flow rate Q. The capillary was 1 mm in diameter and 20 mm in length. For experiments run at the *same* flow rate, what would the pressure drop across the capillary have been if the capillary had been 1 mm diameter but 40 mm in length? What would the pressure drop have been if the capillary had been 20 mm in length, but with a diameter of 0.5 mm? Comment on your answers, and compare to Newtonian predictions (that is, for power-law parameter $n = 1$).

10.10 You are given a sack of polymer and are told to measure viscosity versus shear rate at 200°C on a capillary rheometer in the lab. Describe what experiments you need to run to get an accurate plot of viscosity as a function of shear rate.

10.11 You have air pressure of 90 psig in your lab that you want to use to drive a capillary rheometer. For typical sizes of capillaries ($R = 0.5, 1, 1.5$ mm; $L = 10, 20, 30, 40$ mm), assuming a Newtonian fluid, calculate the maximum viscosities and maximum shear rates you can achieve with lab pressure as your driving force. You estimate that the lowest average exit velocity you can measure is $v_{z,\text{av,min}} = 0.5$ mm/s. Would the proposed capillary rheometer be appropriate for most polymers?

10.12 For the data in Figure 10.10, calculate v_{slip} as a function of wall shear stress.

10.13 For the data in Figure 10.8, calculate the true pressure drop as a function of apparent shear rate.

10.14 The raw data of apparent shear rate versus wall shear stress given in Table 10.7 [131] are for a star-branched polybutadiene ($M_w = 333$ kg/mol) at

379°C. Calculate and plot the true viscosity versus true shear rate at this temperature. What are m and n for a fit of these data to the power-law generalized Newtonian fluid model?

TABLE 10.7
Data for Problem 10.14

$\dfrac{\Delta P R}{2L}$ (dyn/cm^2)	$\dfrac{4Q}{\pi R^3}$ (s^{-1})
0.44	0.021
0.57	0.051
0.66	0.15
0.84	0.21
1.12	0.48
1.28	0.64
1.51	1.01
1.63	1.15
1.80	1.56
1.92	1.77
2.17	2.56
2.22	2.56
2.36	3.13
2.54	3.68

10.15 Derive an expression for the shear rate at the wall for Poiseuille flow between two flat parallel plates. Your expression will be similar to the Rabinowitsch correction discussed for flow in a tube. You may assume that the end and edge effects are negligible. Verify that your solution is correct for the case of a power-law generalized Newtonian fluid (see Problem 7.27).

10.16 The steady-state torque required to rotate the upper plate in the torsional parallel-disk rheometer was calculated for an incompressible Newtonian fluid in Section 3.5.4. Show that this result for the torque is consistent with the viscosity calculated from the Rabinowitsch expression for this geometry, Equation (10.78).

10.17 You are given a sack of polymer and are told to measure viscosity versus shear rate at 200°C on a parallel-disk rheometer in the lab. Describe what experiments you need to run to get an accurate plot of viscosity as a function of shear rate.

10.18 In conducting measurements of G' and G'' on a cone-and-plate rheometer, you notice that the torques that are generated by the fluid are below the minimum sensitivity of the instrument. What can you do to raise the torques generated by the fluid on the cone-and-plate instrument?

10.19 Data were taken with a cone-and-plate rheometer on a fluid of unknown properties. The stress response with time was fit to the following function:

$$\tau_{\theta\phi}|_{\theta=\Theta_0} \text{ (Pa)} = -16(1 - e^{-25t})$$

where t is in seconds; the imposed shear rate $\dot{\gamma}_0 = \dot{\gamma}_{\phi\theta}$ was 10 s^{-1}.

(a) Plot the shear stress as a function of time. It is customary to plot on a log–log plot.

(b) Calculate the stress growth function $\eta^+(t)$ for this fluid at $\dot{\gamma}_0 = 10^{-1}$.

(c) Calculate the steady shear viscosity for this fluid at $\dot{\gamma}_0 = 10^{-1}$.

10.20 You are given a sack of polymer and are told to measure viscosity versus shear rate at 200°C on a cone-and-plate rheometer in the lab. Describe what experiments you need to run to get an accurate plot of viscosity as a function of shear rate.

10.21 The viscosity and first normal-stress coefficient were measured on a cone-and-plate viscometer to be 300 Pa · s and 50 Pa · s², respectively, at a shear rate of 5 s⁻¹. The cone angle was 4°, and the radius of the cone-and-plate fixtures was 25 mm.

(a) What was the torque in this measurement? Give your answer in both N · m and in g · cm.

(b) What was the thrust? Give your answer in both newtons and grams.

(c) What would the torque and thrust have been if fixtures with radius 12.5 mm had been used? Comment on how your calculations would impact your selection of which transducer to use to measure torque and thrust.

10.22 A cone-and-plate rheometer is equipped with a 2000-g · cm transducer to measure torque. What is the maximum viscosity measurable with fixtures having a radius of 12.5 mm and a cone angle of 4°? Give your answer as a function of shear rate. What if the cone angle is 2°? What if the radius is 25 mm?

10.23 Compare and contrast the cone-and-plate and parallel-disk torsional shear geometries. Under what circumstances would each be favored?

10.24 A Couette rheometer is equipped with a 2000-g · cm transducer to measure torque. What is the

maximum viscosity measurable with fixtures having an inner radius of 12.5 mm, a gap of 1 mm (outer − inner radius = gap), and a bob length of 30 mm? Give your answer as a function of shear rate.

10.25 You are given a polymer solution and are told to measure viscosity versus shear rate at room temperature on a Couette rheometer in the lab. Describe what experiments you need to run to get an accurate plot of viscosity as a function of shear rate.

10.26 The viscosity of a fluid was measured on a Couette viscometer to be 6.2 Pa · s. The inner radius was 25 mm, the gap (outer − inner radius) was 1 mm, and the bob length was 30 mm. The shear rate was 10 s⁻¹.

(a) What was the torque in this measurement? Give your answer in both N · m and in g · cm.

(b) What would the torque have been if fixtures with an inner radius of 15 mm and a gap of 1 mm had been used? The length remains the same at 30 mm.

10.27 In experiments with a polymer on the metal-belt extension rheometer (MBER), the force in newtons as a function of time $f(t)$ was measured and empirically fit to the function $f(t) = 8.0 \times 10^{-4}(1 - e^{-0.0010t})^{0.40}$, where t is time in seconds. The initial cross-sectional area was 10.0 mm², and the elongational rate was 0.0010 s⁻¹. What is $\bar{\eta}^+$ for this experimental run? Plot your answer.

10.28 What is the difference between uniaxially and biaxially birefringent materials?

10.29 The refractive index tensor of zircon is given below [81].

$$\underline{\underline{n}} = \begin{pmatrix} 1.923 & 0 & 0 \\ 0 & 1.923 & 0 \\ 0 & 0 & 1.968 \end{pmatrix}_{\xi_1\xi_2\xi_3}$$

(a) What are n_e, n_0, and Δn?

(b) Is zircon uniaxial, biaxial, or isotropic? If uniaxial, is it positively or negatively birefringent?

10.30 For beryl, $n_0 = 1.598$ and $n_e = 1.590$ [81]. What is $\underline{\underline{n}}$ for beryl? What is the birefringence?

10.31 For diamond, $n = 2.417$ and the crystal is isotropic [81]. What is $\underline{\underline{n}}$?

10.32 The refractive index tensor of feldspar is given below [81].

$$\underline{\underline{n}} = \begin{pmatrix} 1.524 & -2.000 \times 10^{-3} & 0 \\ -2.000 \times 10^{-3} & 1.524 & 0 \\ 0 & 0 & 1.530 \end{pmatrix}_{123}$$

Is feldspar uniaxial, biaxial, or isotropic? What is the value of birefringence (or the two values of birefringence if feldspar is biaxial)?

10.33 The light incident on a slab of calcite is characterized by electric field vector $\underline{E}$,

$$\underline{E} = e^{i\omega(\frac{z}{u}-t)} \begin{pmatrix} 2 \\ 0 \\ 1 \end{pmatrix}_{\xi_1\xi_2\xi_3}$$

where $\omega = \frac{2\pi c}{\lambda_v}$ and the light is from a helium-neon laser, $\lambda_v = 0.6328\mu m$. The refractive index tensor for calcite is [81]

$$\underline{\underline{n}} = \begin{pmatrix} 1.658 & 0 & 0 \\ 0 & 1.658 & 0 \\ 0 & 0 & 1.486 \end{pmatrix}_{\xi_1\xi_2\xi_3}$$

Calculate the polarization state of the exiting light. The thickness of the sample is 3 mm, and the light is passing along the $\hat{\xi}_2$-direction.

10.34 Calculate the eigenvalues of the stress tensor $\underline{\underline{\tau}}$ in terms of the coefficients of $\underline{\underline{\tau}}$ in the shear coordinate system [Equations (10.150–10.152)].

10.35 Derive the relationships between the principal stresses σ_i and the shear coordinate system coefficients τ_{21}, τ_{11}, τ_{22}, and τ_{33} [Equation (10.161)].

10.36 Show that the stress-optical law written as

$$\underline{\underline{\tilde{n}}} = C\underline{\underline{\tilde{\tau}}}$$

$$\underline{\underline{\tilde{n}}} \equiv \underline{\underline{n}} - \frac{1}{3}\text{trace}(\underline{\underline{n}})\underline{\underline{I}}$$

$$\underline{\underline{\tilde{\tau}}} \equiv \underline{\underline{\tau}} - \frac{1}{3}\text{trace}(\underline{\underline{\tau}})\underline{\underline{I}}$$

is equivalent to the form discussed in the text. What is B equal to? $\underline{\underline{\tilde{\tau}}}$ is known as the deviatoric stress tensor, that is, the stress tensor written such that the isotropic stress is zero.

APPENDIX

A

Nomenclature

Symbol	Section	Definition
a	2.2	Magnitude of vector $\underline{a}$, $a = |\underline{a}|$
	4.3.2	Length of side of a cube of fluid
	7.2.1	Exponent parameter of Carreau–Yassuda generalized Newtonian fluid constitutive model
$\tilde{a}$	8.4.2	Combination of parameters, $\tilde{a} = \sqrt{-I\omega\rho/\eta^*}$
a_1, a_2, a_3	3.2.2	1-, 2-, 3-components of vector $\underline{a}$
a_T	6.2.1.3	Shift factors for time–temperature superposition of rheological functions
$\underline{a}$	2.2	Arbitrary vector variable
	3.2.2	Stress on 1-plane of a cube of fluid surrounding a point of interest
$\hat{a}$	2.2	Arbitrary unit vector
A	3.2.2	Area
	C.6	3×3 matrix of coefficients of tensor $\underline{\underline{A}}$ in 123 Cartesian coordinate system
$A(t)$	10.2.1.1	Cross-sectional area in elongational rheometry
$A(r)$	10.1.2	Scalar function of r
$\tilde{A}$	8.4.2	Combination of variables in torsional viscometer example, $\tilde{A} = 2\pi R^4 L\rho(a-1)/I$
A_0	10.2.1.1	Initial cross-sectional area in melt stretching experiment
$\mathcal{A}$	8.4.2	Amplitude ratio in torsional viscometer example
$\underline{\underline{A}}$	2.3	Arbitrary tensor variable
A_{11}, A_{12}, A_{13}, etc.	2.3	11, 12, 13, etc. coefficients of tensor $\underline{\underline{A}}$, i.e., A_{23} is the coefficient of the $\hat{e}_2\hat{e}_3$ diad
$\widehat{A}^{11}, \widehat{A}^{12}, \widehat{A}^{13}$, etc.	9.3.2	11, 12, 13, etc. contravariant convected coordinates of tensor $\underline{\underline{A}}$
$\widehat{A}_{11}, \widehat{A}_{12}, \widehat{A}_{13}$, etc.	9.3.2	11, 12, 13, etc. covariant convected coordinates of tensor $\underline{\underline{A}}$
b	4.3.3	Scalar parameter indicating the way streamlines of elongational flow change with rotations around the flow direction
	10.1.1.4	Extrapolation length; used to quantify slip at surfaces
$b(t)$	5.2.2.5	Time-dependent displacement of upper plate in small-amplitude oscillatory shear experiment
b_T	6.2.1.2	Vertical shift factors for time–temperature superposition, $b_T \equiv T_{\text{ref}}\rho_{\text{ref}}/T\rho$
$\underline{b}$	2.3.3	Arbitrary vector variable
	3.2.2	Stress on 2-plane of a cube of fluid surrounding a point of interest
$\underline{b}^{(1)}, \underline{b}^{(2)}, \underline{b}^{(3)}$	9.3.1	nonorthonormal reciprocal basis vectors
$\underline{b}_{(1)}, \underline{b}_{(2)}, \underline{b}_{(3)}$	9.3.1	nonorthonormal basis vectors
b_1, b_2, b_3	2.3.3	1-, 2-, 3-coefficients of vector $\underline{b}$.
B	9.3.1	Volume of parallelepiped formed by vectors $\underline{b}_{(1)}, \underline{b}_{(2)}, \underline{b}_{(3)}$
	C.6	3×3 matrix of coefficients of tensor $\underline{\underline{B}}$ in 123 Cartesian coordinate system

continued

Symbol	Section	Definition
$\mathcal{B}$	8.4.2	Combination of variables in torsional viscometer example, $\mathcal{B} = (a-1)^2 R^2\, K/I$
$B(r)$	10.1.2	Scalar function of r
$\underline{\underline{B}}$	2.3	Arbitrary tensor variable
	D	Finger strain tensor (DPL notation)
	10.3.2	Constant tensor
$\underline{\underline{B}}^{-1}$	D	Cauchy strain tensor (DPL notation)
B_{11}, B_{12}, B_{13}, etc.	2.3	11, 12, 13, etc. coefficients of tensor $\underline{\underline{B}}$, i.e., B_{23} is the coefficient of the $\hat{e}_2\hat{e}_3$ diad
c	2.4	Speed of light in a vacuum
c, c'	C.4	Section on finite deformation tensors in spherical coordinates, $c = \cos\phi, c' = \cos\phi'$
C, C'	C.4	Section on finite deformation tensors in spherical coordinates, $C = \cos\theta, C' = \cos\theta'$
C_1, C_2	4.3.1	Integration constants
c_1^0, c_2^0	6.2.1.3	Constants in WLF equation for shift factor $\log a_T$ versus temperature
C	10.3.2	Stress-optical coefficient
$\underline{\underline{C}}$	9.1.1	Cauchy strain tensor
$\underline{\underline{C}}^{-1}$	9.1.1	Finger strain tensor
$\underline{c}$	3.2.2	Stress on 3-plane of a cube of fluid surrounding a point of interest
c_1, c_2, c_3	3.2.2	1-, 2-, 3-components of vector $\underline{c}$
$\underline{d}$	2.3	Arbitrary vector variable
D	10.1.1.4	Tube diameter
	8.2	Displacement of a spring or dashpot
$\underline{D}$	E.3	Electric displacement vector
De	5	Deborah number, De $\equiv \lambda/t_{\text{flow}}$
d/dt	2	Derivative operator
$\partial/\partial t$	2	Partial derivative operator
D/DT	2.6.3	Substantial derivative operator
e	10.1.1.3	Bagley correction length
$\hat{e}_x, \hat{e}_y, \hat{e}_z$	2.2.1.1	Unit vectors in x-, y-, z-directions of Cartesian coordinate system
	2.2.1.1	Unit vectors in 1-, 2-, 3-directions of Cartesian coordinate system; same as x-, y-, and z-directions
$\hat{e}_r, \hat{e}_\theta, \hat{e}_z$	2.5	Unit vectors in r-, θ-, z-directions of cylindrical coordinate system
$\hat{e}_r, \hat{e}_\theta, \hat{e}_\phi$	2.5	Unit vectors in r-, θ-, ϕ-directions of spherical coordinate system
$\mathcal{E}$	E.3	3×1 matrix of coefficients of vector $\underline{E}$
E	B	Young's modulus
$\underline{E}$	10.3.1	Electric-field vector
$\underline{E}_0$	E.2	Vector prefactor of electric-field vector, $\underline{E} = \underline{E}_0 e^{i(\underline{k}\cdot\underline{r}-\omega t)}$
E_0	E.2	Magnitude of vector $\underline{E}_0$
E_{0x}, E_{0y}, E_{0z}	10.3.1	x-, y-, z-components of $\underline{E}_0$
$\underline{\underline{E}}$	9.3.4	Strain tensor that corresponds to Gordon–Schowalter derivative [138]
$\underline{\underline{E}}^T$	D	Inverse deformation-gradient tensor (DPL notation) [26]
f	8.2	Force on a spring or dashpot
$f(t)$	4.2	Scalar function of time
	10.2.1.1	Force in elongation experiment
	10.2.2	Force on a plate in biaxial stretching experiment
$f(x, t)$	2.6.2	Scalar function of space (x) and time (t)
$f(x_1, x_2, x_3, t)$	2.6.3	Scalar function of three-dimensional space (x) and time (t)
$f(\underline{b})$	2.3.3	Arbitrary vector function
$\underline{f}$	3.2	Force (vector)
$\underline{f}_{(i)}$	3.2	ith force
	9.3.2.2	Force on surface normal to $\underline{g}_{(i)}$
f_1, f_2, f_3	3.2.2	1-, 2-, 3-components of vector $\underline{f}$

continued

Symbol	Section	Definition
$\underline{f}(t)$	10.2.1.1	Vector force as a function of time for melt stretching experiment
F	10.1.1	Force on piston in a capillary rheometer
$\underline{\underline{F}}(t, t')$	9.1.1	Deformation-gradient tensor
$\underline{\underline{F}}^{-1}(t', t)$	9.1.1	Inverse deformation-gradient tensor
$\mathcal{F}$	10.1.3	Axial thrust in cone-and-plate rheometer
g	2.2.1.1	Gravitational force constant (scalar)
	9.3.2	Determinant of matrix g_{ij}
g^{ij}	9.3.2	Contravariant metric coefficients, $g^{ij} = \underline{g}^{(i)} \cdot \underline{g}^{(j)}$
g_{ij}	9.3.2	Covariant metric coefficients, $g_{ij} = \underline{g}_{(i)} \cdot \underline{g}_{(j)}$
$g(t - t')$	8.2	Generalized function that appears in generalized linear viscoelastic constitutive equation
$\tilde{g}_i$	6.2.1.3	Non-temperature-dependent portion of individual relaxation moduli g_i, $g_i = \tilde{g}_i T \rho$
$\underline{g}$	3.2.3	Force due to gravity (vector)
$\underline{g}^{(1)}, \underline{g}^{(2)}, \underline{g}^{(3)}$	9.3.2	Nonorthonormal reciprocal basis vectors of convected coordinate system
$\underline{g}_{(1)}, \underline{g}_{(2)}, \underline{g}_{(3)}$	9.3.2	Nonorthonormal basis vectors of convected coordinate system
G	5.2.2.5	Elastic modulus in Hooke's law for elastic solids
$G(t)$	5.2.2.4	Linear viscoelastic relaxation modulus
$G(t, \gamma_0)$	5.2.2.4	Nonlinear viscoelastic relaxation modulus
G_0	9.2	Modulus parameter in Lodge model
G^0	B	Instantaneous modulus
$G'(\omega), G''(\omega)$	5.2.2.5	Storage and loss moduli measured in small-amplitude oscillatory shear
$G'_r(a_T \omega), G''_r(a_T \omega), G^*_r(a_T \omega)$	5.2.2.5	Time–temperature shifted moduli measured in small-amplitude oscillatory shear
$G^*(\omega), \|G^*(\omega)\|$	5.2.2.6	Complex modulus (complex number, $G^* = G' - iG''$) and magnitude of complex modulus (real number)
$G_{\Psi_1}(t, \gamma_0), G_{\Psi_2}(t, \gamma_0)$	5.2.2.4	First and second normal-stress step shear relaxation moduli
G^0_N	6.1.3	Plateau modulus
G_{sp}	8.2	Force constant of a spring
h	5.4	Gap in small-amplitude oscillatory shear test as defined in problem
	10.3.1	Sample thickness (birefringence)
$h(t)$	10.2.2.2	Height of a sample in lubricated squeezing experiment
H	3.5.1	Slit half-height in Poiseuille flow in an infinite slit
	10.1.2	Gap between parallel plates
H_0	E.2	Magnitude of vector $\underline{H}_0$
$\underline{H}$	E.1	Magnetic-field vector
$\underline{H}_0$	E.2	Vector prefactor of magnetic-field vector, $\underline{H} = \underline{H}_0 e^{i(\underline{k} \cdot \underline{r} - \omega t)}$
$H(t)$	5.2.2.4	Heaviside unit step function
H_{0x}, H_{0y}, H_{0z}	E.2	x-, y-, z-components of $\underline{H}_0$
i	5.2.2.6	$\sqrt{-1}$
$\hat{i}, \hat{j}, \hat{k}$	2.2.1.1	Unit vectors in x; y; z-directions of Cartesian coordinate system
I	8.4.2	Moment of inertia of bob in torsional viscometer example
	C.6	3×3 matrix of coefficients of identity tensor $\underline{\underline{I}}$
I	5.2.2.6	Imaginary operator, $I(a + bi) = b$
$\underline{\underline{I}}$	2.4	Identity tensor, $\underline{\underline{I}} = \delta_{ij} \hat{e}_i \hat{e}_j$
$I_{\underline{\underline{B}}}, II_{\underline{\underline{B}}}, III_{\underline{\underline{B}}}$	2.3.4	First, second, and third invariants of tensor $\underline{\underline{B}}$
$J(t)$	5.2.2.3	Shear creep compliance
$J_1(r)$	8.4.2	First-order Bessel function of the first kind
J_r	5.2.2.3	Recoverable compliance
J^0_s	5.2.2.3	Steady-state compliance
$\hat{J}$	2.6.2	Value of integral as defined in text
k	10.3.1	Magnitude of wave vector $\underline{k}$
$\underline{k}$	E.2	Wave vector of light

continued

Symbol	Section	Definition
$\hat{k}$	E.2	Unit vector in direction of wave vector $\underline{k}$, $\hat{k} = \underline{k}/|\underline{k}|$
k_x, k_y, k_z	E.3	x-, y-, and z-components of wave vector $\underline{k}$
K	E.2	Dielectric constant or relative permittivity
$\underline{\underline{K}}$	E.3.2.1	Dielectric constant tensor in an anisotropic medium
K_m	E.2	Ratio of permeability of medium to that of vacuum
$\mathcal{K}$	8.4.2	Force constant in torque bar in torsional viscometer example
l	4.2	Final separation of two particles of interest
	B	Final length of sample undergoing uniaxial extension
l_0	4.2	Initial separation of two particles of interest
	B	Initial length of sample undergoing uniaxial extension
l_1, l_2	4.2	y-components of two particles of interest
L_{11}, L_{12}, L_{13}, etc.	2.2.1.1	11, 12, 13, etc. elements of 3×3 matrix L
L	3.5.2	Tube length (Poiseuille flow in a tube) or bob length (torsional viscometer)
	9.1.2	3×3 transformation matrix for rotation of orthonormal bases
m	2.2.1.1	Mass (scalar)
	7.2.1	Consistency index of power-law generalized Newtonian fluid constitutive model
$\underline{\underline{M}}$	2.3.3	Arbitrary tensor variable
$\overline{M}$	4.5	Molecular weight of a fluid
$\mathcal{M}$	8.4.2	Combination of variables in torsional viscometer example, $\mathcal{M} \equiv M_1 \sqrt{I/\mathcal{K}}$
M_1	8.4.2	Combination of parameters, $M_1 \equiv \eta^*/\rho R^2 (a - 1)^2$
M_c	6.1.3	Critical molecular weight for entanglement in linear polymers, $M_c \approx 2M_e$
M_e	6.1.3	Entanglement spacing, i.e., molecular weight between entanglements for linear polymers
n	7.2.1	Exponential parameter in power-law and Carreau–Yasuda generalized Newtonian fluid constitutive models
	10.3.1	Refractive index in isotropic medium
$\hat{n}$	2.2	Unit vector perpendicular to vectors $\underline{a}$ and $\underline{b}$ following right-hand rule
	2.6.1	Outwardly pointing unit vector normal to surface S
$\underline{\underline{n}}$	10.3.1	Refractive index tensor
n_1, n_2, n_3	10.3.1	Principal values of refractive index tensor of a material
n_{11}, n_{12}, n_{13}, etc.	2.3	11, 12, 13, etc. coefficients of tensor $\underline{\underline{n}}$, i.e., n_{23} is the coefficient of the $\hat{e}_2\hat{e}_3$ diad
n_{eff}	E.3	Effective refractive index
N_1, N_2	4.5	First and second normal-stress difference, $N_1 \equiv \tau_{11} - \tau_{22}$, $N_2 \equiv \tau_{22} - \tau_{33}$
p	2.4	Thermodynamic pressure (scalar)
P_{atm}	10.2.1.1	Atmospheric pressure
$\underline{P}$	10.3.1	Polarization vector
$\mathcal{P}$	3.5.2	Modified pressure, $\mathcal{P} \equiv p - \rho g z$
P_0, P_L	3.5.1	Pressure at $x_1 = 0$ and $x_1 = L$, see example on Poiseuille flow in a tube
$\mathcal{P}_0, \mathcal{P}_L$	3.5.1	Modified pressure at $x_1 = 0$ and $x_1 = L$, see example on Poiseuille flow in a tube
Q	3.5.1	Volumetric flow rate (scalar)
r	2.5	Coordinate variable in cylindrical or spherical coordinate system; note: definition of r is not the same in the two systems
$\bar{r}$	9.3.2	Radial coordinate variable in cylindrical coordinate system; used when confusion would result from using r
r'	9.1	Particle position at time t'
R	3.5.2	Tube radius
	4.5	Ideal gas constant
$\underline{\underline{R}}$	9.1.2	Rotation tensor

continued

Symbol	Section	Definition
R_b	10.1.1	Radius of barrel in capillary flow
$\mathcal{R}$	10.1.4	Rabinowitsch correction in capillary flow
	10.1.4	Shear-rate correction in parallel-plate flow
$\mathfrak{R}$	5.2.2.6	Real operator, $\mathfrak{R}(a + bi) = a$
s, s'	C.4	Section on finite deformation tensors in spherical coordinates, $s = \sin\phi$; $s' = \sin\phi'$
S	2.6.1	Surface area enclosing arbitrary volume V
$\underline{s}$	9.3.2	Position vector in example
$\widehat{s}^1, \widehat{s}^2, \widehat{s}^3$	9.3.2	Contravariant convected coordinates of vector $\underline{s}$
S, S'	C.4	Section on finite deformation tensors in spherical coordinates, $S = \sin\theta$; $S' = \sin\theta'$
$\underline{S}$	E.2	Poynting vector
t	2.6.2	Time
t'	8.2	Arbitrary time; used as a dummy variable of integration
T	4.5	Absolute temperature
$\mathcal{T}$	10.1.2	Torque
t_0	5.4	Time parameter in model given
t_1	8.2	Intermediate time in strain calculation
t_{flow}	5	Flow time scale; used in definition of De
t_{ref}	5.2.2.4	Reference time for calculation of strain
T_{ref}	6.2.1.3	Reference temperature for time–temperature superposition
T_g	6.2.1.3	Glass-transition temperature
$T_{\text{cylindrical}}$	C.4	3×3 matrix used in calculating deformation gradient tensor $\underline{\underline{F}}$ and its inverse in cylindrical coordinates
$T_{\text{spherical}}$	C.4	3×3 matrix used in calculating deformation gradient tensor $\underline{\underline{F}}$ and its inverse in spherical coordinates
$\tan\delta$	5.2.2.5	Loss tangent; measured in small-amplitude oscillatory shear experiment
u	2.2	Magnitude of vector $\underline{u}$
	E.2	Speed of light in a medium
u	9.1	Quantity in integration by parts
$\underline{u}$	2.2.1.1	Arbitrary vector variable
	8.2	Displacement vector
u_1, u_2, u_3	2.2.1.1	Scalar coefficients of vector $\underline{u}$ in $\hat{e}_1$-, $\hat{e}_2$-, $\hat{e}_3$-directions
$\underline{\underline{U}}$	9.1.2	Right stretch tensor
$\underline{v}$	2.2.1.1	Arbitrary vector
	2.2	Velocity vector
v	9.1	Quantity in integration by parts
$\underline{v}'$	9.3.3	Velocity vector at time t'
$v_{z,\text{av}}$	10.1.1.4	Spatial average velocity
V_{wall}	3.5.1	Velocity of wall
v_1, v_2, v_3	2.2.1.1	Scalar coefficients of vector $\underline{v}$ in $\hat{e}_1$-, $\hat{e}_2$-, $\hat{e}_3$-directions
$\widehat{v}^1, \widehat{v}^2, \widehat{v}^3$	9.3.2	Contravariant convected coordinates of vector $\underline{v}$
$\widehat{v}_1, \widehat{v}_2, \widehat{v}_3$	9.3.2	Covariant convected coordinates of vector $\underline{v}$
$\bar{v}_1, \bar{v}_2, \bar{v}_3$	4.4.1	Scalar coefficients of vector $\underline{v}$ in $\bar{e}_1$-, $\bar{e}_2$-, $\bar{e}_3$-directions
$\tilde{v}_1, \tilde{v}_2, \tilde{v}_3$	2.2.1.1	Scalar coefficients of vector $\underline{v}$ in a particular coordinate system in example
$\tilde{v}_0$	8.4.2	Complex amplitude of fluid velocity in gap, torsional oscillatory viscometer example
$v_{z,\text{slip}}$	10.1.1.4	Slip velocity
$\underline{v}_{\text{frame}}$	8.5	Velocity of frame of reference in turntable example
$\underline{v}_{\text{surface}}$	2.6.2	Velocity of surface element dS
V	2.6.1	Arbitrary fluid volume over which a volume integration is performed
$\hat{V}$	4.5	Specific volume of a gas
$\underline{\underline{V}}$	9.1.2	Left stretch tensor
$\underline{w}$	2.2.1	Arbitrary vector variable

continued

Symbol	Section	Definition
$\underline{w}$	9.3.2	Position vector in example
W	3.5.1	Slit width in Poiseuille flow in infinite slit
$\mathcal{W}$	8.4.2	Combination of variables in torsional viscometer example, $\mathcal{W} = (\tilde{\omega}^2 - 1)/A\tilde{\omega} + \tilde{\omega}/2$
w_1, w_2, w_3	2.2.1.1	Scalar coefficients of vector $\underline{w}$ in $\hat{e}_1$-, $\hat{e}_2$-, $\hat{e}_3$-directions
$\widehat{w}^1, \widehat{w}^2, \widehat{w}^3$	9.3.2	Contravariant convected coordinates of vector $\underline{w}$
$\mathcal{X}$	8.4.2	Combination of parameters, $\mathcal{X} = (r - R)/(aR - R)$
x, y, z	2.6.2	Spatial variables in Cartesian (fixed, orthonormal) coordinate system
x', y', z'	9.1.1	Coordinates for position of particle of interest at time t'
$\widehat{x}^1, \widehat{x}^2, \widehat{x}^3$	9.3.2	Contravariant convected coordinates of position vector $\underline{r}$
$\widehat{x}_1, \widehat{x}_2, \widehat{x}_3$	9.3.2	Covariant convected coordinates of position vector $\underline{r}$
$\bar{x}, \bar{y}, \bar{z}$	4.4.1	Spatial variables in $\bar{e}$ coordinate system
x	2.6.2	Spatial variable
X	E.3	3×3 matrix of coefficients of tensor $\underline{\underline{\chi}}$
y	2.2.1.1	Coordinate variable in Cartesian coordinate system
Y	C.1.1	New variable used in similarity transform solution
$Y_1(x)$	8.4.2	First-order Bessel function of the second kind
y	C.1.1	Dummy variable used in similarity transform solution
Z	2.2.1.1	3×3 matrix of scalar variables
Z_{11}, Z_{12}, Z_{13}, etc.	2.2.1.1	11, 12, 13, etc. elements of matrix Z
α	2.1	Arbitrary scalar variable
α_0	2.6.2	Scalar constant; limit of integration
β	2.1	Arbitrary scalar variable
β_0	2.6.2	Scalar constant; limit of integration
γ	5.2.2.3	Shear strain
γ_0	5.2.2.4	Fixed strain imposed in step-strain experiment
	5.2.2.5	Strain amplitude imposed in small-amplitude oscillatory shear experiment
γ_∞	5.2.2.3	Recoverable shear
$\gamma_{xx}, \gamma_{xy}, \gamma_{xz}$, etc.	4.2	xx, xy, xz, etc. coefficients of tensor $\underline{\underline{\gamma}}$, i.e., γ_{yz} is the coefficient of the $\hat{e}_y\hat{e}_z$ diad
$\underline{\underline{\gamma}}(t_{\text{ref}}, t)$	8.2	Infinitesimal strain tensor between times t_{ref} and t
$\underline{\underline{\gamma}}^{[0]}$	9.1.1	Finite strain tensor, $\underline{\underline{\gamma}}^{[0]} \equiv \underline{\underline{C}} - \underline{\underline{I}}$
$\underline{\underline{\gamma}}_{[0]}$	9.1.1	Finite strain tensor, $\underline{\underline{\gamma}}_{[0]} \equiv \underline{\underline{I}} - \underline{\underline{C}}^{-1}$
$\dot{\gamma}(t)$	4.2	Rate of strain or shear rate; always a positive quantity, $\dot{\gamma} \equiv \|\underline{\underline{\dot{\gamma}}}\|$
$\dot{\gamma}_a$	10.1.1.2	Apparent shear rate, $\dot{\gamma}_a \equiv 4Q/\pi R^3$
$\dot{\gamma}_0$	4.2	Constant rate of strain or shear rate
	5.2.2.5	Strain-rate amplitude imposed in small-amplitude oscillatory shear experiment
$\dot{\gamma}_R$	10.1.1	Wall shear rate in capillary flow
	10.1.2	Rim shear rate in parallel-plate flow
$\dot{\gamma}_{\text{slip-corrected}}$	10.1.1.4	Shear rate in capillary flow corrected for slip
$\underline{\underline{\dot{\gamma}}}$	3.3.2	Rate-of-strain or deformation rate tensor
$\dot{\gamma}_{11}, \dot{\gamma}_{12}, \dot{\gamma}_{13}$, etc.	4.2	11, 12, 13, etc. coefficients of tensor $\underline{\underline{\dot{\gamma}}}$, i.e., $\dot{\gamma}_{23}$ is the coefficient of the $\hat{e}_2\hat{e}_3$ diad
δ	5.2.2.5	Phase difference between strain and stress in small-amplitude oscillatory shear experiment; $\tan \delta$ is known as loss tangent
	10.3.1	Phase difference between x- and y-components of electric-field vector of light
δ_{ij}	2.2.2	Kronecker delta function
δ_m	10.3.4	Time-variable retardation in photoelastic modulator
$\Delta n_{21}, \Delta n_{31}$	10.3.1	Birefringences, $\Delta n_{21} \equiv n_2 - n_1$, $\Delta n_{31} \equiv n_3 - n_1$
$\Delta\sigma_1, \Delta\sigma_2$	10.3.4	Principal stress differences, $\Delta\sigma_1 \equiv \sigma_1 - \sigma_2$, $\Delta\sigma_2 \equiv \sigma_2 - \sigma_3$
$\underline{\underline{\Delta}}^T$	D	Deformation-gradient tensor (DPL notation)

continued

Symbol	Section	Definition
ϵ	4.3.1	Strain in elongational flow; also known as Hencky strain
ε	5.2.2.4	Short time over which a step deformation is imposed on a sample in the step shear strain experiment
	E.2	Permittivity of a medium
ε_{ijk}	2.2.2	Einstein epsilon function
$\dot{\epsilon}(t)$	4.3.1	Strain rate in elongational flow
$\dot{\epsilon}_o$	4.3.1	Constant strain rate in elongational flow
ε_0	10.3.1	Permittivity of free space, $\varepsilon_0 = 8.854 \times 10^{-12}$ F/m
ζ	2.1	Arbitrary scalar variable
	E.3.2.3	Angle between 2-direction of shear flow and wave vector $\underline{k}$
$\hat{\zeta}_1, \hat{\zeta}_2, \hat{\zeta}_3$	9.1.2	Unit eigenvectors of left stretch tensor $\underline{\underline{V}}$
$\eta(\dot{\gamma})$	5.2.1	Non-Newtonian viscosity
$\bar{\eta}(\dot{\epsilon})$ or $\bar{\eta}_1(\dot{\epsilon})$	5.3.1	Steady elongational viscosity or first steady elongational viscosity, $\dot{\epsilon} > 0$
$\bar{\eta}_2(\dot{\epsilon})$	5.3.1	Second steady elongational viscosity
$\bar{\eta}_{B_1}(\dot{\epsilon})$ or $\bar{\eta}_B(\dot{\epsilon})$	5.3.1	Steady biaxial elongational viscosity, $\dot{\epsilon} < 0$
$\bar{\eta}_{P_1}(\dot{\epsilon})$ or $\bar{\eta}_P(\dot{\epsilon})$	5.3.1	First steady planar elongational viscosity, $\dot{\epsilon} > 0$
$\bar{\eta}_{P_2}(\dot{\epsilon})$	5.3.1	Second steady planar elongational viscosity, $\dot{\epsilon} > 0$
$\eta'(\omega), \eta''(\omega)$	5.2.2.5	Linear viscoelastic viscosities
$\eta^*(\omega), |\eta^*(\omega)|$	5.2.2.5	Complex viscosity (complex number, $\eta^* = \eta' - i\eta''$), magnitude of complex viscosity (real number)
$\eta_r'(a_T\omega), \eta_r''(a_T\omega), \eta_r^*(a_T\omega)$	5.2.2.5	Time–temperature-shifted complex viscosities, measured in small-amplitude oscillatory shear
η_0	5.2.1	Zero-shear viscosity, $\eta_0 \equiv \lim_{\dot{\gamma} \to 0} \eta$
η_k	8.2	kth viscosity parameter in generalized Maxwell model
η_∞	7.2.1	Infinite-shear viscosity parameter of Carreau–Yassuda generalized Newtonian fluid constitutive model
$\tilde{\eta}$	5.4	Viscosity parameter in model given
$\eta^+(t, \dot{\gamma})$	5.2.2.1	Shear stress growth coefficient
$\eta^-(t, \dot{\gamma})$	5.2.2.2	Shear stress relaxation coefficient
θ	2.5	Coordinate variable in spherical or cylindrical coordinate systems; note: definition is different in each case
$\theta(t)$	8.4.2	Angular displacement in fluid in torsional viscometer example
θ'	8.4.2	$d\theta/dt$
θ''	8.4.2	$d^2\theta/dt^2$
Θ, Φ, Ψ	C.6	Coefficients of characteristic equation of matrix or tensor
Θ_0	10.1.3	Cone angle in cone-and-plate viscometer
$\Theta_{aR}(t)$	8.4.2	Time-dependent angular displacement of inner wall in torsional oscillatory viscometer example (complex)
$\Theta_{aR}^0(t)$	8.4.2	Real amplitude of driven time-dependent angular displacement of outer wall in torsional oscillatory viscometer example
$\Theta_R(t)$	8.4.2	Driven time-dependent angular displacement of inner wall in torsional oscillatory viscometer example
κ	3.3.1	Dilatational viscosity (scalar)
	3.6	Ratio of inner to outer radius in Couette geometry
λ	5	Material relaxation time
	C.6	Eigenvalue of tensor or matrix
λ_k	8.2	kth relaxation time in generalized Maxwell model
	9.1.2	Eigenvalues
λ_1	6.1.3	Longest relaxation time of a material
$\tilde{\lambda}_i$	6.2.1.3	Non-temperature-dependent portion of individual relaxation time λ_i, $\lambda_i = \tilde{\lambda}_i a_T$
λ_v	E.3.2.2	Wavelength of light in vacuum
Λ	8.5	Retardation time
	C.6	3×3 diagonal matrix of coefficients of a tensor A in principal frame
μ	3.3.1	Newtonian viscosity (scalar constant)

continued

Symbol	Section	Definition
μ	E.2	Permeability of a medium
μ_0	7.2.1	Viscosity parameter of Bingham generalized Newtonian fluid constitutive model
ν	2.3.4	Order of a tensor
ν_1, ν_2, ν_3	9.1.2	Eigenvalues of left stretch tensor $\underline{\underline{V}}$
ξ	9.3.4	Non-affine-motion or slip parameter in Gordon–Schowalter derivative; $\xi = 0$ corresponds to affine motion
	B	Arbitrary variable in definition of error function
	C.6	3×1 column eigenvector
ξ'	B	Dummy variable of integration in definition of error function
$\hat{\xi}_1, \hat{\xi}_2, \hat{\xi}_3$	9.1.2	Unit eigenvectors of symmetric matrix or tensor
$\hat{\xi}_x, \hat{\xi}_y, \hat{\xi}_z$	E	Unit eigenvectors of symmetric susceptibility tensor $\underline{\underline{\chi}}$ and refractive index tensor $\underline{\underline{n}}$; when stress-optic law holds, these are also unit eigenvectors of extra stress tensor $\underline{\underline{\tau}}$
$\underline{\underline{\Pi}}$	3.2.2	Total stress tensor, $\underline{\underline{\Pi}} \equiv \underline{\underline{\tau}} + p\underline{\underline{I}}$
$\Pi_{11}, \Pi_{12}, \Pi_{13}$, etc.	3.2.2	11, 12, 13, etc. coefficients of tensor $\underline{\underline{\Pi}}$, i.e., Π_{23} is the coefficient of the $\hat{e}_2\hat{e}_3$ diad
ρ	3.1	Mass density (scalar)
ρ_{ref}	6.2.1.3	Mass density at reference temperature for time–temperature superposition
$\sigma_1, \sigma_2, \sigma_3$	10.3.4	Principal values or eigenvalues of extra stress tensor
$\underline{\underline{\varsigma}}$	9.2	Extra stress tensor plus an isotropic constant; used in derivation of Lodge equation, $\underline{\underline{\varsigma}} \equiv \underline{\underline{\tau}} + \frac{\eta_o}{\lambda}\underline{\underline{I}}$
$\dot{\varsigma}(t)$	4.2	Shear-rate function defined for simple shear flow
τ_0	5.2.2.3	Constant stress applied in creep experiment
τ_y	7.2.1	Yield stress of Bingham generalized Newtonian fluid constitutive model
τ_R	10.1.1.2	Shear stress at wall of a capillary, $\tau_R \equiv \Delta \mathcal{P}R/2L$
$\tilde{\tau}_0$	5.2.2.6	Complex coefficient of stress wave in small-amplitude oscillatory shear, $\tilde{\tau}_o = i\tau_o e^{i\delta}$
$\underline{\underline{\tau}}$	3.3	Extra stress tensor
$\tau_{11}, \tau_{12}, \tau_{13}$, etc.	3.2.2	11, 12, 13, etc. coefficients of tensor $\underline{\underline{\tau}}$ in 123 coordinate system, i.e., τ_{23} is the coefficient of the $\hat{e}_2\hat{e}_3$ diad
$\tau_{xx}, \tau_{xy}, \tau_{xz}$, etc.	4.5	xx, xy, xz, etc. coefficients of tensor $\underline{\underline{\tau}}$ in xyz coordinate system, i.e., τ_{yz} is the coefficient of the $\hat{e}_y\hat{e}_z$ diad
$\bar{\tau}_{11}, \bar{\tau}_{12}, \bar{\tau}_{13}$, etc.	4.4.1	11, 12, 13, etc. coefficients of tensor $\underline{\underline{\tau}}$ in $\bar{e}$ coordinate system, i.e., $\bar{\tau}_{23}$ is the coefficient of the $\bar{e}_2\bar{e}_3$ diad
$\underline{\underline{\tau}}_k$	8.2	kth stress contribution in generalized Maxwell model
$[\tau]_{123}$	10.3.4	3×3 matrix of coefficients of stress tensor $\underline{\underline{\tau}}$ written in the 123 coordinate system
$[\tau]_{\xi_x\xi_y\xi_z}$	10.3.4	3×3 matrix of coefficients of stress tensor $\underline{\underline{\tau}}$ written in the $\xi_x\xi_y\xi_z$ coordinate system
ϕ	2.2	Coordinate variable in spherical coordinate system
Φ, Ψ, Θ	C.6	Coefficients of characteristic equation of a matrix or a tensor
χ	10.3.1	Susceptibility, a scalar property of an isotropic material
$\underline{\underline{\chi}}$	10.3.1	Susceptibility tensor for anisotropic material
χ_1, χ_2, χ_3	E.3	Principal values of susceptibility tensor
ψ	2.2	Angle between vectors $\underline{a}$ and $\underline{b}$
	2.2.1.1	Angle between a person and the wall in an example
	2.2.1.1	Angle between string and horizontal in example
	E.3.2.3	Angle between 1-direction of shear flow and plane containing wave vector $\underline{k}$ and $\hat{e}_2$
ψ_{uw}	9.1.2	Angle between vectors $\underline{u}$ and $\underline{w}$
Ψ, Θ, Φ	C.6	Coefficients of characteristic equation of matrix or tensor
$\Psi_1(\dot{\gamma}), \Psi_2(\dot{\gamma})$	5.2.1	First and second normal-stress coefficients

continued

Symbol	Section	Definition
$\Psi_1^+(t, \dot{\gamma})$, $\Psi_2^+(t, \dot{\gamma})$	5.2.2.1	First and second normal-stress growth coefficients
$\Psi_1^-(t, \dot{\gamma})$, $\Psi_2^-(t, \dot{\gamma})$	5.2.2.2	First and second normal-stress relaxation coefficients
ω	5.2.2.5	Radian frequency of oscillation in small-amplitude oscillatory shear experiment
	10.3.1	Frequency of light
$\tilde{\omega}$	8.4.2	Combination of variables in torsional viscometer example, $\tilde{\omega} = \omega\sqrt{I/K}$
ω_x	6.1.3	Frequency at which G' and G'' cross at low-frequency end of entanglement plateau; $1/\omega_x$ is often identified as longest relaxation time of a material
ω_{mod}	10.3.4	Resonant frequency of photoelastic modulator
Ω	8.5	Angular velocity of rotating system
∇	2.4	Nabla or del operator; related to differential operations, $\nabla = \partial/\partial x_i \hat{e}_i$ in Cartesian coordinates x_1, y_2, z_3
∇'	9.3.3	Nabla or del operator at position $\underline{r}'$, $\nabla = \partial/\partial x_i' \hat{e}_i$ in Cartesian coordinates x_1', x_2', x_3'
∇^2	2.4	Laplacian operator, $\nabla \cdot \nabla$
$x^{(*)}$	5.2.2.6	Complex conjugate of complex number x
$\underline{\underline{A}}^T$	2.3.4	Transpose of tensor $\underline{\underline{A}}$
$\overset{\triangledown}{\underline{\underline{A}}}$	9.2	Upper convected deriative of $\underline{\underline{A}}$
$\overset{\triangle}{\underline{\underline{A}}}$	9.3.4	Lower convected deriative of $\underline{\underline{A}}$
$\overset{\circ}{\underline{\underline{A}}}$	9.3.4	Corotational deriative of $\underline{\underline{A}}$
$\overset{\square}{\underline{\underline{A}}}$	9.3.4	Gordon–Schowalter deriative of $\underline{\underline{A}}$
$\overset{\frown}{}$	9.3.2	Denotes convected coordinate
$(\cdots)_{123}$	2.3	Indicates that components listed are in the x_1, x_2, x_3 coordinate system with basis vectors $\bar{e}_1$, $\bar{e}_2$, $\bar{e}_3$
$(\cdots)_{\bar{1}\bar{2}\bar{3}}$	4.4.1	Indicates that components listed are in the $\bar{x}_1$, $\bar{x}_2$, $\bar{x}_3$ coordinate system with basis vectors $\bar{e}_1$, $\bar{e}_2$, $\bar{e}_3$
$(\cdots)_{xyz}$	2.5	Indicates that components listed are in the x, y, z coordinate system with basis vectors $\hat{e}_x$, $\hat{e}_y$, $\hat{e}_z$
$(\cdots)_{\bar{x}\bar{y}\bar{z}}$	4.6	Indicates that components listed are in the $\bar{x}$, $\bar{y}$, $\bar{z}$ coordinate system with basis vectors $\hat{e}_{\bar{x}}$, $\hat{e}_{\bar{y}}$, $\hat{e}_{\bar{z}}$
$(\cdots)_{\xi_1\xi_2\xi_3}$	C.6	Indicates that components listed are in the ξ_1, ξ_2, ξ_3 coordinate system with basis vectors $\hat{\xi}_1$, $\hat{\xi}_2$, $\hat{\xi}_3$
$(\cdots)_{\xi_x\xi_y\xi_z}$	E	Indicates that components listed are in the ξ_x, ξ_y, ξ_z coordinate system with basis vectors $\hat{\xi}_x$, $\hat{\xi}_y$, $\hat{\xi}_z$
$[v]_{123}$	C.6	3×1 matrix of coefficients of vector $\underline{v}$ in 123 coordinate system
$[A]_{123}$	C.6	3×3 matrix of coefficients of tensor $\underline{\underline{A}}$ in 123 coordinate system
$(\partial \cdot /\partial \cdot)_{\hat{x}}$	9.3.3	Indicates that derivative is taken at constant convected coordinate $\hat{x}$
J	C.7.1	Jacobian of a transformation
η_s	B	Solvent viscosity
η_{sp}	B	Specific viscosity
$[\eta]$	B	Intrinsic viscosity
$\underline{\underline{\epsilon}}$	E.3.1	Dielectric tensor
$\underline{\underline{\tau}}^p$	B	Polymer contribution to the stress Oldroyd B equation
$\underline{\underline{\tau}}^s$	B	Solvent contribution to the stress in the Oldroyd B equation
ϕ_1, ϕ_2	B	Parameters in the Reiner-Rivlin fluid model
c	B	Concentration
$\lambda_{\mathrm{steady}}$	B	Material time constant in steady flow
We	B	Weissenberg number
Re	10.1	Reynolds number
δ_+	5.2.2.4	Asymmetric delta function
Φ	B	Potential energy function
χ	10.3.4	Orientation angle

Glossary

apparent shear rate   See shear rate, apparent.

apparent viscosity   See viscosity, apparent.

Boger fluid   Experimental fluids that have high viscosity and elasticity. The fluids are designed to give constant, high viscosities (no shear-thinning) but also to have elastic effects such as nonzero first and second normal-stress differences. They are named after David Boger, the rheologist who introduced the study of such fluids. Boger fluids often follow the Oldroyd B constitutive equation.

Cauchy–Green stretch tensors   The right Cauchy–Green stretch tensor $\underline{\underline{U}}$ and the left Cauchy–Green stretch tensor $\underline{\underline{V}}$ are the tensors that result from the polar decomposition of a tensor [9]. The stretch tensors for $\underline{\underline{A}}$ contain all of the stretch information contained in $\underline{\underline{A}}$. For an arbitrary, nonsingular tensor $\underline{\underline{A}}$, the Cauchy–Green stretch tensors are given by (see Section 9.1.2)

$$\underline{\underline{U}} \equiv (\underline{\underline{A}}^T \cdot \underline{\underline{A}})^{\frac{1}{2}}$$

$$\underline{\underline{V}} \equiv (\underline{\underline{A}} \cdot \underline{\underline{A}}^T)^{\frac{1}{2}}$$

See also polar decomposition.

Cayley–Hamilton theorem   For any symmetric three-dimensional tensor $\underline{\underline{A}}$, the Cayley–Hamilton theorem states that

$$\underline{\underline{A}}^3 - \Theta \underline{\underline{A}}^2 + \Phi - \Psi \underline{\underline{I}} = \underline{\underline{0}}$$

where Θ, Φ, and Ψ are the scalar invariants of $\underline{\underline{A}}$ as defined at the end of Section C.6.

Cole–Cole plot   A plot of $G'(\omega)$ versus $G''(\omega)$.

conservative force   A force is conservative if the work the force does on a particle sums to zero over any path that brings the particle back to its initial position [240]. Gravity is an example of a conservative force since the work required to lift a mass to a certain elevation is independent of the precise path followed in arriving at the final position.

Friction is not a conservative force since on any closed path (a path that brings the mass back to its starting place) nonzero work is required to move a mass along the path. See also the discussion of how the conservative nature of gravity is related to the definition of equivalent pressure.

contravariant convected components   The coefficients of a vector or tensor in the convected basis system employing the basis vectors $\underline{g}_{(i)}$ are called the contravariant components of the vector or tensor. For more details on contravariant components, see Section C.7.

covariant convected components   The coefficients of a vector or tensor in the convected basis system employing the reciprocal basis vectors $\underline{g}^{(i)}$ are called the covariant components of the vector or tensor. For more details on covariant components, see Section C.7.

curl of a vector, tensor   The curl of a vector $\underline{v}$ is $\nabla \times \underline{v}$; the curl of a tensor $\underline{\underline{A}}$ is $\nabla \times \underline{\underline{A}}$.

damping function   The damping function $h(\gamma_o)$ is a function that gives the strain dependence of time-strain factorizable constitutive equations. The damping function is related to the relaxation modulus $G(t, \gamma_0)$ as follows for time-strain factorizable fluids:

$$G(t, \gamma_0) = G(t)h(\gamma_0)$$

where $G(t)$ is the linear viscoelastic relaxation modulus. $G(t, \gamma_0)$ is measured directly in the step-strain experiment, and the damping function can be calculated from these data.

Deborah number   The Deborah number De is the ratio of a material relaxation time λ to the time scale of the flow being studied:

$$\text{De} \equiv \frac{\lambda}{t_{\text{flow}}}$$

The relaxation time is calculated from a time-dependent flow such as the step-strain experiment or the cessation of a steady shearing experiment. A related dimensionless number is the Weissenberg number We, which is the ratio of a material relaxation time calculated from a steady flow to the time scale of the flow:

$$\text{We} \equiv \frac{\lambda_{\text{steady}}}{t_{\text{flow}}}$$

The flow time t_{flow} is usually taken to be D/V, where D is a characteristic length scale of the flow and V is a characteristic velocity of the flow [220].

delta function, asymmetric   The delta or impulse functions are symbolic functions rather than real functions. They are used to express discrete events in terms of integrals. The asymmetric delta function is defined as

$$\int_{a+0}^{b} f(\xi)\delta_+(\xi - x)\, d\xi = \begin{cases} 0 & x < a \\ f(x+0) & a \le x < b \\ 0 & x \ge b \end{cases}$$

where $f(x+0)$ means the value of the function $f(x)$ when approached from the positive side. Note also that $a < b$. We choose to represent the asymmetric delta function as follows, but there are other choices [129]:

$$\delta(t) \equiv \lim_{\varepsilon \to 0} \begin{cases} 0 & t < 0 \\ \dfrac{1}{\varepsilon} & 0 \le t < \varepsilon \\ 0 & t \ge \varepsilon \end{cases}$$

Note that the following relations hold for the asymmetric delta function:

$$\int_{-b}^{+a} f(x)\delta_+(x)\,dx = f(0)$$

$$\int_{-b}^{+a} f(x)\frac{d\delta_+}{dt}\,dx = -\frac{df}{dx}(0)$$

where both a and b are positive.

delta function, symmetric   The delta or impulse functions are symbolic functions rather than real functions. They are used to express discrete events in terms of integrals. The symmetric delta function, also known as the Dirac delta function, is defined as [129].

$$\int_{a}^{b} f(\xi)\delta(\xi - x)\,d\xi = \begin{cases} 0 & x < a \\ \dfrac{1}{2}f(x) & x = a \text{ or } x = b \\ f(x) & a < x < b \\ 0 & x > b \end{cases}$$

where a and b may be positive or negative and $a < b$. One way to represent the Dirac delta function with a continuous function is given below [27]; other choices for representing $\delta(x)$ can be found in standard references [129].

$$\delta(x) = \lim_{n \to \infty} \sqrt{\frac{n}{\pi}}\, e^{-nx^2}$$

Note that the following relations hold for the Dirac delta function:

$$\int_{-a}^{+b} f(x)\delta(x)\,dx = f(0)$$

$$\int_{-a}^{+b} f(x)\frac{d\delta(x)}{dt}\,dx = -\frac{df}{dx}(0)$$

where both a and b are positive, and $a < b$.

determinant   For a 2×2 matrix A the determinant $\det|A|$ is defined as

$$A = \begin{bmatrix} A_{11} & A_{12} \\ A_{21} & A_{22} \end{bmatrix}$$

$$\det |A| \equiv A_{11}A_{22} - A_{12}A_{21}$$

For a 3×3 matrix B the determinant $\det|B|$ is defined as

$$B = \begin{bmatrix} B_{11} & B_{12} & B_{13} \\ B_{21} & B_{22} & B_{23} \\ B_{31} & B_{32} & B_{33} \end{bmatrix}$$

$$\det |B| \equiv B_{11}(B_{22}B_{33} - B_{23}B_{32}) - B_{12}(B_{21}B_{33} - B_{23}B_{31})$$
$$+ B_{13}(B_{21}B_{32} - B_{22}B_{31})$$

The term determinant comes from using matrix algebra to solve a system of two equations, two unknowns,

$$A_{11}x + A_{12}y = C_1$$
$$A_{21}x + A_{22}y = C_2$$
$$\begin{bmatrix} A_{11} & A_{12} \\ A_{21} & A_{22} \end{bmatrix} \begin{bmatrix} x \\ y \end{bmatrix} = \begin{bmatrix} C_1 \\ C_2 \end{bmatrix}$$

For this system, the value of $\det|A|$ determines whether the system can be solved for nonzero x and y. For the 3×3 case, the determinant determines whether the system of three equations, three unknowns may be solved [12]. When a matrix is being used to hold the coefficients of a tensor in a chosen coordinate system, the determinant is one of the invariants of the tensor (see Section C.6). For more on determinants of systems of arbitrary size, see Aris [7].

deviatoric stress tensor   The extra stress tensor defined such that its trace is zero [136]. In incompressible fluids the separation between pressure and extra stress in the definition of $\underline{\underline{\tau}}$ becomes arbitrary:

$$\underline{\underline{\Pi}} = \underline{\underline{\tau}} + p\underline{\underline{I}}$$

When the value of p is chosen such that trace $(\underline{\underline{\tau}}) = 0$, $\underline{\underline{\tau}}$ is called the deviatoric stress tensor.

dilatant fluid   A rheologically dilatant fluid is one for which the viscosity increases with increasing shear rate; these are also called shear-thickening fluids. The term dilatant is also used to describe volumetric dilatancy, which is the tendency of concentrated suspensions to expand in volume during flow. The term dilatant fluid is often used to describe both behaviors (volumetric and rheometric dilatancy), causing confusion [178].

Dirac delta function   See delta function, symmetric.

divergence of a vector, tensor   The divergence of a vector $\underline{v}$ is $\nabla \cdot \underline{v}$; the divergence of a tensor $\underline{\underline{A}}$ is $\nabla \cdot \underline{\underline{A}}$.

dummy variable   In integration, as in other mathematical operations, there is sometimes a need for a placeholder variable. This type of variable is called a dummy variable. In the example problem solution below, the limits on the integral are at $x = 0$ and at the variable limit x. To distinguish the variable x in the integral over $g(x)$ and the limit of the integral x, we change the variable in the integral to the dummy variable x'. This change does not affect the result of the equation since it is just a change of a placeholder symbol.

$$\frac{df}{dx} = g(x)$$

$$f(x) = \int_0^x g(x')dx'$$

eigenvalue   See eigenvector.

eigenvector   An eigenvector of a tensor or of a matrix is a vector that stretches but does not rotate when the tensor or matrix acts on it. The magnitude of an eigenvector is not significant, and therefore they are often normalized to have unit length; eigenvectors of unit length are called unit eigenvectors. For a second-order tensor $\underline{\underline{A}}$, the unit eigenvectors $\hat{\xi}_1, \hat{\xi}_2, \hat{\xi}_3$ follow the following equations:

$$\underline{\underline{A}} \cdot \hat{\xi}_1 = \lambda_1 \hat{\xi}_1$$

$$\underline{\underline{A}} \cdot \hat{\xi}_2 = \lambda_2 \hat{\xi}_2$$

$$\underline{\underline{A}} \cdot \hat{\xi}_3 = \lambda_3 \hat{\xi}_3$$

The scalars λ_1, λ_2, and λ_3 are called the eigenvalues, and they indicate the amount of stretching that the eigenvectors undergo in each of the three directions.

Einstein notation   A way of writing vectors and tensors that is most useful in an orthonormal coordinate system. A vector or tensor written in Einstein notation is written as a sum of coefficients and unit vectors (in the case of vectors) or of coefficients and unit dyads (in the case of tensors). The summation signs are not written explicitly in Einstein notation, however; rather a sum is performed over every pair of repeated indices. Three examples of Einstein notation are the last terms in the following:

$$\underline{v} = \sum_{p=1}^{3} v_p \hat{e}_p = v_p \hat{e}_p$$

$$\underline{\underline{A}} = \sum_{s=1}^{3} \sum_{m=1}^{3} A_{sm} \hat{e}_s \hat{e}_m = A_{sm} \hat{e}_s \hat{e}_m$$

$$\underline{v} \cdot \nabla \underline{w} = \sum_{p=1}^{3} v_p \hat{e}_p \cdot \sum_{k=1}^{3} \sum_{j=1}^{3} \frac{\partial w_j}{\partial x_k} \hat{e}_k \hat{e}_j = v_p \hat{e}_p \cdot \frac{\partial w_j}{\partial x_k} \hat{e}_k \hat{e}_j$$

empirical relation   One that results from experimental observation rather than from theoretical derivation.

EOM   equation of motion.

equivalent pressure   The equivalent pressure $\mathcal{P}$ is a function that combines the effects of thermodynamic pressure p and gravity $\underline{g}$. The force due to gravity, $\rho\underline{g}$, is a conservative force, that is, the work done by the force due to gravity is independent of the path taken in the course of doing the work [240]. All conservative forces may be written as the negative of the gradient of a potential energy function,

$$\rho\underline{g} = -\nabla\Phi$$

where Φ is the potential energy due to gravity. Φ is equal to $\rho g h$, where g is the gravitational force constant, and h is the height of a particle of interest above a reference plane. Thus,

$$\rho\underline{g} = -\nabla(\rho g h)$$

and for constant density,

$$\underline{g} = -g\nabla h$$

The equivalent pressure appears when the pressure and gravity terms of the Navier–Stokes equation are combined [246, 64],

$$\rho\left(\frac{\partial \underline{v}}{\partial t} + \underline{v}\cdot\nabla\underline{v}\right) = -\nabla p + \mu\nabla^2\underline{v} + \rho\underline{g}$$

$$= -\nabla p + \mu\nabla^2\underline{v} - \nabla\Phi$$

$$= -\nabla(p + \Phi) + \mu\nabla^2\underline{v}$$

$$= -\nabla\mathcal{P} + \mu\nabla^2\underline{v}$$

where $\mathcal{P} \equiv p + \Phi = p + \rho g h$. To evaluate $\mathcal{P}$, the function h must be correctly expressed in the coordinate system of interest. See also Problem 3.7 for an alternative definition of equivalent pressure that is useful for compressible fluids.

error function   The integral in the definition of the error function erf ξ is not solvable in closed form and has been incorporated into this standard function:

$$\text{erf } \xi \equiv \frac{\int_0^\xi e^{-(\xi')^2}\, d\xi'}{\int_0^\infty e^{-(\xi')^2}\, d\xi'} = \frac{2}{\sqrt{\pi}}\int_0^\xi e^{-(\xi')^2}\, d\xi'$$

Note that erf $(0) = 0$ and erf $(\infty) = 1$.

even/odd permutations   The order of three numbers, such as 123, can be permuted by taking the last digit from the end and adding it to the beginning; this yields a total of three combinations, 123, 312, and 231. These are the even permutations of the three numbers. The odd permutations are obtained by permuting 321: 321, 132, 213.

extra stress tensor   The total stress tensor $\underline{\underline{\Pi}}$ expresses the molecular contributions to stress at a point in a flowing fluid. The extra stress tensor $\underline{\underline{\tau}}$ is the part of the total stress tensor $\underline{\underline{\Pi}}$ that is not the isotropic pressure, that is, $\underline{\underline{\Pi}} = p\underline{\underline{I}} + \underline{\underline{\tau}}$.

extrudate distortion   See melt fracture.

Gibbs notation   The symbolic vector tensor notation used in this book, such as $\underline{\underline{A}}$, $\underline{\underline{\tau}}$, $\underline{v}$, $\underline{s} \times \underline{u}$, $\nabla \underline{v} + (\nabla \underline{v})^T$. Gibbs notation makes no reference to a coordinate system, in contrast to Einstein notation, which includes reference to the specific coefficients of a vector and tensor and which therefore only expresses a vector or tensor in a particular coordinate system.

Heaviside unit step function

$$H(t) \equiv \begin{cases} 0 & t < 0 \\ 1 & t \geq 0 \end{cases}$$

Note that $dH/dt = \delta(t)$, where $\delta(t)$ is the Dirac delta function.

Hencky strain ϵ The strain in an elongational flow experiment:

$$\epsilon = \dot{\epsilon}_0 t = \ln \frac{l}{l_0}$$

The Hencky strain is different from the extension ratio l/l_0, which is called the strain in the study of metals and other solid materials [78]. At the limits of small strains, ϵ(small strains) $= l/l_0 - 1$.

homogeneous/inhomogeneous flow   Homogeneous flow means that the strain rate $\dot{\gamma} \equiv |\dot{\underline{\underline{\gamma}}}|$ is not a function of position. In inhomogeneous flow, the strain rate varies throughout the flow field.

homogeneous material   A material whose physical properties, such as viscosity and density, do not vary with position. Homopolymers and solvents are homogeneous, for example, and suspensions are inhomogeneous. A phase-separated blend is inhomogeneous. A compatible blend, however, is a homogeneous material on length scales that are large enough to include several molecules.

impulse function   See delta function.

invariants of a tensor   For a second-order tensor, there are three scalars associated with the tensor that are invariant to coordinate transformation. These three invariants are defined in one of two alternate ways, as shown below, and the two sets of definitions can be interrelated [see Equations (C.81–C.83)]:

$$I_{\underline{\underline{A}}} \equiv A_{ii} = \text{trace } \underline{\underline{A}}$$

$$II_{\underline{\underline{A}}} \equiv A_{ij} A_{ji} = \underline{\underline{A}} : \underline{\underline{A}}$$

$$III_{\underline{\underline{A}}} \equiv A_{ij} A_{jk} A_{ki}$$

Equivalently,

$$\Theta = I_1 \equiv A_{ii} = \text{trace } \underline{\underline{A}}$$

$$\Phi = I_2 \equiv A_{22} A_{33} - A_{23} A_{32} + A_{33} A_{11} - A_{31} A_{13} + A_{11} A_{22} - A_{12} A_{21}$$

$$\Psi = I_3 \equiv \det \left| \underline{\underline{A}} \right|$$

kinematics   Kinematics refers to all information about the motion in a flow. The kinematics consist of the velocity field $\underline{v}$ or tensors such as $\underline{\underline{\dot{\gamma}}}$ that are a function of $\underline{v}$.

Kronecker delta   The Kronecker delta δ_{pk} is defined as

$$\delta_{pk} \equiv \begin{cases} 1 & \text{for } p = k \\ 0 & \text{for } p \neq k \end{cases}$$

Leonov model   A constitutive equation that is derived from the thermodynamic idea that the stress in polymers should be related to stored elastic strain [151]. It is a nonlinear rheological model that resembles the Giesekus model (Section 9.4.1.2). The Leonov model predicts stress overshoots in both shear stress and first normal-stress coefficient, in agreement with observation [138].

linear viscoelastic limit/region   The term linear viscoelastic limit refers to a range-of-deformation magnitude (strain or strain rate) in which the stress generated is linear in strain, that is, in the linear viscoelastic limit when the strain doubles, the stress doubles. At higher strains, the response is nonlinear, and we say that this is outside of the linear viscoelastic region. All material functions have a linear viscoelastic limit at small strains. The generalized linear viscoelastic (GLVE) constitutive equation is based on the assumption of a linear strain–stress relationship, and therefore the GLVE equation can be used to interrelate material functions in the linear viscoelastic limit.

material objectivity   Constitutive equations must represent material behavior correctly in order for them to be useful. Oldroyd [192, 195] carefully considered the relationship between constitutive equations and material behavior and formulated three rules that all constitutive equations must follow. First, the stresses calculated from constitutive equations must be free of any dependence on reference frame. Second, the value of the stresses calculated for a material element must not change when the particle undergoes a rigid rotation or a simple translation. Third, the value of stresses calculated for a material element must not depend on the rheological history of neighboring fluid elements. These three requirements together are referred to as the requirements of material objectivity [26].

melt-flow index   The melt-flow index or melt index is a rheological measure widely used in the polymer industry to indicate the fluidity of a material. A polymer sample is placed in a barrel, heated to the testing temperature, and subsequently extruded through a capillary die of diameter $D = 2.095$ mm and length-to-diameter ratio $L/D = 7.637$ [162]. A fixed mass is used to drive the flow of the polymer through the capillary, commonly 2.160 kg. The mass of polymer (in grams) that extrudes in 10 minutes is the melt-flow index. High melt-flow index indicates low viscosity and hence low molecular weight.

melt fracture   Melt fracture, or extrudate distortion, is a general term that refers to a family of instabilities that occur in flow in a capillary and in extrusion in general. The main observation during melt fracture is that the exiting polymer is distorted. In capillary flow, when the flow is driven by constant pressure, the spurt instability results; when the flow is driven by a constant flow rate, the pressure oscillates; see Section 6.1.2.

metric coefficients of a basis   There are nine dot products between pairs of basis vectors of a basis. The scalar results of these nine dot products are called the metric coefficients of the basis. The set of metric coefficients of the basis are themselves a tensor by the definition of a tensor in Section C.7. For orthonormal bases the metric coefficients are given by the delta function δ_{ij}. For nonorthonormal bases the metric coefficients are denoted by the symbol $b_{mp} = \underline{b}_{(m)} \cdot \underline{b}_{(p)}$, $g_{mp} = \underline{g}_{(m)} \cdot \underline{g}_{(p)}$, or $g^{ns} \underline{g}^{(n)} \cdot \underline{g}^{(s)}$.

modulus, instantaneous   The instantaneous modulus G^0 is the value of the relaxation modulus function $G(t)$ at $t = 0$ [95].

neo-Hookean model   The neo-Hookean model is another name for the finite-strain Hooke's law, Equation (9.113) [162].

objectivity   See material objectivity.

Oldroyd B constitutive equation   For polymer solutions, both the polymer and the solvent contribute to the stresses generated. The Oldroyd B equation combines a polymer contribution that acts like the upper convected Maxwell equation and a solvent contribution that is Newtonian:

$$\underline{\underline{\tau}} = \underline{\underline{\tau}}^p + \underline{\underline{\tau}}^s$$

where $\underline{\underline{\tau}}^p$ is the polymeric contribution, and $\underline{\underline{\tau}}^s$ is the solvent contribution. The Oldroyd B equation is also the equivalent of the upper convected Jeffreys model (Section 9.4.1.1).

orthogonal tensor   An orthogonal tensor is one for which its transpose is equal to its inverse. For example, a tensor $\underline{\underline{A}}$ is orthogonal if

$$\underline{\underline{A}}^T \cdot \underline{\underline{A}} = \underline{\underline{A}} \cdot \underline{\underline{A}}^T = \underline{\underline{I}}$$

orthonormal basis vectors   As pointed out in Chapter 2, any three nonzero noncoplanar vectors may form a basis in physical (three-dimensional) space. When these three basis vectors are mutually perpendicular and of unit length, they are called orthonormal basis vectors.

physical components   The coefficients of a vector or tensor in a laboratory (usually stationary) basis system, such as the Cartesian, cylindrical, or spherical bases, are called the physical components of the vector or tensor.

polar decomposition   Refers to the separation of a tensor into two parts, one part that contains the stretch information of the original tensor, and a second part that contains rotation information (see Section 9.1.2). For a tensor $\underline{\underline{A}}$, the polar decomposition theorem says that the tensor may be written in the following two ways [9]:

$$\underline{\underline{A}} = \underline{\underline{R}} \cdot \underline{\underline{U}}$$
$$\underline{\underline{A}} = \underline{\underline{V}} \cdot \underline{\underline{R}}$$

where $\underline{\underline{R}} = \underline{\underline{A}} \cdot (\underline{\underline{A}}^T \cdot \underline{\underline{A}})^{-\frac{1}{2}}$ is the rotation tensor, and $\underline{\underline{U}} = (\underline{\underline{A}}^T \cdot \underline{\underline{A}})^{\frac{1}{2}}$ and $\underline{\underline{V}} = (\underline{\underline{A}} \cdot \underline{\underline{A}}^T)^{\frac{1}{2}}$ are the right and left Cauchy–Green stretch tensors. The eigenvectors of $\underline{\underline{U}}$ and $\underline{\underline{V}}$ are the same, and these vectors indicate the directions of pure stretch associated with the tensor $\underline{\underline{A}}$.

positive definite tensors   A tensor $\underline{\underline{A}}$ is positive definite if for any vector $\underline{v}$ the scalar $\underline{v} \cdot \underline{\underline{A}} \cdot \underline{v}$ is positive.

pressure, equivalent   See equivalent pressure.

pseudoplastic fluid   Characterized by a viscosity that decreases with shear rate; also called shear-thinning fluids.

Reiner–Rivlin fluid   A fluid that follows the following constitutive equation [221]:

$$\underline{\underline{\tau}} = -\frac{1}{2}(\phi_1 \underline{\underline{\dot{\gamma}}} + \phi_2 \underline{\underline{\dot{\gamma}}}^2)$$

where ϕ_1 and ϕ_2 are functions of the invariants of $\underline{\underline{\dot{\gamma}}}$. Note that for $\phi_2 = 0$ this family of constitutive equations includes the generalized Newtonian fluids.

relaxation time   Refers to a time scale for the relaxation of stress in a fluid; it is often given the symbol λ. Constitutive equations will often include one or more relaxation times to capture the memory effects observed in polymeric fluids.

retardation time   Refers to a time scale for the buildup of stress in a fluid; it is often given the symbol Λ. Constitutive equations will sometimes include one or more retardation times to capture time scales observed during start-up experiments that are not captured by the relaxation times.

Reynolds' transport theorem   This is another name for the three-dimensional Leibnitz formula, see Equation (2.264).

rheopectic fluid   A fluid that does not exhibit a steady shear viscosity and for which the shear stress generated in a constant shear-rate experiment increases with time [238, 227]; see thixotropic fluid. The time-dependent change in shear stress is reversible for rheopectic fluids.

rheopexy   A type of rheological behavior in which a fluid forms structure under shear (and hence exhibits increased viscosity) that gradually disintegrates when at rest [238, 227].

second moment of a function   A type of function average. See [239].

second-order fluid equation   A simple constitutive equation that predicts a first normal-stress difference. The second-order fluid constitutive equation is given by [162, 26]

$$\underline{\underline{\tau}} = -\left(\eta_0 \underline{\underline{\dot{\gamma}}} - \frac{\Psi_1^0}{2} \overset{\triangledown}{\underline{\underline{\dot{\gamma}}}} + \Psi_2^0 \underline{\underline{\dot{\gamma}}} \cdot \underline{\underline{\dot{\gamma}}}\right)$$

The second-order fluid is only valid at small Deborah numbers [26].

shear rate, apparent   The quantity $4Q/\pi R^3$ obtained in the capillary flow experiment; Q is the flow rate, and R is the radius of the capillary. This quantity is equal to the shear rate at the wall for Newtonian fluids, but for non-Newtonian fluids, correction factors must be applied to obtain the shear rate at the wall; see Section 10.1.1.

shear stress, wall   The quantity $\Delta \mathcal{P} R/2L$ obtained in the capillary flow experiment; $\Delta \mathcal{P}$ is the modified pressure drop across the capillary, R is the radius of the capillary, and L is the length of the capillary. This quantity is equal to the shear stress at the wall for all fluids, Newtonian and non-Newtonian; see Section 10.1.1.

simple fluids   A term due to Noll, Coleman, and Truesdell [191, 50, 247], and it defines a class of materials. Simply put, for a simple fluid, the stress at any point at time t depends only on the deformation experienced by material points in an arbitrarily small neighborhood around that point at all times prior to t [95, 138]. This requirement eliminates the dependence of $\underline{\underline{\tau}}$ on higher spatial derivatives of velocity such as $\partial^2 v_z/\partial r^2$ [238].

singular/nonsingular matrix   A matrix is singular if its determinant is equal to zero. Matrix inverses do not exist for singular matrices.

singular/nonsingular tensor   A tensor is singular if its determinant is equal to zero. The determinant is one of the invariants of a tensor.

slump test   A test for determining slump/workability of concrete, both in the laboratory and in the field. In the slump test the height of a cone of concrete is monitored as a function of time (ASTM C 143-90).

specific interactions   Chemical or electrical interactions between atoms in molecules that are specific to the chemical composition of the molecules in question. For example, if a molecule is charged, it will interact with other charged particles in a particular way. This kind of interaction cannot be captured by general physical arguments about the molecule, that is, by arguments about size, length, flexibility, and so on.

spurt flow   An instability that occurs in flow through a capillary when the flow is being driven such that the driving pressure drop is constant. At a critical value of stress (the equivalent of a critical wall shear stress), the flow rate becomes very high, and material spurts out of the capillary; see Section 6.1.2.

thermorheological complexity   This term is used to describe the situation when a fluid does not exhibit time–temperature superposition (see Section 6.2.1.3). This results when the various relaxation modes of the fluid do not share a common temperature dependence.

thixotropic fluid   A fluid that does not exhibit a steady shear viscosity and for which the shear stress generated in a constant-shear-rate experiment decreases with time [238, 227]; see rheopectic fluid. The time-dependent change in shear stress is reversible for thixotropic fluids.

time-strain factorability   A material is time-strain factorable or time-strain separable if the time and strain dependence of rheological properties can be factored into separate functions. In experimental data this is checked by measuring the large-amplitude step-strain modulus $G(t, \gamma_0)$. If curves of $G(t, \gamma_0)$ all have similar shapes, they can be shifted along the modulus axis to produce a composite curve. The shift factors $h(\gamma_0)$ are known as the damping function.

time–temperature shift factors   For most rheological material functions data at different temperatures can be shifted to produce master curves using the principle of time–temperature superposition. Horizontal shift factors a_T can be calculated from viscosity (see Section 6.2.1.3):

$$a_T \equiv \frac{\eta(T)T_{\text{ref}}\rho_{\text{ref}}}{\eta(T_{\text{ref}})T\rho}$$

Vertical shift factors for time–temperature superposition are sometimes given the symbol b_T:

$$b_T \equiv \frac{T_{\text{ref}}\rho_{\text{ref}}}{T\rho}$$

time–temperature superposition principle   Rheological data taken at many different temperatures can often be shifted to a single reference temperature. The resulting master curve of the property, along with the values of the shift factors, contains the equivalent amount of information as the original, unshifted data. Time–temperature equivalence is predicted for materials whose relaxation times all have the same temperature dependence. Viscosity, small-amplitude oscillatory moduli, and creep compliance are all material functions that are found to follow the time–temperature superposition principle.

traction vector   A vector that represents the tension force per unit area at a point in space.

Trouton ratio   The Trouton ratio is the ratio of a fluid's zero-deformation-rate elongational viscosity to its zero-shear viscosity. Can also refer to the ratio of elongational viscosity to shear viscosity at the same value of $\dot\gamma$ for both flows.

Trouton viscosity   For a Newtonian fluid, the elongational viscosity $\bar\eta$ is equal to three times the shear viscosity, $\bar\eta = 3\mu$. The quantity 3μ or $3\eta_0$ for non-Newtonian fluids is known as the Trouton viscosity. For non-Newtonian fluids, however, $3\eta_0$ may not be equal to the elongational viscosity.

unit vector   A vector whose magnitude is 1 is a unit vector. For example, if for a vector $\underline{s}$,

$$|\underline{s}| = \sqrt{\underline{s}\cdot\underline{s}} = 1$$

then $\underline{s}$ is a unit vector. In the text, unit vectors are written with a caret ($\hat{}$) over the symbol:

$$|\hat{s}| = 1$$

viscosity, apparent η_a The ratio

$$\eta_a = \frac{\Delta P R/2L}{4Q/\pi R^3}$$

measured in capillary flow; ΔP is the pressure drop across the capillary, R is the radius of the capillary, L is the length of the capillary, and Q is the flow rate. The apparent viscosity is equal to the viscosity for a Newtonian fluid. For a non-Newtonian fluid, the Rabinowitsch correction needs to be applied to obtain true viscosity; see Section 10.1.1.2.

viscosity, intrinsic   The intrinsic viscosity $[\eta]$ is a property of polymer solutions. It is defined as

$$[\eta] \equiv \lim_{c \to 0} \left(\frac{\eta_{sp}}{c} \right)$$

$$= \lim_{c \to 0} \left(\frac{\eta_0 - \eta_s}{\eta_s c} \right)$$

where c is the concentration in g/cm^3, η_0 is the zero-shear solution viscosity, η_s is the solvent viscosity, and $\eta_{sp}(c)$ is called the specific viscosity of the solution.

viscosity, specific   The specific viscosity η_{sp} is a property of polymer solutions. It is defined as

$$\eta_{sp} \equiv \frac{\eta_0 - \eta_s}{\eta_s}$$

where η_0 is the zero-shear solution viscosity and η_s is the solvent viscosity.

wall shear stress   See shear stress, wall.

Weissenberg number   See Deborah number.

Young's modulus   The Young's modulus E is a modulus defined in uniaxial extensional flow:

$$E = \lim_{\epsilon \to 0} \frac{\tau_{33} - \tau_{11}}{\epsilon}$$

Here ϵ is the usual Hencky strain, which in this limit is just $l/l_0 - 1$, where l is the deformed length of the sample, and l_0 is the initial length of the sample.

C

Mathematics

C.1 Math Hints

C.1.1 Solving PDEs by Similarity Transformation

Some partial differential equations (PDEs) can be transformed into ordinary differential equations (ODEs) by rearranging the dependent variables. For example, an equation dependent on two variables t and y can sometimes be transformed into an equation dependent on one variable ζ that is a combination of t and y. To find the functional form of $\zeta(t, y)$, we use a standard technique called the *similarity transform*.

To demonstrate the method of combination of variables or similarity transform, we will solve the following partial differential equation subject to the initial and boundary conditions shown:

$$\frac{\partial u}{\partial t} = \nu \frac{\partial^2 u}{\partial y^2} \tag{C.1}$$

$$
\begin{array}{lll}
t \leq 0 & \text{for all } y & u = 0 \\
t > 0 & y = 0 & u = V \\
t > 0 & y = \infty & u = 0
\end{array}
\tag{C.2}
$$

This system describes fluid velocity $u(y, t)$ near a wall suddenly set into motion at time $t = 0$ [28]. We seek a variable ζ that is a combination of t and y raised to some as yet unknown powers, for example, k and m:

$$\zeta = t^k y^m \tag{C.3}$$

The partial differential equation we are seeking to solve, Equation (C.1), places some constraints on the acceptable values of k and m. To find out what these constraints are, we will define two new variables, T and Y, which differ from t and y only in multiplication by a constant λ raised to different as yet undetermined exponents α and β. This parameterization will reveal one constraint that the equation places on how the two independent variables t and y must appear in the combined variable ζ we are seeking:

$$T \equiv \lambda^{\alpha} t \tag{C.4}$$

$$Y \equiv \lambda^{\beta} y \tag{C.5}$$

Substituting T and Y for t and y in Equation (C.1), we obtain

$$\frac{\partial u}{\partial T} \frac{\partial T}{\partial t} = \nu \frac{\partial}{\partial y} \left(\frac{\partial u}{\partial Y} \frac{\partial Y}{\partial y} \right) = \nu \frac{\partial \left(\frac{\partial u}{\partial Y} \lambda^{\beta} \right)}{\partial y} \tag{C.6}$$

$$\frac{\partial u}{\partial T} \lambda^{\alpha} = \nu \lambda^{\beta} \frac{\partial \left(\frac{\partial u}{\partial Y} \right)}{\partial Y} \frac{\partial Y}{\partial y} = \nu \lambda^{2\beta} \frac{\partial^2 u}{\partial Y^2} \tag{C.7}$$

$$\frac{\partial u}{\partial T} = \nu \lambda^{2\beta - \alpha} \frac{\partial^2 u}{\partial Y^2} \tag{C.8}$$

Comparing equations (C.8) and (C.1) we see that they can be made to be similar if we set $\lambda^{2\beta - \alpha}$ equal to 1. Note that if a solution for the exponents α and β can be found at this stage, then a similarity solution does exist for the differential equation in question [269]:

$$\lambda^{2\beta - \alpha} = 1 \tag{C.9}$$

$$\implies \beta = \frac{1}{2}\alpha \tag{C.10}$$

Since Equation (C.10) is the only constraint on α and β in order to obtain similarity between Equations (C.1) and (C.8), we are free to choose a value for one of these two exponents. We will choose $\beta = 1$:

$$\beta = 1 \quad \alpha = 2 \tag{C.11}$$

For the problem expressed in T and Y to be similar in all aspects to the problem expressed in t and y, the boundary conditions must also be similar. For the problem considered, this is automatically achieved:

$$\begin{aligned} T &\leq 0 \quad \text{for all } Y \quad u = 0 \\ T &> 0 \quad Y = 0 \quad u = V \\ T &> 0 \quad Y = \infty \quad u = 0 \end{aligned} \tag{C.12}$$

Similarity of the boundary conditions is easily achieved when the boundary conditions only involve values of the independent variables at zero and infinity, as in the present case. When the boundary conditions involve finite values of the variables (e.g., $u = 0$ at $y = B$), the variables can sometimes be redefined prior to the similarity transform process to produce boundary conditions at zero or infinity (e.g., $u = 0$ at $Y \equiv y - B = 0$).

Returning to our goal of finding a combination of variables, $\zeta = \zeta(t, y)$, for the current problem, we substitute the definitions of T and Y, Equations (C.4), (C.5), into Equation (C.3) using the values for α and β chosen above:

$$T = \lambda^2 t \tag{C.13}$$

$$Y = \lambda^1 y \tag{C.14}$$

$$\zeta = (T^k \lambda^{-2k})(Y^m \lambda^{-m}) \tag{C.15}$$

$$= T^k Y^m \lambda^{-2k-m} \tag{C.16}$$

To eliminate λ from Equation (C.16) we set the expression with λ equal to 1 and solve for the relationship between k and m:

$$k = -\frac{1}{2}m \tag{C.17}$$

Again we have one constraint and two variables, which leaves us with one degree of freedom. If we choose $m = 1$, we obtain the following for ζ:

$$\zeta = t^{-\frac{1}{2}} y^1 = \frac{y}{\sqrt{t}} \tag{C.18}$$

To test the combination of variables thus obtained, we will substitute ζ into Equation (C.1). If the technique is successful, an ordinary differential equation of u as a function of ζ will result:

$$\frac{\partial u}{\partial t} = v \frac{\partial^2 u}{\partial y^2} \tag{C.19}$$

$$\frac{\partial u}{\partial \zeta} \frac{\partial \zeta}{\partial t} = v \frac{\partial}{\partial y} \left(\frac{\partial u}{\partial \zeta} \frac{\partial \zeta}{\partial y} \right) = v \frac{\partial}{\partial y} \left(\frac{\partial u}{\partial \zeta} \frac{1}{\sqrt{t}} \right) \tag{C.20}$$

$$= \frac{v}{\sqrt{t}} \frac{\partial \left(\frac{\partial u}{\partial \zeta} \right)}{\partial \zeta} \frac{\partial \zeta}{\partial y} \tag{C.21}$$

$$\frac{\partial u}{\partial \zeta} \left(-\frac{1}{2} y t^{-\frac{3}{2}} \right) = v\, t^{-1} \frac{\partial^2 u}{\partial \zeta^2} \tag{C.22}$$

$$\frac{-\zeta}{2v} \frac{du}{d\zeta} = \frac{d^2 u}{d\zeta^2} \tag{C.23}$$

The final ordinary differential equation for $u(\zeta)$ is then

$$\boxed{u'' + \frac{\zeta}{2v} u' = 0} \tag{C.24}$$

where prime ($'$) and double prime ($''$) denote first- and second-order ordinary differentiation with respect to ζ. Now we will transform the boundary conditions:

$$
\begin{array}{llll}
t \le 0 & \text{for all } y & u = 0 & \longrightarrow \quad \zeta = \infty \quad u = 0 \\
t > 0 & y = 0 & u = V & \longrightarrow \quad \zeta = 0 \quad u = V \\
t > 0 & y = \infty & u = 0 & \longrightarrow \quad \zeta = \infty \quad u = 0
\end{array} \tag{C.25}
$$

The initial condition and one of the boundary conditions collapse to the same boundary condition on ζ, leaving two unique boundary conditions for the second-order ordinary differential equation for $u(\zeta)$ obtained in Equation (C.24). This ordinary differential equation can then be solved by the usual techniques (integrating-function method [34]).

C.1.2 MISCELLANEOUS MATH HINTS

TABLE C.1
Math Hints for Chapter 7

1. Expression appearing in $\dot{\underline{\gamma}}$ when written in cylindrical coordinates	$-\dfrac{v_\theta}{r} + \dfrac{\partial v_\theta}{\partial r} = r\dfrac{\partial}{\partial r}\left(\dfrac{v_\theta}{r}\right)$
2. l'Hôpital's rule	If $\lim\limits_{x\to 0} f(x) = 0$ and $\lim\limits_{x\to 0} g(x) = 0$, then $$\lim_{x\to 0}\left(\frac{f(x)}{g(x)}\right) = \lim_{x\to 0}\frac{df/dx}{dg/dx}$$
3. Taking the derivative of a constant raised to the power of a variable	What is $\dfrac{d}{dx}(a^x)$? Let $y = a^x$. Then, $\ln y = x\ln a$ $\dfrac{1}{y}\dfrac{dy}{dx} = \ln a$ $$\boxed{\dfrac{dy}{dx} = \dfrac{d(a^x)}{dx} = a^x\ln a}$$

TABLE C.2
Math Hints for Chapter 8

1. Integrals occurring frequently in calculations involving the Maxwell model and other constitutive equations derived from the Maxwell model	$\displaystyle\int_0^\infty e^{-\frac{s}{\lambda}}\cos\omega s\,ds = \dfrac{\lambda}{1+\omega^2\lambda^2}$ $\displaystyle\int_0^\infty e^{-\frac{s}{\lambda}}\sin\omega s\,ds = \dfrac{\omega\lambda^2}{1+\omega^2\lambda^2}$
2. Integration by parts	$\displaystyle\int_a^b u\,dv = (u\,v)\Big|_a^b - \int_a^b v\,du$
3. Solving a type of first-order ordinary differential equations	For differential equations of the following type: $0 = \dfrac{dy}{dx} + y\,a(x) + b$ we can integrate the equation by using an integrating function $u(x)$, where $u(x)$ is defined as $u(x) = e^{\left[\int a(x)\,dx\right]}$ To solve the differential equation, we first multiply through by $u(x)$, and then we factor the left-hand side of the equation: $u(x)\frac{dy}{dx} + u(x)a(x)y = -b\,u(x)$ $\dfrac{d}{dx}[u(x)y] = -b\,u(x)$ The validity of this factorization may be verified by substituting the definition of $u(x)$ into the equation above. The final solution becomes $y(x) = \dfrac{1}{u(x)}\left[-\int b\,u(x')\,dx' + C\right]$ where C is an arbitrary constant of integration.

C.2 Differential Operations in Curvilinear Coordinates

TABLE C.3
Differential Operations in the Cylindrical Coordinate System r, θ, z

$$\underline{w} = \begin{pmatrix} w_r \\ w_\theta \\ w_z \end{pmatrix}_{r\theta z} \tag{C.3-1}$$

$$\nabla = \hat{e}_r \frac{\partial}{\partial r} + \hat{e}_\theta \frac{1}{r}\frac{\partial}{\partial \theta} + \hat{e}_z \frac{\partial}{\partial z} \tag{C.3-2}$$

$$\nabla a = \begin{pmatrix} \dfrac{\partial a}{\partial r} \\[2mm] \dfrac{1}{r}\dfrac{\partial a}{\partial \theta} \\[2mm] \dfrac{\partial a}{\partial z} \end{pmatrix}_{r\theta z} \tag{C.3-3}$$

$$\nabla \cdot \nabla a = \nabla^2 a = \frac{1}{r}\frac{\partial}{\partial r}\left(r\frac{\partial a}{\partial r}\right) + \frac{1}{r^2}\frac{\partial^2 a}{\partial \theta^2} + \frac{\partial^2 a}{\partial z^2} \tag{C.3-4}$$

$$\nabla \cdot \underline{w} = \frac{1}{r}\frac{\partial}{\partial r}(rw_r) + \frac{1}{r}\frac{\partial w_\theta}{\partial \theta} + \frac{\partial w_z}{\partial z} \tag{C.3-5}$$

$$\nabla \times \underline{w} = \begin{pmatrix} \dfrac{1}{r}\dfrac{\partial w_z}{\partial \theta} - \dfrac{\partial w_\theta}{\partial z} \\[3mm] \dfrac{\partial w_r}{\partial z} - \dfrac{\partial w_z}{\partial r} \\[3mm] \dfrac{1}{r}\dfrac{\partial(rw_\theta)}{\partial r} - \dfrac{1}{r}\dfrac{\partial w_r}{\partial \theta} \end{pmatrix}_{r\theta z} \tag{C.3-6}$$

$$\underline{\underline{A}} = \begin{pmatrix} A_{rr} & A_{r\theta} & A_{rz} \\ A_{\theta r} & A_{\theta\theta} & A_{\theta z} \\ A_{zr} & A_{z\theta} & A_{zz} \end{pmatrix}_{r\theta z} \tag{C.3-7}$$

$$\nabla \underline{w} = \begin{pmatrix} \dfrac{\partial w_r}{\partial r} & \dfrac{\partial w_\theta}{\partial r} & \dfrac{\partial w_z}{\partial r} \\[3mm] \dfrac{1}{r}\dfrac{\partial w_r}{\partial \theta} - \dfrac{w_\theta}{r} & \dfrac{1}{r}\dfrac{\partial w_\theta}{\partial \theta} + \dfrac{w_r}{r} & \dfrac{1}{r}\dfrac{\partial w_z}{\partial \theta} \\[3mm] \dfrac{\partial w_r}{\partial z} & \dfrac{\partial w_\theta}{\partial z} & \dfrac{\partial w_z}{\partial z} \end{pmatrix}_{r\theta z} \tag{C.3-8}$$

$$\nabla^2 \underline{w} = \begin{pmatrix} \dfrac{\partial}{\partial r}\left[\dfrac{1}{r}\dfrac{\partial(rw_r)}{\partial r}\right] + \dfrac{1}{r^2}\dfrac{\partial^2 w_r}{\partial \theta^2} + \dfrac{\partial^2 w_r}{\partial z^2} - \dfrac{2}{r^2}\dfrac{\partial w_\theta}{\partial \theta} \\[4mm] \dfrac{\partial}{\partial r}\left[\dfrac{1}{r}\dfrac{\partial(rw_\theta)}{\partial r}\right] + \dfrac{1}{r^2}\dfrac{\partial^2 w_\theta}{\partial \theta^2} + \dfrac{\partial^2 w_\theta}{\partial z^2} + \dfrac{2}{r^2}\dfrac{\partial w_r}{\partial \theta} \\[4mm] \dfrac{1}{r}\dfrac{\partial}{\partial r}\left(r\dfrac{\partial w_z}{\partial r}\right) + \dfrac{1}{r^2}\dfrac{\partial^2 w_z}{\partial \theta^2} + \dfrac{\partial^2 w_z}{\partial z^2} \end{pmatrix}_{r\theta z} \tag{C.3-9}$$

continued

$$\nabla \cdot \underline{\underline{A}} = \begin{pmatrix} \dfrac{1}{r}\dfrac{\partial}{\partial r}(rA_{rr}) + \dfrac{1}{r}\dfrac{\partial A_{\theta r}}{\partial \theta} + \dfrac{\partial A_{zr}}{\partial z} - \dfrac{A_{\theta\theta}}{r} \\[2mm] \dfrac{1}{r^2}\dfrac{\partial}{\partial r}(r^2 A_{r\theta}) + \dfrac{1}{r}\dfrac{\partial A_{\theta\theta}}{\partial \theta} + \dfrac{\partial A_{z\theta}}{\partial z} + \dfrac{A_{\theta r} - A_{r\theta}}{r} \\[2mm] \dfrac{1}{r}\dfrac{\partial}{\partial r}(rA_{rz}) + \dfrac{1}{r}\dfrac{\partial A_{\theta z}}{\partial \theta} + \dfrac{\partial A_{zz}}{\partial z} \end{pmatrix}_{r\theta z}$$ (C.3-10)

$$\underline{u} \cdot \nabla \underline{w} = \begin{pmatrix} u_r\left(\dfrac{\partial w_r}{\partial r}\right) + u_\theta\left(\dfrac{1}{r}\dfrac{\partial w_r}{\partial \theta} - \dfrac{w_\theta}{r}\right) + u_z\left(\dfrac{\partial w_r}{\partial z}\right) \\[2mm] u_r\left(\dfrac{\partial w_\theta}{\partial r}\right) + u_\theta\left(\dfrac{1}{r}\dfrac{\partial w_\theta}{\partial \theta} + \dfrac{w_r}{r}\right) + u_z\left(\dfrac{\partial w_\theta}{\partial z}\right) \\[2mm] u_r\left(\dfrac{\partial w_z}{\partial r}\right) + u_\theta\left(\dfrac{1}{r}\dfrac{\partial w_z}{\partial \theta}\right) + u_z\left(\dfrac{\partial w_z}{\partial z}\right) \end{pmatrix}_{r\theta z}$$ (C.3-11)

TABLE C.4
Differential Operations in the Spherical Coordinate System r, θ, ϕ

$$\underline{w} = \begin{pmatrix} w_r \\ w_\theta \\ w_\phi \end{pmatrix}_{r\theta\phi}$$ (C.4-1)

$$\nabla = \hat{e}_r \frac{\partial}{\partial r} + \hat{e}_\theta \frac{1}{r}\frac{\partial}{\partial \theta} + \hat{e}_\phi \frac{1}{r\sin\theta}\frac{\partial}{\partial \phi}$$ (C.4-2)

$$\nabla a = \begin{pmatrix} \dfrac{\partial a}{\partial r} \\[2mm] \dfrac{1}{r}\dfrac{\partial a}{\partial \theta} \\[2mm] \dfrac{1}{r\sin\theta}\dfrac{\partial a}{\partial \phi} \end{pmatrix}_{r\theta\phi}$$ (C.4-3)

$$\nabla \cdot \nabla a = \nabla^2 a = \frac{1}{r^2}\frac{\partial}{\partial r}\left(r^2\frac{\partial a}{\partial r}\right) + \frac{1}{r^2\sin\theta}\frac{\partial}{\partial \theta}\left(\sin\theta\frac{\partial a}{\partial \theta}\right) + \frac{1}{r^2\sin^2\theta}\frac{\partial^2 a}{\partial \phi^2}$$ (C.4-4)

$$\nabla \cdot \underline{w} = \frac{1}{r^2}\frac{\partial}{\partial r}\left(r^2 w_r\right) + \frac{1}{r\sin\theta}\frac{\partial}{\partial \theta}\left(w_\theta\sin\theta\right) + \frac{1}{r\sin\theta}\frac{\partial w_\phi}{\partial \phi}$$ (C.4-5)

$$\nabla \times \underline{w} = \begin{pmatrix} \dfrac{1}{r\sin\theta}\dfrac{\partial}{\partial \theta}\left(w_\phi\sin\theta\right) - \dfrac{1}{r\sin\theta}\dfrac{\partial w_\theta}{\partial \phi} \\[2mm] \dfrac{1}{r\sin\theta}\dfrac{\partial w_r}{\partial \phi} - \dfrac{1}{r}\dfrac{\partial}{\partial r}(rw_\phi) \\[2mm] \dfrac{1}{r}\dfrac{\partial}{\partial r}(rw_\theta) - \dfrac{1}{r}\dfrac{\partial w_r}{\partial \theta} \end{pmatrix}_{r\theta\phi}$$ (C.4-6)

$$\underline{\underline{A}} = \begin{pmatrix} A_{rr} & A_{r\theta} & A_{r\phi} \\ A_{\theta r} & A_{\theta\theta} & A_{\theta\phi} \\ A_{\phi r} & A_{\phi\theta} & A_{\phi\phi} \end{pmatrix}_{r\theta\phi}$$ (C.4-7)

continued

$$\nabla \underline{w} = \begin{pmatrix} \dfrac{\partial w_r}{\partial r} & \dfrac{\partial w_\theta}{\partial r} & \dfrac{\partial w_\phi}{\partial r} \\[2ex] \dfrac{1}{r}\dfrac{\partial w_r}{\partial \theta} - \dfrac{w_\theta}{r} & \dfrac{1}{r}\dfrac{\partial w_\theta}{\partial \theta} + \dfrac{w_r}{r} & \dfrac{1}{r}\dfrac{\partial w_\phi}{\partial \theta} \\[2ex] \dfrac{1}{r\sin\theta}\dfrac{\partial w_r}{\partial \phi} - \dfrac{w_\phi}{r} & \dfrac{1}{r\sin\theta}\dfrac{\partial w_\theta}{\partial \phi} - \dfrac{w_\phi}{r}\cot\theta & \dfrac{1}{r\sin\theta}\dfrac{\partial w_\phi}{\partial \phi} + \dfrac{w_\phi}{r} + \dfrac{w_\theta}{r}\cot\theta \end{pmatrix}_{r\theta\phi}$$

(C.4-8)

$$\nabla^2 \underline{w} = \begin{pmatrix} \left\{ \dfrac{\partial}{\partial r}\left[\dfrac{1}{r^2}\dfrac{\partial}{\partial r}(r^2 w_r)\right] + \dfrac{1}{r^2 \sin\theta}\dfrac{\partial}{\partial\theta}\left(\sin\theta\dfrac{\partial w_r}{\partial\theta}\right) + \dfrac{1}{r^2\sin^2\theta}\dfrac{\partial^2 w_r}{\partial\phi^2} \right. \\[2ex] \left. - \dfrac{2}{r^2\sin\theta}\dfrac{\partial}{\partial\theta}(w_\theta\sin\theta) - \dfrac{2}{r^2\sin\theta}\dfrac{\partial w_\phi}{\partial\phi} \right\} \\[3ex] \left\{ \dfrac{1}{r^2}\dfrac{\partial}{\partial r}\left(r^2\dfrac{\partial w_\theta}{\partial r}\right) + \dfrac{1}{r^2}\dfrac{\partial}{\partial\theta}\left[\dfrac{1}{\sin\theta}\dfrac{\partial}{\partial\theta}(w_\theta\sin\theta)\right] + \dfrac{1}{r^2\sin^2\theta}\dfrac{\partial^2 w_\theta}{\partial\phi^2} \right. \\[2ex] \left. + \dfrac{2}{r^2}\dfrac{\partial w_r}{\partial\theta} - \dfrac{2\cot\theta}{r^2\sin\theta}\dfrac{\partial w_\phi}{\partial\phi} \right\} \\[3ex] \left(\dfrac{1}{r^2}\dfrac{\partial}{\partial r}\left(r^2\dfrac{\partial w_\phi}{\partial r}\right) + \dfrac{1}{r^2}\dfrac{\partial}{\partial\theta}\left[\dfrac{1}{\sin\theta}\dfrac{\partial}{\partial\theta}(w_\phi\sin\theta)\right] + \dfrac{1}{r^2\sin^2\theta}\dfrac{\partial^2 w_\phi}{\partial\phi^2} \right. \\[2ex] \left. + \dfrac{2}{r^2\sin\theta}\dfrac{\partial w_r}{\partial\phi} + \dfrac{2\cot\theta}{r^2\sin\theta}\dfrac{\partial w_\theta}{\partial\phi} \right) \end{pmatrix}_{r\theta\phi}$$

(C.4-9)

$$\nabla \cdot \underline{A} = \begin{pmatrix} \dfrac{1}{r^2}\dfrac{\partial}{\partial r}(r^2 A_{rr}) + \dfrac{1}{r\sin\theta}\dfrac{\partial}{\partial\theta}(A_{\theta r}\sin\theta) + \dfrac{1}{r\sin\theta}\dfrac{\partial A_{\phi r}}{\partial\phi} - \dfrac{A_{\theta\theta} + A_{\phi\phi}}{r} \\[2ex] \dfrac{1}{r^3}\dfrac{\partial}{\partial r}(r^3 A_{r\theta}) + \dfrac{1}{r\sin\theta}\dfrac{\partial}{\partial\theta}(A_{\theta\theta}\sin\theta) + \dfrac{1}{r\sin\theta}\dfrac{\partial A_{\phi\theta}}{\partial\phi} + \dfrac{(A_{\theta r} - A_{r\theta}) - A_{\phi\phi}\cot\theta}{r} \\[2ex] \dfrac{1}{r^3}\dfrac{\partial}{\partial r}(r^3 A_{r\phi}) + \dfrac{1}{r\sin\theta}\dfrac{\partial}{\partial\theta}(A_{\theta\phi}\sin\theta) + \dfrac{1}{r\sin\theta}\dfrac{\partial A_{\phi\phi}}{\partial\phi} + \dfrac{(A_{\phi r} - A_{r\phi}) + A_{\phi\theta}\cot\theta}{r} \end{pmatrix}_{r\theta\phi}$$

(C.4-10)

$$\underline{u} \cdot \nabla \underline{w} = \begin{pmatrix} u_r\left(\dfrac{\partial w_r}{\partial r}\right) + u_\theta\left(\dfrac{1}{r}\dfrac{\partial w_r}{\partial\theta} - \dfrac{w_\theta}{r}\right) + u_\phi\left(\dfrac{1}{r\sin\theta}\dfrac{\partial w_r}{\partial\phi} - \dfrac{w_\phi}{r}\right) \\[2ex] u_r\left(\dfrac{\partial w_\theta}{\partial r}\right) + u_\theta\left(\dfrac{1}{r}\dfrac{\partial w_\theta}{\partial\theta} + \dfrac{w_r}{r}\right) + u_\phi\left(\dfrac{1}{r\sin\theta}\dfrac{\partial w_\theta}{\partial\phi} - \dfrac{w_\phi}{r}\cot\theta\right) \\[2ex] u_r\left(\dfrac{\partial w_\phi}{\partial r}\right) + u_\theta\left(\dfrac{1}{r}\dfrac{\partial w_\phi}{\partial\theta}\right) + u_\phi\left(\dfrac{1}{r\sin\theta}\dfrac{\partial w_\phi}{\partial\phi} + \dfrac{w_r}{r} + \dfrac{w_\theta}{r}\cot\theta\right) \end{pmatrix}_{r\theta\phi}$$

(C.4-11)

TABLE C.5
Continuity Equation in Three Coordinate Systems

Continuity equation, Cartesian coordinates

$$\frac{\partial \rho}{\partial t} + \left(v_x\frac{\partial \rho}{\partial x} + v_y\frac{\partial \rho}{\partial y} + v_z\frac{\partial \rho}{\partial z}\right) + \rho\left(\frac{\partial v_x}{\partial x} + \frac{\partial v_y}{\partial y} + \frac{\partial v_z}{\partial z}\right) = 0$$

(C.5-1)

continued

Continuity equation, cylindrical coordinates

$$\frac{\partial \rho}{\partial t} + \frac{1}{r}\frac{\partial(\rho r v_r)}{\partial r} + \frac{1}{r}\frac{\partial(\rho v_\theta)}{\partial \theta} + \frac{\partial(\rho v_z)}{\partial z} = 0 \tag{C.5-2}$$

Continuity equation, spherical coordinates

$$\frac{\partial \rho}{\partial t} + \frac{1}{r^2}\frac{\partial(\rho r^2 v_r)}{\partial r} + \frac{1}{r\sin\theta}\frac{\partial(\rho v_\theta \sin\theta)}{\partial \theta} + \frac{1}{r\sin\theta}\frac{\partial(\rho v_\phi)}{\partial \phi} = 0 \tag{C.5-3}$$

TABLE C.6
Equation of Motion for Incompressible Fluids in Three Coordinate Systems

Equation of motion for incompressible fluid, Cartesian coordinates

$$\rho\left(\frac{\partial v_x}{\partial t} + v_x\frac{\partial v_x}{\partial x} + v_y\frac{\partial v_x}{\partial y} + v_z\frac{\partial v_x}{\partial z}\right) = -\frac{\partial p}{\partial x} - \left(\frac{\partial \tau_{xx}}{\partial x} + \frac{\partial \tau_{yx}}{\partial y} + \frac{\partial \tau_{zx}}{\partial z}\right) + \rho g_x \tag{C.6-1}$$

$$\rho\left(\frac{\partial v_y}{\partial t} + v_x\frac{\partial v_y}{\partial x} + v_y\frac{\partial v_y}{\partial y} + v_z\frac{\partial v_y}{\partial z}\right) = -\frac{\partial p}{\partial y} - \left(\frac{\partial \tau_{xy}}{\partial x} + \frac{\partial \tau_{yy}}{\partial y} + \frac{\partial \tau_{zy}}{\partial z}\right) + \rho g_y \tag{C.6-2}$$

$$\rho\left(\frac{\partial v_z}{\partial t} + v_x\frac{\partial v_z}{\partial x} + v_y\frac{\partial v_z}{\partial y} + v_z\frac{\partial v_z}{\partial z}\right) = -\frac{\partial p}{\partial z} - \left(\frac{\partial \tau_{xz}}{\partial x} + \frac{\partial \tau_{yz}}{\partial y} + \frac{\partial \tau_{zz}}{\partial z}\right) + \rho g_z \tag{C.6-3}$$

Equation of motion for incompressible fluid, cylindrical coordinates

$$\rho\left(\frac{\partial v_r}{\partial t} + v_r\frac{\partial v_r}{\partial r} + \frac{v_\theta}{r}\frac{\partial v_r}{\partial \theta} - \frac{v_\theta^2}{r} + v_z\frac{\partial v_r}{\partial z}\right) = -\frac{\partial p}{\partial r} - \left[\frac{1}{r}\frac{\partial(r\tau_{rr})}{\partial r} + \frac{1}{r}\frac{\partial \tau_{r\theta}}{\partial \theta} - \frac{\tau_{\theta\theta}}{r} + \frac{\partial \tau_{rz}}{\partial z}\right] + \rho g_r \tag{C.6-4}$$

$$\rho\left(\frac{\partial v_\theta}{\partial t} + v_r\frac{\partial v_\theta}{\partial r} + \frac{v_\theta}{r}\frac{\partial v_\theta}{\partial \theta} + \frac{v_\theta v_r}{r} + v_z\frac{\partial v_\theta}{\partial z}\right) = -\frac{1}{r}\frac{\partial p}{\partial \theta} - \left[\frac{1}{r^2}\frac{\partial(r^2\tau_{r\theta})}{\partial r} + \frac{1}{r}\frac{\partial \tau_{\theta\theta}}{\partial \theta} + \frac{\partial \tau_{\theta z}}{\partial z}\right] + \rho g_\theta \tag{C.6-5}$$

$$\rho\left(\frac{\partial v_z}{\partial t} + v_r\frac{\partial v_z}{\partial r} + \frac{v_\theta}{r}\frac{\partial v_z}{\partial \theta} + v_z\frac{\partial v_z}{\partial z}\right) = -\frac{\partial p}{\partial z} - \left[\frac{1}{r}\frac{\partial(r\tau_{rz})}{\partial r} + \frac{1}{r}\frac{\partial \tau_{\theta z}}{\partial \theta} + \frac{\partial \tau_{zz}}{\partial z}\right] + \rho g_z \tag{C.6-6}$$

Equation of motion for incompressible fluid, spherical coordinates

$$\rho\left(\frac{\partial v_r}{\partial t} + v_r\frac{\partial v_r}{\partial r} + \frac{v_\theta}{r}\frac{\partial v_r}{\partial \theta} + \frac{v_\phi}{r\sin\theta}\frac{\partial v_r}{\partial \phi} - \frac{v_\theta^2 + v_\phi^2}{r}\right)$$

$$= -\frac{\partial p}{\partial r} - \left[\frac{1}{r^2}\frac{\partial(r^2\tau_{rr})}{\partial r} + \frac{1}{r\sin\theta}\frac{\partial(\tau_{r\theta}\sin\theta)}{\partial \theta} + \frac{1}{r\sin\theta}\frac{\partial \tau_{r\phi}}{\partial \phi} - \frac{\tau_{\theta\theta} + \tau_{\phi\phi}}{r}\right] + \rho g_r \tag{C.6-7}$$

$$\rho\left(\frac{\partial v_\theta}{\partial t} + v_r\frac{\partial v_\theta}{\partial r} + \frac{v_\theta}{r}\frac{\partial v_\theta}{\partial \theta} + \frac{v_\phi}{r\sin\theta}\frac{\partial v_\theta}{\partial \phi} + \frac{v_r v_\theta}{r} - \frac{v_\phi^2\cot\theta}{r}\right)$$

$$= -\frac{1}{r}\frac{\partial p}{\partial \theta} - \left[\frac{1}{r^2}\frac{\partial(r^2\tau_{r\theta})}{\partial r} + \frac{1}{r\sin\theta}\frac{\partial(\tau_{\theta\theta}\sin\theta)}{\partial \theta} + \frac{1}{r\sin\theta}\frac{\partial \tau_{\theta\phi}}{\partial \phi} + \frac{\tau_{r\theta}}{r} - \frac{(\cot\theta)\tau_{\phi\phi}}{r}\right] + \rho g_\theta \tag{C.6-8}$$

$$\rho\left(\frac{\partial v_\phi}{\partial t} + v_r\frac{\partial v_\phi}{\partial r} + \frac{v_\theta}{r}\frac{\partial v_\phi}{\partial \theta} + \frac{v_\phi}{r\sin\theta}\frac{\partial v_\phi}{\partial \phi} + \frac{v_r v_\phi}{r} - \frac{v_\phi v_\theta\cot\theta}{r}\right)$$

$$= -\frac{1}{r\sin\theta}\frac{\partial p}{\partial \phi} - \left[\frac{1}{r^2}\frac{\partial(r^2\tau_{r\phi})}{\partial r} + \frac{1}{r}\frac{\partial \tau_{\theta\phi}}{\partial \theta} + \frac{1}{r\sin\theta}\frac{\partial \tau_{\phi\phi}}{\partial \phi} + \frac{\tau_{r\phi}}{r} - \frac{(2\cot\theta)\tau_{\theta\phi}}{r}\right] + \rho g_\phi \tag{C.6-9}$$

TABLE C.7
Equation of Motion for Incompressible Newtonian Fluids: The Navier–Stokes Equations in Three Coordinate Systems

Equation of motion for incompressible Newtonian fluid (Navier–Stokes equation), Cartesian coordinates

$$\rho\left(\frac{\partial v_x}{\partial t} + v_x\frac{\partial v_x}{\partial x} + v_y\frac{\partial v_x}{\partial y} + v_z\frac{\partial v_x}{\partial z}\right) = -\frac{\partial p}{\partial x} + \mu\left(\frac{\partial^2 v_x}{\partial x^2} + \frac{\partial^2 v_x}{\partial y^2} + \frac{\partial^2 v_x}{\partial z^2}\right) + \rho g_x \tag{C.7-1}$$

$$\rho\left(\frac{\partial v_y}{\partial t} + v_x\frac{\partial v_y}{\partial x} + v_y\frac{\partial v_y}{\partial y} + v_z\frac{\partial v_y}{\partial z}\right) = -\frac{\partial p}{\partial y} + \mu\left(\frac{\partial^2 v_y}{\partial x^2} + \frac{\partial^2 v_y}{\partial y^2} + \frac{\partial^2 v_y}{\partial z^2}\right) + \rho g_y \tag{C.7-2}$$

$$\rho\left(\frac{\partial v_z}{\partial t} + v_x\frac{\partial v_z}{\partial x} + v_y\frac{\partial v_z}{\partial y} + v_z\frac{\partial v_z}{\partial z}\right) = -\frac{\partial p}{\partial z} + \mu\left(\frac{\partial^2 v_z}{\partial x^2} + \frac{\partial^2 v_z}{\partial y^2} + \frac{\partial^2 v_z}{\partial z^2}\right) + \rho g_z \tag{C.7-3}$$

Equation of motion for incompressible Newtonian fluid (Navier–Stokes equation), cylindrical coordinates

$$\rho\left(\frac{\partial v_r}{\partial t} + v_r\frac{\partial v_r}{\partial r} + \frac{v_\theta}{r}\frac{\partial v_r}{\partial \theta} - \frac{v_\theta^2}{r} + v_z\frac{\partial v_r}{\partial z}\right)$$
$$= -\frac{\partial p}{\partial r} + \mu\left[\frac{\partial}{\partial r}\left(\frac{1}{r}\frac{\partial(r v_r)}{\partial r}\right) + \frac{1}{r^2}\frac{\partial^2 v_r}{\partial \theta^2} - \frac{2}{r^2}\frac{\partial v_\theta}{\partial \theta} + \frac{\partial^2 v_r}{\partial z^2}\right] + \rho g_r \tag{C.7-4}$$

$$\rho\left(\frac{\partial v_\theta}{\partial t} + v_r\frac{\partial v_\theta}{\partial r} + \frac{v_\theta}{r}\frac{\partial v_\theta}{\partial \theta} + \frac{v_r v_\theta}{r} + v_z\frac{\partial v_\theta}{\partial z}\right)$$
$$= -\frac{1}{r}\frac{\partial p}{\partial \theta} + \mu\left[\frac{\partial}{\partial r}\left(\frac{1}{r}\frac{\partial(r v_\theta)}{\partial r}\right) + \frac{1}{r^2}\frac{\partial^2 v_\theta}{\partial \theta^2} + \frac{2}{r^2}\frac{\partial v_r}{\partial \theta} + \frac{\partial^2 v_\theta}{\partial z^2}\right] + \rho g_\theta \tag{C.7-5}$$

$$\rho\left(\frac{\partial v_z}{\partial t} + v_r\frac{\partial v_z}{\partial r} + \frac{v_\theta}{r}\frac{\partial v_z}{\partial \theta} + v_z\frac{\partial v_z}{\partial z}\right)$$
$$= -\frac{\partial p}{\partial z} + \mu\left[\frac{1}{r}\frac{\partial}{\partial r}\left(r\frac{\partial v_z}{\partial r}\right) + \frac{1}{r^2}\frac{\partial^2 v_z}{\partial \theta^2} + \frac{\partial^2 v_z}{\partial z^2}\right] + \rho g_z \tag{C.7-6}$$

Equation of motion for incompressible Newtonian fluid (Navier–Stokes equation), spherical coordinates

$$\rho\left(\frac{\partial v_r}{\partial t} + v_r\frac{\partial v_r}{\partial r} + \frac{v_\theta}{r}\frac{\partial v_r}{\partial \theta} + \frac{v_\phi}{r\sin\theta}\frac{\partial v_r}{\partial \phi} - \frac{v_\theta^2 + v_\phi^2}{r}\right)$$
$$= -\frac{\partial p}{\partial r} + \mu\left(\nabla^2 v_r - \frac{2v_r}{r^2} - \frac{2}{r^2}\frac{v_\theta}{\partial\theta} - \frac{2v_\theta\cot\theta}{r^2} - \frac{2}{r^2\sin\theta}\frac{\partial v_\phi}{\partial\phi}\right) + \rho g_r \tag{C.7-7}$$

$$\rho\left(\frac{\partial v_\theta}{\partial t} + v_r\frac{\partial v_\theta}{\partial r} + \frac{v_\theta}{r}\frac{\partial v_\theta}{\partial \theta} + \frac{v_\phi}{r\sin\theta}\frac{\partial v_\theta}{\partial \phi} + \frac{v_r v_\theta}{r} - \frac{v_\phi^2\cot\theta}{r}\right)$$
$$= -\frac{1}{r}\frac{\partial p}{\partial \theta} + \mu\left(\nabla^2 v_\theta + \frac{2}{r^2}\frac{\partial v_r}{\partial\theta} - \frac{v_\theta}{r^2\sin^2\theta} - \frac{2\cos\theta}{r^2\sin^2\theta}\frac{\partial v_\phi}{\partial\phi}\right) + \rho g_\theta \tag{C.7-8}$$

$$\rho\left(\frac{\partial v_\phi}{\partial t} + v_r\frac{\partial v_\phi}{\partial r} + \frac{v_\theta}{r}\frac{\partial v_\phi}{\partial \theta} + \frac{v_\phi}{r\sin\theta}\frac{\partial v_\phi}{\partial \phi} + \frac{v_r v_\phi}{r} + \frac{v_\phi v_\theta\cot\theta}{r}\right)$$
$$= -\frac{1}{r\sin\theta}\frac{\partial p}{\partial \phi} + \mu\left(\nabla^2 v_\phi - \frac{v_\phi}{r^2\sin\theta} + \frac{2}{r^2\sin\theta}\frac{\partial v_r}{\partial\phi} + \frac{2\cos\theta}{r^2\sin^2\theta}\frac{\partial v_\theta}{\partial\phi}\right) + \rho g_\phi \tag{C.7-9}$$

where

$$\nabla \equiv \frac{1}{r^2}\frac{\partial}{\partial r}\left(r^2\frac{\partial}{\partial r}\right) + \frac{1}{r^2\sin\theta}\frac{\partial}{\partial \theta}\left(\sin\theta\frac{\partial}{\partial \theta}\right) + \frac{1}{r^2\sin^2\theta}\left(\frac{\partial^2}{\partial \phi^2}\right)$$

C.3 Projection of a Plane

We seek to calculate the projection of an area dS in the direction $\hat{e}$ [239]. We will analyze this problem in a Cartesian coordinate system such that $\hat{e} = \hat{e}_z$ (Figure C.1). We choose to look at a small differential surface element dS with unit normal $\hat{n}$, where dS is the parallelogram enclosed by the vectors $\underline{a}$ and $\underline{b}$. Note that $\underline{a} \times \underline{b}$ is parallel to $\hat{n}$; we choose $\underline{a}$ and $\underline{b}$ such that $\hat{n}$ and $\underline{a} \times \underline{b}$ are in the same direction.

The area is given by $dS = (a)(b) \sin \theta = |\underline{a} \times \underline{b}|$, where θ is the angle between $\underline{a}$ and $\underline{b}$. The area dS projects down on to the xy-plane as a rectangle included by the vectors $\underline{c}$ and $\underline{d}$, as shown in Figure C.1. Since $\underline{a}$ and $\underline{c}$ both lie in the xz-plane, we can write

$$\underline{a} = \underline{c} + \alpha \hat{e}_z \tag{C.26}$$

where α is an unknown scalar. Likewise since $\underline{b}$ and $\underline{d}$ lie in the yz-plane, we can write

$$\underline{b} = \underline{d} + \beta \hat{e}_z \tag{C.27}$$

where β is also an unknown scalar. We now take the cross product of $\underline{a}$ and $\underline{b}$:

$$\underline{a} \times \underline{b} = (\underline{c} + \alpha \hat{e}_z)(\underline{d} + \beta \hat{e}_z) \tag{C.28}$$

$$= \underline{c} \times \underline{d} + \alpha \hat{e}_z \times \underline{d} + \underline{c} \times \beta \hat{e}_z + \alpha \hat{e}_z \times \beta \hat{e}_z \tag{C.29}$$

$$= \underline{c} \times \underline{d} + \alpha (\hat{e}_z \times \underline{d}) + \beta (\underline{c} \times \hat{e}_z) \tag{C.30}$$

recalling that the cross product of parallel vectors is zero ($\sin 0 = 0$). If we dot this final expression with $\hat{e}_z$, the last two terms on the right-hand side will drop out since they are perpendicular to $\hat{e}_z$. This results in

$$(\underline{a} \times \underline{b}) \cdot \hat{e}_z = (\underline{c} \times \underline{d}) \cdot \hat{e}_z \tag{C.31}$$

Because the dot product is commutative, it does not matter whether the dot product with $\hat{e}_z$ appears on the left or right. Note that $\underline{c} \times \underline{d}$ is parallel to and in the same direction as $\hat{e}_z$ (Figure C.1), and thus $(\underline{c} \times \underline{d}) \cdot \hat{e}_z = cd \cos 0 = cd$, which is the area of the rectangle that is the projection of dS onto the xy-plane:

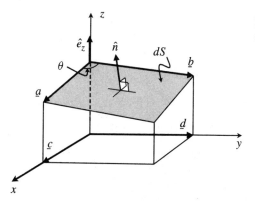

Figure C.1 Geometry for the derivation of the expression for the projection of a plane.

$$\begin{pmatrix} \text{projection of} \\ dS \text{ onto} \\ \text{plane whose unit} \\ \text{normal is } \hat{e}_z \end{pmatrix} = cd = (\underline{a} \times \underline{b}) \cdot \hat{e}_z \tag{C.32}$$

Finally, since $\hat{n}$ is given by

$$\hat{n} = \frac{\underline{a} \times \underline{b}}{|\underline{a} \times \underline{b}|} = \frac{\underline{a} \times \underline{b}}{dS} \tag{C.33}$$

Equation (C.32) becomes

$$\begin{pmatrix} \text{projection of} \\ dS \text{ onto} \\ \text{plane whose unit} \\ \text{normal is } \hat{e}_z \end{pmatrix} = \hat{n} \cdot \hat{e}_z \, dS \tag{C.34}$$

C.4 Finite Deformation Tensors in Curvilinear Coordinates

C.4.1 CYLINDRICAL COORDINATES

In cylindrical coordinates the orthogonal matrix $T_{\text{cylindrical}}$ appears in the expression for the deformation gradient tensor $\underline{\underline{F}}$ [26]:

$$T_{\text{cylindrical}} = \begin{bmatrix} T_{rr} & T_{r\theta} & T_{rz} \\ T_{\theta r} & T_{\theta\theta} & T_{\theta z} \\ T_{zr} & T_{z\theta} & T_{zz} \end{bmatrix}$$

$$= \begin{bmatrix} \cos\theta\cos\theta' + \sin\theta\sin\theta' & \sin\theta\cos\theta' - \cos\theta\sin\theta' & 0 \\ \cos\theta\sin\theta' - \sin\theta\cos\theta' & \sin\theta\sin\theta' + \cos\theta\cos\theta' & 0 \\ 0 & 0 & 1 \end{bmatrix}$$

The coordinates for the deformation gradient tensor $\underline{\underline{F}}$ and its inverse $\underline{\underline{F}}^{-1}$ in cylindrical coordinates may be calculated from these matrix equations:

$$\underline{\underline{F}} = \begin{pmatrix} F_{rr} & F_{r\theta} & F_{rz} \\ F_{\theta r} & F_{\theta\theta} & F_{\theta z} \\ F_{zr} & F_{z\theta} & F_{zz} \end{pmatrix}_{r\theta z} = \begin{bmatrix} \frac{\partial r'}{\partial r} & r'\frac{\partial \theta'}{\partial r} & \frac{\partial z'}{\partial r} \\ \frac{1}{r}\frac{\partial r'}{\partial \theta} & \frac{r'}{r}\frac{\partial \theta'}{\partial \theta} & \frac{1}{r}\frac{\partial z'}{\partial \theta} \\ \frac{\partial r'}{\partial z} & r'\frac{\partial \theta'}{\partial z} & \frac{\partial z'}{\partial z} \end{bmatrix} \begin{bmatrix} T_{rr} & T_{r\theta} & T_{rz} \\ T_{\theta r} & T_{\theta\theta} & T_{\theta z} \\ T_{zr} & T_{z\theta} & T_{zz} \end{bmatrix}^T$$

$$\underline{\underline{F}}^{-1} = \begin{pmatrix} F_{rr}^{-1} & F_{r\theta}^{-1} & F_{rz}^{-1} \\ F_{\theta r}^{-1} & F_{\theta\theta}^{-1} & F_{\theta z}^{-1} \\ F_{zr}^{-1} & F_{z\theta}^{-1} & F_{zz}^{-1} \end{pmatrix}_{r\theta z} = \begin{bmatrix} T_{rr} & T_{r\theta} & T_{rz} \\ T_{\theta r} & T_{\theta\theta} & T_{\theta z} \\ T_{zr} & T_{z\theta} & T_{zz} \end{bmatrix} \begin{bmatrix} \frac{\partial r}{\partial r'} & r\frac{\partial \theta}{\partial r'} & \frac{\partial z}{\partial r'} \\ \frac{1}{r'}\frac{\partial r}{\partial \theta'} & \frac{r}{r'}\frac{\partial \theta}{\partial \theta'} & \frac{1}{r'}\frac{\partial z}{\partial \theta'} \\ \frac{\partial r}{\partial z'} & r\frac{\partial \theta}{\partial z'} & \frac{\partial z}{\partial z'} \end{bmatrix}$$

Note that since $T_{\text{cylindrical}}$ is an orthogonal matrix, that is, $T^T = T^{-1}$, this matrix cancels out when either the Finger tensor $\underline{C}^{-1}$ or the Cauchy tensor $\underline{C}$ is calculated from the deformation gradient tensor or its inverse. To see this, let A be the matrix of derivatives in the preceding equation such that $\underline{\underline{F}}^{-1} = TA$. Then,

$$\underline{\underline{C}}^{-1} = (\underline{\underline{F}}^{-1})^T \cdot \underline{\underline{F}}^{-1}$$

$$= (TA)^T (TA)$$

$$= A^T (T^T T) A$$

$$= A^T A$$

C.4.2 SPHERICAL COORDINATES

In spherical coordinates the orthogonal matrix $T_{\text{spherical}}$ appears in the expression for the deformation gradient tensor $\underline{\underline{F}}$ [26]:

$$T_{\text{spherical}} = \begin{bmatrix} T_{rr} & T_{r\theta} & T_{r\phi} \\ T_{\theta r} & T_{\theta\theta} & T_{\theta\phi} \\ T_{\phi r} & T_{\phi\theta} & T_{\phi\phi} \end{bmatrix}$$

$$= \begin{bmatrix} ScS'c' + SsS's' + CC' & ScC'c' + SsC's' - CS' & -Scs' + Ssc' \\ CcS'c' + CsS's' - SC' & CcC'c' + CsC's' + SS' & -Ccs' + Csc' \\ -sS'c' + cS's' & -sC'c' + cC's' & ss' + cc' \end{bmatrix}$$

where

$$C = \cos\theta \quad S = \sin\theta$$
$$C' = \cos\theta' \quad S' = \sin\theta'$$
$$c = \cos\phi \quad s = \sin\phi$$
$$c' = \cos\phi' \quad s' = \sin\phi'$$

The coordinates for the deformation gradient tensor $\underline{\underline{F}}$ and its inverse $\underline{\underline{F}}^{-1}$ in spherical coordinates may be calculated from the matrix equations

$$\underline{\underline{F}} = \begin{pmatrix} F_{rr} & F_{r\theta} & F_{r\phi} \\ F_{\theta r} & F_{\theta\theta} & F_{\theta\phi} \\ F_{\phi r} & F_{\phi\theta} & F_{\phi\phi} \end{pmatrix}_{r\theta\phi} = \begin{bmatrix} \frac{\partial r'}{\partial r} & r'\frac{\partial\theta'}{\partial r} & r'\sin\theta'\frac{\partial\phi'}{\partial r} \\ \frac{1}{r}\frac{\partial r'}{\partial\theta} & \frac{r'}{r}\frac{\partial\theta'}{\partial\theta} & \frac{r'\sin\theta'}{r}\frac{\partial\phi'}{\partial\theta} \\ \frac{1}{r\sin\theta}\frac{\partial r'}{\partial\phi} & \frac{r'}{r\sin\theta}\frac{\partial\theta'}{\partial\phi} & \frac{r'\sin\theta'}{r\sin\theta}\frac{\partial\phi'}{\partial\phi} \end{bmatrix} \begin{bmatrix} T_{rr} & T_{r\theta} & T_{r\phi} \\ T_{\theta r} & T_{\theta\theta} & T_{\theta\phi} \\ T_{\phi r} & T_{\phi\theta} & T_{\phi\phi} \end{bmatrix}^T$$

$$\underline{\underline{F}}^{-1} = \begin{pmatrix} F_{rr}^{-1} & F_{r\theta}^{-1} & F_{r\phi}^{-1} \\ F_{\theta r}^{-1} & F_{\theta\theta}^{-1} & F_{\theta\phi}^{-1} \\ F_{\phi r}^{-1} & F_{\phi\theta}^{-1} & F_{\phi\phi}^{-1} \end{pmatrix}_{r\theta\phi} = \begin{bmatrix} T_{rr} & T_{r\theta} & T_{r\phi} \\ T_{\theta r} & T_{\theta\theta} & T_{\theta\phi} \\ T_{\phi r} & T_{\phi\theta} & T_{\phi\phi} \end{bmatrix} \begin{bmatrix} \frac{\partial r}{\partial r'} & r\frac{\partial\theta}{\partial r'} & r\sin\theta\frac{\partial\phi}{\partial r'} \\ \frac{1}{r'}\frac{\partial r}{\partial\theta'} & \frac{r}{r'}\frac{\partial\theta}{\partial\theta'} & \frac{r\sin\theta}{r'}\frac{\partial\phi}{\partial\theta'} \\ \frac{1}{r'\sin\theta'}\frac{\partial r}{\partial\phi'} & \frac{r}{r'\sin\theta'}\frac{\partial\theta}{\partial\phi'} & \frac{r\sin\theta}{r'\sin\theta'}\frac{\partial\phi}{\partial\phi'} \end{bmatrix}$$

Since $T_{\text{spherical}}$ is an orthogonal matrix it cancels out when either $\underline{\underline{C}}^{-1}$ or $\underline{\underline{C}}$ is calculated from the deformation gradient tensor or its inverse.

C.5 Coordinate Transformations of Orthonormal Bases

An important property of vectors and tensors is that they transform between coordinate systems while retaining their intrinsic meaning. An example of this is a velocity vector: the magnitude of the speed and the actual direction of the motion of a body are independent of the coordinate system, although the velocity vector may be expressed in any of an infinite number of coordinate systems. It is often desirable to consider the coefficients of vectors and tensors in more than one coordinate system. In this section we describe how to calculate the coefficients of vectors and tensors in an orthonormal coordinate system rotated with respect to the original orthonormal coordinate system [7].

Let $\hat{e}_1$, $\hat{e}_2$, $\hat{e}_3$ be the original coordinate system, and $\bar{e}_1$, $\bar{e}_2$, $\bar{e}_3$ the rotated coordinate system (Figure C.2). Both systems are orthonormal. We begin by writing an arbitrary vector $\underline{v}$ in each of the coordinate systems:

$$\underline{v} = v_1\hat{e}_1 + v_2\hat{e}_2 + v_3\hat{e}_3 \tag{C.35}$$

$$\underline{v} = \bar{v}_1\bar{e}_1 + \bar{v}_2\bar{e}_2 + \bar{v}_3\bar{e}_3 \tag{C.36}$$

We now form the dot product of the original unit vectors $\hat{e}_i$ with each of these two expressions and equate them. We show the case for $i = 1$:

$$\hat{e}_1 \cdot \underline{v} = v_1 = \bar{v}_1(\hat{e}_1 \cdot \bar{e}_1) + \bar{v}_2(\hat{e}_1 \cdot \bar{e}_2) + \bar{v}_3(\hat{e}_1 \cdot \bar{e}_3) \tag{C.37}$$

$$= \bar{v}_1 L_{11} + \bar{v}_2 L_{12} + \bar{v}_3 L_{13} \tag{C.38}$$

where we have defined $L_{ij} \equiv \hat{e}_i \cdot \bar{e}_j$. Similar expressions can be written for $i = 2$ and 3, and we obtain three equations that can be expressed in matrix form:

$$v_1 = \bar{v}_1 L_{11} + \bar{v}_2 L_{12} + \bar{v}_3 L_{13} \tag{C.39}$$

$$v_2 = \bar{v}_1 L_{21} + \bar{v}_2 L_{22} + \bar{v}_3 L_{23} \tag{C.40}$$

$$v_3 = \bar{v}_1 L_{31} + \bar{v}_2 L_{32} + \bar{v}_3 L_{33} \tag{C.41}$$

$$\begin{bmatrix} v_1 \\ v_2 \\ v_3 \end{bmatrix} = \begin{bmatrix} L_{11} & L_{12} & L_{13} \\ L_{21} & L_{22} & L_{23} \\ L_{31} & L_{32} & L_{33} \end{bmatrix} \begin{bmatrix} \bar{v}_1 \\ \bar{v}_2 \\ \bar{v}_3 \end{bmatrix} \tag{C.42}$$

Then

Vector coordinate
transformation
$$\boxed{v = L\bar{v}} \tag{C.43}$$

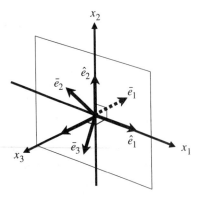

Figure C.2 Schematic of the two different orthonormal coordinate systems considered in Section C.5.

where v and $\bar{v}$ refer to the 3×1 matrices containing the coefficients of $\underline{v}$ in the $\hat{e}$ and $\bar{e}$ coordinate systems, respectively, and L is the 3×3 matrix containing the elements L_{ij}. In these expressions we have used square brackets to enclose the elements of L in order to distinguish this matrix from matrices of tensor coefficients, for which we have been using rounded parentheses. There are no unit vectors $\hat{e}_p \hat{e}_k$ associated with the elements L_{pk} of L. This is an important distinction.

As we see from these equations, the 3×3 matrix L may be used to calculate the coefficients v_i of a vector $\underline{v}$ in one coordinate system from the coefficients $\bar{v}_i$ of $\underline{v}$ in another coordinate system, which is rotated with respect to the first. We derived the expression $L\bar{v} = v$ by taking the dot product of the $\hat{e}_1, \hat{e}_2, \hat{e}_3$ unit vectors with $\underline{v}$ expressed in the two coordinate systems. By taking the dot products of the $\bar{e}_1, \bar{e}_2, \bar{e}_3$ unit vectors with $\underline{v}$ in a similar manner, we can calculate the inverse relationship:

$$\bar{v}_1 = v_1 L_{11} + v_2 L_{21} + v_3 L_{31} \tag{C.44}$$

$$\bar{v}_2 = v_1 L_{12} + v_2 L_{22} + v_3 L_{32} \tag{C.45}$$

$$\bar{v}_3 = v_1 L_{13} + v_2 L_{23} + v_3 L_{33} \tag{C.46}$$

$$\begin{bmatrix} \bar{v}_1 \\ \bar{v}_2 \\ \bar{v}_3 \end{bmatrix} = \begin{bmatrix} L_{11} & L_{21} & L_{31} \\ L_{12} & L_{22} & L_{32} \\ L_{13} & L_{23} & L_{33} \end{bmatrix} \begin{bmatrix} v_1 \\ v_2 \\ v_3 \end{bmatrix} \tag{C.47}$$

Then

$$\text{Inverse vector coordinate transformation} \qquad \boxed{\bar{v} = L^T v} \tag{C.48}$$

where L^T is the transpose of L. We can manipulate the matrix equations for the forward and reverse transformations to conclude the following:

$$v = L\bar{v} \tag{C.49}$$

$$\bar{v} = L^T v \tag{C.50}$$

$$v = LL^T v \tag{C.51}$$

For Equation (C.51) to be true, LL^T must equal the identity matrix I, and thus the transpose of L is also its inverse, $L^T = L^{-1}$. A matrix whose transpose is also its inverse is called an orthogonal matrix.

We can deduce the effect of coordinate rotation on the coefficients of a tensor by an analogous procedure:

$$\underline{\underline{A}} = A_{pm} \hat{e}_p \hat{e}_m = \bar{A}_{ij} \bar{e}_i \bar{e}_j \tag{C.52}$$

$$\hat{e}_k \cdot A_{pm} \hat{e}_p \hat{e}_m = \hat{e}_k \cdot \bar{A}_{ij} \bar{e}_i \bar{e}_j \tag{C.53}$$

$$A_{km} \hat{e}_m = \bar{A}_{ij} (\hat{e}_k \cdot \bar{e}_i) \bar{e}_j = \bar{A}_{ij} L_{ki} \bar{e}_j \tag{C.54}$$

$$\hat{e}_s \cdot A_{km} \hat{e}_m = \bar{A}_{ij} L_{ki} (\hat{e}_s \cdot \bar{e}_j) \tag{C.55}$$

$$A_{ks} = \bar{A}_{ij} L_{ki} L_{sj} \tag{C.56}$$

Then

$$\text{Tensor coordinate transformation} \qquad \boxed{A = L\bar{A}L^T} \qquad (C.57)$$

where A is the matrix of the coefficients of $\underline{\underline{A}}$ in the $\hat{e}$ coordinate system, and $\bar{A}$ is the matrix of the coefficients of $\underline{\underline{A}}$ in the $\bar{e}$ coordinate system. Equation (C.57) was arrived at by requiring that the summations properly follow the rules of matrix multiplication. The inverse transformation is easy to arrive at since $L^{-1} = L^T$:

$$\text{Inverse tensor coordinate transformation} \qquad \boxed{\bar{A} = L^T AL} \qquad (C.58)$$

C.6 Finding Principal Values

Consider a symmetric tensor $\underline{\underline{A}}$ written in Cartesian coordinates x_1, x_2, x_3:

$$\underline{\underline{A}} = \begin{pmatrix} A_{11} & A_{12} & A_{13} \\ A_{21} & A_{22} & A_{23} \\ A_{31} & A_{32} & A_{33} \end{pmatrix}_{123} \qquad (C.59)$$

One important coordinate system for this tensor is that in which the tensor is diagonal, that is, when

$$\underline{\underline{A}} = \begin{pmatrix} \lambda_1 & 0 & 0 \\ 0 & \lambda_2 & 0 \\ 0 & 0 & \lambda_3 \end{pmatrix}_{\xi_1\xi_2\xi_3} \qquad (C.60)$$

The coordinate system in which the only nonzero coefficients of $\underline{\underline{A}}$ are along the diagonal is called the *principal frame*, and the basis vectors of this system, $\hat{\xi}_1, \hat{\xi}_2, \hat{\xi}_3$, are called the *principal axes*. The subscript $\xi_1\xi_2\xi_3$ on the matrix of tensor coefficients in Equation (C.60) indicates that the coefficients of $\underline{\underline{A}}$ are written in the $\hat{\xi}_1, \hat{\xi}_2, \hat{\xi}_3$ coordinate system. Using the methods of Appendix C.5, we seek to find the rotation matrix L that transforms the coefficients A of the tensor $\underline{\underline{A}}$, written in a known Cartesian frame, to the coefficients Λ of $\underline{\underline{A}}$, written in the Cartesian principal frame. Such a transformation can always be found for a symmetric tensor $\underline{\underline{A}}$ [7, 5a]:

$$[A]_{123} \equiv A = \begin{bmatrix} A_{11} & A_{12} & A_{13} \\ A_{21} & A_{22} & A_{23} \\ A_{31} & A_{32} & A_{33} \end{bmatrix} \qquad (C.61)$$

$$[A]_{\xi_1\xi_2\xi_3} \equiv \Lambda = \begin{bmatrix} \lambda_1 & 0 & 0 \\ 0 & \lambda_2 & 0 \\ 0 & 0 & \lambda_3 \end{bmatrix} \qquad (C.62)$$

$$L^T AL = \Lambda \qquad (C.63)$$

Once we know L, we can use Equation (C.43) to calculate the 123 coefficients of the basis vectors $\hat{\xi}_1, \hat{\xi}_2, \hat{\xi}_3$ as follows:

$$[\xi_i]_{123} = L[\xi_i]_{\xi_1\xi_2\xi_3} \tag{C.64}$$

In the new (principal) frame, the coefficients of the basis vectors $\hat{\xi}_i$ are quite simple:

$$\hat{\xi}_1 = \begin{pmatrix} 1 \\ 0 \\ 0 \end{pmatrix}_{\xi_1\xi_2\xi_3} , \qquad \hat{\xi}_2 = \begin{pmatrix} 0 \\ 1 \\ 0 \end{pmatrix}_{\xi_1\xi_2\xi_3} , \qquad \hat{\xi}_3 = \begin{pmatrix} 0 \\ 0 \\ 1 \end{pmatrix}_{\xi_1\xi_2\xi_3} \tag{C.65}$$

Using the coefficients for $\hat{\xi}_i$ given in Equation (C.65) and combining them with Equation (C.64), we see that the coefficients of $\hat{\xi}_1$ in the 123 coordinate system are represented by the first column of L, and similarly, the coefficients of $\hat{\xi}_2$ and $\hat{\xi}_3$ in the 123 coordinate system are contained in the second and third columns of L, respectively.

To solve for Λ, we can left-multiply Equation (C.63) by L to obtain (recall that $L^T = L^{-1}$)

$$AL = L\Lambda \tag{C.66}$$

Written out, this becomes

$$\begin{bmatrix} A_{11} & A_{12} & A_{13} \\ A_{21} & A_{22} & A_{23} \\ A_{31} & A_{32} & A_{33} \end{bmatrix} \begin{bmatrix} L_{11} & L_{12} & L_{13} \\ L_{21} & L_{22} & L_{23} \\ L_{31} & L_{32} & L_{33} \end{bmatrix} = \begin{bmatrix} L_{11} & L_{12} & L_{13} \\ L_{21} & L_{22} & L_{23} \\ L_{31} & L_{32} & L_{33} \end{bmatrix} \begin{bmatrix} \lambda_1 & 0 & 0 \\ 0 & \lambda_2 & 0 \\ 0 & 0 & \lambda_3 \end{bmatrix} \tag{C.67}$$

$$= \begin{bmatrix} \lambda_1 L_{11} & \lambda_2 L_{12} & \lambda_3 L_{13} \\ \lambda_1 L_{21} & \lambda_2 L_{22} & \lambda_3 L_{23} \\ \lambda_1 L_{31} & \lambda_2 L_{32} & \lambda_3 L_{33} \end{bmatrix} \tag{C.68}$$

This is equivalent to the three matrix equations

$$A \begin{bmatrix} L_{11} \\ L_{21} \\ L_{31} \end{bmatrix} = \lambda_1 \begin{bmatrix} L_{11} \\ L_{21} \\ L_{31} \end{bmatrix} \tag{C.69}$$

$$A \begin{bmatrix} L_{12} \\ L_{22} \\ L_{32} \end{bmatrix} = \lambda_2 \begin{bmatrix} L_{12} \\ L_{22} \\ L_{32} \end{bmatrix} \tag{C.70}$$

$$A \begin{bmatrix} L_{13} \\ L_{23} \\ L_{33} \end{bmatrix} = \lambda_3 \begin{bmatrix} L_{13} \\ L_{23} \\ L_{33} \end{bmatrix} \tag{C.71}$$

which can be written more compactly as

$$A[\xi_i]_{123} = \lambda_i [\xi_i]_{123} \tag{C.72}$$

where

$$[\xi_i]_{123} = \begin{bmatrix} L_{1i} \\ L_{2i} \\ L_{3i} \end{bmatrix} \quad \text{for } i = 1, 2, 3$$

Some readers may recognize that this is just the eigenvalue problem of linear algebra. For each column vector $[\xi_i]_{123} = \xi$ (eigenvector) and associated constant $\lambda_i = \lambda$ (eigenvalue), we can write

$$A\xi = \lambda\xi \tag{C.73}$$

$$A\xi = \lambda I \xi \tag{C.74}$$

$$(A - \lambda I)\xi = 0 \tag{C.75}$$

which for nonzero column vector ξ has solutions when the determinant of $A - \lambda I$ equals zero.

To solve for λ_i we solve the characteristic equation, which is a cubic polynomial that results from setting the determinant of $A - \lambda I$ equal to zero:

$$\det|A - \lambda I| = 0 \tag{C.76}$$

$$\Psi - \lambda\Phi + \lambda^2\Theta + \lambda^3 = 0 \tag{C.77}$$

where

$$\Theta \equiv A_{ii} = \text{trace } A \tag{C.78}$$

$$\Phi \equiv A_{22}A_{33} - A_{23}A_{32} + A_{33}A_{11}$$
$$- A_{31}A_{13} + A_{11}A_{22} - A_{12}A_{21} \tag{C.79}$$

$$\Psi \equiv \det|A| \tag{C.80}$$

The roots of the characteristic equation are the three eigenvalues $\lambda_1, \lambda_2, \lambda_3$. The eigenvectors $[\xi_i]_{123}$ can be found from Equation (C.72) and our requirement that $|\hat{\xi}_i| = 1$. The matrix L is assembled from the $[\xi_i]_{123}$. Note that the eigenvalues λ_i are real because we are limiting our discussion to symmetric tensors $\underline{\underline{A}}$ [5a].

The eigenvalues λ_i and eigenvectors $\underline{\xi}_i$ are uniquely associated with the tensor $\underline{\underline{A}}$. The characteristic equation, Equation (C.77), is also uniquely associated with $\underline{\underline{A}}$ and yields the same results, regardless of the coordinate system in which $\underline{\underline{A}}$ is expressed. Thus Θ, Ψ, and Φ are scalar invariants associated with $\underline{\underline{A}}$. The definitions for the invariants of a tensor given in Section 2.3.4 are related to Θ, Ψ, and Φ as shown:

$$\Theta = I_1 = I_{\underline{\underline{A}}} = \text{trace}\underline{\underline{A}} \tag{C.81}$$

$$\Phi = I_2 = \frac{1}{2}\left(I_{\underline{\underline{A}}}^2 - II_{\underline{\underline{A}}}\right) = \frac{1}{2}\left[\left(\text{trace } \underline{\underline{A}}\right)^2 - \text{trace}\left(\underline{\underline{A}}^2\right)\right] \tag{C.82}$$

$$\Psi = I_3 = \frac{1}{6}\left(I_{\underline{\underline{A}}}^3 - 3I_{\underline{\underline{A}}}II_{\underline{\underline{A}}} + 2III_{\underline{\underline{A}}}\right) = \det|\underline{\underline{A}}| \tag{C.83}$$

Θ, Ψ, and Φ are also called I_1, I_2, and I_3.

EXAMPLE

Find the principal coordinate system for the tensor $\underline{\underline{B}}$. Express $\underline{\underline{B}}$ in the principal system.

$$\underline{\underline{B}} = \begin{pmatrix} 1 & 0 & 4 \\ 0 & -1 & 2 \\ 4 & 2 & 0 \end{pmatrix}_{123} \tag{C.84}$$

SOLUTION

To find the eigenvalues, we must solve the characteristic equation:

$$0 = (B - \lambda I)\xi \tag{C.85}$$

$$\det|B - \lambda I| = 0 \tag{C.86}$$

$$= \begin{vmatrix} 1 - \lambda & 0 & 4 \\ 0 & -1 - \lambda & 2 \\ 4 & 2 & -\lambda \end{vmatrix} \tag{C.87}$$

$$= \lambda^3 - 21\lambda - 12 \tag{C.88}$$

and the eigenvalues are given by

$$\lambda_1 = -0.581 \tag{C.89}$$

$$\lambda_2 = -4.264 \tag{C.90}$$

$$\lambda_3 = 4.845 \tag{C.91}$$

The eigenvectors can be found as the solutions to the following equation for each of the eigenvalues:

$$B\xi = \lambda\xi \tag{C.92}$$

$$\xi = \begin{pmatrix} a \\ b \\ c \end{pmatrix}_{123} \tag{C.93}$$

$$\begin{bmatrix} 1 & 0 & 4 \\ 0 & -1 & 2 \\ 4 & 2 & 0 \end{bmatrix} \begin{bmatrix} a \\ b \\ c \end{bmatrix} = \lambda \begin{bmatrix} a \\ b \\ c \end{bmatrix} \tag{C.94}$$

Following this and imposing the condition that $|\hat{\xi}_i| = 1$, we obtain

$$\hat{\xi}_1 = \begin{pmatrix} 0.461 \\ -0.869 \\ -0.182 \end{pmatrix}_{123} , \quad \hat{\xi}_2 = \begin{pmatrix} -0.544 \\ -0.438 \\ 0.716 \end{pmatrix}_{123} , \quad \hat{\xi}_3 = \begin{pmatrix} 0.702 \\ 0.231 \\ 0.674 \end{pmatrix}_{123} \tag{C.95}$$

$\underline{\underline{B}}$ expressed in the principal system is

$$\underline{\underline{B}} = \begin{pmatrix} \lambda_1 & 0 & 0 \\ 0 & \lambda_2 & 0 \\ 0 & 0 & \lambda_3 \end{pmatrix}_{\xi_1\xi_2\xi_3} = \begin{pmatrix} -0.581 & 0 & 0 \\ 0 & -4.264 & 0 \\ 0 & 0 & 4.845 \end{pmatrix}_{\xi_1\xi_2\xi_3} \quad (C.96)$$

This calculation has been programmed into most common math software packages [170, 169, 267].

C.7 Contravariant and Covariant Transformations of Tensors

In this text we have defined tensors with respect to the indeterminate vector product, for example, $\underline{\underline{A}} = \underline{u}\,\underline{v} = A_{ij}\hat{e}_i\hat{e}_j$. Tensors have a wider utility and meaning than was employed here, however [222]. One set of terminology from general tensor analysis that has found its way into this introductory rheology text are the terms *contravariant* and *covariant* coefficients of tensors and vectors (Section 9.3.2). In this section we will explain a bit more about this terminology. You are invited to consult the literature to learn more about tensor analysis, both in rheology [157] and in general [166, 7, 222].

The primary property of a tensor (including first-order tensors, i.e., vectors) is that a tensor is a mathematical quantity that is independent of a coordinate system. Figure C.3 shows a point P as referenced to two different coordinate systems, one an arbitrary one

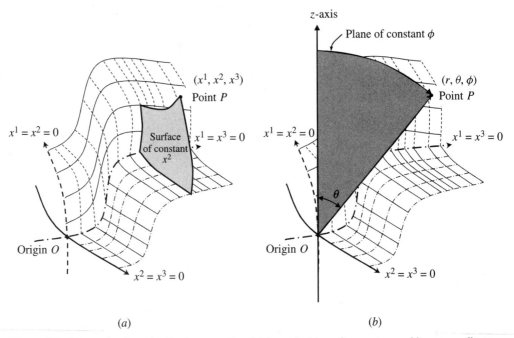

(a) (b)

Figure C.3 Schematic of a point P whose position (a) is marked by reference to an arbitrary coordinate system x^1, x^2, x^3 and also (b) is located independently by reference to a spherical coordinate system with the same origin.

with coordinates x^1, x^2, x^3 and the second the spherical coordinate system with coordinates r, θ, ϕ.[1] A tensor property, by definition, can be expressed in any coordinate system. How the tensor coefficients transform from one coordinate system to another depends on the coordinate systems, however. We saw in Section C.5 how vector and tensor coefficients transform between orthonormal coordinate systems. For nonorthonormal coordinate systems the situation is more complicated, and there are different types of coordinate transformation rules. Two patterns of how coefficients transform will be discussed here: contravariant transformations and covariant transformations.

C.7.1 CONTRAVARIANT TENSORS

Consider a point P with the coordinates x^1, x^2, x^3 in one coordinate system and the coordinates θ^1, θ^2, θ^3 in a second coordinate system with the same origin. Both coordinate systems may be of the most general type, that is, we do not need them to be Cartesian systems.

In order to transform tensor properties from one coordinate system to another, we must find a set of equations that allow us to perform the transformation. The equations that allow us to transform the x^1, x^2, x^3 coordinates into the θ^1, θ^2, θ^3 coordinates can be written as functions as follows:

$$\theta^1 = \theta^1(x^1, x^2, x^3) \tag{C.97}$$

$$\theta^2 = \theta^2(x^1, x^2, x^3) \tag{C.98}$$

$$\theta^3 = \theta^3(x^1, x^2, x^3) \tag{C.99}$$

[For an example of a set of equations of this type involving the Cartesian and cylindrical coordinate systems, see Equations (2.242)–(2.244).] These functions are assumed to be single-valued and differentiable. We would also like to do the reverse transformation, that is, transform properties written in the θ^i system to the x^i system. It is not always possible to find such a unique inversion, since not all sets of three equations in three unknowns [Equations (C.97)–(C.99)] yield unique solutions. The inversion does exist, however, if the following determinant is not zero [166]:

$$J \equiv \begin{vmatrix} \dfrac{\partial \theta^1}{\partial x^1} & \dfrac{\partial \theta^1}{\partial x^2} & \dfrac{\partial \theta^1}{\partial x^3} \\[2mm] \dfrac{\partial \theta^2}{\partial x^1} & \dfrac{\partial \theta^2}{\partial x^2} & \dfrac{\partial \theta^2}{\partial x^3} \\[2mm] \dfrac{\partial \theta^3}{\partial x^1} & \dfrac{\partial \theta^3}{\partial x^2} & \dfrac{\partial \theta^3}{\partial x^3} \end{vmatrix} \neq 0 \tag{C.100}$$

This determinant is denoted by J and is called the *Jacobian* of the transformation.

If the Jacobian is not zero, then the inverse functions to transform from the θ^i system to the x^i system exist:

$$x^1 = x^1(\theta^1, \theta^2, \theta^3) \tag{C.101}$$

$$x^2 = x^2(\theta^1, \theta^2, \theta^3) \tag{C.102}$$

---

[1] The use of superscripts to index contravariant coordinates is standard notation in general tensor analysis. The superscripts do not indicate powers of quantities, which will be designated using parentheses, e.g., $(x^1)^2$. As we will see later, subscripts will be used to designate covariant tensors.

$$x^3 = x^3(\theta^1, \theta^2, \theta^3) \tag{C.103}$$

Now, associated with point P is some property that is composed of a set of three scalars, for example, the velocity components of a fluid passing through that point or the force components in three directions on a particular surface passing through that point. In the x^1, x^2, x^3 coordinate system the property is given by the set of three scalars p^1, p^2, p^3, or collectively p^i. By definition, the property p^i is a first-order contravariant tensor if the coefficients p^i in the x^i system transform to the coefficients π^i in the θ^i system according to the following equations:

$$\pi^1 = \frac{\partial\theta^1}{\partial x^1}p^1 + \frac{\partial\theta^1}{\partial x^2}p^2 + \frac{\partial\theta^1}{\partial x^3}p^3 \tag{C.104}$$

$$\pi^2 = \frac{\partial\theta^2}{\partial x^1}p^1 + \frac{\partial\theta^2}{\partial x^2}p^2 + \frac{\partial\theta^2}{\partial x^3}p^3 \tag{C.105}$$

$$\pi^3 = \frac{\partial\theta^3}{\partial x^1}p^1 + \frac{\partial\theta^3}{\partial x^2}p^2 + \frac{\partial\theta^3}{\partial x^3}p^3 \tag{C.106}$$

In Einstein notation, where a summation from 1 to 3 is implied when subscripts are repeated, these contravariant transformation rules can be written as

$$\pi^1 = \sum_{k=1}^{3} \frac{\partial\theta^1}{\partial x^k}p^k = \frac{\partial\theta^1}{\partial x^k}p^k \tag{C.107}$$

$$\pi^2 = \sum_{k=1}^{3} \frac{\partial\theta^2}{\partial x^k}p^k = \frac{\partial\theta^2}{\partial x^k}p^k \tag{C.108}$$

$$\pi^3 = \sum_{k=1}^{3} \frac{\partial\theta^3}{\partial x^k}p^k = \frac{\partial\theta^3}{\partial x^k}p^k \tag{C.109}$$

or even more compactly as

Coordinate transformation rule for a first-order contravariant tensor p^k
$$\boxed{\pi^j = \frac{\partial\theta^j}{\partial x^k}p^k} \qquad \text{for } j = 1, 2, 3 \tag{C.110}$$

The rule for coordinate transformation for a second-order contravariant tensor P^{km} is defined to be

Coordinate transformation rule for a second-order contravariant tensor P^{km}
$$\boxed{\Pi^{ij} = \frac{\partial\theta^i}{\partial x^k}\frac{\partial\theta^j}{\partial x^m}P^{km}} \qquad \begin{aligned} \text{for } i &= 1, 2, 3 \\ j &= 1, 2, 3 \end{aligned} \tag{C.111}$$

where Π^{ij} are the coefficients of the tensor P^{km} in the $\theta^1, \theta^2, \theta^3$ coordinate system.

EXAMPLE

Are the coefficients of vectors and tensors written with respect to the convected coordinates $\underline{g}_{(1)}, \underline{g}_{(2)}, \underline{g}_{(3)}$ defined in Chapter 9 contravariant tensors as defined by Equations (C.110) and (C.111)? Show why or why not.

SOLUTION

Vectors and tensors are frame-independent quantities that can be written in any coordinate system, for example, as follows:

$$\text{vector} \qquad \underline{v} = \sum_{p=1}^{3} v^p \underline{g}_{(p)} \qquad \text{(C.112)}$$

$$\text{2nd-order tensor} \qquad \underline{\underline{A}} = \sum_{m=1}^{3} \sum_{k=1}^{3} A^{mk} \underline{g}_{(m)} \underline{g}_{(k)} \qquad \text{(C.113)}$$

where $\underline{g}_{(1)}$, $\underline{g}_{(2)}$, and $\underline{g}_{(3)}$ are basis vectors that are not necessarily mutually orthogonal nor of unit length. We have chosen to write the indices on the coefficients in the superscript position, which indicates contravariant coefficients, in anticipation of our findings in this practice problem; the choice is just one of notation. To see if these quantities are contravariant tensors as defined, we need to examine how the coefficients v^p and A^{mk} transform from one coordinate system to another.

Let $\widehat{x}^1, \widehat{x}^2$, and $\widehat{x}^3$ be the coordinate variables associated with the basis vectors $\underline{g}_{(1)}, \underline{g}_{(2)}$, and $\underline{g}_{(3)}$, respectively, and let $\underline{c}_{(1)}, \underline{c}_{(2)}, \underline{c}_{(3)}$ and $\theta^1, \theta^2, \theta^3$ be basis vectors and coordinate variables for another, arbitrary coordinate system with the same origin. We choose that the Jacobian of the transformation between the two chosen coordinate systems be nonzero. Let the vector $\underline{r}$ be a position vector that points from the origin to an arbitrary point. In Chapter 9 [see Equations (9.226)–(9.228)] we defined the convected basis vectors $\underline{g}_{(i)}$ as follows:

$$\underline{g}_{(i)} \equiv \frac{\partial \underline{r}}{\partial \widehat{x}^i} \qquad \text{(C.114)}$$

These basis vectors are defined such that they indicate the direction of the increasing coordinate variable $\widehat{x}^i$ at a point (Figure C.4). The $\underline{c}_{(j)}$ basis vectors are defined similarly with respect to the θ^i-coordinates:

$$\underline{c}_{(j)} \equiv \frac{\partial \underline{r}}{\partial \theta^j} \qquad \text{(C.115)}$$

Substituting the definition for $\underline{g}_{(i)}$ [Equation (C.114)] into Equations (C.112) and (C.113) we obtain

$$\underline{v} = \sum_{p=1}^{3} v^p \frac{\partial \underline{r}}{\partial \widehat{x}^p} \qquad \text{(C.116)}$$

$$\underline{\underline{A}} = \sum_{m=1}^{3} \sum_{k=1}^{3} A^{mk} \frac{\partial \underline{r}}{\partial \widehat{x}^m} \frac{\partial \underline{r}}{\partial \widehat{x}^k} \qquad \text{(C.117)}$$

In these equations we can write the vector $\underline{r}$ in any coordinate system, and we now choose to write $\underline{r}$ in the $\underline{c}_{(j)}$ coordinate system. Note that the $\underline{c}_{(j)}$ are not functions of $\widehat{x}^i$, and they are unaffected by the derivatives with respect to $\widehat{x}^i$:

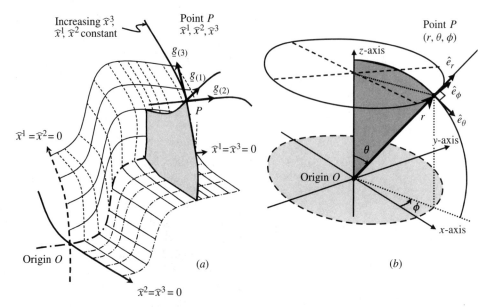

Figure C.4 Schematic showing the relationship between basis vectors and coordinate lines (lines for which two coordinate variables are held constant) for (*a*) the convected coordinate system $\underline{g}_{(1)}, \underline{g}_{(2)}, \underline{g}_{(3)}$ and (*b*) for comparison, for the spherical coordinate system.

$$\underline{r} = \sum_{j=1}^{3} \theta^j \underline{c}_{(j)} \tag{C.118}$$

$$\underline{v} = \sum_{p=1}^{3} \sum_{j=1}^{3} v^p \frac{\partial \theta^j}{\partial \widehat{x}^p} \, \underline{c}_{(j)} \tag{C.119}$$

$$\underline{\underline{A}} = \sum_{m=1}^{3} \sum_{k=1}^{3} \sum_{j=1}^{3} \sum_{s=1}^{3} A^{mk} \frac{\partial \theta^j}{\partial \widehat{x}^m} \frac{\partial \theta^s}{\partial \widehat{x}^k} \, \underline{c}_{(j)} \underline{c}_{(s)} \tag{C.120}$$

We will choose to use overbar notation for the coefficients of $\underline{v}$ and $\underline{\underline{A}}$ in the $\underline{c}_{(1)}, \underline{c}_{(2)}, \underline{c}_{(3)}$ coordinate system, $\bar{v}^j$ and $\bar{A}^{js}$. Thus

$$\underline{v} = \sum_{j=1}^{3} \bar{v}^j \underline{c}_{(j)} \tag{C.121}$$

$$\underline{\underline{A}} = \sum_{j=1}^{3} \sum_{s=1}^{3} \bar{A}^{js} \underline{c}_{(j)} \underline{c}_{(s)} \tag{C.122}$$

Comparing Equations (C.119) and (C.121) we see that the $\underline{g}_{(i)}$ coefficients of $\underline{v}$ (that is, v^i) transform to the $\underline{c}_{(i)}$ coefficients ($\bar{v}^i$) as follows:

$$\bar{v}^j = \sum_{p=1}^{3} v^p \frac{\partial \theta^j}{\partial \widehat{x}^p} \quad \text{for } j = 1, 2, 3 \tag{C.123}$$

which is a first-order contravariant tensor transformation [compare with Equation (C.110)]. For the $\underline{\underline{A}}$ transformation we compare Equations (C.120) and (C.122) and obtain

$$\sum_{m=1}^{3} \sum_{k=1}^{3} A^{mk} \frac{\partial \theta^j}{\partial \widehat{x}^m} \frac{\partial \theta^s}{\partial \widehat{x}^k} = \bar{A}^{js} \tag{C.124}$$

which follows the rule for the transformation of a second-order contravariant tensor [Equation (C.111)].

C.7.2 COVARIANT TENSORS

Another type of transformation law is the covariant transformation. A first-order covariant tensor is a quantity for which the coefficients p_k in the x^k system transform to the coefficients π_j in the θ^j system according to the following transformation rules:

$$\pi_1 = \frac{\partial x^1}{\partial \theta^1} p_1 + \frac{\partial x^2}{\partial \theta^1} p_2 + \frac{\partial x^3}{\partial \theta^1} p_3 \tag{C.125}$$

$$\pi_2 = \frac{\partial x^1}{\partial \theta^2} p_1 + \frac{\partial x^2}{\partial \theta^2} p_2 + \frac{\partial x^3}{\partial \theta^2} p_3 \tag{C.126}$$

$$\pi_3 = \frac{\partial x^1}{\partial \theta^3} p_1 + \frac{\partial x^2}{\partial \theta^3} p_2 + \frac{\partial x^3}{\partial \theta^3} p_3 \tag{C.127}$$

which in Einstein notation is

Coordinate transformation
rule for a first-order
covariant tensor p_k
$$\boxed{\pi_j = \frac{\partial x^k}{\partial \theta^j} p_k} \quad \text{for } j = 1, 2, 3 \tag{C.128}$$

In general tensor theory covariant tensors are indexed with subscripts to differentiate these coefficients from superscripted contravariant tensors. The rule for coordinate transformation for a second-order covariant tensor P_{km} is defined to be

Coordinate transformation
rule for a second-order
covariant tensor P_{km}
$$\boxed{\Pi_{ij} = \frac{\partial x^k}{\partial \theta^i} \frac{\partial x^m}{\partial \theta^j} P_{km}} \quad \begin{array}{l} \text{for } i = 1, 2, 3 \\ j = 1, 2, 3 \end{array} \tag{C.129}$$

where Π_{ij} are the coefficients of the tensor P_{km} in the $\theta^1, \theta^2, \theta^3$ coordinate system.

In Chapter 9 we encountered covariant coefficients of vectors and tensors when we discussed the coefficients of tensors written with respect to the reciprocal basis vectors $\underline{g}^{(1)}, \underline{g}^{(2)}, \underline{g}^{(3)}$. The different transformation rules for the regular and reciprocal bases result from the different relationships between the two sets of basis vectors $\underline{g}_{(i)}, \underline{c}_{(k)}$ and $\underline{g}^{(i)}, \underline{c}^{(k)}$ and the single set of coordinate variables $\widehat{x}^i, \theta^k$. The basis vectors $\underline{g}_{(i)} \equiv \partial \underline{r}/\partial \widehat{x}^i$ and

$c_{(k)} \equiv \partial r / \partial \theta^k$ are tangent to the lines of increasing coordinate variables $\hat{x}^i, \theta^k$ at a point. The reciprocal base vectors, defined as

$$g^{(i)} \equiv \frac{g_{(j)} \times g_{(k)}}{g_{(1)} \cdot g_{(2)} \times g_{(3)}} \qquad \text{for } ijk = 123, 312, 231 \qquad (C.130)$$

$$c^{(i)} \equiv \frac{c_{(j)} \times c_{(k)}}{c_{(1)} \cdot c_{(2)} \times c_{(3)}} \qquad \text{for } ijk = 123, 312, 231 \qquad (C.131)$$

are perpendicular to the surfaces of constant coordinate variables $\hat{x}^1, \hat{x}^2, \hat{x}^3$ for the $g^{(i)}$ and $\theta^1, \theta^2, \theta^3$ for the $c^{(i)}$ (Figure C.5). We saw in the last section that the coefficients of a tensor written with respect to the $g_{(i)}$ transform a coordinate system according to Equation (C.110) and are therefore contravariant. The reciprocal basis vectors $g^{(i)}$ are orthogonal to the $g_{(i)}$, and it is not surprising that vector and tensor coefficients written with respect to the reciprocal basis do not transform in the same way. Using the definitions of the reciprocal bases [Equations (C.130) and (C.131)] and the transformation rules for the $g_{(i)}$, one can

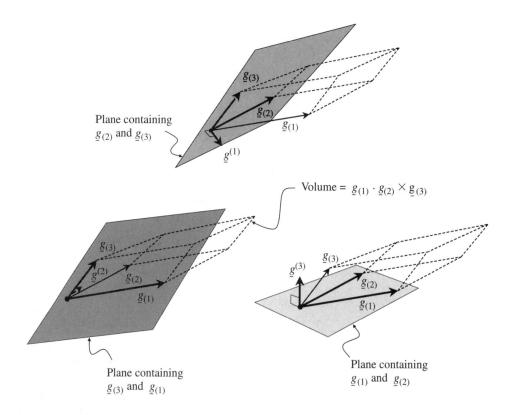

Figure C.5 Relationship between regular $g_{(p)}$ and reciprocal $g^{(p)}$ basis vectors of a coordinate system $\hat{x}^1, \hat{x}^2, \hat{x}^3$.

show that vector and tensor coefficients written with respect to the reciprocal basis vectors transform according to the corvariant transformation laws, Equations (C.128) and (C.129):

$$\underline{v} = \sum_{i=1}^{3} v_i \underline{g}^{(i)} = \sum_{k=1}^{3} \bar{v}_k \underline{c}^{(k)} \tag{C.132}$$

$$\bar{v}_k = v_i \frac{\partial \widehat{x}^i}{\partial \theta^k} \qquad \text{(stated without proof)} \tag{C.133}$$

$$\underline{\underline{A}} = \sum_{i=1}^{3} \sum_{p=1}^{3} A_{ip} \underline{g}^{(i)} \underline{g}^{(p)} = \sum_{k=1}^{3} \sum_{j=1}^{3} \bar{A}_{kj} \underline{c}^{(k)} \underline{c}^{(j)} \tag{C.134}$$

$$\bar{A}_{kj} = A_{ip} \frac{\partial \widehat{x}^i}{\partial \theta^k} \frac{\partial \widehat{x}^p}{\partial \theta^j} \qquad \text{(stated without proof)} \tag{C.135}$$

The proof is omitted.

One set that transforms covariantly is $\partial f/\partial x^1$, $\partial f/\partial x^2$, $\partial f/\partial x^3$, as we will show in the next example.

EXAMPLE

For a scalar function $f = f(x^1, x^2, x^3)$, is the set of three partial differentials of f, $\partial f/\partial x^1$, $\partial f/\partial x^2$, $\partial f/\partial x^3$, a contravariant or a covariant tensor, or neither? Show why or why not.

SOLUTION

We must determine whether this set transforms from

$$\frac{\partial f}{\partial x^1}, \frac{\partial f}{\partial x^2}, \frac{\partial f}{\partial x^3} \quad \text{to} \quad \frac{\partial f}{\partial \theta^1}, \frac{\partial f}{\partial \theta^2}, \frac{\partial f}{\partial \theta^3}$$

following the transformation rules for covariant or for contravariant tensors. To relate the two sets we use the chain rule:

$$\frac{\partial f}{\partial \theta^1} = \frac{\partial f}{\partial x^1} \frac{\partial x^1}{\partial \theta^1} + \frac{\partial f}{\partial x^2} \frac{\partial x^2}{\partial \theta^1} + \frac{\partial f}{\partial x^3} \frac{\partial x^3}{\partial \theta^1} \tag{C.136}$$

$$\frac{\partial f}{\partial \theta^2} = \frac{\partial f}{\partial x^1} \frac{\partial x^1}{\partial \theta^2} + \frac{\partial f}{\partial x^2} \frac{\partial x^2}{\partial \theta^2} + \frac{\partial f}{\partial x^3} \frac{\partial x^3}{\partial \theta^2} \tag{C.137}$$

$$\frac{\partial f}{\partial \theta^3} = \frac{\partial f}{\partial x^1} \frac{\partial x^1}{\partial \theta^3} + \frac{\partial f}{\partial x^2} \frac{\partial x^2}{\partial \theta^3} + \frac{\partial f}{\partial x^3} \frac{\partial x^3}{\partial \theta^3} \tag{C.138}$$

Writing these three relations in Einstein notation,

$$\frac{\partial f}{\partial \theta^i} = \frac{\partial f}{\partial x^m} \frac{\partial x^m}{\partial \theta^i} \quad \text{for } i = 1, 2, 3 \tag{C.139}$$

and we can see by comparison with Equation (C.128) that the set $\partial f/\partial x^1$, $\partial f/\partial x^2$, $\partial f/\partial x^3$ is a first-order covariant tensor.

It is straightforward to show that in the Cartesian coordinate system the basis vectors and the reciprocal basis vectors are identical, that is, the reciprocal vector $\hat{e}^1$, defined analogously to Equation (C.130), is just the familiar $\hat{e}_1$, and so on. Thus there is no distinction between contravariant and covariant coordinates when they are referred to different Cartesian coordinate systems with the same origin. In fact, Cartesian coefficients transform both covariantly and contravariantly, as we demonstrate next.

EXAMPLE

For an arbitrary vector $\underline{v}$, which with respect to a Cartesian basis $\hat{e}_1, \hat{e}_2, \hat{e}_3$ has coefficients v_1, v_2, v_3, calculate the covariant ($\bar{v}_i$) and contravariant ($\bar{v}^i$) coefficients of $\underline{v}$ in the $\bar{e}_1, \bar{e}_2, \bar{e}_3$ coordinate system (see Figure C.6).

SOLUTION

To calculate the contravariant and covariant coefficients, we will use the transformation rules, Equations (C.110) and (C.128). We need then to evaluate the partial derivatives $\partial \bar{x}_j / \partial x_k$ and $\partial x_k / \partial \bar{x}_j$. The two coordinate systems are related as follows:

$$\bar{x}_1 = \cos \psi \, x_1 + \sin \psi \, x_2 \tag{C.140}$$

$$\bar{x}_2 = -\sin \psi \, x_1 + \cos \psi \, x_2 \tag{C.141}$$

$$\bar{x}_3 = x_3 \tag{C.142}$$

Solving these three equations for x_1, x_2, x_3 we obtain

$$x_1 = \cos \psi \, \bar{x}_1 - \sin \psi \, \bar{x}_2 \tag{C.143}$$

$$x_2 = \sin \psi \, \bar{x}_1 + \cos \psi \, \bar{x}_2 \tag{C.144}$$

$$x_3 = \bar{x}_3 \tag{C.145}$$

From these six relations we can calculate all the partial derivatives we need. For the contravariant coefficients we obtain

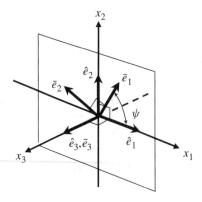

Figure C.6 Schematic of the two bases discussed in the example.

$$\bar{v}^j = \sum_{k=1}^{3} \frac{\partial \bar{x}^k}{\partial \bar{x}^j} v^k \tag{C.146}$$

$$\bar{v}^1 = v_1 \cos \psi + v_2 \sin \psi \tag{C.147}$$

$$\bar{v}^2 = v_2 \cos \psi - v_1 \sin \psi \tag{C.148}$$

$$\bar{v}^3 = v_3 \tag{C.149}$$

and for the covariant coefficients we obtain

$$\bar{v}_j = \sum_{k=1}^{3} \frac{\partial x^k}{\partial \bar{x}^j} v^k \tag{C.150}$$

$$\bar{v}_1 = v_1 \cos \psi + v_2 \sin \psi \tag{C.151}$$

$$\bar{v}_2 = v_2 \cos \psi - v_1 \sin \psi \tag{C.152}$$

$$\bar{v}_3 = v_3 \tag{C.153}$$

We see that the coefficients of $\underline{v}$ in Cartesian systems transform both contravariantly and covariantly, and there is no point in making a distinction between covariant and contravariant coefficients for these systems.

The use of tensors in mathematics and physics has evolved considerably in the latter half of the twentieth century, encompassing far more than the concept of coordinate-invariant transformations discussed here [222]. In fact, the terms contravariant and covariant have fallen out of use in mathematical physics, replaced by more general classifications. In the rheological literature, however, these terms are encountered, as well as terms such as upper convected and lower convected, which derive from the notation used when writing vectors and tensors with respect to contravariant and covariant bases, as discussed in Chapter 9.

C.8 PROBLEMS

C.1 A vector function $\underline{v}$ is written in the Cartesian coordinate system x, y, z as

$$\underline{v} = \begin{pmatrix} \dot{\gamma}_0 y \\ 0 \\ 0 \end{pmatrix}_{xyz}$$

where $\dot{\gamma}_0$ is a scalar constant, and y is the variable in the $\hat{e}_y$-direction. Express that function in the Cartesian coordinate system $\bar{x}$, $\bar{y}$, $\bar{z}$, which is rotated 45° counterclockwise around the $z = \bar{z}$-axis with respect to the original coordinate system.

C.2 For the symmetric tensor $\underline{\underline{A}}$ written in a Cartesian coordinate system,

$$\underline{\underline{A}} = \begin{pmatrix} 1 & 2 & 3 \\ 2 & 0 & 4 \\ 3 & 4 & 1 \end{pmatrix}_{123}$$

find the principal axes and principal values. If possible write $\underline{\underline{A}}$ in a coordinate system in which it is diagonal.

C.3 For the nonsymmetric tensors $\underline{\underline{A}}$ and $\underline{\underline{B}}$, written

$$\underline{\underline{A}} = \begin{pmatrix} 1 & 2 & 3 \\ 0 & -1 & 6 \\ 3 & 2 & 1 \end{pmatrix}_{123}$$

$$B = \begin{pmatrix} -1 & 1 & 0 \\ -1 & 0 & 1 \\ -1 & 0 & 0 \end{pmatrix}_{123}$$

in a Cartesian coordinate system, find the principal axes and principal values. Write each in a coordinate system in which it is diagonal, if possible; if not possible, explain why not.

C.4   Find $\sqrt{\underline{\underline{A}}}$, where $\underline{\underline{A}}$ is the tensor:

$$\underline{\underline{A}} = \begin{pmatrix} 11 & -2 & 5 \\ -2 & 2 & 2 \\ 5 & 2 & 11 \end{pmatrix}_{123}$$

(See discussion in the example in Section 9.1.2).

C.5   Show that the invariants defined in Chapter 2 [Equations (2.179)–(2.181)] and the coefficients of the characteristic Equation (C.77) are related as given in Equations (C.82) and (C.83).

C.6   Show that a necessary condition for the inversion of a coordinate transformation to exist is that the Jacobian be nonzero. *Hint*: Begin by writing the differentials of the two systems, $d\theta$ and dx, in terms of partial derivatives.

C.7   Show that the coordinate transformation from the x, y, z to the $\bar{x}, \bar{y}, \bar{z}$ system is invertible.

$$\bar{x} = 2x + y$$

$$\bar{y} = x - y + 3z$$

$$\bar{z} = 2y + z$$

C.8   Show that the coordinate transformation from the Cartesian x, y, z coordinate system to the r, θ, z cylindrical coordinate system is invertible. Comment on the points where $r = 0$.

C.9   Show that the coordinate transformation from the Cartesian x, y, z, coordinate system to the r, θ, ϕ spherical coordinate system is invertible. Comment on the points where $r = 0$ or $\theta = 0$.

C.10   Calculate the reciprocal basis vectors for the Cartesian coordinate system. Comment on your answer.

C.11   Calculate the reciprocal basis vectors for the cylindrical coordinate system. Comment on your answer.

C.12   Derive a relationship between the reciprocal basis vectors $\underline{g}^{(1)}, \underline{g}^{(2)}, \underline{g}^{(3)}$ and the magnitudes of the basis vectors $\underline{g}_{(1)}, \underline{g}_{(2)}, \underline{g}_{(3)}$. Your answer will include the angles between the vectors ϕ_{ii}, where,

for example, ϕ_{11} is the angle between vectors $\underline{g}^{(1)}$ and $\underline{g}_{(1)}$.

C.13   Using the contravariant and covariant transformation laws, Equations (C.110) and (C.128), calculate the covariant and contravariant coefficients of an arbitrary vector $\underline{v}$ in the $\theta_1, \theta_2, \theta_3$ coordinate system

$$\theta_1 = x_2$$

$$\theta_2 = x_3$$

$$\theta_3 = x_1$$

Express your answer in terms of the Cartesian coefficients of $\underline{v}, v_1, v_2, v_3$. Comment on your result.

C.14   Using the contravariant and covariant transformation laws, Equations (C.110) and (C.128), calculate the covariant and contravariant coefficients of an arbitrary vector $\underline{v}$ in the $\bar{x}_1, \bar{x}_2, \bar{x}_3$ coordinate system, which is an orthonormal coordinate system in which the $\bar{x}_1$-axis is rotated around the x_3 Cartesian axis counterclockwise by an angle ψ from the x_1 Cartesian axis. Express your answer in terms of the Cartesian coefficients of $\underline{v}, v_1, v_2, v_3$. Comment on your result.

C.15   A nonorthonormal basis $\underline{g}_{(1)}, \underline{g}_{(2)}, \underline{g}_{(3)}$ and its reciprocal basis $\underline{g}^{(1)}, \underline{g}^{(2)}, \underline{g}^{(3)}$ are given as

$$\underline{g}_{(1)} = \hat{e}_2 \qquad\qquad \underline{g}^{(1)} = \hat{e}_1 + \hat{e}_2$$

$$\underline{g}_{(2)} = \hat{e}_3 \qquad\qquad \underline{g}^{(2)} = \hat{e}_3$$

$$\underline{g}_{(3)} = \hat{e}_1 - \hat{e}_2 \qquad \underline{g}^{(3)} = \hat{e}_1$$

(a) Verify that $\underline{g}_{(i)} \cdot \underline{g}^{(j)} = \delta_{ij}$.

(b) Calculate the relationships between the Cartesian coordinates x_1, x_2, x_3 and the coordinates $\hat{x}^1, \hat{x}^2, \hat{x}^3$ associated with the basis $\underline{g}_{(1)}, \underline{g}_{(2)}, \underline{g}_{(3)}$. Recall that $\underline{g}_{(i)} = \partial \underline{r}/\partial \hat{x}^i$. Calculate the inverse relationships $\hat{x}^1(x_1, x_2, x_3), \hat{x}^2(x_1, x_2, x_3), \hat{x}^3(x_1, x_2, x_3)$.

(c) Using the contravariant and covariant transformation rules, Equations (C.110) and (C.128), calculate the coefficients $\hat{v}_i$ and $\hat{v}^i$ of an arbitrary vector $\underline{v}$ in the two bases $(\underline{v} = \hat{v}_i \underline{g}^{(i)} = \hat{v}^i \underline{g}_{(i)})$. Write your answers in terms of the Cartesian components v_1, v_2, v_3.

(d) By using the expressions for the $\underline{g}_{(i)}$ and $\underline{g}^{(i)}$ given above, verify that the results you obtained in part (b) are correct.

Predictions of Constitutive Equations

TABLE D.1
Comparison of Nomenclature for Strain Tensors Used in the Literature

Name	This Text	Larson [138]	DPL [26]	Macosko [162]	Middleman [179]
Stress tensor	$\underline{\underline{\Pi}} = \underline{\underline{\tau}} + p\underline{\underline{I}}$	$-\underline{\underline{T}} = -\underline{\underline{\sigma}} + p\underline{\underline{I}}$	$\underline{\underline{\Pi}} = \underline{\underline{\tau}} + p\underline{\underline{I}}$	$-\underline{\underline{T}} = -\underline{\underline{\tau}} + p\underline{\underline{I}}$	$-\underline{\underline{T}} = -\underline{\underline{\tau}} + p\underline{\underline{I}}$
Gradient of a vector	$\nabla\underline{w} = \dfrac{\partial w_p}{\partial x_k}\hat{e}_k\hat{e}_p$	$\nabla\underline{w} = \dfrac{\partial w_p}{\partial x_k}\hat{e}_k\hat{e}_p$	$\nabla\underline{w} = \dfrac{\partial w_p}{\partial x_k}\hat{e}_k\hat{e}_p$	$\tilde{\nabla}\underline{w} = \dfrac{\partial w_p}{\partial x_k}\hat{e}_p\hat{e}_k$	$\tilde{\nabla}\underline{w} = \dfrac{\partial w_p}{\partial x_k}\hat{e}_p\hat{e}_k$
Deformation-gradient tensor	$\underline{\underline{F}}$	$\underline{\underline{F}}$	$\underline{\underline{\Delta}}^T$	$(\underline{\underline{F}}^{-1})^T$	—
Inverse deformation-gradient tensor	$\underline{\underline{F}}^{-1}$	$\underline{\underline{F}}^{-1}$	$\underline{\underline{E}}^T$	$\underline{\underline{F}}^T$	—
Cauchy tensor	$\underline{\underline{C}}$	$\underline{\underline{C}}$	$\underline{\underline{B}}^{-1}$	$\underline{\underline{B}}^{-1}$	—
Finger tensor	$\underline{\underline{C}}^{-1}$	$\underline{\underline{C}}^{-1}$	$\underline{\underline{B}}$	$\underline{\underline{B}}$	—
Finite strain based on Cauchy	$\underline{\underline{\gamma}}^{[0]}$	$\underline{\underline{C}} - \underline{\underline{I}}$	$\underline{\underline{\gamma}}^{[0]}$	$\underline{\underline{B}}^{-1} - \underline{\underline{I}}$	—
Finite strain based on Finger	$\underline{\underline{\gamma}}_{[0]}$	$\underline{\underline{I}} - \underline{\underline{C}}^{-1}$	$\underline{\underline{\gamma}}_{[0]}$	$-\underline{\underline{E}}$	—
Rate-of-strain tensor	$\underline{\underline{\dot{\gamma}}}$	$2\underline{\underline{D}}$	$\underline{\underline{\dot{\gamma}}}$	$2\underline{\underline{D}}$	$\underline{\underline{\Delta}}$
Green tensor	$\underline{\underline{F}}^{-1} \cdot (\underline{\underline{F}}^{-1})^T$	$\underline{\underline{F}}^{-1} \cdot (\underline{\underline{F}}^{-1})^T$	$\underline{\underline{E}}^T \cdot \underline{\underline{E}}$	$\underline{\underline{C}}$	—

TABLE D.2
Predictions of Lodge Equation or Upper Convected Maxwell Model in Shear and Extensional Flows

1. *Shear*

   Startup

   $\eta^+(t, \dot{\gamma})$ $\eta_0 \left(1 - e^{-\frac{t}{\lambda}}\right)$

   $\Psi_1^+(t, \dot{\gamma})$ $2\eta_0\lambda \left[1 - e^{\frac{-t}{\lambda}}\left(1 + \frac{t}{\lambda}\right)\right]$

   $\Psi_2^+(t, \dot{\gamma})$ 0

   Steady

   $\eta(\dot{\gamma})$ $\eta_0 \equiv G_0\lambda$

   $\Psi_1(\dot{\gamma})$ $2G_0\lambda^2 = 2\eta_0\lambda$

   $\Psi_2(\dot{\gamma})$ 0

   Cessation

   $\eta^-(t, \dot{\gamma})$ $\eta_0 e^{\frac{-t}{\lambda}}$

   $\Psi_1^-(t, \dot{\gamma})$ $2\lambda\eta_0 e^{\frac{-t}{\lambda}}$

   $\Psi_2^-(t, \dot{\gamma})$ 0

   Step shear strain

   $G(t, \gamma_0)$ $G_0 e^{-\frac{t}{\lambda}}$

   $G_{\Psi_1}(t, \gamma_0)$ $G_0 e^{-\frac{t}{\lambda}}$

   $G_{\Psi_2}(t, \gamma_0)$ 0

2. *Extension*

   Startup
   Uniaxial ($b = 0, \dot{\epsilon}_0 > 0$)
   or biaxial ($b = 0, \dot{\epsilon}_0 < 0$)

   $\bar{\eta}^+(t, \dot{\epsilon}_0)$
   or $\bar{\eta}_B^+(t, \dot{\epsilon}_0)$ $\dfrac{\eta_0}{\mathcal{AB}}\left(3 - 2\mathcal{B}e^{-\frac{t\mathcal{A}}{\lambda}} - \mathcal{A}e^{-\frac{t\mathcal{B}}{\lambda}}\right)$

   $\mathcal{A} = 1 - 2\dot{\epsilon}_0\lambda$
   $\mathcal{B} = 1 + \dot{\epsilon}_0\lambda$

   Planar ($b = 1, \dot{\epsilon}_0 > 0$)

   $\bar{\eta}_{P_1}^+(t, \dot{\epsilon}_0)$ $\dfrac{2\eta_0}{\mathcal{AC}}\left(2 - \mathcal{A}e^{-\frac{\mathcal{C}t}{\lambda}} - \mathcal{C}e^{-\frac{\mathcal{A}t}{\lambda}}\right)$

   $\mathcal{A} = 1 - 2\dot{\epsilon}_0\lambda$
   $\mathcal{C} = 1 + 2\dot{\epsilon}_0\lambda$

   $\bar{\eta}_{P_2}^+(t, \dot{\epsilon}_0)$ $\dfrac{2\eta_0}{\mathcal{C}}\left(1 - e^{-\frac{\mathcal{C}t}{\lambda}}\right)$

   Steady
   Uniaxial ($b = 0, \dot{\epsilon}_0 > 0$)
   or biaxial ($b = 0, \dot{\epsilon}_0 < 0$)

   $\bar{\eta}(\dot{\epsilon}_0)$
   or $\bar{\eta}_B(\dot{\epsilon}_0)$ $\dfrac{3\eta_0}{(1 - 2\lambda\dot{\epsilon}_0)(1 + \lambda\dot{\epsilon}_0)} = \dfrac{3\eta_0}{\mathcal{AB}}$

   Planar ($b = 1, \dot{\epsilon}_0 > 0$)

   $\bar{\eta}_{P_1}(\dot{\epsilon}_0)$ $\dfrac{4\eta_0}{1 - 4\dot{\epsilon}_0^2\lambda^2} = \dfrac{4\eta_0}{\mathcal{AC}}$

   $\bar{\eta}_{P_2}(\dot{\epsilon}_0)$ $\dfrac{2\eta_0}{1 + 2\dot{\epsilon}_0\lambda} = \dfrac{2\eta_0}{\mathcal{C}}$

TABLE D.3
Predictions of Cauchy–Maxwell Equation or Lower Convected Maxwell Model in Shear and Extensional Flows

1. *Shear*

Startup	$\eta^+(t, \dot\gamma)$	$\eta_0 \left(1 - e^{-\frac{t}{\lambda}}\right)$
	$\Psi_1^+(t, \dot\gamma)$	$2\eta_0\lambda \left[1 - e^{\frac{t}{\lambda}}\left(1 + \frac{t}{\lambda}\right)\right]$
	$\Psi_2^+(t, \dot\gamma)$	$-\Psi_1^+$
Steady	$\eta(\dot\gamma)$	$\eta_0 \equiv G_0\lambda$
	$\Psi_1(\dot\gamma)$	$2G_0\lambda^2 = 2\eta_0\lambda$
	$\Psi_2(\dot\gamma)$	$-\Psi_1$
Cessation	$\eta^-(t, \dot\gamma)$	$\eta_0 e^{\frac{-t}{\lambda}}$
	$\Psi_1^-(t, \dot\gamma)$	$2\lambda\eta_0 e^{\frac{-t}{\lambda}}$
	$\Psi_2^-(t, \dot\gamma)$	$-\Psi_1^-$
Step shear strain	$G(t, \gamma_0)$	$G_0 e^{-\frac{t}{\lambda}}$
	$G_{\Psi_1}(t, \gamma_0)$	$G_0 e^{-\frac{t}{\lambda}}$
	$G_{\Psi_2}(t, \gamma_0)$	$-G_{\Psi_1}$

2. *Extension*

Startup

Uniaxial ($b = 0, \dot\epsilon_0 > 0$) $\bar\eta^+(t, \dot\epsilon_0)$

or biaxial ($b = 0, \dot\epsilon_0 < 0$)   or $\bar\eta_B^+(t, \dot\epsilon_0)$
$$\frac{\eta_0}{C\mathcal{D}}\left(3 - 2\mathcal{D}e^{-\frac{tC}{\lambda}} - Ce^{-\frac{t\mathcal{D}}{\lambda}}\right)$$
$$C = 1 + 2\dot\epsilon_0\lambda$$
$$\mathcal{D} = 1 - \dot\epsilon_0\lambda$$

Planar ($b = 1, \dot\epsilon_0 > 0$) $\bar\eta_{P_1}^+(t, \dot\epsilon_0)$
$$\frac{-2\eta_0}{\mathcal{A}C}\left(2 - \mathcal{A}e^{-\frac{Ct}{\lambda}} - Ce^{-\frac{\mathcal{A}t}{\lambda}}\right)$$
$$\mathcal{A} = 1 - 2\dot\epsilon_0\lambda$$

$\bar\eta_{P_2}^+(t, \dot\epsilon_0)$
$$\frac{-2\eta_0}{\mathcal{A}}\left(1 - e^{-\frac{\mathcal{A}t}{\lambda}}\right)$$

Steady

Uniaxial ($b = 0, \dot\epsilon_0 > 0$) $\bar\eta(\dot\epsilon_0)$

or biaxial ($b = 0, \dot\epsilon_0 < 0$)   or $\bar\eta_B(\dot\epsilon_0)$
$$\frac{3\eta_0}{(1 + 2\lambda\dot\epsilon_0)(1 - \lambda\dot\epsilon_0)} = \frac{3\eta_0}{C\mathcal{D}}$$

Planar ($b = 1, \dot\epsilon_0 > 0$) $\bar\eta_{P_1}(\dot\epsilon_0)$
$$\frac{-4\eta_0}{1 - 4\dot\epsilon_0^2\lambda^2} = \frac{-4\eta_0}{\mathcal{A}C}$$

$\bar\eta_{P_2}(\dot\epsilon_0)$
$$\frac{-2\eta_0}{1 - 2\dot\epsilon_0\lambda} = \frac{-2\eta_0}{\mathcal{A}}$$

TABLE D.4
Predictions of Oldroyd B or Convected Jeffreys Model in Shear and Extensional Flows [26]

1. *Shear*

   Startup

   $\eta^+(t, \dot{\gamma})$ $\qquad \eta_0\left[\dfrac{\lambda_2}{\lambda_1} + \left(1 - \dfrac{\lambda_2}{\lambda_1}\right)\left(1 - e^{-\frac{t}{\lambda_1}}\right)\right]$

   $\Psi_1^+(t, \dot{\gamma})$ $\qquad 2\eta_0(\lambda_1 - \lambda_2)\left[1 - e^{-\frac{t}{\lambda_1}}\left(1 + \dfrac{t}{\lambda_1}\right)\right]$

   $\Psi_2^+(t, \dot{\gamma})$ $\qquad 0$

   Steady

   $\eta(\dot{\gamma})$ $\qquad \eta_0$

   $\Psi_1(\dot{\gamma})$ $\qquad 2\eta_0(\lambda_1 - \lambda_2)$

   $\Psi_2(\dot{\gamma})$ $\qquad 0$

   Cessation

   $\eta^-(t, \dot{\gamma})$ $\qquad \eta_0\left(1 - \dfrac{\lambda_2}{\lambda_1}\right)e^{-\frac{t}{\lambda_1}}$

   $\Psi_1^-(t, \dot{\gamma})$ $\qquad 2\eta_0(\lambda_1 - \lambda_2)\,e^{-\frac{t}{\lambda_1}}$

   $\Psi_2^-(t, \dot{\gamma})$ $\qquad 0$

   SAOS

   $G'(\omega)$ $\qquad \eta_0\dfrac{(\lambda_1 - \lambda_2)\omega^2}{1 + \lambda_1^2\omega^2}$

   $G''(\omega)$ $\qquad \eta_0\omega\dfrac{1 + \lambda_1\lambda_2\omega^2}{1 + \lambda_1^2\omega^2}$

2. *Extension*

   Startup

   Uniaxial ($b = 0, \dot{\epsilon}_0 > 0$)
   or biaxial ($b = 0, \dot{\epsilon}_0 < 0$)

   $\bar{\eta}^+(t, \dot{\epsilon}_0)$
   or $\bar{\eta}_B^+(t, \dot{\epsilon}_0)$ $\qquad 3\eta_0\dfrac{\lambda_2}{\lambda_1} + \dfrac{\eta_0}{\mathcal{A}\mathcal{B}}\left(1 - \dfrac{\lambda_2}{\lambda_1}\right)\left(3 - 2\mathcal{B}e^{-\frac{t\mathcal{A}}{\lambda_1}} - \mathcal{A}e^{-\frac{t\mathcal{B}}{\lambda_1}}\right)$
   $\mathcal{A} = 1 - 2\dot{\epsilon}_0\lambda_1$
   $\mathcal{B} = 1 + \dot{\epsilon}_0\lambda_1$

   Planar ($b = 1, \dot{\epsilon}_0 > 0$)

   $\bar{\eta}_{P_1}^+(t, \dot{\epsilon}_0)$ $\qquad 4\eta_0\dfrac{\lambda_2}{\lambda_1} + \dfrac{2\eta_0}{\mathcal{A}\mathcal{C}}\left(1 - \dfrac{\lambda_2}{\lambda_1}\right)\left(2 - \mathcal{A}e^{-\frac{\mathcal{C}t}{\lambda_1}} - \mathcal{C}e^{-\frac{\mathcal{A}t}{\lambda_1}}\right)$
   $\mathcal{A} = 1 - 2\dot{\epsilon}_0\lambda_1$
   $\mathcal{C} = 1 + 2\dot{\epsilon}_0\lambda_1$

   $\bar{\eta}_{P_2}^+(t, \dot{\epsilon}_0)$ $\qquad 2\eta_0\dfrac{\lambda_2}{\lambda_1} + \dfrac{2\eta_0}{\mathcal{C}}\left(1 - \dfrac{\lambda_2}{\lambda_1}\right)\left(1 - e^{-\frac{\mathcal{C}t}{\lambda_1}}\right)$

   Steady

   Uniaxial ($b = 0, \dot{\epsilon}_0 > 0$)
   or biaxial ($b = 0, \dot{\epsilon}_0 < 0$)

   $\bar{\eta}(\dot{\epsilon}_0)$
   or $\bar{\eta}_B(\dot{\epsilon}_0)$ $\qquad 3\eta_0\left(\dfrac{\lambda_2}{\lambda_1} + \dfrac{1 - \frac{\lambda_2}{\lambda_1}}{\mathcal{A}\mathcal{B}}\right)$

   Planar ($b = 1, \dot{\epsilon}_0 > 0$)

   $\bar{\eta}_{P_1}(\dot{\epsilon}_0)$ $\qquad 4\eta_0\left(\dfrac{\lambda_2}{\lambda_1} + \dfrac{1 - \frac{\lambda_2}{\lambda_1}}{\mathcal{A}\mathcal{C}}\right)$

   $\bar{\eta}_{P_2}(\dot{\epsilon}_0)$ $\qquad 2\eta_0\left(\dfrac{\lambda_2}{\lambda_1} + \dfrac{1 - \frac{\lambda_2}{\lambda_1}}{\mathcal{C}}\right)$

TABLE D.5
Predictions of White–Metzner Equation in Shear and Extensional Flows [26]*

1. *Shear*

   Startup $\eta^+(t, \dot{\gamma})$ $\eta(\dot{\gamma})\left(1 - e^{-\frac{t}{\lambda(\dot{\gamma})}}\right)$

   $\Psi_1^+(t, \dot{\gamma})$ $2\eta(\dot{\gamma})\lambda(\dot{\gamma})\left[1 - e^{-\frac{t}{\lambda(\dot{\gamma})}}\left(1 + \frac{t}{\lambda(\dot{\gamma})}\right)\right]$

   $\Psi_2^+(t, \dot{\gamma})$ 0

   Steady $\eta(\dot{\gamma})$ $\eta(\dot{\gamma})$

   $\Psi_1(\dot{\gamma})$ $2\eta(\dot{\gamma})\lambda(\dot{\gamma})$

   $\Psi_2(\dot{\gamma})$ 0

2. *Extension*

   Steady

   Uniaxial $(b = 0, \dot{\epsilon}_0 > 0)$ $\bar{\eta}(\dot{\epsilon}_0)$ $\dfrac{3\eta(\dot{\gamma})}{\left[1 - 2\lambda(\dot{\gamma})\dot{\epsilon}_0\right]\left[1 + \lambda(\dot{\gamma})\dot{\epsilon}_0\right]} = \dfrac{3\eta(\dot{\gamma})}{\mathcal{A}(\dot{\gamma})\mathcal{B}(\dot{\gamma})}$

   or biaxial $(b = 0, \dot{\epsilon}_0 < 0)$ or $\bar{\eta}_B(\dot{\epsilon}_0)$ $\mathcal{A}(\dot{\gamma}) = 1 - 2\dot{\epsilon}_0\lambda(\dot{\gamma})$
   $\mathcal{B}(\dot{\gamma}) = 1 + \dot{\epsilon}_0\lambda(\dot{\gamma})$

   Planar $(b = 1, \dot{\epsilon}_0 > 0)$ $\bar{\eta}_{P_1}(\dot{\epsilon}_0)$ $\dfrac{4\eta(\dot{\gamma})}{1 - 4\dot{\epsilon}_0^2\lambda(\dot{\gamma})^2} = \dfrac{4\eta(\dot{\gamma})}{\mathcal{A}(\dot{\gamma})C(\dot{\gamma})}$

   $\mathcal{A}(\dot{\gamma}) = 1 - 2\dot{\epsilon}_0\lambda(\dot{\gamma})$
   $C(\dot{\gamma}) = 1 + 2\dot{\epsilon}_0\lambda(\dot{\gamma})$

   $\bar{\eta}_{P_2}(\dot{\epsilon}_0)$ $\dfrac{2\eta(\dot{\gamma})}{1 + 2\dot{\epsilon}_0\lambda(\dot{\gamma})} = \dfrac{2\eta(\dot{\gamma})}{C(\dot{\gamma})}$

*$\lambda(\dot{\gamma}) = \eta(\dot{\gamma})/G_0$ and $\dot{\gamma} = |\underline{\dot{\gamma}}|$.

TABLE D.6
Predictions of Factorized Rivlin–Sawyers Model in Shear and Extensional Flows [26]

1. *Shear*

   Steady

   $\eta(\dot{\gamma})$ $\qquad \displaystyle\int_0^\infty M(s)s(\Phi_1 + \Phi_2)\,ds$

   $\Psi_1(\dot{\gamma})$ $\qquad \displaystyle\int_0^\infty M(s)s^2(\Phi_1 + \Phi_2)\,ds$

   $\Psi_2(\dot{\gamma})$ $\qquad -\displaystyle\int_0^\infty M(s)s^2\Phi_2\,ds$

   SAOS

   $G'(\omega)$ $\qquad \displaystyle\int_0^\infty M(s)(1 - \cos\omega s)\,ds$

   $G''(\omega)$ $\qquad \displaystyle\int_0^\infty M(s)\sin\omega s\,ds$

2. *Extension*

   Steady

   Uniaxial ($b = 0, \dot{\epsilon}_0 > 0$) $\bar{\eta}(\dot{\epsilon}_0)$ $\qquad \dfrac{1}{\dot{\epsilon}_0}\displaystyle\int_0^\infty M(s)\left[\Phi_1\left(e^{2\dot{\epsilon}_0 s} - e^{-\dot{\epsilon}_0 s}\right) + \Phi_2\left(e^{\dot{\epsilon}_0 s} - e^{-2\dot{\epsilon}_0 s}\right)\right]ds$
   or biaxial ($b = 0, \dot{\epsilon}_0 < 0$) or $\bar{\eta}_B(\dot{\epsilon}_0)$

   Planar ($b = 1, \dot{\epsilon}_0 > 0$) $\bar{\eta}_{P_1}(\dot{\epsilon}_0)$ $\qquad \dfrac{1}{\dot{\epsilon}_0}\displaystyle\int_0^\infty M(s)\left[\Phi_1\left(e^{2\dot{\epsilon}_0 s} - e^{-2\dot{\epsilon}_0 s}\right) + \Phi_2\left(e^{2\dot{\epsilon}_0 s} - e^{-2\dot{\epsilon}_0 s}\right)\right]ds$

   $\bar{\eta}_{P_2}(\dot{\epsilon}_0)$ $\qquad \dfrac{1}{\dot{\epsilon}_0}\displaystyle\int_0^\infty M(s)\left[\left(\Phi_1 e^{-\dot{\epsilon}_0 s} + \Phi_2 e^{\dot{\epsilon}_0 s}\right)\left(e^{\dot{\epsilon}_0 s} - e^{-\dot{\epsilon}_0 s}\right)\right]ds$

Optics of Birefringence

The discussion in this appendix draws on the texts by Fowles [81] and Born and Wolf [33].

E.1 Light in a Vacuum

The propagation of light is described by the four Maxwell equations. They describe the interrelation of the electric-field vector $\underline{E}$ and the magnetic-field vector $\underline{H}$. In a vacuum with no charges present, the Maxwell equations are [81][1]

Maxwell equations
(vacuum)

$$\nabla \times \underline{E} = -\mu_0 \frac{\partial \underline{H}}{\partial t}$$
$$\nabla \times \underline{H} = \varepsilon_0 \frac{\partial \underline{E}}{\partial t}$$
$$\nabla \cdot \underline{E} = 0$$
$$\nabla \cdot \underline{H} = 0$$

(E.1)

where ε_0 is the permittivity of vacuum, $\varepsilon_0 = 8.854 \times 10^{-12}$ farads/m, and μ_0 is the permeability of vacuum,[2] $\mu_0 = 4\pi \times 10^{-7}$ henrys/m. By taking the curl of the first equation and the time derivative of the second and combining these expressions, $\underline{H}$ may be eliminated, and we obtain the wave equation for the propagation of light in a vacuum:

Electric-field
wave equation
(vacuum)

$$\nabla^2 \underline{E} = \frac{1}{c^2} \frac{\partial^2 \underline{E}}{\partial t^2}$$

(E.2)

where $c = 1/\sqrt{\mu_0 \varepsilon_0}$ is the speed of light in vacuum, $c = 3.0 \times 10^8$ m/s. An identical equation for $\underline{H}$ may be obtained by eliminating $\underline{E}$ between the first two Maxwell equations:

---

[1] Often the Maxwell equations are written in terms of the vector $\underline{B}$, the magnetic induction. For nonmagnetic media such as we are considering, $\underline{B} = \mu_0 \underline{H}$ [81].

[2] Quantities described in this section look slightly different when expressed in units other than the rationalized MKS units used. For more information see the appendix in [121].

Magnetic-field
wave equation
(vacuum)

$$\nabla^2 \underline{H} = \frac{1}{c^2} \frac{\partial^2 \underline{H}}{\partial t^2}$$

(E.3)

As we see in the next section, similar equations are obtained for the propagation of light in an isotropic medium.

E.2 Light in an Isotropic Medium

When light propagates through an isotropic nonconducting medium rather than vacuum, the Maxwell equations must be modified to include the permittivity ε and the permeability μ of that medium:

Maxwell equations
(isotropic medium)

$$\nabla \times \underline{E} = -\mu \frac{\partial \underline{H}}{\partial t}$$
$$\nabla \times \underline{H} = \varepsilon \frac{\partial \underline{E}}{\partial t}$$
$$\nabla \cdot \underline{E} = 0$$
$$\nabla \cdot \underline{H} = 0$$

(E.4)

The same manipulations used in the last section may be applied to these Maxwell equations, resulting in the following wave equations for light in an isotropic nonconducting medium:

Wave equations
(isotropic medium)

$$\nabla^2 \underline{E} = \frac{1}{u^2} \frac{\partial^2 \underline{E}}{\partial t^2}$$
$$\nabla^2 \underline{H} = \frac{1}{u^2} \frac{\partial^2 \underline{H}}{\partial t^2}$$

(E.5)

where $u = 1/\sqrt{\mu\varepsilon}$ is the speed of light in the medium. The dielectric constant K is defined as the ratio of the permittivity of the medium to the permittivity of vacuum:

$$K \equiv \frac{\varepsilon}{\varepsilon_0} \tag{E.6}$$

This is also called the *relative* permittivity. The relative permeability K_m is the ratio of the permeability of the medium to that of vacuum:

$$K_m \equiv \frac{\mu}{\mu_0} \tag{E.7}$$

The refractive index n is the ratio of the speed of light in vacuum to that in the medium:

Refractive index

$$n \equiv \frac{c}{u} = \sqrt{\frac{\mu\varepsilon}{\mu_0\varepsilon_0}} = \sqrt{K_m K}$$

(E.8)

Note that for nonmagnetic media $K_m = 1$, and therefore $n = \sqrt{K}$. The velocity u is also called the *phase* velocity.

A solution to Equation (E.5) is

$$E = \begin{pmatrix} E_{0x} \\ E_{0y} \\ E_{0z} \end{pmatrix}_{xyz} e^{i(\underline{k}\cdot\underline{r}-\omega t)} \tag{E.9}$$

$$= \underline{E}_0 e^{i(\underline{k}\cdot\underline{r}-\omega t)} \tag{E.10}$$

where t is time, $\underline{r}$ denotes position in space (i.e., $\underline{r} = x\hat{e}_x + y\hat{e}_y + z\hat{e}_z$ in Cartesian coordinates), ω is the radian frequency of the wave, and $\underline{k}$ is called the *wave vector*:

$$\underline{k} = k_x\hat{e}_x + k_y\hat{e}_y + k_z\hat{e}_z \tag{E.11}$$

$$k \equiv |\underline{k}| = \frac{\omega}{u} \tag{E.12}$$

The argument of the exponential, $\underline{k} \cdot \underline{r} - \omega t$, is called the *phase* of the wave, and surfaces where this quantity is constant are called *surfaces of constant phase*. Surfaces of constant phase move in the direction of $\underline{k}$ at a rate equal to the phase velocity u. This solution to the wave equation is called the *solution for plane harmonic waves*. A more general solution is given in Fuller [84].

The magnetic-field vector $\underline{H}$ obeys the same equation as the electric-field vector $\underline{E}$, and thus we may write

$$\underline{H} = \begin{pmatrix} H_{0x} \\ H_{0y} \\ H_{0z} \end{pmatrix}_{xyz} e^{i(\underline{k}\cdot\underline{r}-\omega t)} \tag{E.13}$$

$$= \underline{H}_0 e^{i(\underline{k}\cdot\underline{r}-\omega t)} \tag{E.14}$$

Substituting the solutions for $\underline{E}$ and $\underline{H}$ for plane harmonic waves back into the Maxwell equations for isotropic nonconducting media yields the following equations:

$$\underline{k} \times \underline{E} = \mu\omega\underline{H} \tag{E.15}$$

$$\underline{k} \times \underline{H} = -\varepsilon\omega\underline{E} \tag{E.16}$$

$$\underline{k} \cdot \underline{E} = 0 \tag{E.17}$$

$$\underline{k} \cdot \underline{H} = 0 \tag{E.18}$$

The time rate of flow of electromagnetic energy of the wave per unit area is given by the Poynting vector $\underline{S}$:

$$\text{Poynting vector} \qquad \boxed{\underline{S} = \Re(\underline{E}) \times \Re(\underline{H})} \tag{E.19}$$

The direction of $\underline{S}$ indicates the direction of travel of the wave. From Equations (E.15) and (E.16) we see that for isotropic nonconducting media, $\underline{k}$, $\underline{E}$, and $\underline{H}$ form an orthogonal

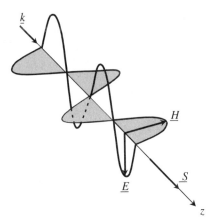

Figure E.1 Schematic of the electric-field $\underline{E}$ and magnetic-field $\underline{H}$ vectors of linearly polarized light propagating in an isotropic medium. The planes of constant phase are propagating in the $\hat{k}$-direction; adapted from Hecht [108]. From *Optics*, 2nd edition, by E. Hecht, Copyright © 1987. Reprinted by permission of Addison-Wesley Educational Publishers, Inc.

triad, and therefore $\underline{k}$ and $\underline{S}$ are parallel (see Figure E.1). Thus for isotropic nonconducting media, $\hat{k} = \underline{k}/k$ is both the direction of propagation of planes of constant phase and the direction of energy transport of the wave.

From Equations (E.17) and (E.18) we see that there is no component of either $\underline{E}$ or $\underline{H}$ in the $\underline{k}$-direction, and therefore both $\underline{E}$ and $\underline{H}$ are transverse waves, as depicted in Figure E.1. If the wave is propagating in the z-direction ($\underline{k} = k\hat{e}_z$, $\underline{S} = S\hat{e}_z$), the expressions for $\underline{E}$ and $\underline{H}$ become:

$$\underline{E} = \begin{pmatrix} E_x \\ E_y \\ E_z \end{pmatrix}_{xyz} = \begin{pmatrix} E_{0x}e^{i(\underline{k}\cdot\underline{r}-\omega t)} \\ E_{0y}e^{i(\underline{k}\cdot\underline{r}-\omega t)} \\ 0 \end{pmatrix}_{xyz} \tag{E.20}$$

$$\underline{H} = \begin{pmatrix} H_x \\ H_y \\ H_z \end{pmatrix}_{xyz} = \begin{pmatrix} H_{0x}e^{i(\underline{k}\cdot\underline{r}-\omega t)} \\ H_{0y}e^{i(\underline{k}\cdot\underline{r}-\omega t)} \\ 0 \end{pmatrix}_{xyz} \tag{E.21}$$

where E_{0x}, E_{0y}, H_{0x}, and H_{0y} are, in general, complex numbers. As shown in Section 5.2.2.6 in discussing small-amplitude oscillatory shear, a complex amplitude is the equivalent of adding a phase difference to the argument of the exponential. Thus E_x and E_y (and by analogy H_x and H_y) need not be in phase in order to satisfy the Maxwell equations. If E_{0x} and E_{0y} are constant and real, however, the x- and y-components of $\underline{E}$ will be in phase, and the wave is said to be linearly polarized (Figure E.2). If E_{0x} and E_{0y} are complex numbers, there is a phase difference between E_x and E_y. When E_{0x} and E_{0y} are constant but complex, the light wave is said to be elliptically polarized. If the phase difference between constant complex E_{0x} and E_{0y} is $\pi/2$ and if $E_{0x} = E_{0y}$, the light is circularly polarized.

The average value of the Poynting flux is called the *irradiance* or *intensity* I of the light:

$$I = |\langle\underline{S}\rangle| = |\langle\Re(\underline{E}) \times \Re(\underline{H})\rangle| \tag{E.22}$$

Using the following identity:

$$\Re(x) = \frac{x + x^{(*)}}{2} \tag{E.23}$$

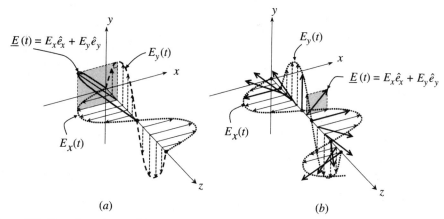

Figure E.2 Time dependence of the x- and y-components of the electric-field vector $\underline{E}$ for (a) linearly and (b) elliptically polarized light. In both cases the phase difference between the two components is constant. For linearly polarized light, the phase difference is zero.

we can show that the intensity is given by (see Problem E.5)

$$I = |\langle \underline{S} \rangle| = \frac{1}{2} \left| \Re \left(\underline{E}_0 \times \underline{H}_0^{(*)} \right) \right| \tag{E.24}$$

$$= \frac{1}{2} E_0 H_0 \tag{E.25}$$

where, as introduced earlier, $\underline{E} = \underline{E}_0 e^{i(\underline{k} \cdot \underline{r} - \omega t)}$, $\underline{H} = \underline{H}_0 e^{i(\underline{k} \cdot \underline{r} - \omega t)}$, $E_0 = |\underline{E}_0|$, and $H_0 = |\underline{H}_0|$. Taking the magnitude of both sides of Equation (E.16), we can see that

$$H_0 = \frac{\varepsilon \omega E_0}{k} = \varepsilon u E_0 \tag{E.26}$$

We thus obtain the familiar expression that the intensity of light is proportional to the square of the electric-field magnitude:

$$I = \frac{1}{2} \varepsilon u E_0^2 \tag{E.27}$$

E.3 Light in an Anisotropic Medium

E.3.1 General Wave Equation

When light propagates in an anisotropic medium, a more general form of the Maxwell equations is needed. This form contains light propagation in a vacuum and in an isotropic medium as special cases. The Maxwell equations for nonconducting nonmagnetic media are [81]

$$\boxed{\begin{aligned}
\nabla \times \underline{E} &= -\mu_0 \frac{\partial \underline{H}}{\partial t} \\
\nabla \times \underline{H} &= \varepsilon_0 \frac{\partial \underline{E}}{\partial t} + \frac{\partial \underline{P}}{\partial t} \\
\nabla \cdot \underline{E} &= -\frac{1}{\varepsilon_0} \nabla \cdot \underline{P} \\
\nabla \cdot \underline{H} &= 0
\end{aligned}}$$

Maxwell equations (anisotropic medium)   (E.28)

These versions of the Maxwell equations contain the polarization vector $\underline{P}$, which describes the volume density of electric dipoles in the medium. Setting $\underline{P}$ to zero recovers the Maxwell equations for a vacuum. Note that the combination $\varepsilon_0 \underline{E} + \underline{P}$ appears in two of the Maxwell equations. This quantity is called the *electric displacement*, and it is denoted by $\underline{D}$:

Electric displacement vector
$$\boxed{\underline{D} \equiv \varepsilon_0 \underline{E} + \underline{P}}$$ (E.29)

The electric displacement vector describes the aggregate response of the bound charges of a medium to the imposed electric field.

The case for an isotropic medium can be recovered by proposing the following simple constitutive equation for $\underline{D}$:

Constitutive equation for $\underline{D}$ (isotropic medium)
$$\boxed{\underline{D} = \varepsilon \underline{E}}$$ (E.30)

where ε is the same scalar permittivity of the medium as discussed in the last section. The polarization vector may then be solved for as

$$\underline{P} = (\varepsilon - \varepsilon_0)\underline{E} \tag{E.31}$$

$$= \varepsilon_0 \chi \underline{E} \tag{E.32}$$

where $\chi = \varepsilon/\varepsilon_0 - 1$ is the electric susceptibility, a scalar.

For anisotropic media, the response of the bound electrons to the light's electric field is not the same in every direction. Thus the electric displacement will not be parallel to the electric field as was assumed in the constitutive equation for $\underline{D}$, Equation (E.30). Rather, the electric displacement vector is a linear vector function of the electric-field vector, and therefore this function may be represented by a tensor:

Constitutive equation for $\underline{D}$ (anisotropic medium)
$$\boxed{\underline{D} = \underline{\underline{\epsilon}} \cdot \underline{E}}$$ (E.33)

The tensor $\underline{\underline{\epsilon}}$ is called the *dielectric tensor*. We may now solve for the polarization vector $\underline{P}$:

$$\underline{D} = \varepsilon_0 \underline{E} + \underline{P} \tag{E.34}$$

$$= \varepsilon_0 \underline{\underline{I}} \cdot \underline{E} + \underline{P} = \underline{\underline{\epsilon}} \cdot \underline{E} \tag{E.35}$$

$$\underline{P} = \varepsilon_0 \left(\frac{1}{\varepsilon_0} \underline{\underline{\epsilon}} - \underline{\underline{I}} \right) \cdot \underline{E} \tag{E.36}$$

$$= \varepsilon_0 \underline{\underline{\chi}} \cdot \underline{E} \tag{E.37}$$

where the susceptibility is now a tensor, $\underline{\underline{\chi}}$ [compare with Equation (E.32)]:

$$\text{Susceptibility tensor} \qquad \boxed{\underline{\underline{\chi}} \equiv \frac{1}{\varepsilon_0} \underline{\underline{\epsilon}} - \underline{\underline{I}}} \tag{E.38}$$

The susceptibility tensor $\underline{\underline{\chi}}$ contains information about how the medium responds to the electric field of the light. For ordinary nonabsorbing solids and liquids, $\underline{\underline{\chi}}$ is symmetric [33].

We can now obtain the wave equation for anisotropic media by substituting Equation (E.37) into the Maxwell equations for anisotropic media and eliminating $\underline{H}$ as before. The result is

$$\text{General wave equation} \qquad \boxed{\nabla \times \nabla \times \underline{E} + \frac{1}{c^2}\frac{\partial^2 \underline{E}}{\partial t^2} = -\frac{1}{c^2}\underline{\underline{\chi}} \cdot \frac{\partial^2 \underline{E}}{\partial t^2}} \tag{E.39}$$

which is the general wave equation for electrically neutral nonmagnetic, nonconducting anisotropic media. Solutions to this equation give $\underline{E}(\underline{r}, t)$ for light traveling in anisotropic media.

E.3.2 HARMONIC WAVES

The equation for harmonic plane waves is a solution to Equation (E.39) under certain circumstances. To find out what circumstances, we can substitute the expression for $\underline{E}$ for harmonic plane waves, $\underline{E} = \underline{E}_0 e^{i(\underline{k}\cdot\underline{r}-\omega t)}$, into Equation (E.39). The result is

$$\underline{k} \times \underline{k} \times \underline{E} + \frac{\omega^2}{c^2}\underline{E} = -\frac{\omega^2}{c^2}\underline{\underline{\chi}} \cdot \underline{E} \tag{E.40}$$

One solution to this equation is $\underline{E} = 0$. To find nontrivial solutions, we first evaluate the $\underline{k} \times \underline{k} \times \underline{E}$ term. This is most easily done using the determinant method of taking cross products. We will carry out this calculation in an arbitrary Cartesian coordinate system, $\hat{e}_x$, $\hat{e}_y$, $\hat{e}_z$:

$$\underline{k} \times \underline{E} = \begin{vmatrix} \hat{e}_x & \hat{e}_y & \hat{e}_z \\ k_x & k_y & k_z \\ E_x & E_y & E_z \end{vmatrix} \tag{E.41}$$

$$= (k_y E_z - k_z E_y)\hat{e}_x - (k_x E_z - k_z E_x)\hat{e}_y + (k_x E_y - k_y E_x)\hat{e}_z \tag{E.42}$$

$$= \begin{pmatrix} k_y E_z - k_z E_y \\ -(k_x E_z - k_z E_x) \\ k_x E_y - k_y E_x \end{pmatrix}_{xyz} \tag{E.43}$$

$$\underline{k} \times \underline{k} \times \underline{E} = \begin{vmatrix} \hat{e}_x & \hat{e}_y & \hat{e}_z \\ k_x & k_y & k_z \\ k_y E_z - k_z E_y & k_z E_x - k_x E_z & k_x E_y - k_y E_x \end{vmatrix} \tag{E.44}$$

Carrying out this final determinant and substituting back into Equation (E.40) results in

$$\begin{pmatrix} \left(-k_y^2 - k_z^2\right) E_x + k_x k_y E_y + k_x k_z E_z \\ k_y k_x E_x + \left(-k_x^2 - k_z^2\right) E_y + k_y k_z E_z \\ k_z k_x E_x + k_z k_y E_y + \left(-k_x^2 - k_y^2\right) E_z \end{pmatrix}_{xyz} + \frac{\omega^2}{c^2} \left(\underline{\underline{I}} + \underline{\underline{\chi}}\right) \cdot \underline{E} = 0 \qquad (E.45)$$

The left-hand side may be expressed as the product of a 3×3 matrix A, with the 3×1 matrix $\mathcal{E}$ that contains the coefficients of $\underline{E}$. If we define X as the 3×3 matrix of coefficients of the tensor $\underline{\underline{\chi}}$ in the same coordinate system, then we obtain the following matrix equation for the coefficients:

$$\left[A + \frac{\omega^2}{c^2}(I + X) \right] \mathcal{E} = 0 \qquad (E.46)$$

where

$$A = \begin{bmatrix} -k_y^2 - k_z^2 & k_x k_y & k_x k_z \\ k_y k_x & -k_x^2 - k_z^2 & k_y k_z \\ k_z k_x & k_z k_y & -k_x^2 - k_y^2 \end{bmatrix} \qquad (E.47)$$

and I is the identity matrix.

Since $\mathcal{E} \neq 0$, the solutions of Equation (E.46) result when values of k_x, k_y, and k_z are found that make the determinant of $A + (\omega^2/c^2)(I + X)$ equal to zero. Recall that $k = \sqrt{k_x^2 + k_y^2 + k_z^2} = \omega/u$. Thus for waves propagating through anisotropic media, the solved-for values of k_x, k_y, and k_z give allowed values of the phase velocity u, and the planes of constant phase travel in directions specified by the vector $\hat{k} = \underline{k}/k$.

We reemphasize that the Poynting vector $\underline{S} = \underline{E} \times \underline{H}$ points in the direction of travel of the light beam, whereas the wave vector $\underline{k}$ points in the direction in which waves of constant phase propagate. In an isotropic material the first two Maxwell equations require that $\underline{k}$ and $\underline{S}$ be parallel. For an anisotropic material, however, the polarizability vector $\underline{P}$ appears in the Maxwell equations, and $\underline{k}$ and $\underline{S}$ need not be parallel. The significance of this is that the planes of constant phase are inclined with respect to the direction of light propagation in an anisotropic material (Figure E.3).

E.3.2.1 Travel along an Optic Axis

Solving for the determinant of $A + (\omega^2/c^2)(I + X)$ can be simplified by taking advantage of the fact that $\underline{\underline{\chi}}$ is symmetric. For symmetric tensors it is possible to find an orthonormal coordinate system in which the tensor takes a diagonal form (see Appendix C.6). The axes of this coordinate system, $\hat{\xi}_x$, $\hat{\xi}_y$, $\hat{\xi}_z$, are called the *principal axes* of the tensor. Thus in its principal frame, the susceptibility tensor may be written as

$$\underline{\underline{\chi}} = \begin{pmatrix} \chi_1 & 0 & 0 \\ 0 & \chi_2 & 0 \\ 0 & 0 & \chi_3 \end{pmatrix}_{\xi_x \xi_y \xi_z} \qquad (E.48)$$

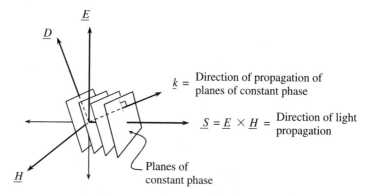

Figure E.3 Schematic showing that when light traverses anisotropic media, the wave vector direction is not necessarily the same as the direction of propagation of energy, the Poynting vector. $\underline{H}$, $\underline{D}$, and $\underline{k}$ are mutually perpendicular, as are $\underline{H}$, $\underline{E}$, and $\underline{S}$. From *Introduction to Modern Optics*, 2nd Edition, by G. R. Fowles (Dover Publications, Inc.: New York, (1989). Copyright © 1975, Grant R. Fowles. Reprinted by permission of Dover Publications, Inc.

The subscript $\xi_x \xi_y \xi_z$ on the tensor coefficients is a reminder that this is the form of $\underline{\underline{\chi}}$ only when this tensor is written in the principal frame. The three nonzero coefficients χ_1, χ_2, and χ_3 are called the *principal susceptibilities*, and it is with respect to these values that the principal dielectric constants K_i and the principal refractive indices n_i are defined:

$$K_i \equiv 1 + \chi_i \qquad (E.49)$$

and

Principal refractive indices
$$\boxed{n_i \equiv \sqrt{1 + \chi_i} = \sqrt{K_i}} \qquad (E.50)$$

From these principal values we can construct the refractive-index ($\underline{n}$) and dielectric-constant ($\underline{K}$) tensors, both of which have the same principal frame as $\underline{\underline{\chi}}$:

$$\underline{\underline{n}} = \begin{pmatrix} n_1 & 0 & 0 \\ 0 & n_2 & 0 \\ 0 & 0 & n_3 \end{pmatrix}_{\xi_x \xi_y \xi_z} \qquad (E.51)$$

$$\underline{\underline{K}} = \begin{pmatrix} K_1 & 0 & 0 \\ 0 & K_2 & 0 \\ 0 & 0 & K_3 \end{pmatrix}_{\xi_x \xi_y \xi_z} \qquad (E.52)$$

Since $\underline{n}$ and $\underline{K}$ are tensors, they can be transformed to other coordinate systems in the usual way (see Appendix C.5). We can see from these definitions that the following relationships hold among the various tensors discussed:

$$\underset{=}{n} \cdot \underset{=}{n} = \underset{=}{I} + \underset{=}{\chi} = \underset{=}{K} \tag{E.53}$$

When $\underset{=}{\chi}$ and $\underset{=}{E}$ are written in the frame of reference formed by the principal axes, the determinant equation $|A + (\omega^2/c^2)(I + X)| = 0$ becomes

$$\begin{vmatrix} \left(\frac{n_1\omega}{c}\right)^2 - k_y^2 - k_z^2 & k_x k_y & k_x k_z \\ k_y k_x & \left(\frac{n_2\omega}{c}\right)^2 - k_x^2 - k_z^2 & k_y k_z \\ k_z k_x & k_z k_y & \left(\frac{n_3\omega}{c}\right)^2 - k_x^2 - k_y^2 \end{vmatrix} = 0 \tag{E.54}$$

Note that we continue to use the notation k_x, k_y, k_z for the coefficients of $\underset{}{k}$, but now they represent the coefficients of $\underset{}{k}$ in the principal frame, $\underset{}{k} = k_x\hat{\xi}_x + k_y\hat{\xi}_y + k_z\hat{\xi}_z$. A notation such as k_{ξ_x}, k_{ξ_y}, k_{ξ_z} would be too awkward and does not conform to standard practice. We continue to use subscripts on the matrix representation of tensor coefficients to emphasize the coordinate system in use.

The solutions to Equation (E.54) fall on a double surface in three-dimensional k_x, k_y, k_z space (Figure E.4). The values of k_x, k_y, and k_z that lie on the surface are solutions to the Maxwell equations for anisotropic media. Since the surface is double, however, for a given direction of wave-front propagation $\hat{k}$, there are two possible values of the magnitude of the wave vector, $k = \omega/u = n\omega/c$, that is, for most directions of wave-front propagation there are two possible values of the phase velocity u or, equivalently, of the refractive index n. How the two solutions affect the travel of a wave depends on the polarization state of the incoming wave, as we will show in the next section.

As seen in Figure E.4, there exist directions for which only one phase velocity is prescribed by the Maxwell equations. These directions are called the *optic axes* of the material. Figure E.5 shows the three cases that can occur, depending on the shapes of the intercepts of the wave-vector surfaces with the k_y-plane. These three cases correspond to uniaxial positively birefringent ($n_1 = n_2 < n_3$), uniaxial negatively birefringent

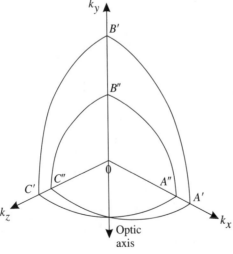

Figure E.4 Three-dimensional surface of solutions of Equation (E.54) [81] for a biaxial crystal. From M. Born and E. Wolf, *Principles of Optics: Electromagnetic Theory of Propagation, Interference and Diffraction of Light*, 6th Edition, Copyright © 1980, Max Born and Emil Wolf. Reprinted with the permission of Cambridge University Press.

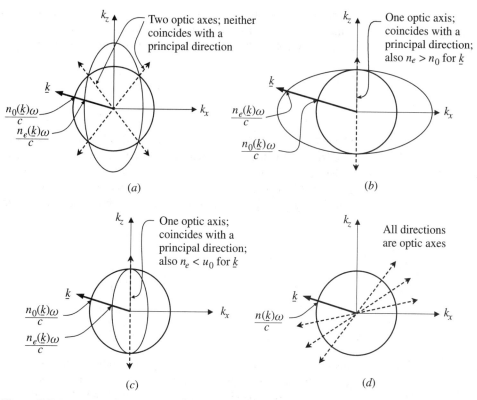

Figure E.5 Intercepts of the wave-vector surfaces in the $k_x k_z$-plane for various materials. (a) Biaxial. (b) Uniaxial positive. (c) Uniaxial negative. (d) Isotropic. An optic axis is a direction in which there is only one possible ray velocity.

($n_1 = n_2 > n_3$), and biaxial materials (n_1, n_2, n_3 all distinct). For uniaxial materials the birefringence Δn is defined as the difference between the principal refractive indices, that is, $\Delta n = n_3 - n_1 = n_3 - n_2$, and for biaxial materials there are two independent birefringences, $\Delta n_{31} = n_3 - n_1$ and $\Delta n_{21} = n_2 - n_1$. For uniaxially birefringent materials, the two equal refractive indices are called *ordinary* indices $n_1 = n_2 \equiv n_0$, and the different index is called the *extraordinary* index, $n_3 \equiv n_e$. In the case where the intercepts of the wave-vector surfaces are circles and coincide exactly, the material is isotropic ($n_1 = n_2 = n_3 \equiv n$). Since the components of $\underline{E}$ propagate with a single common velocity along the optic axes, light passes through a birefringent medium without change in polarization if the direction of propagation corresponds to an optic axis.

E.3.2.2 Travel along a Principal Direction

For light traveling along a direction other than an optic axis, there are two possible phase velocities, u' and u''. It can be shown that these two velocities always correspond to two mutually perpendicular polarizations of the electric-field vector $\underline{E}$ [33]. For example, if the wave fronts of a linearly polarized plane wave are propagating in an anisotropic medium of thickness h along principal direction $\hat{\xi}_z$, the wave vector is $\underline{k} = k_z\hat{\xi}_z = k\hat{\xi}_z$, and Equation (E.54) has two solutions, $k' = n_1\omega/c$ and $k'' = n_2\omega/c$. $\underline{H}$ lies in the $\xi_x\xi_y$-plane, and the incident electric-field vector may be written in the principal frame as

$$\underline{E} = e^{i(kz-\omega t)} \begin{pmatrix} E_{0x} \\ E_{0y} \\ E_{0z} \end{pmatrix}_{\xi_x\xi_y\xi_z} \tag{E.55}$$

where E_{0x}, E_{0y}, and E_{0z} are constant real numbers since the incoming light is linearly polarized. In an anisotropic medium we do not know a priori the direction of propagation of the wave $\underline{S} = \underline{E} \times \underline{H}$, since the vector $\underline{S}$ need not be parallel to $\underline{k}$. We can deduce the direction of $\underline{S}$ by first considering the constraints on $\underline{E}$ presented by the Maxwell equations. Substituting $\underline{k} = k\hat{\xi}_z$ into Equation (E.45), we obtain the following three equations:

$$\left(\frac{\omega^2 n_1^2}{c^2} - k^2 \right) E_x = 0 \tag{E.56}$$

$$\left(\frac{\omega^2 n_2^2}{c^2} - k^2 \right) E_y = 0 \tag{E.57}$$

$$\left(\frac{\omega^2 n_3^2}{c^2} \right) E_z = 0 \tag{E.58}$$

Since $\omega \neq 0$ and $n_3 \neq 0$, the third equation indicates that $E_z = E_{0z}e^{i(kz-\omega t)} = 0$. Thus $\underline{E}$ must lie in the $\xi_x\xi_y$-plane. The direction of $\underline{H}$ is unaffected by travel in the anisotropic sample [see Equation (E.28)], and since both $\underline{E}$ and $\underline{H}$ lie in the $\xi_x\xi_y$-plane, $\underline{S} = \underline{E} \times \underline{H}$ must be parallel to the $\hat{\xi}_z$-direction. Thus for light propagating along a principal direction in anisotropic media, $\underline{S}$ and $\underline{k}$ are once again parallel.

Returning to Equation (E.56), for $E_x \neq 0$, $k = k'$ is a solution whereas $k = k''$ is not. Thus the x-component of $\underline{E}$ travels at the speed $u' = \omega/k' = c/n_1$. In Equation (E.57), for $E_y \neq 0$, $k = k''$ satisfies the equation, and $k = k'$ does not. Thus the y-component of $\underline{E}$ travels at the speed $u'' = \omega/k'' = c/n_2$. These two components of $\underline{E}$ may be treated as two waves traveling independently through the anisotropic sample.

We now see that in an anisotropic sample with light propagating along a principal direction (in this example, $\hat{\xi}_z$) $\underline{k}$ and $\underline{S}$ are parallel, $E_z = 0$, and the ξ_x- and ξ_y-components of $\underline{E}$ travel at different speeds. As a result, upon exiting an anisotropic sample of thickness h, the electric field of the wave will have the following form:

$$\underline{E} = e^{-i\omega t} \begin{pmatrix} E_{0x}e^{\frac{in_1\omega h}{c}} \\ E_{0y}e^{\frac{in_2\omega h}{c}} \\ 0 \end{pmatrix}_{\xi_x\xi_y\xi_z} \tag{E.59}$$

$$= e^{-i\omega t} e^{\frac{in_1\omega h}{c}} \begin{pmatrix} E_{0x} \\ E_{0y}e^{\frac{i\Delta n_{21}\omega h}{c}} \\ 0 \end{pmatrix}_{\xi_x\xi_y\xi_z} \tag{E.60}$$

where $\Delta n_{21} = n_2 - n_1$ is the birefringence of the sample. The effect on the light of having passed through the birefringent sample of thickness h along a principal direction is to have altered the absolute phase of the wave by the factor $n_1 h/c$ and to have introduced a phase difference δ between the ξ_x- and ξ_y-components:

$$\underline{E} = e^{i\left(\frac{2\pi h n_1}{\lambda_v} - \omega t\right)} \begin{pmatrix} E_{0x} \\ E_{0y} e^{i\delta} \\ 0 \end{pmatrix}_{\xi_x \xi_y \xi_z} \tag{E.61}$$

$$\delta \equiv \frac{\Delta n_{21} \omega h}{c} = \frac{2\pi h \Delta n_{21}}{\lambda_v} \tag{E.62}$$

where $\lambda_v = 2\pi c/\omega$ is the wavelength of the light in vacuum.

E.3.2.3 Travel in an Arbitrary Direction

The same procedure as used in the preceding must be followed when the direction of propagation is along a direction other than a principal direction. To calculate the effect of transversing a birefringent medium on the polarization of a beam, allowed phase velocities are calculated from Equation (E.54), and the directions of polarization that are affected are obtained from Equation (E.45). As an example of this more complicated type of calculation, we will derive the polarization change that results when a linearly polarized beam is incident down the 2-direction of shear flow.

The shear-flow coordinate system is not the principal coordinate system for stress or for flow birefringence, as is discussed in Chapter 10. The 3- or neutral direction is a principal direction, but the other two principal axes are rotated counterclockwise from the 1- and 2-directions by an angle χ. We seek to calculate the polarization changes that take place in a linearly polarized beam traversing a birefringent fluid in the $-\hat{e}_2$-direction of shear.

For the incoming light of this example we wish to consider linearly polarized light with waves of constant phase traveling along $\underline{k} = -k\hat{e}_2$ (Figure E.6):

$$\underline{k} = -k\hat{e}_2 = \begin{pmatrix} 0 \\ -k \\ 0 \end{pmatrix}_{123} \tag{E.63}$$

Recall that since $\underline{E}$, $\underline{k}$, and $\underline{S}$ are all perpendicular to $\underline{H}$, they are coplanar. The coordinate system used above is the usual Cartesian shear coordinate system, where $\hat{e}_1$ is the flow direction, $\hat{e}_2$ is the gradient direction, and $\hat{e}_3$ is the neutral direction. Since the incident light is traveling in the $-\hat{e}_2$-direction, both $\underline{E}$ and $\underline{H}$ for the *incident* wave are in the 13-plane, and $\underline{S} = \Re\{\underline{E}\} \times \Re\{\underline{H}\} = -S\hat{e}_2$. Because the medium is anisotropic, $\underline{S}$ for the beam inside the sample is not necessarily parallel to $\underline{k}$, as was discussed in Section E.3.2, and as we will calculate now.

The incoming light is linearly polarized in the 13-plane, and hence $E_{02} = 0$, and E_{01} and E_{03} are constant and real. In the shear coordinate system we can write $\underline{E}$ for the incident light as

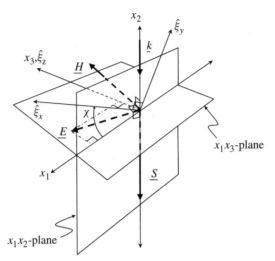

Figure E.6 Relationship between $\underline{E}$, $\underline{H}$, and $\underline{S}$ for linearly polarized incident light about to pass through a shearing cell along the -2-direction of shear flow (not a principal direction). $\underline{E}$ and $\underline{H}$ of the incident light are in the 13-plane and $\underline{S}$ is in the -2-direction. $\hat{\xi}_t, \hat{\xi}_y, \hat{\xi}_z$ are the principal directions of the shear flow.

$$\underline{E}_{\text{incident}} = e^{i(-kx_2-\omega t)} \begin{pmatrix} E_{01} \\ 0 \\ E_{03} \end{pmatrix}_{123} \tag{E.64}$$

$\underline{H}$ is given by

$$\underline{H} = \begin{pmatrix} H_1 \\ H_2 \\ H_3 \end{pmatrix}_{123} = e^{i(-kx_2-\omega t)} \begin{pmatrix} H_{01} \\ 0 \\ H_{03} \end{pmatrix}_{123} \tag{E.65}$$

In order to use Equation (E.54) we must express $\underline{k}$ in the principal frame of the anisotropic medium $\hat{\xi}_x, \hat{\xi}_y, \hat{\xi}_z$. The relationship between the two frames is (see Section 10.3)

$$\begin{aligned}
\hat{e}_1 &= \cos \chi \, \hat{\xi}_x - \sin \chi \, \hat{\xi}_y & \hat{\xi}_x &= \cos \chi \, \hat{e}_1 + \sin \chi \, \hat{e}_2 \\
\hat{e}_2 &= \sin \chi \, \hat{\xi}_x + \cos \chi \, \hat{\xi}_y & \hat{\xi}_y &= -\sin \chi \, \hat{e}_1 + \cos \chi \, \hat{e}_2 \\
\hat{e}_3 &= \hat{\xi}_z & \hat{\xi}_z &= \hat{e}_3
\end{aligned} \tag{E.66}$$

Therefore $\underline{k}$ and $\underline{E}$ are given by

$$\underline{k} = -k\hat{e}_2 = \begin{pmatrix} -k \sin \chi \\ -k \cos \chi \\ 0 \end{pmatrix}_{\xi_x \xi_y \xi_z} \tag{E.67}$$

$$\underline{E}_{\text{incident}} = \begin{pmatrix} E_x \\ E_y \\ E_z \end{pmatrix}_{\xi_x \xi_y \xi_z} \tag{E.68}$$

$$= e^{i[-k(\sin \chi \, \xi_x + \cos \chi \, \xi_y) - \omega t]} \begin{pmatrix} E_{01} \cos \chi \\ -E_{01} \sin \chi \\ E_{03} \end{pmatrix}_{\xi_x \xi_y \xi_z} \tag{E.69}$$

where ξ_x and ξ_y are the coordinate variables in the $\hat{\xi}_x$- and $\hat{\xi}_y$-directions, respectively.

Substituting the principal components of $\underline{k}$ [Equation (E.67)] into Equation (E.54) we arrive at two solutions for k:

$$k' = \frac{\omega n_3}{c} \tag{E.70}$$

$$k'' = \frac{\omega}{c} \left(\frac{\sin^2 \chi}{n_2^2} + \frac{\cos^2 \chi}{n_1^2} \right)^{-\frac{1}{2}} \tag{E.71}$$

which correspond to the two phase velocities u' and u'' that are allowable for the light within the birefringent medium ($k = \omega/u$).

To determine which polarization states exhibit these velocities, we return to Equation (E.45). For $\underline{k}$ given by Equation (E.67), Equation (E.45) becomes[3]

$$\begin{pmatrix} -k^2 \cos^2 \chi \, E_x + k^2 \sin \chi \cos \chi \, E_y \\ k^2 \sin \chi \cos \chi \, E_x - k^2 \sin^2 \chi \, E_y \\ -k^2 E_z \end{pmatrix}_{\xi_x \xi_y \xi_z} + \frac{\omega^2}{c^2} \begin{pmatrix} (1 + \chi_1) E_x \\ (1 + \chi_2) E_y \\ (1 + \chi_3) E_z \end{pmatrix}_{\xi_x \xi_y \xi_z} = 0 \quad \text{(E.72)}$$

Substituting k' into the ξ_z-component of Equation (E.72), we see that the equation is satisfied identically. Thus this speed is allowed for the ξ_z-component of the incoming light, that is, for $E_z \hat{\xi}_z = E_{03} e^{i(-kx_2 - \omega t)} \hat{e}_3$. When k'' is substituted into the z-component, however, $E_z = 0$ is the only solution. Thus k'' is not a possible wave vector for the ξ_z-component of $\underline{E}$, and we conclude that the ξ_z-component of the incoming light must travel with wave number k' only (Figure E.7).[4]

The direction of travel of a ray is the direction of the Poynting vector, $\underline{S} = \Re\{\underline{E}\} \times \Re\{\underline{H}\}$. For the wave that travels with wave number k', $\underline{E} = E_{03} e^{i(-kx_2 - \omega t)} \hat{e}_3$, the corresponding magnetic-field vector is the x_1-component of $\underline{H}$ (since electric and magnetic fields are perpendicular). Thus we can find the direction of the Poynting vector from $\Re\{E_3 \hat{e}_3\} \times \Re\{H_1 \hat{e}_1\}$, and we obtain that $\underline{S}$ for the wave traveling with wave number k' is in the $-\hat{e}_2$-direction ($H_{01} < 0$; Figure E.6).

In summary, $\underline{E}$, $\underline{H}$, and $\underline{S}$ are given for the wave traveling at wave number k' down the $-\hat{e}_2$-direction of shear flow as follows:

Light-related vectors for the wave traveling with wave vector k'

$$\underline{E}(k') = e^{i(-k'x_2 - \omega t)} \begin{pmatrix} 0 \\ 0 \\ E_{03} \end{pmatrix}_{123} \tag{E.73}$$

---

[3] Note that in the field of rheology it is standard to use χ for the angle between the flow direction and the principal direction $\hat{\xi}_x$, whereas in optics it is standard to use χ_1, χ_2, and χ_3 for the principal values of the susceptibility tensor. This is an unfortunate overlap in nomenclature.

[4] Note that the angles in the figure are not the usual spherical coordinate angles (which originate at axis $\hat{e}_3$) but rather angles that define $\underline{S}$ with respect to $\hat{e}_2$.

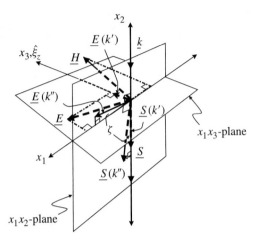

Figure E.7 Relationship between $\underline{E}$, $\underline{H}$, and $\underline{S}$ for light incident down the -2-direction of shear flow (not a principal direction) and these same vectors some time after penetrating an anisotropic material. Initially $\underline{E}$ and $\underline{H}$ are in the 13-plane and $\underline{S}$ is in the -2-direction. The x_3-component of $\underline{E}$ travels at wave number k', and after traveling in the anisotropic medium for some time it still points in the 3-direction $[\underline{E}(k')]$ since $\hat{e}_3$ is a principal direction. The x_1-component of $\underline{E}$ travels at wave number k'', but since the $\hat{e}_2$-direction is not a principal direction, this component of $\underline{E}$ rotates in the 12-plane and becomes vector $\underline{E}(k'')$, as shown. Since $\underline{S} = \Re\{\underline{E}\} \times \Re\{\underline{H}\}$, the Poynting vector $\underline{S}(k')$ for the electric-field vector $\underline{E}(k')$ is parallel to the $-x_2$-axis, but the Poynting vector $\underline{S}(k'')$, for $\underline{E}(k'')$ is rotated away from the $-x_2$-axis by an angle ζ.

$$H(k') = e^{i(-k'x_2 - \omega t)} \begin{pmatrix} H_{01} \\ 0 \\ 0 \end{pmatrix}_{123} \tag{E.74}$$

$$S(k') = \begin{pmatrix} 0 \\ H_{01} E_{03} \cos^2(-k'x_2 - \omega t)\hat{e}_2 \\ 0 \end{pmatrix}_{123} \tag{E.75}$$

Similarly, k' does not satisfy either the ξ_x- or the ξ_y-components of Equation (E.72) for $E_x \neq 0$, $E_y \neq 0$, while substituting k'' into either the ξ_x- or the ξ_y-components of Equation (E.72) results in the following relationship between the components of $\underline{E}$ in the anisotropic medium:

$$\frac{E_y''}{E_x''} = \frac{E_{0y}''}{E_{0x}''} = -\frac{n_1^2}{n_2^2} \tan \chi \tag{E.76}$$

where $\underline{E}(k'')$ can be written as

$$\underline{E}(k'') \equiv \begin{pmatrix} E_x'' \\ E_y'' \\ 0 \end{pmatrix}_{\xi_x \xi_y \xi_z} \tag{E.77}$$

$$= e^{i[-k''(\sin \chi \, \xi_x + \cos \chi \, \xi_y) - \omega t]} \overbrace{\begin{pmatrix} E_{0x}'' \\ E_{0y}'' \\ 0 \end{pmatrix}_{\xi_x \xi_y \xi_z}}^{\equiv \underline{E}_0(k'')} \tag{E.78}$$

The incident light that travels at wave number k'' is just

$$\underline{E}(k'')_{\text{incident}} = e^{i[-k''(\sin \chi\, \xi_x + \cos \chi\, \xi_y) - \omega t]} \begin{pmatrix} E_{01} \cos \chi \\ -E_{01} \sin \chi \\ 0 \end{pmatrix}_{\xi_x \xi_y \xi_z} \tag{E.79}$$

Note that the ratio of the ξ_y-component to the ξ_x-component has changed by an amount n_1^2/n_2^2 as the incident light entered the anisotropic medium [compare Equation (E.76) with Equation (E.79)]. This indicates that the direction of polarization of the beam has rotated.

Referring to Figure E.8 we can see that geometrically the ξ_x- and ξ_y-components of $\underline{E}(k'')$ are related as

$$\frac{-E_y''}{E_x''} = \tan (\chi - \zeta) \tag{E.80}$$

This relation, when combined with Equation (E.76), shows us that $\underline{E}(k'')$ is rotated away from the $\hat{e}_1$-axis in the 12-plane by an angle ζ:

$$\zeta = \chi - \tan^{-1}\left(\frac{n_1^2}{n_2^2} \tan \chi\right) \tag{E.81}$$

Note that $\zeta < 0$ ($\underline{E}(k'')$ rotates clockwise from $\hat{e}_1$) and that when the refractive index is isotropic ($n_1 = n_2$), $\zeta = 0$ as expected. We can also write out $\underline{E}(k'')$ in the 123 (shear) coordinate system. The incident light that travels at k'' is the 1-component of $\underline{E}$:

$$\underline{E}_{\text{incident}}(k'') = e^{i(-k''x_2 - \omega t)} \begin{pmatrix} E_{01} \\ 0 \\ 0 \end{pmatrix}_{123} \tag{E.82}$$

Within the medium, $\underline{E}(k'')$ becomes

$$\underline{E}(k'') = \begin{pmatrix} E_x'' \\ E_y'' \\ 0 \end{pmatrix}_{\xi_x \xi_y \xi_z} \tag{E.83}$$

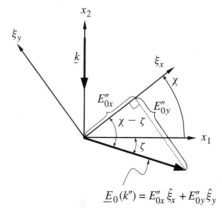

Figure E.8 Relationship defining the angle ζ in the example of light incident down the $-\hat{e}_2$-direction of shear flow.

$$= \begin{pmatrix} E''_x \\ -E''_x \frac{n_1^2}{n_2^2} \tan \chi \\ 0 \end{pmatrix}_{\xi_x \xi_y \xi_z} \tag{E.84}$$

In the 123 coordinate system this becomes

$$\underline{E}(k'') = E''_x \begin{pmatrix} \cos \chi \left(1 + \frac{n_1^2}{n_2^2} \tan^2 \chi \right) \\ \sin \chi \left(1 - \frac{n_1^2}{n_2^2} \right) \\ 0 \end{pmatrix}_{123} \tag{E.85}$$

We see that $\underline{E}(k'')$ has a nonzero component in the 2-direction as concluded earlier. When $n_1 = n_2$, however, the 2-component of $\underline{E}$ becomes zero, as expected. Thus the 1-component of $\underline{E}$ of the incident light rotates by an angle ζ in the 12-plane when it enters the birefringent sample and travels at wave number k''.

We can calculate $\underline{H}(k'')$ in the same manner as we calculated $\underline{H}(k')$. Since $\underline{H}$ is unaffected by travel in an anisotropic medium, it is perpendicular to the incident $\underline{E}(k'')$, which is polarized in the $\hat{e}_1$-direction:

$$\underline{E}_{\text{incident}}(k'') = e^{i(-k''x_2 - \omega t)} \begin{pmatrix} E_{01} \\ 0 \\ 0 \end{pmatrix}_{123} \tag{E.86}$$

$$\underline{H} = e^{i(-k''x_2 - \omega t)} \begin{pmatrix} 0 \\ 0 \\ H_{03} \end{pmatrix}_{123} \tag{E.87}$$

To calculate the Poynting vector we return to its definition and use Equations (E.78) and (E.87):

$$\underline{S} = \Re\{\underline{E}(k'')\} \times \Re\{\underline{H}(k'')\} \tag{E.88}$$

$$= \cos\left[-k''(\sin \chi \, \xi_x + \cos \chi \, \xi_y) - \omega t\right] \begin{pmatrix} E''_{0x} \\ E''_{0y} \\ 0 \end{pmatrix}_{\xi_x \xi_y \xi_z}$$

$$\times \cos\left[-k''(\sin \chi \, \xi_x + \cos \chi \, \xi_y) - \omega t\right] \begin{pmatrix} 0 \\ 0 \\ H_{03} \end{pmatrix}_{\xi_x \xi_y \xi_z} \tag{E.89}$$

$$= \begin{vmatrix} \hat{\xi}_x & \hat{\xi}_y & \hat{\xi}_z \\ E''_{0x} & E''_{0y} & 0 \\ 0 & 0 & H_{03} \end{vmatrix} \cos^2\left(-k''[\sin \chi \, \xi_x + \cos \chi \, \xi_y] - \omega t\right) \tag{E.90}$$

$$= \cos^2 \left[-k''(\sin \chi \, \xi_x + \cos \chi \, \xi_y) - \omega t\right] \begin{pmatrix} E_{0y}'' H_{03} \\ -E_{0x}'' H_{03} \\ 0 \end{pmatrix}_{\xi_x \xi_y \xi_z} \tag{E.91}$$

which shows that $\underline{S}(k'')$ is in the 12-plane. To determine the angle α that $\underline{S}$ makes with $\hat{e}_2$, we can dot $\hat{e}_2$ with $\underline{S}$,[5]

$$\hat{e}_2 \cdot \underline{S} = S \cos \alpha \tag{E.92}$$

$$= \begin{pmatrix} \sin \chi \\ \cos \chi \\ 0 \end{pmatrix}_{\xi_x \xi_y \xi_z} \cdot \cos^2 \left[-k''(\sin \chi \, \xi_x + \cos \chi \, \xi_y) - \omega t\right] \begin{pmatrix} E_{0y}'' H_{03} \\ -E_{0x}'' H_{03} \\ 0 \end{pmatrix}_{\xi_x \xi_y \xi_z} \tag{E.93}$$

and

$$H_{03} |\underline{E}_0(k'')| \cos \alpha = H_{03} (E_{0y}'' \sin \chi - E_{0x}'' \cos \chi) \tag{E.94}$$

$$\cos \alpha = -\cos \zeta \tag{E.95}$$

$$\alpha = \pi \pm \zeta \tag{E.96}$$

Examining Equation (E.91) we see that both the ξ_x- and the ξ_y-components of $\underline{S}$ are negative. Thus we conclude that $\alpha = \pi - \zeta$ (remember that ζ is negative), and $\underline{S}$ points in the direction indicated on Figure E.9. Thus for the wave traveling with wave vector $k = k''$ down the $-\hat{e}_2$-direction of shear flow, $\underline{E}$, $\underline{H}$, and $\underline{S}$ are given as

Light-related vectors
for beam traveling
with wave number k''

$$\underline{E}(k'') = E_x'' \begin{pmatrix} \cos \chi \left(1 + \frac{n_1^2}{n_2^2} \tan^2 \chi\right) \\ \sin \chi \left(1 - \frac{n_1^2}{n_2^2}\right) \\ 0 \end{pmatrix}_{123} \tag{E.97}$$

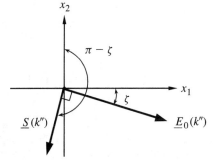

Figure E.9 Direction of Poynting vector for the wave traveling at wave number k''. The angle between $\hat{e}_2$ and $\underline{S}$ was calculated in text to be $\pi - \zeta$.

---

[5] Note that $E_{0y}''/|\underline{E}_0(k'')| = -\sin(\chi - \zeta)$ and $E_{0x}''/|\underline{E}_0(k'')| = \cos(\chi - \zeta)$; see Figure E.8.

$$\underline{H} = e^{i(-k''x_2 - \omega t)} \begin{pmatrix} 0 \\ 0 \\ H_{03} \end{pmatrix}_{123} \tag{E.98}$$

$$\underline{S}(k'') = \cos^2(-k''x_2 - \omega t) H_{03} \begin{pmatrix} E''_{0y} \cos \chi + E''_{0x} \sin \chi \\ E''_{0y} \sin \chi - E''_{0x} \cos \chi \\ 0 \end{pmatrix}_{123} \tag{E.99}$$

In summary, for linearly polarized light incident in the 2-direction of shear flow (not a principal direction), two beams will be produced. One, polarized along $\hat{e}_3$ (the ordinary ray), will travel in the $-\hat{e}_2$-direction, the same direction as the incident light. The other, polarized in the 12-plane (the extraordinary ray), will travel at an angle ζ with respect to $-\hat{e}_2$ [Equation (E.81)].[6]

Recall that to measure birefringence, we need to identify the phase lag between two components of $\underline{E}$ caused by these different components traveling at different speeds in the medium. The splitting of the incident light into two beams that travel in different directions, as calculated before, complicates our measurement. For $\Delta n = n_1 - n_2$ not too large, however, Equation (E.81) shows us that ζ will be small, and the ordinary and extraordinary beams will approximately coincide. For polymer melts for which we wish to measure flow birefringence, Δn is very small. For example, for high-density polyethylene (HDPE), Wales [256] reports a birefringence of 1.6×10^{-5}, which leads to values of ζ of less than 0.02 degree. For such small angles we can neglect the different paths of the ordinary and extraordinary rays and calculate just the changes in polarization experienced by the combined beam as it passes through a birefringent medium of thickness h.

To calculate the phase difference induced by sending light down the -2-direction, we begin with the expression for the electric field of the incident light, written in the shear coordinate system:

$$\underline{E} = \begin{pmatrix} E_1 \\ E_2 \\ E_3 \end{pmatrix}_{123} = e^{i(\underline{k} \cdot \underline{r} - \omega t)} \begin{pmatrix} E_{01} \\ 0 \\ E_{03} \end{pmatrix}_{123} \tag{E.100}$$

We have already established that the 3-component travels in the $-\hat{e}_2$-direction with $\underline{k} = \underline{k}'$,

$$E_3 = E_{03} e^{i(\underline{k}' \cdot \underline{r} - \omega t)} \tag{E.101}$$

For a sample of thickness h and for a wave traveling in the $-\hat{e}_2$-direction, $\underline{r} = -h\hat{e}_2$ at the exit, and $\underline{E}_3$ therefore becomes

---

[6] Upon exiting the medium, the extraordinary ray, which in the medium is polarized according to Equation (E.85), will be polarized in the 1-direction only since in air, which is an isotropic medium, the electric field must be perpendicular to the wave vector, $\underline{k} = -k\hat{e}_2$. The direction of the wave vector, that is, the direction of propagation of planes of constant phase, is not changed by traveling in anisotropic media.

$$E_3 = E_{03} e^{i(hk' - \omega t)} \tag{E.102}$$

The 1-component of $\underline{E}$ travels with $\underline{k} = \underline{k}''$, also along $\underline{S} = -S\hat{e}_2$ (for $\zeta \approx 0$), and analogously,

$$E_1 = E_{01} e^{i(\underline{k}'' \cdot \underline{r} - \omega t)} \tag{E.103}$$

$$= E_{01} e^{i(hk'' - \omega t)} \tag{E.104}$$

Since ζ is small, $\underline{E}$ remains perpendicular to $\underline{k}$, and E_2 is zero. Combining the two results for E_1 and E_3, we obtain

$$\underline{E} \text{ (at exit)} = e^{-i\omega t} \begin{pmatrix} E_{01} e^{ihk''} \\ 0 \\ E_{03} e^{ihk'} \end{pmatrix}_{123} = e^{i(hk' - \omega t)} \begin{pmatrix} E_{01} e^{ih(k'' - k')} \\ 0 \\ E_{03} \end{pmatrix}_{123} \tag{E.105}$$

We can define an effective refractive index for the 12-plane by writing

$$k'' = \frac{\omega n_{\text{eff}}}{c} = \frac{2\pi}{\lambda_v} n_{\text{eff}} \tag{E.106}$$

where $\lambda_v = \frac{2\pi c}{\omega}$ is the wavelength of the light in a vacuum. Therefore, combining with Equation (E.71) we have

$$\frac{1}{n_{\text{eff}}^2} = \frac{\sin^2 \chi}{n_2^2} + \frac{\cos^2 \chi}{n_1^2} \tag{E.107}$$

and the final expression for $\underline{E}$ is

$$\underline{E} = e^{i\left(\frac{2\pi h n_3}{\lambda_v} - \omega t\right)} \begin{pmatrix} E_{01} e^{i\frac{2\pi h}{\lambda_v}(n_{\text{eff}} - n_3)} \\ 0 \\ E_{03} \end{pmatrix}_{123} \tag{E.108}$$

Note the similarity to Equation (E.61). We see that for the values of n_1^2/n_2^2 usually encountered in flow-birefringence measurements, the effect of sending linearly polarized light along a direction that is not a principal direction (in this case, the 2-direction of shear flow) is the same as sending the light along a principal direction, that is, the overall phase changes, and a phase difference is introduced between two nonzero components of $\underline{E}$. For propagation along a nonprincipal direction, however, the measured birefringence ($n_3 - n_{\text{eff}} = \Delta n_{\text{eff},3}$, in this case) is not a principal birefringence, and the direction of propagation of planes of constant phase $\underline{k}$ is only approximately parallel to the direction of energy propagation $\underline{S}$.

The effective refractive index can be related to the components of $\underline{\underline{n}}$ when it is written in the shear coordinate system:

$$\begin{pmatrix} n_{11} & n_{21} & 0 \\ n_{21} & n_{22} & 0 \\ 0 & 0 & n_{33} \end{pmatrix}_{123} \tag{E.109}$$

To arrive at the result, either we carry out the same calculation in the shear frame (rather than in the principal frame, as was done before; see Problem E.7), or n_1, n_2, and χ can be related to n_{ij} in the shear frame through a change-of-frame calculation (see Problem E.6 and below). In either case the result is

$$n_{\text{eff}} = \frac{n_{11}n_{22} - n_{21}^2}{\sqrt{n_{21}^2 + n_{22}^2}} \tag{E.110}$$

The expression for the effective refractive index simplifies when small values of birefringence are considered. To show this, consider $\underline{\underline{n}}$ in the principal frame:

$$\begin{pmatrix} n_1 & 0 & 0 \\ 0 & n_2 & 0 \\ 0 & 0 & n_3 \end{pmatrix}_{\xi_x \xi_y \xi_z} \tag{E.111}$$

We can calculate n_{ij} in the shear frame in terms of n_i and χ by performing a coordinate rotation on $\underline{\underline{n}}$ (see Appendix C.5):

$$[n]_{123} = L[n]_{\xi_x \xi_y \xi_z} L^T \tag{E.112}$$

$$L = \begin{bmatrix} \cos \chi & \sin \chi & 0 \\ -\sin \chi & \cos \chi & 0 \\ 0 & 0 & 1 \end{bmatrix} \tag{E.113}$$

which yields

$$[n]_{123} = \begin{bmatrix} n_1 \cos^2 \chi + n_2 \sin^2 \chi & (n_1 - n_2) \sin \chi \cos \chi & 0 \\ (n_1 - n_2) \sin \chi \cos \chi & n_1 \sin^2 \chi + n_2 \cos^2 \chi & 0 \\ 0 & 0 & n_3 \end{bmatrix} \tag{E.114}$$

$$\underline{\underline{n}} = \begin{pmatrix} n_1 \cos^2 \chi + n_2 \sin^2 \chi & (n_1 - n_2) \sin \chi \cos \chi & 0 \\ (n_1 - n_2) \sin \chi \cos \chi & n_1 \sin^2 \chi + n_2 \cos^2 \chi & 0 \\ 0 & 0 & n_3 \end{pmatrix}_{123} \tag{E.115}$$

Note from Equation (E.115) that n_{11} and n_{22} are both of order n_1 or n_2, the principal refractive indices. By contrast, $n_{21} = n_{12}$ is of order $n_1 - n_2$, which is taken to be a very small quantity. We can therefore conclude that n_{21} is much smaller than either n_{11} or n_{22}. Returning to Equation (E.110) we now see that in the small birefringence limit ($n_{21} \ll n_{11}, n_{22}$)

$$n_{\text{eff}} = n_{11} \tag{E.116}$$

$$n_{11} - n_{33} = \Delta n_{\text{eff},3} \tag{E.117}$$

Thus measurements in the 2-direction of shear flow, in the small-birefringence limit, yield a retardance that corresponds to $n_{11} - n_{33}$, where n_{11} and n_{33} are the 11- and 33-coefficients of $\underline{\underline{n}}$ written in the usual shear coordinate system.

This concludes what turned out to be a lengthy calculation of the effect on electric-field polarization of sending a linearly polarized beam down the $-\hat{e}_2$-direction of shear flow, a direction that is not a principal direction of the refractive index tensor of a material undergoing shear flow. We saw that the beam splits into ordinary and extraordinary beams, with the extraordinary beam traveling at an effective refractive index n_{eff}, which is related to principal refractive indices n_1 and n_2 and the orientation angle χ. Because of the complexity of this type of calculation, it is preferable to send light down a principal direction to measure flow birefringence. When this is not feasible, we showed in the last discussion that meaningful values of birefringence related to the normal stress differences can be measured for light incident down nonprincipal directions when birefringence is not too large.

E.4 Summary

Optics is a complex subject, and this appendix cannot cover all aspects of the field. For further reading on optics in general, Fowles' book [81] is very readable, and it uses the same kind of notation and language that we have used. A more comprehensive and thus more complicated resource is the volume by Born and Wolf [33], which is quite complete. For rheo-optics, Fuller's text [84] goes much further into the details of using all types of optical techniques in rheology. A good resource for some discussion of experimental rheo-optics may be found in Janeschitz-Kriegl [116] and Wales [256].

E.5 PROBLEMS

E.1 For a polarized beam of light you are told that the phase difference δ between the x- and y-components of the electric-field vector is $\pi/4$ and that the amplitudes of the x- and y-components of the electric-field vector are equal, $E_{0x} = E_{0y}$ [see, for example, Equation (E.61) or Equation (10.139)].

(a) How is the beam polarized (linearly, circularly, elliptically)?

(b) For this wave, plot the projection of the tip of the electric-field vector $\underline{E}$ on an $\hat{e}_z$-plane (a plane of unit normal $\hat{e}_z$).

(c) Plot the projection of the tip of the electric-field vector on an $\hat{e}_z$-plane if $\delta = \pi/4$ and $E_{0y} = 2E_{0x}$.

(d) Plot the projection of the tip of the electric-field vector on an $\hat{e}_z$-plane if $\delta = \pi/4$ and $E_{0y} = \frac{1}{2}E_{0x}$.

E.2 For a polarized beam of light with $E_{0x} = E_{0y}$ (amplitudes of the x- and y-components of the electric-field vector equal):

(a) Plot the projection of the tip of the electric-field vector $\underline{E}$ on an $\hat{e}_z$-plane (a plane of unit normal $\hat{e}_z$) for the following phase differences δ:

$$\delta = \frac{\pi}{4}, \frac{\pi}{2}, \frac{3\pi}{4}, \pi, \frac{5\pi}{4}, \frac{3\pi}{2}, \frac{7\pi}{4}, 2\pi$$

See, for example, Equation (E.61) or Equation (10.139).

(b) Indicate the polarization state (linear, circular, elliptical) of each wave.

E.3 Verify that the equation for plane harmonic waves [Equation (E.10)] is a solution to the wave equation for isotropic media [Equation (E.5)].

E.4 Derive the wave equation for plane waves in electrically neutral nonmagnetic nonconducting anisotropic media [Equation (E.40)] from Equation (E.39) and the Maxwell equations.

E.5 The average magnitude of the Poynting flux is called the irradiance or intensity of the light:

$$I = |\langle \underline{S} \rangle| = |\langle \Re(\underline{E}) \times \Re(\underline{H}) \rangle|$$

Using the identity

$$\Re(x) = \frac{1}{2}(x + x^{(*)})$$

show that the intensity is given by

$$I = \frac{1}{2}|\Re(\underline{E}_0 \times \underline{H}_0^{(*)})|$$

E.6   By relating the expressions for $\underline{n}$ in the shear coordi-

nate system (Equation E.109) and in the principal frame (Equation E.111) show that the two equations for n_{eff} in the 13-plane of shear flow [Equations (E.107) and (E.110)] are equivalent.

E.7   For light directed down the -2-direction of shear flow, solve the Maxwell equations in the shear coordinate system (not the principal frame) for k' and k'', that is, derive Equation (E.110).

Data for Problems and Examples

This appendix contains more extensive data tables than were presented in several text examples and chapter practice problems. The data can be converted easily into electronic format by scanning and performing optical character recognition (OCR) on the files.

TABLE F.1
Supplemental Data for Problem 7.18

$a_T \omega \approx \dot{\gamma}_0$ (rad/s)	$\eta_p'(25°C) \approx \eta$ (poise)	$a_T \omega \approx \dot{\gamma}_0$ (rad/s)	$\eta_p'(25°C) \approx \eta$ (poise)
9.97E−01	1.72E+01	4.02E+03	1.94E+00
1.56E+00	1.71E+01	4.99E+03	1.88E+00
2.48E+00	1.70E+01	6.30E+03	1.67E+00
3.89E+00	1.69E+01	8.08E+03	1.60E+00
6.19E+00	1.67E+01	1.00E+04	1.47E+00
9.89E+00	1.62E+01	1.27E+04	1.40E+00
1.58E+01	1.54E+01	1.57E+04	1.30E+00
2.47E+01	1.40E+01	2.02E+04	1.21E+00
3.93E+01	1.20E+01	2.55E+04	1.14E+00
6.26E+01	9.86E+00	3.21E+04	1.12E+00
9.96E+01	7.98E+00	4.05E+04	1.09E+00
1.58E+02	6.54E+00	5.03E+04	1.03E+00
2.49E+02	5.40E+00	6.34E+04	9.99E−01
4.00E+02	4.39E+00	7.99E+04	9.73E−01
6.40E+02	3.68E+00	1.27E+05	9.20E−01
1.01E+03	3.12E+00	2.02E+05	8.67E−01
1.27E+03	3.06E+00	3.17E+05	8.50E−01
1.61E+03	2.67E+00	5.04E+05	7.97E−01
2.03E+03	2.64E+00	8.14E+05	7.81E−01
2.56E+03	2.28E+00	1.27E+06	7.27E−01
3.23E+03	2.15E+00	1.99E+06	7.46E−01
4.01E+03	2.05E+00	3.17E+06	7.30E−01

*Complex viscosity master curve $[\eta'(25°C) = \eta'(T)/a_T(T)]$ at 25°C of a solution of narrow-polydispersity polystyrene ($M_w = 860$ g/mol) in chlorinated diphenyl (concentration = 0.0154 g/cm^3) as a function of $a_T\omega$ (rad/s), where a_T is the time–temperature shift factor (see Section 6.2.1.3) and ω is the oscillation frequency in the small-amplitude oscillatory shear experiment. The data are given in units of poise. The solution viscosity η_s' has been subtracted from the measured η' in order to isolate the polymeric contribution to the complex viscosity, $\eta_p' = \eta' - \eta_s'$.

TABLE F.2
Supplemental Data for Example in Section 8.3, Table 8.1*

ω (rad/s)	G' (Pa)	ω (rad/s)	G'' (Pa)
2.45E+01	9.27E+01	5.56E+00	4.53E+02
3.49E+01	1.73E+02	7.89E+00	6.66E+02
4.98E+01	3.33E+02	1.12E+01	9.44E+02
7.10E+01	6.22E+02	1.59E+01	1.29E+03
1.01E+02	1.16E+03	2.26E+01	1.90E+03
1.44E+02	2.24E+03	3.20E+01	2.60E+03
2.05E+02	3.76E+03	4.55E+01	3.82E+03
2.92E+02	6.54E+03	6.45E+01	5.41E+03
4.16E+02	1.10E+04	9.16E+01	7.68E+03
5.93E+02	1.78E+04	1.30E+02	1.09E+04
8.44E+02	2.79E+04	2.62E+02	2.19E+04
1.20E+03	3.81E+04	3.72E+02	3.10E+04
1.71E+03	5.19E+04	7.48E+02	5.24E+04
2.42E+03	6.60E+04	1.06E+03	6.23E+04
3.44E+03	8.40E+04	1.51E+03	8.24E+04
4.89E+03	9.97E+04	2.13E+03	8.52E+04
6.95E+03	1.18E+05	3.03E+03	8.51E+04
9.86E+03	1.36E+05	4.29E+03	8.81E+04
1.40E+04	1.72E+05	6.08E+03	9.43E+04
1.99E+04	1.84E+05	8.62E+03	1.01E+05
2.83E+04	2.27E+05	1.22E+04	1.12E+05
4.01E+04	2.34E+05	1.73E+04	1.24E+05
5.69E+04	2.50E+05	2.46E+04	1.43E+05
8.08E+04	2.77E+05	3.49E+04	1.76E+05
1.15E+05	3.06E+05	4.95E+04	2.17E+05
1.63E+05	3.51E+05	7.02E+04	2.58E+05
2.31E+05	4.03E+05	9.96E+04	3.18E+05
3.28E+05	4.78E+05	1.41E+05	3.79E+05
4.66E+05	5.67E+05	2.00E+05	4.67E+05
6.62E+05	6.96E+05	2.84E+05	5.76E+05
9.41E+05	8.86E+05	4.03E+05	6.85E+05
1.34E+06	1.17E+06	5.72E+05	8.44E+05

*Small-amplitude oscillatory shear data $G'(\omega)$ and $G''(\omega)$ for a narrowly distributed polystyrene melt of $M_w = 59$ kg/mol, $T = 190°C$ [252]. In the example exercise the relaxation spectrum of the generalized Maxwell model is fit to these data.

TABLE F.3
Supplemental Data for Problem 8.26*

$a_T \omega$ (rad/s)	G^* (Pa)	$\tan \delta$
8.29E+00	8.06E+05	1.54E−01
3.92E+00	7.46E+05	2.82E−01
2.33E+00	6.66E+05	4.06E−01
1.47E+00	6.10E+05	5.83E−01
8.44E−01	5.94E+05	8.79E−01
5.63E−01	3.94E+05	1.28E+00
3.52E−01	2.74E+05	1.96E+00
2.74E−01	2.21E+05	2.54E+00
2.18E−01	1.91E+05	2.77E+00
1.84E−01	1.83E+05	3.00E+00
1.42E−01	1.37E+05	3.86E+00
1.33E−01	1.22E+05	4.30E+00
1.04E−01	9.59E+04	5.55E+00
9.02E−02	8.22E+04	6.21E+00
7.60E−02	7.67E+04	6.64E+00
6.40E−02	6.29E+04	8.27E+00
5.66E−02	5.41E+04	9.05E+00
4.98E−02	5.12E+04	9.74E+00
4.32E−02	4.21E+04	1.14E+01
3.80E−02	3.77E+04	1.13E+01
3.29E−02	3.44E+04	1.40E+01
2.87E−02	2.85E+04	1.65E+01
2.44E−02	2.44E+04	1.89E+01
2.17E−02	2.25E+04	2.16E+01
1.80E−02	1.73E+04	2.75E+01
1.65E−02	1.69E+04	2.79E+01
1.43E−02	1.49E+04	2.93E+01
1.23E−02	1.20E+04	4.55E+01
1.11E−02	1.11E+04	4.42E+01
9.40E−03	9.81E+03	4.72E+01

*Linear viscoelastic data for a generation-4 poly (amidoamine) (PAMAM) dendrimer [248].

TABLE F.4
Supplemental Data for Problem 8.27*

ω (rad/s)	G' (Pa)	ω (rad/s)	G'' (Pa)
6.35E+00	9.81E−01	1.02E+00	1.70E+00
9.92E+00	2.51E+00	1.57E+00	2.68E+00
1.55E+01	5.49E+00	2.54E+00	4.23E+00
2.49E+01	1.16E+01	4.09E+00	6.55E+00
3.99E+01	2.13E+01	6.22E+00	1.01E+01
6.19E+01	3.29E+01	9.83E+00	1.60E+01
9.62E+01	4.81E+01	1.56E+01	2.43E+01
1.47E+02	6.62E+01	2.42E+01	3.41E+01
2.38E+02	8.96E+01	3.91E+01	4.70E+01
3.84E+02	1.19E+02	6.21E+01	6.23E+01
5.99E+02	1.58E+02	9.50E+01	7.94E+01
9.33E+02	2.05E+02	1.54E+02	1.05E+02
1.48E+03	2.72E+02	2.35E+02	1.34E+02
2.31E+03	3.40E+02	3.81E+02	1.75E+02
3.73E+03	4.69E+02	5.93E+02	2.32E+02
4.93E+03	5.40E+02	9.41E+02	3.07E+02
6.05E+03	5.76E+02	1.49E+03	4.24E+02
7.70E+03	6.38E+02	1.90E+03	4.88E+02
9.80E+03	7.21E+02	2.41E+03	5.73E+02
1.22E+04	8.46E+02	3.01E+03	6.60E+02
1.53E+04	9.19E+02	3.90E+03	7.90E+02
1.91E+04	1.10E+03	4.77E+03	9.27E+02
3.08E+04	1.46E+03	6.08E+03	1.03E+03
4.81E+04	1.83E+03	7.72E+03	1.23E+03
7.65E+04	2.24E+03	9.63E+03	1.44E+03
1.22E+05	2.75E+03	1.22E+04	1.66E+03
1.93E+05	3.45E+03	1.53E+04	2.03E+03
3.02E+05	4.23E+03	1.94E+04	2.38E+03
		2.42E+04	2.85E+03
		3.02E+04	3.41E+03
		3.76E+04	4.08E+03
		4.78E+04	4.98E+03
		6.06E+04	6.20E+03
		7.56E+04	7.41E+03
		1.20E+05	1.10E+04
		1.93E+05	1.68E+04
		3.05E+05	2.54E+04
		4.83E+05	3.94E+04
		7.63E+05	6.21E+04
		1.21E+06	9.80E+04
		1.94E+06	1.49E+05
		3.01E+06	2.44E+05

*Linear viscoelastic data for a narrowly distributed solution of polystyrene ($M_w = 860$ kg/mol, $c = 0.015$ g/cm^3 in Aroclor 1248) [167].

TABLE F.5
Supplemental Data for Problem 8.28[*]

ω (rad/s)	G′ (Pa)	ω (rad/s)	G″ (Pa)	ω (rad/s)	G′ (Pa)	ω (rad/s)	G″ (Pa)
2.46E−03	1.18E+01	1.56E−03	8.24E+01	1.44E+00	8.60E+03	1.11E+00	8.73E+03
3.16E−03	2.06E+01	2.27E−03	1.18E+02	1.63E+00	1.03E+04	1.46E+00	9.89E+03
4.59E−03	3.70E+01	4.30E−03	2.12E+02	2.13E+00	1.14E+04	2.01E+00	1.06E+04
7.46E−03	7.13E+01	8.72E−03	3.92E+02	2.51E+00	1.24E+04	2.55E+00	1.18E+04
1.23E−02	1.30E+02	1.06E−02	4.38E+02	3.40E+00	1.40E+04	3.40E+00	1.38E+04
2.20E−02	2.61E+02	1.54E−02	6.30E+02	3.45E+00	1.57E+04	4.02E+00	1.48E+04
3.20E−02	3.30E+02	2.06E−02	7.87E+02	3.91E+00	1.63E+04	5.92E+00	1.70E+04
3.62E−02	3.64E+02	2.27E−02	9.17E+02	4.55E+00	1.77E+04	6.61E+00	1.85E+04
3.83E−02	4.19E+02	3.53E−02	1.15E+03	6.17E+00	2.04E+04	8.14E+00	1.90E+04
4.10E−02	5.02E+02	4.17E−02	1.35E+03	8.14E+00	2.28E+04	1.04E+01	2.15E+04
5.27E−02	6.28E+02	5.07E−02	1.97E+03	9.61E+00	2.55E+04	1.48E+01	2.24E+04
6.13E−02	7.32E+02	6.15E−02	1.74E+03	1.18E+01	2.77E+04	1.74E+01	2.51E+04
7.76E−02	9.14E+02	7.06E−02	1.89E+03	1.40E+01	2.97E+04	2.09E+01	2.65E+04
9.43E−02	1.10E+03	8.45E−02	2.24E+03	1.72E+01	3.32E+04	2.64E+01	2.61E+04
9.96E−02	1.23E+03	9.58E−02	2.91E+03	2.09E+01	3.71E+04	2.60E+01	3.00E+04
1.18E−01	1.41E+03	1.16E−01	2.57E+03	2.50E+01	3.86E+04	3.73E+01	3.26E+04
1.33E−01	1.55E+03	1.35E−01	2.91E+03	2.99E+01	3.97E+04	4.22E+01	3.13E+04
1.69E−01	2.02E+03	1.81E−01	3.35E+03	3.94E+01	4.69E+04	5.65E+01	3.75E+04
2.19E−01	2.26E+03	2.20E−01	3.69E+03	4.79E+01	4.75E+04	6.57E+01	3.85E+04
2.78E−01	2.91E+03	2.17E−01	4.48E+03	5.50E+01	5.39E+04	7.87E+01	4.13E+04
3.15E−01	3.07E+03	2.56E−01	4.12E+03	6.96E+01	5.70E+04	9.82E+01	4.07E+04
3.37E−01	3.53E+03	3.06E−01	4.54E+03	8.68E+01	6.46E+04	1.14E+02	4.55E+04
4.51E−01	3.95E+03	4.58E−01	5.29E+03	1.13E+02	6.82E+04	1.66E+02	4.94E+04
5.55E−01	4.86E+03	4.71E−01	6.09E+03	1.73E+02	8.29E+04	2.38E+02	5.15E+04
6.74E−01	5.36E+03	5.40E−01	6.26E+03	2.38E+02	8.63E+04	2.89E+02	5.52E+04
7.53E−01	6.25E+03	6.84E−01	6.61E+03	2.55E+02	9.52E+04	4.95E+02	5.99E+04
1.01E+00	6.70E+03	6.65E−01	6.89E+03	4.76E+02	1.11E+05	7.40E+02	6.60E+04
1.19E+00	8.26E+03	8.41E−01	7.71E+03	7.62E+02	1.35E+05	1.67E+03	7.68E+04
				2.03E+03	1.73E+05	2.71E+03	7.89E+04

[*] Linear viscoelastic data for a broadly distributed high-density polyethylene [142].

References

1. J. A. vanAken and H. Janeschitz-Kriegl, "New apparatus for the simultaneous measurement of stresses and flow birefringence in biaxial extension of polymer melts," *Rheol. Acta*, **19**, 744–752 (1980).

2. J. A. vanAken and H. Janeschitz-Kriegl, "Simultaneous measurement of transient stress and flow birefringence in one-sided compression (biaxial extension) of a polymer melt," *Rheol. Acta*, **20**, 419–432 (1981).

3. J. J. Aklonis and W. J. MacKnight, *Introduction to Polymer Viscoelasticity,* 2nd ed. (John Wiley & Sons, New York, 1983).

4. D. C. Allport and W. H. Janes, eds., *Block Copolymers* (John Wiley & Sons, New York, 1973).

5a. N. R. Amundson, *Mathematical Methods in Chemical Engineering: Matricies and Their Application* (Prentice-Hall, Englewood Cliffs, NJ, 1966).

5b. S. L. Anna, C. Rogers, and G. H. McKinley, "On controlling the kinematics of a filament stretching rheometer using a real-time active control mechanism," *J. Non-Newt. Fluid Mech.* **87**, 307–335 (1999).

6. L. A. Archer, Y. L. Chen, and R. G. Larson, "Delayed slip after step strains in highly entangled polystyrene solutions," *J. Rheol.*, **39**, 519–525 (1995).

7. R. Aris, *Vectors, Tensors, and the Basic Equations of Fluid Mechanics* (Dover Publications, New York, 1989; unabridged and corrected republication of the work first published by Prentice-Hall, Englewood Cliffs, NJ, 1962).

8. K. R. Arnold and D. J. Meier, "A rheological characterization of SBS block copolymers," *J. Appl. Polym. Sci.*, **14**, 427–440 (1970).

9. G. Astarita and G. Marrucci, *Principles of Non-Newtonian Fluid Mechanics* (McGraw-Hill, New York, 1974).

10. E. B. Bagley, "End corrections in the capillary flow of polyethylene," *J. Appl. Phys.*, **28**, 624–627 (1957).

11. H. A. Barnes, J. F. Hutton, and K. Walters, *An Introduction to Rheology* (Elsevier Science Publishers, New York, 1989).

12. S. Barnett, *Matrices: Methods and Applications* (Clarendon Press, Oxford, UK, 1990).

13. G. K. Batchelor, *An Introduction to Fluid Dynamics* (Cambridge University Press, New York, 1967).

14. F. Bates, "Block copolymers near the microphase separation transition. 2. Linear dynamic mechanical properties," *Macromol.*, **17**, 2607–2613 (1984).

15. M. Baumgaertel, M. E. De Rosa, J. Machado, M. Masse, and H. H. Winter, "The relaxation time spectrum of nearly N. monodisperse polybutadiene melts," *Rheol. Acta*, **31**, 75–82 (1992).

16. S. Berg, R. Kroger, and H. J. Rath, "Measurement of extensional viscosity by stretching large liquid bridges in microgravity," *J. Non-Newt. Fluid Mech.*, **55**, 307–319 (1994).

17. J. W. Berge, P. R. Saunters, and J. D. Ferry, "Mechanical properties of poly-*n*-octyl methacrylate at low frequencies and in creep; entanglements in metacrylate polymers," *J. Colloid Sci.*, **14**, 135–146 (1959).

18. A. Beris, R. C. Armstrong, and R. A. Brown, "Perturbation theory for viscoelastic fluids between eccentric rotating cylinders," *J. Non-Newt. Fluid Mech.*, **13**, 109–148 (1983).

19. A. N. Beris and B. J. Edwards, *Thermodynamics of Flowing Systems with Internal Microstructure* (Oxford University Press, New York, 1994).

20. B. Bernstein, E. Kearsley, and L. Zapas, "A study of stress relaxation with finite strain," *Trans. Soc. Rheol.*, **7**, 391–410 (1963).

21. G. C. Berry and T. G Fox, "The viscosity of polymers and their concentrated solutions," *Adv. Polym. Sci.*, **5**, 261–357 (1968).

22. D. M. Binding, "An approximate analysis for contraction and converging flows," *J. Non-Newt. Fluid Mech.*, **27**, 173–189 (1988).

23. D. M. Binding, "Further considerations of axisymmetric contraction flows," *J. Non-Newt. Fluid Mech.*, **41**, 27–42 (1991).

24. R. J. Binnington and D. V. Boger, "Constant viscosity elastic liquids," *J. Rheol.*, **29**, 887–904 (1985).

25. R. B. Bird, R. C. Armstrong, and O. Hassager, *Dynamics of Polymeric Liquids,* vol. 1: *Fluid Mechanics*, 1st ed. (John Wiley & Sons, New York, 1977).

26. R. B. Bird, R. C. Armstrong, and O. Hassager, *Dynamics of Polymeric Liquids,* vol. 1: *Fluid Mechanics*, 2nd ed. (John Wiley & Sons, New York, 1987).

27. R. B. Bird, C. F. Curtiss, R. C. Armstrong, and O. Hassager, *Dynamics of Polymeric Liquids*, vol. 2: Kinetic Theory, 2nd ed. (John Wiley & Sons, New York, 1987).

28. R. B. Bird, W. Stewart, and E. Lightfoot, *Transport Phenomena* (John Wiley & Sons, New York, 1960).

29. H. Block, J. P. Kelly, A. Qin, and T. Watson, "Materials and mechanisms in electrorheology," *Langmuir*, **6**, 6–14 (1990).

30. L. L. Blyler, Jr., and A. C. Hart, Jr., "Capillary flow instability of ethylene polymer melts," *Polym. Eng. Sci.*, **10**, 193–203 (1970).

31. D. V. Boger and R. J. Binnington, "Separation of elastic and shear thinning effects in the capillary rheometer," *J. Rheol.*, **21**, 515–534 (1977).

32. Bohlin Instruments, Inc., 2540 Route 130, Suite 114, Cranberry, NJ 08512, USA, *www.bohlinusa.com*.

33. M. Born and E. Wolf, *Principles of Optics: Electromagnetic Theory of Propagation, Interference and Diffraction of Light,* 6th (corrected) ed., (Cambridge University Press, New York, 1980).

34. W. E. Boyce and R. C. DiPrima, *Elementary Differential Equations and Boundary Value Problems,* 3rd ed., (John Wiley & Sons, New York, 1977).

35. F. Brochard and P. G. deGennes, "Shear-dependent slippage at a polymer/solid interface," *Langmuir,* **8**, 3033–3037 (1992).

36. Brookfield Engineering, 11 Commerce Boulevard, Middleboro, MA 02346, USA *www.brookfieldengineering.com*.

37. F. Bueche, *Physical Properties of Polymers* (Wiley Interscience, New York, 1962).

38. C-Mold, 9625 Ormsby Station Road, Louisville, KY 40223, USA, tel.: 502-423-4350, *commons.cmold.com*.

39. C. A. Cathey and G. G. Fuller, "The optical and mechanical response of flexible polymer solutions to extensional flow," *J. Non-Newt. Fluid Mech.,* **34**, 63–88 (1990).

40. F. M. Chapman and T. S. Lee, "Effect of talc filler on the melt rheology of polypropylene," *SPE J.,* **26**, 37–40 (1970).

41. S. Chatraei, C. W. Macosko, and H. H. Winter, "Lubricated squeezing flow: A new biaxial extensional rheometer,"*J. Rheol.,* **25**, 433–443 (1981).

42. I. J. Chen, G. E. Hagler, L. E. Abbott, D. C. Bogue, and J. L. White, "Interpretation of tensile and melt spinning experiments on low density and high density polyethylene," *Trans. Soc. Rheol.,* **16**, 473 (1973).

43. N. P. Cheremisinoff, *An Introduction to Polymer Rheology and Processing* (CRC Press, Boca Raton, FL, 1993).

44. R. P. Chhabra and J. F. Richardson, *Non-Newtonian Flow in the Process Industries* (Butterworth-Heinemann, Oxford, UK, 1999).

45. E. B. Christiansen and W. R. Leppard, "Steady-state and oscillatory flow properties of polymer solutions," *Trans. Soc. Rheol.,* **18**, 65–86 (1974).

46. H.-K Chuang and C. D. Han, "Rheological behavior of blends of nylon with a chemically modified polyolefin," *Advances in Chemistry: Polymer Blends and Composites in Multiphase Systems,* **206**, 171–183 (1984).

47. C. I. Chung and J. C. Gale, "Newtonian behavior of a styrene-butadiene-styrene block copolymer," *J. Polym. Sci.,* **14**, 1149–1156 (1976).

48. F. N. Cogswell, "Converging flow of polymer melts in extrusion dies," *Polym. Eng. Sci.,* **12**, 64–73 (1972).

49. F. N. Cogswell, "Measuring the extensional rheology of polymer melts," *Trans. Soc. Rheol.,* **16**, 383–403 (1972).

50. B. D. Coleman and W. Noll, "Recent results in the continuum theory of viscoelastic fluids," *Ann. NY Acad. Sci.,* **89**, 672–714 (1961).

51. A. A. Collyer and D. W. Clegg, eds., *Rheological Measurement* (Elsevier, Essex, UK, 1988).

52. S. L. Cooper and A. V. Tobolsky, "Properties of linear elastomeric polyurethanes," *J. Appl. Polym. Sci.,* **10**, 1837–1844 (1966).

53. W. P. Cox and E. H. Merz, "Correlation of dynamic and steady flow viscosities,"*J. Polym. Sci.*, **28**, 619–622 (1958).

54. M. J. Crochet, A. R. Davies, and K. Walters, *Numerical Simulation of Non-Newtonian Flow* (Elsevier, New York, 1984).

55. J. S. Dahler and L. E. Scriven, "Angular momentum of continua," *Nature*, **192**, 36–37 (1961).

56. W. Dannhauser, W. C. Child, Jr., and J. D. Ferry, "Dynamic mechanical properties of poly-*n*-octyl methacrylate," *J. Colloid Sci.*, **13**, 103–113 (1958).

57. J. M. Dealy, "On the relationship between extensional viscosities for uniaxial and biaxial extension," *Trans. Soc. Rheol.*, **17**, 255–258 (1973).

58. J. M. Dealy, *Rheometers for Molten Plastic: A Practical Guide to Testing and Property Measurement* (Van Nostrand Reinhold, New York, 1982).

59. J. M. Dealy, U. S. Patent 4,463,928 (1984).

60. J. M. Dealy, "Official nomenclature for material functions describing the response of a viscoelastic fluid to various shearing and extensional deformations," *J. Rheol.*, **37**, 179–191 (1994).

61. J. M. Dealy and K. F. Wissbrun, *Melt Rheology and Its Role in Plastics Processing* (Van Nostrand Reinhold, New York, 1990).

62. P. G. de Gennes, "Reptation of a polymer chain in the presence of fixed obstacles," *J. Chem. Phys*, **55**, 572–579 (1971).

63. DeGroot, University of Minnesota internal report; cited in Macosko [162].

64. M. M. Denn, *Process Fluid Mechanics* (Prentice-Hall, Englewood Cliffs, NJ, 1980).

65. M. M. Denn, "Issues in viscoelastic fluid mechanics," *Ann. Rev. Fluid Mech.*, **22**, 13–34 (1990).

66. M. Doi and S. F. Edwards,"Dynamics of concentrated polymer systems, Part 1—Brownian motion in the equilibrium state," *J. Chem. Soc., Faraday Trans. II*, **74**, 1789–1801 (1978).

67. M. Doi and S. F. Edwards, "Dynamics of concentrated polymer systems, Part 2—Molecular motion under flow," *J. Chem. Soc., Faraday Trans. II*, **74**, 1802–1817 (1978).

68. M. Doi and S. F. Edwards, "Dynamics of concentrated polymer systems, Part 3—The constitutive equation," *J. Chem. Soc., Faraday Trans. II*, **74**, 1818–1829 (1978).

69. M. Doi and S. F. Edwards, "Dynamics of concentrated polymer systems, Part 4—Rheological properties," *J. Chem. Soc., Faraday Trans. II*, **75**, 38–54 (1979).

70. M. Doi and S. F. Edwards, *The Theory of Polymer Dynamics* (Clarendon Press, Oxford, UK, 1986).

71. P. Dvornic and D. A. Tomalia, "Molecules that grow like trees," *Sci. Spectra*, **5**, 36–41 (1996).

72. Y. Einaga, K. Osaki, M. Kurata, S. Kimura, and M. Tamura, "Stress relaxation of polymer solutions under large strain," *Polym. J.*, **2**, 550–552 (1971).

73. A. Einstein, "Eine neue Bestimmung der Moleküldimensionen," *Ann. Phy.*, **19**, 289–306 (1906).

74. A. Einstein, "Berichtigung zu meiner Arbeit: 'Eine neue Bestimmung der Moleküldi-mensionen'," *Ann. Phys.*, **34**, 591–593 (1911).

75. J. D. Ferry, *Viscoelastic Properties of Polymers* (John Wiley & Sons, New York, 1980).

76. J. D. Ferry, W. C. Child, Jr., R. Zand, D. M. Stern, M. L. Williams, and R. F. Landel, "Dynamic mechanical properties of polyethyl methacrylate," *J. Colloid Sci.*, **12**, 53–67 (1957).

77. L. J. Fetters, D. J. Lohse, D. Richter, T. A. Witten, and A. Zirkel, "Connections between polymer molecular weight, density, chain dimensions, and melt viscoelastic properties," *Macromol.*, **27**, 4639–4647 (1994).

78. W. N. Findley, J. S. Lai, and K. Onaran, *Creep and Relaxation of Nonlinear Viscoelastic Materials* (Dover Publications, New York, 1989; unabridged, corrected republication of the work originally published by North-Holland Publishing Company, Amsterdam, 1976).

79. P. F. Flory and J. Rehner, "Statistical mechanics of cross-linked polymer networks. I. Rubberlike elasticity," *J. Chem. Phys.*, **11**, 512–520 (1943).

80. R. W. Flumerfelt, M. W. Pierick, S. L. Cooper, and R. B. Bird, "Generalized plane Couette flow of a non-Newtonian fluid," *Ind. Eng. Chem. Fundam.*, **8**, 354–357 (1969).

81. G. R. Fowles, *Introduction to Modern Optics*, 2nd ed., (Dover Publications, New York, 1989; unabridged, corrected republication of the 2nd edition (1975) of the work originally published by Holt, Rinehart, and Winston, New York, 1968).

82. P. L. Frattini and G. G. Fuller, "A note on phase-modulated flow birefringence: A promising rheo-optical method," *J. Rheol.*, **28**, 61–70 (1984).

83. M. K. Fukuda, K. Osaki, and M. Kurata, "Nonlinear viscoelasticity of polystyrene solutions. I. Strain-dependent relaxation modulus," *J. Polym. Sci., Polym. Phys. Ed.*, **13**, 1563–1576 (1975).

84. G. G. Fuller, *Optical Rheometry of Complex Fluids* (Oxford University Press, New York, 1995).

85. C. J. Geankoplis, *Transport Processes and Unit Operations*, 3rd ed. (Prentice Hall, Englewood Cliffs, NJ, 1993).

86. P. G. de Gennes, "Reptation of a polymer chain in the presence of fixed obstacles," *J. Chem. Phys*, **55**, 572–579 (1971).

87. A. Ghijsels and J. Raadsen, "A collaborative study on the melt rheology of a styrene-butadiene-styrene block copolymer," *Pure Appl. Chem.*, **52**, 1359 (1980).

88. A. J. Giacomin and J. M. Dealy, "A new rheometer for molten plastics," *SPE (ANTEC) Tech. Papers*, **32**, 711–714 (1986).

89. A. J. Giacomin, T. Samurkas, and J. M. Dealy, "A novel sliding plate rheometer for molten plastics," *Polym. Eng. Sci.*, **29**, 499–504 (1989).

90. H. Giesekus, "Die Elastizität von Flüssigkeiten," *Rheol. Acta*, **5**, 29–35 (1966).

91. H. Giesekus, "A simple constitutive equation for polymer fluids based on the concept of deformation-dependent tensoride mobility," *J. Non-Newt. Fluid Mech.*, **11**, 69–109 (1982).

92. Göttfert Werkstoff - Prüfmaschinen GmbH, Siemensstrasse 2, D–74722 Buchen/ Odenwald, Germany; Goettfert-USA, 488 Lakeshore Parkway, Rock Hill, SC 29730, USA, tel.: 803-324-3883, *www.goettfert.com*.

93. R. J. Gordon and W. R. Schowalter, "Anisotropic fluid theory: A different approach to the dumbbell theory of dilute polymer solutions" *Trans. Soc. Rheol.*, **16**, 79 (1972).

94. E. V. Gouinlock and R. S. Porter, "Linear dynamic mechanical properties of an SBS block copolymer," *Polym. Eng. Sci.*, **17**, 535–542 (1977).

95. W. W. Graessley, "The entanglement concept in polymer rheology," *Adv. Polym. Sci.*, **16**, 1–179 (1974).

96. W. W. Graessley, "Effect of long branches on the flow properties of polymers," *Acc. Chem. Res.*, **10**, 332–339 (1977).

97. W. W. Graessley, "Viscoelasticity and flow in polymer melts and concentrated solutions," *Phys. Properties Polym.*, **2**, 97–143 (1993).

98. R. G. Green and R. G. Griskey, "Rheological behavior of dilatant (shean-thickening) fluids. Part I. Experimental and data," *Trans. Soc. Rheol.*, **12**, 13–25 (1968).

99. M. S. Green and A. V. Tobolsky, "A new approach to the theory of relaxing polymeric media," *J. Chem. Phys.*, **14**, 80–96 (1946).

100. M. D. Greenberg, *Foundations of Applied Mathematics* (Prentice-Hall, Englewood Cliffs, NJ, 1978).

101. J. Greener and J. R. G. Evans, "Uniaxial elongational flow of particle-filled polymer melts," *J. Rheol.*, **42**, 697–709 (1998).

102. J. T. Gruver and G. Kraus, "Rheological properties of polybutadienes prepared by *n*-butyllithium initiation," *J. Polym. Sci. A*, **2**, 797–810 (1964).

103. M. Gupta, "Effect of elongational viscosity on die design for plastic extrusion," *SPE ANTEC Tech. Papers*, **45**, 1254 (1999).

104. M. Gupta, "Effect of elongational viscosity on axisymmetric entrance flow of polymers," *Polym. Eng. Sci.*, **40**, 23 (2000).

105. R. K. Gupta, D. A. Nguyen, and T. Sridhar, "Extensional viscosity of dilute polystyrene solutions: Effect of concentration and molecular weight," *Phys. Fluids* (2000), as cited in Li et al. [153].

106. N. H. Hartshornt and A. Stuart, *Crystals and the polarising microscope,* 4th ed. (Edward Arnold Publishers, London, UK, 1970).

107. J. W. S. Hearle, *Polymers and Their Properties* (John Wiley & Sons, New York, 1982).

108. E. Hecht, *Optics*, 2nd ed. (Addison-Wesley Publishing Company, Reading, MA, 1987).

109. C. G. Hermansky and D. V. Boger, "Opposing-jet viscometry of fluids with viscosity approaching that of water," *J. Non-Newt. Fluid Mech.*, **56**, 1–14 (1995).

110. J. O. Hirschfelder, C. F. Curtiss, and R. B. Bird, *Molecular Theory of Gasses and Liquids* (John Wiley & Sons, New York, 1964).

111. G. Holden, E. T. Bishop, and N. R. Legge, "Thermoplastic elastomers," *J. Polym. Sci.*, **C26**, 37–57 (1969).

112. R. J. Hunter, *Foundations of Colloid Science*, vol. 1 (Clarendon Press, Oxford, UK, 1986).

113. J. F. Hutton, "Fracture and secondary flow of elastic liquids," *Rheol. Acta*, **8**, 54–59 (1969).

113a. N. J. Inkson, T. C. B. McLeish, O. G. Harlen, and D. J. Groves, "Predicting low density polyethylene melt rheology in elongational and shear flows with 'pom-pom' constitutive equations," *J. Rheol.*, **43**, 873–896 (1999).

114. Instron, Inc., 100 Royall St., Canton, MA 02021-1089, USA; tel.: 800-564-8378, *www.instron.com*.

115. H. M. James and E. Guth, "Theory of the elastic properties of rubber," *J. Chem. Phys.*, **11**, 455–481 (1943).

116. H. Janeschitz-Kriegl, *Polymer Melt Rheology and Flow Birefringence* (Springer-Verlag, New York, 1983).

117. S. H. Johnson, P. L. Frattini, and G. G. Fuller, "Simultaneous dichroism and birefringence measurements in transient shear flow," *J. Colloid Inter. Sci.*, **104**, 440 (1985).

118. D. D. Joseph, *Fluid Dynamics of Viscoelastic Liquids* (Springer-Verlag, New York, 1989).

119. D. Kalika and M. M. Denn, "Wall slip and extrudate distortion in linear low-density polyethylene," *J. Rheol.*, **31**, 815–834 (1987).

120. R. M. Kannan and J. A. Kornfield, "Stress-optical manifestations of molecular and microstructural dynamics in complex polymer melts," *J. Rheol.*, **38**, 1127–1150 (1994).

121. M. Karplus and R. N. Porter, *Atoms and Molecules: An Introduction for Students of Physical Chemistry* (Benjamin/Cummings Publishing, Reading, MA, 1970).

122. A. Kaye, College of Aeronautics, Cranfield, Note 134 (1962).

123. A. Keller and J. A. Odell, "The extensibility of macromolecules in solution: A new focus for macromolecular science," *Colloid Polym. Sci.*, **42**, 49–64 (1992).

124. S. A. Khan and R. G. Larson, "Comparison of simple constitutive equations for polymer melts in shear and biaxial and uniaxial extensions," *J. Rheol.*, **31**, 207–234 (1987).

125. S. A. Khan, R. K. Prud'homme, and R. G. Larson, "Comparison of the rheology of polymer melts in shear, and biaxial and uniaxial extensions," *Rheol. Acta*, **26**, 144–151 (1987).

126. J. Kim, C. D. Han, and S. G. Chu, "Viscoelastic behavior and order–disorder transition in mixtures of a block copolymer and a midblock-associating resin," *J. Polym. Sci., Polym. Phys. Ed.*, **26**, 677–701 (1988).

127. S. Kimura, K. Osaki, and M. Kurata, "Stress relaxation of polybutadiene at large deformation. Measurements of stress and birefringence in shear and elongation,"*J. Polym. Sci.*, **19**, 151–163 (1981).

128. M. Kompani and D. C. Venerus, "Equibiaxial extensional flow of polymer melts via lubricated squeezing flow. Part I: Experimental analysis," *Rheologica Acta*, in press (2000).

129. G. A. Korn and T. M. Korn, *Mathematical Handbook for Scientists and Engineers* (McGraw-Hill, New York, 1968).

130. J. A. Kornfield, G. G. Fuller, and D. S. Pearson, "Infrared dichroism measurements of molecular relaxation in binary blend melt rheology," *Macromol.*, **22**, 1334–1345 (1989).

131. G. Kraus and J. T. Gruver, "Rheological properties of multichain polybutadienes," *J. Polym. Sci. A*, **3**, 105–122 (1965).

132. I. M. Krieger and S. H. Maron, "Direct determination of the flow curves of non-Newtonian fluids. III. Standardized treatment of viscometric data," *J. Appl. Phys.*, **25**, 72–75 (1954).

133. R. S. Kroger and H. J. Rath, "Velocity and elongational rate distributions in stretched polymeric and Newtonian liquid bridges," *J. Non-Newt. Fluid Mech.*, **57**, 137–153 (1995).

134. W. M. Kulicke, H. E. Jeberien, H. Kiss, and R. S. Porter, "Visual observation of flow irregularities in polymer solutions at theta-conditions," *Rheol. Acta*, **18**, 711–716 (1979).

135. S. Kurzbeck, F. Oster, H. Munstedt, T. Q. Nguyen, and R. Gensler, "Rheological properties of two polypropylenes with different molecular structure," *J. Rheol.*, **43**, 359–374 (1999).

136. W. M. Lai, D. Rubin, and E. Krempl, *Introduction to Continuum Mechanics* (Pergamon Press, New York, 1978).

137. R. G. Larson, "Instabilities in viscoelastic flows," *Rheol. Acta*, **31**, 213–263 (1992).

138. R. G. Larson, *Constitutive Equations for Polymer Melts and Solutions* (Butterworths, Boston, 1988).

139. R. G. Larson, *The Structure and Rheology of Complex Fluids* (Oxford University Press, New York, 1999).

140. R. G. Larson, S. A. Khan, and V. R. Raju, "Relaxation of stress and birefringence in polymers of high molecular weight," *J. Rheol.*, **32**, 145–161 (1988).

141. R. G. Larson and K. Monroe, "The BKZ as an alternative to the Wagner model for fitting shear and elongational flow data of an LDPE melt," *Rheol. Acta*, **23**, 10–13 (1984).

142. H. M. Laun, "Description of the non-linear shear behavior of a low density polyethylene melt by means of an experimentally determined strain dependent memory function," *Rheol. Acta*, **17**, 1–5 (1978).

143. H. M. Laun, "Prediction of elastic strains of polymer melts in shear and elongation," *J. Rheol.*, **30**, 459–501 (1986).

144. H. M. Laun, S. T. E. Aldhouse, H. Coster, D. Constantin, J. Meissner, J. M. Starita, M. Fleissner, D. Frank, D. J. Groves, G. Ajroldi, L. A. Utracki, J. L. White, H. Yamane, A. Ghijsels, and H. H. Winter, "A collaborative study of the stability of extrusion, melt spinning and tubular film extrusion of some high-, low- and linear-low density polyethylene samples," *Pure Appl. Chem.*, **59**, 193–216 (1987).

145. H. M. Laun, and H. Schuch, "Transient elongational viscosities and drawability of polymer melts," *J. Rheol.*, **33**, 119–175 (1989).

146. M. N. Layec-Raphalen and C. Wolff, "On the shear-thickening behavior of dilute solutions of chain macromolecules," *J. Non-Newt. Fluid Mech.*, **1**, 159–173 (1976).

147. H. Leaderman, "Textile materials and the time factor. I. Mechanical behavior of textile fibers and plastics," *Textile Res. J.*, **11**, 171–193 (1941).

148. G. L. Leal, *Laminar Flow and Convective Transport Processes: Scaling Principles and Asymptotic Analysis* (Butterworth-Heinemann, Boston, 1992).

149. C. S. Lee, B. C. Tripp, and J. J. Magda, "Does N_1 or N_2 control the onset of melt fracture," *Rheol. Acta*, **31**, 306–308 (1992).

150. J. S. Lee and G. G. Fuller, "The spatial development of transient Couette flow and shear wave propagation in polymeric liquids by flow birefringence," *J. Non-Newt. Fluid Mech.*, **26**, 57–76 (1987).

151. A. I. Leonov, "Nonequilibrium thermodynamics and rheology of viscoelastic polymer melts," *Rheol. Acta*, **15**, 85–98 (1976).

152. O. Levenspiel, *Engineering Flow and Heat Exchange* (Plenum Press, New York, 1984).

153. L. Li, R. G. Larson, and T. Sridhar, "Brownian dynamics simulations of dilute polystyrene solutions," *J. Rheol.*, **44**, 291–322 (2000).

154. A. S. Lodge, "A network theory of flow birefringence and stress in concentrated polymer solutions," *Trans. Faraday Soc.*, **52**, 120–130 (1956).

155. A. S. Lodge, *Elastic Liquids* (Academic Press, New York, 1964).

156. A. S. Lodge, *Body Tensor Fields in Continuum Mechanics* (Academic Press, New York, 1974).

157. A. S. Lodge, "Stress relaxation after a sudden shear strain," *Rheol. Acta*, **14**, 664–665 (1975).

158. A. S. Lodge, R. C. Armstrong, M. H. Wagner, and H. H. Winter, "Constitutive equations from Gaussian molecular network theories in polymer rheology," *Pure Appl. Chem.*, **54**, 1350–1359 (1982).

159. A. S. Lodge and J. Meissner, "On the use of instantaneous strains, superposed on shear and elongational flows of polymeric liquids, to test the Gaussian network hypothesis and to estimate the segment concentration and its variation during flow," *Rheol. Acta*, **11**, 351–352 (1972).

160. J. M. Lupton and J. W. Regester, "Melt flow of polyethylene at high rates," *Polym. Eng. Sci.*, **5**, 235–245 (1965).

161. M. E. Mackay and D. V. Boger, "Flow visualization in rheometry," chap. 14 in *Rheological Measurement,* A. A. Collyer and D. W. Clegg, eds. (Elsevier, Essex, UK, 1988), pp. 433–477.

162. C. W. Macosko, *Rheology Principles, Measurements, and Applications* (VCH Publishers, New York, 1994).

163. J. J. Magda, C. S. Lee, S. J. Muller, and R. G. Larson, "Rheology, flow instabilities, and shear-induced diffusion in polystyrene solutions," *Macromol.*, **26**, 1696–1706 (1993).

164. H. Markovitz, "A property of Bessel functions and its application to the theory of two rheometers," *J. Appl. Phys.*, **23**, 1070–1077 (1952).

165. H. Markovitz, *Rheological Behavior of Fluids* (Educational Services (Video), Watertown, MA, 1965).

166. G. E. Mase, *Schaum's Outline of Theory and Problems of Continuum Mechanics* (McGraw-Hill; New York, 1970).

167. D. J. Massa, J. L. Schrag, and J. D. Ferry, "Dynamic viscoelastic properties of polystyrene in high-viscosity solvents. Extrapolation to infinite dilution and high-frequency behavior," *Macromol.*, **4**, 210–214 (1971).

168. T. Masuda, K. Kitagawa, T. Inoue, and S. Onogi, "Rheological properties of anionic polystyrenes. II. Dynamic viscoelasticity of blends of narrow-distribution polysty-renes," *Macromol.*, **3**, 116–125 (1970).

169. The Math Works, Inc., 24 Prime Park Way, Natick, MA 01760, USA, tel.: 508-647-7000, *www.matlab.com*.

170. Mathsoft, Inc., 101 Main St., Cambridge, MA 02142–1521, USA, *www.mathsoft.com*.

171. J. E. Matta and R. P. Tytus, "Liquid stretching using a falling cylinder," *J. Non-Newt. Fluid Mech.*, **45**, 215–229 (1990).

172. B. Maxwell and A. Jung, "Hydrostatic pressure effect on polymer melt viscosity," *Mod. Plast.*, **35**, 174–182, 276 (1957).

173. J. Meissner, "Development of a universal extensional rheometer for the uniaxial extension of polymer melts," *Trans. Soc. Rheol.*, **16**, 405–420 (1972).

174. J. Meissner, "Experimental aspects in polymer melt elongational rheometry," *Chem. Eng. Commun.*, **33**, 159–180 (1985).

175. J. Meissner, "Elongation of polymer melts—experimental methods and recent results," in *Proc. 12th Int. Congr. on Rheology*, A. Ait-Kadi, J. M. Dealy, D. F. James, and M. C. Williams, eds. (August 18–23, 1996), p. 7.

176. J. Meissner and J. Hostettler, "A new elongational rheometer for polymer melts and other highly viscoelastic liquids," *Rheol. Acta*, **33**, 1–21 (1994).

177. E. V. Menezes and W. W. Graessley, "Nonlinear rheological behavior of polymer systems for several shear-flow histories," *J. Polym. Sci., Polym. Phys.*, **20**, 1817–1833 (1982).

178. A. B. Metzner and M. Whitlock, "Flow behavior of concentrated (dilatant) suspen-sions," *Trans. Soc. Rheol.*, **2**, 239–254 (1958).

179. S. Middleman, *Fundamentals of Polymer Processing* (McGraw-Hill, New York, 1977).

180. S. Middleman, *An Introduction to Fluid Dynamics* (John Wiley & Sons, New York, 1998).

181. M. J. Miller and E. B. Christiansen, "The stress state of elastic fluids in viscometric flow," *AIChE J.*, **18**, 600–608 (1972).

182. Moldflow Corporation, 91 Hartwell Avenue, Lexington, MA 02173, USA, tel.: 617-674-0085, *www.moldflow.com*.

183. M. Mooney, "Explicit formulas for slip and fluidity," *Trans. Soc. Rheol.*, **2**, 210–222 (1931).

184. F. A. Morrison and R. G. Larson, "A study of shear-stress relaxation anomalies in binary mixtures of monodisperse polystyrenes," *J. Polym. Sci., Polym. Phys. Ed.*, **30**, 943–950 (1992).

185. F. A. Morrison, P. Manjeshwar, and J. W. Mays, "Anomalous shear behavior of polybutadiene melts," unpublished manuscript, 2000.

186. S. J. Muller, R. G. Larson, and E. S. G. Shaqfeh, "A purely elastic transition in Taylor–Couette flow," *Rheol. Acta*, **28**, 499–503 (1989).

187. H. Münstedt, "Dependence of the elongational behavior of polystyrene melts on molecular weight and molecular weight distribution," *J. Rheol.*, **24**, 847–867 (1980).

188. K. A. Narh and A. Keller, "The effect of counterions on the chain conformation of polyelectrolytes, as assessed by extensibility in elongational flow: The influence of multiple valency," *J. Polym Sci. B*, **32**, 1697–1706 (1994).

189. P. Neogi, ed., *Diffusion in Polymers* (Marcel Dekker, New York, 1996).

190. I. S. Newton, *Principia Mathematica* (1687); as cited in Macosko [162].

191. W. Noll, "A mathematical theory of the mechanical behavior of continuous media," *Arch. Rational Mech. Anal.*, **2**, 197–226 (1958).

192. J. G. Oldroyd, "On the formulation of rheological equations of state," *Proc. Roy. Soc. A*, **200**, 523–541 (1950).

193. J. G. Oldroyd, "Finite strains in an anisotropic elastic continuum," *Proc. Roy. Soc. A*, **202**, 345–358 (1950).

194. J. G. Oldroyd, "Non-Newtonian effects in steady motion of some idealized elastico-viscous liquids," *Proc. Roy. Soc. A*, **245**, 278–297 (1958).

195. J. G. Oldroyd, "An approach to non-Newtonian fluid mechanics," *J. Non-Newt. Fluid Mech.*, **5**, 9–46 (1984).

196. S. Onogi, T. Masuda, and K. Kitagawa, "Rheological properties of anionic polystyrenes. I. Dynamic viscoelasticity of narrow-distribution polystyrenes," *Macromol.*, **3**, 109–116 (1970).

197. N. V. Orr and T. Sridhar, "Probing the dynamics of polymer solutions in extensional flow using step strain rate experiments," *J. Non-Newt. Fluid Mech.*, **82**, 203–232 (1999).

198. K. Osaki, N. Bessho, T. Kojimoto, and M. Kurata, "Flow birefringence of polymer solutions in time-dependent field. Relation between normal and shear stresses on application of step-shear strain," *J. Rheol.*, **23**, 617–624 (1979).

199. K. Osaki, S. Kimura, and M. Kurata, "Relaxation of shear and normal stresses in step shear deformation of a polystyrene solution. Comparison with the predictions of the Doi–Edwards theory," *J. Polym. Sci., Polym. Pys. Ed.*, **19**, 517–527 (1981).

200. H. C. Öttinger, *Stochastic Processes in Polymeric Fluids* (Springer-Verlag, Berlin, 1996).

201. L. I. Palade, V. Verney, and P. Attane, "Time–temperature superposition and linear viscoelasticity of polybutadienes," *Macromol.*, **28**, 7051–7057 (1995).

202. J. Parnaby and R. A. Worth, "Die variator mandrel forces encountered in blow moulding parison control systems–computer-aided design," *Proc. Inst. Mech. Eng.*, **188**, 357–364 (1974).

203. R. C. Penwell, W. W. Graessley, and A. Kovacs, "Temperature dependence of viscosity-shear rate behavior in undiluted polystyrene," *J. Polym. Sci., Polym. Phys. Ed.*, **12**, 1771–1783 (1974).

204. Perkin-Elmer Corporation, 761 Main Avenue, Norwalk, CT 06859, USA, tel.: 800-762-4000, *www.perkin-elmer.com*.

205. C. J. S. Petrie, *Elongational Flows. Aspects of the Behaviour of Model Elasticoviscous Fluids* (Pitman, London, UK, 1979).

206. C. J. S. Petrie and M. M. Denn, "Instabilities in polymer processing," *AIChE J.*, **22**, 209 (1976).

207. J. M. Piau, N. El Kissi, and B. Tremblay, "Low Reynolds number flow visualization of linear and branched silicones upstream of orifice dies," *J. Non-Newt. Fluid Mech.*, **30**, 197–232 (1988).

208. J. M. Piau, N. El Kissi, and B. Tremblay, "Influence of upstream instabilities and wall slip on melt fracture and sharkskin phenomena during silicones extrusion through orifice dies," *J. Non-Newt. Fluid Mech.*, **34**, 145–180 (1990).

209. D. J. Plazek, "Temperature dependence of the viscoelastic behavior of polystyrene," *J. Phys. Chem.*, **69**, 3480–3487 (1965).

210. D. J. Plazek, "The temperature dependence of viscoelastic behavior of poly(vinyl acetate)," *Polym. J.*, **12**, 43–53 (1980).

211. D. J. Plazek, "The creep behavior of amorphous polymers," in *Relaxations in Complex Systems,* K. L. Ngai and G. B. Wright, eds. (National Technical Information Service, U.S. Department of Commerce, Springfield, VA, Oct. 1984).

212. Polyflow S. A., 16, Place de l'Université, B–1348 Louvain-la-Neuve, Belgium, tel.: 32-10-452861, *www.polyflow.be*.

213. G. Pomar, S. J. Muller, and M. M. Denn, "Extrudate distortions in linear low-density polyethylene solutions and melt," *J. Non-Newt. Fluid Mech.*, **54**, 143–151 (1994).

214. A. V. Ramamurthy, "Wall slip in viscous fluids and influence of materials of construction," *J. Rheol.*, **30**, 337–357 (1986).

215. M. Reiner, "The Deborah number," *Phys. Today*, **17**, 62 (1964).

216. Rheometrics Scientific, One Possumtown Road, Piscataway, NJ 08854, USA, *www.rheosci.com*.

217. R. S. Rivlin and D. W. Saunders, "Large elastic deformations of isotropic materials. VII. Experiments on the deformation of rubber," *Phil. Trans. R. Soc.*, **A243**, 251–288 (1951).

218. W. B. Russel, D. A. Saville, and W. R. Schowalter, *Colloidal Dispersions* (Cambridge University Press, New York, 1989).

219. D. Satas, ed., *Handbook of Pressure-Sensitive Adhesive Technology,* 2nd ed. (Van Nostrand Reinhold, New York, 1989).

220. W. R. Schowalter, *Mechanics of Non-Newtonian Fluids*, (Oxford University Press, New York, 1978).

221. R. M. Schulken, R. H. Cox, and L. A. Minnick, "Dynamic and steady-state rheological measurements on polymer melts," *J. Appl. Polym. Sci.*, **25**, 1341–1353 (1980).

222. B. Schutz, *Geometrical Methods of Mathematical Physics* (Cambridge University Press, New York, 1980).

223. T. Schweizer, K. Mikkelsen, C. A. Cathey, and G. G. Fuller, "Mechanical and optical responses of the M1 fluid subject to stagnation point flow," *J. Non-Newt. Fluid Mech.*, **35**, 277–286 (1990).

224. R. B. Secor, "Operability of extensional rheometry by stagnation, squeezing, and fiber-drawing flows: Computer-aided-analysis, viscoelastic characterization, and experimental analysis," Ph.D. dissertation, University of Minnesota (1988).

225. R. B. Secor, C. W. Macosko, and L. W. Scriven, "Analysis of lubricated planar stagnation die flow,"*J. Non-Newt. Fluid Mech.*, **23**, 355–381 (1987).

226. R. W. G. Shipman, M. M. Denn, and R. Keunings, "Mechanics of the 'falling plate' extensional rheometer," *J. Non-Newt. Fluid Mech.*, **40**, 281–288 (1991).

227. A. H. P. Skelland, *Non-Newtonian Flow and Heat Transfer* (John Wiley & Sons, New York, 1967).

228. M. J. Solomon and S. J. Muller, "The transient extensional behavior of polystyrene-based Boger fluids of varying solvent quality and molecular weight," *J. Rheol.*, **40**, 837–856 (1996).

229. P. R. Soskey and H. H. Winter, "Large-step shear strain experiments with parallel-disk rotational rheometers," *J. Rheol.*, **28**, 625–645 (1984).

230. P. R. Soskey and H. H. Winter, "Equibiaxial extension of two polymer melts: Polystyrene and low density polyethylene," *J. Rheol.*, **29**, 493–517 (1985).

231. R. S. Spencer and R. E. Dillon, "The viscous flow of molten polystyrene. II," *J. Colloid Sci.*, **4**, 241–255 (1956).

232. M. R. Spiegel, *Schaum's Outline of Theory and Problems of Advanced Mathematics for Engineers and Scientists* (McGraw-Hill: New York, 1971).

233. S. H. Spiegelberg, D. C. Ables, and G. H. McKinley, "Extensional rheometry of viscous polymer solutions," presented at the 67th Annual Meeting of the Society of Rheology (Sacramento, CA, Oct. 1995), Paper T4.

234. T. Sridhar, V. Tirtaatmadja, D. A. Nguyen, and R. K. Gupta, "Measurement of extensional viscosity of polymer solutions," *J. Non-Newt. Fluid Mech.*, **40**, 271–280 (1991).

235. Z. Tadmor, "Non-Newtonian tangential flow in cylindrical annuli," *Polym. Eng. Sci.*, **6**, 203–212 (1966).

236. Z. Tadmor and C. G. Gogos, *Principles of Polymer Processing* (John Wiley & Sons, New York, 1979).

237. H. Tanaka and J. L. White, "Experimental investigations of shear and elongational flow properties of polystyrene melts reinforced with calcium carbonate, titanium dioxide, and carbon black," *Polym. Eng. Sci.*, **20**, 949–956 (1980).

238. R. I. Tanner, *Engineering Rheology,* rev. ed. (Clarendon Press, Oxford, UK, 1988).

239. G. B. Thomas and R. L. Finney, *Calculus and Analytic Geometry* 5th ed. (Addison-Wesley, Reading, MA, 1979).

240. P. A. Tipler, *Physics* (Worth Publishers, New York, 1976).

241. V. Tirtaatmadja and T. Sridhar, "A filament stretching device for measurement of extensional viscosity," *J. Rheol.*, **37**, 1081–1102 (1993).

242. D. A. Tomalia, "Dendrimer molecules," *Sci. Amer.*, **272**, 62–66 (1995).

243. L. R. G. Treloar, "The elasticity of a network of long-chain molecules—II," *Trans. Faraday Soc.*, **39**, 241–246 (1943).

244. L. R. G. Treloar, *The Physics of Rubber Elasticity*, 3rd ed. (Oxford University Press, London, UK, 1975).

245. B. Tremblay, "Estimation of the elongational viscosity of polyethylene blends at high deformation rates," *J. Non-Newtonian Fluid Mech.* **33** 137–164 (1989).

246. D. J. Tritton, *Physical Fluid Dynamics* (Oxford University Press, Oxford, UK, 1988).

247. C. Truesdell and W. Noll, *The Nonlinear Field Theories of Mechanics* (Springer-Verlag, Berlin, 1965).

248. S. Uppuluri, "Structure–property relationships of poly(amidoamine)dendrimers: Evaluation of the skeletal macromolecular isomerism principle," Ph.D. dissertation, Michigan Technological University, Houghon, MI (1997).

249. L. A. Utracki, "Pressure oscillation during extrusion of polyethylenes. II," *J. Rheol.*, **28**, 601–623 (1984).

250a. W. C. Uy and W. W. Graessley, "Viscosity and normal stresses in poly(vinyl acetate) systems," *Macromol.*, **4**, 458–463 (1971).

250b. D. C. Venerus, M. Kompani and B. Bernstein, "Equibiaxial Extensional Flow of Polymer Melts via Lubricated Squeezing Flow. Part II: Flow Modeling," *Rheologica Acta,* in press (2000).

251. S. Venkatraman, M. Okano, and A. Nixon, "A comparison of torsional and capillary rheometry for polymer melts: The Cox–Merz rule revisited," *Polym. Eng. Sci.*, **30**, 308–313 (1990).

252. G. V. Vinogradov, "Critical regimes of deformation of liquid polymeric systems," *Rheol. Acta*, **12**, 357–373 (1973).

253. G. V. Vinogradov, A. Ya. Malkin, Yu. G. Yanovskii, E. K. Borisenkova, B. V. Yarlykov, and G. V. Berezhnaya, "Viscoelastic properties and flow of narrow distribution polybutadienes and polyisoprenes," *J. Polym. Sci. A-2*, **10**, 1061–1084 (1972).

254. C. M. Vrentas and W. W. Graessley, "Relaxation of shear and normal stress components in step-strain experiments," *J. Non-Newt. Fluid Mech.*, **9**, 339–355 (1981).

255. C. M. Vrentas and W. W. Graessley, "Study of shear stress relaxation in well-characterized polymer liquids," *J. Rheol.*, **26**, 359–371 (1982).

256. J. L. S. Wales, *The Application of Flow Birefringence to Rheological Studies of Polymer Melts* (Delft University Press, Rotterdam, The Netherlands, 1976).

257. F. T. Wall, "Statistical thermodynamics of rubber. II," *J. Chem. Phys.*, **10**, 485–488 (1942).

258. K. Walters, *Rheometry* (Chapman & Hall, London, UK, 1975).

259. K. Walters, *Rheometry, Industrial Applications* (Research Studies Press, New York, 1980).

260. Waterloo Maple, Inc., 450 Phillip St., Waterloo, Ont. N2L 5J2, Canada, tel.: 519-747-2373, *www.maplesoft.com*.

261. K. Weissenberg, as cited by B. Rabinowitsch, "Über die Viskosität und Elastizität von Solen," *Z. Phys. Chem.*, **A145**, 1–26 (1929).

262. R. F. Westover, "Effect of hydrostatic pressure on polyethylene melt rheology," *SPE Trans.*, **1**, 14–20 (1961).

263. F. M. White, *Viscous Fluid Flow* (McGraw-Hill, Inc.: New York, 1974).

264. J. L. White and A. B. Metzner, "Development of constitutive equations for polymeric melts and solutions," *J. Appl. Polym. Sci.*, **7**, 1867–1889 (1963).

265. M. L. Williams, R. F. Landel, and J. D. Ferry, "The temperature dependence of relaxation mechanisms in amorphous polymers and other glass-forming liquids," *J. Am. Chem. Soc.*, **77**, 3701–3707 (1955).

266. H. H. Winter, C. W. Macosko, and K. E. Bennet, "Orthogonal stagnation flow, a framework for steady extensional flow experiments," *Rheol. Acta*, **18**, 323–334 (1979).

267. Wolfram Research, Inc., 100 Trade Center Drive, Champaign, IL 61820-7237, USA, *www.mathematica.com*.

268. Yao, M. W., personal communication (1997).

269. D. Zwillinger, *Handbook of Differential Equations*, 3rd ed. (Academic Press, New York, 1989).

Index